T0190311

CAMBRIDGE LIBRARY COLLECTION

Books of enduring scholarly value

Astronomy

From ancient times, humans have tried to understand the workings of the world around them. The roots of modern physical science go back to the very earliest mechanical devices such as levers and rollers, the mixing of paints and dyes, and the importance of the heavenly bodies in early religious observance and navigation. The physical sciences as we know them today began to emerge as independent academic subjects during the early modern period, in the work of Newton and other 'natural philosophers', and numerous sub-disciplines developed during the centuries that followed. This part of the Cambridge Library Collection is devoted to landmark publications in this area which will be of interest to historians of science concerned with individual scientists, particular discoveries, and advances in scientific method, or with the establishment and development of scientific institutions around the world.

The Scientific Papers of Sir William Herschel

By the time of his death, William Herschel (1738–1822) had built revolutionary telescopes, identified hundreds of binary stars, and published astronomical papers in over forty volumes of the Royal Society's *Philosophical Transactions*. This two-volume collection, which originally appeared in 1912, was the first to gather together his scattered publications. It draws also on a wealth of previously unpublished material, from personal letters to numerous papers presented to the Philosophical Society of Bath. Although Herschel is best known for his discovery of Uranus, this collection highlights the true range of his observations and interests. Focusing on his early work, Volume 1 includes notes on the discovery of Uranus, unpublished papers on electricity, and studies of the lunar mountains and the poles of Mars – both of which he believed to be inhabited. It also features a biographical account by the historian of astronomy J.L.E. Dreyer.

Cambridge University Press has long been a pioneer in the reissuing of out-of-print titles from its own backlist, producing digital reprints of books that are still sought after by scholars and students but could not be reprinted economically using traditional technology. The Cambridge Library Collection extends this activity to a wider range of books which are still of importance to researchers and professionals, either for the source material they contain, or as landmarks in the history of their academic discipline.

Drawing from the world-renowned collections in the Cambridge University Library and other partner libraries, and guided by the advice of experts in each subject area, Cambridge University Press is using state-of-the-art scanning machines in its own Printing House to capture the content of each book selected for inclusion. The files are processed to give a consistently clear, crisp image, and the books finished to the high quality standard for which the Press is recognised around the world. The latest print-on-demand technology ensures that the books will remain available indefinitely, and that orders for single or multiple copies can quickly be supplied.

The Cambridge Library Collection brings back to life books of enduring scholarly value (including out-of-copyright works originally issued by other publishers) across a wide range of disciplines in the humanities and social sciences and in science and technology.

The Scientific Papers of Sir William Herschel

Including Early Papers Hitherto Unpublished

VOLUME 1

EDITED BY
JOHN LOUIS EMIL DREYER

CAMBRIDGE
UNIVERSITY PRESS

University Printing House, Cambridge, CB2 8BS, United Kingdom

Published in the United States of America by Cambridge University Press, New York

Cambridge University Press is part of the University of Cambridge.
It furthers the University's mission by disseminating knowledge in the pursuit of
education, learning and research at the highest international levels of excellence.

www.cambridge.org
Information on this title: www.cambridge.org/9781108064620

This edition first published 1912
This digitally printed version 2013

ISBN 978-1-108-06462-0 Paperback

THE SCIENTIFIC PAPERS

OF

SIR WILLIAM HERSCHEL

KNT. GUELP., LL.D., F.R.S.

Wm: Herschel

From a pastel by J. Russell. R.A. 1794.
In the possession of Sir W. J. Herschel. Bart.

THE SCIENTIFIC PAPERS

OF

SIR WILLIAM HERSCHEL

KNT. GUELP., LL.D., F.R.S.

INCLUDING EARLY PAPERS HITHERTO UNPUBLISHED

COLLECTED AND EDITED UNDER THE DIRECTION
OF A JOINT COMMITTEE OF THE ROYAL SOCIETY
AND THE ROYAL ASTRONOMICAL SOCIETY

WITH A BIOGRAPHICAL INTRODUCTION COMPILED MAINLY FROM
UNPUBLISHED MATERIAL BY J. L. E. DREYER

VOL. I

LONDON: PUBLISHED BY
THE ROYAL SOCIETY AND THE ROYAL ASTRONOMICAL SOCIETY
AND SOLD BY DULAU & CO., LTD.
37 SOHO SQUARE, LONDON, W.
1912

PREFACE

SOON after the death of Sir WILLIAM HERSCHEL in 1822, his distinguished son, Sir JOHN F. W. HERSCHEL, F.R.S., formed the plan of republishing his father's papers; but he found on inquiry that no publisher would be willing to undertake the risk of so extensive a work. He therefore resigned the idea, considering that he might contribute more effectually towards a monument to his father's memory by devoting himself to extending and carrying out, with his own instruments and after his own manner, his father's processes of observation. A German translation was commenced by Professor J. W. PFAFF of Erlangen, but only the first volume was published (*W. Herschel's Sämmtliche Schriften.* Erster Band: "Ueber den Bau des Himmels." Dresden und Leipzig, 1826, 8°).

The papers of W. HERSCHEL are scattered over about forty volumes of the *Philosophical Transactions*, and they have become difficult of access to many to whom their study is of importance. A useful summary of their contents was published by Professors HASTINGS and HOLDEN in the Smithsonian Report for 1880, but a collected edition of the papers has always been considered a *desideratum* in astronomical literature, and from time to time it has been strongly urged that a complete reprint should be made available.

Early in the year 1910 the matter was taken in hand, and a Committee was appointed by the Royal Society and the Royal Astronomical Society to prepare an edition of WILLIAM HERSCHEL'S works at the joint expense of the two Societies. It consisted of Sir WILLIAM HUGGINS (who was prominent in promoting the work, but was only able to preside over two meetings of the Committee before his death), Sir JOSEPH LARMOR and Professor R. A. SAMPSON, representing the Royal Society, and Sir DAVID GILL, Mr J. A. HARDCASTLE and Professor H. H. TURNER, representing the Royal Astronomical Society. Soon afterwards Dr. J. L. E. DREYER and Major E. H. HILLS, and finally Mr. F. W. DYSON, Astronomer Royal, were added to the Committee.

It seemed desirable to reprint all HERSCHEL'S published papers exactly as they had been issued by him, without omissions or alterations,—except that actual errors of observation, or identification of the observed objects, should be pointed out. Some errors of this kind in the observations of Double Stars had been found, chiefly by the late Mr. H. SADLER, from an examination of the MS. sheets belonging to the Royal Astronomical Society; and these have now been checked where necessary by reference to the original observing Journal. But it appeared to be specially important to take this opportunity to make a revision of the three catalogues of Nebulæ in order to clear up many difficulties in reconciling HERSCHEL'S results with those of later observers. This has been done by an examination of the original "sweeps," which, together with CAROLINE HERSCHEL'S Zone Catalogue and the observations of the objects in MESSIER'S catalogue, were lent for this purpose by the Royal Society. It is hoped that this revision has resulted in improving the accuracy of the three catalogues; but at the same time it should be stated that the revision has furnished additional proofs of the very great care with which the observations were both made and reduced.

The reprint has been made complete as regards the published scientific work, though this has involved the insertion in Vol. II. of three papers on NEWTON'S coloured rings, which have not more than a personal interest (see Introduction, pp. lvii-lviii).

The Committee have been under great obligations to Sir WILLIAM

J. HERSCHEL, Bart., who generously placed at their disposal his grandfather's letters, his observing Journal, the record of the polishing of mirrors, and also autobiographical memoranda and unpublished papers. Everything of importance that could be extracted from these valuable materials and brought within reasonable limits of space, has been given in the Introduction to the present volumes.

Many incidents of HERSCHEL'S career have thus been presented in a new light, especially as regards his early life; while the papers read before the Bath Philosophical Society form an interesting record of his early modes of thought and of the versatility of his mind, in addition to affording an illustration of the remarkable activity of genuine physical speculation in England, at a time when formal mathematical analysis was but slightly cultivated.

The joint Committee aforesaid of the Royal Society and the Royal Astronomical Society are responsible for the general plan of the work. But they desire to record their obligation, and that of the astronomers who will use the reprint, to one of their number, Dr. J. L. E. DREYER, who very generously undertook both the collation of the Manuscripts and the preparation of the Introductory Memoir.

LONDON, *February* 1912.

CONTENTS OF VOL. I

INTRODUCTION

A SHORT ACCOUNT OF SIR WILLIAM HERSCHEL'S LIFE AND WORK, chiefly from unpublished sources. By J. L. E. DREYER.

UNPUBLISHED PAPERS.

I. *Papers read before the Philosophical Society of Bath*, 1780–1781.

 b

PAPERS PUBLISHED IN THE PHILOSOPHICAL TRANSACTIONS OF THE ROYAL SOCIETY AND ELSEWHERE

PLATES IN VOL. I

CORRIGENDA IN VOL. I

Page 64. *Under* II. 2, *for* 30° 35′ *read* 21° 28′ (see Vol. II. p. 281).

„ 81. *To* V. 44 *add this footnote*: This observation is not in the *Journal*; 1783, Sept. 17, only one companion mentioned, following.

„ 202, line 1. *For* V. 70 *read* IV. 70.

„ 356, line 1. *For* II. *read* III. (first column).

„ 366. *Under* II. 679–680, *for* Sw. 709 *read* Sw. 729 (Apr. 15, 1787).

„ 369, last line but one. The] should be at the end of the last line.

„ 373, line 17. *Add* this footnote: This was not the seventh satellite; see below, p. 405, under August 29.

A Short Account of
SIR WILLIAM HERSCHEL'S LIFE AND WORK

CHIEFLY FROM UNPUBLISHED SOURCES

I.—Early Years, 1738–1782

THE following short account of William Herschel's life is almost exclusively founded on his unpublished memoranda, journals of observations, and letters. A few necessary particulars have been taken from the *Memoir and Correspondence of Caroline Herschel* (London, 1876), but as a rule readily accessible, printed material has only been referred to, but not quoted at length. In 1783 Herschel sent to Lichtenberg a short account of his life, written in English, and dated in his letter-book the 15th February. This was printed in German in Lichtenberg and Forster's *Göttinger Magazin der Wissenschaften und Literatur*, iii. 4.* There is a slightly longer MS. account, written in November 1784 at the request of the editor of the *European Magazine*, and headed, "In a letter to Dr Hutton." But the principal source of information about Herschel's early life is a MS. book in 4to, written much later,† entitled "Biographical Memorandums," which we proceed to give almost complete, omitting only unimportant names of pupils and such dates of musical engagements as were not essential to illustrate the chronological sequence of events and his wonderful activity.

MEMORANDUMS FROM WHICH AN HISTORICAL ACCOUNT OF MY LIFE MAY BE DRAWN

My Great-Grand-Father, Hans Herschel, was a citizen and brewer in Pirna, a town two [German] miles from Dresden.‡

My Grandfather, Abraham, second son of Hans Herschel, was born in 1651. He learned gardening in the Elector's gardens at Dresden, and had also a good knowledge of arithmetic,

* Reprinted in the first (and only) volume of J. W. Pfaff's *W. Herschel's Sämmtliche Schriften*, Dresden und Leipzig, 1826, pp. 10–12. Translated from this back into English in Holden's *William Herschel, his Life and Works*, London, 1881, pp. 3–6. There is a similar very short sketch in vol. i. of *Public Characters* (1799), translated by Zach in his *Monatliche Correspondenz*, Bd. v. p. 70 (1802).

† Later than 1793, which date occurs in it. It is a fasciculus of sixty pages in small quarto, stitched at home in a cover of thick grey paper like all the other MS. books of Herschel. The last twenty-four pages are blank. The first three pages being copied from a memorandum by Isaac Herschel (still extant), are written in German. They have here been slightly condensed. Use has also been made of some stray notes by Sir John Herschel.

‡ About the origin of the family, Carl Christian Herschel wrote to William Herschel in 1790 that his own father had learned that "Drei Brüder wendeten sich, da sie der protestantischen Religion wegen aus Mähren vertrieben wurden, nach Sachsen und kauften sich an." The same writer wrote in 1819, when he was "Königl. Sächs. Oberhof-Gerichts Actuarius," the following: "Wegen Religionsverhältnissen wanderten drey Brüder Herschel aus Mähren nach Sachsen aus. Einer blieb in Schmilka, der ander in Postelnitz, welche beyde Grenzdörfer an der Elbe nach Böhmen sind; der dritte Bruder ging nach Pirna, wo er Gärtnerey trieb, sich aber von Pirna nach Hannover begab. Von diesem dritten Bruder stammen Sie ab; ich hingegen von dem Schmilkaischen." That the third brother went to Hanover, is a mistake. There is nothing about these three brothers in Isaac Herschel's MS., which is still in existence.

writing, drawing, and music. He was employed at the country-seat of Hohentziatz, in the principality of Anhalt-Zerbst, 4 [German] miles from Magdeburg, as a landscape gardener.* His eldest son, Eusebius, learned gardening at Gotha, but afterwards took to farming. He married and had five sons.

My Father, Isaac Herschel, third son of Abraham, † was born at Hohentziatz on the 14th January 1707. His own short account of himself is as follows:

"My parents wished me to devote myself to gardening, but my Father died in 1718 before I could take it up. My brother (Eusebius) took up my Father's work for some years, and as I had picked up some knowledge of gardening, he found a place for me at Zerbst as a gardener. But as I had already at Hohentziatz procured a violin and learned to play it by ear, I took proper lessons at Zerbst from an hautboy-player in the court-band. I also bought an hautboy and was never so happy as when I could occupy myself with music. My desire was to become a member of a band, for I had lost all interest in gardening. When I had reached the age of 21 I thought I knew enough, I gave up my situation and went to Berlin. But as I found the Prussian service as a bandsman very bad and slavish,‡ I went to Potsdam and took lessons for a whole year in music and especially in playing the hautboy from the old Prussian conductor Päbush. My dear brother and my dear sister § supplied me with the necessary funds, although my brother often urged me to come to him and learn agriculture (economics) from him. I took his advice, but I could not resist my love of music and travelling, and in July 1731 I went to Brunswick, and finding the service too Prussian there, I went on to Hanover. There it was my destiny to stay, and on the 7th August 1731 I was engaged as hautboy-player in the Foot-Guards. The following year, 1732, I married Anna Ilse Moritzen, daughter of a citizen of Neustadt on the Rübenberg near Hanover."

My Mother had ten children, four of them died early, the remaining six were:

Sophia Elisabeth, born April 12, 1733. She was married to a musician named Griesbach and had a numerous family.||

Heinrich Anton Jacob, born November 20, 1734.¶

Friederich Wilhelm, born November 15, 1738. I was naturalized April 30, 1793, by the name of William Herschel.

Johann Alexander Herschel, born November 13, 1745. His name is mentioned in the memorandums taken from some of my pocket books.**

Caroline Lucretia, born March 16, 1750. She became my assistant when I was appointed Astronomer to His Majesty.††

Johann Dieterich, born September 13, 1755. He is an eminent musician and well known as a member of several Academies.‡‡

My father's great attachment to music determined him to endeavour to make all his sons complete musicians, and my eldest brother Jacob, having already made great progress in music, my father taught me to play on the violin as soon as I was able to hold a small one made on purpose for me. §§ Being also desirous to give all his children as good an education as his very limited circumstances would allow, I was at a proper time sent to a school where, besides religious instruction, all the boys received lessons in reading, writing, and arithmetic; and as I very readily learned every task assigned me, I soon arrived at such a degree of

* His wife was Eva Meves of Coburg, according to Isaac Herschel.

† The second son died in infancy.

‡ This was during the reign of Frederic William I.

§ Appollonia, second child of Abraham, married in a rather novel-like manner ("auf eine etwas romanhafte Art") a country gentleman, Herr von Thümen.

|| She lost her husband early (1772); her children settled in England.

¶ Died in 1792. He was in the Court orchestra at Hanover.

** Died in 1821. He lived at Bath for many years from 1770, but retired to Hanover in 1816.

†† Died at Hanover 9th January 1848.

‡‡ Died in 1827. He was also in the Court orchestra at Hanover.

§§ "I have heard my Father say that at four years of age he was set on a table with a little fiddle to play a solo." (Note by John Herschel.)

perfection, especially in arithmetic, that the master of the school made use of me to hear younger boys say their lessons and to examine their arithmetical calculations.

On the 1st May 1753 I was engaged as musician in the Hanoverian Guards, being then 1753. fourteen years and some months of age, and I remember playing for this purpose on the hautboy and on the violin before General Sommerfeld, who approved my performance. This engagement furnished the means for my improvement not only in music, which was my profession, but also in acquiring a knowledge of the French language, with the advantage 1754. of studying above two years under a very well informed teacher, who, taking a great liking to me, did not confine his instruction to language only, but encouraged the taste he found in his pupil for the study of philosophy, especially logic, ethics, and metaphysics, which were his own favourite pursuits.*

Soon after the great earthquake which on the 1st November destroyed Lisbon, the 1755. Guards marched out of Hanover, and about the end of March 1756 we were at Ritzbüttel 1756. and embarked at Cuxhafen for England, where after a passage of 16 days we arrived in April. We disembarked at Chatham and marched to Maidstone in Kent and were quartered in that town. Here I applied myself to learn the English language and soon was enabled to read Locke on Human Understanding. From Maidstone we marched to Coxheath, where the Hanoverian troops were encamped. Here as well as at Maidstone my father, my eldest brother and myself made several valuable acquaintances with families that were fond of music, and which on mine and my brother's return to England proved of great service to us. During our stay in camp we took leave of absence for a short visit to view London. While we were encamped on Coxheath my brother obtained his wished for dismission from the regiment and returned to Hanover.

In the autumn when the camp broke up, the Guards marched to Rochester, where again I made some valuable acquaintances, but our stay there was not of long duration, as we soon after received orders to embark again for Germany.

Early in the spring my father and I went with the regiment into a campaign which proved 1757. very harassing by many forced marches and bad accommodations. We were many times obliged after a fatiguing day to erect our tents in a plowed field, the furrows of which were full of water.† About the time of the battle of Hastenbeck (July 26) we were so near the field of action as to be within the reach of gunshot; when this happened my father advised me to look to my own safety.‡ Accordingly I left the engagement and took the road to Hanover, but when arrived there I found that having no passport I was in danger of being pressed for a soldier; it was therefore thought proper for me to return to the army.§ When I had rejoined the regiment I found that nobody had time to look after the musicians, they did not seem to be wanted. The weather was uncommonly hot and the continual marches

* This teacher's name was Hofschläger. He afterwards obtained a good appointment at Hamburg, and lived to a very advanced age. In the letter to Hutton, Herschel adds here:—"To this fortunate circumstance it was undoubtedly owing, that altho' I loved music to excess and made considerable progress in it, I yet determined with a sort of enthusiasm to devote every moment I could spare to the pursuit of knowledge, which I regarded as the sovereign good, and in which I resolved to place all my future views of happiness in life." In her *Memoir*, Caroline Herschel says that her two eldest brothers were about this time often introduced as solo performers or assistants in the Court orchestra, and on returning home kept her from going to sleep by their talk on music or on philosophical subjects, in which her father joined. He also assisted William by contriving self-made instruments, among which was a neatly turned 4-inch celestial globe (*Memoir*, p. 7).

† Isaac Herschel had had a similar experience the night after the battle of Dettingen (1743); his daughter says that it left him with an impaired constitution and an asthmatic affection which he never got over.

‡ "During the battle with balls flying over his head he walked behind a hedge spouting speeches, rhetoric being then his favourite study. Many heads of speeches he has lately burnt." (Note by J. Herschel.)

§ Caroline Herschel writes (*Memoir*, p. 10): "I can now comprehend the reason why we little ones were continually sent out of the way and why I had only by chance a passing glimpse of my brother as I was sitting at our street door, when he glided along like a shadow, wrapped in a great coat, followed by my mother with a parcel containing his accoutrements. After he had succeeded in passing unnoticed beyond the last sentinel at Herrenhausen, he changed his dress. . . . My brother's keeping himself so carefully from all notice was undoubtedly to avoid the danger of being pressed, for all unengaged young men were forced into the service. Even the clergy, unless they had livings, were not exempted."

were very harassing. At last (in September) my father's opinion was, that as on account of my youth I had not been sworn in when I was admitted in the Guards I might leave the military service; indeed he had no doubt but that he could obtain my dismission, and this he after some time actually procured (in 1762) from General Spörcken who succeeded General Sommerfeld.

[We interrupt Herschel's narrative to point out that the existence of this formal discharge puts an end to the legend, too long and too readily believed, that he deserted from the army and that he received a formal pardon for this offence from George III. on the occasion of his first audience in 1782. The fact that Herschel visited Hanover quite openly in 1764 and 1772 ought to have been sufficient to refute the legend. We give here the full text of the discharge; it is a printed form with name and other particulars inserted in writing (here in italics), and signed and sealed by the Colonel of the Foot-Guards with his private seal showing his armorial bearings.

"Des Allerdurchlauchtigsten Grossmächtigsten Fürsten und Herrn, Herrn Georg des Dritten, Königs von Grossbritannien, Franckreich und Irrland, Beschützers des Glaubens, Herzogen zu Braunschweig und Lüneburg, des Heiligen Römischen Reichs Ertz-Schatzmeister und Churfürsten, bestalter General der Infanterie, Obrister über die Garde zu Fuesse

ICH AUGUST FRIEDERICH VON SPÖRCKEN

Füge hiemit jedermänniglich zu wissen, wasgestalt Vorzeiger dieses *Friederich Wilhelm Herschell* gebürtig von *Hannover*, *25* Jahre alt, *blondt* von Haaren, *Langer* Statur, bey dem mir Allergnädigst anvertrauten Garde-Regimente, und zwar bey *dem 1ten Battaillon vier* Jahre *sechs* Monate *als Hautboiste* treu und ehrlich gedienet, auch sich bey allen vorgefallenen Kriegs-Expeditionen, Zug und Wachten solchermassen bezeiget, als es einem rechtschaffenen *Hautboisten* wohl anstehet und gebühret, so dass man allenthalben eine völlige Zufriedenheit darab nehmen mögen. Nachdem aber derselbe nunmehro *um seine Fortun auf ander Art zu suchen gebührend angehalten* dimittiret worden, so ist ihme darüber dieser Abschied ertheilet. Es gelanget auch an alle und jede Hohe und Niedrige Militär- und Civil-Obrigkeiten und Bediente, mein respective unter-dienst- und dienstfreundliches Anersuchen, obbesagten *Friederich Wilhelm Herschell* nicht nur aller Orts sicher und ohngehindert pass- und repassiren zu lassen, sondern demselbem auch wegen seines Wohlverhaltens, nach Nothdurfft und Gelegenheit, allen geneigten Willen zu gönnen. Welches in dergleichen und ähnlichen Fällen zu erwiedern, ich jederzeit bereit und willig seyn werde. Gegeben *im Winter Quart: Hameln d. 29ten Mertz 1762.*

"[Signed] *A. F. v. Spörcken.*"

Isaac Herschel's discharge is also still in existence. It is dated at Paderborn the 11th May 1760, issued by General Sommerfeld, and Isaac Herschel is described as "Hautboiste," so that he was not a bandmaster, as stated by some writers. His surname is spelt with a single l. He was discharged " wegen Engbrüstigkeit."]

I found no difficulty to leave the army, as nobody seemed to mind whether the musicians were present or absent. My intention being to go to England I took the road to Hamburg; and the French having taken Hanover, my brother Jacob left the place and came to me at Hamburg about the end of October and soon after went with me to England. When we arrived in London, we made use of the recommendation of some of the families we had been known to when we were in England before. We were introduced to some private concerts, my brother attended some scholars and I copied music, by which means we contrived to live pretty comfortably in the winter,* and in the summer we visited some families in and near Maidstone and Rochester and had a concert at Tunbridge Wells.

1758

London was so overstocked with musicians that we had but little chance of any great

* When he arrived in England "he had not half a Guinea in the world. He went to a music shop in town and asked to write music. An opera was put into his hands to copy, and he returned it with such dispatch as amazed his employer, who from that time kept him in full employ till he found a more profitable line of labour." (Note by John Herschel.)

success. My brother having a promise of a place in the Hanoverian orchestra received notice of a vacancy in September 1759, and accordingly left me to go and, as was the custom, to play for the place, which he did on the 15th October, and his performance being approved of, his name was on the 9th January 1760 inscribed in the list of Court Musicians. 1759.

Having given everything I could possibly spare to my brother when he left me,* I found myself involved in great difficulties, and seeing no likelihood of doing well in London I intended to try for better success in the country. Very opportunely I had an offer of going into Yorkshire, where the Earl of Darlington wanted a good musician to be at the head of a small band for a regiment of Militia of which he was the Colonel. The engagement being upon a liberal plan and not binding for any stipulated time, I gladly accepted it.† As the regiment was quartered at Richmond in Yorkshire, I took a place in the stage-coach, and on my arrival found that our musical band consisted only of two hautboys and two French horns. The latter being excellent performers I composed military music on purpose to show off our instruments. 1760.

Before my brother left England we made an agreement to carry on a regular correspondence by means of a friend in London, who being in a public office could send or receive our letters twice a week to and from Hanover free of postage. My brother being an excellent musician and eminent composer and also fond of intelligent disquisitions, my letters to him were chiefly relating to music; they also contained an account of my situation and circumstances, with occasional moral and metaphysical dissertations. He preserved most of my letters, and when I visited our family at Hanover in 1764, as he knew I was then about writing a treatise on music, he allowed me on account of their musical contents to take them back again.‡

The want of some account during this and other intervals is partly supplied by dates that are affixed to many of my musical compositions, by which the time and place of my temporary residence are ascertained.

[Herschel here gives a list of the contents of three volumes, each containing six symphonies, stating merely the date and place of composition of each. Next comes an index of his letters to Jacob, noting merely whether each letter was "metaphysical," "musical," "moral" or "historical," but in a few cases adding some particulars about his own doings, all of which are given below. These two sources we have interwoven, so as to form the following short chronological summary of the years 1760–1764, which contains everything known to us of this period of his life.]

June and September. Two symphonies written at Richmond, Yorkshire. 1760.
October. One, written at Sunderland.
Nov. 5 and 15. Two, written at Halnaby near Darlington.
Nov. 30. A symphony written at Sunderland.
Jan. 30. A symphony written at Sunderland. 1761.

Feb. 4. I went from Sunderland to Newcastle where on the 5th February I attended the regular subscription concert of Mr Avison the organist, in which I was engaged as first violin and solo player. From Newcastle I set out the next day on a journey to Edinburgh, where I had been informed it was probable that there would be a vacancy in the musical department, the manager of the concerts intending to leave that place. On my arrival there I was intro-

* In an undated note addressed to "My dear Nephew," Caroline Herschel says, that Jacob was one of the first violinists and oboe-players in Germany, and therefore would not condescend to earn his own living by teaching, but let his brother be his fag, allowing him to pay his tailor's bills, and to part with his last farthing for travelling expenses.

† "My father was engaged month by month to *teach* the Durham militia band, but was not himself one of that band." (Note by J. H. of a conversation with his father, 1st November 1818.)

‡ These letters, which extend to some 400 pages, are still extant but have not been at our disposal. We are informed that Herschel in them interweaves his philosophy and even his musical studies with references of an earnest kind to the Creator as a beneficent Deity, expressing his gratitude and addressing him in a prayerful spirit.

duced to M[r] Hume, the metaphysician, and a few days after, at one of their regular concerts, I was appointed to lead the band of musicians, while some of my symphonies and solo concertos were performed. M[r] Hume, who patronized my performance, asked me to dine with him, and accepting of his invitation I met a considerable company, all of whom were pleased to express their approbation of my musical talents, so that there would have been no doubt of my success, had not their established manager agreed to stay.

March 11. [In a letter of this date] it is mentioned that I had already asked to be dismissed from my engagement with Lord Darlington, and was in hopes in a few days to go to Edinburgh.*

April 20 and June 20. Two symphonies dated from Sunderland.

June 29 to August 3. Letters alternately from Richmond and from Sunderland.

August 12. I spent a week at Halnaby, the seat of Sir Ralph and Lady Milbank. I had been there often before and in November 1760 I made a long stay there to accompany Lady Milbank who was an excellent performer on the Harpsicord. I had now the honour of being of the musical party with M[r] Avison and M[r] Garth to entertain the Duke of York, the King's brother, who played the violoncello and accompanyed me in several solos which I had the honour of playing before him.

August 14. A weekly garden concert was established at Newcastle of which I was appointed to direct the music.

August 20. A symphony dated from Sunderland.

September 22 being the King's coronation day, I directed a band of 30 musicians at a concert given at Newcastle in honour of the day.

Sept. 27. Letter written from Pontefract.

Nov. 1 and Dec. 1. Symphonies composed at Pontefract.

1762. March 24. Symphony composed at Pontefract.

April 14. Symphony dated from Leeds.

April 25. [In a letter] a circumstantial account of my being approved of as a performer and composer so far as to induce a set of Gentlemen to wish to retain me at Leeds as a manager of their concerts. This situation, with the certain prospect of a number of scholars, appeared to me so desirable that I immediately determined to endeavour to obtain it, and in this I completely succeeded. Accordingly I left Pontefract and settled at Leeds, in which place I remained till the prospect of being chosen organist at Halifax brought on a change.

June 3, 16, July 1, 31. Four symphonies dated from Pontefract.

Sept. 16. Among the concertos in score, No. 7 is dated Barnard Castle in the county of Durham. I remember being a very short time at that place to settle some accounts relating to my quitting the Militia service.

1763 January 20. Violino Solo No. 6, composed at Leeds.

April 9. End of the winter concerts.

1764. In the spring of the year, after the winter concerts, I paid a visit to my relations at Hanover.†

[These are the only notes from the years 1763 and 1764, and there are none from 1765.

* In a note headed "Mr. Herschel" in the *Pocket Magazine of Classic and Polite Literature*, vol. ii. p. 215 (1818), it is stated that a Dr. Miller, organist of Doncaster, claimed to have been the means of drawing H. from a state of obscurity. M. tells how he met H. at Pontefract about the year 1760 when dining with the officers of the Durham militia, that he was astonished at the performance of the young German on the violin, how he suggested to H. to leave the militia and come and live with himself, that the offer was accepted, that he introduced H. at Mr. Copley's concerts, but that he soon lost his companion, whose fame was presently spread abroad. The name of Miller does not occur once in Herschel's Memoranda. When the above story appeared in print, John Herschel asked his father about it and noted the following: "The paragraph in question is founded on truth, but inaccurate in particulars. D[r] Miller, it seems, assumes credit to himself more than his due. My Father was never living with D[r] M. in his house as the Doctor would insinuate, though it is true that he was there occasionally for a night or so on a visit, and repeatedly passing through Doncaster where M. was organist often called on him. Yet it is not unlikely that M. introduced Father at Copley's concerts."

† He arrived on the 2nd April, and seems only to have stayed a couple of weeks. This was the last time he saw his father, who died 22nd March 1767, aged 60. (*Memoir of C. H.*, pp. 16–19.)

These years were spent principally at Leeds, but Herschel no doubt travelled about a good deal in the same way as he did in 1766. Speaking of the years which elapsed between his arrival in England and his appointment as organist at Halifax, Herschel writes in the letter to Hutton: "During all this time, tho' it afforded not much leisure for study, I had not forgot my former plan, but had given all my leisure hours to the study of languages. After I had improved myself sufficiently in English I soon acquired the Italian, which I looked upon as necessary to my business; I proceeded next to Latin, and having also made considerable progress in that language I made an attempt of the Greek, but soon after dropt the pursuit of that as leading me too far from my other favourite studies by taking up too much of my leisure. The theory of music being connected with mathematics had induced me very early to read in Germany all what had been written upon the subject of harmony; and when not long after my arrival in England the valuable book of Dr Smith's Harmonics came into my hands, I perceived my ignorance and had recourse to other authors for information, by which I was drawn on from one branch of mathematics to another."

The "Memorandums" now proceed to give a kind of rough diary for 1766 and following years. Many of the entries are very laconic, giving merely a date and the name of a town, and after his arrival at Bath names of pupils; nearly all of these are here omitted.]

MEMORANDUMS TAKEN FROM SOME POCKET-BOOKS *

As I at this time attended many musical families in the neighbourhood of Leeds, Halifax, Pontefract, Doncaster &c., the memorandums of the places will partly show how I spent my time, a great deal of which was taken up with travelling on horseback.†

1766. January 1. Wheatley. This was the country seat of Sir Bryan Cook, where every fortnight I used to spend two or three days. Sir Bryan played the violin, and some of his relations generally came from Doncaster to make up morning concerts. Our music was chiefly Corelli, Geminiani &c. Lady Cook loved music, and I gave her lessons on the guitar, which was then a fashionable instrument. Sir Bryan being an invalid, Lady Cook, an elderly Miss Wood and I generally passed the evening in playing at Tredrille.

Jan. 8. Having this time spent a whole week at Wheatley, my mare, standing idly in the stable and being overfed by Sir Bryan's grooms, died.

Jan. 12. Halifax. I received letters from Hanover. From this it appears that I lived there [*i.e.* at Halifax] at this time.

Feb. 19. Wheatley. Observation of Venus.‡

Feb. 24. Eclipse of the moon at 7 o'clock A.M. Kirby.‡

March 4. Sir Bryan Cook died.

March 7. Halifax. The Messiah. This oratorio was performed at a private club of chorus singers held at the Rev. Mr Bates', the clerk of the parish church and father of the well-known musical Mr Joc [Joah] Bates, where it was agreed to rehearse the same oratorio every other Friday, in order to perform it in the church at the opening of a new organ erecting there. I was the leader of the orchestra, and Mr Joc Bates, who played a chamber organ, directed the performance; his brother played the violoncello. I was a candidate for the place of organist, which by the interest of the Messrs Bates and many musical families I attended I had great hopes to obtain.§ About this time I was an inhabitant of Halifax.

* Judging by the writing, the following extracts seem to have been copied about 1818. Compare *C. H.*, p. 218.

† "Once (I have often heard him tell the story) he was reading on horseback, and found himself standing before the horse with book in his hand, being tossed over horse's head." (Note by J. H., 10th September 1816.)

‡ These are the first (and for a long time to come the only) notes on astronomical subjects.

§ "A dispute and law suit took place about the organ—one party maintaining that it was a heathenish thing, and meanwhile the organ was not permitted to be played. Now he had agreed with a parcel of chorus singers and others to meet once a week to rehearse the Messiah till it was settled, which they did, but it continued a whole twelvemonth. When the organ was at last suffered to be played, the treble voice fell ill (the Messiah was to be played at the opening). Allot (afterwards Dean of Raphoe) undertook the treble part, which he did capitally—a very short time afterwards (a fortnight says Father) he sung bass ∴ his voice was iust on the turn." (Note by John Herschel.)

My leisure time was employed in reading mathematical books, such as the works of Emerson, Maclaurin, Hodgson, Smith's Harmonics &c. This happened to be noticed by one of the Mess[rs] Bates, who told his brother: "M[r] Herschel reads Fluxions!"

March 21. Messiah at Halifax. The members of the club paid 6[d] each.

March 30. I composed the first song in an intended Oratorio which I called The Success of Satan against man. The words taken from Milton's P.L.

March 31. Second song.

April 2. Composed the 2nd chorus: "Better to reign in Hell than serve in Heaven."

April 18. Messiah rehearsed. My 2nd chorus was rehearsed. Composed a 6th and 7th catch.

April 28. Liverpool, rehearsal of the Messiah; 29th, of Judas Maccabæus.

30. The Messiah at Peter's Church.

May 1. Judas Maccabæus at do., concert at the Playhouse.

2. Jud. Mac. and Händel's Coronation Anthem. Warrington.

July 13. Played the organ during the service at Leeds.

22 and 24. Composed Preludiums Nos. 13–14.

27. Played the organ Prel. 15 at Wakefield for the organist.

28 &c. Organ every day by way of practice at Leeds. Prel. 16.

August 28. Oratorio at the opening of the organ by M[r] Bates.

August 29. Oratorio. Letter from M[r] Dechair in which I was nominated as the intended organist of an organ to be erected in the Octagon Chapel at Bath.

August 30. Another candidate for the organist's place played first, after which I also played. The Mess[rs] Bates and principal Gentlemen of the town were in the body of the church, and it was unanimously decided that I was to be their organist.*

August 31. I played the Halifax organ during service time.

October 18. Delivered a note to signify my intention to give up my place of organist at the end of the quarter.

Oct. 22. Dined at M[r] Walker's. He communicated to me a proposal from the gentlemen at Halifax to increase my salary if it would induce me to stay.

November 30. The last Sunday of playing the organ. For the 13 Sundays of my being organist I was paid 13 Guineas.

December 1–5. Halifax to London.

Dec. 8. Set off for Bath.

Dec. 9. Arrived at Bath.

Dec. 10. Was introduced to the Primate of Ireland.†

1767. January 1. A benefit concert at Rooms, the music chiefly of my composition. I had but little company but it was select. I performed a solo concert on the violin, one on the hautboy, and a sonata on the harpsichord.

Jan. 3. I set off for Leeds, Halifax &c. to do some business that required my presence and arrived again at Bath the 18th. I performed the journey on horseback.

Jan. 21. I began to take scholars.

* This does not quite agree with the account of the competition given in the paragraph in the *Pocket Magazine* already quoted. According to this there were seven competitors, and the performance of the second was so rapid, that Dr. Miller asked Herschel what chance he could have after that. "He replied, 'I don't know, I am sure fingers will not do.' On which he ascended the organ loft and produced from the organ so uncommon a fulness, such a volume of slow, solemn harmony, that I could by no means account for the effect. After this short extempore effusion he finished with the Old Hundredth psalm-tune which he played better than his opponent. . . . Having afterwards asked M[r] H. by what means, in the beginning of his performance, he produced so uncommon an effect, he replied, 'I told you fingers would not do!' and producing two pieces of lead from his waistcoat pocket, 'One of these,' said he, 'I placed on the lowest key of the organ, and the other upon the octave above; thus, by accommodating the harmony, I produced the effect of four hands instead of two. However, as my leading the concert on the violin is their principal object, they will give me the preference to a better performer on the organ; but I shall not stay long here, for I have the offer of a superior situation at Bath, which offer I shall accept.'" To this paragraph John Herschel added: "This anecdote is told by D[r] Miller. This gentleman perhaps speaks too strongly of the assistance he rendered my Father—but on the whole tells truth."

† Richard Robinson. Founded and endowed the Armagh Observatory in 1790.

Jan. 23. Letter from Mr Derrick, the Master of the Ceremonies, offering me a situation in the established band of musicians that played at the public subscription concerts, the Pump Room, the balls, the Play House &c. This I at first refused but some time after accepted, when I found that Mr Lindley the first musician in the place was one of this band and that like him I might be allowed to send a deputy when not convenient to attend personally.

March 28. Taken a house from the 25th March to Sept. 29 in Beauford Square.

April 5. Went into mourning for the death of my Father.

April 6. Went into the new house.

— 16. Mr Bullman, the appointed clerk of the Octagon Chapel, arrived from Leeds, at whose house I lived when I resided in that town.

June 24. My brother Jacob arrived at Bath.

— 29. The organ began to be put up.

July 15. I removed to another house in Beauford Square.

October 4. The octagon chapel was opened.

— 28. Oratorio in the morning, concert at rooms in the evening.

Oct. 29. Oratorio. In these performances I was the leader of the instrumental band and my brother Jacob was at the organ. The organ being thus opened, I attended it regularly every Sunday. Dr Dechair intending to introduce Cathedral Service, I had prepared a choir of singers and composed the required music for the purpose, which on account of its simplicity was generally approved of.

Nov. 17. The pump room music began, and concerts soon followed. I had many scholars.

Dec. 25. An anthem of my composition. I had instructed a set of singers.

1768. I was in full employment with public business and private scholars.

July 22. J. Dieterich arrived from Hanover. In the summer I attended resident scholars.

Sept. 24. The plays began and soon after all public business. Full of business to the end of the year.

1769. Besides public business I attended generally four, five or six private scholars every day.

June 1. Took a house in New King Street No. 7.

July. John [Dieterich] returned with Jacob to Hanover. In the summer I attended plays at Bristol.

December. My this year's receipts amounted to £316.

1770. The business as usual.

April. Paid half a year's rent, 15 Guineas, for a house at the north side of New King Street, No. 7, from Michaelmas 1769.

July 9. Alexander Herschel arrived at Bath, came to England with Jacob, who soon after returned to Hanover.

November. Mr Bullman lived at my house, he paid £10 per annum.

December. My this year's receipts amounted to £352.

1771. New furnished my house. Sometimes seven scholars per day.

Dec. My receipts, cast up at the end of the year, amounted to nearly £400.

1772. March 28. This week gave 39 lessons to ladies, the following week 38, other weeks little less.

May 13. Under my direction, a concerto spirituale was performed at the Octagon Chapel. A printed copy of the performance is preserved.

July 6. I set off upon a tour to Paris, Nancy en Loraine and Hanover.

July 11. Paris. At the Grand Opera I was much surprized to hear all the recitatives chanted like Cathedral music. The composition was by Lulli. There was an excellent orchestra of 60 musicians.

July 21. Nancy. I came to this place on a short visit to D^r and Mrs Dechair.

August 2. Hanover. I found all the members of my family in good health.

Aug. 16. Set off on my return to England in company with my sister.

Aug. 27. Arrived at Bath.

Sept. 1. Began again to teach my resident scholars.

1773. April 19. Bought a quadrant and Emerson's Trigonometry.*

May 10. Bought a book of astronomy and one of astronomical tables.

May 24. Bought an object glass of 10 feet focal length.

June 1. Bought many eye glasses, and tin tubes made.

— 7. Glasses paid for and the use of a small reflector paid for.

— 14. Boxes for glasses paid for. The hire of a 2 feet reflecting telescope for 3 months paid for.

June 21 to Aug. 23. Many glasses, tubes.

Sept. 15. Hired a 2 feet reflector.

— 22. Bought tools for making a reflector. Had a metal cast.

Oct. 2. Bought a 20 feet object glass and nine eye-glasses &c. Emerson's Optics. Attended private scholars as usual.

November 8. Attended 40 scholars this week. Public business as usual.

— 15. Attended 46 private scholars; nearly 8 per day.

1774. January. Gave 6, 7 and 8 private lessons every day.

March. Nearly the same number of scholars. Astronomical journal begun.

April. Specula grinding. Mirrors cast. Tools bought.

Midsummer. House at Walcot.†

August. Maps, glasses, putty &c. An astronomical time-piece.

September. Attended 6, 7, or 8 scholars every day. At night I made astronomical observations with telescopes of my own construction.

December 27. Bought an astronomical clock made by Field.

1775. By arrangement of the Marchioness of Lothian her own private concerts and those that occasionally were given by ladies of her party were modelled into a regular succession, so that one of them was to be held every Saturday evening at the house of one of the ladies who joined the party. As the music was chiefly to consist of the singing and harpsichord playing of my scholars I engaged only a sufficient accompanyment to make up a quartetto. Twenty of these concerts were given on so many successive Saturdays from November 1775 to March 1776.

1776. May 1. I observed Saturn with a new 7 feet reflector.

July 13. I viewed Saturn with a new 20 feet reflector I had erected in my garden.

1777. August 1. Went to Germany about my brother Joh. Dieterich.‡

Sept. 29. Moved to No. 19 at the south side of New King Street. Musical business carried on as usual. All my leisure time was given to preparing telescopes and contriving proper stands for them. I kept a regular account of any experiments of polishing.

1778. January. We had weekly concerts at Bristol, at the lower rooms at Bath and also at the New Rooms.

1779. January. I gave up so much of my time to astronomical preparations that I reduced the number of my scholars so as seldom to attend more than three or four per day.

February. A musical prize question being proposed in the Ladies Diary for this year, I sent a solution of it directed as required. My solution of the question was published in the Ladies Diary for 1780.

* The first note showing scientific tastes since February 1766.

† That is, he moved into a house near Walcot turnpike. (*Memoir of C. H.*, p. 37.)

‡ See *Memoir of C. H.*, p. 38. Dieterich ran away from home (W. H. missed him on the way), came to Bath, where he got musical employment and stayed there till 1779.

May. In summer we had concerts at Spring Gardens for which I composed glees, madrigals, songs and duets. Among them was the Echo Catch, which had a great run. All my leisure time was employed in grinding and polishing 7, 10 and 20 feet mirrors and making observations with them.

August 17. I began a regular review of the heavens with a very good 7 ft. reflector.

December. I moved to a house in River Street, and having no room for my 20 feet telescope I hired a convenient garden for it on the rising ground at the back of the Crescent.

About the latter end of this month I happened to be engaged in a series of observations on the lunar mountains, and the moon being in front of my house late in the evening, I brought my 7 feet reflector into the street and directed it to the object of my observations. While I was looking into the telescope a gentleman coming by the place where I was stationed stopped to look at the instrument. When I took my eye off the telescope he very politely asked if he might be permitted to look in, and this being immediately conceded, he expressed great satisfaction at the view. Next morning the gentleman, who proved to be D[r] Watson jun. (now Sir William), called at my house to thank me for my civility in showing him the moon, and told me that there was a Literary Society then forming at Bath, and invited me to become a member of it, to which I readily consented.

The Literary Society of Bath began their weekly Friday meetings about the middle of January. Papers were given in to be read by the Secretary as at the Royal Society of London. In the course of these meetings, during the time I attended the Society, 31 papers of mine were read, of which the copies are still in my possession. Public business as usual; I had not time to attend many scholars. 1780.

Feb. 16. In lent we had oratorios, on Wednesdays at Bath, and the same oratorios on Fridays at Bristol; the choruses were under my direction.

September. Public business as usual.

In the beginning of March I moved again to No. 19 in New King Street. 1781.

March 13. Discovered the Georgian planet.*

— 30. Two oratorios the 21st and 23rd. I still attended the meeting of the Bath Society.

April 10. In passion week we had four oratorios, two at Bath and two at Bristol.

May 3. I was at the Royal Society. I had many scholars and was engaged with preparation for casting a 30 feet mirror.

August 11. I cast the great metal.

October. Public business as usual.

Jan. I gave up much time to astronomy and also attended many scholars. Some of them made me give them astronomical instead of musical lessons. 1782.

March 20. In passion week we had four oratorios, two at Bath and the same two at Bristol.

April. Attending scholars by day and astronomical observations at night. I was informed by several gentlemen that the King expected to see me, and by my journal it appears that about the end of April I made out a list of celestial objects that I might show the King.†

May 19. This being Whit-Sunday one of my Anthems was sung at St Margaret's Chapel, when for the last time I performed on the Organ. ‡

* Inserted afterwards.

† "For the King. With 250, γ Virginis, π Bootis, 54 Leonis, Castor, α Herculis rather obscure and difficult, β Cygni, γ Andromedæ. With 500, γ Leonis. Compound, γ Virginis."

‡ There are three more pages of occasional notes (up to February 1787) which will be quoted in the next two chapters.

II.—Scientific Work at Bath

The indomitable energy and perseverance which had made the youthful Isaac Herschel overcome all obstacles and succeed in becoming a musician, was inherited by his brilliant son and enabled him to make good the deficiencies of his education, and to acquire the knowledge for which his mind longed. We have seen from his autobiographical notes, how he had worked for years at mathematics, physics and the theory of music, but the turning point of his life came during the first winter after his sister's arrival at Bath, when the study of Smith's *Optics* roused in him the desire to see for himself the wonders of the heavens described in that book.* In the spring of 1773, when the heavy, musical work was over for a while, he began to divide his leisure hours between attempts at making telescopes and continuing his reading, often, as his sister tells us, "retiring to bed with a basin of milk or glass of water and Smith's *Harmonics* or *Optics*, Ferguson's *Astronomy* &c., and so went to sleep buried under his favourite authors." He soon commenced keeping a careful record of his work in making telescopes, which he continued throughout his life till December 1818. It enables us to follow him step by step in the work which culminated in the 40-foot reflector and also made a comfortable income for him.† He describes the small beginning in the following words :—

"In the spring of the year 1773 I began to provide myself with materials for astronomical purposes. The 19th of April I bought an Hadley's quadrant, and soon after Ferguson's *Astronomy*. In May I procured some short object glasses and had tubes made for them, beginning with a 4 feet one of the Huyghenian construction. With this I began to look at the planets and stars. It magnified about 40 times. In the next place I attempted a 12 feet one and contrived a stand for it. I saw Jupiter and its satellites with it. After this I made a 15 feet and also a 30 feet refractor and observed with them. The great trouble occasioned by such long tubes, which I found it almost impossible to manage, induced me to turn my thoughts to reflectors, and about the 8th September I hired a two feet Gregorian one. This was so much more convenient than my long glasses that I soon resolved to try whether I could not make myself such another, with the assistance of Dr Smith's popular treatise on *Optics*.‡ I was, however, informed that there lived in Bath a person who amused himself with repolishing and making reflecting mirrors. Having found him out he offered to let me have all his tools and some half-finished mirrors, as he did not intend to do any more work of that kind. The 22nd September when I bought his apparatus, it was agreed that he should also show me the manner in which he had proceeded with grinding and polishing his mirrors, and going to work with these tools I found no difficulty to do in a few days all what he could show me, his knowledge indeed being very confined. About the 21st October I had some mirrors cast for a two feet reflector, the mixture of the metal was according to a receipt I had obtained with the tools. It was at the rate of 32 copper, 13 tin and one of Regulus of Antimony, and I found it to make a very good, sound, white metal. In the beginning of November I had other mirrors cast, among them was one intended for a 5½ feet Gregorian reflector, and as soon as they were ground and figured as well as I could do them, I proceeded to the work of polishing. About the middle of December I got so far as to give a tolerable gloss to some metals, and having advanced considerably in this work it became necessary to think of mounting these mirrors."

* "A compleat System of Opticks in four books, by Robert Smith, Professor of Astronomy and Experimental Philosophy at Cambridge and Master of Mechanicks to his Majesty," Cambridge, 1738, 2 vols. 4to.

† This record is in four volumes in folio, written in Herschel's own hand. As his sister made a complete copy of vol. i. (148 pp., 1773–1790), the original volume i. was in 1893, by his grandson Sir W. J. Herschel, presented to the University Observatory, Oxford.

‡ Caroline says that her brother wrote to inquire the price of a mirror for (she thinks) a five- or six-foot telescope. The answer was, that there were none of so large a size, but a person offered to make one at a price much above what H. could afford to give. (*Memoir*, p. 35.)

In the beginning of January 1774 Herschel put his 5½-foot mirror into a square wooden tube, but found the adjustment of a Gregorian telescope very troublesome. He therefore had recourse to the Newtonian form, and on the 1st March he commenced his astronomical journal by noting that he had viewed Saturn's ring with a power of 40, appearing "like two slender arms," and also "the lucid spot in Orion's sword belt." This was the modest beginning of the vast number of observations he was to make. The success of this instrument encouraged him to go on using the Newtonian construction, and he made a 7-foot telescope "with many different object mirrors, keeping always the best of them for use and working on the rest at leisure." * On the 1st May 1776 he "saw with a new one Saturn's ring and two belts in great perfection." At the same time he had worked at a 10-foot telescope, for which purpose he prepared several mirrors of 9 inches aperture, one with a hole in the centre, as he still intended to take up the Gregorian form. Having no concave tool he ground the convex tool with the speculum, using fine emery. To prevent the speculum from running "loosely and unevenly about" (as he ground without weight on the metal), he filed a pretty deep double cross in the brass tool, which made the speculum move much more evenly and take a proper hold of the tool. A pitch polisher 14 inches in diameter was next prepared and figured with the speculum, but it was soon found to be too large and reduced to 11 inches, though he shortly afterwards decided that a proportion of 4 to 3 would be better. Keeping the pitch soft during the polishing gave some trouble; he tried first laying the polisher in warm water, but this did not turn out a success, as the temperature was liable to change considerably; the addition of a little grease answered better. On 28th May 1776 he was able to use a 10-foot reflector with a power of 240 to view minute parts of the moon. In view of his preference of the "front view" in after years it is interesting to read under the same date: "I tried a 10 feet mirror without the small one, looking in at the front of the tube and holding the eye glass in my hand. I liked the method very well."

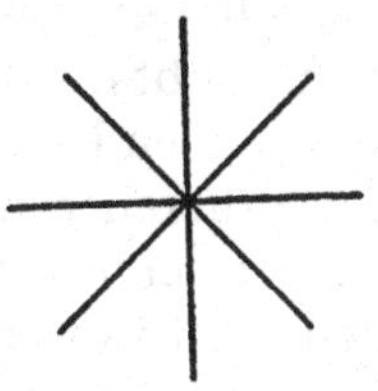

At the same time he was at work on a 20-foot telescope, and made three object mirrors for it. "My contrivances for stands and apparatus kept pace with the optical work. Indeed, with the assistance of my mathematical knowledge, the Optics of Dr Smith and Mechanics of Emerson, I found no difficulties but what I could get the better of." The observing Journal records an observation of Saturn made with a new 20-foot on 13th July 1776. This was at a house near Walcot turnpike, where Herschel had taken up his abode in the summer of 1774. There was more room there for work on telescopes, a place for observing on the roof, and a grass plot behind the house where the 20-foot was erected.

In 1777 the experiments on mirrors were resumed, and it is stated in the record (7th June) that Herschel had, "after numberless attempts," at last succeeded in polishing a 10-foot Gregorian telescope, using a tool with "star-cut gutters" like this. Having obtained "a tolerable gloss," and finding on trial that the figure was very good, he went on polishing. But this spoiled the figure, so that by the time he had a good polish an object gave three or four images. This he attributed to some circular strokes made to restore the polisher on the edges, and the figure was mended by very long straight strokes, so as always to bring the whole perforation of the mirror over the edges. This operation was in its turn found to have been continued too long, and it was concluded that "a circular motion ought sometimes to be used in order to unite all the straight cross strokes and thereby to take off a certain glare or streams of light that seem to run in a sort of glory from a luminous object." But the most valuable experience gained on that occasion was the following, which turned out to be of fundamental importance: "Being often at a

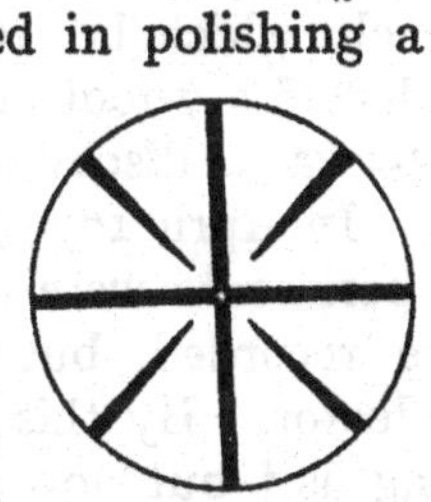

* See Caroline Herschel's *Memoir*, p. 36, for an account of how the house was turned upside down during all this work on mirrors and mountings. There is no record of telescope-making between March 1774 and April 1776.

loss for something to moisten the polisher to prevent its growing too bare, I found that Crocus mixed with water might be used at all times without making scratches and would prevent the speculum's catching hold of the pitch, which it will do if nothing but water be used."

In 1778 the records are very lengthy, extending from March to the end of November. He had now come across Mudge's paper on the making of reflecting telescopes (*Phil. Trans.*, vol. lxvii. p. 296), and he "prepared everything necessary for trying his method, of the success of which, from my own experience, I had great reason to doubt." In order better to ascertain the effect of different strokes he divided a 6-inch mirror by diaphragms into five rings, leaving a circle an inch in diameter in the middle, covered by the small mirror. He made innumerable trials to obtain the parabolical figure by using round strokes according to Mudge's directions. In the end he became convinced "that Mr Mudge's method of polishing will not succeed with Newtonian specula and that beginning with round strokes immediately destroys the figure of the mirror. Perhaps when a complete spherical figure is once obtained, a few round strokes may produce the so much wished for parabola. But to what end shall we attempt a parabolical figure before we can polish a spherical one? The difference in the focal length of the outside and inside rays of the metal which I am going to polish, if it came out spherical would be less than ·015; and if I had a speculum so far true it would make a most capital instrument. I have already too often found that when the central rays are shortened in focus, it is not easy to restore them to their proper length, whereas by a few round strokes they may soon be shortened when too long."

The experiments were therefore continued with redoubled energy, and Herschel found that the grinding with the brass tool should be persisted in until the figure had become as true as possible. When thus prepared, the mirror should be polished upon a polisher cut into squares and of a diameter not above 5 : 4 in proportion to the mirror. In July he polished a new 10-foot Newtonian with great success, using only strokes like these.

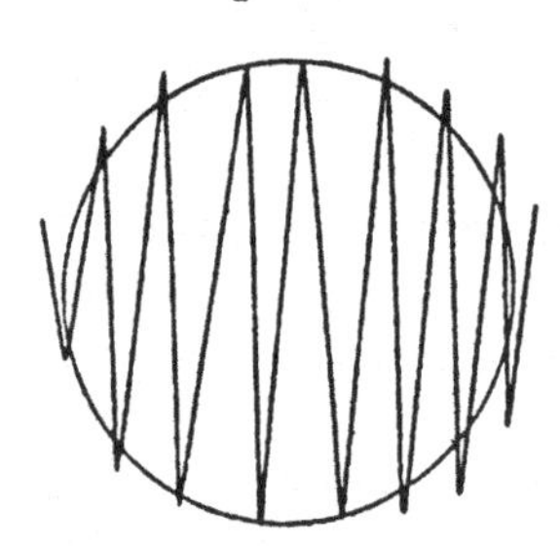

The speculum was not turned, but Herschel changed his place now and then, regularly, in order to go over the polisher in all directions. But every speculum seemed to require different treatment, and the work went on throughout the summer and autumn. Every possible combination of straight and round strokes was tried; the "glory stroke," directing the strokes over the centre of the polisher to different points of the compass like the spokes of a wheel, the "excentric stroke," avoiding the middle of the polisher, etc. It is not to be wondered at that he did not succeed in laying down any hard and fast rule for polishing mirrors by hand, since the half unconsciously acquired skill of an operator must accomplish much, which he would find it impossible to define in words. In November he finally re-polished a 7-foot mirror, and made it "a most capital speculum." The planet Uranus was afterwards discovered with it.*

In April 1779 Herschel started work again after the end of the Bath season, but the experiments were not so numerous this year as in the previous one. In 1780 even fewer are recorded, but in January 1781 he began preparations for constructing a 30-foot reflector. By this time he had made a good many mirrors of various apertures and focal lengths,† but now he thought that the experience thus gained justified him in attempting to make a very large instrument, and believed that he could manage a speculum 4 feet in diameter. The consideration of the expense of the undertaking, and particularly the brittleness of the speculum metal at that time generally employed, induced him to try if some

* For an engraving of Herschel's mountings for telescopes of this size, see Pearson's *Practical Astronomy*, pl. vi. fig. 6.

† In 1795 Herschel, speaking of his work at Bath, wrote that he had made "not less than 200 7 feet, 150 10 feet, and about 80 20 feet mirrors" (below, p. 485). But all these can hardly have been made at Bath; at least there is no sign of such a great number of mirrors in the polishing record.

other metal might not be suitable. The very high polish often given to domestic articles of iron or steel suggested experiments on these materials, and he got a 6-inch iron mirror cast, but it turned out to be very porous and would not take a very high polish. He procured a plate of wrought iron of the same size, brought it nearly to the proper figure but gave up the idea of polishing it, as it had not been well forged. Perhaps it was the failure of these experiments which decided him to be content for a while with a speculum 3 feet in diameter, and he next proceeded to experiment himself on various compositions of metal, as he found on inquiry that there was no chance of his getting a large mirror cast either in Bath or in Bristol. A furnace and melting oven were therefore built in a room on the ground floor of his house. The most successful of his experiments was one in which there was a percentage of 29 37 of tin, and he therefore decided on a proportion of 5 lbs. of tin to 12 lbs. of copper, the same which he afterwards continued to use for moderate sized mirrors (29·41 per cent.).

While this preliminary work was paving the way for the casting of the great mirror, a stand for the telescope was also being prepared. It differed from that afterwards adopted by Herschel for his 20- and 40-foot telescopes, and we therefore insert his description of it here :—" Its construction was singular but very simple. It would enable me in a long but narrow garden behind my house, the south view of which was clear from buildings, to observe the heavens in the meridian from about 10° above the horizon up to the zenith, and to view in general all the stars that were at a great altitude. With my 20-feet telescope I use a long pole, to the top of which is fastened a very short arm holding a set of pulleys, and when the telescope is elevated by them I use a moveable ladder with a back to mount up to the eye-piece.* Instead of this pole and ladder I here unite both in one contrivance. It consists of three stout long poles, well fastened in the ground by brick-work, at the distance of four feet from each other, so as to make an equilateral triangle. They are joined at the top by a strong circular cap, in the centre of which is an iron pivot ; on this moves an arm that comes out beyond the circle, and at the end of the arm is a set of pulleys by which the front of the telescope may be brought up to any altitude. Within the three poles is a moveable platform properly railed in, which may be raised up and fixed at a proper height, so as to bring me opposite the eye-piece. The tube is intended to rest below in the same manner as the 20 feet telescope, and to have the side motion brought to my hands by cords in a similar way. The fine vertical motion is also to be brought under the hand, as I have it in the 20 feet instrument."

As the mirror would be too heavy to be lifted on to the polisher or off it by hand, arrangements had to be made for doing this by a crane. On the back of the mirror there was therefore cast a ring concentric with it, 24 inches in diameter, 1½ inches high and 3 inches broad, which would also help to strengthen the mirror ; it was provided with twenty-four holes through which ropes might be passed, meeting a little distance above and fastened together in an iron ring, by which the mirror might be lifted. The entire weight of the metal was to be 469 lbs. 15 ozs. A mould was made of loam, annealed by filling it with charcoal which was burned in it. On the 11th August 1781, the casting took place. The metal ran into the mould " very quietly," but when nearly full, a small crack near the bottom of the mould began to let some of the metal out, which caused a great deficiency in one side of the mirror. The disappointment, however, did not last long, as the mirror cracked in two or three places while cooling.

Nothing daunted, the indefatigable worker immediately began to prepare to cast another mirror. Believing the metal of the first one to have been too brittle, he made the percentage of tin somewhat smaller, 5·82 ozs. to a pound of copper (or 26·67 per cent.). The total amount of metal was 537·9 lbs. But before the metal was sufficiently fluid for casting, some of it began to drop through the bottom of the furnace into the fire. " The crack soon increased and the metal came out so fast that it ran out of the ash hole which was not lower

* This was the mounting of his earliest 20-foot telescope of 12 inches aperture.

than the stone floor of the room; when it came upon the pavement the flags began to crack and some of them to blow up, so that we found it necessary to keep at a proper distance and suffer the metal to take its own course." *

As already mentioned, Herschel began on the 1st March 1774 to keep a record of what he saw in the heavens. But if he always noted down what he looked at, it would seem that he must for some years have spent more time in making telescopes than in using them. In several of his later papers he quotes some of these early observations, *e.g.* in his paper of 1789 on Saturn, where observations of the ring and the surface markings from 1774 and following years are quoted.† In May 1776 he tried a new 10-feet telescope (with a power of 240) on the moon and was at once "struck with the appearance of something I had never taken notice of before and which I immediately took to be woods or large quantities of growing substances in the moon." The Mare Humorum he thought was a forest, and he even believed he could see the shadow cast by the trees at the edge of the wood. The so-called seas, he says, have hitherto been supposed to be a kind of soil that reflected the light less copiously; but "my observation proves them all to be forests." The next night he noted that as he could not use the 10-foot telescope nor any power higher than 150, he could not see any woods; and after that there is no further mention of them, and he does not seem to have devoted much time to the moon until he commenced to measure the heights of lunar mountains in 1779. In April 1777 he began to make sketches in pen and ink of the surface of Mars, and the following year of Jupiter, all of which were utilised in his paper of 1781 on the rotation of these planets.‡ The sketches are rather rough and show very little detail; Herschel was not a draughtsman and made no pretence of being one, but his sketches were quite sufficient for the use he made of them. The "lucid spot in Orion" he examined frequently, but here his want of skill in drawing prevented him from making observations of any importance. Gradually he settled down to the steady pursuit of sidereal astronomy, which was soon to become so fruitful: in October 1777 he turned his attention to the much-neglected variable star Mira Ceti, and three months later he got the idea of determining the annual parallax of stars by measures of close companions, which gradually led him to search for and measure hundreds of double stars.§ He must at first have thought that parallactic displacement might be visible even without measures, for on the 12th of March 1778 he writes: "To my great disappointment I found the stars in the tail of Ursa major just as I saw them three months ago, at least not visibly different." There is, however, not any record of what he had seen three months earlier, though he had in January commenced looking for companions to bright stars.

Hard at work as a musician as Herschel was during the greater part of the year, instructing numerous pupils and from 1776 directing the public concerts at Bath, and hard as he worked in his leisure hours at polishing mirrors for telescopes, and using them occasionally at night, he now began to be active in yet another field, as a writer. His first attempt was an answer to the following prize question proposed by Landen in *The Ladies' Diary* for 1779:

"The Prize Question, by Peter Puzzlem.

"The length, tension, and weight of a musical string being given, it is required to find how many vibrations it will make in a given time, when a small given weight is fastened to its middle and vibrates with it."

* The above account differs in several particulars from that written from memory by Caroline Herschel many years later. According to her, it was the second mirror which was cracked in the cooling, and she seems to think that the casting took place in the spring of 1782. She writes: "The stone flooring flew about in all directions, as high as the ceiling; my poor brother fell, exhausted with heat and exertion, on a heap of brickbats." (*Memoir*, p. 44.) Compare below, p. 486, at top.

† This edition, *infra*, pp. 370 and 377.

‡ *Infra*, pp. 20 *sq*.

§ A diagram and investigation similar to those given below on p. 53 are found in the Journal between 26th and 28th January 1778.

March 12. 5h. 45' in the morning

Mars seems to be all over bright but the air is so frosty & undulating that it is possible there may be spots without my being able to distinguish them. Nr 4. 20ft.

53' I am pretty sure there is no spot on Mars

---

the Shadow of Saturn ~~sig~~ lays at the left upon the ring

Tuesday March 13

Pollux is follow'd by 3 small stars at abt 2' and 3' distance.

♂ as usual. ρ H

---

in the quartile near ζ Tauri the lowest of two is a curious either Nebulous star or perhaps a Comet.

~~preceding the star that preceeds ν Geminorum double about 30".~~

a small star follows the Comet at 2/3 of the field's distance

~~Σ 7. 33.~~

FROM HERSCHEL'S JOURNAL.
Slightly reduced in size.

[*To face page* xxix.

The question was "Answered by M^r Wm. Herschel" in the same publication for 1780. The result is only a very rough approximation; the increase of period (expressed in modern notation) is stated to be in the ratio $\left(1 + \frac{\pi}{2}\frac{\delta M}{M}\right) : 1$, but the short note finishes with the reservation that "the above solution is not to be considered as mathematically true, but as a practical solution approaching near the truth, since the chord loaded with a single weight cannot vibrate exactly in the same curve as when unloaded, nor as when uniformly loaded throughout."

The same year in which this first slight attempt at authorship was made, was that in which Herschel made the acquaintance of Dr. William Watson, jun., of Bath,* which again led to his joining the new Philosophical Society of Bath. The weekly meetings of this Society gave him the chance of meeting men interested in science, and encouraged him to prepare papers giving accounts of various physical experiments and of some of his astronomical observations. The Society only lasted a couple of years and did not publish any Memoirs or Proceedings; but in the history of English science it deserves to be remembered with gratitude, as it undoubtedly was of great use to Herschel at the opening of his scientific career. In all, he communicated thirty-one papers to the Bath Society, none of which have ever been printed till now, except the few which, in 1780–81, were also communicated to the Royal Society at the instance of Dr. Watson. Most of the others are given in the present volume, and will be read with interest, as giving an insight into the versatile mind of Herschel before he had quite settled down to the work of his life.

In the beginning of 1781, Herschel had thus commenced to make a name in the scientific world, as the three papers which he had sent to the Royal Society showed him to be a careful observer with instruments made altogether by his own hands. He was now to make a discovery which drew the attention of astronomers all over the world to the amateur astronomer at Bath. And this was not a lucky accident, but a discovery which was bound to be made sooner or later by an observer who searched the heavens as systematically as he did. He had already made a review of the stars of the first four magnitudes with a 7-foot Newtonian of 4½ inches aperture,† when he started on the 17th of August 1779 to make a second review of the heavens. This was made with a 7-foot telescope of 6·2 inches aperture and power 227; it extended to all the stars in Harris' maps and telescopic ones near them as far as the eighth magnitude. The main purpose of this review was the registering of double stars, but it led to a result which had never been dreamt of by anybody. It is recorded in his Journal in the following manner:—‡

"Tuesday, March 13.

"In the quartile near ζ Tauri the lowest of two is a curious either nebulous star or perhaps a comet. A small star follows the comet at ⅔ of the field's distance."

At that time Herschel had never come across a planetary nebula, which was perhaps fortunate, as he might have taken the stranger for one of these bodies, noted its place, and not looked it up again for some time; in which case it might have figured for years as a lost nebula, like the lost stars of Flamsteed. He observed Mars and Saturn on Thursday morning between five and six o'clock, but the sky had probably not been clear early enough to allow him to look for the comet. But next we read:—

* See above, p. xxiii.

† See his paper on the proper motion of the sun, *infra*, p. 109. This review is not mentioned in his Journal as a systematic undertaking, and it only led to his finding a few double stars.

‡ Reproduced in facsimile on the plate opposite (slightly reduced in size). The two vertical lines down the middle means that the observations have been copied into the separate books for fixed stars, planets, etc.

"Saturday, 17th March 1781. 11^h.

"I looked for the Comet or Nebulous Star and found that it is a Comet, for it has changed its place. I took a superficial measure 1 Rev. 6 parts and found also that the small star ran along the other wire. Exactly measured 1 Rev. 12 pts. both dia^{rs} included

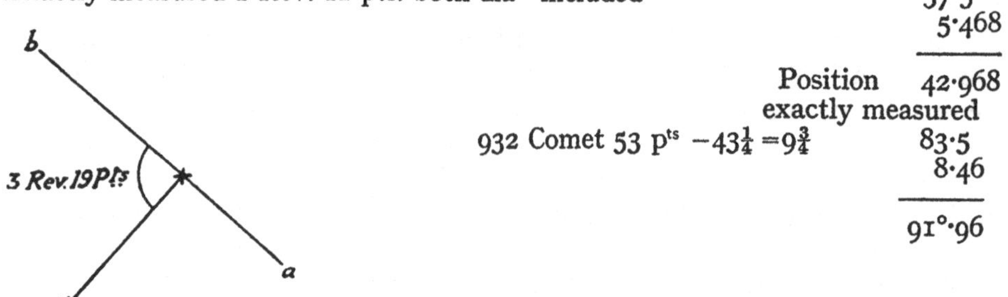

"When I took off the rays, by a strong light 51 parts & this is probably too large." *

This is the first determination of the "Comet's" position made by Herschel. He now followed it regularly, and his observations are given in the paper which was read before the Royal Society on the 26th April.†

The discovery was very soon communicated to the only public observatories then existing in England, at Greenwich and Oxford. Maskelyne wrote on the 4th April to Dr. Watson that he had for the last three nights observed stars near the position pointed out by Mr. Herschel, whereby he was enabled on the 3rd to discern a motion in one of them, which convinced him that "it is a comet or new planet, but very different from any comet I ever read any description of or saw." On the 23rd April he wrote to Herschel: "It is as likely to be a regular planet moving in an orbit nearly circular round the sun as a comet moving in a very excentric ellipsis. I have not yet seen any coma or tail to it." Hornsby wrote on the 24th that he had searched for the comet, but could not find anything like a comet except an object which turned out to be a small cluster; he asked for further particulars. A week later he had not yet found it, but on the 14th April he wrote that he had found it immediately after receiving Herschel's last letter, and had in fact observed it on the 29th and 30th March "unknowingly." He adds: "I do not in the least question but this is the comet of 1770, but whether it has passed its Perihelion or has not yet come to it, is more than I can say at present. I will very soon try to construct its orbit."

It is not necessary in this place to give an account of the attempts made to calculate a parabolic orbit for the new "comet." That the new star was at a very great distance, became evident after some time, though Méchain had at first computed a parabolic orbit with a perihelion-distance of 0·46, the comet being about to pass the perihelion on the 23rd May. These elements were communicated to "Monsieur Hertsthel à Bath" by Messier, who wrote to the discoverer to express his wonder as to how he had found this stellar object, the motion of which could not be recognised in the course of a night. Herschel replied at once, explaining that it was by its appearance that he had distinguished the comet from a star. Later on Lexell computed a perihelion distance of 16, and announced that the perihelion would not be reached till April 10, 1789. It is difficult to decide who was the first to announce publicly, that the star moved in an orbit of small excentricity and at a distance about twice that of Saturn; but it appears that Saron, Lexell, and Laplace found this independently. Herschel had himself imagined that his observations made in March and April showed a considerable parallax. His paper as printed in the *Philosophical Transactions* contains a paragraph "Remarks on the path of the comet," ‡ in which he points out that the apparent distortions

* The diagram is exactly like those which accompany most of Herschel's measures of double stars; *ba* is always the parallel.

† Below, p. 30.

‡ *Infra*, pp. 36, 37.

in his figures of the comet's path were simply caused by the inevitable errors of his measures.* But this had not been his opinion at first. His paper as originally presented to the Royal Society did not contain the paragraph just mentioned, but instead of it another and longer one entitled "Remarks on the diurnal parallax of the Comet." † Looking over the delineations he had made of the comet's path, Herschel says he had found a certain irregularity for which he could not account. He shows first that it could not arise from refraction and next considers whether it might be caused by parallax. On March 18 "we find the comet elevated in the early observation and depressed in the late measurement of the same evening. The difference of parallax in about 3½ hours seems here to amount to about 4″, to which we must add nearly 1″ for the contrary effect of refraction being taken off." On the 25th the distance in the late measure was greater than it ought to have been by 5 or 6″. "For in this situation the whole effect of parallax would fall on the distance alone; an addition of about ¾″ should be made for the effect of refraction. . . . March 26 was a late and March 28 an early observation, this accounts for the apparent deviation of the comet from the regular path. For we find the distance too great on the 26th and the angle $A\beta$ comet too small on the 28th, as upon the principle of parallax they should be." He therefore took great care on the 16th April to obtain accurate measures as early and as late as possible, and the result was in his opinion decisive. "I cannot hesitate to say that no other cause but parallax could give that evident deviation from a regular path which in this observation amounted to no less than 11″." Finally he says that the observed diameters of the comet agree with the observations of its place in pointing to its having a considerable parallax.

In a letter to Hornsby of May 21, 1781, Herschel writes that, without having entered into calculations, he supposes that a parallax of not less than 10″ nor much more than 20″ would follow from his observations. This was very soon disproved by more accurate observations made elsewhere, the cause of the discrepancies was found, and the paragraph in question was not printed. ‡

At the time when Herschel was laying the foundation of the great skill which he gradually acquired in making specula, and enlarged the solar system by the first discovery of a planet since the beginning of history, there were two other amateurs in England who were engaged in telescope-making, Michell and Edwards, and as it has been asserted that from the former "Herschel received his first lessons in speculum-grinding," § it seems desirable to show that there is no foundation whatever for this statement.

On the 12th August 1780, seven years after Herschel had commenced to make telescopes, Priestley wrote to him as follows :—

"Dear Sir,—I do not keep up a regular correspondence with Mr Michell, but if you will be so good as to send me an account of the *construction* and *effects* of the telescopes you have made, and what farther views you have in the same way, I will not fail to take an early opportunity of writing to him to procure an account of what *he* is doing, and when I have his answer, shall lose no time in transmitting it to you."

Herschel was (not unnaturally) curious to know something of the methods adopted by a distinguished man like Michell. Probably he failed to hear anything through Priestley, as there are no other letters from him to Herschel in existence. But on the 21st January 1781 Michell wrote to Herschel's friend, Dr. Watson, on the same subject :—

"I look upon myself as very much obliged to you for your favour from Bath, and particularly for the very interesting account, both of what Mr Herschel has done and what he has seen, both of

* Compare *infra*, p. 171, where he explains that the errors were caused by a fault in the micrometer.

† Vol. ii. of the "Archives" Series of the Royal Society, No. 12.

‡ In a letter of February 26, 1782, Hornsby expressed his admiration of Herschel's diligence in measuring double stars, but threw some doubt on the value of his micrometers, since his measures had led him to place the new comet below the orbit of Mars. Compare below, p. xxxv.

§ *Dict. of National Biography*, article "Michell," quoted by several writers. Michell left Cambridge in 1767 to become Rector of Thornhill in Yorkshire, while Herschel had left Yorkshire the year before.

which seem to be very important. I shall be very happy if I should be able to succeed as well, or near as well, as from your account he seems to have done, and I shall be very glad of the favour of his correspondence; at the same time I think it very probable that I may be more likely to learn from him what may be useful to myself, than he to learn anything from me."

Michell then remarks that almost all theoretical telescope-makers have supposed that the figure of their specula would come out perfectly spherical; but this is not the case, as all concave specula have a tendency to recede from the spherical form the way they ought to do in order to make them parabolical, as they grind and polish away a little more towards the edges than they do in the intermediate parts. He mentions that he has lately had a 29½-inch mirror cast, the focal length of which, when finished, will be only 12½ or 13 feet. Neither in this letter nor in a second one to Watson, dated the 23rd February 1781, is there anything about grinding or polishing. The second letter ends, "With compliments to Mr Herschel, though I have not yet the pleasure of a personal acquaintance with him."

Up to that time, therefore, Michell and Herschel had never met, nor can they have exchanged a letter; and this sufficiently disposes of the story of Michell having given Herschel any information about making specula. There is one letter extant, written by Michell to Herschel, and dated from Thornhill, 12th April 1781, in reply to two letters written by the latter. As Herschel was in the habit of preserving all letters dealing more or less with scientific matters, it seems certain that he never received another letter from Michell. The letter is of considerable length (3½ pp. foolscap, closely written); it deals with the forms of Gregorian and Cassegrain mirrors and their relative merits for large and small apertures. It is written in a slightly superior tone, and it is possible that Herschel did not care to continue the correspondence. This letter also gives no information about grinding or polishing.*

At the same period of his life Herschel also got into correspondence with the Rev. John Edwards, the author of several tracts on making and using reflecting telescopes. Herschel had tried the composition of metal recommended by Edwards, and when the latter heard it from Maskelyne, with the addition that Herschel had found it very white and beautiful but wished that it was not so brittle, he wrote in April 1782: "If you will be pleased to add a small portion of brass and silver, you will find it will be still whiter and not so brittle. Copper 32 lbs., tin 15 to 16 lbs. (according to the purity of the copper), brass, silver, and arsenic, each 1 oz." Herschel failed in casting two 6½-inch specula of this composition, and Edwards suggested that perhaps he had let them cool too fast. He had himself never had a failure, but on the other hand he had never cast one larger than 4½ inches. He did not succeed in persuading Herschel to adopt the additional ingredients.

In November 1781 Herschel received the Copley medal, and on the 6th December he was elected a Fellow of the Royal Society.† His paper on the parallax of the fixed stars was laid before the Society in December, and his first Catalogue of Double Stars a month later, while he continued his observations of these objects throughout the winter. But much as his discoveries of a new planet and of numerous double stars were admired, there were some particulars in his papers which scientific men in London found it hard to accept.

* After Michell's death in 1793 his son-in-law wrote to Herschel that he wished to dispose of his largest telescope. The tube was 12 feet long, made of rolled iron, the diameter of the speculum 29 inches, focal length 10 feet, "it is now cracked." There were also eight small concave mirrors of 3¼ to 5 inches aperture. Herschel bought the telescope and also a polishing tool for a 10-foot mirror, and used it in 1799. Probably it is this trivial circumstance which gave rise to the myth about Michell having given him lessons. (See below, pp. lv and lx.) The 29-inch speculum is no doubt the one alluded to by Dr. Watson in two letters in 1785; he had heard from different people that it was a success "and performs extremely well upon day-objects." Herschel saw the telescope at Thornhill in 1792, and bought it during a short visit there in 1793.

† By resolution of the Council he was exempted from payment of the usual admission fee and annual subscription.

In several letters written from London in December 1781, Watson told Herschel that people did not believe in the enormously high magnifying powers he had used, and he gave some sound advice about the necessity of stating how they were determined. Maskelyne and Aubert (a well-known amateur astronomer) had said that they never saw fixed stars round and well defined. In reply Herschel wrote to Watson, on the 7th January 1782, as follows :—

" From the contents of your letter I begin to have a much better opinion of my own observations than I had before. I thought what I have seen had been within the reach of many a good telescope, and am surprised that neither Mr Aubert nor Dr Maskelyne have seen the stars round and well defined. I do not say without the least aberration, for so far I will not go even with Jupiter or Saturn ; but that I have a thousand and a thousand times seen them (with 460) as well defined as I ever saw Jupiter (with 227) I am very well convinced of to myself. And I believe till these gentlemen can see them so, they will not be able to find that ζ Cancri, *h* Draconis &c. are double (I mean the preceding of the two stars of ζ Cancri), for the aberration of one of the stars will efface the other star, or make them appear as one. I make no doubt of soon being able to mention a good list of respectable names to ascertain a great many of the facts I have reported ; but there will always be a few that cannot be otherwise than so difficult to prove to another person that we must not be surprised to see them doubted, if my own observations are not sufficient to make them believed. I do not suppose there are many persons who could even find a star with my power of 6450, much less keep it, if they had found it. Seeing is in some respect an art which must be learnt. To make a person see with such a power is nearly the same as if I were asked to make him play one of Handel's fugues upon the organ. Many a night have I been practising to see, and it would be strange if one did not acquire a certain dexterity by such constant practice.

" When I said that *I surmised* the diameter of α Lyræ, &c.,* I did not speak without a very deliberate consideration, of which I will give you a little abstract. I have seen α Lyræ with the naked eye and with a power of 200. The naked eye may be taken to have an aperture of 2 tenths of an inch (in the dark) if not more. Now if the diameter of a speculum that magnifies 200 times be 6 inches, and if we suppose that no light at all were lost in the reflection, then we shall have just 30 times the diameter of the aperture of the eye for the diameter of the speculum, and the light in proportion as 900 to one, which being scattered over an apparent disk 40000 times as large ought to make the star appear very dim. Now *by experience* I find the contrary to be true. The star appears considerably brighter than with the naked eye. Again, I have seen the star magnified 200 times and have also seen it magnified 32 times as much, that is 6400 times, with the very same aperture. Therefore in the latter case the same quantity of rays were diffused over a surface full 1000 times as large as in the former, yet did I not see the least sensible deficiency of light ; I do not know whether it did not appear rather brighter than before. Now the step from 6450 to 100000 is (by far) less than from 200 to 6450. Therefore *I surmise* (and to you, Sir, who are a philosopher, I will say *I am sure*) this star would bear a power of 100000 with no more aperture than 6 inches ; had I a telescope that was capable of such a power, and could I have my telescope made to keep pace with the diurnal motion of the earth, so as to keep the star in view. However, if you think it better not to mention this circumstance I shall not have the least objection to any alteration. That *surmise* may appear strange, when my reasons for it are not known.

" When I said that I had always found the diameters of the stars to measure less and less the more I magnified, I mean it literally as I expressed it, that is, they really measured less and not proportionally. For instance (star 30, 6th class) α Aurigæ with 227 measured 2″·5 ; (star 66), α Tauri with 460 measured 1″ 46‴, with 932 it measured 1″ 12‴. α Lyræ with 6450 measured no more than 0″·3553. This proves that they actually measured less, and so indeed they ought ; for the real diameter I should suppose (I speak merely at random) must measure much less than 1‴, and most probably, would we magnify 100000 times, we should then reach the *real* diameter, which would put an end to the changeable measures, and any still higher power we should apply would always give the same real measure."

To Aubert, Herschel wrote two days later, sending him two maps with the first five classes of double stars marked, and asked him to confirm his observations of them. He continued :—

" It would be hard to be condemned because I have tried to improve telescopes and practised continually to see with them. These instruments have played me so many tricks that I have at last found them out in many of their humours and have made them confess to me what they would have

* See below, p. 81.

concealed, if I had not with such perseverance and patience courted them. I have tortured them with powers, flattered them with attendance to find out the critical moments when they would act, tried them with specula of a short and of a long focus, a large aperture and a narrow one; it would be hard if they had not been kind to me at last."

After urging Aubert to get a set of eye-pieces made for his $3\frac{1}{2}$-foot refractor, magnifying 200, 250 . . . 500 times, and assuring him that he would see the stars round and (on fine evenings) well defined, he added:—

"If I say too much or am too particular, pray excuse it, Sir, but judging from myself I know that no person could give me more real pleasure than by entering into such minutiæ as I know to be generally of considerable consequence when we come to refinements and want to *screw an instrument up to its utmost pitch* (as you are a harmonist you will pardon the musical phrase). . . . You will soon hear the description I am now drawing up of the last new Micrometer, which occasion'd you obligingly to exclaim that 'we would go to Bedlam together.' I shall have no objection to any place in such astronomical company, but pray, Sir, do not let me be sent there by myself, which I fear will be the case, unless you procure me a reprieve by verifying some of the facts I have mentioned, and by that means make it appear that they may *all* be true."

Again, on the 28th January, Herschel wrote to Aubert:—

"When you want to practise seeing (for believe me, Sir—to use a musical phrase—you must not expect to see at sight or *à livre ouvert*) apply a power something higher than what you can see well with, and go on increasing it after you have used it some time. These practices I have repeated upon Castor, ϵ Bootis &c., increasing the power by degrees till I fairly could not distinguish the least appearance of the two stars, and they had run perfectly into each other. The consequence of this was that every time I tried again, my eyes acquired more practice, and I can now see with powers that I used to reject for a long time."

The outcome of the scepticism with which Herschel's accounts of the high powers he had used had been received, was a letter to Sir Joseph Banks, dated the 28th March 1782, and printed in the *Philosophical Transactions*, in which he described how the magnifying powers of his eye-pieces had been determined.* But though he did not quite escape criticism, his achievements were so considerable as to attract a great deal of attention even among the general public. We have seen from his autobiographical memoranda that he gradually sought to release himself from some of his musical engagements; and the time was now coming when he was to be enabled to devote himself altogether to science. He says that he was "informed by several gentlemen that the King expected to see him." One of these intimations is preserved among his letters, and deserves to be inserted here on account of its historical importance.†

"Chesterfield Street, 10th May 1782.

"Dear Sir,—In a conversation I had the honour to hold with His Majesty the 30th ult° concerning you and your memorable Discovery of a new Planet, I took occasion to mention that you had a twofold claim as a Native of Hanover and a Resident of Great Britain, where the Discovery was made, to be permitted to name the Planet from His Majesty. His Majesty has since been pleased to ask me when you would be in Town, to which I could not certainly answer. I am now setting out for Taunton, where the Worcestershire Regiment at present is quartered and shall endeavour to see you in my way. I yesterday discoursed with S[r] Joseph Banks on this subject, who has the same sentiments with me on the matter. I remain Dear Sir your very obedient humble servant,

"John Walsh.

"M[r] Herschel."

* See below, p. 97.

† The writer is mentioned by Caroline Herschel as Colonel Walsh (*Memoir*, p. 43). He was evidently in the Worcestershire Militia.

This looks as if the initiative was taken by Herschel's friends and not by the King, although there would have been nothing remarkable in the King's having desired to see Herschel and one of his telescopes, as the King was (or at least had been) much interested in astronomy, and had, in 1768–69, built an observatory for his own use in the Deer Park at Richmond, and equipped it with a transit instrument, a mural quadrant, a zenith sector, and other instruments.* What happened next is told thus in Herschel's Memoranda :—

" Having packed up my 7 feet telescope, I went the 20th of May to London on a visit to Dr Watson sen. who lived in Lincoln's Inn Fields, and on the Saturday following I had an audience of His Majesty, who received me very graciously. The King said that my telescope in three weeks was to go to Richmond and mean while to be put up at Greenwich. Some letters to my sister at Bath which are still in her possession contain an account of what passed at this and other audiences I had of the King." †

" May 29. My telescope having been carried to Greenwich, I prepared it for Dr Maskelyne's observations."

" June. Not only Dr Maskelyne, but every gentleman acquainted with astronomical telescopes who observed with mine during the course of all this month that it remained at Greenwich declared it to exceed in distinctness and magnifying power all they had seen before.‡ My telescope having undergone this kind of examination, the King desired it to be brought to the Queen's Lodge at Windsor."

" July 2. I had the honour of showing the King and Queen and Royal Family the planets Jupiter and Saturn and other objects." §

When Herschel had been some weeks in London, Dr Watson wrote to him from Bath,‖ reminding him that Banks had promised to " do his best endeavours with the King." But if he should be prevented from doing so, Herschel ought to speak to the King himself, to show that he would like to take the post at the Kew Observatory, vacant by the death of Dr Demainbray (Watson did not know that Demainbray's son had already been appointed to the post). Otherwise the King might think that he was in flourishing circumstances at Bath, as he might have heard " of musicians gaining very high profits, and much above what he would chuse to confer on his Astronomer."

On the 23rd June Watson wrote that he rejoiced to hear that Herschel's affairs were " in the most flourishing train," and that he hoped H. had informed the King of the trial his instrument had undergone.

" I think you should take great pains to get at Mr Hornsby as soon as ever he comes to Town. I take it for granted that at present he lies under strong prepossessions against you—not only with respect to your high powers, but thinks you have been guilty of a blunder with respect to your measurement of the distance of the new star from the fixed stars near it. You should as soon as possible get at him and inform him what has passed, as likewise why you are sure that you are not so much out in your measurement as appeared to him. It would not be amiss that Mr Aubert be desired likewise to set him aright in these particulars as soon as possible. As the King seems inclined to abide by his report, it is of the highest importance that Mr Hornsby should have all his prejudices removed. This is the only circumstance that can now stand in your way, and it behoves you to remove it by your exertions. I am surprized not to hear that the interview at Kew is settled, at the same time I cannot but be glad it is delayed, and I could wish it had not to take place till Mr Hornsby has been made a convert of."

* See "History of the Kew Observatory," by R. H. Scott, *Proc. Roy. Soc.*, vol. xxxix. p. 37 (1885).

† *Memoir of C. Herschel*, pp. 45–49.

‡ The journal shows that H. observed at Greenwich on 29th May, 1st, 2nd, 11th, 14th, 15th, 16th, and 19th June. A number of double stars were found, and these and others measured on these occasions. (See below, p. 174 *sq.*) On the first three evenings H., Maskelyne and the assistant, Mr. Linley, compared H.'s telescope (on double stars) with an achromatic of 46 inches focal length, and a reflector by Short of 6 feet focal length, very much to the disadvantage of the two latter instruments. On 15th June Mr. Playfair, Mr. Aubert and others took part in the observations.

§ Double Stars (I. 36, 37, IV. 61, V. 75, 76, VI. 93, 94) were found and measured at the Queen's Lodge, showing how Herschel never lost a moment.

‖ The letter is only dated "Sunday, June 1782."

On the 29th June Watson wrote that he was glad to see that Herschel recognised the necessity of applying to the King, as the King could not be expected to offer before he knew that his offer would be accepted. He was sure the King must be disappointed at not having received an application.

Herschel must have followed his friend's advice soon afterwards, as Watson in the next letter (of 14th July) expressed his satisfaction at hearing that the King would make him "independent of music," and hoped soon to hear how it was to be done. He had learned from Herschel's letter that "the business of the dedication of the new star is next to be done," and he thought novum sidus Georginum a very good name, though perhaps sidus Georginum was sufficient. About this matter he wrote six days later the following letter :—

"Bath, July 20, 1782.

"My dear Friend,—I will now tell you the result of my consultation with our friends M[r] Collings and M[r] Webb. In the first place we think the star should be called not Georginum Sidus, but Georgium Sidus, in the same manner as Horace Liber I, Ode XII,

"Micat inter omnes
Julium Sidus—

"Mr Webb recommends that either in the print or at the bottom with some mark referring to the star the words Georgium Sidus should be written, and under it these words, 'jam nunc assuesce vocari.' The quotation is taken from the first book of the Georgics, line 42, where Virgil after invoking Cæsar as a future God among other things tells him he must now accustom himself to be call'd upon with vows, or, as Dryden has it, 'And use thyself betimes to hear our prayer.'* You see that the words as we apply them simply imply 'to be call'd'—but you know that upon such occasions it is a common and allow'd of practice to make a quotation and apply it in a different sense. M[r] Collings further recommends that instead of saying more at the bottom of the print, you write at the top of the paper which is to accompany the print :—

"'An Account of the new Star, discover'd . . . . . . and in honor of the best of Princes, Patron of Arts and Sciences nam'd Georgium Sidus by W. H., F.R.S.'

"With respect to what is usually called a dedication, I believe they agree with me in not understanding what the word can mean when applied to a star. † Such are the hints which I have collected from my friends, you are at liberty to make just what use you please of them either to use them in whole or in part or reject them altogether. I confess the quotation from Virgil pleases me and is as good a one as could be expected on the subject.

"I hope to hear very soon of your establishment. I think it is a long time about. But the King must now, I think, be much harrassed by the confusion things are in, since the new ministry have done us the favour to take our affairs into their hands.

"I am Dear Sir yours most sincerely W. WATSON.

"I wrote this according to the date, since which I had the pleasure of receiving yours of the 18th—and am sorry to find that you are not likely to be settled at Kew. I am likewise a little fearful that your appointment may not be so considerable as your friends could have wished it. On the other hand if we had known that Demainbray's son had succeeded his father, our expectations would not have reached even this arrangement, especially if we consider the King's present situation. With respect to Mr Webb's quotation I should have observed that this was said by Virgil on a somewhat similar occasion, viz. making a new constellation for Cæsar and doubtless to be call'd by his name, telling him that the Scorpion will contract his claws to give him room." ‡

The only reference to these events in Herschel's Memoranda is the following, which is written immediately after the entry of 2nd July, quoted above :—

"About the last week of this month it was settled by His Majesty that I should give up my musical profession and, settling somewhere in the neighbourhood of Windsor, devote

* [Ignarosque viæ mecum miseratus agrestes
Ingredere et votis jam nunc assuesce vocari.]

† Herschel had apparently intended to publish a chart of the place in the heavens where he found the planet. Inside this letter there is a loose piece of paper on which he has written : "To Geo. 3 King of great Britain &c. is humbly dedicated this plate representing the situation of the Georgium Sidus in which it was discovered at Bath March 13 1781 by His Majesties most loyal Subject and devoted Servant W[m] Herschel."

‡ Georgics, I. 32. Compare Ideler's *Sternnamen*, p. 175.

THE 20-FOOT TELESCOPE.

From a drawing made either at Datchet or at Clay Hall.

[*To face page* xxxvii.

my time to astronomy, in consequence of which I took a house at Datchet with a convenient garden in which my 20 feet reflector might be placed. I went then to Bath to pack up my telescopes and furniture, to be sent to Datchet, to which place I returned immediately."

The salary granted by the King was £200 a year, not a large amount certainly. But in fairness to the King it should be remembered, not only that the purchasing power of money was much greater then than now,* but that no definite duties were attached to the post except that Herschel should occasionally attend in the evening to show something of interest to members of the Royal Family, an obligation which seems gradually to have dropped into oblivion. Besides, within the next five years the King gave £4000 for the 40-foot telescope, a salary of £50 a year to Caroline Herschel, and an annual allowance for the upkeep of the telescope. Herschel always expressed great gratitude for the King's liberality.

III.—Observatory Work after 1782

"August 2. This day the wagon which brought my astronomical apparatus &c. arrived very safely and was delivered at my house, of which I took possession and where in a few days I began to erect my 20 feet telescope.† My brother and sister were with me, the former on a visit, ‡ the latter to be my assistant in astronomy, in which capacity she had already acted at Bath. I employed myself now so intirely in astronomical observations, as not to miss a single hour of star-light weather, for which I used either to watch myself or to keep up somebody to watch; and my leisure hours in the day time were spent in preparing and improving telescopes."

At Datchet, about a mile and a half from Windsor Castle, on the left bank of the Thames, Herschel remained for three years. The low situation of the place rendered it subject to frequent floods, in one of which he caught a dangerous ague which nearly proved fatal, by continuing his work at night in the open air while the country round his house was under water. § He therefore, in the beginning of June 1785, moved to a place called Clay Hall, at Old Windsor, where observations were commenced on the 10th June. || But here unlooked-for troubles arose in consequence of the landlady being "a litigious woman who refused to be bound to reasonable terms," and finally he took a house at Slough, where the first observations were taken on the 3rd April 1786, and where he remained for the rest of his life. ¶ Though a not unimportant part of his work (most of his observations of double stars and two-fifths of his observations of nebulæ) had been done before he went to Slough, the name of this place will always be associated with that of Herschel to the exclusion of any other, and he conferred immortality on a hitherto obscure spot, as another great observer, Tycho Brahe, had done two hundred years previously.

The surprising activity which Herschel had displayed at Bath was continued without interruption during nearly the whole of the forty years which he spent in the neighbourhood of Windsor. Of the yearly volumes of the *Philosophical Transactions* for the years 1780 to

* We may also mention that Maskelyne's salary was only £300 a year.

† The observations of double stars were resumed on the 3rd August. On the 5th Herschel believed he had found a comet and watched it for some nights to see if it moved. It was the globular cluster Messier 5, found under similar circumstances by Gotfried Kirch in 1702. On the 20th August the 20-foot was first used.

‡ Alexander, who continued to live at Bath.

§ Compare vol. ii. p. 241. "He used to rub himself all over face, hands &c. with a raw onion to keep off infection of the ague; however, he caught it at last." (Note by Sir J. H.)

|| "A time piece of John Shelton's used for the first time; but it is not yet properly fixed, nor have I got my time. All the apparatus hardly settled."

¶ "The last night at Clay Hall was spent in sweeping till daylight, and by the next evening the telescope stood ready for observation at Slough" (*Memoir and Corresp. of C. H.*, p. 58). According to the sweep-book two sweeps were observed on the 28th March from $8^h\ 38^m$ to $13^h\ 45^m$, S.T. The next night is 3rd April, when a short sweep containing only three stars and no nebulæ was observed. It begins with the note: "The telescope removed to Slough, and neither my time nor meridian ascertained."

1818 inclusive, only those for 1813 and 1816 contain nothing by him, while not a few volumes contain several papers from his hand. All his writings (except one from the last year of his life) were published in that series, and he remained deaf to the entreaties of Bode, who begged him as a native of Germany to send some of his results to the *Astronomisches Jahrbuch*, but who had to be content with making and inserting abstracts of some of the papers, not always to Herschel's satisfaction. Occasionally unauthorised accounts of his work appeared in that periodical.* Thus von Zach, who spent some years in England, sent in November 1784 an account of a supposed volcano in the Moon, seen by Herschel on the 4th May 1783.† Though Herschel in his short account of what he saw in 1787 ‡ promised to communicate his observation of 1783 to the Royal Society, he never did so, and it will perhaps be well to give here his own account of it, from a letter to his Portuguese acquaintance, Magellan, who had asked for it:—

"May 4, 1783. I perceived in the dark part of the moon a luminous spot. It had the appearance of a red star of about the 4th magnitude. It was situated in the place of Hevelii Mons Porphyrites, the instrument with which I saw it was a 10 feet Newtonian Reflector of 9 inches aperture. Dr Lind's lady who looked in the telescope immediately saw it, tho' no person had mentioned it, and compared it to a star. Dr Lind tried to see it in an achromatic of 3½ feet of Dollond's but could not perceive it, tho' he easily saw it in my reflector. However, I could also tho' with difficulty perceive it in the refractor.

"May 13, 1783. I saw two small conical mountains, which I suppose to have been thrown up in the last eruption of the volcano. They are situated just by a third much larger which I have often seen before and remarked, tho' the two small ones were never before perceived in that place nor expressed in a drawing I had made of the spot.

"This is all I can give you at present. Believe me, Sir, I have not the least desire of keeping such observation to myself, but have so many subjects (which I think of greater consequence in astronomy) in hand at present that I had postponed giving an account of them to some other opportunity."

In April 1787, when Herschel saw something of the same appearance, he published his observation. On 20th May he wrote to Sir Joseph Banks:—

"Last night I had an opportunity to view the moon in a favourable situation and found that the volcano of which I saw the eruption last month was still considerably luminous. The crater seemed to glow with a degree of brightness, which I should not have been able to account for, if I had not seen the eruption last month. It appeared to me as if the crater was nearly doubled in its dimension since last month. It would not be difficult to account for this; for instance, the actual eruption might be much less than the real crater, and much of the ignited matter may remain lodged within its boundaries and thus produce a larger luminous circle: or part of the crater might fall in, and thus occasion an increase of its dimension, as these craters are generally much larger at the base than at the upper part. However, I leave this to be discussed by those who are better acquainted with volcanic phenomena than I am." §

In the following October, Herschel wrote to Dr. Blagden, Sec. R.S., that "the volcanoes in the Moon are in a quiescent state," and he saw them still in March 1788. In May 1788 Lalande wrote to Herschel that the volcano had been well seen during the last few days; but there were astronomers at Paris who were tempted to think that the mountain of Aristarchus, which is naturally very bright, might well return the light from the earth so

* When speaking of unauthorised announcements, it is hardly necessary to allude to the absurd weather-prognostications attributed to him, which he did his best to protest against and to refute. A facsimile copy of a letter from him on this subject is given in Smyth's *Cycle of Celestial Objects continued at the Hartwell Observatory* (1860), p. 400.

† *Astr. Jahrbuch*, 1788, p. 144.

‡ Below, p. 316.

§ On the same day Herschel wrote to some official at Windsor Castle asking him to inform the King that the crater was still very luminous, and adding that he would himself be at Windsor in the evening in good time to see the King's 10-foot telescope brought out and prepared, if it should please His Majesty to have it done. The severe illness of the King in the following year probably put an end to his star-gazing and to Herschel's duties in connection with it.

as to produce the luminous appearance. As Lalande was at Slough two months later, he may have persuaded Herschel of the truth of this suggestion.

When Herschel commenced his career as a professional astronomer in August 1782, his principal instruments were a 20-foot telescope of 12 inches aperture * and the 7-foot of 6·2 inches aperture. The latter had served him in his "second review" of the heavens, which resulted in the first catalogue of double stars and the discovery of Uranus. At the end of December 1781 he had started his third review, using the same instrument but a higher power (460 instead of 227), and extending the review to all Flamsteed's stars and all small stars near them. This work was now pushed on at Datchet with redoubled energy, the observations being often continued for ten or twelve hours at a time, some four hundred stars being examined and measures taken of many of them in the course of a night. The series was closed in January 1784, and was published as the second catalogue of double stars, but, in the course of later years, 145 other doubles were picked up during other work and supplied the materials of the last paper published by him. But, even before the close of the third review, Herschel had planned and commenced a much more extensive piece of work. In his paper dated the 1st February 1783 he announced the discovery of the motion of the solar system towards λ Herculis, and great views on the construction of the heavens were beginning to form in his mind. He must also have seen that his hopes of determining the distance of the fixed stars could never be realised by means of the comparatively rough micrometer measures which his instruments allowed him to make. The publication of Messier's short list of nebulæ and clusters of stars in the *Connaissance des Temps* for 1783 (extended in the volume for 1784 to 103 objects) at once attracted his attention, and he soon began to examine them with his 12-inch reflector. Believing, as a result of this examination, that all nebulæ would be resolved into clusters of stars, if only they could be viewed with a sufficiently large telescope, he decided to carry out a systematic search for these objects and at the same time to "gauge" the depth of the sidereal system by counting the number of stars visible in the field of view in various parts of the heavens.† On the 28th October 1783 he began therefore to "sweep" the heavens with his new 20-foot reflector of 18·7 inches aperture, with which he had commenced observing on the 23rd, still using the Newtonian form. The instrument was pointed to the meridian, but there was a lateral motion under the point of support of the speculum, allowing a change of 15° in azimuth, while the observer, standing on a gallery about 9 feet long, could also draw the tube along to the same extent, so that a range of 30° could be had without moving the stand. Drawing the telescope along, Herschel let it perform slow oscillations of 12° or 14°; and having noted if anything remarkable was seen (and very often there was nothing to note), the telescope was raised or lowered 8′ or 10′ and another oscillation made. From 10 to 20 such were called a "sweep," and the telescope was then re-set to a slightly different P.D. and another series or sweep commenced. But working in this way without any assistance was very fatiguing; sketching the positions of stars in the finder and the field of the telescope, whenever an object of interest appeared, spoiled the sensitiveness of the eye for faint objects, and very few results of any value were obtained.‡ This plan of operation was therefore discarded after forty-one sweeps had been made, and it was decided to have recourse to vertical sweeps, to employ a workman to raise and lower the telescope, and to appeal to the loving devotion of his sister to write down the observations, so that Herschel had not to take his eye from the telescope nor to expose it to any artificial light. A few sweeps (42–45) were made by way of experiment, and with Sweep 46 on December 18, 1783 the long series of observations was commenced,

* Known as "the small 20 ft." to distinguish it from "the large 20 ft." of 18·7 inches aperture used in the sweeps and for observations of the satellites of Uranus and Saturn, etc.

† See his paper of June 1784, this edition, p. 157 *sq*.

‡ The following interesting experiment occurs in Sweep 25, 19th November: "Syrius. I can count the lines of a page in the *Transactions* by its light. I refracted it thro' a prism & saw all the prismatic colours in the telescope very distinctly."

which was not closed till the 30th September 1802 (Sweep 1112). Beginning with this 46th sweep, the observations were now entered in separate "sweep-books," at first by himself, afterwards always by the indefatigable Caroline.*

A description of the method of observation adopted, which was gradually modified and improved by the addition of various pieces of apparatus, was given in the introduction to the first catalogue of nebulæ and in the account of the 40-foot telescope; † we shall here supplement it with a few more details. The telescope being placed in the meridian, it was used as a transit-instrument, the time of transit over a vertical wire being noted at first by a watch to the nearest minute, afterwards (from 24th December) by a sidereal clock.‡ Until the 30th December no attempt was made to fix the P.D. directly; when a nebula or cluster was found, a sketch was made of the field of the finder or of several consecutive fields to connect it with a known star, and also of the field of the telescope when the nebula was in the centre. These sketches cease from the middle of March 1784, when better arrangements had been made for determining the P.D. by dividing the board which indicated it more accurately and bringing it into the room close to the clock. We may here remark that Herschel never attempted to make elaborate and careful drawings of any objects in the heavens. His sketches of planets show very little detail, and those of nebulæ were always made with pen and ink, and were solely intended to show the relative positions of a nebula and neighbouring stars or (before the sweeping began) to give an idea of the general character of the object. The places obtainable by these sketches were necessarily very rough, but, with very few exceptions, these objects were re-observed later. The sweeps were from 2° to 3° broad in P.D.; whenever an interruption from clouds occurred or the telescope was re-set in P.D., a new sweep was commenced, so that often three or four sweeps were recorded during a night.§ In this way the systematic exploration of the part of the heavens north of about 122° P.D. went on, and in the spring of 1789 the second catalogue of a thousand nebulæ and clusters was finished. After that the work progressed more slowly, not only owing to the many and long interruptions caused by other series of observations, but also because the regions north of the zenith had now frequently to be searched, which took more time.

In every sweep from three to five standard stars were observed. It was a great misfortune that the only extensive star-catalogue available was that of Flamsteed; and as Herschel purposely avoided the brightest stars in order not to make it difficult for his eye to see faint objects, the choice was often a limited one. Very frequently, particularly north of the zenith, stars of the sixth or seventh magnitude were observed, but a note was afterwards added: "Not in Fl.," or "Not in Woll.," so that the star could not be made use of. From 1790 the catalogue compiled by Francis Wollaston was adopted, but Lalande's Zones of Circumpolar Stars, published in the *Mémoires de l'Académie* for 1789 and 1790, were not made use of, though they would have been specially valuable by often enabling Herschel to connect a nebula with a star nearly on the same parallel, which in small P.D. would have counteracted the increased influence of the errors of adjustment of the instrument.|| The correction of the clock and of the dial showing P.D. (the P.D. clock, erected 7th August 1784) was not deduced for the whole sweep from all the Flamsteed stars observed, but each nebula

* There are three volumes in quarto and four in folio, stitched in covers of limp cardboard or thick paper by Caroline Herschel. They were presented to the Royal Society by Sir John Herschel in 1863.

† This edition, vol. i. p. 262 *sq.* and pp. 513–518. The 40-foot was never used for "sweeping."

‡ Magellan (a Portuguese who had settled in London, where he died in 1790) wrote to Bode in 1785 (*Astron. Jahrbuch*, 1788, p. 162) that he had been present at the observations on the 6th January, and that Herschel let every object pass three times through the field of view in order to examine it thoroughly. This can only refer to a sliding eyepiece, as the telescope was certainly only moved in P.D. during the sweeping.

§ During the first month or two the sweeps had been very short, and a nebula had sometimes to be determined by means of a star observed in another sweep. This was never done afterwards.

|| These stars observed but not used have often been found very useful during the revision of the catalogues of nebulæ for the purpose of the present edition.

or cluster was referred to the star nearest to it in R.A., and in this way the $\Delta\alpha$ and ΔP.D. given in the three catalogues were obtained.

As to the accuracy of the resulting positions, Herschel states * that the places of the few objects observed before the 13th December 1783 may be in error as much as 1^m and 8′ or 10′, while the errors gradually became smaller till the end of 1784, when they should seldom exceed half those quantities. Between that time and the 24th September 1785 they should rarely amount to 10^s or 12^s and 3′ or 4′. On that day the new contrivance came into use, showing the P.D. of the tube at any moment in degrees and minutes, after it had been set to the P.D. of the zone at the beginning of each sweep; and after that the errors ought to be only 4^s or 6^s in R.A. and $1\frac{1}{2}'$ or 2′ in P.D., and often less. Auwers, after reducing all W. Herschel's published observations, came to a slightly different result.† By comparison with the places of Sir John Herschel, the probable errors of which, according to d'Arrest, are $\pm 1^s{\cdot}02 \sec\delta$ and $\pm 0'{\cdot}34$, and neglecting the seven objects observed before 13th December 1783, he found the probable errors to be :—

In 1784	$\pm 8^s{\cdot}3$	$\pm 1'{\cdot}45$.
January to September 1785	$\pm 5{\cdot}9$	$\pm 1{\cdot}24$.
After September 1785	$\pm 4{\cdot}8$	$\pm 1{\cdot}05$.

The errors in R.A. were found to increase from about P.D. 40° considerably more rapidly than $\sec\delta$, and as the circumpolar nebulæ were nearly all observed after 1785, Auwers concluded that the increase in accuracy was not so great as assumed by Herschel. But Auwers was not aware that the $\Delta\alpha$ and ΔP.D. given in the three catalogues do not always correspond to the date of the first observation (in the second column), but are often the results of a later and better observation. The accuracy attained in 1784 was therefore in reality much less than he supposed. But after Herschel had completed his arrangements, the accuracy of his results was really remarkable, considering the nature of his instrument (moved by ropes, etc.), and the few star-places he could employ; and they compare favourably with many of the positions of new nebulæ found a hundred years later by observers using equatorially mounted refractors, and having plenty of excellent star-places at their disposal.

The general appearance of the record of a sweep may be seen from the following specimen (No. 729), which is shorter and contains fewer objects than many other sweeps, but is otherwise quite representative (see next page).

The results of each observation, *i.e.* the description and the $\Delta\alpha$ and ΔP.D. from the nearest star, were copied by Caroline Herschel on single sheets of foolscap (the "Register Sheets"), each object having a sheet to itself, so that there are 2508 in all, in addition to a similar set of sheets for Messier's nebulæ and clusters.‡ Her perseverance in doing the enormous amount of copying required of her was equalled by her carefulness and accuracy in performing her tasks; § the books and sheets written in her clear and most legible hand seem completely free from slips, either clerical or arithmetical. In fact, the only mistakes she ever made in her work seem to have been two or three slips of 1° in P.D. in the *Zone Catalogue* of her brother's nebulæ, which she made in her old age.

The watch for nebulæ or clusters was occasionally varied by counting the stars in some consecutive fields for his "star-gauges"; but this was always done when there seemed to be no risk of losing a nebula in this way.

Until September 1786, Herschel used the 20-foot telescope in the Newtonian form, although he had already, in 1776, tried to do without the small mirror.|| In one of his earliest

* Below, p. 264.

† *Königsberger Beobachtungen*, xxxiv. p. 157.

‡ There is a similar series of register-sheets of all the double star observations, now belonging to the Royal Astronomical Society, while those of the nebulæ belong to the Royal Society.

§ She also copied out the observations of sun, moon, and planets into separate books.

|| See above, p. xxv.

	729.	Sweep.	Breadth 2° 10′.
			April 15, 1787.
12	19	90 5	Zero.* Began.
	26 2	90 17	29 (γ) Virginis . . . . . . Fl. 12 26 1 89 45
			+ 4 56 + 32
			12 30 57 90 17
			26 2 17
			+ 4 55 + 0
	31 32	91 34	cB. E. pL. from ab^t. 70° sp. to nf. 29 (γ) Virginis f. 5′ 30″ s. 1° 17′.
	32 28	92 12	pB. S. E. in the parallel, mbM. 29 (γ) Virginis f. 6′ 26″ s. 1° 55′.
	37 0	90 2	vB. elliptical vgmbM. cL. 29 (γ) Virginis f. 10′ 30″ n. 0° 15′.
	−28†		
13	14 27	90 33	Two, the place is that of the most south, cL. pB. bM. The smallest ab^t 6″ preceding & 4′ south. F. S. iF. 90 (p) Virginis p. 24′ 28″ s. 0° 5′.
	36 18	91 12	vF. S. iR. following two stars & in the par. with them. 90 (p) Virginis p. 2′ 37″ s. 0° 44′.
	37 7	90 5	cF. cL. vlbM. 90 (p) Virginis p. 1′ 48″ n. 0° 23′.
	38 18	90 24	vvS. mbM. 90 (p) Virginis p. 0′ 37″ n. 0° 4′.
	39 25	90 28	6m. 90 (p) Virginis . . . . . Fl. 13 38 51 89 58
	−30†		+ 4 56 + 29
	40	90 5	Left off.‡
			13 43 47 90 27
			38 55 28
			+ 4 52 − 1

sweeps, No. 29, 20th November 1783, he notes: "I tried the front-view of the telescope and found it incomparably better and brighter tho' under the greatest disadvantages of situation." Nothing further seems to have been done in this direction till 22nd September 1786, when we find in Sweep 600 the following:—

"Mem. I repeated some former experiments by looking into the telescope at the front without the small reflecting mirror, and found the image as good as at the side; the light is incomparably more brilliant, & I thought sometimes that the stars were, if not better, at least full as well defined as in the Newtonian way, so that it seems I have heretofore too hastily laid it aside. In high sweeps the position of looking is a very convenient one; and in no other situation can it be a very bad one. On a mature consideration we find that writing and often reading are generally done by looking nearly either more or less downwards and even perpendicularly on the paper or book; and yet these are things that can be done for many hours together without great fatigue." §

The "front-view," as he now called it, was after that night adopted altogether.||

* Zero means the upper limit of the sweep. On September 1787 there is the note: "The word Zero, being no longer proper, ought to have been changed when I introduced polar distances instead of numbers beginning from 0. We shall for the future call it *Bolt*, to signify that the telescope is rested there or bolted."

† Correction to the meridian (estimated); see *infra*, p. 264.

‡ The next sweep began two minutes later, at 91° 5′.

§ He had hitherto had the eyepiece mounted so as to look down at an angle of 45° with the vertical, when the telescope was horizontal (below, p. 260).

|| The note to the nebula V. 15 (pp. 287 and 294) means that Herschel had compared two observations of that object made in 1784 with the Newtonian telescope with one made in October 1786 with the front view. The slide which carried the eyepiece was placed at one of the corners of the octagon tube. "I have another on the opposite side in case I should want the left eye any long time, but then of course the inclination of the speculum will require an alteration" (Letter to Schröter, 4th January 1794).

Another experiment which does not seem to have been tried more than a few times, was that of binocular vision. Under date 18th October 1787, Herschel notes: "This evening I tried to sweep with two eye glasses, one for each eye, at once. The images of the stars are as distinct as in a single eye glass. I expect several considerable advantages from this construction. I tried the same scheme under a different form about two years ago; but it would not answer my end at that time. It consisted then of an eye-glass-pavement, as I called it, being several small eye glasses joined in one box to serve the same eye, by way of extending the field of view." He notes, however, at the end of the short sweep: "All objects were taken with the left eye glass." On 3rd May 1788, "I tried the double eye glass better executed than before, by having made room for the nose between, so as easily to come at the proper focus of pencils with each eye, and find it to act so well that I suppose I shall continue it." For a few nights there is the note at the beginning, "double eye glass," but on the 4th June there is again "single eye glass," after which there is no further mention of the double one.

To the front-view Herschel attributed the discovery of the two brightest satellites of Uranus on the 11th January 1787, and from that time the sweeps for nebulæ were generally interrupted to observe this planet at its culmination, when not too far from the opposition. Encouraged by this discovery, he turned his attention to Saturn in the following August, and for some years devoted a great deal of time to this planet and its satellites around the time of opposition, especially in 1789, when he was able to announce the discovery of two satellites, now known as Enceladus and Mimas, though the former had already been seen two years before. These observations are so fully described and discussed in his two papers on Saturn of 1789 and 1790, that no further account is required here. Naturally the work on nebulæ was completely suspended from July 1789 to February 1790, while the heavy work went on of following Saturn for hours by night and working out the results by day; and the planet and its satellites continued for some years to claim a great deal of attention, especially in 1793, when he determined the period of rotation anew.

In 1794 a new field of work presented itself to Herschel's mind: the determination of the relative brightness of the stars in order to secure materials for settling the question of the secular variation of their light. The laborious observations, on which his six catalogues of comparative brightness were founded, were nearly all made in the years 1795–97, and there are therefore no sweeps between the 18th October 1794 and the 22nd November 1797. The work was hardly estimated at its full value until towards the end of the following century, when it was reduced and discussed by Professor E. C. Pickering, who gives the following estimation of the importance of the work: * "Herschel furnished observations of nearly 3000 stars, from which their magnitudes a hundred years ago can now be determined with an accuracy approaching that of the best modern catalogues. The average difference from the photometric catalogues is only $\pm 0{\cdot}16$, which includes the actual variations of the stars during a century, as well as the errors of both catalogues. The error of a single comparison but little exceeds a tenth of a magnitude."

The sweeps for nebulæ and clusters finally came to an end on the 30th September 1802 with Sweep 1112,† and the third and last catalogue of Herschel's discoveries of these objects came out in the *Phil. Trans.* for the same year. From time to time during the next twelve or thirteen years he viewed some of the more interesting of the bright objects catalogued by Messier with instruments of different apertures, from his 40-foot reflector downwards; but old age gradually put a stop to the marvellous activity he had shown as an observer both in middle age and later. His opinions about the nature of nebulæ had greatly altered

* *Annals of the Observatory of Harvard College*, xxiii. p. 231; compare *ibid.*, l. p. 10, and lxiv. pp. 104 and 125.

† An attempt to try this work again was apparently made on the 31st May 1813, when Sweep 1113 was begun at $14^h\ 35^m$ in twilight, and stopped at $14^h\ 59^m$ by clouds; nothing but a star 9 mag. being observed (P.D. 79°). "New 20 ft. mirror, the stars are very distinct."

in the course of the long series of "sweeps." He had started with the idea that all nebulæ were composed of stars, and he therefore included clusters, even rather scattered ones, in his observations, as representing with dense clusters and nebulæ the different gradations of the same class of bodies. But the discovery of some indubitably nebulous stars, or stars with atmospheres (notably IV. 69 in November 1790), compelled him to recognise that there must be some kind of luminous fluid here and there in space. From thence he lost interest in the scattered clusters, and after 1790 he only catalogued nine of them.

During the last four months of 1801 Herschel devoted a good many morning hours of twenty-two nights to searching for the planet found by Piazzi on the 1st January, but as he expected to recognise it by its appearance, it is not strange that he failed to find Ceres.* The following is a specimen of his notes.

"Oct. 9, 1801. The trapezium γ η 42 50 Leonis. I examined this place with the 7 ft. telescope, power 460; the air was very clear and I saw γ Leonis double in high perfection; but there was not one of the small stars, though I examined at least 60 in that space, which had more than a stellar diameter. A planet of one tenth the diameter of Mars (or a thousand times less) would have been instantly perceived."

The regular search for nebulæ had hardly been stopped, before Herschel turned back to the double stars, which he had been the first to observe on a large scale twenty years previously. The brilliant discovery of binary systems made him re-observe many of his double stars in 1802 and following years.

We have here made no allusion to Herschel's observations of solar phenomena and other shorter series of observations of various celestial objects, nor to his numerous laboratory experiments on heat and light. His discovery of the infra-red part of the solar spectrum was certainly one of his greatest achievements, while his researches on the heat-rays emanating from heated terrestrial objects marked an epoch in the study of this subject. All these observations and experiments were fully described in the papers which he published almost as soon as each piece of work was finished, and as these are reprinted in chronological order, they form a continuous record of his activity from year to year, and render a detailed review of it in this place unnecessary.

But no account of the work of the Slough Observatory would be complete which did not mention Caroline Herschel's discoveries of comets. Immediately after leaving Bath her brother suggested to her to search for comets, and she began to do so on the 22nd August 1782, using a small refractor.† In July 1783 she began to use a small Newtonian "sweeper" of 2 feet focal length, an aperture of 4·2 inches, and a power of 30.‡ From time to time she picked up a bright nebula, but from December 1783 she "became entirely attached to the writing-desk, and had seldom an opportunity after that time of using the newly acquired instrument." It was therefore not till the summer of 1786 that she found her first comet, during her brother's absence in Germany. The following is a list of all the comets found by her. Numbers 5, 6, 7 were found with the larger sweeper of 5 feet focal length and 9·2 inches aperture.

1. Comet 1786 II., discovered 1st August; see below, p. 309, and *Memoir*, p. 64 *sq.*
2. Comet 1788 II., discovered 21st December; below, p. 327, *Memoir*, p. 80.
3. Comet 1790 I., discovered 7th January, only observed on four nights, 9th to 21st January.
4. Comet 1790 III., discovered 17th April; *Memoir*, p. 85, Maskelyne's *Greenwich Observations*, iii. p. 56.
5. Comet 1792 I., discovered 15th December 1791; below, p. 438.
6. Comet 1793 I., found 7th October; below, p. 451. But it had already been found by Messier on 27th September.

* The "Review of the ecliptic with a very high magnifying power" (1792–95, see vol. ii. p. 614) was doubtless also carried out in the hope of finding a planet.

† *Memoir*, pp. 52 and 54.

‡ *Infra*, p. 294, and vol. ii. p. 44.

7. Comet 1795, discovered 7th November (second apparition of Encke's Comet); below, p. 528, *Memoir*, p. 93.

8. Comet 1797, found 14th August; visible to the naked eye, and also found the same evening by Bouvard at Paris, and (at ten o'clock) by Lee at Hackney. *Memoir*, pp. 94 and 147; * Maskelyne, iii. p. 123. Lee wrote that it was "very visible to the eye without a glass."

Caroline Herschel's observing station was on the roof of a small detached building to the north of the dwelling house, used as a library.† The 20-foot telescope stood close to and east or south-east of it. Her nephew records that she "never observed with the 40-feet, being afraid to mount the ladders." ‡

IV.—The 40-foot and other Telescopes

In the course of the first winter he spent at Datchet, Herschel polished several smaller mirrors. In the autumn of 1783 he got two 18¾-inch mirrors cast for a 20-foot telescope, and began to observe with it on the 23rd October. This was "the large 20 feet" with which all his "sweeps" and most of his other important observations were henceforth made. In addition to telescopes for his own use, he had hitherto only made one or two for personal friends, but he now by degrees made a regular business of this work. One of the last entries in his book of "Memorandums" says: "The goodness of my telescopes being generally known, I was desired by the King to get some made for those who wished to have them. Getting the woodwork done by His Majesty's cabinet maker, I fitted up five 10 feet telescopes for the King, and very soon found a great demand for 7 feet reflectors. This business in the end not only proved very lucrative, but also enabled me to make expensive experiments for polishing mirrors by machinery." These mirrors for the King were of 9 inches aperture or about 8·8 inches polished surface when finished. They were cast in a small furnace erected in Herschel's house.

In the summer of 1785 Herschel entered in his polishing record the following "Memorandum of the method I have hitherto used to make my polishers":—

"1. The foundation on which the pitch is cast ought to be metal (lead, cast iron or brass) and not less than an inch thick for any diameter not exceeding 18 inches; if more, proportionally thicker, and it should, in a coarse way, have the figure of the intended polisher.

"2. Put double paper round the foundation just high enough to hold the required thickness of the pitch, from ¼ to ⅜ in.; about ½ inch or more is best. Cut the paper by measure, as it will serve to cast the pitch by, so that it may become of an even thickness. When you have cast the pitch or are casting it, elevate or depress the polisher, so that the pitch may fill the paper everywhere and be equally thick, which is very essential.

"3. Before the pitch is quite cold, tear off the paper and begin to figure it with the speculum (or concave tool if there be any). To do this, let the ridge of pitch at the circumference be cut away with a knife or (if warm enough) be pressed down by the hand.

"4. To prevent sticking or scratching in giving the figure to the polisher, apply paper or linen over the pitch with soap and water upon it; and to hasten the work let the speculum or tool be heated in water to about 90 to 100°. As it cools, restore the heat by dipping it again in the hot water. It will be necessary to bring the speculum by degrees to this temperature, especially in winter, though there is not much danger while the heat comes on; but take care that no cold water comes suddenly near it when it is heated. A difference of 20° F. is what hardly any metal will stand; I mean cold

* The comet was not found on the 6th, as stated in the *Memoir*, p. 94, but on the 14th, as correctly stated on p. 147. C. H. writes: "Tuesday morning the 15th I went to Greenwich and carried my memorandums to Dr Maskelyne. When I arrived he had had no intelligence yet of the comet. In the evening Dr M. received a letter from Stephen Lee Esq. who had also seen it on the 14th. I had the pleasure to find that my observations agreed perfectly with Mr Lee's." The packet containing eight paper-cases, in which are letters and papers connected with the eight comets, bears the superscription, "This is what I call the Bills & Recds of my Comets." Recds probably means Records.

† Compare *Memoir of C. H.*, p. 135.

‡ MS. note by Sir John Herschel.

applied to heat. It is even dangerous to put a very hot speculum upon a cold polisher; but this could hardly answer any end; for the polisher ought to be at least as warm as the speculum, which will bring it to a proper form in a very short time.

"5. When the polisher is nearly of the required form, take away the paper or linen cloth and go on with the bare speculum, but as the polisher grows dry, use either a little soap and water or colcothar, to prevent too great an adhesion. This operation ought to be continued till both the polisher and the speculum have lost their heat nearly as far as to have the same temperature with the air or not to exceed it more than 5 or 8°. If you want to give over and find the change of temperature to come on too slowly, try to accelerate it by applying a towel dipt in cool water to the back of the speculum or place the polisher upon cooling substances, but it must be done gradually. Supposing the polisher now to have an uniform face, let it be cooled in pump water or otherwise till its temperature be less than 50°.

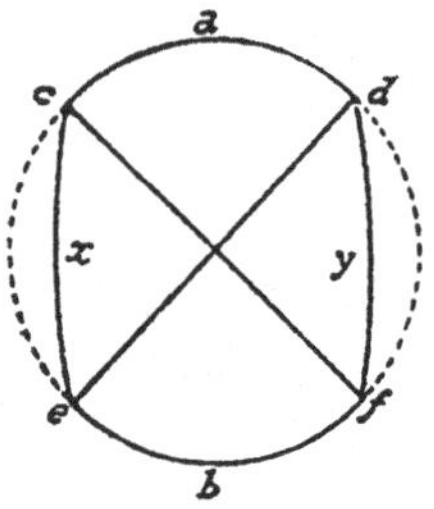

"Size and shape of cuts. When a single cross is to be made, let it be directly over the center. If the polisher is round, mark it with the point of a pen knife or proper cutter over any part of it, but if it is oval, let the cross be diagonally placed. The cut must have sloping sides like the letter V and may be from 2 to 6 or 8 tenths of an inch broad at the surface of the polisher, but join at the bottom. The breadth of the cut will be in some coarse proportion to the size of the polisher, but in the largest I have ever made it exceeded not 1 inch, nor have I used it less than about 0·2.

"Shape of the polisher. When the cross has been cut and the polisher is to be oval, mark the length *ab* and breadth *xy*. From the center of the cross draw two arches *cd* and *ef*; then with a proper radius draw two other arches to pass through *cxe* and *dyf*, which will bring the polisher to the required size."

The hope of making a very large telescope had been frustrated at Bath, but it had never been abandoned. Herschel now writes:—

"Hitherto I had looked upon all my polishing work only as a preparation for making a much larger mirror; but having already been at a very great expense with my new 20 feet mirrors, machinery and optical instruments, and my present situation being much more limited with regard to income than my former one at Bath, I thought it prudent to request the favour of the President of the R.S. to make an application to the King. His Majesty most graciously granted my petition, which was for the amount of the expense of the materials and of such work as carpenters, bricklayers, and smiths can only do; intending not only to execute all the optical parts myself, but also to contrive, direct and complete the whole construction of a large telescope. It remained now only to fix upon the size of it, and having proposed to the King either a 30 or a 40 feet telescope, His Majesty fixed upon the largest."

It was in September 1785 that this grant of £2000 was made, and preparations for the work were at once commenced. In August 1787 a second sum of £2000 was granted, with an allowance of £200 a year for the upkeep of the telescope and for paying the men who attended at night.* At the same time a salary of £50 a year was bestowed on Caroline Herschel.

The great mirror was cast in London, gauges for the mirror and tools having been made by Herschel. When he went to town on the 31st October 1785 to supervise the casting, the founder told him that he had procured the metal ready mixed as being much the better way, and showed some large slabs of what he called hard metal and others of white metal, and though this was not the metal Herschel wanted, he gave way on being assured that many mirrors had been cast of the white metal for opticians, and that it was much whiter and tougher than the proposed mixture of copper and tin. Fourteen slabs of hard and seven of white were used, and the casting was made. The next day the mould was opened and everything looked promising. The weight was 1023 lbs. But it turned out that the back of the mirror was depressed in the centre instead of being uniformly convex, so that it was far thinner (0·9 inch) than intended, although Herschel had reduced its thickness to

* *Memoir of C. H.*, p. 75. We are not aware how long this allowance was continued.

only 2⅛ inches to make it as light as possible. Still, as he had always meant to have two mirrors, he thought it worth while to finish this thin one to gain experience thereby. The mirror had been cast with an offset which fitted into an iron ring,* in which the mirror was secured by four flat bars, and it could be suspended by a crane in a horizontal position, face downwards, over the grinding or polishing tool. The rough irregularities of the surface having been filed or sawn off, handles were fixed to the ring for ten men to grind the metal, and this work was now proceeded with, coarse emery being used. The surface when ground appeared "uncommonly beautiful," so that great hopes were entertained that an excellent mirror would be the result.

The iron foundation provided for the polisher, cast to the curvature of the mirror, 49½ inches in diameter and 2⅛ inches thick at the sides, was next covered with pitch about ¾ inch thick. The mirror, suspended by the crane, was then dipped in a shallow tub of hot water and laid on the polisher, the latter being covered with a strong linen cloth. This was repeatedly done, the mirror being set on different parts of the polisher, so that the surface of the latter might acquire the proper curvature. The cloth was then taken off, the polisher covered with strong soap and water, and the ten men moved the polisher gently about. When the polisher was believed to be of the proper figure, a deep cross-gutter was cut in the pitch, and the polishing began in earnest. But the size and weight of the mirror and the number of men at work made the operation very troublesome, as ten men could not be made to "agree in those delicate attentions which are required in polishing mirrors, and those indications of the state of the mirror and of the polisher which were obtained by the touch when polishing by hand were entirely lost, so that former experience became almost useless." As neither the tube nor the rest of the mounting were ready, the work of polishing was not hurried on, as there was no way of testing the mirror properly; yet, after a great expenditure of time and trouble, a very fine evenly polished surface was produced. At last, on the 19th February 1787, Herschel writes:

"The apparatus for the 40 feet telescope was by this time so far completed, that I could put the mirror into the tube and direct it to a celestial object; but having no eye-glass fixed, nor being acquainted with the focal length which was to be tried, I went into the tube, and laying down near the mouth of it I held the eye-glass in my hand and soon found the place of the focus. The object I viewed was the nebula in the belt of Orion, and I found the figure of the mirror, though far from perfect, better than I had expected. It showed the four small stars in the Nebula and many more. The nebula was extremely bright."

This experiment was made a little more than a month after the discovery of two satellites of Uranus, which had convinced Herschel of the great value of the "front-view" construction, and this only was therefore adopted for the 40-foot telescope.

The work of polishing the speculum was continued during the spring and summer of 1787. Some trouble was caused by the pitch of the polisher adhering to the warm speculum round the circumference, and to prevent that a linen cover was put on the polisher and the speculum laid on it for four or five minutes. Various trials were made of additional gutters and holes cut in the pitch to promote the even distribution of the rouge, which was thrown into them with a long spoon when they were exposed by the motion of the speculum. Several times the speculum was put into the tube and tried on stars. The casting of the second speculum was delayed in order to gain as much experience as possible with the first one, but finally Herschel proceeded in December to consider the question of the composition of the second mirror. The proportion of tin hitherto employed, 29·4 or 29·5 per cent., gave a very white surface, but the mixture was of considerable brittleness. Still, Herschel thought it worth risking, and a speculum of this composition (1946 lbs. of copper and 812 of tin) was cast on the 26th January 1788. Every precaution was taken, and an overflow cup or "feeder" was fitted on the middle of the cover, into which the metal might rise after filling the mould,

* See below, p. 524, fig. 46.

But when the mould was opened three days after the casting, the mirror was found to be cracked in four places. Another mould was therefore prepared, and the proportion of tin was reduced to 25 per cent. by adding 430 lbs. of copper. The second casting was made on the 16th February and was successful; the weight of the mirror, after removing the feeder from its back, was 2118 lbs., and it was $3\frac{1}{2}$ inches thick, while the surface to be polished was 48 inches in diameter. In the following July it was ground, and the operation was repeated in September to reduce the focal length a little. On the 1st October the polishing began, but the work of directing twenty men (for so many were now employed) was found so wearisome that Herschel made up his mind to polish specula by machinery in future.* On the 12th January 1789 and following days a 20-foot mirror was polished by a machine, the essential parts of which were much the same as those of the machine he finally adopted. Like Cortez, who burned his ships, Herschel now destroyed the apparatus for polishing 20-foot mirrors by hand, which had served him so long and so well, to prevent himself from being tempted to polish again by hand. After ten days' work with the machine an additional motion was introduced, by which the speculum at every other stroke was turned the 126th part of its circumference upon its centre, so as to come once round in 252 strokes, or, at the rate of 60 to a minute, once round in $4^{m}\cdot2$. For a whole month Herschel worked almost every day, part of the time on a 7-foot mirror, the rest of the time continuing to practise on the 20-foot mirror, while a machine for the 40-foot specula was being prepared. This was tried early in March, but it was not till the 1st and 2nd June that the thin speculum was polished, and found to be greatly improved. In July and August the thicker speculum was polished by the machine. On the 24th August polishing began at ten o'clock and went on till six o'clock; "stroke 36 inches, 8 in 1′, colcothar laid on pretty dry, side motion pretty frequently but never far, that is about 3, 4, 5 inches from the centre each way. At 6^{h} we took off the speculum and the polish was greatly advanced towards the sides; by the trial on the Castle it appears to be of a very excellent shape. It acts well without any limiting diaphragm. The polish is still very far from being complete, but I shall try the speculum on celestial objects before I polish any more." Everything was now got ready, and Herschel dated the completion of the 40-foot telescope from the 28th August 1789, when he with that instrument discovered a sixth satellite of Saturn, or, rather, confirmed the discovery made two years previously but not fully realised.

The great telescope was described in a paper read before the Royal Society in 1795, and it was indeed worthy of a detailed description. The whole of it had been designed by the astronomer who was to use it, and every part of it had been made on the spot under his own immediate and never-ceasing supervision, while the amount of personal, mental and manual, labour he performed must have been immense.† But though the published account goes into the most minute details as to the tube, the mounting, and their accessories, there is not one word in it about the composition of the speculum or how it was ground and polished. These matters Herschel apparently considered as trade secrets not to be revealed; but he continued year after year to keep a careful record of his work as a maker of telescopes—a record which must have been of great value to his son, and would have been the same to Lord Rosse and Mr. Lassell, could they have seen it. It would be difficult to compress within reasonable limits of space an epitome of Herschel's experience in polishing specula, but it seems sufficient to give a summary of the procedure finally adopted by him taken from the account of it written by Sir John Herschel,‡ who, in the last year of his father's life and under

* See the paper "On polishing Specula by a Machine," sent to the Royal Society in March 1789, but not published (*infra*, p. cviii). It will be noticed that it does not give any description of the machine. On the 24th October 1788 the second speculum was tried on Saturn, but the focal length was 14 or 15 inches longer than convenient and the polishing was not finished.

† "One of my Father's workmen observing his dexterity in turning and at the same time in forging iron asked which he had been brought up to. 'To fiddling,' was the answer." (Note by Sir J. H.)

‡ *Encyclopædia Britannica*, 8th edition, article "Telescope."

his continual guidance, made and figured the 18¾-inch mirror of which he afterwards made such excellent use.*

The speculum was worked face downwards by straight or nearly straight strokes on a polisher covered with pitch, the consistency of which was tested by a " pitch-gauge" designed in August 1789. This consisted of a weight of about half a pound, resting at one end of a wooden lever on a supporting blunt edge of a piece of brass, by the depth of whose impression, in one minute, the hardness was judged. The polisher was carefully covered with rouge and water. If the polisher was circular, its diameter was very little larger than that of the speculum (about 1·06), or, if oval (diameters 1·12 and 0·97), the larger diameter was in the direction of the stroke. When an oval polisher was used, the gutters, by which the polisher was divided into a number of small squares,† were cut at angles of 45° with the longer axis, so that the stroke carried the mirror across and in no case along them. The machine was adapted to give the following movements to the speculum and the polisher—viz., 1st, the stroke, being a reciprocating movement by which, acting alone, the centre of the speculum would describe a straight or nearly straight line to and fro in a nearly invariable direction ; 2nd, the side-motion, by which the track of the centre was shifted laterally at every stroke (or every alternate one) through a short interval, so as to carry it backwards and forwards by regular steps to a certain distance on either side of the centre of the polisher ; 3rd, the rotation of the speculum, by which it was turned at each stroke or alternate stroke, through a certain angle ; and 4th, the rotation of the polisher (when a circular one was used), by which the gutters were presented at every angle to the direction of the stroke.

The speculum was surrounded by a " polishing ring " laid upon it, within which, loosely fitting and held in its place by three pins above and below, there was a thin flat ring on which was screwed a ratchet ring, and which carried three cocks, resting on the speculum by flanges covered with felt and adjustable, so as to hold the speculum without pinching it. By the action of an arm on the notches of the ratchet (against which it was pressed by a spring) the inner ring and the speculum with it could be carried round on its centre. To the outer ring was attached a claw (ending at C in the figure) and a connecting pin G, on which could be hooked a tailpiece for communicating the side-motion. The loop C was pinned on a fixed point of the lever AB, movable to and fro by a force applied at A round a firm centre B.

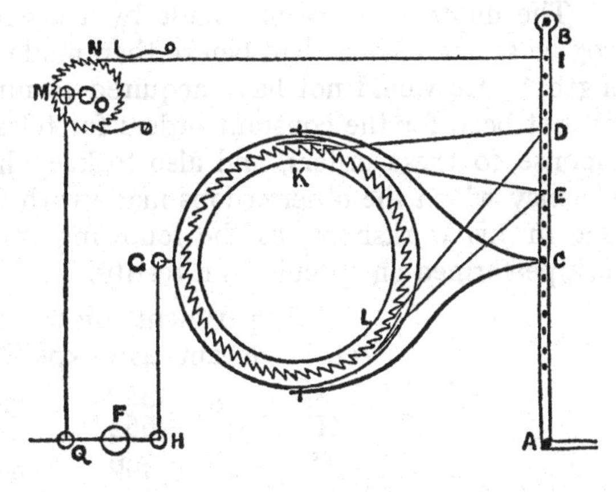

This would carry the centre of the speculum over the same space which C describes, which is the "length of the stroke" and can be measured on a scale. To prevent the centre from wandering, a tailpiece GH was looped on to G, either working at the other end H on a fixed centre F or on a stud-joint in a lever HF, which gave the side-motion. By proper adjustment of the length of the arm GH a very nearly rectilinear motion in any one stroke could be given to the centre of the speculum. The round motion was given to the speculum by two arms DL and DK, one pushing, the other pulling, acting on the ratchet wheel by claws and held lightly against it by springs. The other ends of these arms worked on a pin D attached to the lever BA, and the difference of motion of A and D caused the claws at K and L to work along the ratchet and turn the inner ring and the mirror round through an arc determined by the

* "The interest he took in this work and the clearness and precision of his directions during its execution showed a mind unbroken by age and still capable of turning all the resources of former experience to the best account." (MS. by Sir J. H.)

† For years he had only used a few gutters, but their number was gradually increased.

number of teeth of the ratchet brought into action at each stroke. The round motion of the polisher (when a circular polisher was used) was similarly communicated by placing it on a bed, revolving horizontally on rollers on a solid iron ring, ratched at the circumference the contrary way to the speculum's round motion, and worked round by arms attached at E to the under side of the lever. The angular motion of both these rotations could be varied by shifting the pins D and E.

A similar arm attached to a pin at I gave its impulse to the side-motion, by acting on a ratchet-wheel N, carrying an arm OM ; but as it was necessary that the tailpiece GH and the lever arm BC should be on contrary sides of the line of stroke, the revolving motion of the crank-arm OM was communicated by a rod MQ acting on the opposite arm of the lever FH, and thus giving a reciprocating motion to H, and therefore to G, at right angles to the line of stroke. The extent of the side-motion was varied by altering either the length of the crank-arm OM or of the lever-arm FH, or both.

All these movements were therefore adjustable in their relative extent. The stroke or the side-motion might also be made excentric or non-symmetrical with respect to the polisher, by varying its situation or that of the pin C, or the length of the arm GH. Every change in these dispositions was found by experience to have its peculiar influence on the figure ; and by a long induction from an immense number of experiments, Herschel was able to communicate at pleasure an elliptic, parabolic, or hyperbolic form to his specula. The length of the stroke and the extent of the side-motion were apparently of most importance, and 0·47 was assigned for a good working length of stroke on a round polisher without side-motion, and 0·29 stroke and 0·19 total amount of side-motion (from side to side) with it, as good average adjustments, the diameter of the speculum being 1.

The number of specula made by Herschel in the course of years was very great. In 1795 he stated that he had before then made 200 of 7 feet, 150 of 10 feet, 80 of 20 feet focal length.* He would not have acquired so much experience in the polishing of specula if it had not been for the constant orders for telescopes which continued to be sent to him. In response to these orders, and also to keep his own telescopes up to the high standard of efficiency which the observations made with them demanded, a very great deal of work was done in his workshops, as the following list of "experiments," *i.e.* of separate pieces of work performed on specula, can testify.

"Experiments on the construction of specula."

Vol. I.	contains	472	experiments	up to	Nov. 28, 1790.
II.	,,	530	,,	,,	Dec. 16, 1793.
III.	,,	662	,,	,,	Jan. 4, 1804.
IV.	,,	496	,,	,,	Dec. 5, 1818.

In all 2160 recorded operations. The number of telescopes made to order and the prices paid for them may be seen from the following autograph list:—

		£	s.	d.			£	s.	d.
25 feet	King of Spain . . .	£3150	0	0	7 ft.	Duke of Saxe-Gotha . .	£105		
20 feet	Prince Usemoff . . .	500	0	0		Dijon, Pere Fabard . .	105		
7 ft.	Baron Hahn . . .	157	10	0		Duke of Tuscany . .	105		
10 ft.	5 for the King † . .	787	10	0		Mr Bode, Berlin . .	105		
	Sir Fr. Drake . . .	210	0	0		Mr Pond	105		
	Col. Martin . . .	232	2	0		Capt. Thomas, Thetis .	105		
	Mr Edwards . . .	210	0	0		Dr Blair	71		
	Mr Wilson . . .	80	0	0		Mr Nairne	105		
7 ft. large,	King of Spain . .	157	10	0		Duke of Richmond . .	117	12	0
7 ft.	Greenwich . . .	105	0	0		Spanish Ambassador .	117	12	0
	Denmark, Ct Brühl . .	105				Count Brühl . . .	31	10	0

* Compare above, p. xxvi, footnote †.

† One of these mirrors (of 9 inches aperture) is at Armagh since the dissolution of the King's Observatory at Kew in 1840. It is still remarkably bright.

Size	Name	£	s.	d.
7 ft.	Stockholm . . .	£31	10	0
	Count Knühl [?] . .	31	10	0
	Mr Schröter . . .	65	0	0
10 ft.	Dr Hamilton* . . .	210	0	0
7 ft.	Mr Piazzi	105	0	0
	Dr van Marum . . .	105	0	0
10 ft.	Empress of Russia . .	210	0	0
	Hesse Darmstadt . .	63	0	0
	Mr Muller	8	8	0
	Duke of Richmond, repair	10	10	0
6 ft.	Mr Shard	31	10	0
	Mr Pond, repolishing .	31	10	0
7 ft.	Milan, Mr Oriani . .	105	0	0
	Mr Lloyd	105	0	0
	Count Brühl . . .	105	0	0
	Mr Newenham . . .	105	0	0
	Mr Trudaine . . .	107	2	0
	Sir Wm. Watson . .	0	0	0
	Clock	80	0	0
	Mr Aubert, By Quad. .	30	0	0
	Mr Mendoza, Platina Spec.	0	0	0
	Count Gesler . . .	105	8	0
	Mr Pitt, 1799 . . .	105	0	0
	Duke of Tuscany . .	106	11	6
7 ft.	Princess Roskow . .	£168	0	0
	Mr Paugh Düren . .	31	10	0
	Mr Vogel, Danzig . .	105	0	0
	Capt. Huddart . . .	106	1	0
	Mr Ferguson . . .	106	11	6
	Empress of Russia . .	525	0	0
	King of Spain, 7 feet .	150	0	0
	Glasgow	400	0	0
	Mr Brehm, Gloster . .	210	0	0
	London Society . .	210	0	0
	Dr Dauney, Aberdeen .	105	0	0
	Chevallier Forteguerri .	111	16	6
	Honble Fitzmaurice . .	105	0	0
	Oxford, 10 ft. . . .	315	0	0
	Dr Belcher	105	0	0
	Dr Dinvidee	26	5	0
	Mr Trimmer . . .	31	10	0
	Rev. Mr Deane . . .	157	10	0
	Greenwich	315	0	0
	Prince Canino, x ft. & 7 ft.	2310	0	0
	To China	105	0	0
	Emperor of Austria . .	210	0	0
	Col. Martin, 7 ft. . .	105	0	0
	Abbé Fontana, 7 ft. .	105	0	0

"The above list of persons for whom my telescopes were made, is partly put down from memory and cannot be supposed to contain a complete account of them." †

One cannot help remarking, how very little work of any value was done with all these telescopes, scattered broadcast over Europe. Probably next to none, except with Schröter's.

The most celebrated of all the instruments made by Herschel was, of course, the 40-foot telescope, erected close behind the house at Slough, south of the 20-foot telescope.‡ We have seen how Herschel already at Bath tried to make a large telescope, and how he in 1785 succeeded in obtaining the means of realising the strong wish he had entertained so long. As nearly all his work continued to be made with smaller instruments, it is of interest to consider why the ardently desired great telescope was so little used.

In a postscript to his second catalogue of nebulæ the discovery was announced of a sixth satellite of Saturn "with the forty-feet reflector." But though the discovery of this object (now known as Enceladus), as well as that of the seventh satellite (Mimas), was verified with the 40-foot telescope, both objects had previously been found with the 20-foot telescope. Enceladus was really found on the 19th August 1787, though the discovery was not followed up.§ At the opposition of Saturn in 1789 Enceladus was seen on the 18th July (but taken for the first or Tethys) and again on the 27th, on both occasions with the 20-foot.|| On the 28th August the 40-foot was directed to Saturn with a power of 189, and the following observation was recorded:—

* For Armagh Observatory (sold in 1835); it was accompanied by a paper giving detailed instructions for putting the telescope together and adjusting it.

† The total amount of the prices given above is £14,743. In 1785 Herschel wrote to Bode (in reply to an inquiry) that his brother made telescopes of all sizes, and that a 7-foot one, complete, and magnifying up to 3000 times, came to 100 guineas, but that the mirror and eyepieces could be made for 30 guineas. In January 1788 he informed a correspondent in London that a 10-foot Newtonian cost 200 guineas; with high powers, micrometer, etc., 275 to 350 guineas; a 20-foot front view reflector, capable of showing the satellites of the Georgian and faint nebulæ, 1000 guineas. A 10-foot Newtonian of 9-inch diameter, giving excellent images, is now in the possession of H.'s great-grandson, Mr J. A. Hardcastle.

‡ The nearest point of the circular track of the 40-foot was 40 feet east by south of the house.

§ *Infra*, p. 375. When Herschel remarked this, he wrote on 4th September 1789 to Sir J. Banks that the P.S. to the catalogue of Nebulæ ought to run thus: "Saturn has six satellites. An account of its discovery, its revolution and orbit will be given in the next vol. of the *Ph. Tr.*"

|| *Infra*, p. 397.

"Saturn with 5 stars in a line, very beautiful. The nearest of these five is probably a satellite, which has hitherto escaped observation. It is less bright than the others. What makes me take it immediately for a satellite is its exactly ranging with the other 4 and with the ring. The ring is very bright but extremely slender. I see very extended dark spots upon Saturn, much stronger than I have ever seen them before. The new satellite is nearer

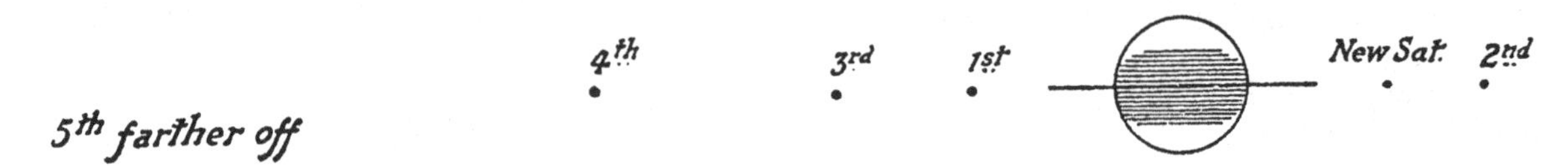

to the end of the ring than the length of the projecting part of the ring is beyond the body of Saturn. It seems to be bright enough, now it is discovered, I may perceive and follow it in the 20 feet telescope. Having no side-motion on, I can follow it no longer than the distance of my ladders will permit."

The 20-foot was therefore immediately turned on Saturn and the new satellite observed with it.

The seventh satellite (Mimas) was discovered with the 20-foot on the 8th September 1789, and was probably also seen on the 14th.* On the 17th September at 21^h it was seen just preceding the ring, between it and the first satellite, and definitely recognised as a satellite. Following the planet there was another which he first took for a new one and showed to Mrs. Herschel, and to his brother and sister; but it turned out afterwards that it was Enceladus. At $23^h\ 31^m$ he verified the six satellites with the 40-foot † and the discovery of Mimas was thus finally settled.

After that night Herschel continued till the end of the year to follow the satellites of Saturn, and there is no doubt whatever that all the observations (with the exceptions given above) published in his paper on their motions, including those of 16th October, when he followed the new satellites up to the very limb of the planet, were made with the 20-foot telescope, which instrument must indeed have been in the highest state of efficiency. In the description of the 40-foot he says that he had in 1789 "many times taken up Saturn two or three hours before the meridian passage and kept it in view with the greatest facility till two or three hours after the passage." ‡ But it is unlikely that anything of value can have been recorded on these occasions, as no observations with the 40-foot are found in the two MS. volumes devoted to Saturn except what we have already mentioned, and the following.§ On October 15 he used the 20-foot in observing Enceladus, || and only looked at Saturn with the 40-foot for ten minutes about $23^h\ 20^m$ and noted that the ring could hardly be seen. On October 20 he also tried the 40-foot during an interval of observations with the 20-foot and found that it would not act well, though he could see all the satellites (the speculum was subsequently found to be covered with thick dust). On December 2 the 40-foot was tried on Saturn, but the speculum was found "materially injured" by condensed moisture.

In 1790 Herschel appears only to have looked once (9th October) at Saturn with the great instrument; he noted which satellites were preceding and which following, after which he observed them with the 20-foot. In 1791 Saturn was viewed four times with the 40-foot and some measures of the ring taken ¶; on the 5th November (after using the 20-foot)

* *Infra*, p. 401. It was *not* seen on 29th August, as might be inferred from p. 373; compare p. 405.

† *Infra*, p. 401. This observation is not in the MS. books on Saturn; there is only at the beginning of the night the reference: "See obs. 40 feet." The same reference occurs on 11th, 16th, and 24th September. On these nights there are intervals of an hour or more between observations with the 20-foot, during which the 40-foot was probably used.

‡ *Infra*, p. 508.

§ Compare what Caroline Herschel wrote to her nephew in 1827 about observations made with the 40-foot under unfavourable circumstances, being written on slates or loose papers and not entered in the books afterwards. (*Memoir*, p. 210.)

|| *Infra*, p. 373.

¶ See Vol. I. p. 432.

he noted that the 40-foot showed the fifth and sixth satellites much larger than the 20-foot did.* After that we find nothing till the 7th February 1798, when with the 40-foot and a power of 800 he noticed that the shadow of the body on the ring was gibbous, *i.e.* " of a much shorter radius than that of Saturn." On the 1st April 1801: " The speculum is much injured by time, I see the phenomena, however, of the ring, the belts, the satellites &c. very well." On the 5th May 1805 Herschel made use of the 40-foot to verify his observations of the " singular figure " of Saturn.† Eight years passed before the old astronomer again looked at Saturn with the great telescope, which had grown old like himself :—

" July 29, 1813. 40 feet. I viewed Saturn, it was very bright and I saw many of the satellites, but the stars of the milky way being scattered over the neighbourhood the satellites could not be identified unless their situation had been calculated before. The mirror is so much tarnished that the image of Saturn was very imperfect. At all events the light collected is much more than is required for viewing this planet even with a high power, for which reason the delicacy of a 10-feet mirror in perfection is fully adequate to critical observations of Saturn's phenomena."

In August 1815 (date not given), Herschel, then in his seventy-seventh year, had his last look at Saturn with the 40-foot: " Saturn was very bright and considerably well defined. The 4th satellite appears like a star of the first magnitude to the naked eye. The mirror is extremely tarnished." With the 20-foot he found Saturn " well defined, but the great light is rather offensive, and so large an instrument is not wanted except for the purpose of observing the small satellite." This is his last recorded observation of Saturn.

If we turn to the observations of the satellites (real and supposed) of Uranus, we find the same disinclination to use the 40-foot. Of the observations of the planet and small stars around it, given in the paper on the four additional satellites, many are made with 7-foot and 10-foot telescopes, most with the 20-foot, but none with the 40-foot, although a large aperture might be supposed to be particularly useful in an inquiry of that kind. In his papers of 1814 and 1815 Herschel seems to have thought that this required some explanation, and points out that he has made it a rule never to use a larger instrument when a smaller one would answer, and that it took time to get the 40-foot ready for action, and it required two workmen to move it.‡ The second paper on the satellites of Uranus does, however, contain three observations made with the 40-foot,§ and Herschel seems to have enjoyed examining large globular clusters with it.

But it is most remarkable that nebulæ, a class of objects which played a very important part in Herschel's life-work, and with which his name is more associated than with any other celestial bodies except perhaps double stars, that these striking-looking objects should have been so very little looked at with the great telescope.|| Herschel was apparently not skilful in drawing, but still the study of the forms of the various kinds of nebulæ with a large aperture might have given him many hints as to their nature ; and in particular he might have discovered the spiral form of some of the nebulæ, which ought to have been within the reach of a 48-inch mirror. It is also strange that in the few recorded observations of the nebula in Orion there is no distinct mention of the fifth and sixth star of the trapezium, which alter the whole character of this miniature constellation, and simply obtrude themselves on the attention of an observer with a large telescope.

No one doubts that Herschel was capable of grinding and polishing even a speculum four feet in diameter, and thereby giving it both an excellent figure and a brilliantly polished surface. But all the same it is likely enough that the instrument did not generally perform

* Vol. II. p. 47.—A Mr. Greathead and his brother were shown Saturn in the 40-foot. This is deserving of notice, as Herschel is supposed not to have allowed anyone else (except Prof. Vince) to use this telescope.

† See Vol. II. p. 335. The speculum had not been polished since November 1798.

‡ Vol. II. pp. 536 and 543.

§ 8th March 1790, 13th February 1792, and 25th May 1810 (Vol. II. p. 553 *et seq.*). We have not been able to find any others in the MSS.

|| See the observations of Messier's nebulæ and clusters, Vol. II. p. 651 *et seq.*

well. The speculum was supported in an iron ring, resting there at its lowest point and confined there by an iron cross over its back. It would seem that a speculum weighing a ton and supported in this simple manner *must* have been subject to considerable flexure and cannot as a rule have done justice to the skill of its maker.* Besides, whenever it was polished, it became tarnished very rapidly, a fault for which the low composition of the metal was undoubtedly responsible.† The work of polishing it went on from time to time, but it was not till October 1796 that the foci agreed perfectly at all distances from the centre of the thick mirror, while the polish was excellent and some " blotches " had been got rid of. But even if the speculum was good and the effect of flexure not troublesome, the mounting of the telescope must have been very unwieldy, and after a few years it was probably exceedingly hard to move. What particular kind of observations Herschel had mainly intended the great instrument for is nowhere stated. But as he had been for years of the opinion that nebulæ were nothing but distant clusters, it is likely enough that he expected a large aperture to prove this by resolving many of them. Once he had recognised (in his paper on nebulous stars) that many nebulæ were not composed of stars, he may have become less desirous of using a large aperture, particularly when it involved a considerable loss of time. This is only conjecture, but for one reason or another he very soon gave up the hope of making extensive use of the 40-foot telescope. But though he only used it very rarely, the polish of the thick mirror was occasionally renewed. In 1797 it was polished in June, in 1798 in May, June, October, and November, the mirror being tested by looking at letters put up on a tree. Then it does not seem to have been touched till October 1806 : ‡ " We polished 1½ hour, I used six men at the beam. The mirror appeared by trial on a tree of a good figure. I ought to have gone on but found it too fatiguing to do more in one day." In 1807 polishing of the same mirror (the thin one seems to have been quite discarded §) was done on four days (October to December) ; the figure was considered good, but some blotches remained, which he attributed to the polisher not being exactly of the same figure as the mirror. He tried to remedy this by letting the mirror rest on the polisher in various positions for 7½ hours.‖ Again, on four days in January and February 1808, with much the same result, a few blotches remaining. The same was the case on the 24th March 1809, when the great mirror was polished for the last time; still Herschel was evidently pleased with the trial on letters " of a very small size," using three apertures and the whole mirror, as he finishes by saying: " If I had two mirrors this should not go on again till the second were better." ¶

The second largest telescope made by Herschel was one of 25 feet focal length and 24 inches aperture, for the King of Spain. Two mirrors were ground in September 1796 and polished in the following winter and a few times in the following years, the last time in December 1800. It was erected close to the 20-foot, and seems to have been a success; it is mentioned several times in the observing books as defining very well. The last time it was used was on the 1st April 1801: " it shews Saturn very well; by limiting the aperture to half its

* From the first, Lord Rosse made use of the system of uniform support devised by Thomas Grubb. *Phil. Trans.*, 1840, p. 524.

† Compare the remark that a smaller mirror may be made of a metal which reflects more light (Vol. II. p. 543, last line).

‡ It was not polished in September 1800, as one might think from C. H.'s *Memoir*, p. 106.

§ It seems to have been polished for the last time in April 1797.

‖ Herschel wrote to Dr. Watson on October 20, 1807: " My 40 feet thick speculum is at present in a state of being repolished. It has already recovered the brightness it had lost by tarnish, and I can read very small letters at 802 feet distance with it; but I do not mean to stop at this. Had I two mirrors of equal thickness, I should certainly keep this as it is at present till the second mirror excelled it. Just when a good mirror has been made, then is the time to make one still better, because the first has then put all the apparatus in the highest order, but with a single mirror one is unwilling to go on when it is in a very promising condition."

¶ C. H. says (*Memoir*, p. 123) that the 40-foot was polished in June 1814. There is no mention of it in the Polishing Record after 1809, and in June 1814 only a 7-foot mirror was polished. Perhaps preparations were made but given up again.

diameter it shews the planet in higher perfection of distinctness than my 10 feet." It was taken down a day or two after this observation * and sent off to Spain.

Another remarkable instrument was the "large 10 feet," also described as the "X feet" to distinguish it from the ordinary 10-foot telescopes of 9 inches aperture. Its aperture was 24 inches. It appears first in the Polishing Record as having been ground in April and May 1799, during a visit to Bath, the tool being 33·6 inches in diameter with a 10-inch hole in it; "this tool I bought at Thorpe [*sic*] after Mr Michell's death." † Having made a polisher with a foundation of yellow deal, Herschel polished it (still at Bath), and it "showed the stars very well defined." It was used in the Newtonian form, and was employed a good deal, as it was a convenient instrument for an old man, owing to its short focal length; it seems to have performed well.‡ It was sold to Lucien Bonaparte in 1814.

The following list of instruments at Slough bears no date, but was doubtless written not very long before Herschel's death. It is in Caroline Herschel's hand, and is found in a small book in which she also entered a copy of the first half of her brother's autobiographical notes and a list of his MSS. :—

Account of Telescopes.

40 feet Reflector. The woodwork is fast decaying and cannot be effectually mended, and, considering that even my 20 feet Telescope has not been brought into general use, I cannot recommend the 40 feet to be kept up.

The difficulty of repolishing its mirror, which is tarnished, and preserving or restoring its figure when lost, is so great that if a larger telescope than a 20 ft. should ever be wanting, I am of opinion that one of 25 ft. with a mirror of 2 feet in diameter, such as I have made § and which acted uncommonly well, should be a step between the 20 and 40 feet Instruments.

20 feet Reflector. The woodwork of it is also decaying, but the whole apparatus may be renewed at a moderate expense, and the method of repolishing and preserving or restoring the figure of the mirror when tarnished is not attended with such critical difficulties as will occur when the weight of the mirror requires a crane to lift it on and off the polisher.

10 ft. Newtonian Reflector, fitted up in light common wood for my own use.

Another 10 ft. Newtonian. Mahogany stand, some of the motions are not yet put on.

A 7 ft. Newtonian Reflector fitted up commodiously for my own use.

Another 7 ft. Newtonian in mahogany with all its motions complete.

A 7 ft. skeleton Newtonian telescope. The tube and stand consisting of bars to be screwed together and taken asunder so as to lie in a small compass for convenient carriage.

A large 5 ft. Newtonian Sweeper with a mirror of 9¼ inches in diameter fitted up for the purpose of discovering comets.||

Account of Mirrors.

There are two mirrors of 40 feet focal length; one of them is much too thin to keep its figure when put into the telescope. The other which is in the tube is of a proper thickness but is so much tarnished that it no longer gives a distinct image of celestial objects.

There are two 20 feet mirrors, one of them is in a close metal box, the other is loosely placed into an open case with two handles; the box is lined with thick woollen cloth; upon

* In the sweep book we find under date 5th April 1801: "On examination I find that by an accident of taking down the 25 feet telescope, my instrument [20 ft.] has been drawn out of the meridian towards the east at least 5 or 6 degrees in azimuth, perhaps more." Compare Vol. II. p. 234, under I. 282.

† Michell died in 1793. Herschel must have bought other things of his, as he worked on a mirror of 4·9 inches, 15 feet focal length, in October 1798: "It had a very bad figure, being one, polished by Mr Michell's directions, on the principle of springing the polisher by suspending a weight to a hook under the centre, the polisher being supported at the rim by an iron ring . . . . some of the emery marks remaining still visible, as Mr Michell's workman had not gone on long enough to bring them out."

‡ See for instance Vol. II. pp. 314 and 537, and observations of the nebula of Orion, Vol. II. p. 656.

§ See *Phil. Trans.*, 1800, p. 75, and 1815, p. 336 [Vol. II. p. 46 and p. 560].

|| See *Phil. Trans.*, 1800, p. 71 [Vol. II. p. 44†].

the mirror is laid a concave circular plate of tin, and an outward cover incloses the box with all its contents. Both mirrors are highly polished and of an excellent figure. Either of them may be put into the tube when observations are to be made, but should not be left in to be kept there.

There are eight 10 feet mirrors, one of them is in a brass box with adjustments and belongs to my 10 feet Newtonian telescope. Four are without cases. . . .* One is in a brass case; it was cast hollow as an experiment to reduce the weight of the mirror. Two are of glass.

There are eight seven feet mirrors. . . . The 6th and 7th are of glass. An eighth mirror is of white glass or Tussi's compound. By experiment I find it reflects double the quantity of light reflected by common glass.

A mirror for the 5 feet Newtonian Sweeper of 9¼ inches diameter, cast with a hole in the center so that it might also serve for a Gregorian construction.

Slider and plain mirrors with arms belonging to the 20 feet telescope, when it was used in the Newtonian form.

Plain mirrors belonging to my 10 ft. Telescope; two of them are mounted on arms fitted to the slider for alternate trials of their action.

Plain mirrors belonging to my 7 ft. Telescope; two of them are mounted on arms fitted to the slider for alternate trials of their reciprocal action.

A collection of mirrors in eight parcels contains several sorts of mirrors, that have served for making experiments relating to their figure and polish. Among them, set in cocoa-wood, are two mirrors intended for a trial of Dr. Smith's microscope, the large one being concave and the small one of an equal convexity.

Account of polishing machines, grinding tools and polishers.

There are polishing machines for mirrors of the focal length of 40 feet, 25 feet, 20 feet large, 20 feet small, 14 feet, 10 feet large, 10 feet common, 7 feet, 6 feet Gregorian, 5 feet Sweeper, 2½ feet Sweeper, 2 feet Gregorian. There are also polishing machines for plain mirrors for Newtonian telescopes of various sizes, and a polishing jack for 7 feet plain mirrors &c.

There are tools for giving the proper focal length to the above-mentioned mirrors by grinding.

There are iron or leaden foundations for polishers of a proper size for the above-mentioned mirrors, many of those foundations still having the polishing substance remaining on them, as they were set by, after having been used. There are also iron and brass gauges for mirrors of 40, 25, 20, 14, 10 and 7 feet focal length &c. &c.

V.—Later Years

In the course of years Herschel naturally came to exchange letters with most contemporary astronomers, both in England and on the Continent. As shown by his Letter Book, which contains drafts or copies of about 190 letters (1781–1817), he never wrote in German to his German colleagues, but always either in English or in French. All letters received by him were carefully preserved; they show the universal respect in which he was held, and on the part of those who had had the good fortune to meet him, many signs of warm friendship. Among his German correspondents one of the most frequent was Schröter, who sometimes sent lengthy extracts from his own observations and was as verbose in his letters as in his books. There is a letter of November 1793 about the appearance of Venus; he had evidently not heard of the sharp attack on his own observations made by Herschel, but

* [Some particulars as to cases and packing are omitted.—Ed.]

probably at the time not yet in print. Neither does he allude to it in the next letter (of September 1796), in which he thanks for the paper on the 40-foot telescope.

Among French astronomers Lalande was Herschel's principal correspondent; and there are thirty letters extant in his almost microscopic but very neat and distinct handwriting. He paid a visit to Slough in the summer of 1788, and was delighted with everything. He never fails to send his respects to "la savante Miss" or "l'aimable Caroline," and caused a niece of his to be called after her, as she was born on the day when Caroline Herschel's comet of 1790 was first observed in Paris. Herschel's replies (*i.e.* those of which copies are preserved) are generally written in English; only in the years 1797–1800, when intercourse between England and France was extremely difficult and letters must have been frequently opened on the way, Herschel wrote in French, probably to show the more readily how harmless the contents of the letters were. Most of these letters from foreign correspondents are written to thank him for copies of papers received, or to inquire about or to order telescopes; whenever the contents of Herschel's papers are alluded to, it is rarely at any great length.

Of English astronomers Alexander Aubert was one of the earliest to recognise the genius of Herschel, and he remained till his death in 1805 a very warm friend of his, many letters giving evidence of his strong affection for the great astronomer. In 1786 he presented Herschel with a quadrant by Bird and a clock by Shelton, the work of which he said was as fine as watchwork. Both these instruments were made good use of in the sweeps with the 20-foot telescope.*

The friendship with Doctor (afterwards Sir William) Watson, jun., formed at Bath, was continued till Herschel's death, and brought about a frequent intercourse by letters and many exchanges of visits.† This was also the case with Patrick Wilson, the son and successor of Professor Alexander Wilson of Glasgow. He corresponded with Herschel from 1783; in 1795 he expressed some regret that his father's name and his views on the constitution of the sun had not been mentioned in Herschel's paper on the same subject; but Herschel's reply to this letter seems to have satisfied him perfectly. Wilson resigned his professorship at Glasgow at the end of 1798, and settled in London (from 1805 at Windmill Hill, Hampstead), and he and his sister were on the most cordial terms with Herschel and his wife and sister, and often paid visits to Slough. His letters are always very lengthy (there are more than eighty of them extant); in the years 1807–1810 there is a good deal in them on coloured rings, on which subject he sided with Herschel against the Royal Society. He died in 1811.

With Maskelyne Herschel also corresponded frequently, about his papers and about new comets, as to which Maskelyne often sent him early information. Comets also form the subject of many letters from Pigott and from Sir Henry Englefield, who sent both elements and observations. Information about current astronomical events, discoveries of comets, minor planets, etc., was often sent by Sir Joseph Banks, who was President of the Royal Society during nearly the whole period of Herschel's scientific activity (1778–1820), as also by Blagden, whose position as Secretary to the Royal Society enabled him to be well posted as to scientific publications and similar news.

Banks was a valuable ally in 1809, when Herschel's second paper on coloured rings met with a very strong opposition from the Committee of Publication of the Royal Society. Banks wrote to him on the 25th May that the Committee was "unanimous in thinking you have somehow deceived yourself in your experiments and have in consequence deduced incorrect results." Herschel replied at once that he would come to town, in the hope that he might be "admitted to a conference." He saw Wollaston (who was one of the secretaries), and the

* The Shelton clock is still in the possession of Herschel's descendants, and performing well. Aubert was born in 1730 and died in 1805. His observatory was first at Loampit Hill, near Deptford, from 1788 at Highbury House, Islington.

† William Watson, M.D., born 1744, knighted 1796. He was a son of Sir William Watson (1715–1787), known by his researches in electricity. He died in 1825.

objections were given to him in writing. To these Herschel replied at considerable length (eleven pages folio in his Letter Book), and announced his intention of preparing a supplement to his two papers. His reply is written with great heat. To the charge of having confined his study to what was written by Newton, he answers that it was highly proper that he should do so, as he only wished to investigate the coloured rings discovered by Newton.* To the remark that the phenomena exhibited by scattered hair-powder did not prove that it possessed any properties not common to other bodies, he replies that they prove that specular rings are not due to alternate fits of easy transmission and reflection, and that "Newton's elaborate calculation as given in the 8th observation of his *Optics* p. 276 cannot possibly reach them." That colours shown by mica and other thin plates have anything to do with the "rings between object-glasses" he stoutly denies. He does not yield on a single point, but merely reiterates the statements in his paper.

On the 3rd July Banks wrote: "Your paper was ordered to be printed, I believe unanimously. I wish I could say that the opinion of the committee was as favourable to your deductions as their ballot was. You tell me, however, that you have now placed your opinions on a sound basis, which I hear with great pleasure."

It may be recalled that it was in 1801 that Thomas Young communicated to the Royal Society his explanation of the Newtonian fits as involved in his physical principle of the interference of trains of undulations, which gained no public recognition till it was revived by Fresnel in 1815.† Herschel must have been acquainted with Young's writings, though there is no direct reference to them, neither in his papers nor in his reply to the objections.‡ As Foreign Secretary to the Royal Society from 1804 to 1830, Young would be cognisant of the discussion relating to papers submitted for publication.§

This was the only occasion on which Herschel published a paper on an important subject which did not throw out ideas destined to influence the trend of scientific progress. It shows that his mind did not recognise its own limitations; but this very circumstance was an important factor in encouraging him to set forth views on the construction of the heavens which continued to prevail during the following century. A more cautious mind would have perceived that questions connected with the nature of light lay outside its proper sphere; but it might also have recoiled from venturing without any guidance from previous investigators to attack the deepest problems of sidereal astronomy. These problems had been examined by Kant and Lambert mainly by mere speculation, after the manner of the early Greek philosophers; but the genius of Herschel, which acknowledged no boundaries to its explorations, planned work on original lines and found untrodden paths.

After the presentation of the Copley medal in 1781, the next public recognition which Herschel received in Great Britain was the degree of LL.D., conferred on him by the University of Edinburgh in 1786; an example followed six years later by the University of

* The report of the Royal Society's meetings in the *Phil. Mag.* (xxvi. p. 364 and xxvii. p. 84) speaks of Herschel's paper as a "curious" one, and adds: "It may not be improper to remark here, that many of the Author's observations on Newton's theory of colours have been anticipated both by Dr Bancroft and by an anonymous writer (*Phil. Mag.*, viii. p. 78) on that subject. Perhaps it was a knowledge of this circumstance that induced him to mention so pointedly, that most of his experiments were made several years ago."

† There is only one letter from Young in Herschel's collection, written in 1807, at the request of Olbers, to announce the discovery of Vesta.

‡ He used the word "interference" in a fragment of a paper giving a quite different train of ideas (below, p. cxiv).

§ In a letter to a correspondent (Wackerbarth of Upsala), who seems to have suggested that William Herschel's papers ought to be reprinted, Sir John Herschel wrote in 1866: "I doubt whether it would be desirable to include in any collection of his papers those of 1809 and 1810, and perhaps also that of 1807, on the Newtonian coloured rings, the views of the subject there taken having found no concurrence on the part of later photologists." In his analysis of W. Herschel's works (*Annuaire du Bureau de Longitude*, 1842, p. 598), Arago says that these papers have not contributed to the progress of the theory of these curious phenomena: "J'apprends de bonne source que le grand astronome en portait le même jugement. C'était, disait-il, la seule fois qu'il dût regretter d'avoir, suivant sa constante méthode, publié des travaux au fur et à mesure de leur exécution." We have not found anything in Herschel's papers or letters which supports this statement; but it is, of course, possible that Arago had heard it from Sir John Herschel.—See Preface and Vol. II. p. 440.

Glasgow, while gradually nearly all the learned Societies and Academies of Europe enrolled him among their members. At Datchet, Clay Hall, and Slough the work went on day and night, only varied by occasional visits to London to attend the meetings of the Royal Society, when strong moonlight prevented observations of nebulæ. Only in the summer of 1786 did Herschel allow himself to take a longer holiday, when he paid his last visit to Germany (the third after settling in England), saw his mother for the last time, and delivered a telescope, presented by the King to the Göttingen Observatory.

From 1772 Caroline Herschel had been her brother's constant companion, his housekeeper as well as his associate, at first in his musical, afterwards in his scientific, work. When he gave up his musical profession, she could doubtless, like her brother Alexander, have continued to follow it at Bath, but she preferred to give it up and to devote herself to an entirely new sphere of work, simply because her beloved brother William wished it, and she felt she could be very useful to him. She had her reward, not only in the consciousness of being indispensable to him, but also in the universal respect and admiration which her devotion to scientific work, so unusual in a woman at that time, won her from astronomers and other men of science; though she sometimes resented this homage, when she imagined that it took something from her brother's great renown.

But a day came when she had, to some extent, to yield her place to another. On the 8th May 1788, Herschel married Mary, daughter of Mr. Adee Baldwin, a merchant of the City of London,* and widow of John Pitt, Esq., by whom she had one child, a son, who died in early youth towards the end of 1792, the same year in which Herschel's and her only child, John Frederick William, was born on the 7th March. She is said to have been a lady of singular amiability and gentleness of character, and she made Herschel's home-life thoroughly happy, while her jointure made a considerable addition to his income. Certainly she did not in the smallest degree distract his attention from his work; nor did the altered circumstances prevent Caroline Herschel from continuing her work as Astronomical Assistant, though she preferred to live in lodgings at Slough. The hope expressed by Lalande, when he offered his congratulations, that the marriage would not prevent Miss Caroline from discovering comets, was amply fulfilled. And her letters show that she was on the best possible terms with her amiable sister-in-law, as well as with the latter's niece, Miss Baldwin, who soon became a more or less constant inmate of Herschel's house till her marriage in 1819.

The easy circumstances, as regards his finances, in which Herschel now found himself, enabled him to seek recreation every year in shorter or longer holiday journeys. A complete record of these is left, extending from 1791 to 1817. Most of these accounts are mere skeletons, giving the dates, the names of the places he stopped at, and the number of miles between them. But of the tours made in 1792 and 1802 there are lengthy and detailed descriptions. The tour in 1792 was made in company with General Komarzewski, a Pole, whose acquaintance Herschel had made a couple of years earlier, and who seems to have been a man of considerable culture, specially interested in chemistry and mineralogy.† It was probably on his account that this trip took the special form of a tour of inspection of the principal factories of the middle and the north of England, at every one of which Herschel made detailed notes of the machinery he saw, illustrating them by numerous sketches of everything that interested him. He had a 7-foot telescope with him, and often treated his hosts to a view of Jupiter, etc., in the evening. Starting from Slough on the 29th May, the two friends went to Coventry, where they visited a spinning mill, thence to Birmingham, where they first spent three days at Soho ("dined with Mr Watt"), and devoted nearly a fortnight to

* His name was *not* James. He was a son of "Mr Baldwin of Slough who sold the estate"; his mother was born in 1667, daughter of Nicholas Adee, inducted in the Vicarage of Redbourne Cheney 1671, died 1712. Mrs Herschel's mother was Elisabeth Brooker, died 1798.

† He and Watson were the godfathers of John Herschel. He had large estates in Poland, and in 1798 he mentions in a letter that the Russian Emperor had made him a Privy Councillor and granted him a pension. He was the owner of copper mines in Galicia.

examining various iron-works in the vicinity. Next to Bangor and Carnarvon, then to Manchester, where there were cotton mills in plenty and also glassworks to be seen; then *via* Liverpool, Preston, and Carlisle to Glasgow, arriving there on the 29th June.

At Glasgow the municipal and University authorities vied with each other in doing honour to the distinguished travellers. On the 3rd July the Lord Provost presented them both with the freedom of the city, and the same day the Principal and Professors gave a dinner in their honour. "After dinner the Principal addressing himself to me desired to drink my health as Doctor of Laws of the University of Glasgow and read a diploma to that purpose, which he afterwards presented to me." * On the way to Edinburgh a visit was paid to Mr. Bruce, the Abyssinian traveller. At Edinburgh the usual sight-seeing was done and the acquaintance of the leading scientific and literary men was made.† The only astronomical matter mentioned is:—

"The Observatory contains many of Mr Short's Gregorian reflectors, a 2 feet, 2·6 feet, and a 10 or 12 feet one. I viewed several land objects thro' them. The large one which has the reputation of being a very bad instrument appeared to me not to be very defective, but the presumption seems much to be in favour of its being good. I could only see the top of a steeple, and even there great part of the light was lost by being confined in the opening of the roof, which is like our equatorial ones at Greenwich and Cambridge. . . . I saw again the large reflector at the observatory and measured its power, which was said to be 800, I found it 130, the aperture is 12 inches."

Turning southward, the travellers passed by Sunderland and Durham into Yorkshire, where Herschel at Richmond met some acquaintances from the time of his youth. Passing through Thornhill, where Michell had been the rector since 1767, they naturally called on him, and Herschel notes:—

"We saw Mr Michell's telescope, it is on an equatorial stand, being without cover behind. I put my hand into the opening and felt the face of the object speculum so wet as to moisten my fingers. Mr Michell was very indifferent in health."

Slough was reached on the 19th July, many notes and sketches of machinery having been made on the way (often without stating where); but already three days later Herschel and his friend started on a fortnight's tour on a somewhat similar plan through Devonshire and Cornwall. The following year he and his wife, in the course of a holiday trip, spent a couple of hours at Thornhill, where Michell had died on the 21st April, and Herschel notes that he "bought Mr Michell's great telescope and paid Mr Turton 30 pound." He does not mention that the speculum was "cracked," as Mr. Turton had informed him some months previously.‡ Possibly the stand was utilised for Herschel's 10-foot telescope of 24 inches aperture, made six years later.

The records of the holiday tours during the following years are generally very laconic, except that notes are occasionally made as to the geological formation of the country visited, showing Herschel's wide interest in science. But in 1802, when the Peace of Amiens again for a short time opened France to English travellers, Herschel and his wife and son and Miss Baldwin, like many others, paid a visit to Paris, where he had only been once before, in 1772, when he was an unknown musician, while he now came as a renowned man of science and Foreign Associate of the Institute of France. They arrived in Paris on the 24th July; the following is a complete copy of the notes made during the stay there:—

"July 25. Hotel Montorgueil, rue Montorgueil, bien bon. This day I saw M. Méchain, M. Delambre, M. Carrochet, M. Bouvet. I saw the great telescope in an unfinished state. The apparatus is certainly the worst contrived that can possibly be imagined. From what

* The official diploma was, however, not sent till March 1793; it was written with a quill from an eagle's wing, as being most appropriate.

† Among these were Principal Robertson, Dr. Hutton, "Mr Clark on naval tactics" [the celebrated Clerk of Eldin], etc.

‡ See above, p. xxxii, note.

I already know of M. Carrochet's work, I suppose the mirror may be a very good one, but as things are arranged, I believe it impossible that it can have a fair chance of shewing objects as it ought to do. I saw through it both in the Newtonian way and by the front view, but objects appeared very ill defined. This, however, was by no means a proper trial, for in sunshine terrestrial objects viewed horizontally can very seldom be seen to any advantage.

July 28. Dined with M. de la Place. There were at dinner M. Chaptal, Ministre de l'Intérieur, M. Berthollet, Sir Charles Blagden and other persons of eminence in literature. We went to the Institute at 6 o'clock, where I saw M. Haüy, the President, M. Messier, M. Lacroix, Secretary, M. Fourcroy, who read a memoir on Chemistry, M. Jeaurat, the old astronomer. In the morning I saw also M. Legendre.

July 29. We all dined with Madame Gautier at Passy. It is a beautiful place, and there was a very elegant entertainment.

July 30. I breakfasted with M. le Sénateur La Place. His lady received company a bed, which to those who are not used to it appears very remarkable. Mad. La Place, however, is a very elegant and well informed woman.

July 31. We saw the paintings and statues. I got my passport signed by the Minister. I also reviewed the Observatory with Sir Charles Blagden.

August 1. I dined with the Sénateur M. Berthollet, where I also met Dr Kapp from Leipzig and M. Vauquelin the chemist. M. Berthollet also is a very sensible man. The house is pleasantly situated at Arcueil about 4 leagues from Paris. In going to his house we saw the formerly famous waterworks at Marly. After dinner we walked to see the Aqueduct.

August 2. We saw the Jardin des Plantes, Museum of French monuments, Temple, the place where the Bastille has been. In the evening to one of the Gardens.

August 3. We saw the Parade after having breakfasted with Madame Laplace. In the evening we saw the opera Semiramis and the ballet of La Dansomanie.

August 4. I dined with M. La Place. I went to the meeting of the Institute and afterwards to the play.

August 5. We dined with M. le Sénateur Sieyès at his house one league beyond Marly. I had much conversation with M. Sieyès and found him a very excellent philosopher.

August 6. I went with General Komarzewski to a bookseller opposite the Tuileries, who is a member of the Institute. He is blind but a very agreable man and seems to be well informed.

August 7. We went with General Komarzewski to see M. Charles, the aeronaut and lecturer. He has a very magnificent apparatus for experiments in natural philosophy. We dined with the General at the Gardens of the Tuileries, and he spent the evening with us.

August 8, Sunday. We breakfasted with Madame La Place to take leave. I dined with the Minister of the Interior, M. Chaptal. His house is a magnificent palace, most elegantly finished. About 7 o'clock the Minister conducted M. La Place, Count Rumford and me to Malmaison, the palace of the First Consul, in order to introduce me to him. We saw Madame Bonaparte, who, after the Minister had introduced me was pleased to accost us with great politeness. After a considerable walk with her in the garden we met the First Consul, who was engaged in making improvements in the garden by directing the persons employed how to conduct water for the irrigation of the plants. The Minister introduced me and Count Rumford. The First Consul addressed himself to me in politely asking some questions relating to astronomical subjects, and after a considerable conversation he also addressed Count Rumford by questions of a nature which related to the character of that philosopher, well known as a man that has much contributed to the comforts of the poor at Munich &c. The First Consul afterwards entered into a general conversation with the Minister of the Interior, with M. La Place, Count Rumford and myself on common subjects.

After about half an hour's walk he led us towards the house, but stopping short, on meeting with some other Gentlemen, he entered into conversation with them on the subject of a canal which is to be made in France. The Consul seemed to be perfectly acquainted with the subject. He now led us to a room where after a short time spent in conversing, he seated himself in a chair and politely desired me to sit down. As the same invitation was not given to the rest of the company, nor any of them took seats, I only bowed my thanks for the First Consul's great civility and kept standing with the rest. The First Consul then asked a few questions relating to astronomy and the construction of the heavens, to which I made such answers as seemed to give him great satisfaction. He also addressed himself to M. La Place on the same subject and held a considerable argument with him, in which he differed from that eminent mathematician. The difference was occasioned by an exclamation of the First Consul's, who asked in a tone of exclamation or admiration (when we were speaking of the extent of the sidereal heavens) 'and who is the author of all this.' M. de La Place wished to shew that a chain of natural causes would account for the construction and preservation of the wonderful system; this the First Consul rather opposed. Much may be said on the subject; by joining the arguments of both we shall be led to 'Nature and Nature's God.'

Soon after, the conversation took another turn, which was the breeding of horses in England; in this the Minister of the Interior and Count Rumford joined. I remarked that the Second and Third Consuls who were now also present hardly ever joined in the conversation. Madame Bonaparte and her daughter, who came in a short time after the First Consul had led us into the room, were seated on a sofa near the Consul. She is a very elegant lady and seems hardly old enough to be the mother of her daughter, who, I hear, is married to one of the First Consul's brothers. It appeared to me pretty remarkable that neither the Second nor Third Consul sat down in the presence of Bonaparte, but remained intermixed with the company: Cambacérès standing close by me, and Le Brun between some General and M. La Place. The conversation turned also upon the police, which it was remarked is very badly kept in England compared to what it is in Paris. The great licence of English newspapers was also reprobated by some; the First Consul however did not join in expressions of that nature.

Between the conversation ices were handed about, which were of an excellent flavour, and the weather being extremely hot were very refreshing. Bonaparte said that the thermometer had been 28° in the shade, that is 95° of Fahrenheit's scale. After some not particularly interesting conversation the First Consul quitted the room. The Minister of the Interior addressed Madame Bonaparte, and after some short conversation on the subject of the opera &c. he took his leave, and we followed his example. We then returned to the house of the Minister, where we rested a little while, and where we found Madame Chaptal and her family, and also Madame La Place and her family, who had spent the day at M. Berthollet's. They said that the thermometer there had been observed by M. Berthollet at 31°, that is $101\frac{3}{4}$° of Fahrenheit's scale.

As M. La Place and I went and returned in a carriage by ourselves, I led the conversation upon the subject of my last paper, of which I gave him some of the outlines. I mentioned the various possible combinations of revolving stars united in double or triple systems. When I mentioned three stars at an equal distance revolving round a center, he remarked that he had shewn in, I believe, his *Mécanique Céleste*, that six stars could turn round in a ring, about their common center of gravity.

A few days ago I saw M. Messier at his lodgings. He complained of having suffered much from his accident of falling into an ice-cellar. He is still very assiduous in observing, and regretted that he had not interest enough to get the windows mended in a kind of tower where his instruments are, but keeps up his spirits; he appeared to be a very sensible man in conversation. Merit is not always rewarded as it ought to be.

August 9, Chantilly. . . . Aug. 15, Ramsgate. . . . Aug. 25, Slough."

The annual holiday trips were often varied by visits to Sir William Watson at Bath or at Dawlish near Exeter.* In 1810 he spent two days (July 17–18) with James Watt at Heathfield near Birmingham. Glimpses of Herschel's home life are afforded us by the memoirs of Madame D'Arblay and Dr. Burney and the letters of the poet Campbell,† while the diary of his sister shows us the inevitable but very gradual advance of the infirmities of old age; though these were not visible to the outer world, which could only continue to admire the memoirs still flowing from his pen. Two tardy official recognitions were granted him in 1816, when he received the third class of the Hanoverian Guelphic Order on the 5th April, and was knighted by the Prince Regent a month later.‡ When the Astronomical Society of London (now the Royal Astronomical Society) was founded in 1820, Herschel was elected its first President. Though he was too feeble to come to town to attend the meetings of the Society, he showed his interest in it by allowing his last paper to appear in its *Memoirs*.

On the 25th August 1822 Sir William Herschel died at his house at Slough, in his eighty-fourth year.§ He was buried in the church of St. Laurence at Upton, close to Slough; his grave is marked by a stone tablet on the floor under the tower, bearing a long Latin inscription, composed by Dr. Goodall, Provost of Eton.||

The features of William Herschel are known to posterity through a number of portraits, of which the following are the principal ones:—

A Wedgwood medallion executed in 1783 by Flaxman.

A painting by Abbott in the National Portrait Gallery (date about 1788); an engraving from it is given in the *Memoir of Caroline Herschel* and in Holden's *William Herschel*.

A crayon by J. Russell, dated 1794 (frontispiece to Vol. I.).

A small engraving by C. Westermayr, in von Zach's *Allgemeine Geographische Ephemeriden* (Bd. I., 1798).

A large engraving by F. Rehberg, date 1814. Reproduced in Newcomb-Engelmann's *Populäre Astronomie* (ed. of 1881).

A painting by Artaud of 1819 (frontispiece to Vol. II.).

How Caroline Herschel at once left England, where she had spent exactly fifty years, and returned to Hanover, where she had become a perfect stranger, and how she spent the remaining twenty-five years of her long life in looking back to the many happy years she had spent as her brother's assistant, has been vividly told in her *Memoir and Correspondence*. But in this place it seems desirable to say a few words of the labour to which she devoted the first years of her retirement, crowning the work she had done in England by preparing a Catalogue of the 2500 Nebulæ and Clusters found by her brother, since this work has remained unpublished and can only have been seen by very few people. It was very properly rewarded by the Astronomical Society by the bestowal of their gold medal in 1828.

The work forms a folio-volume of 104 pages, all written in the same clear and distinct handwriting (extremely like her brother's), in which she had recorded most of the "sweeps," and written out the "register-sheets" of the observations of each object separately. The title is: "A Catalogue of the Nebulæ which have been observed by Wm Herschel in a Series of Sweeps; brought into Zones of N.P. Distance and order of R.A. for the year 1800,

* In 1799 he made a stay of some months at 6 Sion Hill, Bath, and perhaps again in 1800, when he gave up the house there at midsummer.

† Holden, *William Herschel*, pp. 100 *sq.*

‡ In 1784 Lalande, seeing the appellation *Esq.* in the title of one of his papers, congratulated him on his "new dignity," adding: "C'est une justice de votre digne Monarque."

§ Though he had been very weak for a long time, his death must have been unexpected, as his son had started for the Continent on the 8th August, and was unable to get back till after the funeral, at which his friend Babbage acted for him. For an account of Herschel's will, see Holden, p. 114. Consols stood at 81 at the time.

|| Printed in *The Observatory*, iv. p. 274. This is the correct text, while that given in the *Astronomisches Jahrbuch* for 1826, p. 222 (there stated to have been composed by J. Herschel), differs somewhat in several places. Among a number of newspaper cuttings, obituary notices, etc., collected by Sir J. Herschel, we have found a MS. copy of the inscription, agreeing exactly with that of the *Jahrbuch*. Probably it was first written by J. H. and then improved by Dr. Goodall.

by applying to the determining stars the variations given in Wollaston's or Bode's Catalogues." First come the few "circumpolar nebulæ to 9 degrees N.P. Distance"; then the Zones 10°–14° and 15°–16°, after which follow separately every degree of Polar Distance from 17° to 121° inclusive. The last two pages give errata in the three printed catalogues (most of them in the first catalogue of 1786), and corrections to her catalogue of omitted stars and to her index to Flamsteed's observations. In every zone there are eleven columns, containing:—the general number (in order of discovery, being the numbers of the register-sheets); the class and number; the date of discovery; the Flamsteed number and constellation of the star; the letters *p* or *f*; the difference of R.A.; the letters *n* or *s*; the difference of N.P.D.; the resulting Right Ascension and Polar Distance; the number of the sweep. Every observation of the objects observed more than once is given separately. The whole work is carried out with the same scrupulous care evinced in her work on the current sweeps. Of course the resulting positions are not quite as accurate as those afterwards derived by Auwers, since Caroline Herschel did not pay any attention to the difference of precession of nebula and comparison star, and also neglected the influence of nutation and aberration; but still, on the whole, the results may be said to do justice to the original observations.

The catalogue was primarily intended for the use of Sir John Herschel, whom his aunt knew to be engaged in the revision and extension of his father's work by independent observations; and Sir John, in more than one place, bears testimony to the immense value of the catalogue to his own work. But he never published it; partly, no doubt, because he shared the universal opinion at the time, that very few of his father's nebulæ could be seen, or at least usefully observed with any but the largest telescopes; but chiefly because he always intended to bring out a General Catalogue of all known Nebulæ and Clusters, a task which the vast amount of valuable work he carried out did not allow him to complete till 1864. But it certainly is to be regretted that Caroline Herschel's catalogue was not at once given to the public, since the publication of the separate results of her brother's observations would have cleared up many difficulties of identification, and prevented the continuance of many an error. Even in the revision of the three printed lists of 1786, 1789, and 1802, for the purpose of the present edition, though it was founded on the original sweeps, the *Zone Catalogue* was often found useful.

The house in Windsor Road, Slough, is still standing, since 1888 again occupied by some of Herschel's grandchildren, who with loving hands guard the manuscript treasures of their great ancestor. Of the 40-foot telescope only a small fragment remains, but the 4-foot speculum hangs on the wall of the entrance hall. The telescope stood for over fifty years as a monument to the ambition of its maker to explore the heavens to the farthest possible extent; it had then, in December 1839, to be taken down, as the woodwork was dangerously decayed,* but the tube was left on the site, extended horizontally on low piers.† Thus it lay for many years, until a tree was blown down on it in a storm, destroying all but about ten feet of the tube. This small remainder now lies in the garden, somewhat farther from the house, the original cover of the speculum being kept inside it. But no earthly monument is wanted to perpetuate the memory of the man who, solely guided by his genius and his indefatigable energy, rose to the greatest eminence in a hitherto almost unexplored field of science, and widened the horizon of man to an extent no one had imagined possible before his time.

* Immediately before it was taken down, a photograph was secured of it, a glass negative having been taken by Sir John Herschel, from which paper prints were successfully made many years later.

† Compare *Astronomische Nachrichten*, No. 405 (Bd. xvii.); Weld, *Hist. of the Royal Society*, vol. ii. p. 193.

UNPUBLISHED PAPERS

I.—*Papers read before the Philosophical Society of Bath, 1780–1781*

A List of my Papers that were read at the Bath Philosophical Society by Mr. Rack the Secretary *

I. Observations on the Growth and Measurement of Corallines.—*Read January* 28, 1780.
II. Observations on Dr. Priestley's Optical Desideratum—" What Becomes of Light ? " —*Read February* 4, 1780.
III. Addition to Observations on Dr. Priestley's Desideratum, &c.—*Read February* 11, 1780.
IV. Observations on Corallines (*continued*).—*Read February* 18, 1780.†
V. On the Central Powers of the Particles of Matter.—*Read February* 18, 1780.
VI. Observations on the Mountains of the Moon.‡
VII. Continuation of the Observations on the Height of the Lunar Mountains.‡
VIII. Astronomical Observations on the Periodical Star in Collo Ceti.‡
IX. Proposition and Queries.—*Read March* 23, 1780.
X. Experiments in Electricity made by the Committee.—*Read April* 7, 1780.
XI. The Question, whether a Knowledge of the Classic Authors in their own Language is essentially necessary to one who would write English with Elegance and Perspicuity, considered.—*Read April* 14, 1780.§
XII. Experiments on Prince Rupert's Glass Drops.—*Read April* 14, 1780.
XIII. On the Utility of Speculative Inquiries.—*Read April* 14, 1780.
XIV. Experiments in Electricity continued.—*Read April* 21, 1780.
XV. On the Electrical Fluid.—*Read May* 5, 1780.
XVI. On the Existence of Space.—*Read May* 12, 1780.
XVII. Continuation of the Experiments on Glass Drops.—*Read May* 19, 1780.
XVIII. Communication of my Letter to the Reverend Doctor Maskelyne, On the Measurement of the Lunar Mountains.‖
XIX. Observation of the Occultation of Gamma Virginis, made with a view to determine whether any Effect of a Lunar Atmosphere could be perceived.—*Read August* 17, 1780.

* The papers that were read at this Society were copied into a book by the Secretary with a view of an intended publication ; but as the Society a few years after its establishment was dissolved, the copies in the handwriting of the Secretary were taken out of the books and returned to their respective authors.

† [Not reprinted.—Ed.]

‡ These three papers were read at the Bath Society, but were not inserted by the Secretary in his collection, as they were soon after presented to the Royal Society at London, where they were read and honoured with a publication in the *Philosophical Transactions* for 1780.

§ [Not reprinted.—Ed.]

‖ *Phil. Trans.*, vol. lxx., 1780 [below, p. 12. Some remarks about lunar inhabitants omitted from the *Phil. Trans.* are given below, p. xc].

XX. On Liberty and Necessity.—*Read September* 1, 1780.

XXI. Measurement of the Height of Landsdown and of Barrow Hill, by Dr. Watson and Mr. Herschel.*

XXII. A Short Account of some Curious Experiments on Glass Bottles taken from the Last Publication of the Bologna Society. (*Bonon. Scient. Inst.*, p. 97.)—*Read November* 24, 1780.

XXIII. Short Account of some Experiments upon Light that have been made by Zanottus, with a few Remarks upon them. (*Bonon. Scient. Inst.*, p. 105, De luce.)—*Read December* 1, 1780.

XXIV. Remarks on a Passage in Dr. Priestley's *History of Electricity*.—*Read December* 22, 1780.

XXV. Answer to the Question whether the Electrical Fire contains any Sensible Degree of Heat.—*Read December* 22, 1780.

XXVI. On the Heat of the Electrical Fire (*continued*).—*Read December* 29, 1780.

XXVII. Appendix to the Electrical Experiments on the Thermometer.—*Read January* 6, 1781.

XXVIII. On Central Powers.—*Read January* . ., 1781.

XXIX. On the Periodical Star in Collo Ceti.—*Read February* 2, 1781.

XXX. Della Torre's Method of Making Crystal Globules to be used instead of Lenses for Single Microscopes. Translated from the Italian Scelta di Opusculi.—*Read February* 9, 1781.

XXXI. Account of a Comet.—*Read March* . ., 1781.†

Observations on the Growth and Measurement of Corallines.

By Mr. Herschel.

Read January 28, 1780.

January 22^{d}. 7^{h} *morning.* I examined a particular branch of the Coralline. In one of them was a black spot, seemingly contained within the hollow capacity of the branch. The diameter of the spot was, by estimation, about $\frac{1}{4}$ by $\frac{1}{8}$ of the dia^{r} of the branch. It appeared to be surrounded by another clear spot nearly resembling the lucid part which is about a grain of sago not much boiled. The diameter of this latter was about $\frac{5}{8}$ of the thickness of the Branch.

Jan. 24. 9^{h} *morning.* I examined the same branch and found that several little arms were sprouted out in places where I had not observed any before. The little black spot was now considerably larger, and almost circular.—That I might the better ascertain the growth of the little arms, I thought it necessary to measure the lengths of some of them. One of the longest just near the black spot measured ·0038^{inch} or 38 ten thousands of an inch.

At 6^{h} *in the afternoon.* I measured the same branch again and found it ·0036^{inch} therefore I concluded that there was no sensible alteration nor could I perceive any difference in the shape of the black spot.—However I could hardly expect much change in so few hours. I measured the longest little arm I found near that place which was upon a neighbouring Branch its length was ·0152^{inch}. The longest upon the same Branch where the dark spot is situated measures ·0106^{inch}.

* [Not reprinted.—Ed.]

† This paper is dated 22nd March 1781, but there is no copy of it among the Secretary's returned papers. [It is the account of the discovery of Uranus, also sent to the Royal Society and published in the *Phil. Trans.* and below, p. 30.—Ed.]

Those who are not acquainted with the manner of measuring Microscopic Objects will wonder by what instrument I could determine a magnitude less than the 4 thousands part of an inch; and still more so when I shall say that nothing but a common ruler divided into inches and tenths, and a good pair of compasses is necessary.

It will not be amiss to describe my Method of measuring, tho' it is not quite new. In the first place we must find the power of the Microscope. This is done by placing a ruler finely divided into quarters of tenths, or any other small known measure under the Microscope, and observing with one Eye how much of it fills the whole field of View, while the other Eye observes how much of the same ruler will fill the same field of View in its natural or unmagnifyed Condition. In my Microscope ·09 inch when magnify'd, filled the field of View; and 6 inches not magnifyed will also fill it. This gives us the power of the Instrument. For to make ·09 inch cover a space of 6 inches it is required to magnify 66⅔ times, or neglecting fractions I shall say 66 times only.

The power of the Instrument being thus known, the measuring of small Objects is easy enough.

Place the Object under the Microscope and bring it near the side of the field of View. While one Eye is looking at the Object, let the other with a pair of fine Compasses measure the magnifyed image. This requires a little practice, but may easily enough be learned.

In this manner I found that the little arm which was grown out of the branch of the Coralline measured 25^{inch}.

Now as ·25^{inch} was the Size of it when magnifyed 66 times, it is easy enough to imagine that the 66th part of ·25^{inch} viz: ·0038^{inch} must be the real Size of it.

Jan. 25. 8^{h} *morning.* I now found the black Spot again as it was the 22^{d} for which reason I ascribe the apparent change I found in its Size on the 24th to a different view of it tho' apparently in the same Situation. Perhaps it is loose in the branch and therefore liable to be somewhat displaced. I could not perceive any sensible difference in the length of the arms.

Jan. 26. 7^{h} *morning.* The little arms were now very much encreased; most of them measured ·0045^{inch} several ·009^{inch} ·01 ·013^{inch} and one of them ·023^{inch}.

Jan. 27. 8^{h} *morning.* The black spot in the Branch is no larger to-day than it was at first. It measures ·0026^{inch}. The branch ·01^{inch}. One of the little arms measured ·05^{inch} and there were several of this Size, and perhaps larger, but this was most conveniently situated for measuring.

Jan. 28. 8^{h} *morning.* I found now a great number of long arms. I measured five of the following lengths:—·015 ·018 ·022 ·048 ·059 or ·06 Inch.

Remarks.

That vegetation is carried on in these Corallines is without doubt, nor is there anything surprising in it. It is also evident that no signs of life have hitherto appeared. Now, if these Corallines are the habitation of Polypes of a shape resembling, and adapted to, their Houses, it would almost look something like an instance of Leibnitz's pre-establish'd Harmony. But without having recourse to so extraordinary a doctrine, we need but reflect on the nature of Polypes, (as far as we are already acquainted with them by experiments) to solve the difficulty.

Does it not appear that any part of a polype which is cut off will grow again? or when the whole is cut into several pieces, each piece will form itself into a compleat Polype? Does it not then follow that a polype will grow in that part of its body which, by its situation and circumstances seems to be wanted. And consequently, if a Polype should grow in a Branch of a Coralline, it is no longer surprising that it should grow in such a manner as to have its body adapted to the shape of its habitation.

Wm HERSCHEL.

January 28, 1780.

Observations on Dr. Priestley's Optical Desideratum—" What becomes of Light?"

Jan. 28.

THE following desideratum by D[r] Priestley was propos'd as a subject curious in itself, & proper for the Consideration of this Society, by M[r] Herschel.

The Cause of the extinction of Light in Bodies is a subject of very great Difficulty. If Light was subject to no Laws but those of reflection and refraction, no place into which Light was admitted, would ever be dark and the Light of the Universe would be continually encreased.

If the same Cause produces all these Effects it might be expected that it would sometimes produce effects of an intermediate kind: in consequence of which Light might not always be reflected or refracted with equal velocity. M. Bouguer supposes that the power which absorbs or extinguishes the Light is confind to the surfaces of bodies; and that it operates chiefly when the rays fall upon it with a certain degree of Obliquity: whereas Newton supposes that a Ray is never stopd but when it impinges on some solid parts of bodies.

To whatever Cause it be owing, that Light is stopd in bodies, the question is " What becomes of it?" In consequence of which the following Letter was sent by M[r] Herschel.

The difficulty of the subject makes it necessary to proceed with the utmost caution; but as every different point of view, in which we can consider the question, may serve to throw light upon the subject, let us first see in what manner the more palpable effects of the extinction of light can be explained. We find for instance, that light is gone the moment a candle is extinguished.

From a great number of experiments we are assured that no kind of objects, how bright soever, reflect so much as *nine tenths* of the *incident* light. Certainly not more. Let us then suppose a candle to be placed in the center of a hollow sphere of polished metal of about 20 f[t] in diameter, and examine how long by the law of reflexion, under these circumstances light will continue to be visible after the candle is extinguished. In the present case, rays proceed from the center to the concave surface, and therefore will be reflected thro' the center to a point diametrically opposite, then back again, and so on.

Now, if nine tenths of the incident light are reflected from the first surface, we may admit that this will also be the case when the light so reflected reaches the opposite part of the hollow globe; that is to say, nine-tenths, out of nine-tenths, or the square of $\frac{9}{10}$ will be returned at the second reflexion. And at the third again nine-tenths of the light incident from the second reflexion, or the cube of $\frac{9}{10}$ and so on. Therefore at the nth reflexion, the nth power of $\frac{9}{10}$, or $\overline{\frac{9}{10}}|^{n}$ will be returned.

To proceed, let us suppose now that the light of a candle, if it continued but one quarter of a second after the candle had been put out, might be observed not to be lost instantaneously; And to find the number of reflexions made by the reverberating light in that time, from which we shall obtain the value of n, we have the velocity of light from the Sun to the earth, at the rate of about 92 Millions of miles in 8 minutes and 13 seconds. This gives us 187000 miles in one second, or 46 750 in a quarter of a second. If we take a mile to contain 5280 feet, light will be found to move thro' 236,840,000 feet in a quarter of a second, and therefore will make very near 12 millions of reflexions within the above mention'd globe of about 20 feet dia[r], in that time.

Having now found the value of n to be 12 millions we may determine the expression $\overline{\frac{9}{10}}|^{n}$; And here it appears that the 12 millionth power of $\frac{9}{10}$ is a fraction whose numerator is *one*, and whose denominator is a one with five hundred and forty nine thousand and eighty nine cyphers annexed to it, or $\frac{1}{1000000000000\ \&c.}$ to 549089 cyphers. (At the rate that my cyphers are wrote here, the denominator of this fraction would just be about a mile and a half long.) Now this quantity is infinitely little indeed, and may safely be called nothing;

but this is all the light there can remain in a quarter of a second after the extinction of the candle.

What is more, if 999 parts out of a thousand were reflected, the remainder, in a quarter of a second, would be no more than $\overline{\frac{999}{1000}}\vert^{n}$, or $\frac{1}{1000\ \&c.}$ to 5213 cyphers; which again is so small a quantity that we may call it a *real nothing*.

From this calculation it appears that to account for the extinction of light in all possible cases of this kind, we need but account for the extinction of a single particle. It also appears that if there be not one single particle lost at the first reflexion, there can be no reason to suppose that any one will be lost at the 2^{d}, 3^{d} or n^{th} reflexion; and therefore, as Dr Priestley says, "if light was only subject to reflexion and refraction" (and not to extinction also) "the light of the whole universe" (at least in those parts where there *are* reflecting and refracting mediums) "would be continually increased."

Let us now examine the hypothesis of M. Bouguer. He supposes "that the power which absorbs or extinguishes the light is confined to the surfaces of bodies." (Pr. optics page 779). I believe this to be contrary to experience, which proves beyond a doubt that light is lost even within the most transparent substances. If it were true that light is only lost at the surfaces, a very thin, and a very thick lens should be equally transparent, which we find otherwise. Nay he observes himself that "Sea water would be perfectly opaque at the thickness of 679 f^t^., and the air of our atmosphere would cease to be transparent, if the light had 518385 toises of it to traverse." (ib page 426). It seems then that M. Bouguer does not give us any great satisfaction upon the subject of our present inquiry.

The supposition of Newton seems also to have its difficulties. He says that light "is never stopped but when it impinges on some of the solid parts of bodies" (ib. page 779). I have before remarked that the light is lost not only at the surfaces of bodies but also within them; To this however I must now add, that much more is lost at the surfaces than there is absorbed within. If this was not the case a piece of glass one inch in thickness and ten pieces of the same, each one tenth of an inch thick, ought to transmit an equal quantity of light; but this is contrary to experience, the ten pieces stopping incomparably more than the one. Now, can it be supposed that the light should meet with parts sufficient to impinge upon, at the surfaces of bodies, and find hardly any at all within them? If light be sensibly stopped at the surface of water, for instance, by impinging on some of its solid particles, how can it possibly traverse a considerable quantity of it without being totally stopped?

Having thus shown the difficulties, which the above recited opinions are liable to, it will be expected that I should give my own upon this subject. I shall therefore attempt to solve the question in a manner perhaps not intirely new, yet as far as I can find never before put in that light, in which it appears to me to obtain in Nature. It will be proper to see what has been said upon the manner in which bodies are supposed to act upon the rays of light. And here I need but mention a very few authors. S^r^. Is. Newton says that reflexion and refraction may be caused by powers of *repulsion and attraction* belonging to bodies, and extending to certain distances beyond their surfaces.

M^r^. Boscovich goes a little farther and maintains that matter consists of phisical points only, endowed with powers of attraction and repulsion taking place at different distances; that is, surrounded with various spheres of attraction and repulsion, in the same manner as solid matter is generally supposed to be. And that it acts upon light by these powers.

M^r^. Michell, also, is of this latter opinion which he gathered from M^r^. Baxter's treatise on the immateriallity of the Soul.

Now, it appears to me that this Hypothesis of powers of attraction and repulsion taking place at different distances, is not philosophical, nor consistent with that regularity and systematical manner in which Nature seems to act. To suppose a power of attraction to extend to a certain distance, there to cease at once, and at that place another power totally opposite, namely repulsion to begin, (which also is to have its limits),—and to suppose perhaps many more such alternate hollow spheres of powers one within the other, according as there shall be occasion for them, has too much the appearance of an arbitrary Hypothesis to give satisfaction. Besides it realy seems to be demonstrable that these different spheres of powers can not exist in such a manner. Is it not absurd to suppose that one of the outermost coats, or hollow spheres of powers, that has not the least connection with its center, should move along with it, and never become excentric?—and that several such spheres should all remain as it were stuck together like different coats upon a globe, when it is evident one cannot be connected with the other except we would maintain that one power shall be the support or substratum of an other power, which is absurd.

However, I will not insist longer upon the absurdity of this Hypothesis as we shall find that my own is not very different from it, but only establishes the same effects upon more unexceptionable principles. S[r] I. Newton has already proved that every particle of matter as far as we hitherto have been able to make experiments, is endowed with an attractive power which diminishes as the square of the distance encreases. That such a power existed was already supposed by Kepler, Hook and several other philosophers before Newton proved the laws by which it acted. My Hypothesis, then, is, that every particle of matter, is endowed with at least four more *central powers*, each of them acting under very different Laws; but none of them under the law of that power, established by the newtonian philosophy.

It will presently appear that this Hypothesis can solve every difficulty of our question; And if, by a careful investigation of nature, and attention to experiments, we could arrive at the knowledge of the laws by which these powers act, we should be able to show, *a priori*, in what manner a ray of light is stopped. But, to proceed, I shall beg leave to call these powers by the following names.

1)—The attraction of cohesion.

This will be found to be in the inverse ratio of some very high power of the distance.

2)—The repulsion of emission or reflexion of light. (I do not mean that this power is only to serve for the purpose of emitting or reflecting light but only that it is adequate to it.) This power must also be in the inverse ratio of some high involution of the distance, but far inferior to the former

3)—The attraction of refraction and inflection.

This again, must be in the inverse ratio of some considerable power of the distance

4)—The repulsion of deflection.

Still in the inverse ratio of some power higher than the square of the distance.

5) The Newtonian attraction of gravity.

Known to be in the inverse ratio of the square of the distance.

As I observed before, every one of these central powers must act according to different laws. For should any one repulsive power act under the same law with an attractive force, they would exactly balance each other's effects, and thereby destroy one another. And should two powers of the same kind act under the same law, it would be no more than doubling the force; that is to say no more than one power.

It is evident at first sight that this Hypothesis is fruitful beyond all bounds, since from the different number and connections of particles tho' possessed of no other central powers (which however I would by no means affirm) such very different effects, in regard to the extent of the circles or spheres of action must arise as will evidently carry us into an endless field of speculation and experiments. And for this very reason the investigation of the laws under which these central powers act will be attended with the utmost difficulty, and require all the assistance that mathematics joined to experimental philosophy can afford. In M[r] Boscovich's Hypothesis all the spheres of powers must be (I do not say equidistant but) at an equal distance in all sorts of bodies in general, except he would indeed be very arbitrary in his suppositions. But in the present, every different number of particles that are held together by cohesion will produce a different arrangement of the extent of the spheres of

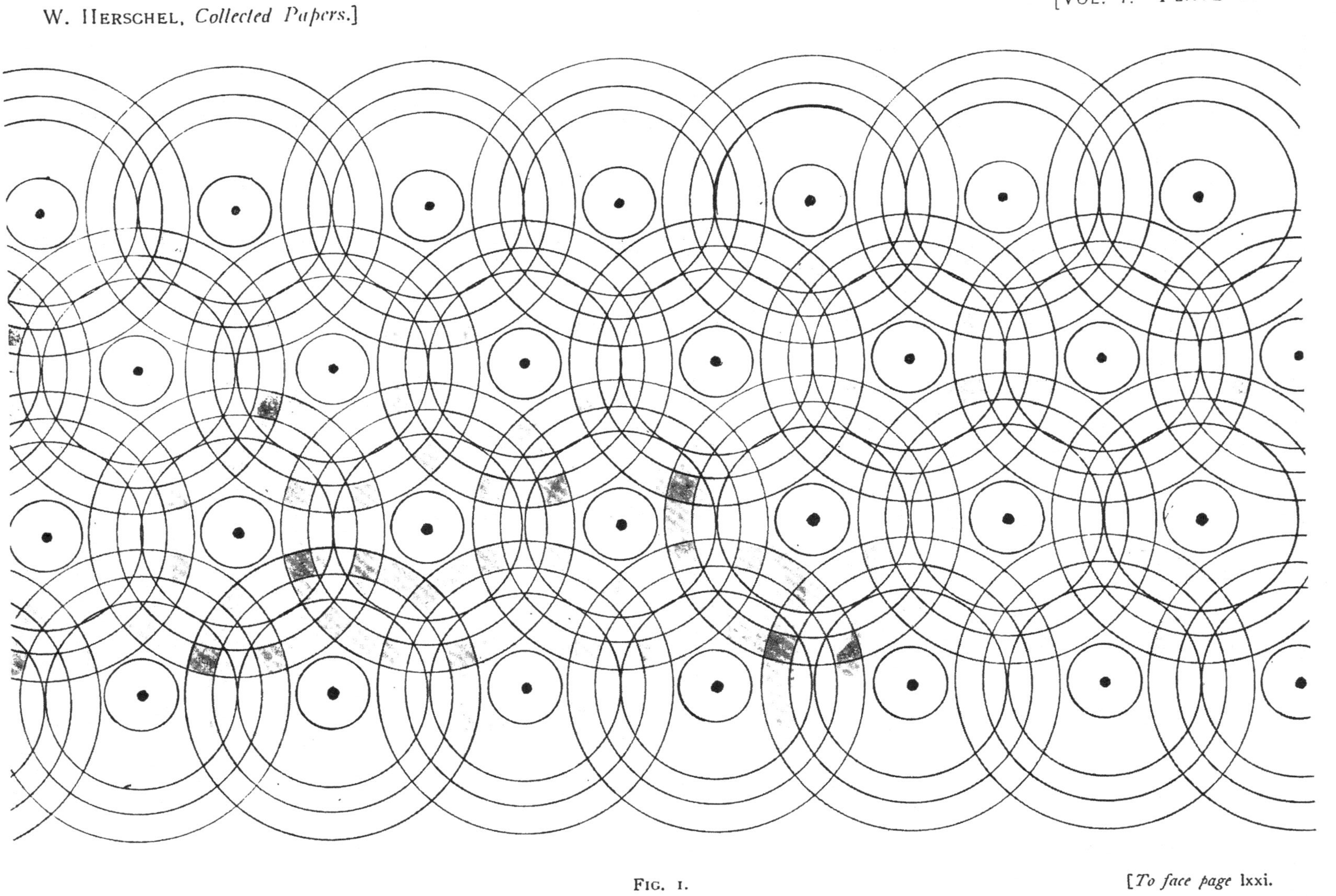

Fig. i.

[*To face page* lxxi.

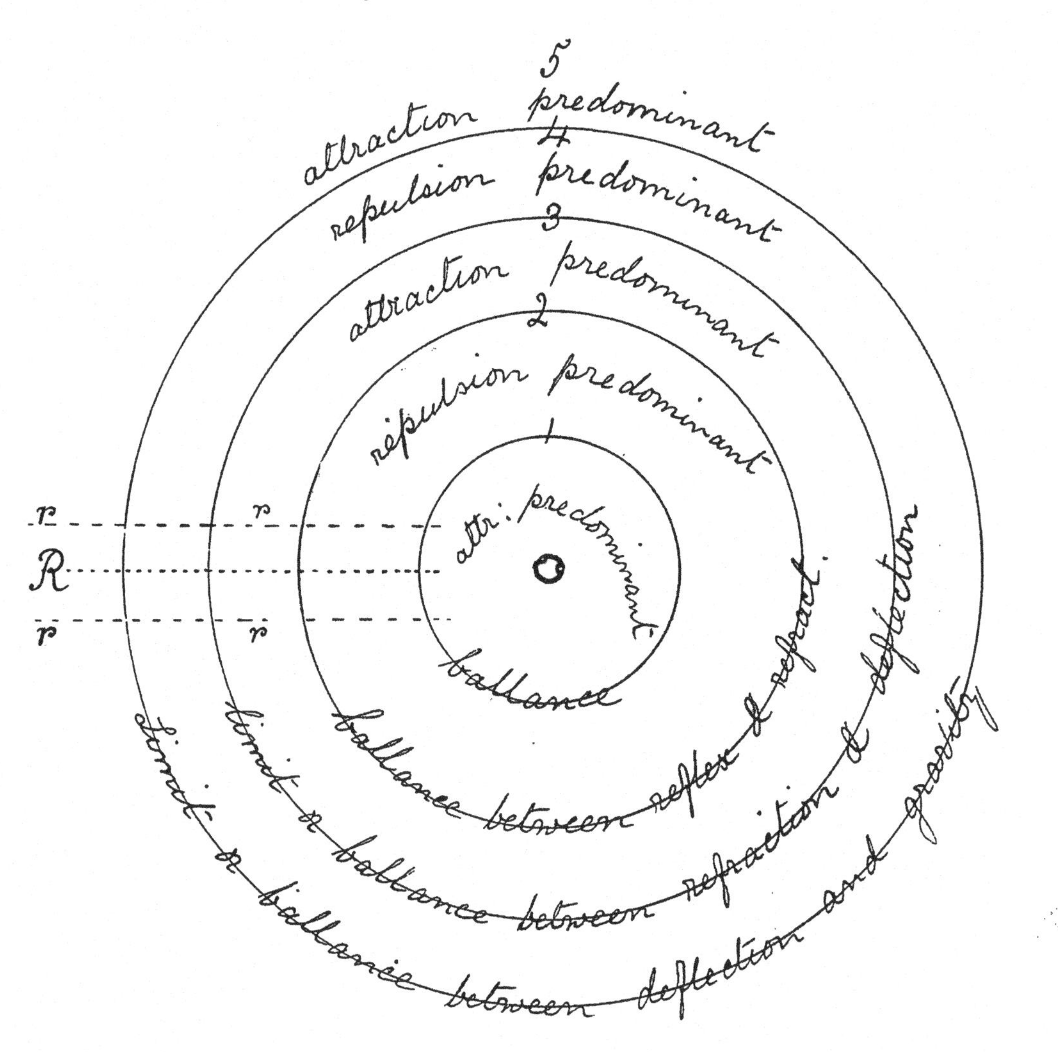

FIG. 2.

[*To face page* lxxi.

action, and by that means afford such an infinite variety as, I fear, will for ever elude our most careful attention.

But as my present design is not to pursue this subject farther than as it concerns the question we are endeavouring to resolve, I shall just hint my thoughts in hopes that other philosophers better qualifyed and more happily situated in regard to experimental assistance, may pursue this subject to its fullest extent.

In the anexed figure, let the black centers represent particles of glass which I suppose to be made up of particles of matter endowed with such powers as I have been describing, and which are held together by the attraction of cohesion. Let the surrounding circles represent the various distances at which the several powers begin, or cease, to be predominant. But here let it be noted that the figure is not intended to convey the least Idea of the real proportion of the extent of each circle, as that must depend intirely upon the Laws, and intenseness * of the centrifugal and centripetal forces and also on the number of particles that are united together in the center to make up a constituent or specific particle of Glass.

To instance in a single particle of Glass, the figure Nº 2 [Plate D], will explain by inspection what I would represent by the collouring of fig: 1 [Plate C].

For the attraction of cohesion which holds together the particles of matter in the center will extend but a little way beyond it where it will be ballanced at 1, and soon after overpower'd by the repulsion of reflexion; this again will be ballanced at 2, and a little farther off, overcome by the attr: of refraction; this also will meet with its limit at 3, where it will begin to be overpowered by the repuls. of deflection; and lastly this will also be ballanced at 4, where the attr: of gravity will predomine over all other powers, and continue at 5, and as far as we know, on to infinity.

I shall now have but little to add; it will immediately appear that upon this Hypothesis of central forces, both M. Bouguer's and Sr I. Newton's opinions agree very well together and will now bid fair for resolving the Question. It is evident that a ray R, falling in a direction, perpendicular to the center, or nearly so as r. r. fig. 2. if its velocity be sufficient, (as we suppose it to be) to pierce thro' the circles of deflection, refraction and reflexion will arrive at the sphere of attraction of cohesion, and will there be stopped. This agrees well enough with Sr I. Newton's rays impinging upon solid particles. As I objected to Newton's Hypothesis, that rays should find particles enough to impinge upon at the surface and hardly any within to stop them, our present explanation will solve that difficulty. Glass being a regular body, consisting nearly of homogeneous particles, we may suppose that they are arranged, by the powers I have mention'd into a regular order, perhaps not unlike what I have drawn, fig. 1, where they are supposed to be held together by the ballance between reflexion and refraction, so that they could not approach without being repulsed nor recede without being attracted. In this case every particle of light that escapes the sphere of the attraction of cohesion of a few of the first rows, will have a good chance to get thro' the whole substance. But at the same time, we know that glass is very far from consisting of particles so intirely homogeneous, or arrang'd with that precise order as never after to afford a single particle that might chance to be situated directly in the way of a ray of light. And this will account for the loss of a great many rays within the body of glass. This Hypothesis therefore, differs from the Newtonian no otherwise than as it ascribes to a power of attraction what Newton ascribes to impinging; and if we consider the number of particles of light that are lost, the Hypo: of impinging would hardly account for so many; For, the excessive smallness of the particles of light would make it next to impossible that they should ever meet with any of the particles of glass at all; but if we suppose a small circle of attraction of cohesion it will effectually stop every particle that comes within its compass.

* I have before observed that the laws of the decrease of the central forces must be different; to which I ought to have added that the forces themselves must also be different in intenseness, as otherwise attraction would prevail at every imaginable distance.

Our present Hypothesis agrees also with Mr. Bouguer's supposition *so far*, that the power which absorbs the light is *chiefly*, tho' not intirely confined to the surfaces of bodies. For, it is plain that in all transparent Mediums one of the constituent particles will assist another to carry a ray of light thro'; because every particle of the Medium having a repulsive power of reflex: will throw off any ray of light that comes in its way, and thereby help it thro' rather than stop it, after it has passed the first surface, where the powers, not being assisted or ballanced by those of other particles, must produce very different effects.

The exact way a ray of light must take that falls either perpendicularly or obliquely, on a plane surface of glass and is not stopped in its passage, can not be delineated except we could trace the laws according to which these powers act, and even *then*, the size of the particles of light and of the medium would also be required. And where so many data are wanting we have but little hopes of arriving at any mathematical certainty.

That light passes through any transparent Medium not in straight lines, but in waving Curves of contrary flexure and in various kinds of such curves, I may venture to surmise; and perhaps there may arrive a time when even those curves shall become the subject of calculation; But that period we can hardly expect to see very soon. I am apt to believe that Newton had some views upon the investigation of the laws according to which that power, which I have called the repulsion of deflection, acts, when he was so particular as to mark the spaces taken up by the fringes of light that are seen about shadows.

That some happy second Newton may arise and succeed in such discoveries must be the wish of every philosopher.

Wm. HERSCHEL.

February 1st. 1780.
Bath.

Addition to Observations on Dr. Priestley's Desideratum, &c.

Read February 11, 1780.

At the last meeting of the Society an experiment was mention'd by Mr. Collins, of bodies continuing to be visible in a dark room, several seconds after they were taken out of the light. Now, as I had before shown, in my Observations on Dr. Priestley's desideratum that the light of a candle could not be visible any sensible time after the candle was extinguished, Mr. Collins thought, that this experiment seemed to be against my argument. Dr. Stamer also inclined to that opinion. It will therefore not be amiss if I observe that my demonstration only regards the reflexion of light from the *surfaces* of bodies, where the loss of it is found to be so considerable.

It is well known that several bodies of the phosphoreal kind, (and by Mr. Collins's experiment it should seem to be probable that perhaps *all*,) have a power of absorbing light in different degrees, so as afterwards to emit some of it again, and by that means to become visible in dark places. This is so curious a subject that I beg leave to see whether it can be accounted for by the powers I have ascribed to the particles of matter in my last observations.

Any kind of compound body, such as a hand, which was what Mr. Collins mention'd, must consist of very heterogeneous particles; and for that reason their arrangement can not be so regular as that of the particles of glass which I delineated in my first Figure. Since, then, the inward texture of the particles near the surface must be exceedingly complicated, and the powers of the attractive and repulsive spheres be render'd very unequal, it may occasion an *internal* reflexion of the particles of light very near the surface, which is not liable to the same law of diminishing, as the light, which in my example of the candle, was reverberated from *surface* to *surface*. And the reason why light is not extinguished or stopped *within* bodies so soon as at their *surfaces* is obvious; because *there* we find a continual ballance of the central powers, which is so wanting at the surface.

After a great number of these *internal* reflexions it is easy enough to imagine that several particles of light must be disengaged again. For, out of no less than, at least two billions of reflexions very near the surface that must be made in one second, it would be very extraordinary, if some one reflexion or other should not happen to be made upon such a sphere of repulsive power, and in such a direction as to be thrown outwards again.

These kind of experiments might be curiously varied; and if any Member of this Society could have an opportunity to make a very dark room I would propose the following tryals, as very much contributing to illustrate the subject.

Exp: 1.

Let an opake object be selected which is known to give a tollerable strong light in the dark after it has been exposed in the manner mention'd by M[r] Collins; but let it be as thin as it can possibly be, without being transparent when held to the light of a candle at a distance. This I would have exposed to bright daylight, or sunshine, on *one* side, then immediately cover'd, and the *other* side observed in the dark room.

By this means we should find how far these internal reflexions of light pervade the pores of solid bodies.

Exp: 2.

The same should be tryed with transparent objects of considerable thickness. They should be covered one side when exposed to the light, and on the contrary side when viewed in the dark room.

By this we might find out how far internal reflexions reach in transparent bodies, which I suppose would be a considerable way.

Exp: 3.

If glass will return light in the dark, a prism should be exposed with two sides cover'd, to try if it would colour the light which it emits again in the dark, where those two sides should be uncovered and the side which was exposed to the light covered.

I am inclined to suppose that there would be no colours since the light returned must have undergone at least 4,800,000,000,000 reflexions in about two seconds of time, and therefore could hardly be affected with the refrangibility of its incidence.

P: S:

I beg leave to correct a mistake in my last paper where I observed that the different number of particles united together by the force of cohesion would occasion a different arrangement in the *extent* of the spheres of the powers. I find it would only affect the *forces* of the spheres, but not their *extent*, which would always remain the same; * In consequence of this mistake I observed that it would be next to impossible ever to arrive at the knowledge of the Laws whereby these powers act; but this difficulty is so far lessen'd by this circumstance that I believe experimental philosophers may flatter themselves with the greatest hopes of success in their attempts to investigate them. A great number of experim[ts] have already been made that may assist us. For instance, the curve that is formed by Milk or any other colourd liquor between two glass plates, set at a very small angle (most probably owing to that power which I have called the attraction of refraction) may be observed and

* *Demonstration.*

Let
F = Force of repulsion of reflexion
f = force of gravity
D = Distance where these two powers ballance each other
n = number of particles

And let the law of the force of repulsion of reflexion be as $\frac{1}{D^x}$. that of gravity as $\frac{1}{DD}$.

Then we shall have $\frac{nF}{D^x} = \frac{nf}{DD}$. which will always be true whatever be n.

has already been the subject of calculation. The rising of Water in capillary tubes and many more experiments of the same nature, may all be taken into consideration.

Dr. Priestley's Desideratum mentions also, that, "if the same cause produce all these effects, it might be expected that it would sometimes produce effects of an intermediate kind; in consequence of which light might not always be reflected or refracted with equal velocity."

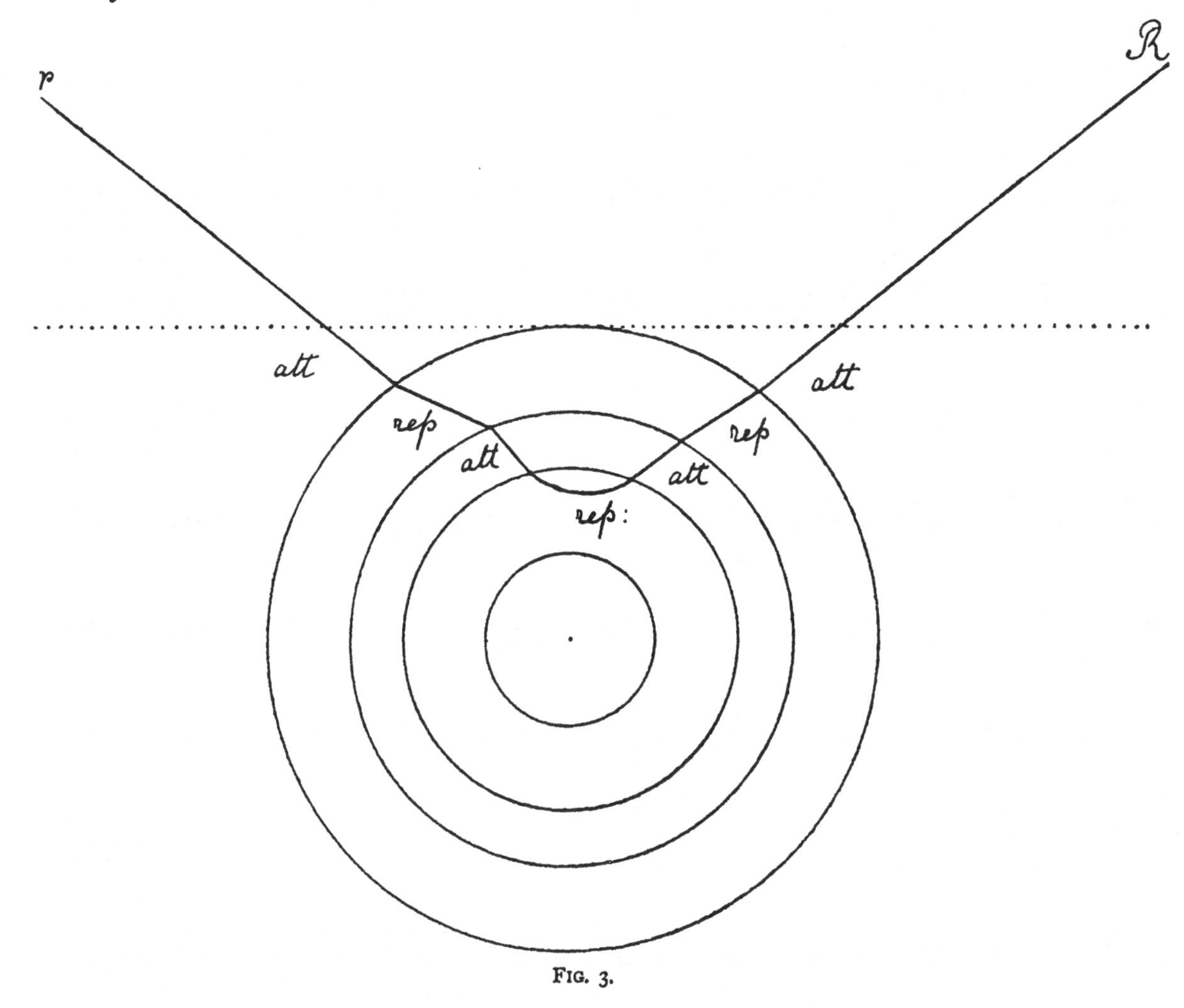

Fig. 3.

In answer to this, it follows most evidently from my hypothesis of central attracting and repelling powers that no such intermediate effect can ever take place. For if a particle of light can possibly escape at all either by reflexion or refraction, it must come off with the same velocity it enter'd the medium, or came within the sphere of its active powers; because there will always be a perfect ballance of attraction to attraction, and repulsion to repulsion in the same sphere. The bare inspection of the annexed figure will be a sufficient proof of what I advance, which however I could easily give in a mathematical form if it were required.

Wm. HERSCHEL.

On the Central Powers of the Particles of Matter.

Dr. Priestleys desideratum has insensibly led me to say something more on a subject which on account of its intricacy I had reserved for much more contemplation: I mean the central powers of the particles of matter. However, as I look upon every member of this Society to be a real lover of truth, I shall not hesitate to lay my thoughts before them, even in their present imperfect state, although many experiments are still wanting for the more particular, and proper establishment of the bold Hypothesis I have hazarded to advance. I flatter myself that I am animated with an equal Zeal for truth, and therefore it will at all events help to advance it; since a discussion of the subject will lead to a farther *developement* (if I may use that expression) of thoughts. By attempting to explain, I shall instruct myself; and if experiment should prove to be against me, or convince me that I have stept farther into the mazes of philosophy than I ought to have done, I shall with pleasure confess either my error or my inability, and sit down contented with the assurance that my endeavours at least were laudable, tho' not successful.

Dr. Watson, very philosophically observed to me at the last meeting of the Society, that he thought the powers of deflection and reflexion might be one.—It is true, that a good philosopher never assigns different causes for effects that may be explained by the same. But how shall I perhaps surprize him, when so far from retracting, I confess that I am partly sure I have not named *all* the central powers there are reasons to suspect may exist. Perhaps I may soon prove the existence of *four more*; But let us see what reasons I can assign for having already named five.

The attraction of cohesion I believe will easily be granted me. This power has been so often mention'd by our best philosophers, that I need not scruple to affirm it to be a *general* and *central* property of every particle of matter.—But how?—of every *particle*?—What can I mean by that word? I have all along used it in preference to *atom*.—It is evident the word particle does not imply an indivisible point. Nor would I have it to mean it. I am firmly of opinion that this very particle, which I suppose to be endowed with those central powers is divisible *ad infinitum*; and that, if it be divided in two parts for instance, and these set at a distance, each half will carry along with it the same spheres of central powers the whole was possessed of, but with half the energy only. Nor can this be more difficult to comprehend than it is, that, when another particle equal to the former, and possessed of the same powers, is added to it, the whole will become doubly powerful: This law is known to obtain in *Gravity*;—where is the difficulty to admit it in *cohesion*? To return, then, to the central powers—

I take it for granted that gravity is admitted as the outermost of them all; Tho' this is perhaps not so absolutely proved by experiment, as one might imagine. Its extent reaches undoubtedly far beyond the orbit of Saturn; and several very excentric comets have carried the proofs of its existence a long way into the truly immense space that lies between us and the nearest fixed Stars. But, that a very weak repulsion, in the inverse ratio of some power of the distance less than two, perhaps only as the distance, does not balance gravity somewhere between the confines of our Solar System and that of other fixed Stars, and thereby prevent the effects of their mutual attraction, which must otherwise in time, tho' perhaps not in many thousand ages, destroy that equilibrium, which seems to be kept up with wonderful order,—I say, that such a power does *not* exist, is by no means clear to me.

Let us, in the next place, examine what proofs we can bring for the existence of those other powers I have mention'd. Sr. I. Newtons experiments in optics are so well known that nobody can doubt the deflection of rays of light; and whoever has attended to the circumstances will easily admit this effect to be owing to a central power of repulsion. Is it not generally admitted that the attraction of gravity will explain the phenomena of nature, such as the retention of planets in their orbits, the ebbing and flowing of the sea, the descent of heavy bodies, &c. much better than any other hypothesis? I maintain therefore, and have the opinion of Newton, Boscovich, Michell and Priestley on my side, that the deflection of light is occasioned by a repulsive power; but I add, that this power also, like gravity, is central and acts without cessation or interruption, by one constant law of decrease, from the center on to infinity. And I flatter myself that the manner in which I suppose it to exist, will find admittance rather than any other, as being the most simple, most consistent, most easily comprehended, and at the same time producive of the greatest variety of effects; and let me add, that every demonstration which proves gravity to be such a central power, makes it at least probable, by analogy, that *this* is also established upon such principles.

The actual existence of a power of refraction is also very obvious from numberless instances, where the rays of light are found to bend towards the perpendicular of the medium into which they enter, when it is denser than that out of which they come; from which we may infer that this effect is owing to attraction. But what I shall endeavour to show at present is that the attractive power of refraction takes place *within* the sphere of deflection. For this purpose I shall once more appeal to an experiment of Sr I. Newton's, where the light that is made to pass between two knives is first of all deflected from each side to the opposite, but *afterwards*, on approach of the knives, is attracted, and separates in the middle, as may be seen by a dark shadow, which is left there. So far, then, our proofs of the actual existence of these powers are established upon the firm foundation of experiments, not only made by our great philosopher, but repeated after him, by various opticians and lovers of science.

I am now come to the power of reflexion, which is what Dr Watson observed might perhaps be owing to the same impulse which occasions deflection; or which at least ought not to be ascribed to another cause without sufficient reason.

That there is such a power, is so generally known and acknowledged, that I need only proceed to show that the effects which I ascribe to it are not occasion'd by the repulsion of deflection. It will soon appear that the sphere of its predominance is situated much nearer to the center than the sphere where deflection is known to act: nay even much within the last mention'd attraction of refraction.

From the above quoted experiment of Newton we learn that the power of deflection takes place at the 320th part of an inch from the knives; and as the next attractive power is also perceived to begin at the 800th part of an inch, it is thereby proved that the extent of action of the deflective power is somewhere between these extremes. Now I shall find it easy enough to prove that reflexion must be occasion'd by a power situated much nearer the center. The microscope affords us instances without number, not only of points, but whole animals, with all their parts, visible by reflexion, who do not make up the size of such a sphere; and yet the rays of light can reach them and by reflexion make them visible. This proves that light pierces far thro' the power of deflection, which consequently must be much too weak, and by its situation, inadequate to the purpose of reflexion. I have myself measured the side of the hexagon of the eye of a fly, when magnifyed by a good solar microscope, where I found it to be between 7 and 8 [thousandth parts of an] inch in length. Now if we say that the naked eye can perceive a point of the 100th part of an inch in diameter, we shall certainly not exceed the truth; but the real magnitude of one of the sides of these hexagons, is only, in round numbers, about the 7000^ths^ part of an inch, and therefore it will follow from this estimation that I saw particles of matter less than the seven hundred thousands part of an inch in diameter, when I looked upon some small visible point in these magnifyed sides. This one experiment only, which is, perhaps, very far from assigning the smallest particle of matter that may be render'd visible by the assistance of the Microscope, is sufficient to show that the sphere, where the repulsion of reflexion prevails, lays much closer to the particles of matter, than that which I have called the repulsion of deflection, and that therefore they must be very different powers. Moreover, from the newtonian experiment I have already proved, that between these two repulsive powers at least one attraction prevails, namely that of refraction.

Thus I think I have made it appear from undoubted experiments that *no less* than these five powers can be admitted. I might now proceed to mention, what reasons I have for suspecting four more, (besides a sixth which I have hinted at above); but I confess that the experiments, which indicate them, are at present involved in such mystery to my conceptions that I dare not pronounce them decisive, till I shall have had an opportunity to repeat them more at large, with corroberating circumstances.

I must now beg leave to answer an observation of Mr Parson's.—He seemed to think that my Hypothesis of central powers was no other, than that of Mr Boscovich and Mr Michell. Let us see what Dr Priestley reports their opinion to be; for I confess that the want of a scientific library at Bath, has put it out of my power, ever to see more on this subject than what I have been able to gather from that author, who seems to be sufficiently clear and remarkably particular, in the description of Father Boscovich's and Mr Michell's hypothesis. The words that are most of all to the purpose are these following: (see *Optics* p. 390)

" Mr Boscovich supposes that matter is not impenetrable, as has been, perhaps, universally taken for granted, but that it consists of physical points only, endued with powers of attraction and repulsion, taking place at different distances; that is, surrounded with various spheres of attraction and repulsion in the same maner as solid matter is generally supposed to be."

(And, page 392) " Mr Michell on reading *Baxter on the immateriality of the Soul*, found that this Authors Idea of matter was, that it consisted as it were of bricks, cemented together by an immaterial mortar. These bricks, if he would be consistent to his own reasoning, were again composed of lesser bricks, cemented, likewise, by an immaterial mortar, and so on *ad infinitum*. This putting Mr Michell upon the consideration of the several appearances of nature, he began to perceive that the bricks were so covered with this immaterial mortar, that if they had any existence at all it could not possibly be perceived, &c."

In the first place I must observe that the various spheres of attraction and repulsion, mention'd by Mr Boscovich, Mr Michell, or Dr Priestley, are never called central powers; and tho' they might surround physical points as a center, it does not appear that they were meant to act all along *from the center on to infinity* according to one universal law, as *I* would be understood to mean when I call them central powers. Nor, indeed, have I proved them to act in this manner, rather than in any other. At present I maintain no more than that it strikes me as being the most eligible supposition, for the support of which, I partly foresee, I shall soon be able to bring experimental demonstrations.

The next very obvious difference between the Boscovichean Hypothesis and mine, is, that his spheres are, in all their extent, of an equal force; so that the sphere of attraction, for instance, where ever it begins and ends is equally powerful. For, as no particular exceptions or remarks are made to the contrary it follows of course that we have no right to imagine anything of that kind was meant when spheres of attraction or repulsion were named. Mr Michell's Idea being taken from Baxter's, is even more expressive of that uniformity in the power of spheres, surrounding what he compares to *bricks*; for by alluding to an *immaterial mortar*, every Idea of that increase and decrease of the predominant power which must happen, near the limits, seems to be intirely excluded.

Again, in my Hypothesis, there are several places where the attractive and repulsive forces balance each other in such a manner that they are in a perfect equipoise, or as it were destroy each other's effects; next to which they begin respectively to exert their predominant powers by infinitely small degrees, so that in reality there is not any one assignable space that agrees in power with the space just adjoining on either side.

I might now observe another difference in the Hypotheses, which, however, as it is foreign to our present subject I will but just name.—It is, that Mr Boscovich, Michell & Priestley, all, in fact, maintain—(it will sound like a paradox when I say it)—the *immateriality of matter*. Boscovich ascribes the *surrounding spheres* to *physical points*. And Michell thinks that the immaterial *mortar* will do *without the bricks*; whereas I ascribe the *central powers* to *particles of matter*: divisible *ad infinitum*; *impenetrable*; in short—material, in the usual acceptation of the word.—But this would at present lead me to another subject, which I may perhaps, some time or other lay before this Society, in a little treatise, called "Remarks on Dr Priestley's Disquisition on Matter and Spirit," which I wrote about a twelvemonth ago, partly with a view to publish; but which I never had sufficient leisure to review and correct so far as to make it fit for *public* view, tho' I should not hesitate to submit it to the inspection of such *indulgent judges*.

Wm HERSCHEL.

Proposition and Queries.

Read March 23, 1780.

Mr Chairman,

Not having it in my power to be personally with the Society this evening, I beg leave to ask, if it will be agreable to any of the members to form themselves into a committee for resolving such questions and problems in Electricity as are either already pointed out by Dr Priestley or may hereafter occur to the committee in the course of their meetings The method of doing this might perhaps be the following, viz. to meet at the Societies experimental Room every day at a certain hour and spend a short time (not less than one hour) on the subject under consideration. If either of the hours, one, two, four, or five should be thought to be least liable to interfere with other engagements, any one of them might perhaps be fixt upon for the time of meeting. Or, whether appointing the hour from time to time should be thought a properer method I leave to be considered.—The committee might report the result of their experiments to the Society at every meeting, and if any conclusions of consequence sufficient to deserve a general inspection should be obtained, a general meeting might then be appointed for that purpose.

The following Queries of Dr Priestley's, I believe might very probably be soon resolved.

"Is the tone of a glass vessel, made in the form of a bell, the same when it is charged as when it is uncharged? or would the ringing of it make it more liable to break in those circumstances?"

To preserve the tone of the bell while it is charging a string of steel or Brass wire may be hung upon a pin with a certain weight at the end, and tuned to the tone of the bell. Or the number of vibrations which the bell makes in a given time before it is charged may be calculated and the same may be done afterwards.

"Is the refractive power of glass the same when it is charged or excited?"

This may be determined by charging the object glass of a Refracting telescope, or the eye glass of a reflector. I have a 15 feet refractor very proper for the purpose; because any alteration in refraction would be soon perceived in so great a focal length.

As Dr Priestley proposes to see the effect of electricity upon the vibrations of an electric namely a *glass* bell, the same experiment properly varied might be made upon a bell of metal; or upon another conducting medium thus:

"Is the tone of a brass string affected by an electric shock sent thro' it at the time that it is sounding."

This might be tryed upon a monochord with a pretty large string, and the discharge of a small battery.

The intention of a committee for the above purpose is by no means to exclude other members that would like to assist *now and then*, as the experimental Room at the time appointed for the committee's meeting might also be open to any Gentleman of the Society; but only to find which of the members have time and leisure sufficient and like the subject well enough to devote a *certain fixed hour* to it.

I should suppose that any number *not less* than four would be sufficient to form a committee, any two whereof being met at the time appointed might proceed to business. But these are only meant as hints. Perhaps an afternoon hour might on this account be preferable, that in case of any *very engaging* experiment, the interruption of dinner would not prevent the Gentlemen of the Committee to stay a little longer than usual if not otherwise engaged.

I am &c:

March 23, 1780. Wm HERSCHEL.

Electrical Experiments.

Read April 7, 1780.

April 1, 1780.

1/ A Phial charged *Plus*, put into a wine and water glass (which had been rubbed with a silk handkerchief, and shewed signs of being excited by attraction) repelled at the coating, do when charged *minus*.

2/ The same phial charged plus and put into the same glass had a silk string fastened to the knob, which string being held in one hand while the other held the glass; attracted at the coating. Whenever the string was quitted the coating repelled,—but Not minus.

3/ The same phial charged plus, and put into the same glass, shewed three spheres about the coating, viz. attraction, repulsion and attraction.—but not minus.

THOs CURTIS—Wm HERSCHEL

4/ The same phial charged plus was put into a glass. All the three spheres were visible, but there is a suspicion that the outermost attraction might be owing to the upper part of the glass of the phial and the brass knob, which attract strongly.

5/ The lower part of the glass of the phial just above the coating half way upwards has a stronger repulsive power than the coating. Wm HERSCHEL.

April 1. *Afternoon.*

6/ The same phial charged plus, put into the glass and the glass put down to stand sometime insulated upon the table, it was observed that the spheres of attraction about the coating, and glass just above the coating, were more extended than they were before.

7/ Charged plus as before, the knob and chain were thrown out; the phial put into the glass. The same two spheres of repulsion and attraction were perceived about the coating, but extremely weak.

8/ The phial and glass being set down upon a plate of glass for about a quarter of an hour, the knob and chain were put into it again. And the two spheres of repulsion and attraction were found to be stronger than before.

9/ A third sphere of attraction surrounding the former two was also observed, but it was almost evident that it proceeded from the attraction of the top of the phial.

10/ The lower part of the glass of the phial repelled and the upper part attracted.

11/ When the phial was discharged over a calcined oister shell, it showed a very fine sea green spot near the middle. THO^s CURTIS Wm HERSCHEL

12/ *April* 4th

The same phial charged plus was presented with the bottom part, which is concave, to the ballance, by being held in the hand. No kind of effect could be perceived, at 10, 9, 8, 7, 6 and every lesser distance up to the very phial. The bottle was discharged and found to contain a very good full charge. The distances were measured by a glass ruler not excited but perfectly dry. The room had not been electrified before, this being the first charge.

13/ At the convex side of the coating there was found repulsion at 3¼ inches and attraction at 1 inch, when the phial was presented to the ballance in the glass.

14/ By letting the phial and glass stand some time insulated the spheres were enlarged. THOMAS CURTIS. Wm HERSCHEL.

April 5th

15/ A Newtonian reflector of 7 f^t was placed so as to have an object in view, and when the great Speculum was electrified, it occasion'd not the least sensible difference in the distinctness or perfection of vision. Power 222.

16/ A shock being thrown through the speculum it occasioned no alteration in the vision.

17/ An eye lens being electrified by means of two small circular plates of tin foil, the reflector shewed no signs of any alteration in the refractive power of the glass. Power 30.

18/ A Thermometer placed upon the conductor when the cylinder was turned, was not sensibly affected. THOMAS CURTIS
Wm HERSCHEL

Experiments on Prince Rupert's Glass Drops.

THE satisfactory manner in which Mr Arden explained the breaking of those glass drops generally called Prince Rupert's drops, if I remember right is, " that the particles of glass are in a state of repulsion when they are taken out of the furnace, and by being dropt into cold water the outward coating becomes hard very suddenly, and thereby these outward particles are put into a state of attraction, while the inner still remain in their repulsive condition; therefore as soon as the end is broken off, there is an opening for the repulsion of the inner particles to exert its power and rend the drop to pieces." So far I think we may rest satisfyed with this rational hypothesis. On the remaining part of the explanation, wherein Mr Arden ascribes the breaking of the neck of a glass bottle when it is filled with water and the end of the glass drop broken off when the drop is immersed in it, to the density of the water, as more able to transmit the shock to the bottle than atmospheric air, I would beg leave to make an observation and report some experiments I have made relative to that very singular circumstance.

I conjectured that the breaking of the neck of the bottle, might perhaps be owing to the elasticity of glass, and that the drop when it exploded might probably communicate a certain tremor or vibration to the particles of water, which being conveyed to the glass might occasion a similar vibration in the neck of the glass bottle and by that means break it.

To put this to an experiment I imagined that if Mr Arden's explanation was right the explosion of the glass drop in water ought to break through any other confinement of equal strength with the glass bottle, the medium and distance being the same; but, that on the contrary, if the elasticity of glass was concerned as I conjectured, the drop would not destroy a much weaker obstacle than the neck of a glass bottle. I will now relate the experiments,

Exp: 1.

In a tube of stiff, but thin pasteboard of exactly the size of the neck of a glass bottle, filled with water, I broke one of these drops. It made a very sensible report, but the pasteboard tube was not in the least injured.

Exp: 2.

In a wooden tube of about the thickness of the neck of a glass quart bottle, filled with water, I broke another drop ; but the tube remained unhurt.

Exp: 3.

In a small thin stone vessel, but which was full twice as large in diameter as the neck of a quart bottle, filled with water I broke the drop ; it occasioned a very smart shock and I thought at first the little pot had been cracked, but on examination, found it quite sound. This vessel having a glazed coat, both inside and outside was partly elastic, and I suppose had it been a little thinner or less in diameter would have been broken.

Exp: 4.

Having but four of these glass drops, I ventured now, *very heroically*, to break the last in my hand, under water, and found that the sensation was in no respect different from what I suppose it would have been, if I had broken it in the open air.

Wm HERSCHEL.

April 8. 1780.

On the Utility of Speculative Inquiries.

Read April 14, 1780.

THE following reflexions were occasioned by the accidental debate on the admission of the question proposed at the last meeting of the Society for Speculative inquiry, "Whether Space be anything actually existing."

One of our members, observed that speculations and metaphysics were of little use to mankind ; that it was better to keep only to experimental philosophy, and as such even doubted the propriety of introducing the above-mentioned question.

When I mentioned the admirable essay on human understanding wrote by that excellent Metaphysican, Mr. Locke, he said that this book had done no good, and that Locke himself had allowed it to be of little use.

It was alledged that metaphysics served to instruct the mind and guide it in the search of those truths which could not be ascertained by experiments; this he answered by comparing the Metaphysician to a dancing master who came in with a fine bow, and after having amused you a while with his various hops and skips, left you with another very fine bow, no wiser than you were before.

When mathematical demonstrations were urged as instances of truths that depended not upon experiments ; and the well known proposition, of the three angles of any triangle being equal to two right angles was mentioned, this Gentleman seemed to derive the certainty of that Theorem from its being known to be matter of fact.

All these doctrines appearing new to me, I beg leave to give my opinion upon them with that openness and freedom which every member should maintain, who would wish to deserve the honour of belonging to a philosophical society. But before I proceed, I also beg leave to assure this gentleman, that, if I have in any respect misunderstood his meaning, I shall be very glad to have it explained better, and in that case, whatever I say, as contrary to his

l

supposed opinion, can not at all concern him ; but may be looked upon as so many general observations in answer to any one who should maintain Speculative knowledge to be of little use.

To begin with the propriety of the question " whether space be anything actually existing," I must observe that it is not so ill expressed as this Gentleman intimated ; for it was given in that plain concise manner with a particular view to avoid circumlocution and obscurity. And I shall be very glad to be shewn how it could have been expressed better. But this being foreign to my present purpose I proceed. The Sixth rule of our Society mentions very particularly, that " the members shall be at liberty to discuss with moderation and freedom any subject within the circle of the arts and *Sciences.*" From which I conclude that a metaphysical inquiry can not be called an improper subject.

Names indeed are not very material, when the meaning is understood. However I beg leave to say that the authorities of the best latin writers are not wanting to translate metaphysical into *sublime* or *elevated*-physical. And in that sense I shall for the future continue to use it, for the credit of the science which is called Meta-physics or the most elevated part of physics. And in that sense I always thought it was used by every metaphysician I have had the honor to be acquainted with.*

It was said that speculation and metaphysics were of little use to mankind. This I deny. The perfection of our nature is evidently to be looked for in the superior powers of reason and speculation. What would all experiments avail if we should stop there, and not argue upon them so as to draw general conclusions ? And how can we argue and draw conclusions if the superior intellectual powers are not improved by frequent exercise in speculative researches ? Half a dozen experiments made with judgment by a person who reasons well, are worth a thousand random observations of insignificant matters of fact. But setting aside the very obvious consequences of improved faculties, the subjects of mere speculative knowledge are of the highest concern to those who love wisdom. By Metaphysics we are enabled to prove the existence of a first cause, the infinite Author of all dependent beings. By mathematics we come to have a just Idea of the superlative perfection of His works. By Logic, we can prove them to others ; By Ethics we are made sensible of our duty towards the Author of our existence, and to our fellow-creatures ; . . . But, were I to go on to enumerate the endless advantages that we derive from a speculative turn of mind I should never have done. By what I have already said it appears to me that the works of a Locke, or of any other good Metaphysician, can not with any propriety be compared to the performance of a dancing master who enters with a fine bow and leaves you with another.

When I place Ethics among the *speculative* sciences, I do not take this term in a sense opposite to *practical*, but only mean to denote that the precepts it inculcates do not owe the force of their conviction to experiments ; thus, it is not necessary to have seen murder committed in order to be convinced that it is wrong to kill our neighbour.

As to mathematics, their use is so extensive and generally known that it would be impertinent to say anything in their praise ; but that it should be imagined they owe the least degree of their evidence to experiments, is what I never heard advanced before ; and what I suppose will very easily appear to be a mistake. Should any one doubt whether the three angles of a triangle are equal to two rectangles. let him by no means apply to an experiment. For, if he does, he will surely never believe it. In all probability there never existed a triangle in the world (at least not such a one as could fall under our senses) wherein this proposition would not have been found to fail. If the most exact triangular figure that art can make be examined with a microscope and its angles measured by a good micrometer, we shall soon give up the equality of the three angles to two right ones.

To say *more* upon the utility of speculative inquiries, would be useless, to a society, by

* An english Dictionary says, Metaphysical, above physical. Altieri renders Metafisica, scienza divina, o prima filosofia.

whose fundamental rules it appears that one principal end of its very institution, was the investigation of speculative truths; and that I have *said so much*, is owing to an apprehension that the remarks which here I have animadverted upon, might have had some effect in depreciating the value of that branch of knowledge which, in my opinion, must for ever remain the glory of rational beings.

WILLIAM HERSCHEL.

April 12th. 1780.

Experiments in Electricity (continued).

By W. HERSCHEL and T. CURTIS.

Read April 21.

Exp^t 19th The usual size Phial charged plus shewed not the least repulsion when presented by the hand to the ballance with the concave bottom but plainly repelled with the convex coating on the side quite near the bottom.

20th The Phial charged plus and put into the usual Glass, when charg'd only a single turn of the Cylinder repell'd from ab^t 3½ inches to the very point of contact; charged two turns of the Cylinder the Coating attracted at 5 inches but repell'd at 1½ up to the point of contact.

21st. *April* 24. The same Phial being charged plus and insulated attracted at 4 Inches from the coating repell'd at 2 inches and attracted at 1/8 of an inch.

N.B.—When the attraction was observ'd at 4 inches the glass rod being presented between the brass knob of the ballance and the Phial at ½ an inch from the knob stop'd that motion and in several instances repell'd.

T. CURTIS.
W^{M} HERSCHEL.

[Copied from Mr. Rack's copy, W. H.'s own MS. being missing.—ED.]

On the Electrical Fluid.

Read May 5, 1780.

THE Idea of positive and negative electricity has always appeared to me to labour under very great difficulties; and I believe the following experiments, which were made with a view to throw some light upon the subject will prove very unfavourable to the supposition of *plus* and *minus*.

The argument is this, if an increase of the quantity of the electrical fluid, be found to accelerate the pulse when a person is electrified *plus*, we ought to conclude that a diminution of the same should retard the circulation of a person electrified *negatively*.

To ascertain the matter of fact I took an assistant to the experimental room of the Society, and having prepared everything proper for the purpose, I rested a while to let my pulse settle to its natural temperament, and found it to beat 75 times in a minute. Then I insulated myself by standing upon four glass bottles, well rubbed with a silk hankerchief, having no better contrivance at hand; and lapping one end of a chain round the negative conductor, I put the other into my bosom. Having stood for a few minutes in this manner while the assistant was working the cylinder, I began then to examine the pulse and found it to beat between 80 and 90 times in a minute. I continued the same experiment for 10 or 12 minutes, but, perhaps on account of the imperfect insulation or other causes not known, could not increase the pulse any farther, tho' I changed the chain for a brass rod with a ball at each end, placing one of them upon the negative conductor while I took the other into my mouth.

I was however convinced that my pulse had been accelerated by what is called being electrified *minus*; that is, according to the common explication, by the diminution or absence of the electrical fluid.

To see how far I might rely upon this experiment, I now received the electricity from the positive conductor exactly in the same manner, but could not by any means bring the pulse to beat oftener in a minute than it had done when I was electrified negatively. In both situations I was sufficiently electrified to make sparks visible that were drawn from my hands by the touch of the assistant.

I had now recourse to the electrical syphon, which is an experiment well known to every electrician; when I suspended the vessel containing the water, at the negative conductor I found that from single drops the discharge was immediately changed to a small regular stream, when the Cylinder was worked, and return'd again to drops when I left off turning.

The very same thing happen'd also when the vessel was suspended at the positive conductor.

These experiments being thus evidently contrary to what we ought to have expected according to the usual division of electricity into *plus* and *minus*, must be look'd upon as very unfavourable to that opinion.

I would however recommend a repetition of the first experiment with all the required accuracy, that it might appear, not only that both *plus* & *minus* do actually accelerate the pulsation, but also whether they do so in an equal degree. If any difference should be discovered when the operation is continued for any length of time, such as half an hour or more, it might open a new view of the electrical power when applied to medicinal purposes.

W^m. HERSCHEL.

March 21. 1780.

On the Existence of Space.

Read May 12, 1780.

THE subject I am going to consider, namely the existence of Space, has been the occasion of great debate and so much has been said for and against the opinion which I shall here endeavour to support, that it will require a method somewhat different from that which is generally taken by those who either maintain or deny the existence of space, in order to arrive at a fair view of the question. With this intention I shall go back to the very foundation of our knowledge. We shall perhaps find that one of the principal sources of it, has either been intirely overlooked, or at least has not been sufficiently attended to. Knowledge, according to Mr. Locke, consists in the perception of the agreement or disagreement of our Ideas. This may be obtained three different ways, viz. by Intuition, by demonstration and by sensation.

The perception by Intuition, in my opinion, should be distinguished into two classes; one of which we might call Consciousness, the other Intuition. The reason for this distinction appears to me very considerable and deserving of some attention. When we perceive by what I now call Intuition we have two Ideas in our mind, and perceive their agreement or disagreement; but when we perceive by what I shall call Consciousness, we have but a single perception, which is not only full as evident and clear as Intuition, but moreover still less compounded, not requiring two Ideas as Intuition does. I shall endeavour to explain this more at large.

Des Cartes says we perceive our existence by demonstration, and his argument runs thus.

" Whatever thinks exists.

" I think;

" Therefore, I exist."

Locke maintains we perceive our own existence by Intuition: That is, by only reflecting on the terms *I am.*

I shall say that we perceive our existence by Consciousness; To justify this distinction, when we examine the operations of our own minds, it will be found that in all intuitive perceptions there are these three things concerned. viz. two Ideas and the relation between them; As, the *whole* is *greater* than a *part. Whole, greater, part.* But in what I call consciousness, there is but a single Idea or perception. Thus when we say, *I am,* the truth of it is not perceived by reflecting on the Idea annexed to *I*, and that of *am*; nor do we wait to see the agreement between those two Ideas to perceive our existence. Consciousness is still more simple. The very word *I*, means something that exists, and *am* is only as it were a repetition of the same Idea. We find even that several languages, as the greek, the latin, the italian, do not so much as express the personal pronoun, but simply say, *AM*, which is perfectly well understood.

If it should be observed that in the proposition I am, there may be an Idea formed of what we call *I*, and another of *being*, or existing; and these two may be affirmed of each other: I will by no means deny it; but this is only saying that we may also perceive by Intuition what we can perceive by consciousness, which I not only grant but likewise add that we may also perceive it by demonstration. For I believe no one can deny the truth of Des Cartes's demonstration, " whatsoever thinks must exist; I think, therefore, I exist." All I would observe is, that, where a superior way of perception can be had, an inferior is of little use. And what is more, when the nobler way of knowledge is wanting, inferior methods will, most generally be found to fail. I question whether any kind of intuitive reflexion, much more so, whether a demonstration would convince me of my own existence, if consciousness were wanting.

It seems then to be one of the distinguishing characteristicks of Consciousness, that no other kind of perception but consciousness can be imployed as a proof in behalf of a proposition that belongs to its province. It is not so with Intuition; where that is wanting demonstration may often supply the defect.

Again, Consciousness is not communicable like Intuition or Demonstration. Thus, that I am is very evident to me because I perceive it by Consciousness; but it may not be equally evident to another person. Nay, if any one will be so sceptical as to deny it, it may not be in my power to convince him. For, since we can only employ consciousness as a proof, the only way to argue with a person upon the truth of that proposition, is to put him into such a Situation that his own consciousness may convince him that I am. To this end, I may try several means: place the argument in several different points of view, and in any one, which I think favourable to my design, may ask as a postulatum *grant that I am.* For instance let him say " I have the Idea of such a person as you affirm yourself to be, I see you, I hear you, and so forth; but I know that all these, what you call proofs are but Ideas in my own mind that have no necessary connection with existence;—it is possible I might have all these Ideas and you not exist. Therefore your existence is not proved to me."

What can we say to answer this argument? Neither Intuition, demonstration nor sensation can be of service. We shall soon find that where Consciousness is required no inferior perception will serve the turn. An argument taken from either of those will only involve us in endless cavils, misapprehensions and ambiguities. In this instance I should answer, " As I know that a certain succession of Ideas will be in your mind when I speak to you, the existence of which Ideas you neither can nor will doubt, because you are *conscious* you have them, I say unto you:—Imagine that in your turn you wanted to convince me that you exist. All the arguments you have used to deny my existence you are conscious will equally hold good if I make use of them against yours; and yet you are *conscious* that, if I do make use of them I am in the wrong. Does not this, by Consciousness strike you as an argument that those reasons by which you would confute my existence are fallacious? Therefore, *grant that I exist.*" In vain will he say to me that this is no proof, but a downright petitio principii. I grant that if he did not perceive by consciousness the truth of my argument, it will, and must appear to him a mere petitio principii. For, my reasonings in putting home the question to him in the above manner evidently do *suppose* that I exist, (which was to be proved) as otherwise the case would not be similar. After all I have no great objection to the terms. A postulatum amounts nearly to the same thing as a petitio principii; And if we can make an argument to prove what it was required to prove, by placing it into such a point of view that the principles it is built upon shall be easily grantable, why may we not add this seemingly new way of proving a point in question, by a petitio principii, to those different methods already introduced in the schools. We have arguments ad hominem, ad ignorantiam, ad absurdum; let me add this ad petitionem principii to it.

It is time now to apply what I have said to my purpose.

It will not be amiss, in some manner to define what I mean by the word space; and that I may avoid the scholastic terms of substance and accident or mode, which I am apprehensive might lead me into difficulties it is not my business to enter into, I shall call Space a simple, infinitely-extended Existence. When it is mention'd in reference to the absence of matter it is often called a *vacuum*; or when considered in relation to bodies not in contact, such a part of it as lies between them is called *distance*. When conceived as occupied by a body it is called its *place*. Metaphysicians also have stiled it sometimes the *Inane*. Several negative properties may be affirmed of it such as actual indivisibility and immobility; but admitting the Idea of it which it was my present business only to point out, to be sufficiently clear and well determined, I shall immediately proceed to endeavour to make it appear that Space is a real Existence and not merely an abstract Idea of the understanding.

I do not see that our Ideas of things that exist have any other connexion with existence than by consciousness; therefore all proofs relative to existence must be taken from that principle. The very evident proposition, *I exist*, and that equally true one, *several other things exist*, as we have seen above can only be proved, or rather illustrated by consciousness. All that is incumbent upon me to do, therefore, is to produce such instances, in nature or in Idea, in which the existence of space may be perceived by consciousness: in which it may be reasonable for me to say grant that Space exists: in which it may appear that the existence of Space is, if a petitio principii, at least a very rational one.

First, an instance taken from Nature.

Let there be two spherical bodies placed so that there may be one foot distance between them. Observe that the distance or space between them will remain the same whether it be filled with any matter such as Iron, Stone, Mercury, Water, Ether, Air, Flame, or any other still more subtle medium. This suggests to us by Consciousness that distance is not owing to the intervention of any kind of matter, but is itself something that exists. Let there be two other bodies of the same spherical figure placed at only half a foot's distance. Do we not plainly perceive by evident consciousness that they are differently placed from the former. If the space between them may be filled with anything you please without changing the distance, it appears from the same consideration that the medium which surrounds the two bodies on every other side is out of question and can not be concerned in their distance. Here then we have two bodies placed at one foot's distance, two others at half a foot; If neither the foot nor the half a foot's distance, have any thing to do with matter, *i.e.* may be filled with one sort or another, or be perfectly void, who does not see that Distance or Space must be something actually existing? Is not one Distance twice as long as the other? Who will affirm that one nothing can be twice as long as another nothing? Therefore grant that Space exists.

Another instance taken from Nature.

Suppose a man to be sitting in a chair, and while he remains there he fills or takes up a certain portion of Space which we call his place. Suppose him now to rise, and leave that place, and let a child sit down upon the same chair; we perceive presently that the child does not take up all the man's place but only some part of it. The man's place then, it seems, is greater than the child's place? But how could this be, if neither of these places were anything?

An Instance from Ideas.

Imagine a globe of iron, such as a cannon ball, to exist and every other kind of material thing to be removed out of being. To me it appears evident that the ball would be moveable. Suppose it then to be in motion, and let it move on in a straight line. After having moved sometime, does it not appear reasonable to think that it is in another place than it was before it begun to move? If the ball be in another place I believe we shall perceive by consciousness that both places, and all the places it passed thro' in its motion must be something really existing. Could it move out of one nothing into another nothing? Can the Space thro' which it moved be called nothing?

Huygens said that it was possible some of the fixed Stars might be so far off from us that their light tho' it travelled ever since the creation at the inconceivable rate of 12 Millions of Miles per Minute, was not yet arrived to us. The thought is noble and worthy of a Philosopher. But shall we call this immense distance a mere imagination? Can it be an abstract Idea? Is there no such thing as Space? *

It is to me the most inconceivable thing, that the very name of Space does not convey to every one who hears, and understands the meaning of it, the Idea of something really existing. Inane, vacuum, plenum, room, place, distance; call it what you please; How could we exist but in space? How could I stretch out my hand if there was no room! Could we get up and walk if we did not leave the place we were in! If we go sometime straight forward are we not at a distance from the place we set out? How idle it would be for philosophers to talk of a plenum or vacuum if neither of these terms meant anything really existing. Nor is there any great refinement required to perceive that space is something besides a mere Idea. Consciousness and plain common sense are sufficient. Nay it is by ill timed refinements we are mislead. We do often reason common sense out of doors.

If it be thought absurd to deny the existence of matter, how much more so must it be to deny the existence of Space. Every sense furnishes us with the Idea of it; and every particle of matter which can not exist but in space: can not move but out of one part of it into another, is an evident witness of its actual existence.

It is not my present business to give a full definition or description of what Space really is; suffice it that its existence is made clear; A fuller account of it I may perhaps hereafter find an opportunity to delineate. However as I have in the beginning of this paper called it . . . . a simple *infinitely extended* existence, I will just observe that as Space is the stage on which Omnipotence displays its wonders, it can not be thought extraordinary that I have ascribed infinity to it.

W^m. HERSCHEL.

* The Newtonian law of the attraction of gravity says that bodies attract in the inverse ratio of the square of the *distance*. By this is meant that according to the *space* between the attracting and attracted body this law takes place; which would be an absolute absurdity if space did not exist. It may be proved by a thousand experiments that this law does not depend on any medium that may be between the attracting and attracted body but only on their distance. Therefore every experiment that proves the Newtonian law must evidently suppose as its foundation, that space exists.

Continuation of the Experiments on Glass Drops.

HAVING now by the favour of Mr Parsons obtained more glass drops I resolved to make some experiments in order to examine the first part of Mr Arden's explication. For, upon a more mature consideration I was rather inclined to think that the explosion of the drop when the point was broken off was owing to the very great elasticity of the drop, and to no other cause. By the manner of their being made it is obvious that the drops must be as elastic as any glass can possibly be, and that for this reason any violent tremor must instantly communicate itself throughout the whole and by that means rend the drop to pieces. This led one to imagine that if any of these drops could be broken in such a manner as to prevent that fatal vibration, the remainder would perhaps escape unhurt; and with this view I tryed the following experiments

EXPERIMENT V.

Upon a small, but coarse grindstone I ground the thick end of one of the drops and took off a considerable part without any injury to the rest, which I preserved. While I was grinding, I perceived that notwithstanding I held the drop very much pressed between my fingers to prevent any violent vibration the small end of it trembled extremly, so as to become almost invisible. I judged that the extend of the vibrations was about $\frac{1}{8}$ of an inch, both sides taken together.

Exp: 6.

Another drop which had a very thick point seemed to me to be proper for a tryal to grind that part of it down. I succeeded so well in this, that I was convinced I could have ground away the whole drop without danger. In this opinion I begun to grind the same drop upon the side of the narrow end and ground a good part of it away. Imagining now, there was no kind of danger I was rather careless in not holding the drop sufficiently hard, to prevent the vibrating motion it received by grinding violently, when it exploded, and gave me a very smart blow.

Ex 7.

When the drop exploded in the last experim^t I had just before observed that I was near one of those bubbles in the inside of the drop.

Experiment 8.

I took another and ground till I came to one of the bubbles, which I laid open to its middle without any accident. It appears that these bubbles are little cavities.

Exp 9.

One of the drops I ground almost all around and at the bottom as far as I could conveniently come at it without breaking it.

Exp. 10.

I attempted to grind down the point of another but it exploded.

Exp 11.

I put the small end of a drop thro' a hole drilled in a piece of brass into which it fitted very firmly. On breaking the point near the end the drop remained intire. I broke it again pretty near the middle and the drop was not hurt. I broke it a third time quite close to the brass, then the drop exploded. I ascribed this to some little motion of the drop in the brass hole which I could not help making when I broke it as it was exceedingly hard.

Exp 12.

I tried the same experiment again and broke the point of the drop twice very safely. A similar drop left unconfined exploded with the smallest touch of a rivetting hammer at a place very near the extremest edge of the point.

Exp 13.

Upon a very fine brass tool with polishing emery I ground away, sideways one half of the thin end of a drop till it fell off and the drop remained unhurt. Attempting to grind off another piece it exploded. Wm HERSCHEL.

The Specimens of the drops mentioned in the 5th 8th 9th 12th Experiments were produced at the Meeting of the Society and, in the presence of the Members.

Exp: 14.

The drop mentioned in the 9th Exp: was broken off at the point upon which it exploded in the usual manner.

Exp: 15.

On the Brass tool with fine emery I ground away in two places, part of the thin end of a glass drop. See the specimen.

Exp 16.

When I ground the thick part of the drops in the 5th 8th and 9th experiments, I observed every now and then, as it were, small explosions or crackling noises, like taking sparks from the Cylinder of an electrical apparatus weakly excited ; And upon viewing the part that was ground always found the surface uneven, being bright in some places as if the particles had been broken off and dull in others as if they were ground away. I ascribed this to partial and internal vibrations occasioned by the coarseness of the grindstone, tearing off small parts ; Therefore upon the brass tool with fine emery I ground away a good part of the thick end of a drop, and in the operation found no perceptible crackling ; nor were there any bright places to be seen on the drop as before except very small ones, such as are perhaps in proportion to the former as the fineness of this manner of grinding is to that which was used in those experiments. (See the Specimen)

Exp 17.

I began now to suspect an internal vibration of the particles of glass, which might perhaps occasion a drop to explode tho' the outward vibration should be intirely stopped. Therefore I cast one of them into wax, leaving the small end to stand out. When the wax was perfectly cooled I broke off the end, without the least precaution two different times ; But breaking it again a third time, the internal vibration reduced the drop to pieces. However I thought they were not so small as usual.

Exp: 18.

I tried the same experiment again and broke the end twice very safely. When I broke it the third time the drop exploded and the pieces were plainly not so small as usual ; the bottom part of the drop being partly left whole, tho' in all appearance very much shivered within. The vibrations seem to have affected the drop in a certain regular manner. As soon as I can procure a fresh supply of drops many other experiments shall be tried which may perhaps throw more light upon this subject, which, it seems is very far from being exhausted.

W^{M} HERSCHEL.

Letter from Mr. Herschel to the Reverend Dr. Maskelyne, Astronomer Royal.

SOME time ago I delivered to this Society a series of Observations upon the height of the lunar Mountains, which paper, having since been presented to the Royal Society by Dr Watson, I have been desired, in a Letter from the Revd Dr. Maskelyne, Astronomer Royal, to Dr Watson, to deliver a further explanation of my method of measuring the projection of the moon's Mountains, and in answer to that Letter, I have drawn up the inclosed Memorandum in a Letter to Dr Maskelyne. Now, as the contents of that Memorandum, are in some respect necessary to compleat the Theory of my former paper, I take the liberty of presenting a copy of the Letter to this Society.*

I beg leave to observe Sir, that my saying there is almost an absolute certainty of the Moon's being inhabited, may perhaps be ascribed to a certain Enthusiasm which an observer, but young in the Science of Astronomy can hardly divest himself of when he sees such wonders before him; And if you will promise not to call me a Lunatic I will transcribe a passage (from a series of observations on the Moon of a different nature I begun about 18 months ago) which will shew my real sentiments on the subject.

"Perhaps conclusions from the analogy of things may be exceedingly different from truth; but as in things beyond the reach of Observation we have no other way to come at knowledge, the imperfection of these Arguments may in some measure be excused; And I may venture to say that if we do not go so far as to conclude a perfect resemblance, we must allow great weight to inferences taken from this source. For instance, seeing that our Earth is inhabited and comparing the Moon with this planet: finding that in such a satellite there is a provision of light and heat: also, in all appearance a soil proper for habitation full as good as ours, if not perhaps better—who can say that it is not extremely probable, nay beyond doubt, that there must be inhabitants on the Moon of some kind or other? Moreover it is perhaps not altogether so certain that the moon *is* out of the reach of observation in this respect. I hope, and am convinced, that some time or other very evident signs of life will be discovered on the moon."

"When we call the Earth by way of distinction a planet and the moon a satellite or attendant, we should consider whether we do not perhaps, *in a certain sense*, mistake the matter. Perhaps—and not unlikely—the moon is the planet and the earth the satellite! Are we not a larger moon to the moon than she is to us? Does it not appear that there is a much more uninterrupted, even temperature there than here? What a glorious View of the heavens from the moon! How beautifully diversified with hills and valleys! No large oceans to take up immense plains, fit for pasture &c: Uninterrupted day on one half, and on the other a day and night of a noble length, equal to many of ours! Do not all the elements seem at war here when we compare the earth with the moon? Air, Water, Fire, Clouds, Tempests, Vulcanos &c: all these are either not on the moon, or at least kept in much greater subjection than here. If as a prerogative we assign the size of the earth, and its motion since it carries the moon along with it; I answer, And is not the sun still larger than the earth, yet we allow him to be only the servant as he yields us light and heat. And tho' by his attraction the earth is made to revolve round him, yet we allow the earth to be the nobler of the two. Even so I say it is with moon. The Earth acts the part of a Carriage, a heavenly waggon to carry about the more delicate moon, to whom it is destined to give a glorious light in the absence of the sun; whereas we as it were travel on foot and have but a small lamp to give us light in our dark nights and that too, often enough extinguished by clouds. For my part, were I to chuse between the Earth and Moon I should not hesitate a moment to fix upon the moon for my habitation."

I am very much flatter'd with Dr. Maskelyne's wish that I might repeat and continue

* [Printed in the *Phil. Trans.*, vol. lxx. (below, pp. 12–15), but the following remarks, at the end of the letter, about inhabitants of the moon were omitted.—ED.]

my observations, which I shall not fail to do under all the circumstances that are mention'd.

An object glass Micrometer has undoubtedly very eminent advantages, but as that could not possibly be used without introducing *refraction*, I fear that the noble simplicity and distinctness of a Newtonian *reflector* would suffer very much by such an addition I have once or twice intended to set about cutting an object speculum and contriving the same kind of sliding motion for it, which is used in the Object-glass Micrometer, but the difficulty of the execution has made me lay aside all thoughts of it, tho' I believe in the hands of a good optician it would be nothing.

There is a way of doing the same thing by only cutting the eye glass, which will answer upon several of the moon's mountains, and upon all such objects as have a sensible parallel diameter. I have attempted it, but tho' I did not succeed in the execution I saw enough to find that it could be reduced to practise. Since the above mention'd attempts, I have constructed another what may be called a wire Micrometer, tho' instead of two wires there are extended two single threads of a silkworm. I have a series of observations in hand that require the utmost nicety of measuring, which has been the occasion of my constructing no less than five different wire Micrometers. If I could suppose you would think a perusal of their description worth your while I should not hesitate to give it at length; two of them I believe are new.

Inclosed I send some observations that were made with a view to find whether any effects of a lunar Atmosphere could be perceived. They may perhaps be acceptable.

I ought now to beg your pardon for the extraordinary length of this letter; but from what Dr Watson says, to whom I am under the highest obligation for introducing me to the honour of your correspondance, I flatter myself that what I have said will be received as having been wrote with no other view than to be laid before you in order to be examined; And if it should meet with your approbation, either the whole, or any part of it, which you may think proper or necessary, may be added to the paper which contains my Observations; All which I leave intirely to your superior judgment and remain

Revd Sir
Your most obedient
humble Servant
Wm HERSCHEL.

June 12th 1780
Rivers Street. Bath.

Observation of the Occultation of Gamma Virginis, made with a view to determine whether any effect of a Lunar Atmosphere could be perceived.

Read Augt 17, 1780.

March 20, 1780. 11^h 48′. I looked with a compound eye piece that takes in a very large field of view and soon found γ ♍ at some distance from the moon. This glass made the apparent distance of the two little stars not much more than a diameter.

12^h 6′. With a power of 222. I measured under very unfavourable circumstances the distance of the two stars and found it, both diameters included 8″·75; intending to measure them again as they approached the moon's limb, and by that means to judge whether their apparent place suffered any change from a refraction of a lunar Atmosphere; but in measuring I found that it would be impossible, in so short a time as I should have, to be sufficiently accurate, as I already knew, the measure I had thus hastily taken to be much too large, and as the weather was not favourable for that purpose.

12^h 15′.* I saw the occultation of the stars one after another, without the least visible

* As I had no other view in observing this occultation than that which I have mentioned, my time is not at all to be considered as true, having taken no pains to gain it, and my time pieces having lately been removed from another house.

change in their place with regard to each other. I did not chuse to measure the distance of the stars again when they were very near the moon, as that would have taken off my attention to the occultation. Besides, their distance with a distinct power of 222 times, was so well defined and so small, not exceeding two diameters of the smallest of them, that I thought it best to trust to the eye.

On Liberty and Necessity.

Read September 1, 1780.

DOCTOR PRIESTLEY'S doctrine of Philosophical Necessity has given occasion to the following reflections on the subject of Liberty and Necessity. To avoid confusion in a discourse which is to express thoughts which are somewhat abstruse, I shall define the word Necessity as I mean to use it in the following reflections.

Necessity when applied to Actions, or rather effects, denotes that kind of connection between them and the Causes from whence they proceed, whereby they are the *Unavoidable* and *Mechanical* consequences of them. I do not affirm that there is a contradiction in the *Nature* of things that such and such causes might not have had different effects if the Laws of Nature had been otherwise ordained; it is sufficient to constitute a natural necessity if by the present established laws of Nature certain effects unavoidably and mechanically flow from certain causes. This definition which I take to be Agreeable to the general as well as Philosophical use of the Word, will in my Opinion assist us greatly in having a clear view of the subject.

In the first place it is evident that actions considered as necessary effects of certain causes or motives, are thereby said to be *unavoidable* and *mechanical*; and it appears to me that this is the sense in which Dr. Priestley looks upon all our Actions to be *Necessary*. A weight not supported, necessarily, that is to say unavoidably and mechanically falls to the ground.

Give me leave to draw a few Consequences from this Instance. If the weight by its fall should crush a man to pieces, it is not guilty of Murder; we do not blame it for what was *Unavoidable*; it deserves no reproof for what was *Mechanical.*

Here then we have the criterion of Morality established at once by a very simple investigation. No Action can be said to be moral, that is Unavoidable, that is mechanical. In Short, that is necessary. I shall now also, in the next place define what I mean by Liberty.

Actions are called free when they are not the Unavoidable (though perhaps certain consequences of) Motives that Occasion them; when they are not the mechanical but *voluntary* though regular effects of the causes that determine the Agent.

This also appears to me to be the common as well as philosophical acceptation of the word Liberty. For it is by no means requisite to make an Agent free, that He should be able to Act without any Reason or Motive at all; on the contrary it is evidently against the nature of a rational Being to Act without a Reason.

The Ideas of *Certainty* and *Necessity* are very different. An Action may very *certainly* flow from such and such motives, which in no Respect can be said to be the *necessary* consequence of them. A man walking in an Unknown Path which is crossed by a Precipice will *ccrtainly* turn aside when he comes to that place, supposing him to see it. If he is of a humane disposition he will *certainly* relieve a person in great distress when He has it in his Power. These Actions we Approve.

It follows, then, in general that *certainty* does not exclude Morality though Necessity does. If an Action be not unavoidable; not mechanical; but voluntary or self-determined from proper motives, it is a moral Action, and cannot with any sort of propriety be called necessary though it may be very certain.

The distinction between Mechanism and Morality is so obvious and evident, that no man ought to make use of the term necessary instead of certain when he means to speak philosophically of a Moral Action.

It is not absurd to say that a good and wise man will *certainly* Act in Such and such manner upon such and such an Occasion; but it would be highly improper to say that he will or rather *must* unavoidably and mechanically or necessarily act in that manner.

Perhaps a Necessarian will say that there is but little difference between certainty and Necessity. I answer, in this case the difference between the two terms amounts to as much as the whole difference of the two Modes of Action of conscious or Unconscious Beings.

Well regulated moral Actions we call certain; Unavoidable mechanical effects we call Necessary. When an intelligent moral Being acts very uniformly from proper motives and reasons, we ought certainly not to use the same expression to denote this voluntary and moral regularity that we employ to signify the Stubborn mechanical effects of unintelligent causes. And surely to act regularly, with consciousness and intelligence, upon proper motives, is not the same thing as to Act (or rather to be Acted upon) without sense or perception by a mere Mechanism and necessity of Nature.

We may grant all the force that can be required by the strongest Necessarian to proper and sufficient motives. We may even allow (would it were true!) that Rational Beings are always certainly determined by them. This is the very circumstance which distinguishes a conscious being from inanimate Nature. A Stone is incapable of being determined by a Motive; nothing but a Necessity of Nature can determine its effects. The excellence of an intelligent creature consists in this Superior way of being determined in its Actions.

And thus we see that certainty is not necessity, and that we are not Necessarians when we admit the regular and certain influence of motives.

It remains only to take notice of an Objection which has been made to Liberty in general; that is, if motives *certainly* determine our Actions then they are no longer free.

I should be glad to hear a Proof of this; for I cannot by any means admit it. On the contrary it appears to me, that may most surely and certainly determine Actions which notwithstanding shall be free in the fullest sense of the word. If they who make the Objection suppose that to be free I must be Able to Act without any motive at all, they affirm a fancy of their own invention; and Dr. Priestley's Son B. is a chimera that never existed, which therefore is very easily confuted.

We know but two sorts of finite Beings or principles; one intelligent or cogitative, the other unintelligible or Uncogitative. The Unintelligent principle acts by unavoidable mechanical operations, unconscious, and *Un*willing of its own Actions.

The intelligent principle, on the contrary, acts according to motives, is influenced by considerations, thoughts, and judgment, is conscious of its own determinations thus influenced, and exerts voluntarily its powers accordingly.

If these two methods are not evidently different, and deserve a distinction, I know not what can deserve to be distinguished. And if the general consent of mankind has introduced terms for this purpose, calling the Actions of the former necessary and Mechanical, those of the latter free and spontaneous, I do not see why we should not still continue to use them as being very proper to express things in the manner they are found to be in Nature.

August 25. 1780.

Short Account of some curious Experiments on Glass Bottles, taken from the last Publication of the Bologna Society. De Bonon: Scient: Inst: Comment: p. 97.

Read Nov[r] 24. 1780.

IT is often found that those peculiar qualities which we ascribe to some particular sort of things are common to the whole kind; that this is the case with a certain extraordinary property of some glass bottles, was accidentally discovered by Casalius, who found that all common bottles in general have the same wonderful quality. The discovery was made as follows.

In a place where diamonds, gold and other valuable things were publickly kept as Pawns * it was custumary to wrap up the Diamonds in papers. But to save the trouble of opening so many papers when the pawns were looked over by the Governors, it was ordered, that for the future the Diamonds should be placed open, ready for inspection, in glass vessels. When this order was complyed with it was observed after a few days that several of the glasses were cracked. This was immediately reported and the broken glasses shewn to the Directors.

Cavalius who was one of them reflected upon this singular circumstance and made the following experiments.

He placed a Diamond weighing 16 grains in a small glass vessel, which remained two days unhurt, but the third day it fell to pieces. Another glass broke in 15 hours. These glasses were placed upright. Two more also placed upright contained, one 4 Diamonds weighing 11 grains, the other 5, weighing 8 grains. These glasses remained six days sound. Being inclined sideways, one of them broke in three days time, the other soon after. Another glass vessel in which were 5 Diam[ds] was laid down, and it broke after a month's time. One containing some very small Diam[ds] remained whole 10 weeks tho' its position was often chang'd.

These glasses were much of the shape of our Apothecary's Phials but larger and thicker, and so strong that they might be suffered to fall from a height of 3 or 4 feet upon a stone floor without any sort of damage.

Experiments were also made with Agates and Crystals. Six and thirty grains weight of pieces of Agate were placed in 4 upright glasses; they all broke at different times, the soonest in 2 hours the latest in 5 days. In another upright vessel was thrown a single piece of Agate weighing but little more than 2 grains and the glass was broke in 5 hours, so that there seems not the least proportion between the weight and the time.

When the glasses were inclined they all broke sooner; there were even some that remained whole when upright, but broke soon after they were laid down upon the side.

Some large pieces of Mountain Crystal were placed into a glass vessel, which remained unhurt tho' they were kept there a good while, but when some of the Crystal was bruised, a smaller weight put into the vessel broke it in 3 hours. The most surprizing thing that happened was that about 33 grains of bruised Crystal were put into a glass that had not been touched by any kind of stone; After some days when no fracture appeared, the position of the glass was changed several times but no crack was found; A whole month passed away and all hopes of the glasses breaking being now given up the Crystal powder was taken out of the glass, when, beyond all expectation the forty first day after the Crystal had been taken out of the glass it fell to pieces.

* In several parts of Germany and it seems also in Bologna, no private Pawnbroker is allowed; in lieu of which the Governm[t] has appointed an Office where money is lent at a very reasonable rate on valuable Deposits.

Short Account of some Experiments upon Light that have been made by Zanottus, with a few remarks upon them.

(*De Bonon: Scient: Instit: Comment:* page 105, de Luce.)

Read Decr 1. 1780.

THE Newtonian doctrine of Light teaches us that it consists of innumerable different shades or gradations of colours such as are expressed in the Rain-bow, and that every one of them when separated from the rest is of so fixed and unalterable a nature that no power of refraction or reflexion can produce the least change therein.

Zanottus, who knew that this was sufficiently proved by a number of experiments made upon the rays of light with opaque and transparent Objects, thought that to these might be added some tryals with lucid or burning Mediums, which in his opinion were more likely to operate any change than those that had been tried.*

Accordingly he separated sometimes one and sometimes another of the colours by the usual prismatic method, and made it to pass through the flame of a candle; but always found that it went on to the paper which was placed to receive it, without any change either of colour or brightness. Some little cloudiness which was occasionally perceived could easily be accounted for from the smoke or thicker part of the flame, which did not always burn equally pure, and therefore, acted sometimes as an opaque body would have done, by intercepting some of the rays.

As a pure flame had no effect upon any particular colour, he next had recourse to a green flame thro' which he occasioned a red ray to pass, and found as before that it suffered not the least alteration. Being satisfyed with these experiments he proceeded to some of a different nature.

He wanted to know whether light is derived from the Surface of a flame only or whether it also proceeds from the innermost parts of it. To investigate this by experiments he placed two lights upon a table at a small distance and observed what degree of brightness they threw upon a paper with small Characters so placed as to be but just legible. While he kept observing the paper, the two flames were several times made to join into one and separated again, and he could perceive no alteration in brightness. The same experiment he also tryed with three flames and the result was always the same.† From which he concludes that the same degree of light may remain tho' the surface of the flame be diminished.

The next experiment he made upon this subject was to let a flame shine thro' a hole made in a board or screen and placing a paper so as hardly to be able to read it, he caused the flame to be encreased by the addition of other flames, when he found that he could read with much more ease than before, tho' in this case the surface of the flame, (confined by the hole in the board) was not encreased. From these Experiments Zanottus concludes that the light of

* Three years ago I made an experiment upon the rays of light that bears some resemblance to those that were made by Zanottus; but is rather a more forcible argument of the extreme subtlety and unchangeable qualities of light than even those are which Zanottus reports. Having fixed up a page of a Newspaper against a Wall at some distance, I placed a seven f^t. Newtonian Reflector so as to read it. While I was reading, (which was about noon in the Summer of the year 1777) the Sun shining very bright, the focus of the Sun's rays collected by a concave Mirror of 6 inches diameter was contrived to be thrown directly into the place where the first image is formed in the telescope. My apparatus was disposed so that I could at pleasure cover or uncover the Mirror which threw the collected rays of the Sun into this image; and by several tryals I found that not the least change could be perceived in the effect of the Telescope, whether the Solar focus was thrown upon the Image or not.

† A wellknown experiment which I have often tryed with two candles is to place them so as to unite their flames or separate them occasionally, and the result always is that the united flame gives a much stronger light than the two separate flames. However this may be accounted for upon principles that do not contradict Zanottus's Theory. For by the approach of the second candle the heat being considerably encreased must dispose the wax or tallow of the candles to inflame much faster than before.

a flame is not in proportion to the surface of flames but rather in proportion to their whole mass.*

In consequence of these conclusions the Author attempts to solve a difficulty. He says that it has been observed in an eclipse of the Sun that the degree of light which is taken away by the interposition of the moon does not at all answer to the surface which is covered; but that in the beginning only a little obscuration takes place, and towards the middle a much greater degree is found to come on, than what can be explained from the surface which is covered. This he says is easily to be accounted for if we admit that light is in proportion to the whole mass and not the surface, since towards the margin of the Sun a part of his body contains but little mass in comparason to what is contained under the same surface towards the Center; and therefore the obscuration must in the beginning of the eclipse be much less in proportion to the cover'd part of the disk than afterwards.†

If the fact, which I think I have also read elsewhere, is well ascertained it is extreamly curious and some proper method ought to be found out to discover with certainty the real ratio which the Obscuration in an Eclipse of the Sun, bears to the part of the disk which is hid from us. Every Eclipse of the Sun should be observed with this view, as such a discovery would perhaps lead us to many others, in regard to the Nature of this glorious body of light.

Perhaps a method like the following might not be amiss. Suppose a person to be placed in a darkened room with a book before him and sitting in any certain posture, the book remaining all the time in the same place. In a window shutter or board let there be a long thin opening, *e.g.* 2 feet long by $\frac{1}{4}$ of an inch broad. Let this be covered by a slider that may be gradually withdrawn by means of a wheel and pinion, the handle of which should carry or be itself an index to denote the length of the part uncovered by the slider. Now let this be set so as to admit only light sufficient that the person placed in the dark room may just be able to read. In an Eclipse of the Sun, as light is gradually lost, our Observer must open the slider in such a manner as always to preserve just enough to read with equal ease; then will the quantity of the Aperture observed and compared by a clock beating Seconds (which should be done by an assistant,) give us the ratio of light which is lost. If the Assistant with the Clock were placed outside of the room he might very easily note the times, and from a Scale on the slider read off the corresponding parts uncovered by the person in the room, who would not be interrupted in his attention. Moreover, as the room would be rather too dark

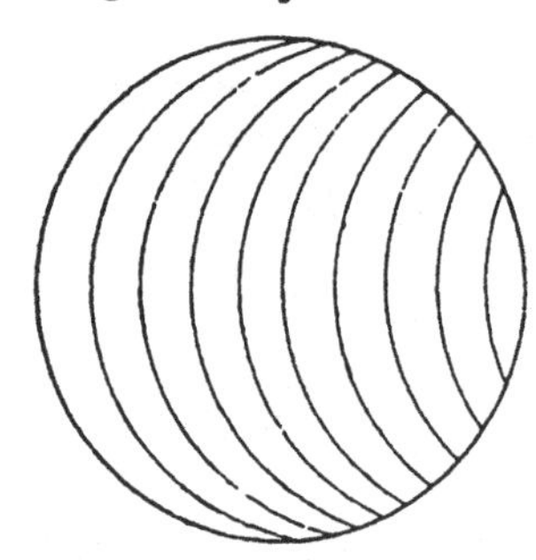

* It is a wellknown property of light that it will penetrate other light, and therefore the inner parts of a flame will certainly penetrate the outer parts and thus encrease the light; but how far this may hold good when a flame becomes very considerable in its dimension has not been investigated by proper Experiments; and in my opinion would perhaps be found to fail. This is a subject that might be tryed by our Experimentalists as soon as a room is fixed upon. I can hardly suppose that a ray of light would traverse a clear flame of a yard diameter without being totally stopped, much less, that the same would hold good when we suppose a flame of several thousand miles in depth. It will not be easy to find a certain method to discover how far light can penetrate light. For if we imploy flames for the purpose it is evident that when they become large their own light, by mixing with the ray which is to traverse them, will prevent our discovering whether they either have stopped or any way changed the ray. In the focus of a burning mirror this inconvenience is prevented because every ray there has a particular known direction and can not mix with another ray; For this reason it seems that an Experiment with several strong burning Mirrors could be much more depended upon, than those that may be tryed with flames; however these latter also ought not to be neglected.

† To this I object, in the first place that it is not at all probable that the whole mass of the Sun should consist of one intire clear flame. Nay, the contrary is proved by the observation of so many spots, which indicate plainly that under the outward flaming Surface is most probably contained a solid globe of unignited Matter. The gravitation of all the planets to the Sun indeed is alone a sufficient demonstration of his solidity.—Now, if we admit that the sun is only a superficial fire tho' perhaps covered with flames that may rise to an altitude of several thousand miles, Zanottus's solution of the difficulty to explain the Phenomena of a Solar Eclipse will not hold good at all, but on the contrary we shall find it still more unaccountable than ever For in this case, just contrary to his supposition, the margin of the Sun being covered ought to deprive us of the greatest light; since there, the flame must be deepest of all, when taken in a perpendicular direction to an Observer placed on Earth.

within, for reading off the divisions on the slider or handle of it, with perfect ease, it would on that account also be better to have the assistant without.

When I consider that it is not without some trouble to calculate what proportion of the Sun's disk is realy cover'd by the Moon in any given equal portions of time such as for instance every five minutes; (As will be sufficiently clear by inspection of the annexed figure) I am very much inclined to doubt the matter of fact. And this is an additional reason why it ought to be investigated the first opportunity that may offer.

Remark on a passage in Dr. Priestley's History of Electricity.

Read Decr 22, 1780.

Dr. PRIESTLEY mentions the charging a plate of air in the same manner as plates of glass had usually been charged. (page 300 &c.) As the experiment is very well known I will only remark that there does not seem to me the least resemblance between what I think is very improperly called charging a plate of air, and charging a plate of glass.

When a plate of glass is charged by means of moveable coatings, we may take off those coatings and find no kind of electric power in them. Instead of those we may apply other coatings, which we are sure contain no charge, to the same plate of glass; and on making a proper communication there will be an explosion. This shews that the plate of glass contained the charge.

In the experiment called charging a plate of air if we remove the coatings the power is found to reside in them where ever they are carried; and if in their stead other, not charged, coatings are substituted and a communication made, there will not be the least explosion. It is therefore evident that the plate of air does not contain any charge. If we make a communication between the positive and negative conductors of an electrical apparatus there will be an explosion; yet we do not say we have charged the air or space that separates the two conductors. The truth is, that air in any of those experiments neither does contain nor is capable of containing a charge.*

Answer to the question whether the Electrical Fire contains any sensible degree of Heat?

Read Decr 22, 1780.†

NOTWITHSTANDING the many plausible arguments that have been urged to prove that the electrical fire contains no sensible degree of heat I have always found myself inclined to maintain the contrary opinion. A worthy Member of this Society who is an excellent Electrician alledged as a proof of his opinion that a needle might be made blue by an electrical discharge and yet not discover the least warmth to the touch. I found a way to reconcile this experiment (tho' very much in favor of those who deny the heat of electrical fire) to my own Ideas on the subject, by supposing that the electrical fluid may find so easy and direct a passage thro' the needle that no sensible effect of its presence can be left, tho' at the same time it may be accompanyed with a very great heat; in the same manner as the astoñishing heat of the focus of a large burning Speculum, as soon as it is covered or turned aside, leaves not the least trace of heat in the place where the moment before it would have melted copper in a few seconds of time.‡

How far this opinion is founded upon truth may be judged from the following Experiments.

* [It is interesting in connection with these vague remarks to note that Henry Cavendish's discovery (only recently published) of specific dielectric influence dates from 1773. See Cavendish's *Electrical Researches*, edited by Clerk Maxwell, p. xlviii.—ED.]

† [The heating effects of disruptive discharges were developed by similar methods in exact detail by Riess in 1838.—ED.]

‡ See Experiments with Mr. Vilette's Mirour. *Phil. Trans:*

I communicated my design of making some tryals upon Thermometers in order to discover the heat of electrical fire to Mr Smith the Optician, who told me that he expected a very capital Apparatus. He had by him a little Instrument called a luminous Conductor, consisting of a glass tube which may be exhausted by an air pump, and containing a Ball on one end and point at the other, which becomes luminous when ever it is held to the prime conductor of an excited electrical apparatus, whence it has obtained its name. It appeared extremely probable that a thermometer inclosed in such a luminous Conductor might perhaps be as good a method as any to try whether a sensible heat could be discovered in the electrical fluid ; because a thermom^{r} thus placed would be exposed to a continual surrounding flash of the electrical fire. Mr Smith's electrical machine being arrived and proving to be a very excellent Cylinder that yielded an admirable quantity of fire, I hastened to try the experiment. With an extempore apparatus of a small Thermom^{r} tyed to a wooden, ill divided scale to which it did not belong I resolved with his assistance to make the tryal, with an intention at an other time to repeat the same with more exactness in case the success of this tryal should give encouragement to pursue it further.

Having placed the Thermom^{r} into the luminous conductor and cemented the ends properly we exhausted it as well as we could by means of an air pump, tho' in all probability we did not succeed quite so well as we could have wished because we had some little suspicion of the ends not being perfectly safe, nor did the screw of the luminous conductor fit well to the screw of the air pump. Everything being prepared, we placed that end of the luminous conductor which contain'd the brass Ball and the bulb of the Therm^{r} over the positive prime Conductor at such a distance from the same as to receive a continual succession of sparks. To the upper end of the luminous Conductor was fastened a chain to carry off the electrical fluid after its passage thro' the vacuum which inclosed the Thermom^{r}.

When everything was thus adjusted the whole was left a sufficient time for the mercury of the Therm. to subside ; for by handling the tube of the lumin^{s} Conduc^{r} as well as warming the ends when we applyed the cement the mercury had been considerably expanded. As soon as we found it sufficiently settled the Cylinder was put in motion.

The very first appearance was extremly remarkable and if the experiment had answer'd no other end would have sufficiently rewarded us for our trouble. A stream of very bright electrical fire seemed to issue from the Mercury within the Thermo^{r} and to fill all the vacuum within it, to the very end, where it vanished.

This beautiful illumination gave me the greatest hopes of success as it evidently shewed that the agitation of the electrical fluid did reach the very innermost parts of the Thermometer. Nor were my hopes disapointed.

In about five minutes of time the mercury was found to have rose visibly, and continued rising the next 5 minutes still more perceptibly We kept the Cylinder in motion about a quarter of an hour, when we found the Mercury to stand a division and a half above the place where it had been before the Cylinder was put in Motion.

In order to be assured that the rising of the Mercury was not owing to a change of temperature in the room (for I thought it possible that the heat of the person who turned the Cylinder, which we did by turns, might affect the thermom^{r}) I thought it proper to stop, And while we remained in the same situation as before the Mercury immediately began to fall and in about 5 or 6 minutes time was found to have returned to its former position.

In order to obviate every possible objection, such as the nearness of the person who turned the Cylinder ; the approaching of a candle when the rising was observed ; I had the experiment repeated with the following alterations. The luminous conductor was placed at a considerable distance from the apparatus, and the electrical fire, by means of a long chain carried from the positive prime Conductor to an assistant, from which the luminous conductor received a continual spark as before. A candle was fixed at a proper distance and every thing so contrived that an observer could in a few moments read off the division where the mercury stood and return again immediately without any fear of disordering the Thermom^{r} by his presence. After a sufficient time allowed for the Mercury to settle ; the Cylinder was put in Motion and in about five minutes time we perceived that the Mercury was certainly rising. It rose still faster the next five minutes, after which it still rose tho' but slowly for the next five minutes which we kept the Machine at work ; and at the end of the quarter of an hour we found the Mercury to be a *full* division above the point where it had rested before we begun this Experiment. The Cylinder was now left at rest and the Mercury immediately begun to fall and in a few Minutes was observed to be something below the place from whence it had rose by the working of the Machine.

I was extremly well satisfyed with the Situation of everything during this 2^{d} Experi^{t}, yet, as the air of a room where there is a fire on one hand, door and windows on the other, can not be in a very settled degree of temperature for a length of time, I proposed to repeat the same once more tho' it was not at present in our power to make some intended improvements in the apparatus which I shall mention hereafter.

It will not be amiss to take notice that as it grew late the fire had been suffered to go down, and was nearly extinct, the door also by some interruption had been opened several times, so that the Thermometer seemed hardly to be stationary when we begun to make the third experim^{t}.

However being unwilling to give it up I begun to turn the Cylinder, and in a few minutes the Mercury was found to be rising ; at the end of the first five minutes I resigned my place to Mr Smith and saw it full half a division above the mark where it stood before. At the expiration of ten Minutes we could not perceive that it had rose any higher ; on the contrary I rather suspected it in a falling condition and at the end of the quarter of an hour we found the Mercury actually returned to the place where it stood before we begun to work the Machine. To find whether the reason to which I ascribed this difference was realy the alteration of the temperature of the room, we gave over working the Apparatus when the Mercury fell very quickly down lower than it had been when we begun this 3^{d} Experiment, and in about 6 or 7 minutes was no less than 4 divisions lower than it had been at any time the whole evening.

The Cylinder acted incomparably well tho' the day was by no means favorable for electrical purposes.

On examining the Thermometer which had been inclosed in the luminous conductor it was found that a division on the wooden scale above mention'd answered to 1,56 Degrees of Farenheits, and that consequently the greatest effect we had perceived to arise from the electrical fire answered to 2$\frac{1}{3}$ degrees of the same Measure.

These experiments seem to amount to a demonstration that electrical fire is accompanyed with heat. However I shall take an early opportunity to repeat them with the following necessary improvements.

Two equal Thermometers to be provided of a proper scale adjusted to each other and equally sensible of the least alteration of heat and cold. The degrees of one of them to be marked upon the glass that there may be no occasion to introduce wood or metal into the luminous conductor. The other to be placed just by the outside of the luminous conductor, that it may appear whether there is any change in the temperature of the room while the inclosed Thermomet^{r} rises and falls. Instead of a chain, to connect the assistant by a wire of a sufficient thickness, to the prime conductor. And if, under these circumstances the Experim^{t} should answer as I fully expect it will, I think we may call it decisive.

The appearance of the luminous vacuum within the Thermometer is a very curious circumstance by no means to be overlooked, and may lead to several new researches. Those who maintain glass to be impervious to the electrical fluid may ascribe it to the negative fire within being disturbed by the positive electricity induced into the luminous conductor and acting upon the outside of the Thermometer ; I shall however at present wave all sorts of argum[ts] upon that subject, confessing myself intirely ignorant of the propriety of the terms of positive and negative when applyed to Electricity. And very much doubting whether anything exists at all in the nature of the electrical fluid, that can deserve those appellations.

Dec[r] 14. 1780. W[M] HERSCHEL.

On the Heat of the Electrical Fire (continued).*

Read Dec. 29. 1780.

I HAVE now the pleasure of laying before this Society two experiments on the Thermometer, wherein the heat of the electrical fire shewed itself so incontestibly that there can be no room left for any farther doubt on the subject.

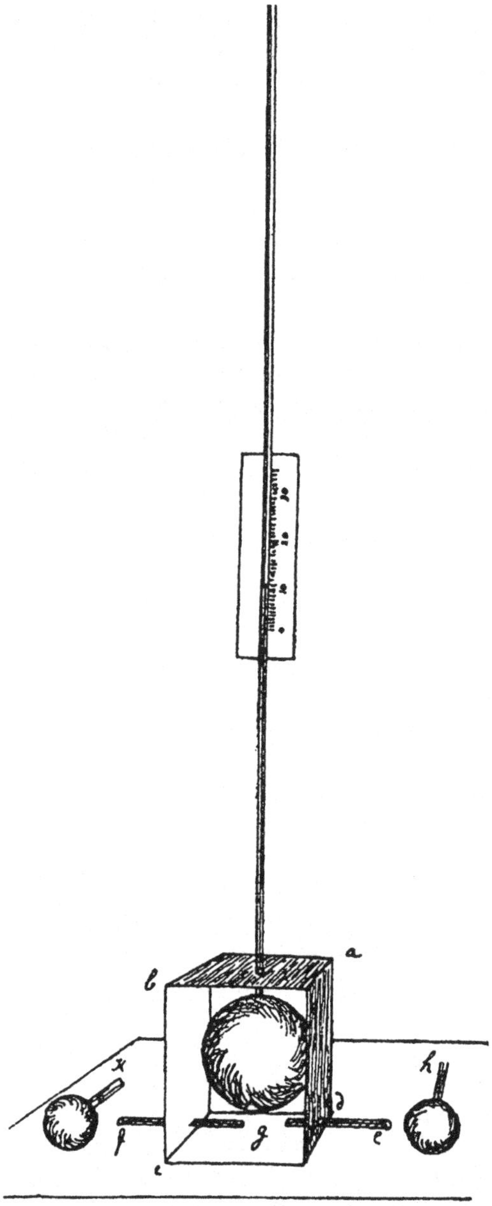

D[r] Watson, ever ready in the cause of Philosophical pursuits, assisted at the execution, and took a Memorandum of the results, of which the following is a Copy.

Before I relate the facts it will not be amiss to give an Account of the Methods that were taken in these experiments. When the question was debated at the Meeting on Fryday the 10th of November, several methods of trying the Thermometer were suggested by M[r] Atwood, M[r] Bryant, M[r] Parsons and myself; Such as putting interupted coatings upon the bulb of the Thermometer, in Order to surround it with a continual succession of Sparks. When I came to reflect upon several circumstances that might contribute to produce the desird effect, I perceivd that keeping together what heat might be collected would be of great service, & therefore conjectured that if the ball of the Thermometer with the Sparks given to it could be inclosed in some nonconducting covering, the effect would probably answer so much the better; Accordingly I prepared the following apparatus.

With six pieces of Glass, fastend together by sealing wax I formed a Cube (*a.b.c.d.*) not much larger than the Thermometer I wanted to inclose. Into two opposite sides I introducd near the bottom 2 pieces of Wire (*e.f.*) rounded at the ends, and kept at the distance of about four tenths of an inch. Over the middle (*g*) I suspended the Ball of the Thermometer by passing the tube through the upper surface (*a.b.*) of the Glass cube, to which it was fastend with Sealing Wax.

A Temporary Scale well divided into small divisions, was fastend to the Tube of the Thermometer. The Apparatus being thus adjusted, was placd upon a plate of Glass, and two Wires terminating in balls (*h.k.*) the one served to convey the electrical fire to the end of the Wire at *e*, while the other receivd it after its passage through *e.g.f.* and carried it off to the ground.

* Exists only in the Secretary's copy.

An excellent Thermometer of Nairne's was placed about 4 Inches from the Apparatus to indicate any change in the temperature of the Room that might happen during the experiment.

As soon as the Thermometers were perfectly settled, we noted the divisions pointed out by the Mercury and then began to work the electrical Machine.

In the following table the first Column contains the time of working the machine. The second contains the temperature of the room, pointed out by the Standard Thermometer, and the third contains the rising of the Mercury in the Thermometer which was electrified.

Time.	Stand: Ther:	Elec: Ther:
o Minutes.	49½ degrees	o Divisions
3.	49½.	rising
5.	49½.	rising fast
8.	49½.	5 above o
12.	49½.	6 d°
16.	49½.	7 d°

When the Machine ceasd acting the electrified Thermometer descended, and in 30 Minutes was again where it begun; the Standard Thermometer being at 49¼.

2nd Experiment.

As soon as the Machine was set at work, the electrifyd Thermometer begun rising. When it had rose 4 divisions we recollected our intention of trying the experiment with the negative fire, and removd the Apparatus to the Negative Conductor, and then continued the work

Time.	Standard Thermometer.	Electrified Thermometer.
o Min:	49¼ Degrees.	o Division.
7.	49¼.	4. . . . . . .
	Removed to the Negative Conductor.	
8.	49¼.	4
21.	49	9

By a former very Accurate determination of the freezing & boiling water points, I know one degree of Farenheits of the scale of the electrifyed Thermometer to be equal to 567 ten thousandths of an inch: And find also that one of the divisions affixed to it during the tryal is equal to 322 ten thousandths, consequently 9 Divisions which the Thermometer rose in the 2nd Experiment answer to five degrees and one tenth of Farenheits Measure; to which if we add a quarter of a degree which the Standard Therm: fell during the course of the same experiment, we shall have 5,35 degrees for the total effect of the electrical fire.

I cannot at present recollect where I have read some experiments made formerly on the Thermometer which proved unsuccessful; but one Article relating to the heat of electrical Fire, I find in Dr Priestleys History of Electricity, vol. i. P. 392.

He says "that there is really no such thing as cold fusion, either by electricity or lightning, was most clearly demonstrated by Mr Kinnersley in a letter to Dr Franklin dated Philadelphia March 12 1761. He suspended a piece of Small brass wire about 24 Inches long, with a pound weight at the lower end; and by sending through it the charge of a case of bottles, containing above 30 feet of coated Glass, he discoverd what he calls a new method of wire drawing. The wire was redhot, the whole length well anneal'd, and above an inch longer than before. A Second Charge melted it so that it parted near the middle, & measurd when the ends were put together four inches longer than at first."

I beg leave to remark that in my opinion there is a circumstance wanting to make this experiment a proof that there is no such thing as cold fusion in electricity; for unless Mr Kinnersley had put his finger to the Wire while He saw it red, and found it warm, I cannot see where the force of the demonstration lies. May not those who admit cold fusion, bring this very instance against Mr Kinnersley & say, if He had put his hand upon the Wire he would have felt it cold in confirmation of their Hypothesis?

However, to do justice to Mr Kinnersley I confess that a cold fusion appears to me a great Absurdity; and on acct of those several instances where Metals have been melted both by Lightning & electricity, I have always maintaind the Opinion that the electrical Fire containd heat; though it has not, as far as I know, been hitherto brought to any certain proof till (on Occasion of the debate arising in this Society) the subject has been investigated in the manner above related.

Appendix to the Electrical Experiments on the Thermometer.

Read January 6, 1781.

On looking over again Dr Priestley's History of Electricity I find that the heat of electric fire has already been discovered in the following instance.

"That even artificial electricity, says Dr Watson, in a paper read at the Royal Society June 28th, 1764, when in too great a quantity, and hurried on too fast, through a fine iron wire, has a remarkable effect upon the wire, appears from a very curious experiment of Mr Kinnersley. This gentleman, in the presence of Dr Franklin, made a large case of bottles explode at once through a fine iron wire. The wire at first appeared red hot, and then fell into drops, which burned themselves into the surface of his table or floor." (page 341)

Experiments have also been made, on the elasticity of air, from some of which the heat of electric fire may be deduced, tho' perhaps not in so direct and unexceptionable a manner as it seems to follow from those I have delivered. The first which was made by Dr Franklin did not succeed, and is as follows.

He made an experiment with a small glass syphon, one leg passing thro' the cork into a bottle, wherein there was a round thick wire inserted at the bottom. The other leg of the Syphon had in it a drop of red ink, which readily moved on the least change of heat or cold in the air contained in the phial; but not at all on the airs being electrified by means of the inserted wire. (page 223)

"Mr Kinnersley contrived an instrument. It consisted of a glass tube about 11 inches long and one inch in diameter, made air tight, closed with brass caps at each end, and a small tube, open at both ends, let down through the upper plate, into some water at the bottom of the wider tube. Within this vessel he placed two wires, one descending from the brass cap at the upper end, and the other ascending from the brass cap at the lower end. &c: &c:

"The charge of a jar which contained about five gallons & a half, darting from wire to wire, would cause a prodigious expansion in the air; and the charge of his battery of thirty square feet of coated glass would raise the water in the small tube quite to the top. Upon the coalescing of the air, the column of water, by its gravity, instantly subsided, till it was in equilibrio with the rarified air. It then gradually descended, as the air cooled, and settled where it stood before. &c: (page 255, 256, 257).

This experiment is intended to prove that the electrical fire rarifies air by its heat. It seems however to be liable to the following objection. It is known that electricity raises water into imperceptible steam or exhalation. Steam is known to be very elastic, for which reason we can not certainly determine how far the elasticity of the air in this experiment made with water may be owing to the heat of electrical fire; or whether the effect may not be partly owing to the elastic vapour arising from the water, which, being afterwards attracted by the sides of the glass or condensing on the surface of the water may occasion the sinking of the *electrical air thermometer*. There are also other, and perhaps unknown, causes that may affect the elasticity of air independent of heat. However as from the experiments I have delivered it appears that the heat of a succession of sparks alone was found sufficient to raise the mercurial Thermometer above 5 degrees of Farenheits measure, it is exceedingly probable that so great an explosion, as that of thirty square feet of coated glass must have been accompany'd with a considerable heat so as to rarify the air very sensibly.

On Central Powers.

Read January 1781.

Some former papers I delivered on the subject of the central powers of the particles of matter contained experiments by which the existence of such powers may be proved. This is so extensive a subject that I have proposed to consider it more at large and in several different points of view.

I shall, at present take it for granted that there are particles of matter invested with such various central powers of attraction and repulsion as I have described; for the intention of this paper is not to investigate the existence of these powers nor to trace out the circumstances or laws under which they act; but only to facilitate the calculation of them, provided sufficient data can be found on which we may superstruct a theory that shall agree with the appearances we find in nature.

Many hypotheses may often appear plausible at first sight, when on a nearer examination, they are found to contain evident absurdities. In matters that will admit of calculation the surest test we can put them to is the rigorous application of such analytical methods as mathematics can furnish us with.

I have asserted in the former papers that opposite central powers might exist in the same particle of matter, in such a manner as by their natural exertion to form several concentric circles or rather hollow spheres of alternate attractions and repulsions. By this theory we shall be cleared at once of all those arbitrary hypotheses, we might otherwise be induced to adopt to account for attractions at certain distances and repulsions at others. I hope to make it appear that what Chimists have called elective attraction is also to be explained upon these principles; whereas the very terms of *elective* attraction contain something that either revolts against common sense or at least is very much involved in obscurity and mystery, and may truely be called *occult*. To explain polarity by election or choice is the worst of all refuges, and if this can not be admitted in the more palpable instances of the magnetical virtue, it can give as little satisfaction to an inquiring mind, in the more minute but more powerful effects of chimical operations.

Let us then take a view of the conditions that may be required in central Forces in order to answer the purposes of our Theory.

That a certain indefinitely small point, which at present for want of a better name we may call *matter*, should be endowed with a power of attracting such another *material* point, when placed at a suitable distance, is by no means difficult to conceive; since we have something analogous with larger particles every day before us. Thus, two drops of water suspended on the points of two needles, when brought within a certain distance will run into each other. It is of little consequence, if we are told that they are not attracted but pushed together. For, as I hope to shew hereafter, they who will maintain that attraction is owing to the impulse of a surrounding elastic medium, run into far greater difficulties than those they would wish to avoid. We shall therefore, in this instance take it as the best and shortest, as well as clearest way, to reason from things as if they realy *were* as they *appear* to be.

If in the next place we affirm the same indefinitely small point to be also possessed of a power of repulsion, we can, without difficulty, find instances enough, where such a power appears to be exerted, either before or after that of attraction takes place. I have many times seen drops of water, when falling in a particular manner on the surface of the same water, run a great way upon it before they were absorbed. The same may be seen more frequently with drops of quicksilver. Nay, in vinegar it is not possible to make drops of quicksilver unite, though their spherical figure still denotes that they retain the usual attraction of particles within. However to say no more on the possibility of the co-existence of these powers, I shall as I observed before take all this for granted and proceed to a mathematical examination of the conditions necessary to central Forces, that they may produce the required effects.

[Some calculations are here omitted.]

And thus I think I have made it appear, that a supposition of central Forces existing in the manner here explained, and shewn to be consistent with mathematical reasoning will account for all phenomena of alternate attractions and repulsions taking place at different distances from the centers of material particles, without any necessity of admitting arbitrary hypotheses of rings or hollow spheres of powers; the idea of which seems to revolt against that beautiful simplicity we find in all the principles of Nature. But the more extensive use and application of this Theory I leave to a future opportunity.

January 14, 1781. WILLIAM HERSCHEL.

On the Periodical Star in Collo Ceti.

Read Feb. 2, 1781.

LAST year I presented to this Society a Memorandum of the uncommon lustre of the periodical Star in Collo Ceti; It is somewhat remarkable that the succeeding Period should be distinguish'd by the very reverse of the former. The subject is so involved in obscurity that I shall attempt little more than to deliver my observations accompanyed with a few conjectures, leaving it to future Observers to frame some plausible theory, when many succeeding periods may have furnished means to direct the thoughts in this pursuit.

A star, (a *Sun* I should say,) perhaps surrounded with a system of Planets depending upon it, undergoes a change, which, were it to happen to *our* Sun, would probably be the total destruction of every living creature! What an amazing alteration from the first magnitude down to the 6th, 7th or 8th! But let me not take up a time in admiration which may perhaps be more philosophically imployed in reciting plain matters of fact.

The observations I have been able to make upon this wonderful star are this time fewer than I could wish to have made; yet are sufficient to deserve to be mentioned as they are so far connected with the former, that we may presume to conjecture the remarkable want of brightness in this period to be some natural consequence of the super-abundant light in that immediately preceeding.

Observations in 1780.

August 3^d 2^h 35′ in the morning

The periodical star is not to be found with the naked eye. The night uncommonly fine.

Augt 8. 2^h in the morning.

I looked with a compound eye piece that takes in a very large field of view, and examined every very small star near the place of the periodical star but could not find it. I also looked with the power of 222 without success.

Septr 8. I could just discover the periodical star with the naked eye, tho' with some doubts; but on applying the Telescope I saw it perfectly well and the small star which follows it. I measured the distance between them with great accuracy.

1st measure. 1′ 44″·062
2^d measure. 1′ 44″·374.

The colour was very remarkable, being a darker red, (or rather garnet colour) than any I remember to have seen before among the fixt Stars.

Septr 19. The periodical Star is considerably encreased, being nearly equal to δ Ceti.

Septr 20. I took a measure of the distance, which I found 1′ 43″·688.

Septr 24. *o* Ceti equal to δ.

Septr 30. *o* Ceti not sensibly encreased being still about the apparent size of δ.

Here is a vacancy in my Memorandum of more than a month, during which time I wrote down no observation on this star, because I found nothing remarkable, and had no suspicion of what I afterwards found. However it is very evident that the periodical star never exceeded δ in brightness. For tho' I have no memorandum on *o* Ceti, my journal contains observations upon some of the Stars near that place which were made Octr 3. 4. 10. 12. 23. 27. and as they do not mention any changes in the periodical star which I could not but observe, I may take it for granted there were none.

Novr 7. The periodical star is hardly so large as δ.

Novr 24. The periodical Star is less than it was, instead of being encreased as I expected. 11 o'Clock.

Decr 15. *o* Ceti is diminished since I saw it last.

Decr 17. *o* Ceti is hardly visible to the naked eye tho δ is bright enough.

Decr 23. I can not find the periodical star.

The opinion of Keill who supposes the greatest part of the surface of this Star to be cover'd with dark spots seems to be the most likely of all; For if, according to Maupertuis, the periodical changes in the appearances were owing to some cause like that which occasions the phases of Saturn's ring, there would probably be much more regularity in the times and gradations of light than has been found by observation.

We have seen that Keill's report " Non enim singulis annis eandem obtinet stella magnitudinem, quandoque secundi ordinis fixas superat magnitudine, aliquando inter tertium ordinem vix consistere videtur; " is fully verifyed in two periods immediately succeeding each other, whence it should seem probable that some of the dark spots on this Star are occasionally consumed, as they are found to be on the Sun. If I am not mistaken the place where a dark spot has been is for some time afterwards more luminous and of a paler colour'd light more resembling flames or melted metal than other parts of the Sun. It has also been observed that the dark spots on the Sun are much deeper than the luminous parts which may be seen protuberant in shelving sides leading down to the dark places. In the year 1779 in April, I observed a very remarkable spot on the Sun, the length of which when I measured it, (which was not till after it was considerably decreased) was 34476 english miles. I saw every day the flames, or vivid, bright-yellow light, encroach upon it, first dividing it in the middle and afterwards crossing it in several other places. But on the periodical Stars the changes must be much more sudden and extensive; involving the greatest part of that globe in one general Chaos of confusion and destruction.

Wm HERSCHEL.

Jan: 30. 1781.

Della Torre's Method of making crystal globules to be used instead of lenses for single Microscopes.

Translated from the italian Scelta di Opusculi.

Read Feb 9. 1781.

THREE things are required to make little globules of Crystal for single Microscopes. A pair of small perpetual Bellows. Some good Tripoli. And some solid Cylinders of Crystal of different thicknesses.

The bellows must move easily and be kept full so as to blow the flame equally, thro' a little pipe, against the cristal. The point of the pipe must enter a little way into the lower part of the flame, and its diameter should be a line. Care must be taken not to melt it.

The blast of the flame should be directed horizontally and the flame will be found to consist of two different parts. Two thirds from the base of it are white, from thence to the point is transparent and without colour. With this latter part the cristal is to be melted. There is no danger of smoking the cristal in this part of the flame, which, on the contrary soon happens when the white part of the flame is touched:

The crystal must be well cleaned before it is exposed to the flame and never touched with the finger. Paper is the best for cleaning the cristal or any part of the apparatus.

The tripoli should be in a piece of 4 or 5 inches long and 4 broad; and made smooth on one part. The best Tripoli is whitish of a fine grain and heavy; and which after calcination turns reddish. To calcine the tripoli put it in the midst of charcoal not much set on fire and let it remain there till all the charcoal has taken fire, is burnt out and extinguished of itself. When all is cold take out the tripoli, which is calcined and fit for use. On the smoothed side of the Tripoli make several little holes of various sizes, in which the globules are to be placed. Take care not to touch them with your fingers. If it wants cleaning let it be done with paper.

The Crystal should be round; for, if square the globules will be apt to be smoked.

The thickness of the Cylinders of Crystal should be different such as half a line, a line &c:

Having all this ready the work is as follows. Take two Cylinders; put them into the flame, end to end; when they begin to melt draw them out (holding one in each hand,) to as fine a thread as you please and separate them in the fire. Put by one of them. Hold the extremity of the other in the point of the flame and it will run up into a semiglobular form, which may be made as large as you please. These semi-globules must be broke or cut off and put by in papers. When you have a sufficient number of them place them into the little holes of the Tripoli. Blow the flame, not against them (for they would be blown away) but against the tripoli which will soon be red hot. Go on in that manner till the little globules are of a white heat. Then they are finished. Take them out of the fire to cool. When you want to make large globules, it will be well to shake the tripoli a little that the globules may the better assume a spherical form. But this must be left to experience which soon will teach many little circumstances that can not so well be described.

In this manner globules have been made that magnified from 42 times up to 1280. 1920. 2560. 3840. 5120. 10240 times.

II.—Papers communicated to the Royal Society

Observations upon Algol.

Read May 8th 1783.

THE most extraordinary Phenomenon of the occultation of Algol is so interesting a subject, that we cannot too soon collect every observation that may serve to ascertain its period, or assist us to find what quantity of light the star looses on these occasions. The wonderful train of consequences that may be drawn from such regular occultations, engage our utmost attention. That stars are Suns has long been inferred from the intensity of their light at such great distances; and that these Suns may have Planets around them, has also been concluded from analogy; It has even been surmised that the change in the appearance of periodical stars might be owing either to spots revolving on their surfaces or to dark clouds swimming in their atmospheres, and performing regular gyrations at some distances. The Idea of a small Sun revolving round a large opake body has also been mentioned in the list of such conjectures.* But the present observations seem to lead us much farther, and will probably furnish us with the strongest arguments & facts to verify former conjectures, of a plurality of solar and planetary systems.

Observations.

August 17. 1779. I observed Algol with a 7 feet reflector, aperture 4 inches; power 222; and found it a single star, with nothing remarkable about it.

Octr 19. 1779. I examined π, ρ, ω, p Persei, and had Algol been less than its usual size should have perceived it, as these stars are just about it.

Augt. 2. 1780. With the same instrument & power I saw Algol distinctly single, without perceiving anything to attract further notice.

Sept. 14. 1781. I observed Algol with an improved 7 feet reflector, 6·1 inches aperture & power 460. It was perfectly round and well defined & I marked down its colour *white*.

Sept. 4. 1782. I examined Algol with 460 looking at it for several minutes successively & found it round and well defined.† Its colour was again marked down *white*.

April 27. 1783. Having been informed by the Revd Dr. Maskelyne that Algol suffered a regular diminution of light every 2 days & 21 hours I begun a regular series of observations upon this star, and examined it the first opportunity I could have, which was this evening at 9 o'clock when I found it of its usual magnitude. It was the same at 10 o'clock. The star being so low afforded no further opportunity for observation.

April 29. 9^h $30'$ Algol of its usual size.

April 30 9^h $10'$ As usual.

May 3. 1783. 8^h $53'$. The wonderful Phenomenon of the star's occultation seems to be now happening. Algol is but very little brighter than ρ Persei, to which it may very conveniently be compared.

May 3. 1783. 9^h $15'$ The same appearance continues; ρ according to Flamsteed is of the 4th magnitude and from the present comparative brightness of Algol I conclude it to be a small star of the 3d or large one of the 4th magnitude.

9^h $40'$ Still as much eclipsed, but not more.

* See explanatory Note to p. 17 of my paper on the motion of the solar system delivered to Dr. Maskelyne, April 1st in answer to some letters concerning that paper. [Below, p. 115.]

† The reason of my continuing to look at this star so long was that I had formerly missed some double stars of the first class for want of this particular attention in this review of the stars, therefore I always looked at every star of the 1st 2d & 3d magnitudes for a considerable time together.

$10^h\ 10'$ The star seems not to have regained any light but is now too low to judge well of its magnitude, as ρ is no longer visible to compare it with.

$15^h\ 15'$ It has now regained its full light being extremely bright notwithstanding day light is already very strong. ρ is but just visible.

[This paper was not intended for publication. In a letter to Pigott of May 20, Herschel explained that he handed it to Sir Joseph Banks at the Royal Society's dinner on being asked if he had seen the phenomenon, but that he was not aware that Goodricke's paper on the same subject had not been read.—ED.]

---

*On polishing Specula by a Machine.**

(1789)

ABOUT six years ago, I made various attempts to polish Mirrors for telescopes by a machine; but after many trials, with Specula of ten feet focal length, and nine inches diameter, I found that notwithstanding I had provided the same kind of movements in my apparatus, which I generally used to give by hand, the figure of a Speculum always suffered under the operation. So far indeed was the machine from giving a proper shape to the mirror, that it came out deformed, notwithstanding I had compleatly figured, and polished it before I placed it on the machine.†

This bad success made me lay aside all thoughts of any further attempt; and had it not been for the particular situation in which I found myself last summer, when I was obliged to employ twenty men to polish my second forty feet Speculum, I might perhaps never have made another trial to do such work by mechanical powers. The idea of a machine however was now again as it were forced upon me; when I considered, that all the essential part, I had formerly taken in the construction of a Speculum, was fairly excluded in the present operation. The enormous weight, of about five and twenty hundred pounds, to be moved upon the polisher, would not permit the use of those delicate touches of the hand, by which I had been accustomed to form small mirrors; and I found myself reduced to the situation of merely directing the unwieldy manœuvres of a set of men, who when they did their best, could only act like a very imperfect machine.

As soon as I perceived that I was, in fact, already working with a machine, there wanted not much to convince me that twenty men made a very bad one; and that I should find no manner of difficulty in contriving another, that would do the work much more to my satisfaction. This point being brought home to me with such forcible arguments, I caused all my apparatus for polishing with the twenty-men-machine to be pulled to pieces that I might never be tempted to use it again; and began now to consult the very compleat theory of polishing, which long experience had furnished me with. I perceived at once the error of my former mechanical constructions, and could explain the cause of my disappointment six years ago. A few days sufficed to plan an apparatus; and having got it executed, of such a size as would polish a Speculum for my 20 feet telescope, I began to work one of the mirrors with it, by way of trying the performance of the machine. A few hours pointed out what still was wanting; and when such little defects, as appeared to require an alteration, were corrected, I found myself in a condition to begin a series of experiments, on the effects of working mirrors mechanically.

My former principles, of certain strokes having certain effects, upon the figure of a Speculum, were now very strictly re-examined; that I might see whether every mode of operation would have the same result, when a machine was used. Most of the rules I had laid down were confirmed; one way of working however proved to be of a different import

* [Sent to the Royal Society, but apparently not read at a meeting.—ED.]
† [These experiments are not mentioned in the Polishing Record.—ED.]

from what had been ascribed to it; but this I could easily explain. For, in the first place, the considerable heat of the hand, which is always of a much higher temperature than the place, and materials which are employed in the work, was now removed; and, next to this, I comprehended very well, that experiments made with a machine, which for any given time will truely perform a certain stroke, with mechanical accuracy, must be more conclusive, than when the same operation is executed by hands, which it is not so easy, nor perhaps possible, to direct, for an hour or two together, without some considerable deviations from the proposed movements.

These experiments had taken me up about five weeks. At last, by the 14th of February, I had not only settled my plan of working, but had prepared everything requisite for the purpose; and that now my trial might be decisive, I totally destroyed the figure and polish of one of my 20 feet Specula, before I laid it on the machine; so that, besides lateral defects, the focus of the outside rays was above an inch shorter than that of the inside. In this condition I set the machine to work.

The day was pretty far advanced when we began; but as my share of the business was only to feed the machine from time to time with polishing stuff, I could now have the advantage of making it continue the work without interruption, while the usual calls, of supplying our own machines with necessary refreshments, took me away from the polishing apparatus. A little before nine in the evening I took off the Speculum, with an intention to proceed next day; but on examination it appeared, that, in little more than 8 hours, the Speculum had acquired a most beautiful, glossy polish; and this occasioned an alteration of my design, which now was to expose the Speculum to optical trials.

By a day object, of some letters, fixed up at a convenient distance, I found that there was no dispersion of rays. I could read them with more facility than with the best of my 20 feet mirrors, which hitherto I had preserved with the greatest care. It is remarkable that this machine-polished Speculum, which now excelled the former, had frustrated my endeavours for near two years past; in which time I could never succeed to make it perform like the other. This failure I used to ascribe to certain cristallizations, that have taken place in cooling, when the metal was cast; and which indeed are very detrimental to Specula, from a difference in the *grain*, if I may so call it, which is occasioned by them. When I directed the telescope to celestial objects, I was still more convinced of the perfection of this new mirror. The small stars were all brought to so precise a focus that their light was gathered into a point; and I saw the satellites of the Georgian planet with greater ease than I could ever see them before.

Thus it appears that we have now obtained an addition to our mechanisms, which certainly must be valuable, as it will secure to us the perfection of astronomical telescopes, by reducing the art of making Specula for them to a certainty.

An apparatus for polishing a forty feet Speculum is nearly finished, and by some trials I have already made, I find that its construction will conveniently answer all the movements I require. The simplicity of the machine, indeed, is such, that there can hardly be any difficulty in its performance.

Slough near Windsor, March 4th 1789.

Observations of the new planet.

Read February 18, 1802.

THE discovery of an additional planet of the solar system by Mr. Piazzi of Palermo, must undoubtedly be highly interesting to all astronomers. Before the elements of its orbit could be well settled, the planet was lost for some time, and when I was upon the look out for it about the place where by calculation it was likely to be, and where we now are assured it really was, I could perceive no star with any visible disk, whereby I might have distinguished it from the rest. Hence I surmised that it would require fixed instruments to rediscover it; and not being in possession of any, I requested my much esteemed friend Dr. Maskelyne, to give me the earliest notice of its situation, as soon as he should have observed it at Greenwich. Accordingly, the 5th. of this month, I received his account of the place where he had seen it early in the morning of the 4th, and, by directing my telescopes to the star thus pointed out I obtained the following observations.

Feb. 7. 1802. 13h.

With a ten-feet reflector and a magnifying power of 600 I viewed the place where I expected the new planet to be, in hopes of distinguishing it from the neighbouring stars by its visible disk. Being sufficiently used to direct my telescope to any given part of the heavens, I immediately perceived a star which appeared sufficiently different from another at no great distance, to occasion a surmise that it was the planet. In order to verify my suspicion I put on a magnifier of 1200, and comparing the supposed planet with the same fixed star, I found a doubt still remaining that there might be a mistake.

The 20 feet telescope with a power of 300 and of 600, would not resolve the doubt; but the supposed planet being still too low for very distinct vision, with such high powers, I intended to examine it when in, or near the meridian, as soon as the air should be sufficiently pure.

The following days, though cloudy, afforded every now and then an opportunity of ascertaining, by its change of place, that the star I had examined must be the new planet.

Feb. 13. 1802. 5 o'Clock in the morning.

Having long been disappointed by cloudy weather, a favourable change enabled me at last to view the same star again at a sufficient altitude to see it with great distinctness.

When I examined it with a magnifying power of 600, I found, by comparing it alternately many times with the star I had chosen as a standard, that there was a sufficient difference in their appearance, and that a very minute planetary disk might be perceived in the one, which was not to be seen in the other. After having clearly satisfied myself of the planetary nature of the new star, I wished to ascertain its magnitude. The advanced time of the morning, and an apprehension of clouds coming on, would not permit me to apply the lamp and lucid disk micrometers.* I therefore had recourse to a comparative view of the Georgian planet, and the newly discovered one, as their situation was such that I could easily change the direction of my telescope from one to the other.

When I turned from the new planet to the Georgium Sidus, and compared its diameter with that of the former, I judged it to be apparently from four to six times as large. Immediately after this I directed the telescope again to the new planet, and as the last of luminous objects in succession is apt to make proportionally the strongest impression, its diameter appeared to me now to be nearly one fourth of the diameter of the Georgian planet. On viewing again our known planet, in order to compare it once more with the new one, I estimated its diameter to be not less than 5 or 6 times as large as that of which I was desirous to ascertain the magnitude.

Apprehensive of not having soon again so fair an opportunity, I examined the new planet with an attention to its appearance, and found its colour is faintly ruddy. Perhaps it appeared rather the more so, on account of my viewing it after the Georgian planet which is of a mild bluish tint.

There was no appearance, nor indeed the least suspicion, of any ring surrounding it; its disk, though very minute, being perfectly well defined all round.

* For a description of these micrometers see *Phil. Trans.*, Vol. 72 page 163, and Vol. 73 page 5. [Below, pp. 91 and 103.]

It would be premature to give a decisive result of the magnitude of the new planet, for which purpose many repeated accurate measures will be required. But we may even now, from these few observations, draw a very important conclusion, with regard to the remarkable smallness of this primary planet, when compared with the rest. For, admitting the diameter of the Georgian planet at the time of the observation to have been 4″; and, without entering into minute investigations, estimating the diameter of the new one at one fourth of that quantity, we obtain the following result.

Suppose the distance of the new planet from the Sun, till we have more accurate calculations, to be to that of the earth as 3 to 1. Then, at the time of observation, its distance from us would be 2.28066. Now, had the moon been at that distance in the place of the new planet, we should have seen it under an angle of 1″,673 whereas by observation the diameter of the planet was estimated only at one second.

Hence it appears that the real diameter of the new planet, remarkable as it may appear, is less than five eighths of the diameter of our moon.

WM. HERSCHEL.

Slough near Windsor
February 14—1802

Announcement of a Comet.

Read Dec. 12, 1805.

Sir

At 5^h 25′ While I was examining the Constellation of Aquarius I perceived a Comet.* It is of considerable brightness and easily to be seen by the naked eye. The extent of its coma is about 20 minutes, and its present situation nearly 4 degrees south following Flamsteed's 100^dth^ Aquarii, making almost an equilateral triangle with the 100^dth^ and 107th of this Constellation.

I have the honour to remain, Sir,

your most obedt and most humble Servt

WM. HERSCHEL

Slough, Sunday evening, Dec 8. 1805.

III.—Miscellaneous Papers of unknown date

On Central Powers.†

In my introductory remarks on the construction of the heavens, given with the catalogue of a second thousand of new nebulæ and clusters of stars, I have hinted at a possibility that other central powers besides the Newtonian attraction might be concerned in the formation of these clusters.‡ My expressions were given with sufficient caution not to make it incumbent upon me to prove that other powers, besides the attraction of gravity which is already known to us, have actually any share in the formation or rather preservation of clusters of stars; much less is it required that I should shew that these other powers exist, and to point out the laws by which they act. I shall therefore in this paper, confine myself merely to prove the possibility of various central powers being united and acting together according to settled laws; and to shew what results will arise from such combinations.

* [Biela's Comet 1806 I. Discovered by Pons 10th November 1805.—Ed.]

† [This remarkable speculation is an expansion of the paper written at Bath. In a list of his MSS., drawn up towards the end of his life, Herschel wrote of this paper: "It was written many years ago with an intention of being presented to the Royal Society. I had shewn it to Professor Vince, who after reading it said that its contents were new to him and were different from Boscovich's theory; in contradiction to Mr Cavendish's opinion, to whom I also had shewn it."—Ed.]

‡ *Phil. Trans.*, 1789. [Below, p. 334.]

When I call the Newtonian attraction a central Power rather than a centripetal force, it is done with a design to convey the idea of a power residing in a center, point, or particle.

Without entering into a metaphisical disquisition relating to the nature of matter, I may suppose myself authorised to argue the existence of this power, in the manner here admitted from the most obvious and common facts ; no particle of matter, how small soever, as far as we know being without it. Thus, for instance, it is well known that weight, which is another name for attraction, is in proportion to the mass. And that when a certain quantity of this mass weighing one pound is divided into a thousand parts, each of those parts will weigh the one thousandth part of a pound. Now as we can with as much facility, in idea, subdivide this small particle as we can divide the pound, it will follow that the notion of an attractive point, center, or particle, in the sense I take it, is by no means an absurd one. Nor is it necessary that such divisions should be carried on ad infinitum before we can come to so small a particle as that which I am to take for a center ; the indivisibility of the particle not being a condition that enters into my account.

If a particle of matter, therefore may be admitted to possess attraction, I can find no difficulty in admitting another particle to be possessed of repulsion. And if an attractive particle may be united to another attractive particle, I see no reason, why it may not as well be united to a repulsive one ; provided the powers of attraction and repulsion be so adjusted to each other as to permit the union. Or, which is the same thing, no reason can be assigned why the same particle may not be possessed of both these powers of attraction and repulsion at the same time.

I would willingly clear this subject from all objections that might be brought against it, and therefore it may not be amiss to mention what the great discoverer of the law of the attraction of gravity says in his preface to the Principia. " Multa me movent ut nonnihil suspicor cætera naturæ phenomena ex viribus quibusdam pendere posse, quibus corporum particulæ per causas nondum cognitas vel in se mutuo impelluntur & secundum figuras regulares cohærent, vel *ab invicem fugantur &* recedunt." By this it appears that Newton puts repulsive powers upon the same ground with attractive ones ; and therefore those who are inclined to believe that attraction is not a primary power of nature, but merely the result of a certain mechanism, which some have been ingenious enough to invent, may also, if they please try how far they can succeed in assigning another mechanism that may explain the phenomena of repulsion. We can do no better than to follow the example that is set us by the excellent philosopher whose words I have quoted and leave to others to settle whether attraction and repulsion should stand as qualities immediately inherent in matter, or whether they ought to be explained upon mechanical principles derived from other known laws of matter and motion.

To those who wish not to admit attractive and repulsive forces as primary qualities, the terms central powers may only stand for general words expressing centripetal and centrifugal forces from whatever cause they may proceed, and in this signification these words will equally denote what will serve for the foundation of a mathematical theory ; tho' I should myself be inclined to use them in the sense pointed out before as denoting powers that were lodged in the center or particle under consideration.

We are now to consider the result of the union of various central powers ; and in order to apply mathematical strictness it will be necessary to lay down a theorem from which calculations may be formed, when observation has furnished us with proper data.

[Here follows a Mathematical Theorem.]

Thus we have shewn the mathematical possibility of the existence of a power which may prove a barrier to the destructive approximation of the fixed stars, or of siderial Systems.

It is certain that Sirius, on the supposition of its distance being as I have stated it before for the sake of an example, and taking it to be a body of an equal size with our sun, would be near 100 millions of years falling into the sun; and though this is a time beyond conception great, yet it does not seem to be suitable to the magnificent construction of the universe that a principle should exist which tends to the final destruction of it. It is true that the contrary attractions of other stars will in a great measure counteract this destructive cause and therefore we are not sufficiently acquainted with the great fabric of sidereal systems to judge what is necessary to their preservation. Moreover as by means of projectile forces the several parts of the solar system are preserved from the decay that would ensue if they were given up to the powerful effects of gravitation alone, so there is a sufficient possibility that the same provision may have been made to guard the great parts of the universe from destruction.

If however we admit the existence of a moderating power of repulsion all difficulties that may arise in our minds are done away. For though on the other hand it might be suggested that the stars would be dispersed by this power, it must be remembered that not only the force of repulsion will be extremely weakened by the immense distance, but that there is a striking difference in the exertion of the two powers: The longer attraction is at work upon a system of bodies the stronger will grow its efficacy to destroy that system, while on the contrary the tendency to danger of repulsion is gradually weakened by the continuance of its exertion. For the approach of the bodies in the first case will encrease the central powers, but their retreat in the latter, will have the contrary effect; nor will it be necessary to observe that the distance at which it may be required to place the neutral boundary is probably not so near as the nearest fixed star. For if the great bodies of the universe by their mutual contrary attractions keep each other partly in a state of rest, the moderating principle ought to be less powerful, and by our theorem we see that it may take place at any required limit; for instead of investigating a single star with it, the united exertion of sidereal systems may become its residence.

It should after all be remembered that we do not assert the actual existence of the above moderating, repulsive power; much less do we assign the law of its decrease to be as has been assumed in the solution of the foregoing problem, which has been given merely as an illustration of the subject, and as a proof of the possibility that such a power may take place.

The discovery of the laws by which various central powers may act, will be attended with the greatest difficulties. The extensive range of the exertion of gravity has given us such data that the law of its action could no longer remain concealed from the penetration of its great discoverer. Now since we find by experience that attraction ceases to be predominant at certain small distances, it is highly to be presumed, or I may venture to say, it is evidently proved that other powers also have existence; but the space in which their action is performed is so small and the influence of the central powers of neighbouring particles so great; also, in the case of the existence of the moderating power before alluded to, the regions in which it is exerted are so remote, that we must not hope to make any hasty advances in the discovery of the laws by which such powers are guided.

Having now sufficiently shewn the possibility of the existence of such powers as have been hinted at in my former paper, I have only to add that the foregoing theory, though these powers should not be necessary for the preservation of the great systems of the universe, may yet be of considerable service in the investigation of the more confined phenomena of nature. There are many optical as well as chemical facts which indicate the exertion of powers that act at various distances. Now if powers of attraction and repulsion are to be admitted how can we conceive their existence in a better way than by referring them to certain centers as the seat of their energy? And as we have seen that this may be done I hope that the foregoing investigation of the phenomena that must happen when they are considered in this manner will not be unacceptable.

I ought to remark that several eminent philosophers have already supposed that there are spheres of attraction and repulsion about bodies, and have even ascribed such spheres to particles of matter, but they have hitherto been considered in a view that must render them inadmissible. An uniform layer or coat of attraction placed at a certain distance about a body or particle, and followed by such another layer of repulsion at some other distance is a notion that revolts against our clearest conceptions: It is giving us an effect without an efficient. For, no doubt, attraction as well as repulsion are effects; and that one effect at a certain distance should leave off and another quite contrary effect should begin is certainly too arbitrary an hypothesis to be admitted.

Our attractive and repulsive spheres with intermediate neutral boundaries as given in this paper are of a very different kind. They are the natural result of the uninterrupted exertions of united central powers, and are as evidently consistent with sound mathematical reasoning as the effects of the power of gravitation.

Interference of Light.—A Fragment.

REPEATING this several times I could never find the least difference neither in distinctness nor light. I look upon this as one of the greatest proofs of the smallness of the particles of light, but we need certainly not have recourse to penetrability to explain it. If two Guns both charged with a good quantity of small shot, were placed at a distance of 20 or 30 yards and discharged both at the same time at rectangles to each other so that the shot of both might cross in the Air at the angular point, I question whether a single grain would ever touch another tho' this were repeated a hundred times. Now if the Velocity is encreased as well as the smallness of the particles to an almost inconceivable degree, it will make it almost impossible they should ever interfere

When I made the above experiment I could have wished to have had a much stronger reflecting Mirror. Perhaps if such a one as Mr Vilette's mentiond in the *Transactions* were used some small effect might have been discovered.

Since then those experiments that are most in favor of penetrability which are the Phenomena of light are so easily shown to prove nothing on that side of the question and that even some, as the loss of light in passing thro' transparent Mediums, are plainly against it, I may at least be allow'd to put aside experience as being neutral in the Case. To return then to our argum^t^. I must observe that Dr Priestley says we have the Idea of *Impenetrability* only from resistance, but it appears to me rather that we derive it from that of *Solidity*, which term I can not allow to have no other signification than *impenetrability* tho' Dr P. seems to put them indifferently each for the other.

For I appeal to the reader whether the Idea of resistance arising from any power of repulsion how great soever & from no other cause would not rather give us the idea of *penetrability* than its contrary. It is but imploying a *greater* power and the struggle will cease, we shall *penetrate*. On the contrary when a substance is solid no power can make another of the same kind penetrate it, as we find by reflecting only on the Ideas represented by the terms of that Proposition, and not by putting our hand on a table or any other experim^t^ whatsoever. We perceive immediately that when there is an apparent penetration, it is a mere deception. The parts were *removed* that opposed and not penetrated, for *Solidity is impenetrable.*

Now Resistance and impenetrability are clearly two very different things; for according to Dr P. we may have resistance by a power of repulsion without impenetrability, and according to my Ideas, which I hope will agree with most of my readers, there may be impenetrability without resistance; for let but the parts be so detached that they may yield and be easily removed; and the most impenetrable Substances will resist but very little or not at all.

Thus Iron when in fusion will not resist the motion of a stick, tho' certainly not less Solid and impenetrable than before. Also, I may dip my finger into a Bason of Mercury without great resistance and yet shall meet with a very considerable one if I press it against the table. Nevertheless I maintain that the Mercury contains more solid and impenetrable particles than the wood.

Ice is something lighter than Water, from which we infer that it contains more pores than the former, *i.e.* contains fewer Solid particles under equal Bulk, yet it resists much more than water.—I might bring a thousand other instances that would evidently prove resistance is not the foundation of our Idea of Impenetrability. But these I hope will suffice.

Having now I hope made it appear that Matter is an extended Substance possessed of Solidity, impenetrability to other Solid Substances, Moveability and Divisibility, I propose now to take a short view of Dr Priestley's Definition of its being endowed with the powers of attraction or repulsion.

That such Powers can not be essential to Matter we have already seen.

Let us now endeavour to find what reasons we have to admit their existence at all.

The Cohesion of solid Bodies, the emision, reflexion and refraction of light, the Revolution of Planets and comets in elliptical Orbits, the Elasticity of Air, electrical Experiments in short a thousand other Phenomena will here crowd upon us and force us to confess that Matter is possessed of powers that are capable to produce the most astonishing effects; if we attempt to reduce them to some order, calling that power by which the Solid parts of Bodies cohere, planets are retained in their orbits, tides ebb & flow, heavy Bodies fall to the Earth, &c, Attraction; the Emision & reflexion of light, the resistance we find in bringing Bodies into perfect contact and a number of electrical Phenomena, Repulsion. I shall not object to these Names provided nothing else be meant by them than the very Effects we have observ'd to exist. But as soon as those terms are to stand for the very powers themselves whose effects they can only denote, I must look upon it as no more than an ingenious and perhaps useful Hypothesis which nevertheless I shall always suspect as deviating from truth, and which I believe there are some good reasons to believe has no real foundation in Nature. It will be expected that I should explain myself more at large.

In the first place let me see how we are to understand Dr P. when he speaks of the powers of attraction or repulsion. As he has no where expressly declared himself we must gather his meaning from his writings in the definition of Matter (page 38 Introd) we find that he calls it a Substance possessed of the property of extension and the powers of attraction or repulsion.*

Let us first of all begin with attraction, and since Matter is an extended substance possessed of the power of attr: which surrounds it in a sphere, I should be glad to know whether that whole sphere of attraction be what is called an extended Substance or Matter, or whether it be only the Center of it. If it is the Center, I ask what can there be in the Center to deserve the distinction of extended Substance? or by what effects do we find that these centers can deserve such a distinction. Besides if they are not absolutely *Mathematical* or *imaginary* points but possessed of the property of extension, they will in fact become little globes and why they should not rather be Globes of the size of the Sphere of attraction I can not see.

* [Here follow footnotes:—]

Page 4, resistance is occasion'd by a *power of repulsion* always acting at a real & in general an assignable distance from what we call the Body itself

Page 5. When we suppose bodies to be divested of a *power* of *attraction* they become to be *nothing at all.*

Page 6. take away the power of attraction and the solidity of the atom intirely disapears.

Ibid—in consequence of taking away attr: which is a *power, solidity* itself vanishes

Ibid. without a foreign power every particle would fall from each other & be dispersed.

Page 7. These powers are not *self-existent* in Matter

Page 16. *that it* (matter) is possessed of powers of attraction and repulsion, and of several spheres of them one within another, I know;

Page 17. Matter has in fact no properties but those of attraction and repulsion.

These passages will be sufficient for my purpose

If that is allow'd it follows that on the Newtonian Hypothesis of Attraction they must be infinitely great.

Introduce we next a sphere of repulsion to the same particle. I ask what binds these two spheres together that they do not become excentric? Is there another attraction that holds them together? Nay it is said that there are several of these spheres. And are they all inseparable, what connects them to each other? *

Does not this Hypothesis involve us in the most perplexing difficulties?

Is every particle of Matter or rather sphere of attraction and repulsion exactly like another?

How are these spheres to act upon each other? Is it the Center that acts upon Center? That is to say an imaginary point, or is it one sphere that acts upon another? If it is, how shall we account for the various densities and qualities of different Bodies? Will not every sphere attract or be attracted till it comes as near as it can to that which is next to it, which must therefore make all Matter equally dense? Are the particles of light such spheres also? If a piece of Iron is nothing but a collection of such particles, why will a ray of light not dart thro' it at once? If this be ascribed to the density of Iron, why will a deal Board stop light when Glass will transmit it, which is known to be more than four times as dense?

Besides, it is said that Matter has in fact no other properties but those of attraction and repulsion. If this be the Case, will those powers suffice to explain an infinity of the mysterious operations of Nature, such as the Mechanism of organised Bodies and the formation of an embrio?

If it be said that one is more regularly porous than the other, I ask how can a Substance want pores that is perfectly penetrable, or how can there be irregular pores in a collection of such regular spheres?

In short, we are not one bit the wiser for such Hypotheses as very plainly they appear to be. Insuperable difficulties occur which they can not solve. But it will perhaps be asked whether I deny those Matters of fact upon which they are founded. I answer, by no means, but the manner of resolving them appears not Satisfactory. If we go no farther than they point out the way, I will not object to the terms attraction or repulsion; that is *e.g.* if I see little parcels of Quicksilver form themselves into small spheres when dispersed upon a table, let me be told it is owing to attraction. If this means that they are moved towards a center by some power essential and inherent in Quicksilver, I believe it not, but if by that word attraction is only meant the *effect* of the particles going together which is obvious to sense without an intention to instruct me by acquainting me with the *cause* of it, I readily agree to it and am satisfy'd.

Again if I see a stone fall to the ground and find when it rains that Water drops from the Clouds I may go so far as to believe both to proceed from the same cause Attraction, or Gravity & so far the word is of use as from the analogy of things we . . . .†

The next experiment I shall mention is one I made to burn the image formed in the focus of a Newtonian reflector; as I do not recollect that this was ever tryed before I shall give a description of the manner it was done.

The principal pencil which is reflected sideways out of the great Tube on its return from the Object Speculum was made to pass thro' a small one, about 4 inches long and 1·5 in Diar. It was adjusted in such a manner that the focus might fall within half an inch of that end of the small tube which is towards the Eye.

In general all experiments upon light must be of great use.

The shadow between Newton's two knives is attraction.

The fringes are repulsion.

The sticking together of two speculum-tools is perhaps not altogether owing to the pressure of the air. This should be tryed in a very perfect vacuo.

* [*Cf.* the modern theory of the deformation of the electrical fields of force of a molecule due to its motion through the æther.—ED.]

† [MS. fails, and there follow some sheets which we here omit.]

The angle where reflexion begins to be strongest in different Mediums may be of use.

A Ray deflected by a knife should be deflected again by other knives. A triangle of knives or square may lead to discoveries, as by that means the interference of powers may be examined as between thin plates. It will be well to try if the red rays come sooner in an Emersion or the violet later in an Immersion of a Satellite, as from that time if found a calculation of the different Velocities of rays might be made.

How much longer the spots of the Sun are hid than appear.

To account for it by the dia^r of the Sun or their depth (P. Trans. 1767 page 398) The figure and Theory of Mr Horsley is wrong besides being founded on false data.

Experiments for finding the central powers.

The rising of fluids in capillary tubes of different diameters.—Sugar held over water may perhaps by a Microscope be seen to act at some distance and tho' by the naked eye it seems only to attract when in contact. The rising of quicksilver at the edge of the vessel which holds it may perhaps be owing to repulsion tho' generally ascribed to the superior attraction of the quicksilver.

The curve between two Glass plates may be investigated. The fringes of colours about shadows. The colours between thin glass plates are very capital experiments in this case, . .*

On the Imperfection of language as an instrument of thought.

In what follows I have used the words Sensation and reflection to express the sources from which all our ideas are drawn, but some little explanation of the manner in which I believe that we receive ideas these two different ways seems to be necessary.

By sensation the ideas are forcibly produced in us by some foreign cause (of which we will at present suppose that we know nothing) and are clearly and fully before us in all their parts and bearings.

Ideas by reflection do not come to us in this manner ; they are not forced upon us from without and by a foreign cause : When we cast an attentive intellectual view upon ideas we have received by sensations, whether they be actually present or only remain in the mind by recollection, ideas by reflection will spring up spontaneously and establish themselves in the mind as firmly as the ideas we have received by sensation.

Memorandum of the defect of language.

I see a candle, I feel the table, &c and if these are called sensations we want a term to express the capacity of the mind to have such sensations. Sensibility will not do. Receptivity might express a power of receiving the sensations conveyed by the senses. A new coined word would be sensativity.

When the candle is absent and I see it with the mind's eye, this may be called imagination but I do not like the term ; and here again we want a word to express the capacity of the mind to imagine the candle, such as imaginability, which is not a word that can properly express the meaning.

When we have an actual sensation, or the imagination of it in the mind, we can fix an attentive intellectual view upon them, this I believe is called reflection. The term is not quite satisfactory, intellectual inspection seems to be more expressive.

At the time that the mind is in this manner inspecting its own sensations or imagina-

* [End of MS.—Ed.]

tions new ideas will spring up relating to the objects of its intellectual view; these have been called ideas by reflection. The expression is very defective.

I see a triangle coarsely drawn upon a white wall with charcoal. Suppose the three angles of the triangle to be about 30, 50 and 100 degrees each, but it will be of no consequence if the three angles should together make a few degrees more or less than 180. When I shut my eyes I see the triangle and wall no longer but I still can imagine them. When my eyes are opened again I see the wall and triangle as before, and with proper attention I perceive by intellectual inspection, or, if you please by reflection that the wall and triangle are conceptions of the mind.

I also perceive by intellectual inspection, or reflection that when my eyes are opened I cannot help seeing these objects, and that the causes which make me see them are not conceptions of the mind but something without me not depending upon my perceptions.

What are these causes? What is their nature? how do they act? My answer is I do not know.

But do such causes exist? I cannot doubt it; the mental view I have of their effect assures me of it.

Now the real causes of the perception of the mental wall and mental triangle being perfectly unknown, if we want to converse about them must have names assigned to them. Like unknown quantities in algebra, we might call the *real* wall x, and the *real* triangle y; but this would at once cut off all investigation by the confusion that must be introduced in the use of thousands and thousands of letters and characters to express unknown causes. The *real* wall, the *real* triangle, therefore appear to me the properest expressions, provided it be always understood that these are things of which we either know nothing, or at least not much, whereas on the contrary the *ideal* wall and *ideal* triangle are perfectly well known.

Having often intellectually viewed the black triangle on the wall when I actually saw it, or when looking another way, I only had it by recollection in my mind, it occurred to me in one of these mental inspections that two of the sides of the triangle howsoever taken must be bigger than the remaining side.

This is one of those ideas that spring up spontaneously in the mind by reflection or rather in consequence of intellectual inspection. I now review the triangle again by actually seeing it on the wall, and find it miserably ill drawn; but reflection tells me that this is of no consequence, nor do I wish to see any other triangle to confirm the truth of the idea which by reflection has sprung up in my mind. Reflection itself tells me that this is not a matter of experience.

Does this certainty belong to all ideas that spring up in the mind by reflection? I answer no! For instance I take up a stone and opening my hand I see it drop to the ground. That there is a cause for this falling I am sure from one single experiment.

Will the next stone I let go out of my hand drop to the ground?

I cannot tell; but after having tried some thousands and found them all to fall to the ground I begin to expect that the next will do the same; but by reflection I perceive that I can never come to any complete certainty in this argument which is by induction, and can only be strengthened by the number of instances I may try.

By an intellectual inspection of the various objects of the sight and the touch, that is by much reflection on them, an idea of extensions in three dimensions is springing up in the mind. Or in the language of Locke, by attending to my sensations, reflection has furnished me with the idea of extension.

It is a property of ideas by reflection that at the time we receive them we are convinced that they are as intirely in our own mind, as those by Sensation were found to be.

To the question what is *real* extension I answer I know not.

But let us examine our *mental* extension. We have already seen that ideas by reflection have appeared variously conditioned; the property of the triangle I have described, left no doubt in the mind ; the cause of the falling of a stone remained under all those doubts which arguments by induction can never intirely remove. How stands the case with extension ?

I have before me a cube of a foot in diameter ; reflection convinces me that it is extended in three dimensions, and when after having seen and handled it, I take it up and put it in another place I find that this extension is gone with it and remains still in the Cube. It is therefore something belonging to it.

I also find that every visible and tangible object is extended though in very different dimensions and that when they are moved their extensions go along with them.

But in this examination reflection raises up a new idea. In moving the cube out of one place which we will call Nº 1 to another place at 4 feet distance Nº 2, I find that although the extension has gone with the Cube yet the space where the Cube has been remains in its place. The same thing happens when I take the Cube from Nº 2 to put it again in Nº 1. The extensions of the cube go with the cube, but the space where the cube is taken from remains in the same place.

What is this space ? I say the ideal space perceived by reflection is perfectly well understood and comprehended and all its properties are known.

But to the question what is *real* space ? I answer of this I know nothing.

In these examinations a very remarkable simple idea is perceived by sensation. It is that of motion. The Cube by being taken from the place Nº 1 and put into Nº 2 had a motion of 4 feet towards the right. In answering the question what is motion, those who did not attend to the simplicity of the idea have given all sorts of unsatisfactory definitions. If the word motion should not be understood we can only explain it by a synonymous term. My answer therefore to the question what is motion will be, it is the *passing* of a body from one place of space into another place of space. Thus the Cube in *going* or *passing* from the place of space Nº 2 into the place Nº 1 will have a motion of 4 feet to the left. It is evident that in this account of what is motion I have only used a term which like the translation of a word into another language expresses the same thing, for I might have said that motion happens when a body *moves* from one place of space into another place of space.

END OF UNPUBLISHED PAPERS

PAPERS

PUBLISHED IN THE

Philosophical Transactions of the Royal Society

AND ELSEWHERE

I.

Astronomical Observations on the periodical Star in Collo Ceti. By Mr.* WILLIAM HERSCHEL, *of* Bath; *communicated by Dr.* WATSON, *Jun. of* Bath, *F.R.S.*

[*Phil. Trans.*, vol. lxx., 1780, pp. 338–344.]

Read May 11, 1780.

THIS remarkable star, we are told,† "was first observed by DAVID FABRICIUS, the 13th of August, 1596, who called it the stella mira, or wonderful star; which has been since found to appear and disappear, periodically, seven times in six years, continuing in the greatest lustre for fifteen days together, and is never quite extinguished."

My own observations on this wonderful star are but few, yet sufficiently verify the surprizing appearances that have been ascribed to it. I shall transcribe them from my astronomical journal in the order they were made.

October 20, 1777, I looked out for the periodical star in Collo Ceti, but it was not visible. If its period is 312 days I may expect to see it about Christmas, not being visible at present.

Dec. 18, 1777, I saw the periodical star in Collo Ceti. It appeared in the very place where, about a fortnight ago, I imagined (but was not sure) there was a faint appearance of it. It was in magnitude about equal to ζ, but not so large as δ. ‡

* BAYER's character for this star is *o*. † See FERGUSON's *Astronomy*. § 366.

‡ ζ is marked by BAYER as a star of the fourth magnitude, and δ of the third. [In the Journal the name "Batenketos" is given instead of ζ. Therefore ζ is not a mistake for ξ, as supposed by Argelander, *Bonner Beob.*, vii. p. 324.—ED.]

Jan. 26, 1778, The periodical star was larger than δ, but less than γ. Being taken up with other observations, I paid no more attention to it during the rest of this period.

Sept. 18, 1779, The periodical star was visible to the naked eye, when I first looked for it.

Oct. 6, 1779, The periodical star was exceedingly bright this evening. It exceeded α and β Ceti; which latter, I must here observe, is considerably larger than the former, and affords a proof of the change in the magnitude of the fixed stars; as we can hardly suppose BAYER should have made a mistake in the magnitude of the two first stars of this constellation.

The apparent magnitude of o Ceti was not round but elliptical, when observed through the telescope, and not very well defined; but as it was too near the horizon, this shape might arise from that cause, for other low stars were also irregular in their forms, yet Bellatrix was exceedingly fine and quite round.

Oct. 7, 1779, The periodical star was perfectly round in the telescopes, and its apparent diameter well defined, full, and very large, for a star of that magnitude.

When I speak of the apparent diameter, I would be understood to distinguish it not only from the real diameter, but also (if I may be allowed the expression) from the real apparent diameter. To explain this a little more at large: the body of the sun, for instance, is of a certain dimension which we call his real diameter, and this remains always the same. His apparent diameter (which I here call real apparent) is changeable, according as we approach to, or recede from, him, and is between 31′ 33″ and 32′ 39″; but were he removed to the distance of one of the nearest fixed stars, neither his real, nor real apparent diameter, could then be known to us by any method we have hitherto been acquainted with: for at the distance of at least 20 billions of miles, his real apparent diameter could not much exceed thirty fourths of a degree*; and a telescope must magnify above fourteen thousand times to make him appear of only two minutes in diameter, which still is hardly sufficiently large to distinguish a square from a circle: and yet I doubt not, but that we should observe some apparent diameter or other of the sun thus removed from us; and this is what I here have called the apparent diameter. This must be owing to some optical deception. DE LA LANDE explains it thus: "Si l'on voit dans les lunettes une lumière éparse qui environne les étoiles, qui les amplifie et les fait paroitre come si elles avoient 5 a 6″ de diametre, on doit attribuer cette apparence à la vivacité de leur lumière, à l'air environnant et illuminé, à l'aberration des verres, à l'impression trop vive qui se fait sur la rétine."

Oct. 19, 1779, The periodical star preceded a very obscure telescopic star at the distance of 1′ 45″·16. This measure was extremely difficult to take, the small

* [That is, $\frac{111}{120}$″.—ED.]

star being so obscure as hardly to bear the field of view to be sufficiently enlightened: both also having nearly the same declination, they therefore pass the parallel hairs of the micrometer almost at rectangles. However, I should suppose, the measure must be true to 3 or 4″.

I measured the distance once more, with a better light and more satisfaction to myself, and found it 1′ 50″·47.

12 o'clock, The periodical star is now about the meridian, and brighter than α Arietis.

Oct. 30, 1779, 9 o'clock, The periodical star is still increased, and visibly larger than α Arietis, though not in the meridian at present.

— o'clock, o Ceti being now in the meridian, is almost of a middle size between Aldebaran and α Arietis. Its apparent diameter by the telescope is also increased.

Nov. 2, 1779, The lustre of the periodical star is still increased. The body is very full and round in the telescope. I magnified it 449 times very distinctly, the evening being so fine; but my usual power is only 222.

Nov. 20, 1779, The periodical star seemed to be as bright as before, but no brighter.

Nov. 30, 1779, o Ceti is considerably decreased. Its magnitude is less than β but greater than α. I have before remarked, that α is less than β.

Dec. 4, 1779, The lustre of o Ceti is only equal to α. I measured the distance of this star from the obscure one which follows it:

First measure, 1′ 53″·437
Second measure, 1′ 50″·625

The weather was so cold, that I could hardly finish this last measure; but I believe it to be too small. The disagreement is owing to the difficulty, and not to want of accuracy in the micrometer.

Jan. 4, 1780, The periodical star is very much diminished. I took the distance, but the evening was very indifferent, and therefore the small star so faint as to allow but little light. It measured 1′ 45″·937. I thought the measure too little when I took it, but could not succeed better.

Feb. 7, 1780, The periodical star was invisible to the naked eye. I was but little prepared to look a long time for it with the telescope; but suppose I shall be able to find it another time.

MAUPERTUIS accounts for the periodical appearances of changeable stars, by supposing that they may be of a flat form, like Saturn's ring, which becomes invisible when the edge is presented to us.

As the periodical star in Collo Ceti appeared always full and round when I viewed it with a telescope, this might at first appear to contradict the supposition of MAUPERTUIS; but, upon proper consideration, will be found not to be at all against it: for, suppose the real apparent diameter of this star to be one-third of a degree *; then, since it appeared to me (I did not measure it) at least of one second, when at the full, it will follow, that there was an aberration, whatever might be the cause of it, which amounted to 59‴, by which its real apparent diameter was increased from 1‴ to 1″. Now, if this star, in one certain position, should present its circular disk of 1‴ in diameter, and in another situation only its flat edge which would appear as a line of 1‴ in length, both appearances, with the aberration included will still remain of a circular form: for, adding the aberration 59‴ to the length 1‴, the whole becomes 1″; and adding 59‴ to scarce any breadth at all, the whole breadth will also become nearly 1″.

KEILL says, "It is probable, that the greatest part of this star is covered with spots and dark bodies, some part thereof remaining lucid; and while it turns about its axis, does sometimes shew its bright part, sometimes it turns its dark side to us; but the very spots themselves of this star are liable to changes, for it does not every year appear with the same lustre. Sometimes it resembles a star of the second magnitude; in other years it can scarcely be reckoned among stars of the third order; nor are the times of its visiting us always of the same duration."†

* [That is, $\frac{1}{60}''$.—ED.]

† [For further observations of *o* Ceti see above, No. xxix. of the papers read before the Bath Philosophical Society.—ED.]

II.

Astronomical Observations relating to the Mountains of the Moon. By Mr. HERSCHEL *of* Bath. *Communicated by Dr.* WATSON, *Jun. of* Bath, *F.R.S.*

[*Phil. Trans.*, vol. lxx., 1780, pp. 507–526.]

Read May 11, 1780.

AT the time when the telescope was first invented this noble instrument was immediately applied to astronomical observations with the most surprising success. Several very eminent persons have given us an account of their discoveries; and, notwithstanding the imperfect state of telescopes in those times, we still owe a great deal of our knowledge of the heavenly bodies to the observations that were made by those first telescopic observers, who made amends for the deficiencies of their instruments by their uncommon diligence and attention.

It may, perhaps, be esteemed to be a mere matter of curiosity to search after the height of the lunar mountains. I grant that there are more necessary and more useful objects of inquiry in the science of astronomy; but when we consider that the knowledge of the construction of the Moon leads us insensibly to several consequences, which might not appear at first; such as the great probability, not to say almost absolute certainty, of her being inhabited, we shall soon agree, that these researches are far from being trifling.

My reason for repeating observations that have been made by very good astronomers was not that I doubted either their veracity or diligence. The names of GALILEO, HEVELIUS, KIRCHER, and several more, will always deserve to be mentioned with particular respect for the eminent services they have rendered to astronomy; but as we know that their instruments were far from being arrived to that degree of perfection we have now obtained, I thought it by no means improper or useless to repeat their observations on the lunar mountains, and to extend them to other parts of the Moon's visible hemisphere, and thereby to establish this theory on the firmest evidence of a survey taken by a very excellent instrument.

The method used by HEVELIUS and others to find the height of a mountain in the Moon is this. Let a ray of light SLM (fig. 1) proceeding from the Sun, pass by the Moon at L, and touch the top of a mountain at M: then the space between L and M will appear dark, and the top of the mountain will be seen to stand at some distance from the illuminated part of the Moon's disk. With a good micrometer let the distance LM be taken by observation.* Draw LC perpendicular to LM; draw also MC from the top of the lunar mountain to the center of the Moon: then in the triangle MLC, rectangled at L, we have given the side LC, which is the Moon's radius, and the side LM taken by observation. Therefore, by trigonometry, we can find the hypothenuse MC†, from which, subtracting the part *p*C or radius, there remains the perpendicular height of the mountain M*p*. I have followed the same method, as being the least liable to error.

GALILEO takes the distance of the top of a lunar mountain from the line that divides the illuminated part of the disk from that which is in the shade to be equal

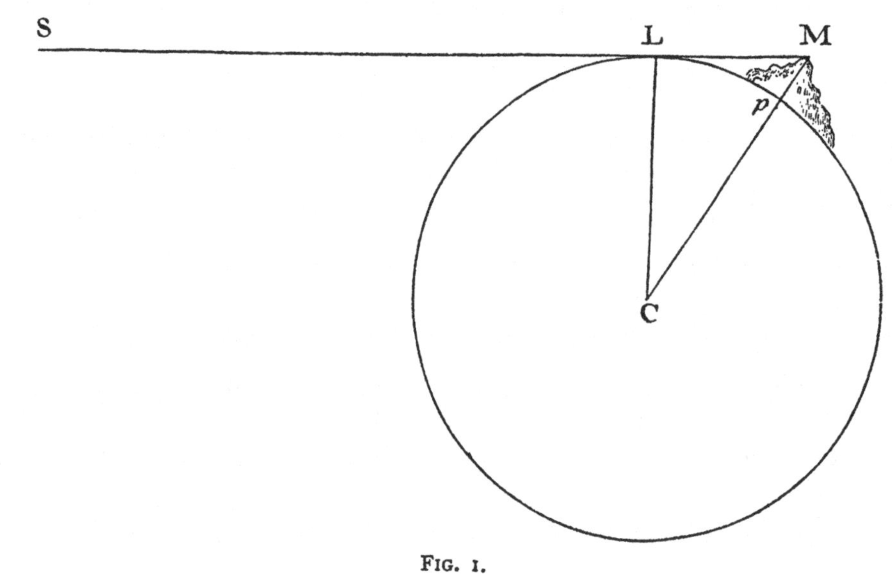

FIG. 1.

to a 20th part of the Moon's diameter; but HEVELIUS affirms, that it is only the 26th part of the same.

When we calculate from thence the height of such a mountain it will be found, in English measure, according to GALILEO, almost 5½ miles; and, according to HEVELIUS, something more than 3¼ miles, admitting the Moon's diameter to be 2180 miles.

He says, in his *Selenography*, p. 266: "Vera distantia illustratarum cuspidum, à confinio luminis et umbræ, præsertim tempore quadraturæ, invenitur, unâ vigesimâ sextâ parte, totius Lunæ dimetientis constare; quando nimirum sunt remotissimæ: quemadmodum hoc ex phasi trigesimâ secundâ, monteque Apennino; ex phasi trigesima prima, monteque Didyme; et trigesimâ phasi, monteque Tauro et Antitauro, manifestissimè demonstratur." Having afterwards mentioned that GALILEO makes the distance LM to be the 20th part of the Moon's diameter, HEVELIUS proceeds: "Quamobrem, cum distantiæ a nobis designatæ, paululum

* I do not recollect that HEVELIUS mentions in what manner he took the distance LM; but I am apt to believe it was by a micrometer.

† $\sqrt{\overline{LC}^2+\overline{LM}^2}=MC$.

sint minores, idcirco et montes aliquantulum depressiores inveniuntur, quàm GALILÆUS æstimavit: neque non tamen illi terrenis nostris montibus, quoad altitudinem, non solum æquiparari possunt meritissimò; sed et multo certe sunt excelsiores, quàm nostri omnium maximi; prout confestim, ex adjecto diagrammate patebit." He gives us then his calculation according to German and Italian measure; and having found, in the manner above mentioned, the hypothenuse MC, he adds: "Semidiameter Lunæ erit 1976 octav. part. Si igitur hæc à totâ hypothenusa auferetur, restabunt adhuc sex, hoc est sex octavæ unius milliaris, vel tres quartæ unius milliaris Germanici, sive tria milliaria Italica: quæ est vera, et genuina altitudo istius montis." As a German mile in the time of HEVELIUS was a very uncertain measure, we may suppose that he meant geographical miles, 15 of which make a degree of latitude. The observations of HEVELIUS have always been held in great esteem; and this is most probably the reason why later astronomers have not repeated them. M. DE LA LANDE, who is one of our most eminent modern astronomers, agrees to the sentiments above cited.

In his *Abrégé d'Astronomie*, p. 435, he says, "Je terminerai ce qui concerne la sélénographie, en disant un mot de la hauteur des montagnes de la lune, qui étoient quelquefois éclairées, quoiqu' éloignées de la ligne de lumière de la treizième partie du rayon de la Lune; de la on peut conclure que ces montagnes ont de hauteur la 338[ieme] partie du rayon Lunaire, ou une lieue de France." He then gives us a particular calculation, and the result is: "Avec ces données on trouve la hauteur de 2643 toises, c'est à dire, plus d'une lieue commune."

He also mentions the opinion of GALILEO, and adds: "Mais on doit préférer à cet égard les observations d'HEVELIUS, qui ont été plus répétées, plus détaillées et plus exactes."

Mr. FERGUSON says (*Astronomy explained*, § 252.) "Some of her mountains, by comparing their height with her diameter, are found to be three times higher than the highest hills on our earth."

KEILL, in his *Astronomical Lectures*, has calculated the height of St Katherine's hill, according to the observations of RICCIOLUS, and finds it nine miles.

Before I report my own observations, it will be necessary to explain by what method I have found the height of a lunar mountain from observations that were made when the Moon was not in her quadrature; for the method laid down by HEVELIUS will only do in that one particular case: in all other positions the projection of the hills must appear much shorter than it really is. Let SLM, or *slm* (fig. 2) be a line drawn from the Sun to the mountain, touching the Moon at L or *l*, and the mountain at M or *m*. Then, to an observer at E or *e* the lines LM, *lm*, will not appear of the same length, though the mountains should be of an equal height; for LM will be projected into *on*, and *lm* into ON. But these are the quantities that are taken by the micrometer when we observe a mountain to project from the line of illumination. From the observed quantity *on*, when the Moon is not in her quadrature, to find LM we have the following analogy. The triangles oOL, rML, are similar; therefore, Lo : LO :: Lr : LM, or $\frac{\text{LO} \times on}{\text{L}o} = \text{LM}$; but LO is the radius of the Moon, and Lr, or on, is the observed distance of the mountain's projection; and Lo is the sine of the angle ROL $=o$LS, which we may take to be the distance of the Sun from the Moon without any material error, and which therefore we may find at any given time from an ephemeris.

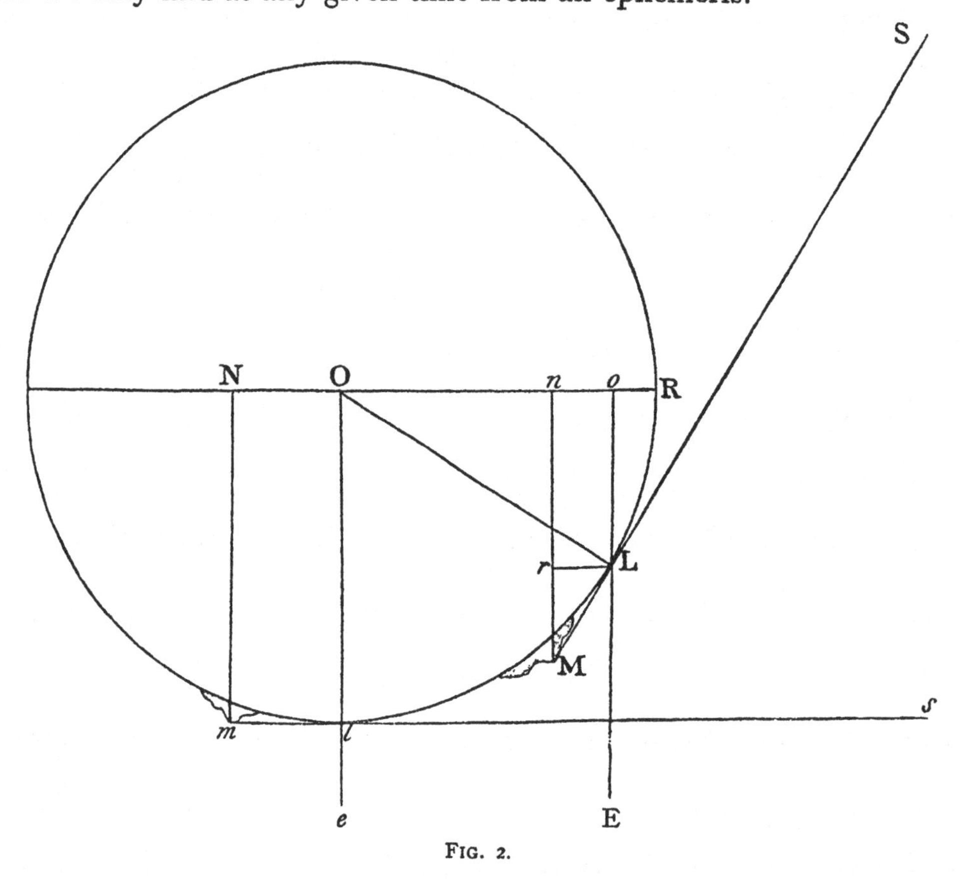

FIG. 2.

I will now give an account of my own observations relating to the mountains in the Moon; but, perhaps, it may not be amiss to mention the instrument they were made with, and a few of the circumstances, that it may appear how far their accuracy may be depended upon.

The telescope I used was a Newtonian reflector of six feet eight inches focal length, to which a micrometer was adapted consisting of two parallel hairs, one of which was moveable by means of a fine screw. The value of the parts shewn by the index was determined by a trigonometrical observation of a known object at a known distance, and was verified by several trials. The power I always used, except when another is mentioned, was 222 times, also determined by experiment, which I have often found to differ somewhat from theory, on account of some little errors in the *data*, hardly to be avoided. The moon having sufficient light,

I used no more aperture of the object speculum than four inches; and, I believe, that for distinctness of vision this instrument is perhaps equal to any that was ever made.

OBSERVATIONS.

November 30, 1779, six o'clock in the morning, a rock, situated near what HEVELIUS calls *Lacus niger major*, was measured to project 41″·56. To reduce this quantity into miles, put R for the semi-diameter of the Moon in seconds, as given by the *Nautical Almanac* at the time of observation, and Q for the observed quantity, also in seconds and centesimals; then it will be in general $R : 1090 :: Q : \frac{1090\ Q}{R} = on$, in miles. Thus it is found that 41″·56 is 46·79 miles. This distance of the Sun and Moon at the same time was, by the *Nautical Almanac*, about 93° 57′½. The sine of which to the radius 1 is ·9985, &c. and $\frac{on}{LO}$ in this case, is LM = 46·85 miles. Then, by HEVELIUS's method the perpendicular height of the rock is found to be about one mile.

The same morning, a great many rocks, situated about the middle of the disk, projected from 25″·93 to 26″·56. This gives *on* about 29·3 miles, and these rocks are all less than half a mile high.

January 13, 1780, 7 o'clock, I examined the mountains in the Moon; but there was not one of them that was fairly placed on level ground, which is a condition very necessary for an exact measurement of the projection. If there should be a declivity on the Moon before the mountains, or a tract of hills placed so as to cast a shadow on that part before them which would otherwise be illuminated, it is plain that the projection would appear too large; and, on the contrary, should there be a rising ground before them, it would appear too little.

As far as I was able to judge of the direction of the line of illumination, the highest hill projected 25″·31, or 30·36 miles: from thence we find as before, that the perpendicular height is (·42 mile) less than half a mile.

January 14, 11 o'clock, I took the projection of the highest mountain which was situated at the Western edge. It measured 24″·68, or about 27 miles; and the perpendicular height comes out less than half a mile. There was not one mountain in the edge of the disk so high as this.

January 17, 7 o'clock, a very high mountain projected no less than 40″·625. Its situation is in the South-east quadrant. The Moon's semi-diameter, at the time of observation, by the *Nautical Almanac*, was 16′ 2″·6; therefore, $\frac{1090\ Q}{R} = 45{\cdot}98$ miles $= on$.

Sun's longitude at 7 h.*	9^s	27°	39′	0″
Moon's longitude at 7 h.	2	2	46	52
Their nearest distance,	4	5	7	52

* [9^s means nine signs of the Zodiac, of 30° each.—ED.]

or about 125° 8′; the sine of which is ·8104: thence we find LM 56·73 miles; and the perpendicular height of the mountain is 1m. 47, or less than a mile and a half.

January 22, 8 h. 20′ the highest mountain, situated near Snell or Petavius, projected 11″·437, which is 12·34; and LM comes out to be 35·3 miles: therefore the perpendicular height is ·57 mile.

Another, just behind Mare Crisium, measured only 7″, therefore is less than half a mile high.

January 25, 7 h. 30′ in the morning, a mountain near Aristoteles measured 18″·59 which gives 20·6 miles; and LM is found 28·53 miles; the perpendicular height is therefore only ·37 mile.

Other mountains about Mare Nectaris measured about 23″·5; but they had hills before them, and their situation was not so proper for my purpose. However, it is evident they were of no considerable height.

January 28, 6 o'clock in the morning, the highest mountain in the disk measured 30″·937; the Moon's semi-diameter at that time 15′ 40″; and *on* therefore equal 31·37 miles: but as the Moon is within four hours of her quadrature we may be assured that this mountain is less than half a mile high.

February 19, Mons Sinopium projected 5″·781; therefore *on* = 6·26 miles, and the quantity LM 56·54 miles; and consequently the height of this mountain, which it seems proves to be a very high one, is not much less than a mile and a half. However, my journal observes, that the measure was very full; therefore the mountain in all probability does not exceed a mile and a quarter. Moreover, I think that observations made so near the full or new Moon are less to be depended upon, because a small error in measuring will produce a great one in the height of a mountain.

From these observations I believe it is evident, that the height of the lunar mountains in general is greatly over-rated; and that, when we have excepted a few, the generality do not exceed half a mile in their perpendicular elevation. It is not so easy to find any certain mountain exactly in the same situation it has been measured in before; therefore some little difference must be expected in these measures. Hitherto I have not had an opportunity of particularly observing the three mountains mentioned by HEVELIUS; nor that which RICCIOLUS found to project a sixteenth part of the Moon's diameter. If KEILL had calculated the height of this last mentioned hill according to the theorem I have given, he would have found (supposing the observation to have been made, as he says, on the fourth day after new Moon) that its perpendicular could not well be less than between eleven and twelve miles.

I shall not fail to take the first opportunity of observing these four, and every other mountain of any eminence; and if other persons, who are furnished with good telescopes and micrometers, would take the quantity of the projection of the lunar mountains, I make no doubt, but that we should be nearly as well acquainted with their heights as we are with the elevation of our own. One caution I would beg leave to mention to those who may use the excellent $3\frac{1}{2}$ feet refractors of Mr. DOLLOND. The admirable quantity of light, which on most occasions is so desirable, will probably give the measure of the projection somewhat larger than the true, if not guarded against by proper limitations placed before the object glass. I have taken no notice of any allowance to be made for the refraction a ray of light must suffer in passing through the atmosphere of the Moon, when it illuminates the top of the mountain, whereby its apparent height will be lessened, as we are too little acquainted with that atmosphere to take it into consideration. It is also to be observed, that this would equally affect the conclusions of HEVELIUS, and therefore the difference in our inferences would still remain the same.

Bath, February 28, 1780.

Continuation of the same observations.

March 11, 1780, 7 h. Promontorium Acherusia projected 17″·187. It is very properly situated for measuring. By a proper deduction from the Moon's semi-diameter, as given by the *Nautical Almanac*, at the time of observation, we find the quantity *on* = 20·1 miles, and LM 22·6 miles; from which it appears, that the perpendicular height of this mountain is a little less than a quarter of a mile.

Antitaurus, the mountain measured by HEVELIUS was badly situated, because Mount Moschus and its neighbouring hills cast a deep shadow, which may be mistaken for the natural convexity of the Moon. A good, full, but just measure, 25″·105; in miles 29·27: therefore, LM 31·7 miles, and the perpendicular height not quite half a mile.

7 h. 45′. I was desirous of being very exact in this measure, therefore I repeated it. I took two different observations. A narrow measure 21″·562; quite full enough 24″·062. These measures give the perpendicular height less than half a mile.

8 h. I measured Lipulus, 19″·063. It is also badly situated, though rather better than Antitaurus. I found that the projection increased, therefore concluded that this was not the highest part of the mountain, and waited some time when I measured it again.

9 h. Lipulus now projected 28″·75.

10 h. It measured 28″·75: this gives *on* = 33·64 miles. Distance of Sun and Moon about 63° 23′: therefore, LM 37·54 miles. From hence we find the perpendicular height ·64 mile, or very near two-thirds of a mile.

March 12, 1780, 7 h. One of the Apennine mountains, between Lacus Trasimenus and Pontus Euxinus projected 44″·062. This gives us *on* = 51·11 miles; and LM = 52·9 miles: therefore the perpendicular height of these mountains, which I know to be very high, comes out to be $1\frac{1}{4}$ mile.

Mons Armenia (near Taurus) projected 31″·406 = 36·43 miles; LM = 38 miles nearly, and the height two-thirds of a mile.

Mons Leucopetra 34″·479 or 40 miles; LM = 41·4 miles, and the perpendicular height three quarters of a mile.

There was a very fine shade of a high rock near it, which shewed the direction of the illuminating ray, and thereby assisted me in measuring to a great exactness; but the mountain itself is not very favourably situated.

March 16, 10 h. 30′. Mons Lacer projected 45″·625; but I am almost certain that there are two very considerable cavities or places where the ground descends below the level of the convexity, just before these mountains, so that these measures must of course be a good deal too large: but supposing them to be just, it follows, that *on* is 50·193 miles, LM = 64 miles, and the perpendicular height above $1\frac{3}{4}$ miles.

Another of the same mountains situated on the borders of S. Sirbonis measured 41″·875. This ridge of mountains is the same of which I measured one on January the 17th, which was then found to be 1·47 miles high.

The following additional Memoranda of the Manner in which Mr. HERSCHEL *made his Observations are taken from a Letter of his to the Rev. Dr.* MASKELYNE, *Astronomer Royal.*

IN the second figure of my observations, the points L, S, E, *r*, are all supposed to be in one plane; and as the illuminating ray SL is also in this plane, it follows, that the line L*r* (= *on*) will always be perpendicular to the right line which joins the cusps of the Moon *; and the truth of the theorem there delivered depends upon this circumstance.

For this reason I have taken care in all my observations to measure the line, which in fig. 3 (taken from your letter to Dr. WATSON) is marked *on*, parallel to the line CD, or perpendicular to AB, and not the line *rn*, perpendicular to the elliptical curve A*ro*B.

The manner of taking it is easy enough: however, I have occasionally used three different methods, and will describe them all, which I should have done in the paper delivered by Dr. WATSON, had I not feared to be too particular.

The first method I used was to set the immoveable hair *hh* (fig. 4) of my micrometer parallel to a line AB, joining the cusps of the Moon; then, by opening the moveable parallel hair till it included the projection *on*, intended to be taken, I marked that down as the measure of *on*. As this method required some attention (that part of the ellipsis of illumination AVB which is the vertex V of the lesser axis may serve as a direction) and took up some time, on account of the small field of view of my telescope, I used occasionally these two following ways.

* It is here supposed, that rays from the Sun S, and the eye of the observer E, to any part of the Moon L, may be taken for parallel; and therefore, that different planes, made by several sections of the Moon, according as the point L is taken North or South of the diameter of the Moon, which is at rectangles to the line joining the cusps, may also be taken to be parallel to that diameter.

When there was any remarkable figure on the disk of the Moon near the line of illumination, I put on a compound eye-piece whose magnified field of view is full 40°, and power about 90 times, so that it takes in the greatest part of the whole Moon; by this means I was enabled to view the projection intended for measuring at the same time with the rest of the Moon, and to fix upon some mark in the disk very near to its edge towards which I judged the line *on* should be directed; then putting on the eye-piece which carries the micrometer I took the distance according to this judgment as well as I could.

The third method I took was the following, which, indeed, I look upon as the best of all, and which I therefore most frequently put in practice. I took a view of some neighbouring shades of rocks or mountains, if there happened to be

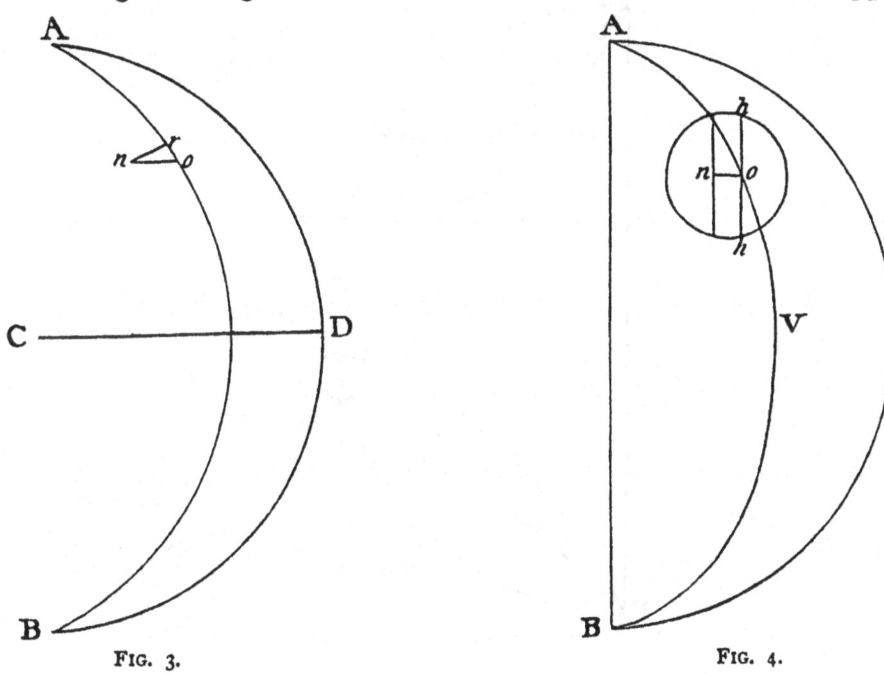

FIG. 3. FIG. 4.

any near, and directed the measure of the micrometer by them, as they plainly pointed out the direction of the illuminating ray; or, which is the same thing, indicated the line perpendicular to a line joining the cusps.

Mons Leucopetra was measured by this last method, which circumstance I have mentioned in my observations of the 12th of March, where I saw the whole rock and its highest point, as well as the whole shade, and its last termination, upon very even ground, at the same time that I directed my micrometer in that line, to take the projection *on*, of the above mentioned mountain.

Sometimes I compared together a measure taken in the direction *on*, and one taken in the direction *rn*; but as most of my observations were made upon mountains not situated near the cusps or limb of the Moon, I never found so much difference between these two measures, that it could have occasioned any very material error, if I had intirely neglected it.

By the nature of the ellipsis it will appear, that, when we do not come too near the limb or cusps of the Moon, a tangent drawn to a point in the curve of illumination will seldom make with the subtangent an angle that exceeds (or is so much as) 26° ; and in all such cases the error that can arise from taking the line *rn* instead of *on* will be less than the tenth part of the whole measure : but, if the angle the tangent makes with the subtangent is only about 18°, the error will be less than a 20th part ; and all the measures I have taken, I believe, will be found to be much within these last-mentioned limits. From this consideration it will appear, that if I had not been aware of this circumstance, my observations would still be sufficiently accurate to disprove the usually assigned great height

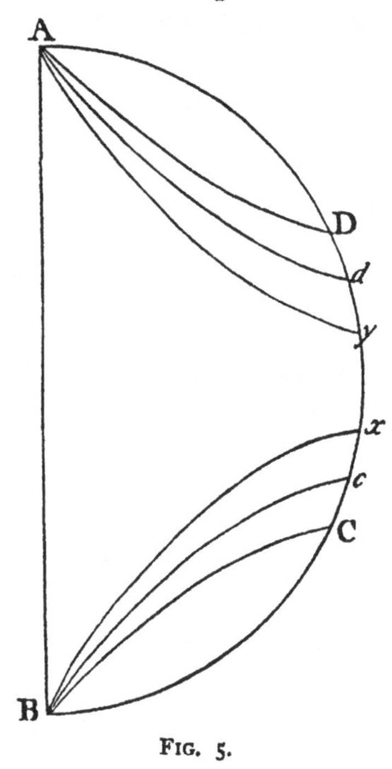

FIG. 5.

of the lunar mountains ; but as I took all the precaution the situation of each mountain would afford, by using any one of the above mentioned three methods, which suited best, I believe there can hardly be a possibility of any error that should amount to a 40th part of the whole height of any mountain I have measured.

The figure ABCD (fig. 5) contained by the diameter AB, the arch CD, and the two curves AD, BC, shews in what portion of the Moon's semi-disk we may safely measure the line *rn*, instead of *on*, without being liable to so great an error as one tenth part of the whole, and the figure AB*cd* contains that part wherein the measure *rn* being taken instead of *on*, the error will be less than the 20th part of the whole measure. In a portion something more confined the error will soon vanish, so that the difference may be safely neglected intirely. Thus in the space AB*xy* the error cannot amount to a hundredth part. These figures may be constructed by taking the several points D, *d*, *y*, and C, *c*, *x*, 26°, 18°, 8°, respectively, from the vertex, the curves AD, A*d*, A*y*, BC, B*c*, B*x*, being the loci of those points of the tangents which touch the several ellipses of illumination that may be contained in the semi-disk of the Moon, when these tangents make those several angles of 26°, 18°, 8°, with their subtangents.*

* [Here ends this paper as printed in the *Phil. Trans.* For the continuation, about lunar inhabitants, see above under No. xviii. of the Bath papers.—ED.]

III.

Astronomical Observations on the Rotation of the Planets round their Axes, made with a View to determine whether the Earth's diurnal Motion is perfectly equable. In a Letter from Mr. WILLIAM HERSCHEL *of* Bath *to* WILLIAM WATSON, *M.D., F.R.S.*

[*Phil. Trans.*, vol. lxxi., 1781, pp. 115–138.]

Read January 11, 1781.

Bath, October 18, 1780.

SIR,—The various motions of the planet we inhabit; the annual revolution in its orbit; the diurnal rotation round its axis; the menstrual motion round the common center of gravity of the moon and earth; the precession of the equinoctial points; the diminution of the obliquity of the ecliptic; the nutation of the earth's axis: in short, every one of the motions that arise from the actions of the sun, moon, and planets, combined with the spheroidical figure of the earth, and the projectile and rotatory motions first impressed upon it, have all been considered by astronomers, and their real and apparent inequalities investigated. And to the great honour of modern astronomers it must be confessed, that no science has ever made such considerable strides towards perfection in so short a time as astronomy has done since the invention of the telescope.

There is one of the motions of the earth however which, it seems, has hitherto escaped the scrutiny of observers; I mean the diurnal rotation round its axis. The principal reason why this has not been looked into, is probably the difficulty of finding a proper standard to measure it by; since it is itself used as the standard by which we measure all the other motions. We have, indeed, no cause to suspect any very material periodical irregularity, either diurnal, menstrual, or annual; for the great perfection of our present time-pieces would have discovered any considerable deviation from that equability which we have hitherto ascribed to the diurnal motion of the earth. And yet, it is not perhaps altogether impossible but that inequalities may exist in this motion which, in an age where observations are carried to such a degree of refinement, may be of some consequence.

To shew how far time-keepers, though ever so perfect, are from being a proper, or at least a sufficient, standard to examine the diurnal motion of the earth by, I may ask, whether it is probable, that any clock would have discovered to us the aberration of the fixed stars? And yet that aberration produces a change in longitude, and of consequence in right ascension, which causes an annual irregularity in a star's coming to the meridian, which a time-piece, were it a sufficient standard, would soon have discovered, and which we might have attributed to an inequality of the earth's diurnal motion, had we not been acquainted with its real cause. And if we were to find out any apparent irregularity, acceleration, or retardation, should we not much rather suspect the clock than the diurnal motion ? I may therefore venture to say, that the aberration of the fixed stars, though attended with the above mentioned consequence, would for ever have remained a secret to us, if it had not been found out by other methods than time-keepers.

Now, if time-pieces do fail us in this critical case, where we stand in the greatest need of their assistance, it is almost in vain to expect any help from another quarter ; for what mechanical movement on earth, or motion of the heavens, is there that can measure out such equal portions of time as we require to compare the diurnal motion of the earth to ? However, to proceed, since we have already great proofs that the diurnal motion of the earth is, if not perfectly equable, at least more so than any other motion we are acquainted with, it will not appear absurd to suppose the diurnal rotation of the other planets to be so likewise. This suggested to me the thoughts of estimating the diurnal motion of one planet very exactly by that of another, making each the standard of the other. In this manner we may obtain a comparative view, by which future astronomers, if they shall hereafter be inclined to pursue the subject, may be enabled to make some estimate of the general equability of the rotatory motions of the planets. For if in length of time they should perceive some small retardation in the diurnal motion of a planet occasioned by some resistance of a very subtle medium in which the heavenly bodies perhaps move ; or, on the other hand, if there should be found an acceleration from some cause or other, they might then ascribe the alteration either to the diurnal motion of the earth, or to the gyration of the other planet, according as circumstances, or observed phænomena, should make one or the other of these opinions most probable.

Now, this method of comparing together different rotations of several planets, simple as it may appear, was not without some difficulties. In the first place it was evident, that the common account of their diurnal motions,* which makes that of Jupiter 9 h. 56′, of Mars 24 h. 40′, how true soever it may be in a general way, was much too inaccurate for this critical purpose. The gyration of Venus was still less to be depended upon, being only noted to the hour without the minutes : it became, therefore, necessary to proceed to observations of a more determinate kind. From what I had already seen of the rotation of the planets,

* Venus spatio 23 horarum gyrationem circa axem ab occidente in orientem perficit. Mars similem rotationem, horis 24 min. 40 absolvit. Macularum revolutionibus sæpius observatis, Ds. CASSINI comperuit periodum Jovis circa proprium Axem esse horarum 9, minutorum 56. KEILL, *Ast. Lect.* V.

I concluded, that Mars on several accounts would be the most eligible planet for my purpose: for the spots on Jupiter change so often that it is not easy, if at all possible, to ascertain the identity of the same appearance, for any considerable length of time. Nor do the dark spots only change their place, which may be supposed to be large black congeries of vapours and clouds swimming in the atmosphere of Jupiter; but also the bright spots, though they may adhere firmly to the body of Jupiter, may undergo some apparent change of situation by being differently covered or uncovered on one side or the other, by alterations in the belts. It will be seen hereafter, that I have observed the revolution of a very bright spot, not suspected of any change of situation, to be first, by one set of observations, at the rate of 9 h. 51′ 45″·6; and afterwards, by another set immediately following at the rate of 9 h. 50′ 48″.

As the principal belts on Jupiter are equatorial, and as we have certain constant winds upon our planet, especially near the equator, that regularly, for certain periods, blow the same way,* it is easily supposed, that they may form equatorial belts by gathering together the vapours which swim in our atmosphere, and carrying them about in the same direction. This will, by analogy, account for all the irregularities of Jupiter's revolutions, deduced from spots on his disk that may have changed their situation; for if we suppose the rotation of Jupiter, according to CASSINI, to be 9 h. 56′, then some spots that I have observed must have been carried through about 60° of Jupiter's equator in 22 of his revolutions or days. This would certainly be a very great velocity in the clouds, which is, however, not unparalleled by what has happened in our own atmosphere.

But to return to my purpose: on the planet Mars we see spots of a different nature; their constant and determined shape, as well as remarkable colour, shew them to be permanent and fastened to the body of the planet. These will give the revolution of his equator to a great certainty, and by a great number of revolutions, to a very great exactness also. Supposing then, that, by a method I shall hereafter describe, we can determine whether a spot on the disk of Mars is, or is not, in the line which joins the center of the earth and the center of that planet, to half an hour's time with certainty (I believe ten or twelve minutes will be found sufficient for that purpose), in this case we shall in 30 days have the revolution true to a minute; and, by continuing these observations for three months, we shall have it to 20″. When we are so far certain, we can easily arrive to a much greater degree of exactness; for as we now can no longer mistake a whole revolution, if we take the time of any particular spot's being in the line which joins the centers of the planets during one opposition of Mars, and take the same again at or near the next opposition, we shall have an interval of about 780 days, which will give the diurnal motion of that planet true to about 2″. The next opposition

* See *Acta Eruditorum*, 1687. Dr. HALLEY's Account of periodical Winds.

will give it to one, and so forth; by which means, and by taking a proper number of such periods, we may determine the rotation of Mars to as great an exactness as we shall think necessary for the purpose of our comparative view.

Had such observations as these been made two thousand, or perhaps only so many hundred years ago, we might now, by repeating them, most probably become acquainted with some curious minute changes of the solar system that have hitherto passed unnoticed.

There is a certain circumstance which would almost create a suspicion that there has been some retardation in the diurnal motion of the earth. The difference between the equatorial and polar diameters of the earth, by actual measurement, has been found to be about 36 English miles and 9 tenths; but, by a calculation wherein the present rotation is made use of, it will only amount to about 33 miles and 8 tenths: from which it should seem probable, that when the earth assumed the present form, the diurnal rotation was somewhat quicker than it is at present, by which means the centrifugal force bore a greater proportion to the force of gravity to which it is contrary, and thus occasioned a higher elevation of the equatorial parts. But I would not lay much stress upon this argument; for, in the calculation, it has been supposed, that the earth is nearly of an equal density at the surface and towards the center, which it seems is not agreeable to some late curious experiments and calculations that have been made under the conduct of the Astronomer Royal upon the attraction of a mountain,* the result of which ought now to be taken into consideration, and the calculation repeated. If all the *data* could be exactly depended upon, it would be practicable enough from the laws of gravity, and the present rotation and given form of the earth, to find the centrifugal force required to produce that form, and thence to shew what must have been its diurnal motion when it assumed the same. However, these are researches that in my present situation I neither have opportunity nor perhaps ability enough to investigate properly; and which, therefore, I hope some of our excellent mathematicians will think worth while to look into.

I shall now relate my observations on Jupiter and Mars. The telescopes I used are of my own construction; and are, a twenty-feet Newtonian reflector, a ten-feet reflector of the same form, and a seven-feet reflector already mentioned in my paper on the mountains of the moon. My time I gained by equal altitudes taken with a brass quadrant of two-feet radius, carrying a telescope which magnifies about 40 times; for the correction of altitudes taken of the sun I used DE LA LANDE's tables. I kept my time by two very good pieces; one having a deal pendulum-rod, the other a compounded one of brass and iron, both having a proper contrivance not to stop when winding up. The rate of going of my clocks I determined by the transit of stars.

*** See Mr. HUTTON's Account of the Calculations made from the Survey and Measures taken at Schehallien, in order to ascertain the mean Density of the Earth. *Phil. Trans.* 1778.**

Observations on Jupiter in the year 1778.

February 24. Clock 1′ 10″ too soon. About 9 o'clock I saw a bright belt on one part of the disk of Jupiter, see Plate I. fig. 1.

About 10 o'clock it was advanced as far as the center, fig. 2.

11 h. The white belt still more advanced, fig. 3.

11 h. 25′. It approached towards the edge of the disk; and at 12 h. was extended all over, as in fig. 4.

February 25. 8 h. The same bright belt I observed yesterday extends all over.

8 h. 45′. It is divided by a darkish spot, situated at some distance from the center, as in fig. 5.

9 h. 5′. The small dark division is advanced a little farther than the center, as in fig. 6.

9 h. 23′. The spot is visibly advanced a considerable deal farther.

March 2. 8 h. 2′. The darkish spot, with some alteration in its shape, is now in the middle of the disk, see fig. 7.

March 3. 10 h. 34′. The bright belt on the south of the equator is now in the middle; that is to say, if a line be drawn perpendicular to the equatorial belt, and through the center, the end of the equatorial belt now touches it; fig. 8.

13 h. 49′. The darkish spot, in which there has been some alteration since yesterday, seems now to be in the center; fig. 9.

March 14. The clock altered to true equated time; but the rate of going not changed, being well regulated.

7 h. 35′. The spot is now in the center, but does not seem quite to fill the white belt; nor is it so large and distinct as it was before; fig. 10.

April 7. 9 h. 31′. There are three dark spots in the equatorial belt nearly in the center, see fig. 11.

April 12. 7 h. 50′. The three dark spots are in the center. The southernmost of the three is nearly quite vanished; the other two are also much fainter. They are, however, distinct enough to be known; fig. 12.

Observations on Jupiter in 1779.

April 14. Clock 52″ too late. 8 h. 48′. A remarkable bright spot in the equatorial belt towards the north is in the center, see fig. 13.

8 h. 58′. The spot is a little past the center.

April 19. Clock true mean time. 7 h. 10′. There is a bright spot just now in the center, which, from its shape, I take to be the same that was there April 14th.

7 h. 20′. The spot is visibly past the center.

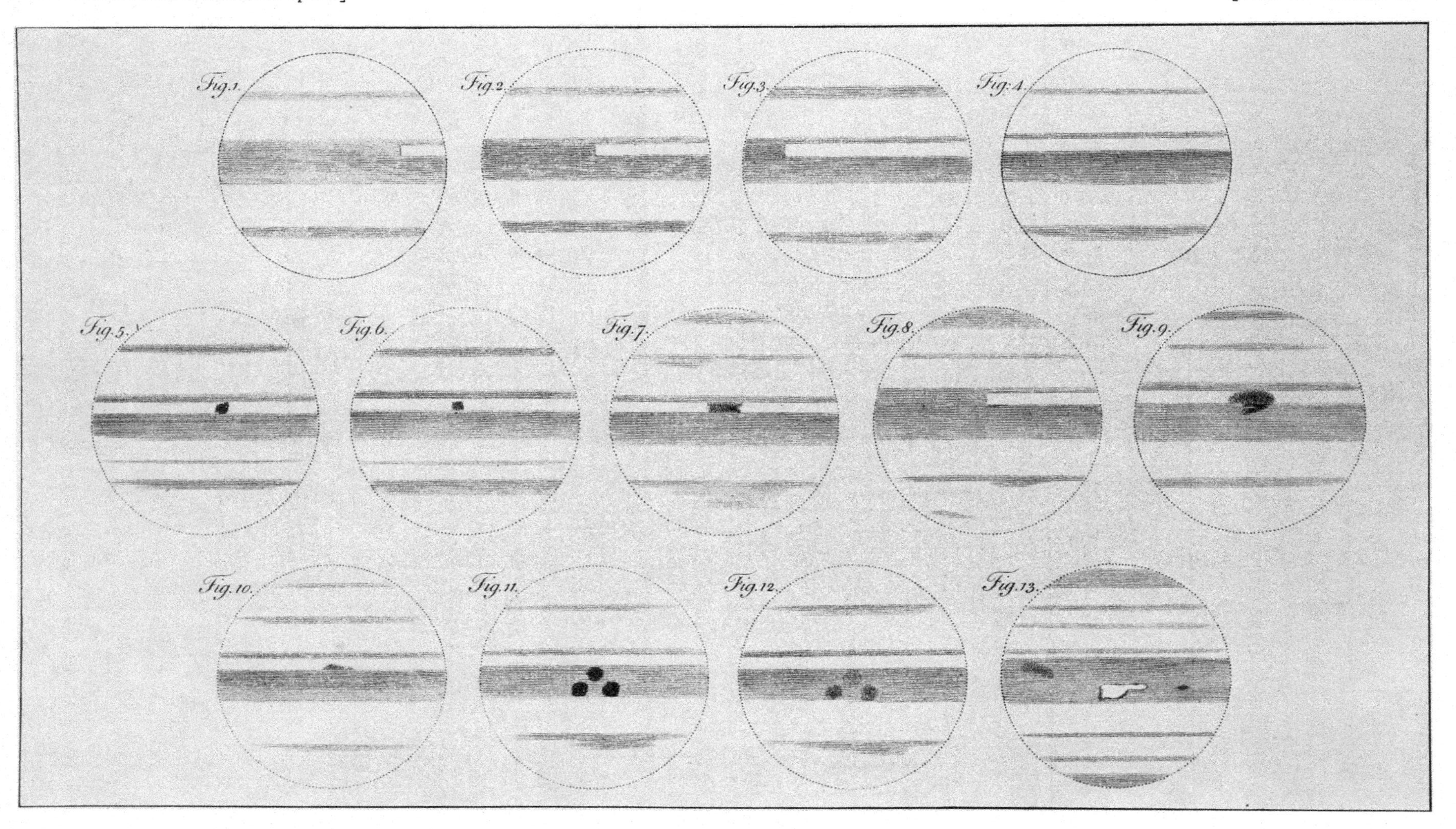
Fig. 1.
Fig. 2.
Fig. 3.
Fig. 4.
Fig. 5.
Fig. 6.
Fig. 7.
Fig. 8.
Fig. 9.
Fig. 10.
Fig. 11.
Fig. 12.
Fig. 13.

[*To face p.* 20.

April 23. Clock shews true time. 9 h. 38′. The same bright spot is in the center.

9 h. 43′. It is past the center. *Memorandum,* my time-piece may be depended upon to a few seconds.

It will not be amiss to observe, that the spots, as well as a great many other phænomena, were watched as they came on, passed over the center, and went off the disk of Jupiter ; but I have only selected those observations that were necessary to my present purpose.

Comparing together the observations that were made in the year 1778, February 24th and March 3d, we obtain an interval of 7 days 34 minutes, which being divided by 17 revolutions made by Jupiter on his axis, we have the time of one synodical revolution equal to 9 h. 54′ 56″·4.

The dark spot on February 25 was observed some time before, and also just after it was past the center ; therefore I have supposed it to be in the center about 8 h. 58′ : and we have,

From				To				Interval.				No. rev.	Value of 1 rev.		
d	h	′		d	h	′	″	d	h	′	″		h	′	″
Feb. 25	8	58		Mar. 2	8	2		4	23	4		12	9	55	20
25	8	58		3	13	49		6	4	51		15	9	55	24
25	8	58	0	14	7	36	10*	16	22	38	10	41	9	55	4·6
Mar. 2	8	2		3	13	49		1	5	47		3	9	55	40
2	8	2	0	14	7	36	10	11	23	34	10	29	9	54	58·2
3	13	49	0	14	7	36	10	10	17	47	10	26	9	54	53·4
Apr. 7	9	31		12	7	50		4	22	29		12	9	51	35

Again, comparing together the observations of 1779, which were made with the utmost attention to time, we have,

From				To				Interval				No. rev.	Value of 1 rev.		
Apr. 14	8	48	52	Apr. 19	7	10	0	4	22	21	8	12	9	51	45·6
19	7	10		23	9	38		4	2	28		10	9	50	48

And taking both together,

From				To				Interval				No. rev.	Value of 1 rev.		
Apr. 14	8	48	52	Apr. 23	9	38	0	9	0	49	8	22	9	51	19·4

These several results are so exceedingly various, that it is evident Jupiter is not a proper planet for the critical purpose of a comparative view of the diurnal motions ; nor can this great variety proceed from any inaccuracy in the observations : for, in my opinion, it is not well possible to make a mistake in the situation of a spot that shall amount to so much as five minutes of time. The observation

* Allowing 1m 10s for the alteration of the clock.

of April 23, 1779, was made with a view to ascertain this point, when it was found that five minutes of time made a sensible difference in the situation of a spot when near the center.

If we reduce the synodical revolutions to sydereal ones, the result will be so little different from the above, that I have not thought it worth while to do it in this place. By a comparison of the different periods it appears, that a spot which is carried about in the atmosphere of Jupiter generally suffers an acceleration, or, which is the same thing, performs its revolutions by degrees in less time than it did at first; for the spot observed in 1778 moved at the following rates. From February 25 to March 2 in 9 h. 55′ 20″; to March 3 nearly the same; to March 14 in 9 h. 55′ 4″; from March 2 to March 3 in 9 h. 55′ 40″; to March 14 in 9 h. 54′ 58″; from March 3 to March 14 in 9 h. 54′ 53″. In 1779 a spot moved from April 14 to April 19 at the rate of 9 h. 51′ 45″; to April 23 in 9 h. 51′ 24″; and from April 19 to April 24 in 9 h. 50′ 48″; all which is agreeable enough to the theory of equatorial winds, since it may probably take up sometime before a spot can acquire a sufficient velocity to go as fast as those winds may blow. And, by the by, if Jupiter's spots should be observed in different parts of his year, and be found in some to be accelerated, in others to be retarded, it would almost amount to a demonstration of his monsoons and their periodical changes; but if his axis should not be inclined enough to his orbit, to occasion such a change, they may probably always blow in the same direction.

Observations on Mars in the year 1777.

Twenty-feet Newtonian reflector; power 300.

April 8. 7 h. 30′. I observed two spots upon Mars, with a bright belt or partition between them. The belt was not very well defined, see Plate II. fig. 14.

9 h. 30′. The spots are advanced, and more spotted parts are visible; fig. 15.

10 h. The revolution of Mars on his axis is now very evident; fig. 16.

April 17. Ten-feet Newtonian reflector; power about 211. 7 h. 50′. Mars appeared as in fig. 17. At *a* and *b* there were two bright spots, so luminous that they seemed to project beyond the disk. At *c* and *d* there were two very dark spots, joined by a lesser black line in the middle, which however was crossed at *e* and *f* by a very faint whitish partition.

April 26. Ten-feet reflector; power 211.

9 h. 5′. The spots on the planet are very faint, and much about as in fig. 18.

April 27. Ten-feet reflector; power 324.

8 h. 40′. The evening very fine: my telescope in compleat order. The spots as in fig. 19.

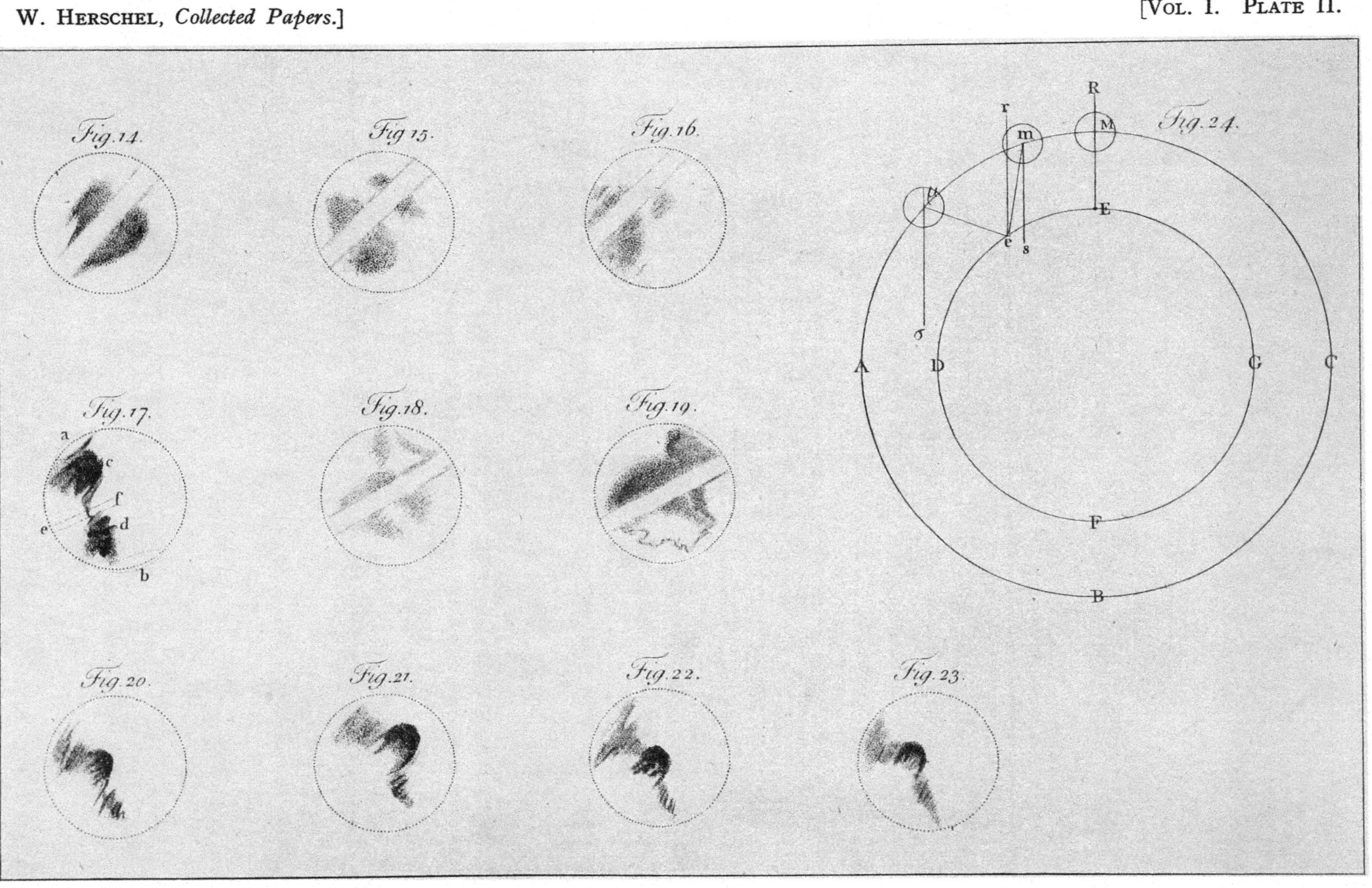
Fig. 14.
Fig 15.
Fig. 16.
Fig. 24.
R
r
m
M
E
e
s
σ
A
D
G
C
F
B
Fig. 17.
a
c
f
d
e
b
Fig. 18.
Fig. 19.
Fig. 20.
Fig. 21.
Fig. 22.
Fig. 23.

[*To face p.* 22.

Observations on Mars in the year 1779.

May 9th. Clock 15″ too fast; by equal altitudes on the 14th of April, and by the transit of a star, is found to lose 1″·45 *per* day.

11 h. 1′ by the clock, I found the situation of the spots on Mars as in figure 20; there is a very remarkable dark spot not far from the center.

11 h. 30′. The figures are gone from the center.

May 11. Clock 12″ too fast.

10 h. 18′. The same spot that was visible May 9 is on the disk, the darkest place being intirely south-east of the center, see fig. 21.

11 h. 43′. The darkest part is almost arrived at the center, fig. 22.

12 h. 17′. The dark spot is with its edge just near the center, as in fig. 23.

May 13. Seven-feet reflector; power 222. Clock 9″ too fast.

11 h. 26′. Mars seems now to be in the same situation he was the 11th, at 10 h. 8′.

May 22. Clock 4″ too slow.

12 h. 5′. The figure of May 11th is not on the disk; but some other fainter spots are visible. The air is full of vapours.

June 6. The clock set by ten equal altitudes taken to-day, and by the transit of δ Scorpii loses 1″·9 *per* day. What I have perhaps improperly called a transit is the occultation of a star passing behind the perpendicular edge of a high building at about 40 yards distance, observed with a fixed telescope directed to the place where it vanishes.

10 h. 10′. The same figure is upon the disk of Mars which was there April 8, 1777, at 7 h. 30′.

June 15. Clock 17″ too slow. 9 h. 45′. The same figure is upon Mars that was there May 9 at 11 h. 1′; but it is more advanced. I suppose it to be the same, and in the same situation, as April 17, 1777, at 7 h. 50′.

June 17. Clock 20″ slow.

9 h. 12′. The dark spot on Mars is rather more advanced than it was May 11th, at 10 h. 18′.

10 h. The spot is visibly advanced: I suppose it will take near an hour to come to the center.

10 h. 15′. A very thick fog obscures the sky.

11 h. 15′. The same darkness.

June 19. Clock 22″ too slow by the transit of δ Scorpii observed this evening.

8 h. 40′. The figure on the disk of Mars appears now to be as it was April 26, 1777, at 9 h. 5′, see fig. 18.

11 h. 30′. The figure of May 11 which I have been hitherto watching, is not come to the position it was then at 11 h. 43′, but cannot be far from it. I fear,

as Mars approaches the horizon, I shall not be able to follow him till the figure comes to the center.

11 h. 47′ The state of the air near the horizon is very unfavourable. With much difficulty I can but just see that the figure is not quite so far advanced as it was May 11th, at 11 h. 43′, but can certainly not be above two or three minutes from it.

11 h. 51′. The undulation of the air prevents all further observation.

Let us now examine the result of the above mentioned observations: comparing together the two following short intervals of the year 1779, we have,

	D.	H.	M.	S.
From May	9	11	0	45
to May	11	12	16	48
Divided by 2 revol.	2	1	16	3

Gives 1 revolution = 24 h. 38′ 1″·5.

A second small interval.

	D.	H.	M.	S.
May	11	10	17	48
	13	11	25	51
2 revol.	2	1	8	3

1 revolution = 24 h. 34′ 1″·5.

Here we have two very short intervals that agree to 4′, which is more than we could have expected in such short periods of time.

Comparing together observations that were made at a greater distance, we find,

First monthly period,

	D.	H.	M.	S.	
May	11	10	17	48	
June	17	9	9	20	allowing 3′ because the obs. says the spot was rather more advanced.
36 revol.	36	22	51	32	

1 revolution = 24 h. 38′ 5″·9.

Second monthly period,

May	11	11	42	48	
June	19	11	50	22	allowing 3′ for the time the spot would have taken to come to the place mentioned.
38 revol.	39	0	7	34	

1 revolution = 24 h. 38′ 5″·4.

Third monthly period,

	D.	H.	M.	S.
May	13	11	25	51
June	17	9	9	20
34 revol.	34	21	43	29

1 revolution = 24 h. 38′ 20″·3.

This last is, perhaps, as likely to be near the truth as any, since the same spot was here observed for the third time, and therefore its motion become more familiar.

Here we have three longer periods that agree to fifteen seconds, which is quite sufficient for extending the interval of time to those observations that were made in the year 1777. But as these are the synodical revolutions, it will be necessary first to reduce them to sydereal rotations.

In figure 24 let us suppose the orbit of Mars, MABC, to be in the same plane with the orbit of the earth, EDFG; and the axis of Mars to be perpendicular to his orbit. Let M, E, *m*, *e*, be the situations of Mars and the earth on the 13th of May and 17th of June; then will the line EM, that connects the centers of Mars and the earth, point out the geocentric place of Mars on the 13th of May; and the line *em*, the geocentric place of the same planet on the 17th of June. Draw *er* and *ms* parallel to ER; then will *er* point out the geocentric place of Mars on the 13th of May; and the angle *sme* is equal to the angle *mer*. Now, by an ephemeris * the geocentric place of Mars, May 13 at 11 h. 26′ was $7^s\ 20°\ 59'\ 21''$; and on the 17th of June, at 9 h. 9′, it was $7^s\ 12°\ 27'\ 22''$, by which we obtain the difference or angle $rem = ems = 8°\ 31'\ 59''$.

Now a spot on Mars, situated in the direction ME, will have made a sydereal revolution when it returns to the same, or a parallel direction *ms*. From which we gather, that the spot on the 17th of June, after coming to the line *me*, where it finishes the synodical revolution, will have to go through an arch of 8° 31′ 59″, in order to arrive into the direction of the line *ms*, where it finishes the sydereal rotation. The time it will take to go through this arch, at the sydereal rate of 24 h. 39′ 20″ to 360 degrees, or 4″·109 *per* minute of a degree, will be 35′ 3″·8; this being divided by the numbers of revolution 34, gives 1′ 1″·8; which, added to 24 h. 38′ 20″·3, gives us 24 h. 39′ 22″·1 for the sydereal revolution of Mars, as found by the third of the monthly periods. This quantity will help us to find a proper divisor for the three following long biennial periods.†

It is to be observed, that Mars has been retrograde in the above example, for which reason the measure of the angle *ems* was to be added to the synodical revolution when we wanted to find the sydereal rotation; but if he had been direct, or if his place had been more advanced in the ecliptic than that to which we compared it, as at μ, then the line $\mu\sigma$ parallel to EM would be the direction to which the spot should return, in order to accomplish a sydereal revolution, and therefore the quantity of the angle $\sigma\mu e = \mu er$, or difference of the geocentric places, ought to be subtracted from the synodical revolution to obtain the sydereal one.

* The *Nautical Almanac* gives the geocentric place of Mars only to every sixth day; for which reason I used WHITE's Ephemeris, where it is given for every day, though perhaps not with so much exactness as I could wish.

† [It was pointed out by Mädler, *Astr. Nachr.*, xvi. p. 361, that Herschel neglected to correct for phase and aberration, and that the result ought to have been $24^h\ 38^m\ 44^s$ or more probably somewhat less.—ED.]

	D.	M.	S.
First biennial period, 1777, April	8	7	30
1779, June	6	10	10
	789	2	40

The geocentric places of Mars at those times were,

S.	°	′	″
6	6	31	26
7	13	48	30
1	7	17	4

turned into time at 4″·109 *per* minute of a degree and subtracted, because Mars is more advanced in the ecliptic, is

	789	2	40	0
	—	2	33	11·8
Divided by 768 rev.	789	0	6	48·2

1 revolution = 24 h. 39′ 23″·03.

	D.	H.	M.	S.
Second biennial period, 1777, April	17	7	50	0
1779, June	15	9	45	17
	789	1	55	17

	S.	°	′	″
Geocentric places	6	3	31	27
	7	12	40	23
	1	9	8	56

Turned into time	789	1	55	17
and subtracted	—	2	40	52
768 revol.	788	23	14	25

1 revolution = 24 h. 39′ 18″·94.

	D.	H.	M.	S.
Third biennial period, 1777, April	26	9	5	0
1779, June	19	8	40	22
	783	23	35	22

	s.	°	′	″
Geocentric places	6	1	24	36
	7	12	31	48
	1	11	7	12
Turned into time	783	23	35	22
and subtracted	—	2	45	15·6
763 revol.	783	20	50	6·4

1 revolution = 24 h. 39′ 23″·04.

As these three periods are supported by observations of equal validity, I shall take a mean of them all for the nearest approximation to the true sydereal revolution of Mars on his axis, which therefore is 24 h. 39′ 21″·67.*

It remains now only to see how far we may depend upon this determination of Mars's diurnal rotation as coming near the truth ; and looking over those causes which may possibly produce any errors, we find, first of all, that in the long biennial periods a mistake in the number of revolutions would produce a considerable deviation from truth. Secondly, in the observations of a spot which moves so slow, we are also liable to some considerable mistake in estimating the time when it comes to a certain place ; and the more so, if that place is not the center. Lastly, the time itself is liable to inaccuracy.

As to the first, it appears from the three monthly periods observed in the year 1779, when the proper allowances for the geocentric places are made, that the sydereal revolution of Mars cannot well be less than 24 h. 39′ 5″, nor more than 24 h. 39′ 22″ ; but if we should divide any one of the three biennial periods by a supposed number of revolutions, only one more or one less than we have done, the difference would be so considerable, that nothing but a mistake in every one of the three monthly periods, of at least one whole hour, could justify such a supposition ; and that such a mistake in the situation of a spot on Mars cannot have been made in those observations, I think, is evident enough from the exactness with which they were made, and from their agreement with each other.

The second cause of error, which is the uncertainty in assigning the exact time when a spot comes to the center, is of some force. But it seems to me highly probable, from the manner in which I have seen the spots on Mars pass over the disk of that planet, that there can hardly be so great an error as 10′ in an observation of any remarkable spot's coming to the center. However, not being willing to trust more to the eye than I ought to do, I had recourse to the following experiment. I

* [The number of revolutions assumed between 1777 and 1779, being based on the erroneous result found from the observations of 1779, should be increased by one, so that the correct result comes out $24^{h}\ 37^{m}\ 26^{s}{\cdot}3$ in good accordance with more recent determinations (Mädler, *l.c.*).—Ed.]

drew several circles of one inch radius, taking care to make no visible impression of a center; and placed in each a fine point at the several distances of ·0424, ·0636, 0848, in ten thousands of an inch from the real center; some to the right, others to the left. These measures are the sines to radius one, of 2° 26′, 3° 39′, and 4° 52′, which are the arches a spot on Mars passes over in 10, 15, 20 minutes respectively. I exposed them to several persons unacquainted with my designs, and found, that not one of them made a single mistake in saying whether the point was, or was not, in the center of the circle, and which way it deviated from it. As the direction of the motion of a spot on Mars is known, I thought the persons who were to judge of the place of the points were intitled to be acquainted with the line in which they were placed, which for that reason was always to the right and left only. The points that answer to the excentricity of 15 and 20′ are indeed so visibly out of the center, that I believe we may safely say, that any mistake, in estimating the time of a spot on Mars coming to the center, cannot well exceed a quarter of an hour at the outside.

As for the third and last occasion of error, the time itself, I believe my manner of obtaining and keeping it in the year 1779 will appear satisfactory, and may, I think, be depended upon to a few seconds; but the observations of the year 1777, indeed, are far from having the same advantage. I was not then provided with an altitude instrument, therefore set my clock by a good sun-dial, with the equation of time contained in the *Nautical Almanac*, and found it to agree generally to a minute or two with the time calculated for the eclipses of Jupiter's first satellite, as I deduced it for Bath from the *Nautical Almanac*. However, it was certainly liable to an error of several minutes; therefore, allowing no less than 10′ for the clock in 1777, and 20′ for an error in estimating the situation of a spot in 1779, it will both amount to half an hour: then, if we take a mean of the three numbers, whereby we have divided the three biennial periods, we have $766\frac{1}{3}$; and half an hour, divided by $766\frac{1}{3}$, will therefore give us the quantity to which, it seems, can amount, all the uncertainty in the sydereal diurnal rotation of Mars, which is 2″·34.

A nearer approximation to truth I hope to obtain at the next opposition, which will happen about the middle of July 1781.

I have ventured to calculate the times for that opposition, when the edge of the remarkable dark spot will be seen near the center, as it is in figure 23, or, which is the same thing, as it was the 11th of May 1779, at 12 h. 17′. The spot not being visible at the time of the opposition, I have taken the nearest period, before and after, in which it will pass over the disk. There is, however, a circumstance which may make the appearance of the spot not quite similar to the figure I have drawn, even though the rotations should perfectly answer as to the times; for the position of the axis of Mars being still in some measure unknown, I could make no allowance for a change, which a difference in the situation of no less than two signs may occasion, though in all probability it will not be very considerable.

Those who are provided with proper telescopes will have an opportunity to see how far the calculated times agree with the spot's appearance; and it is by this means I also hope to correct and improve the tables I have drawn up for this purpose, and further to approximate to a true theory of the gyration of this planet.

Not knowing the exact difference of meridians between Greenwich and this place, I have calculated the spot's appearance for the meridian of Bath. From an eclipse or two of Jupiter's satellites, of which, by the favour of the Rev. Mr. HORNSBY, I have seen correspondent observations, I suppose the difference cannot be much less than 9′ west of Greenwich; and at the same time I join an account of the solar eclipse of the 24th of June 1778, which may be depended upon as a very compleat observation, and may serve to ascertain the longitude of this place.

Eclipse of the sun observed at Bath.

June 24, 1778.

	H.	M.	S.
Beginning by equated or mean time,	3	30	10·7
End, ...	5	18	7·7

Calculations, or (as the principles on which they are founded are still established upon a few observations only, and require some time for mature confirmation) I would rather, if I might be allowed the expression, call them calculated conjectures of the times when the remarkable dark spot will be seen near the center of the disk of Mars.

For June, July, and August, of the year 1781.

D.	H.	M.	S.	D.	H.	M.	S.
June 28	9	48	39	July 8	16	13	23
29	10	27	10	August 3	8	11	6
30	11	5	42	4	8	49	57
July 1	11	44	13	5	9	28	48
2	12	22	32	6	10	7	39
3	13	0	51	7	10	46	31
4	13	39	10	8	11	25	44
5	14	17	29	9	12	4	56
6	14	55	46	10	12	44	8
7	15	34	4				

I have the honour to be, &c.

IV.

Account of a Comet. By Mr. HERSCHEL, *F.R.S.; communicated by Dr.* WATSON, *Jun. of* Bath, *F.R.S.*

[*Phil. Trans.*, vol. lxxi., 1781, pp. 492–501.]

Read April 26, 1781.

ON Tuesday the 13th of March, between ten and eleven in the evening, while I was examining the small stars in the neighbourhood of H Geminorum, I perceived one that appeared visibly larger than the rest: being struck with its uncommon magnitude, I compared it to H Geminorum and the small star in the quartile between Auriga and Gemini, and finding it so much larger than either of them, suspected it to be a comet.

I was then engaged in a series of observations on the parallax of the fixed stars, which I hope soon to have the honour of laying before the Royal Society; and those observations requiring very high powers, I had ready at hand the several magnifiers of 227, 460, 932, 1536, 2010, &c. all which I have successfully used upon that occasion. The power I had on when I first saw the comet was 227. From experience I knew that the diameters of the fixed stars are not proportionally magnified with higher powers, as the planets are; therefore I now put on the powers of 460 and 932, and found the diameter of the comet increased in proportion to the power, as it ought to be, on a supposition of its not being a fixed star, while the diameters of the stars to which I compared it were not increased in the same ratio. Moreover, the comet being magnified much beyond what its light would admit of, appeared hazy and ill-defined with these great powers, while the stars preserved that lustre and distinctness which from many thousand observations I knew they would retain. The sequel has shewn that my surmises were well founded, this proving to be the Comet we have lately observed.

I have reduced all my observations upon this Comet to the following tables. The first contains the measures of the gradual increase of the Comet's diameter. The micrometers I used, when every circumstance is favourable, will measure extremely small angles, such as do not exceed a few seconds, true to 6, 8, or 10 thirds at most; and in the worst situations true to 20 or 30 thirds: I have therefore given the measures of the Comet's diameter in seconds and thirds. And the parts of my micrometer being thus reduced, I have also given all the rest of the measures in the same manner; though in large distances, such as one, two, or three minutes, so great an exactness, for several reasons, is not pretended to.

TABLE I. Measures of the Comet's diameter.*

Days.	″	‴	Powers.
March 17	2	53	932. 460.
19	2	59	932. 460.
21	3	38	460.
28	4	7	932 } these measures agree to 9‴.
—	3	58	227 }
29	4	7	227 rather too small a measure.
—	4	25	227 seems right.
April 2	4	25	227
6	4	53	227
15	5	11	227
—	5	20	227 very good; not liable to half a second of error.
18	5	2	227 true to 12‴ or 18‴ at most.

Having measured the diameter of the Comet with such high power as 932 and 460, it may not be amiss to make one observation on this subject, lest it should be misapprehended that I pretend to a distinct power of such magnitude upon all celestial objects in general. By experience I have found, that the aberration or indistinctness occasioned by magnifying much, provided the object be still left sufficiently distinct, is rather to be put up with, than the power to be reduced, when the angles to be measured are extremely small. The reason of this may, perhaps, be that a small error of judgement, to which we are always liable, is of great consequence with a low power, as bearing a considerable proportion to the diameter of the object; whereas with a higher power the proportion of this error to the whole becomes much less, and the measure more exact, even after we have made allowance for a small additional error occasioned by the want of that perfect distinctness which is required for other purposes. However, to enter deeply into an explanation of this would lead me to speak of the causes of the aberration of rays in the focus of an object speculum, of which there are some that are seldom taken into consideration by opticians, and indeed are such as cannot be calculated; but this not being my present purpose, suffice it to observe, that the method is justified by experience.

* There are several optical deceptions which may affect the measures of objects that subtend extremely small angles. Thus I have found, by experience, that a very small object will appear something less in a telescope when we see it first than when we become familiar with it. There is also a deflection of light upon the wires when they are nearly shut; but as none of these deceptions are well enough understood to apply a correction, I leave them affected with them.

When the diameter of the Comet was increased to about 4″, I thought it advisable to lessen the power with which I measured; and, as I made use of two different micrometers, as well as eye-glasses, I took a measure with both of them. The agreement of the micrometers to 9‴ is no small proof of the goodness of the observations of the 28th of March, and very properly connects the measures of the high powers with those that were made with 227.

TABLE II. Distance of the Comet from certain telescopic fixed stars which I have marked α, β, γ, δ, ϵ, ζ.

	D.	H.	M.		′	″	‴	
Mar.	13	10	30	from α,	2	48	0	by pretty exact estimation true to 20″.
	17	11	0	fig. 1.*	0	41	58	by the micrometer and power 227.
	18	7	20		1	0	35	
	—	9	16		1	6	59	
	—	10	55		1	10	40	
	19	7	4		1	46	40	
	—	10	42		1	51	23	
	21	10	0		3	39	46	
	24	8	12	from β,	2	55	39	true to 4 or 5″, an indifferent observation.
	—	10	58	fig. 2.	2	53	4	true to 4 or 5″.
	25	7	24		2	12	46	true to 2 or 3″.
	—	9	47		2	14	18	
	26	10	43		1	48	3	true to 2 or 3″.
	28	7	46		2	55	49	true to 4 or 5″.
	29	8	50	from γ,	2	20	51	true to 2″.
	30	7	55	fig. 3.	1	28	48	true to 2 or 3″.
Apr.	1	7	45		2	39	20	
	6	8	50	from δ, fig. 4.	2	51	23	
	15	10	18	from ϵ,	4	27	57	estimated by the field, true to 5 or 6″.
	16	7	50	fig. 5.	3	9	14	by the micrometer, true to 3 or 4″.
	—	10	47		2	50	56	true to 3 or 4″.
	18	8	18		3	18	4	} mean 3′ 17″, true to 1″ or 1½.
	—	—	—		3	15	57	}
	—	8	50	from ζ,	2	24	57	
	19	8	38	fig. 6.	3	2	5	true to 3 or 4″.

* The figures are drawn upon a scale of 80 seconds to one inch.

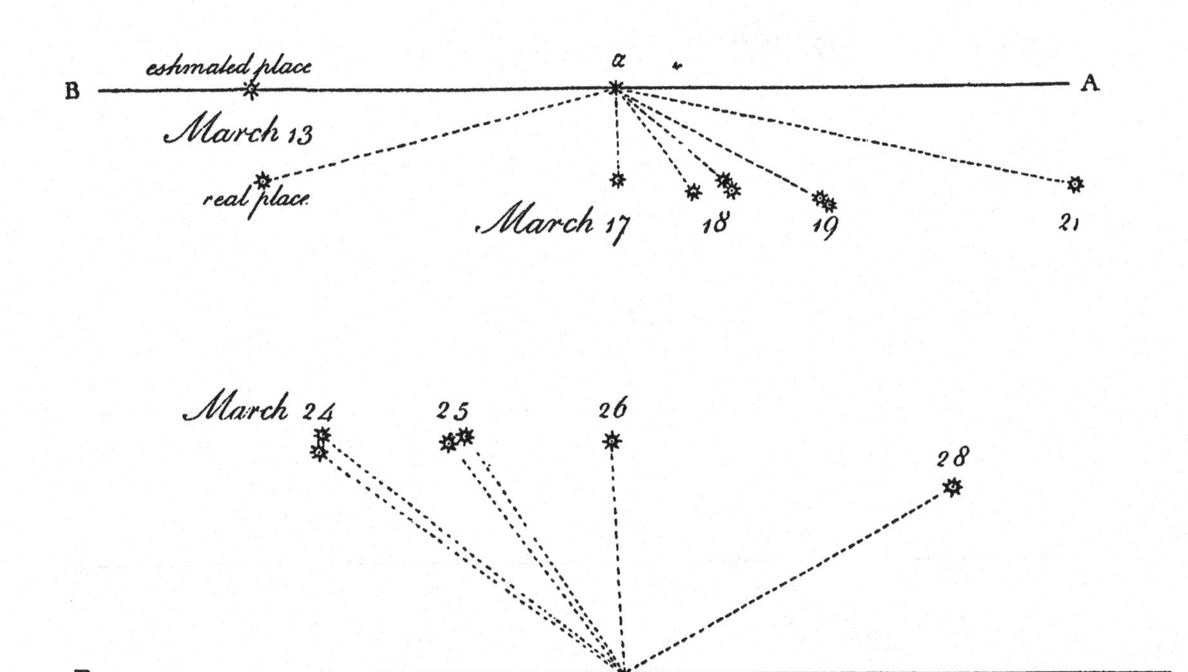
estimated place
α
B
A
March 13
real place
March 17
18
19
21
March 24
25
26
28
B
A
β

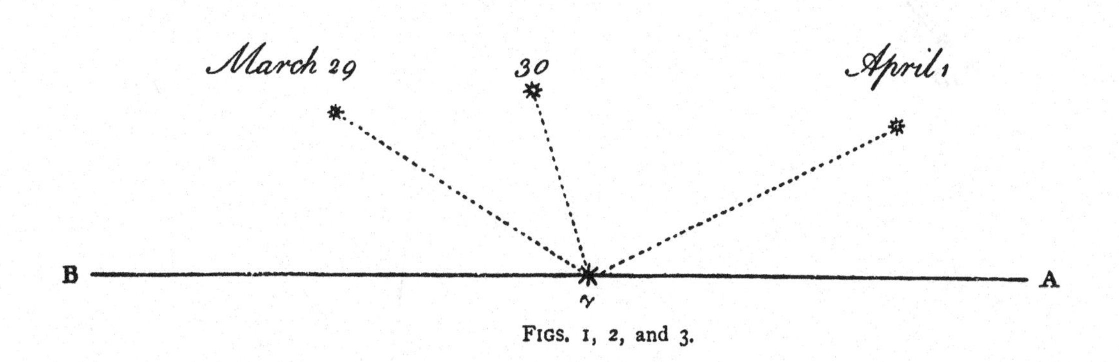
March 29
30
April 1
B
A
γ

FIGS. 1, 2, and 3.

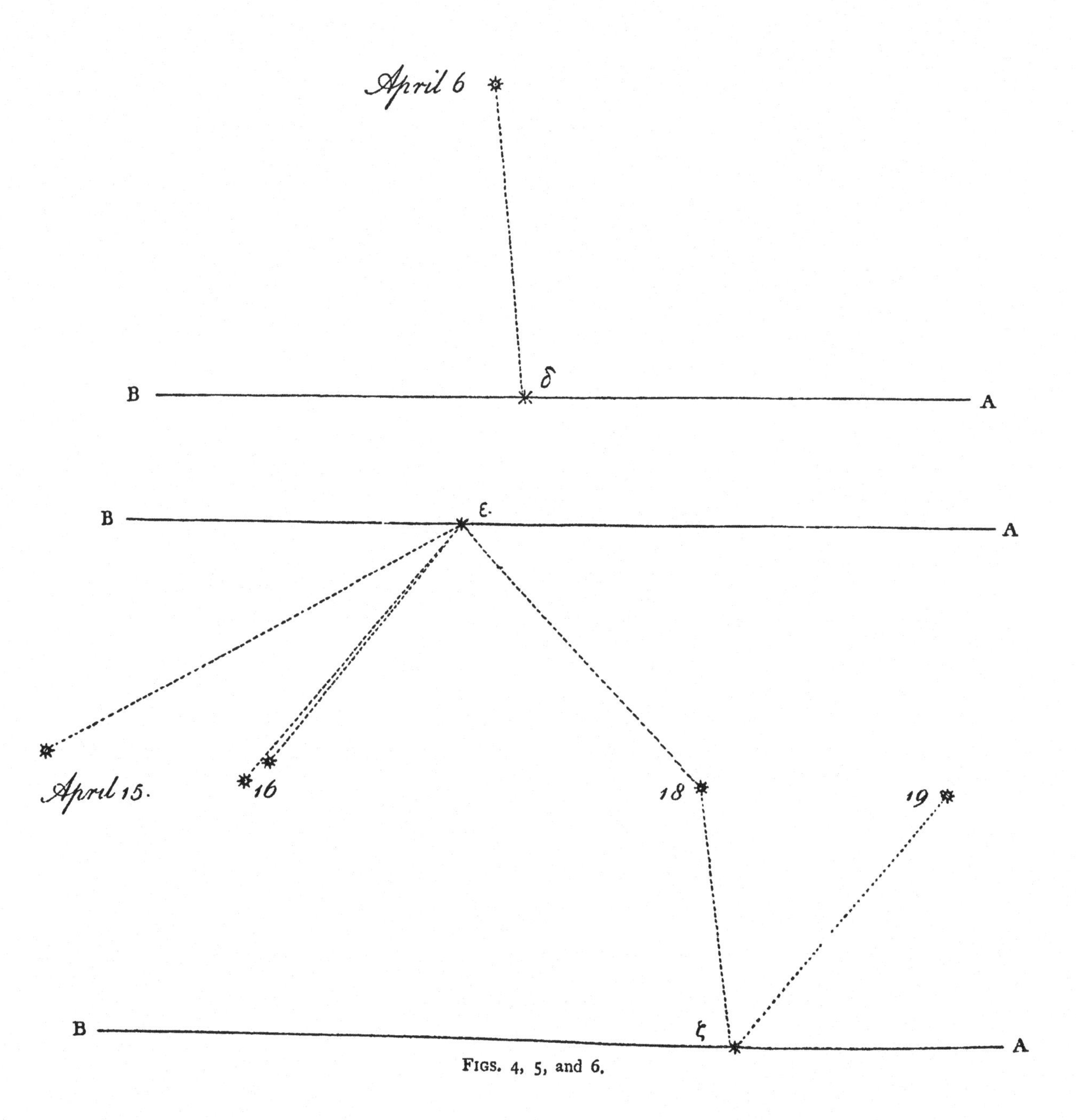

FIGS. 4, 5, and 6.

TABLE III. Angle of position of the Comet with regard to the parallel of declination of the same telescopic fixed stars measured by a micrometer, of which I have given the description, and a magnifying power of 278. See fig. 1. 2. 3. 4. 5. 6.

	D.	H.	M.		°	′	
Mar.	13	10	30	B α Comet.	0	0	by superficial estimation, liable to an error of 10 or 12 degrees.
	17	11	0	A α Comet, fig. 1.*	89	56	by the micrometer.
	18	8	20		56	39	
	—	9	24		41	33	true to 1°.
	19	7	23		29	47	true to 1°.
	21	10	10		11	46	true to 4 or 5°.
	—	11	48		12	14	
	24	8	23	B β Comet, fig. 2.	38	39	true to 2 or 3°.
	—	11	4		36	14	true to 3 or 4°, air very tremulous.
	25	7	33		53	18	
	—	9	55		56	32	liable to a considerable error.
	26	10	55	A β Comet.	87	0	true to 2 or 3°.
	28	7	58		28	51	true to 3 or 4°.
	29	9	25	B γ Comet, fig. 3.	32	19	true to 1 or 2°.
	30	8	25		72	14	true to 3 or 4°.
Apr.	1	7	55	A γ Comet.	28	51	well taken, } 27° 46′, true to 1°.
	—	—	—		27	14	more exact, }
	6	8	28	B δ Comet, fig. 4.	84	42	true to less than 2°.
	15	10	27	B ε Comet, fig. 5.	29	9	true to 2 or 3°.
	16	8	1		49	11	true to 1°.
	—	10	55		50	47	true to 1½ or 2°½.
	18	8	31	A ε Comet.	47	9	very well taken, } 47°, true to less than 1°.
	—	—	—		46	35	pretty well, }
	—	9	8	B ζ Comet.	82	39	
	19	8	56	A ζ Comet, fig. 6.	48	18	} 49° 3′, true to 1°.
	—	—	—		49	48	}
	—	10	45		47	30	true to 2 or 3°.

* The angles are drawn true to the measure, without allowing for errors.

Miscellaneous observations and remarks.

March 19. The Comet's apparent motion is at present 2¼ seconds *per* hour. It moves according to the order of the signs, and its orbit declines but very little from the ecliptic.

March 25. The apparent motion of the Comet is accelerating, and its apparent diameter seems to be increasing.

March 28. The diameter is certainly increased, from which we may conclude that the Comet approaches to us.

April 2. This evening at 8 h. 15′ the Comet was a little above the line drawn from η to θ in fig. 7. This figure is only delineated by the eye, so that no very great exactness in the distances of the stars is to be expected; but I shall take the first opportunity of measuring their respective situations by the micrometer.

April 6. With a magnifying power of 278 times the Comet appeared perfectly sharp upon the edges, and extremely well defined, without the least appearance of any beard or tail.

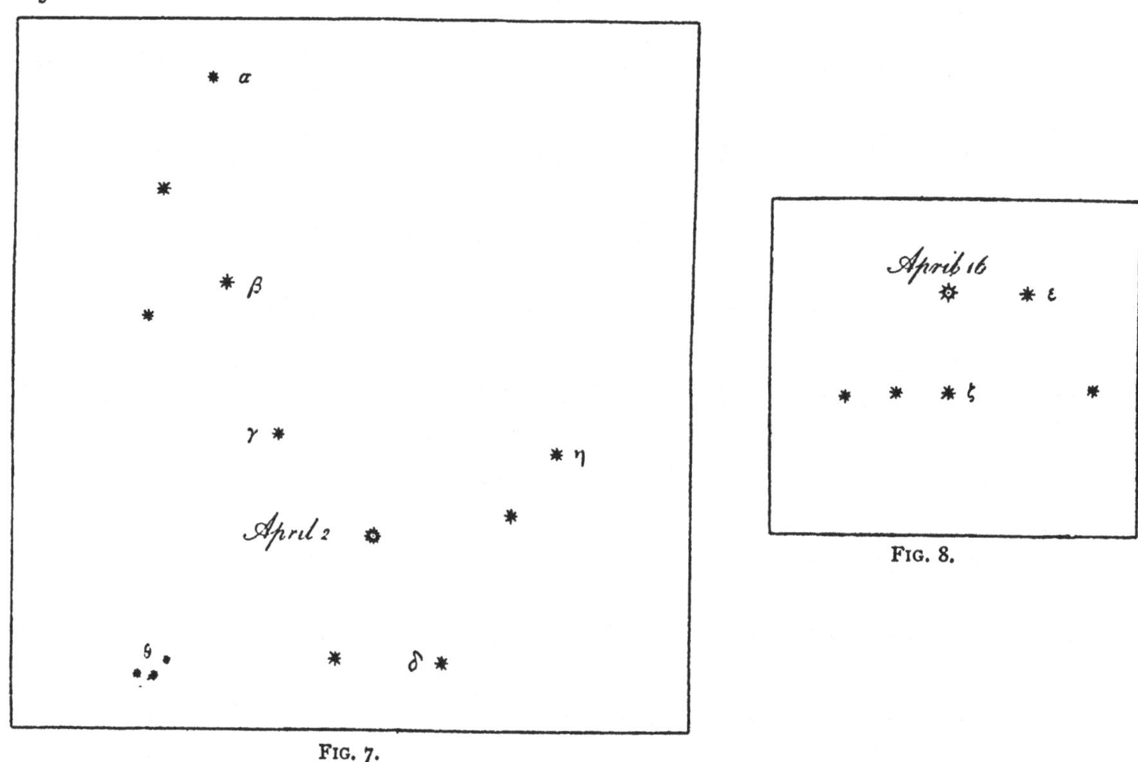

Fig. 7.

Fig. 8.

April 16. Fig. 8 represents the situation of the Comet this evening about nine o'clock, and is only an eye-draught of the telescopic stars.

Remarks on the path of the Comet.

We may observe, that the method of tracing out the path of a celestial body by taking its distance from certain stars, and the angle of position with regard to them, cannot be expected to give us a compleatly just representation of the tract it describes, since even the most careful observations are liable to little errors, both from the remaining imperfections of instruments, though they should be the most accurate that can be had, and from the difficulty of taking angles and positions of objects in motion. Add to this a third cause of error, namely, the obscurity of very small telescopic stars that will not permit the field of view so well to be enlightened as we could wish, in order to see the threads of the micrometer perfectly distinct.

This will account for the apparent distortions to be observed in my figures of the Comet's path. Some little irregularity therein may also proceed from different refractions, as they have not been taken into account, though the observations have been made at very different altitudes, where consequently the refractions must have been very different. But though this method may be liable to great inconveniences, the principal of which is, that many parts of the heavens are not sufficiently stored with small stars to give us an opportunity to measure from them, yet the advantages are not less remarkable. Thus we see that it enabled me to distinguish the quantity and direction of the motion of this Comet in a single day (from the 18th to the 19th of March) to a much greater degree of exactness than could have been done in so short a time by a sector or transit instrument; nay even an hour or two, we see, were intervals long enough to shew that it was a moving body, and consequently, had its size not pointed it out as a Comet, the change of place, though so trifling as 2¼ seconds *per* hour, would have been sufficient to occasion the discovery. A gentleman very well known for his remarkable success in detecting Comets * seems to be well aware of the difficulty to discover a motion in a heavenly body by the common methods when it is so very small; for in a letter he favoured me with, speaking of the Comet, he says: "Rien n'etoit plus difficile que de la reconnoître et je ne puis pas concevoir comment vous avés pu revenir plusieurs fois sur cette étoile ou Comête; car absolument il a fallu l'observer plusieurs jours de suîte pour s'apperçevoir qu'elle avoit un mouvement."

I need not say that I merely point this out as a temporary advantage in the method I have taken; for as soon as we can have regular, constant, and long continued observations by fixed instruments, the excellence of them is too well known to say any thing upon that subject: for which reason I failed not to give immediate notice of this moving star, and was happy to surrender it to the care of the Astronomer Royal and others, as soon as I found they had begun their observations upon it.

Description of a micrometer for taking the angle of position.

Fig. I. Represents the micrometer inclosed in a turned case of wood, as it is put together, ready to be used with the telescope. A is a little box which holds the eyeglass. B is the piece which covers the inside work, and the box A is screwed into it. C is the body of the micrometer containing the brass work, shewing the index plate *a* projecting at one side, where the case is cut away to receive it. D is a piece, having a screw *b* at the bottom, by means of which the micrometer is fastened to the telescope. To the piece C is given a circular motion, in the manner the horizontal motion is generally given to Gregorian reflectors, by the lower part going through the piece D, where it is held by the screw E, which keeps the two pieces C and D together, but leaves them at liberty to turn upon each other.

* Mons. MESSIER.

Fig. II. Is a section of the case containing the brass work, where may be observed the piece B hollowed out to receive the box A, which consists of two parts inclosing the eye lens. This figure also shews how the piece C passes through D, and is held by the ring E: the brass work, consisting of a hollow cylinder, a wheel and pinion, and index plate, is there represented in its place. F is the body of the brass work, being a hollow cylinder with a broad rim *c* at the upper end; this rim is partly turned away to make a bed for the wheel *d*. The pinion *e* turns the wheel *d*, and carries the index plate *a*. One of its pivots moves in the arm *f*, screwed upon the upper part of *c*, which arm serves also to confine the wheel *d* to its place upon *c*. The other pivot is held by the arm *g* fastened to F.

Fig. III. Is a plan of the brass work. The wheel *d*, which is in the form of a ring, is laid upon the upper part of F or *c*, and held by two small arms *f* and *h*, screwed down to *e* with the screws *i*, *i*.

Fig. IV. Is a plan of the brass work. *d*, *d*, is the wheel placed upon the bed or socket of the rim of the cylinder *c*, *c*, and is held down by the two pieces *f*, *h*, which are screwed upon *c*, *c*. The piece *f* projects over the center of the index plate to receive the upper pivot of the pinion. *m*, *n* is the fixed wire fastened to *c*, *c*. *o*, *p*, the moveable wire fastened to the annular wheel *d*, *d*. The index plate *a* is divided into 60 parts, each sub-divided into two, and milled upon the edge. When the finger is drawn over the milled edge of the index plate from *q* towards *r*, the angle *m*, *s*, *o*, will open, and if drawn from *r* towards *q*, it will shut again. The case C, C, must have a sharp corner *t*, which serves as a hand to point out the division on the index plate.

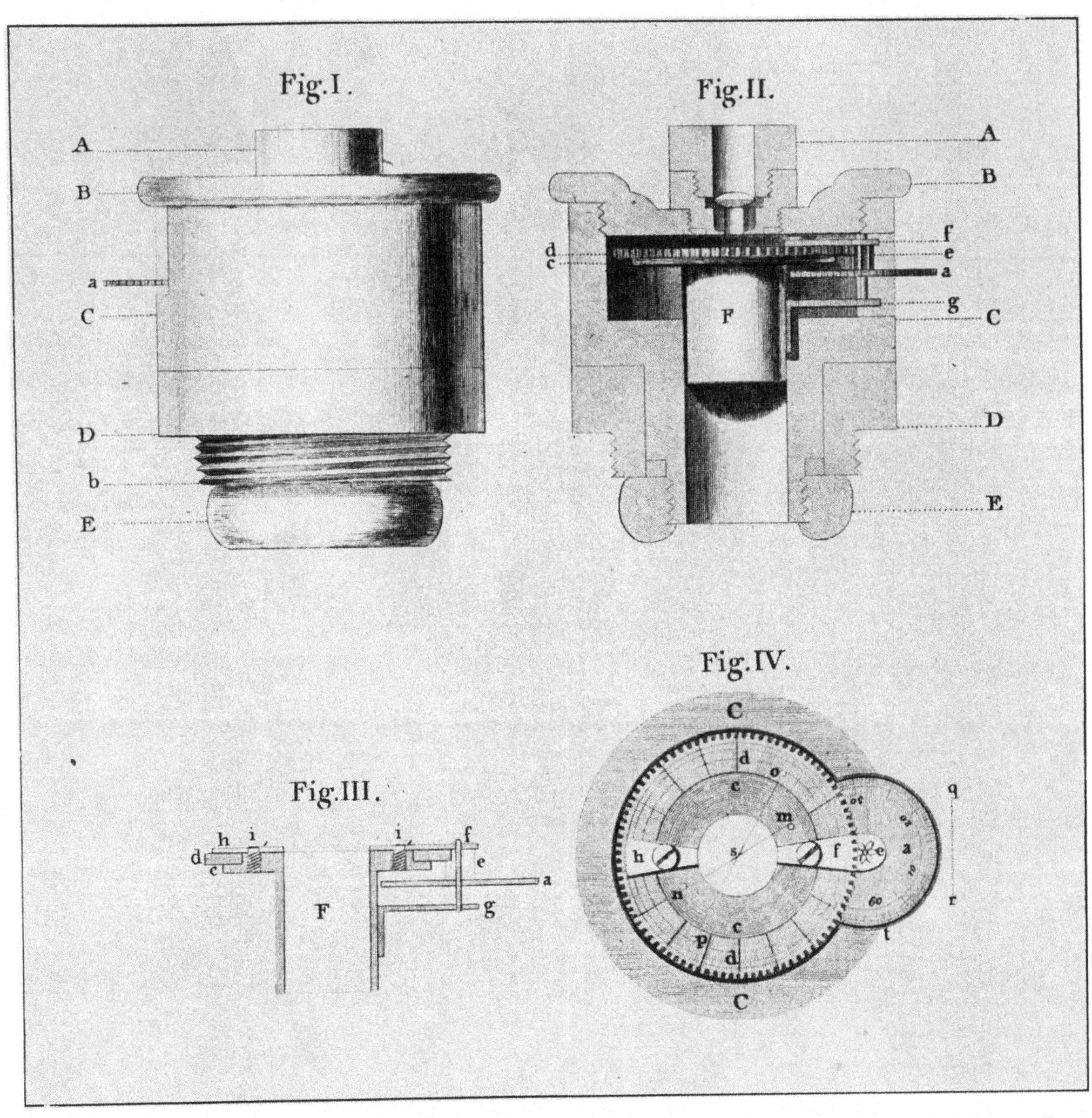
Fig.I.
A
B
a
C
D
b
E
Fig.II.
A
B
d
c
f
e
a
F
g
C
D
E
Fig.III.
h
i
d
c
i
f
e
a
F
g
Fig.IV.
C
d
o
c
m
h
s
f
e
a
n
c
p
d
C
q
r
t
60

[*To face p.* 38.

V.

On the Parallax of the Fixed Stars. By Mr. Herschel, *F.R.S.; communicated by Sir* Joseph Banks, *Bart., P.R.S.*

[*Phil. Trans.*, vol. lxxii., 1782, pp. 82–111.]

Read December 6, 1781.

To find the distance of the fixed stars has been a problem which many eminent astronomers have attempted to solve; but about which, after all, we remain in a great measure still in the dark. Various methods have been pursued without success, and the result of the finest observations has hardly given us more than a distant approximation, from which we may conclude, that the nearest of the fixed stars cannot be less than forty thousand diameters of the whole annual orbit of the earth distant from us. Trigonometry, by whose powerful assistance the mathematician has boldly ascended into the planetary regions, and measured the diameters and orbits of the heavenly bodies, for want of a proper base, can here be but of little service; for the whole diameter of the annual orbit of the earth is a mere point when compared to the immense distance of the stars. Now, as it is not in our power to enlarge this base, we can only endeavour to improve the instruments by which we measure its parallax.

There are two things requisite for measuring extremely small angles with accuracy. First, that the instrument we use for this purpose, be it quadrant, sector, or micrometer, should be divided and executed with sufficient exactness; and, secondly, that the telescope, by which the observations are to be made, should have an adequate power and distinctness. Upon the first head, the great improvements of mathematical instrument-makers have hardly left us any thing to desire: we can now measure seconds with almost as much facility and truth as former observers could measure minutes; nor do I think it impossible to go still further, and divide instruments that would shew thirds with sufficient accuracy. It is in the latter, or optical part, we find the greatest difficulty. To see a single second of a degree with precision requires a telescope of very great perfection; therefore, supposing the mechanical part of an apparatus well executed, it will still be necessary to try how far the power of our telescope will enable us to ascertain with confidence the division or number of seconds it points out. If upon trial we find that our instrument will give us the same measure within the second, every time the experiment is repeated, we may pronounce it capable of measuring seconds ; if otherwise, it will remain to be examined, whether the fault lies in the mechanical or optical part.

Let us now suppose that the parallax of the fixed stars does not amount to a single second, yet still the case is by no means desperate ; and though the difficulty of measuring seconds will soon suggest to us what extraordinary powers and distinctness of the telescope, and accuracy of the micrometer, are required to measure thirds ; this ought by no means to discourage us in the attempt. Could we measure angles, much smaller than seconds, might we not hope to find the parallax of some of the fixed stars at least to amount to several thirds ? On the other hand, if it should appear, indeed, that even with such improved methods of measurement we could not reach the remote situation of such almost infinitely distant suns, we might still derive a valuable approximation towards truth from such repeated observations, even though they should not be attended with all the success we expected from them. On this assurance, I endeavoured to take such a method for attempting the investigation of the parallax of the stars as to avail myself of the improvements I had already made, and was still in hopes of making, in my telescopes.

The next thing that was necessary to consider in this undertaking was, the manner of putting it into execution. The method pointed out by GALILEO, and first attempted by HOOK, FLAMSTEED, MOLINEUX, and BRADLEY, of taking distances of stars from the zenith that pass very near it, though it failed with regard to parallax, has been productive of the most noble discoveries of another nature. At the same time it has given us a much juster idea of the immense distance of the stars, and furnished us with an approximation to the knowledge of their parallax that is much nearer the truth than we ever had before. Dr. BRADLEY, in a letter to Dr. HALLEY on the subject of a new discovered motion of the fixed stars, says, " I believe I may venture to say, that in either of the two stars last mentioned (γ Draconis and η Ursæ majoris) it (the annual parallax) does not amount to 2″. I am of opinion, that if it were 1″ I should have perceived it in the great number of observations that I made, especially upon γ Draconis ; which agreeing with the hypothesis (without allowing any thing for parallax) nearly as well when the sun was in conjunction with, as in opposition to, this star, it seems very probable, that the parallax of it is not so great as one single second." *Phil. Trans.* n. 406. p. 637. Dec. 1728. As I do not know that any thing more decisive has been done upon the subject, it will not be amiss to see how far this method of finding the parallax has really been successful. The instrument that was used upon this occasion was the same as the present zenith sectors, which can hardly be allowed sufficient to shew an angle of one or even two seconds with accuracy ; yet, on account of the great number of observations, and above all the great sagacity of the observer, we will admit that if the parallax had amounted to two seconds he would have perceived it. The star on which these observations were made is marked of the third magnitude in the catalogue of PTOLEMY ; in TYCHO BRAHE's of the third ; in the Prince of HESSE's of the third ;

in HEVELIUS's between the third and second; in FLAMSTEED's of the second; and now appears as a very bright star of the third, or small star of the second magnitude; therefore its parallax is probably considerably less than that of a star of the first magnitude. Several authors who have touched upon this subject seem to have overlooked this distinction; and from Dr. BRADLEY's account of the parallax of γ Draconis, have concluded the parallax of the stars in general not to exceed 1″; but this appears to me by no means to follow from the doctor's observations. It is rather evident that, for aught we know to the contrary, the stars of the first magnitude may still have a parallax of several seconds; and I believe this to be as accurate a result as that method is capable of giving, at least in latitudes where there is not a star of the first magnitude that passes directly through the zenith.*

In general, the method of zenith distances labours under the following considerable difficulties. In the first place, all these distances, though they should not exceed a few degrees, are liable to refractions; and I hope to be pardoned when I say that the real quantities of these refractions, and their differences, are very far from being perfectly known. Secondly, the change of position of the earth's axis arising from nutation, precession of the equinoxes, and other causes, is so far from being completely settled, that it would not be very easy to say what it exactly is

* DE LA LANDE, in his excellent book of Astronomy, says, that the parallax of the fixed stars has been proved to be absolutely insensible (*Ast.* liv. XVI. § 2782). He reports the observations of TYCHO BRAHE, PICARD, HOOK, and FLAMSTEED, and concludes (§ 2778) from the discovery of the aberration by Dr. BRADLEY (which it seems he also allows to be the most decisive upon the subject) that now the question about parallax is resolved. In giving us the opinion which the doctor had of the result of his own observations with regard to the annual parallax, DE LA LANDE only mentions " M. BRADLEY pense que si elle (la parallaxe) eût été seulement de 1″ il l'auroit apperçue dans le grand nombre d'observations qu'il avoit faites, surtout de γ du Dragon." But if we also take in those lines upon which Dr. BRADLEY seems to lay the greatest stress, viz. " I believe I may venture to say, that in either of the two stars last mentioned it does not amount to two seconds;" and if we allow for the magnitude of the stars upon which the observations were made, I think I have fairly stated the full amount of all the actual proofs we have of the smallness of the annual parallax. Now, since it has escaped the finest observations of BRADLEY, it is not likely that it should come up to the full quantity to which it might amount without being perceived; and therefore the doctor might think it highly probable, " that it is not so great as one single second;" and his opinion, as well as DE LA LANDE's, who believes it to be absolutely insensible, are perfectly consistent with all the observations that have hitherto been made; though the *actual proofs*, which are the subject of our present inquiry, do not extend so far. Against the parallax of Sirius DE LA LANDE (§ 2781) mentions " forty-five meridian altitudes taken by Dr. BEVIS [a], with the eight-feet mural quadrant of the Royal Observatory at Greenwich, none of which differed 3 or 4″ from the mean altitude." Now, if they differed 3 or 4″ from the mean, we may suppose they differed 6 or 8″ from each other; and that observations, subject to so many causes of error as I shall presently enumerate, and which differed so much from each other, cannot give the least evidence either for or against a parallax, will need no proof. Refraction alone, which is liable to such changes at the meridian altitude of Sirius, notwithstanding the most careful observations of the barometer and thermometer should be made to ascertain its quantity, would, with me, remain an unanswerable argument against the validity of such observations in a subject of this critical nicety.

[a] These observations were not made by Dr. BEVIS, but extracted from the registers of the Royal Observatory at my desire, and calculated by myself, and sent in a letter by Dr. BEVIS to Paris. NEVIL MASKELYNE.

at any given time. In the third place, the aberration of light, though best known of all, may also be liable to some small errors, since the observations from which it was deduced laboured under all the foregoing difficulties. I do not mean to say, that our theories of all these causes of error are defective; on the contrary, I grant that we are for most astronomical purposes sufficiently furnished with excellent tables to correct our observations from the above mentioned errors. But when we are upon so delicate a point as the parallax of the stars; when we are investigating angles that may, perhaps, not amount to a single second, we must endeavour to keep clear of every possibility of being involved in uncertainties; even the hundredth part of a second becomes a quantity to be taken into consideration.

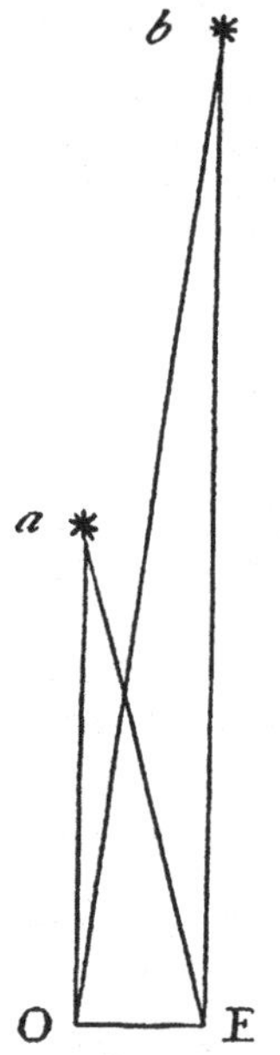

I shall now deliver the method I have taken, and shew that it is free from every error to which the former is liable, and is still capable of every improvement the telescope and mechanism of micrometers can furnish.

Let OE (fig. 1) be two opposite points of the annual orbit, taken in the same plane with two stars *a*, *b*, of unequal magnitudes. Let the angle *a*O*b* be observed when the earth is at O: and let the angle *a*E*b* be also observed when the earth is at E. From the difference of these angles, if any should be found, we may calculate the parallax of the stars, according to a theory that will be delivered hereafter. These two stars, for reasons that will soon appear, ought to be as near each other as possible, and also to differ as much in magnitude as we can find them.

FIG. 1.

GALILEO, I believe, was the first who suggested this method; but in the manner he mentions it in his third dialogue of the *Systema Cosmicum*, it would be exposed to all the difficulties we have enumerated, and would wish to avoid; for he does not observe, that the two stars should be so near each other as thereby to preclude the influence of every cause of error.

This method has also been mentioned by other authors; and we find that Dr. LONG observed the double star which is the first of Aries in PTOLEMY'S catalogue; that in the head of Castor; the middle one in the sword of Orion; and that in the breast of Virgo, with telescopes of fourteen and seventeen feet, and "was persuaded they would be found always to appear the same." But when the theory of parallax will be explained, it will be seen that every one of these stars are totally improper for the purpose; for the stars of γ Arietis are near 10″ distant from each other, and moreover equal in magnitude. In α Geminorum the stars, though near enough, do not sufficiently differ in magnitude to shew any parallax. The stars in the Nebula of Orion, on account of their extreme smallness or distance, are still more improper than any; and those of γ Virginis are equal in magnitude.

I do not find that any thing else has been done upon the subject. GALILEO justly remarks, that such observations ought to be made with the best telescopes, and upon this occasion mentions the power of his own, which enlarged the disk of the sun a thousand times, from which we find it magnified about thirty-two times; but we can hardly think his nor even Dr. LONG's, whose power might probably be sixty or seventy, sufficient for the purpose. What would GALILEO say, if he were told that our present opticians make instruments that enlarge the disk of the sun above forty thousand times? What would even CASSINI say, if he were to view the first star of Aries, which appeared to him as split in two, through a telescope that will shew η Coronæ borealis and *h* Draconis to be double stars?

But to proceed, I shall now prove that this method, if stars properly situated (such as I have found) are taken, is free from all the errors occasioned by refraction, nutation, precession of the equinoxes, changes of the obliquity of the ecliptic, and aberration of light; and that the annual parallax, if it even should not exceed the tenth part of a second, may still become visible, and be ascertained at least to a much greater degree of approximation than it ever has been done.

It will also appear, from the great number of observations I have already made upon several double stars, especially ϵ Bootis, that we can now with much greater certainty affirm the annual parallax to be exceedingly small indeed; and that there is a great probability of succeeding still farther in this laborious but delightful research, so as to be able at last to say, not only how much the annual parallax *is not*, but how much it really *is*.

Let there be two stars at a distance from each other, not exceeding five seconds; suppose them to be observed at an altitude of 20°; and let them be so situated with respect to each other, that one of them may be 20°, and the other 20° and 5″ high: then the whole effect of mean refraction at that altitude, by Dr. MASKELYNE's excellent tables, will be 2′ 35″·5 for 20°, and 2′ 35″·4888 for 20° 5″. The difference is 0″·0111. Now, in the first place, we have nothing to do with the refraction itself, since the real altitude of the stars is not in question. In the next place, we also have no concern with the difference of refraction between the two stars, though no more than the ·0111th part of a second, because the real distance between the two stars is not required. It follows then, that these observations can only be affected by the difference of the difference; that is, by an alteration in the quantity of refraction occasioned by the change of heat and cold, or weight of the atmosphere, and pointed out to us by the rise and fall of the barometer and thermometer. Let us then see what this difference of the difference may amount to. Suppose a change of 22° of FAHRENHEIT's thermometer, that is, from the freezing point to the moderate air of a summer's night, and a difference of an inch in the height of the barometer; these two causes both conspiring, which does not often happen, may occasion an alteration of ·00096th part of a second in five, at an altitude of 20°; but this being less than the thousandth part of a second may safely be rejected as a quantity altogether insensible.

Since it may not be always convenient to view those stars at the altitude of 20°, it remains to see what effect different altitudes may have: let us then make the most unfavourable supposition, that they may one time be seen in a horizontal position, having before been seen vertical. In this case, as the whole difference of refraction in a difference of 5″ of altitude is no more than ·0111, provided they are observed not lower than 20°, and the whole difference of the difference of refraction is only ·0009; the sum ·012, when both conspire, not exceeding much the hundredth part of a second, may still be rejected as insensible. Let us also examine how near the horizon it may be safe to observe such stars. At 10°, for instance, the refraction is 5′ 14″·6; the difference for 5″ is ·0388; the joint effect of the changes in the barometer and thermometer is ·0034; the sum of the whole together amounts to ·0422, which is less than half the tenth of a second: now this may either be taken into consideration, or such low observations may be avoided, as being by no means necessary, and but ill suiting the high powers a telescope proper for this purpose ought to bear.

The change of position of the earth's axis I look upon as an unsurmountable obstacle to taking the parallax of stars by the method of zenith distances: for though refraction is much reduced in the zenith, this change is there no less sensible than in other parts of the heavens; but as this will always affect our two stars exactly alike, we are entirely freed from this embarrassment.

The aberration of light can have no influence of the least consideration upon our two stars, as a mere inspection of the tables will shew. In a whole degree, its effects, when greatest, amount but to four-tenths of a second, and consequently in 5″ to no more than ·0005, or the two thousandth part of a second.

Observations of the relative distance of the two stars that make up a double star, being thus cleared of every impediment, are capable of being continually improved by every degree of perfection the telescope may acquire: we can chuse stars that may be viewed sufficiently high to be clear of the vapours that swim near the horizon, and consequently employ the greatest powers our instruments are capable of. From experience I can also affirm, that the stars will bear a much higher degree of magnifying than other celestial objects. Too much has hitherto been taken for granted in optics: every natural philosopher is ready enough to allow the necessity of making experiments, and tracing out the steps of nature; why this method should not be more pursued in the art of seeing does not appear. Theories are only to be used when proper data are assigned; but the data are carefully to be re-examined, when new improvements may widely alter the result of former experiments. Thus, we are told, that we gain nothing by magnifying *too much*. I grant it; but shall never believe I magnify too much till by experience I find, that I can see better with a lower power. Nor is even that sufficient: a lower power may shew more of the object; it may shew it brighter, nay even distincter, and therefore upon the whole better; and yet the greater power may, in a particular case, be preferable: for if the object is so small as not to be at all visible with the lower power, and I can, by magnifying more, obtain a view of it, though neither so bright nor distinct as I could wish, is it not evident, that here this power is preferable to the former?

The naturalist does not think himself obliged to account for all the phænomena he may observe; the astronomer and optician may claim the same privilege. When we increase the power we lessen the light in the inverse ratio of the square of the power; and telescopes will, in general, discover more small stars the more light they collect; yet with a power of 227 I cannot see the small star near the star following o Aquilæ, when, by the same telescope, it appears very plainly with the power of 460: now, in the latter case, the power being more than double, the light is less than the fourth part of the former. In such particular cases I generally suspect my own eyes, and have recourse to those of my friends. I had the pleasure of shewing this star to Dr. WATSON junior, who soon discovered the small star, which accompanies the other, with the power of 460; but saw nothing of it with 227, though the place where to look for it had been pointed out to him by the higher power. The experiment has been too often repeated to be doubtful, and has also been confirmed by others of nearly the same nature: for instance, the smallest of the two that accompany the star near k Aquilæ, the small star near μ Herculis, and the small star near α Lyræ, are invisible with my power of 227, and visible with the same aperture when the power is 460. Also the small stars near FLAMSTEED'S 24th of Aquila, the smallest of two near σ Coronæ, the small star near the star south of ϵ Aquilæ, the small star near the second o Persei, the small star near the star which accompanies FLAMSTEED'S 10th sub pede et scapula dextra Tauri, the small star near β Delphini, and the small star near the pole star, are all much brighter and stronger, and therefore much sooner seen with 460 than with 227.

Great power may also, in particular circumstances, be favourable, even with an excess of aberration. When two stars are so close together as to make the scale for measuring the distance of their centers too small, if, by magnifying much, we can enlarge that distance, we may gain a considerable advantage, provided the centers or apparent bodies of the stars remain distinct enough for the purpose of these measures. The appearance of α Lyræ in my Newtonian reflector with a power of 460 is represented in fig. 2; with 2010 in fig. 3; with 3168 in fig. 4; and with 6450 in fig. 5. Now in all these figures we see, that the centers are still distinct enough to measure their distances with sufficient truth; or if any little error should be introduced by the magnitude of the central point, it will be more than sufficiently balanced by the largeness of the scale. In this manner, with a power of 3168, I have obtained a scale of no less than ten inches six tenths for the distance of the centers of the two stars of α Geminorum; and as we know these centers to be but a few seconds distant, it is plain how great an advantage we gain by such an enlarged scale.

FIG. 2.

These experiments have but very lately pointed out to me a method of making a new micrometer, upon a construction entirely different from any that are now in use, which I have been successful enough to put in practice, and by which I have already begun to determine the distance of the centers of some of the most remarkable double stars to a very great degree of accuracy.*

The powers that may be used upon various double stars are different, according to their relative magnitudes : ϵ Bootis, for instance, will not bear the same power as α Geminorum, nor would it be difficult to assign a reason for it ; but as I here shall merely confine myself to facts, it will be sufficient in general to mention, that two stars, which are equal, or nearly so, will bear a very high power : with α Geminorum I have gone as far as 3168 ; but with the former only to 2010. The difficulty of

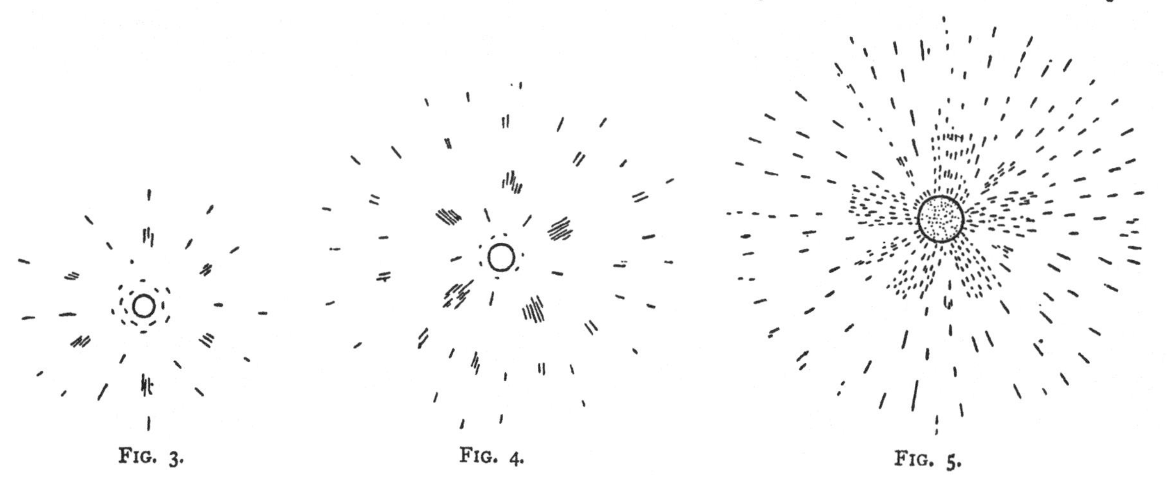

FIG. 3. FIG. 4. FIG. 5.

using high powers is exceedingly great ; for the field of view takes in less than the diameter of the hair or wire in the finder, and the effect of the earth's diurnal motion is so great, that it requires a great deal of practice to find the object, and manage the instrument. It appears to me very probable, that the diurnal motion of the earth will be the greatest obstacle to our progress in magnifying, except we can introduce a proper mechanism to carry our telescopes in a contrary motion.

Notwithstanding opticians have proved that two eye-glasses will give a more correct image than one, I have always (from experience) persisted in refusing the assistance of a second glass, which is sure to introduce errors greater than those we would correct. Let us resign the double eye-glass to those who view objects merely for entertainment, and must have an exorbitant field of view. To a philosopher this is an unpardonable indulgence. I have tried both the single and double eye-glass of equal powers, and always found that the single eye-glass had much the superiority in point of light and distinctness. With the double eye-glass I could not see the *belts on Saturn*, which I very plainly saw with the single one. I would, however, except all those cases where a large field is absolutely necessary, and where power joined to distinctness is not the sole object of our view.

* For a description of this micrometer see a subsequent paper.

The application of the different powers of a telescope in general is of some consequence; and in answer to those who may think I have strained or over-charged mine, I must observe, that a single glance at the subsequent *h* Draconis, η Coronæ, and the star near μ Bootis, with a power of 460, shewed them to me as double stars; when, in two former reviews of the heavens, I had twice set them down in my journal as single stars, where I used only the power of 222 and 227, and in all probability should never have found them double, had I not looked with a higher power.

We are to remember, that it is much easier to see an object when it is pointed out to us than when it falls in our way unexpectedly, especially if of such a nature as to require some attention to be seen at all; but to say no more of other advantages of high powers, it is evident, that in the research of the parallax of the fixed stars they are absolutely necessary. If we would distinctly perceive and measure or estimate extremely small quantities, such as a tenth of a second, it appears, that when we use a power of 460, this tenth of a second will be no more in appearance than 46″, and even with a power of 1500 will be but 2′ 30″, which is a quantity not much more than sufficient to judge well of objects and distinguish them from each other, such as a circle from a square, triangle, or polygon.*

It has been observed, that objects grow indistinct when the principal optic pencil at the eye becomes less than the 40th or 50th part of an inch in diameter. In the experiments that have been made upon this subject it appears to me, that the indistinctness which is ascribed to the smallness of the optical pencil may be owing to very different causes: at least it will be easy to bring contrary experiments of extremely small pencils, not at all affected by this inconvenience; for instance, it is well known, that microscopes, consisting of a single lens or globule, are remarkable for distinctness. We also know, that they have been made so small as to magnify above 10,000 times.† From this we may infer that their apertures, and consequently the diameters of the optic pencil at the eye could not exceed the 2500th part of an inch. I am therefore inclined to believe, that we must look for distinctness in the perfection of the object-speculum or object-glass of a telescope; and if we can make the first image in the focus of a speculum almost as perfect as the real object, what should hinder our magnifying but the want of light? Now, if the object has light sufficient, as the stars most undoubtedly have, I see no reason why we should limit the powers of our instruments by any theory. Is it not best to have recourse to experiments to find how far our endeavours to render the first image perfect have been successful?

As soon as I was fully satisfied that in the investigation of parallax the method of double stars would have many advantages above any other, it became necessary to look out for proper stars. This introduced a new series of observations. I

* By a set of experiments, made in the year 1774, I found, that I could discover or perceive a bright object, such as white paper, against the sky-light, when it subtended an angle of 35″; but could only distinguish it to be a circle, and no other figure, when it appeared under an angle of 2′ 24″.

† See Padre DELLA TORRE'S Method, &c., *Scelta di Opuscuh.*

resolved to examine every star in the heavens with the utmost attention and a very high power, that I might collect such materials for this research as would enable me to fix my observations upon those that would best answer my end. The subject has already proved so extensive, and still promises so rich a harvest to those who are inclined to be diligent in the pursuit, that I cannot help inviting every lover of astronomy to join with me in observations that must inevitably lead to new discoveries. I took some pains to find out what double stars had been recorded by astronomers; but my situation permitted me not to consult extensive libraries, nor indeed was it very material: for as I intended to view the heavens myself, Nature, that great volume, appeared to me to contain the best catalogue upon this occasion. However, I remembered that the star in the head of Castor, that in the breast of the Virgin, and the first star in Aries, had been mentioned by CASSINI as double stars. I also found the Nebula in Orion was marked in HUGEN'S *Systema Saturnium* as containing seven stars, three of which (now known to be four) are very near together. With this small stock I begun, and in the course of a few years observations have collected the stars contained in my catalogue. I find, with great pleasure, that a very excellent observer, whom I have the honour to call my friend,* has also, though unknown to me, met with three of those stars that will be found in my catalogue: and upon this occasion I also beg leave to observe, that the Astronomer Royal, when I was at Greenwich last May, with his usual politeness, shewed me, among other objects, α Herculis as a double star, which he had discovered some years ago. The Rev. Mr. HORNSBY also, when I had the pleasure of seeing him at Oxford, in a conversation on the subject of the stars of the first magnitude that have a proper motion, mentioned π Bootis as a double star. It is a little hard upon young astronomers to be obliged to discover *over-again* what has already been discovered; however, the pleasure that attended the view when I first saw these stars has made some amends for not knowing they had been seen before me.

If I should mention in my list of observations a few that may be found difficult to be verified by other telescopes, I must beg the indulgence of the observers. I hope it will sufficiently appear, that I have guarded against optical delusions; and every astronomer, I make no doubt, will find, by those observations that fall within the compass of his instruments, and attention to circumstances necessary to the right management of them, that I have had all along truth and reality in view, as the sole object of my endeavours; and therefore he will be inclined to give some credit to what he does not immediately perceive, when he finds himself successful where he takes the proper precautions so necessary in delicate observations, even with the best instruments.

I have been in some doubt in what manner to communicate these observations. My first view was to have methodized them properly; but I find them so extensive

* *Phil. Trans.* for the year 1781, part II. double stars discovered in 1779, at Frampton-house, Glamorganshire, by NAT. PIGOTT, Esq. F.R.S. &c.

that there is but little probability that one person should be able to bring them to a conclusion, for which reason I have now resolved to give them unfinished as they are, that every person who is inclined to engage in this pursuit may become a fellow-labourer.

In settling the distances of double stars I have occasionally used two different ways. Those that are extremely near each other may be estimated by the eye, in measures of their own apparent diameters. For this purpose their distance should not much exceed two diameters of the largest, as the eye cannot so well make a good estimation when the interval between them is greater. This method has often the preference to that of the micrometer: for instance, when the diameter of a small star, perhaps not equal to half a second, is double the vacancy between the two stars. Here a micrometer ought to measure tenths of seconds at least, otherwise we could not, with any degree of confidence, rely on its measures; nay, even then, if the stars are situated in the same parallel of declination and near the equator, their quick motion across the micrometer makes it extremely difficult to measure them, and in that case an estimation by the eye is preferable to any other measure; but this requires not a little practice, precaution, and time, and yet with proper care it will be found that this method is capable of great exactness. Let two small circles be drawn either equal or unequal, at a distance not exceeding twice the diameter of the largest; let these be shewn to several persons in the same light and point of view. Then, if every one of them will separately and carefully write down his estimation of the interval between them, in the proportion of either of their diameters, it will be found upon a comparison that there will seldom be so much as a quarter of a diameter difference between all the estimations. If this agreement takes place with so many different eyes, much more may we expect it in the estimations of the same eye when accustomed to this kind of judgement.

I have divided the double stars into several different classes. In the first I have placed all those which require indeed a very superior telescope, the utmost clearness of air, and every other favourable circumstance to be seen at all, or well enough to judge of them. They seemed to me on that account to deserve a separate place, that an observer might not condemn his instrument or his eye if he should not be successful in distinguishing them.

As these are some of the finest, most minute, and most delicate objects of vision I ever beheld, I shall be happy to hear that my observations have been verified by other persons, which I make no doubt the curious in astronomy will soon undertake. I should observe, that since it will require no common stretch of power and distinctness to see these double stars, it will therefore not be amiss to go gradually through a few preparatory steps of vision, such as the following: when η Coronæ borealis (one of the most minute double stars) is proposed to be viewed, let the telescope be some time before directed to α Geminorum, or if not in view to either of the following stars, ζ Aquarii, μ Draconis, ρ Herculis, α Piscium, or the curious double-double star ε Lyræ. These should be kept in view for a considerable time, that the eye may acquire the habit of seeing such objects well and distinctly. The observer may next proceed to ξ Ursæ majoris, and the beautiful treble star in Monoceros's right fore-foot; after these to *i* Bootis, which is a fine miniature of α Geminorum, to the star preceding α Orionis, and to *n* Orionis. By this time both the eye and the telescope will be prepared for a still finer picture, which is η Coronæ borealis. It will be in vain to attempt this latter if all the former, at least *i* Bootis, cannot be distinctly perceived to be fairly separated, because it is almost as fine a miniature of *i* Bootis as that is of α Geminorum. If the observer has been successful in all these, he may then, at the same time, try *h* Draconis, though I question whether any power less than 4 or 500 will shew it to be double; but the former I have all seen very well with 227.

To try the stars of unequal magnitudes it will be expedient to take them in some such order as the following: α Herculis, ω Aurigæ, δ Geminorum, *k* Cygni, ε Persei, and *b* Draconis; from these the observer may proceed to a most beautiful object, ε Bootis, which I have closely attended these two years as very proper for the investigation of the parallax of the fixed stars.

It appears, from what has been said, that these double stars are a most excellent way of trying a telescope; and as the foregoing remarks have suggested the method of seeing how far the power and distinctness of our instruments will reach, I shall add the way of finding how much light we have. The observer may begin with the pole-star and α Lyræ; then go to the star south of ε Aquilæ, the treble star near *k* Aquilæ, and last of all to the star following *o* Aquilæ. Now, if his telescope has not a great deal of good distinct light, he will not be able to see some of the small stars that accompany them.

In the second class of double stars I have put all those that are proper for estimations by the eye or very delicate measures of the micrometer. To compare the distances with the apparent diameters the power of the telescope should not be much less than 200, as they will otherwise be too close for the purpose. The instrument ought, moreover, to be as much as possible free from rays that surround a star in common telescopes, and should give the apparent diameters of a double star perfectly round and well-defined, with a deep black division between them, as in fig. 6, which represents α Geminorum as I have often seen it with a power of 460. It will be necessary here to take notice, that the estimations made with one telescope cannot be applied to those made with another: nor can the estimations made with different powers, though with the same telescope, be applied to each other. Whatever may be the cause of the apparent diameters of the stars, they are certainly not of equal magnitude with the same powers in different telescopes, nor of proportional magnitude with different powers in the same telescope. In my instruments I have ever found less diameter in proportion the higher I was

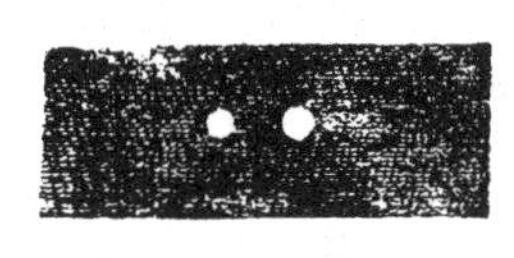

Fig. 6.

able to go in power, and never have I found so small a proportional diameter as when I magnified 6450 times;* therefore if we would wish to compare any such observations together, with a view to see whether a change in the distance has taken place, it should be done with the very same telescope and power, even with the very same eye-glass or glasses; for others, though of equal power and goodness, would most probably give different proportional diameters of the stars.

In the third class I have placed all those double stars that are more than five but less than 15″ asunder; and for that reason, if they should be used for observations on the parallax of the fixed stars, they ought not to be looked upon as quite free from the effects of refraction, &c. In the same manner that the stars in the first and second classes will serve to try the goodness of the most capital instruments, these will afford objects for telescopes of inferior power, such as magnify from 40 to 100 times. The observer may take them in this or the like order: ζ Ursæ majoris, γ Delphini, γ Arietis, π Bootis, γ Virginis, ι Cassiopeæ, μ Cygni. And if he can see all these, he may pass over into the second class, and direct his instrument to some of those that were pointed out as objects for the very best telescopes, where, I suppose, he will soon find the want of superior power.

The fourth, fifth, and sixth classes contain double stars that are from 15 to 30″, from 30″ to 1′, and from 1′ to 2′ or more asunder. Though these will hardly be of any service for the purpose of parallax, I thought it not amiss to give an account of such as I have observed; they may, perhaps, answer another very important end, which also requires a great deal of accuracy, though not quite so much as the investigation of the parallax of the fixed stars. I will just mention it, though foreign to my present purpose. Several stars of the first magnitude have already been observed, and others suspected, to have a proper motion of their own: hence we may surmise, that our sun, with all its planets and comets, may also have a motion towards some particular part of the heavens, on account of a greater quantity of matter collected in a number of stars and their surrounding planets there situated, which may perhaps occasion a gravitation of our whole solar system towards it. If this surmise should have any foundation, it will shew itself in a series of some years; as from that motion will arise another kind of hitherto unknown parallax,† the investigation of which may account for some part of the motions already observed in some of the principal stars; and for the purpose of determining the direction and quantity of such a motion, accurate observations of the distance of stars that are near enough to be measured with a micrometer, and a very high power of telescopes may be of considerable use, as

* See the measures of the diameter of α Lyræ, Catalogue of double stars, 5th class.

† See the note in the Rev. Mr. MITCHELL's paper on the Parallax of the Fixed Stars, *Phil. Trans.* vol. LVII. p. 252.

they will undoubtedly give us the relative places of those stars to a much greater degree of accuracy than they can be had by transit instruments or sectors, and thereby much sooner enable us to discover any apparent change in their situation occasioned by this new kind of systematical parallax, if I may be allowed to use that expression, for signifying the change arising from the motion of the whole solar system.

I shall now endeavour to deliver a theory of the annual parallax of double stars, with the method of computing from thence what is generally called the parallax of the fixed stars, or of single stars of the first magnitude, such as are nearest to us. It may be observed, that the principles upon which I have founded the following theory are of such a nature, that they cannot be strictly demonstrated, in consequence of which they are only proposed as postulata, which have so great a probability in their favour, that they will hardly be objected to by those who are in the least acquainted with the doctrine of chances.

GENERAL POSTULATA.

1. Let the stars be supposed, one with another, to be about the size of the sun.*

2. Let the difference of their apparent magnitudes be owing to their different distances, so that a star of the second, third, or fourth magnitude is two, three, or four times as far off as one of the first.†

In fig. 7 let OE be the whole diameter of the earth's annual orbit; and let a, b, c, be three stars situated in the ecliptic, in such a manner that they may be seen all in one line Oabc, when the earth is at O. Let the line Oabc be perpendicular to OE, and draw PE parallel to cO. Then, if Oa, ab, bc, are equal to each other, a will be a star of the first magnitude, b of the second, and c of the third. Let us now suppose the angle OaE, or parallax of the whole orbit of the earth, to be $1''$ of a degree: then we have $\text{PE}a = \text{O}a\text{E} = 1''$; and, because very small angles,

* See Mr. MITCHELL's Inquiry into the probable Parallax and Magnitude of the Fixed Stars, *Phil. Trans.* vol. LVII. p. 234. 236. 237. 240. and Dr. HALLEY on the Number, Order, and Light, of the Fixed Stars, *Phil. Trans.* vol. XXXI.

† The apparent magnitude is here taken in a stricter sense than is generally used; and by it is rather meant the order into which the stars *ought to be* distinguished than that into which they *are* commonly divided: for as the order of the magnitudes is here to denote the different relative distances, we are to examine carefully the degree of light each star is accurately found to have: and considering then that light diminishes in the inverse ratio of the squares of the distances, we ought to class the stars accordingly. An allowance ought also perhaps to be made for some loss that may happen to the light of very remote stars in its passage through immense tracts of space, most probably not quite destitute of some very subtle medium. This conjecture is suggested to us by the colour of the very small telescopic stars, for I have generally found them red, or inclining to red; which seems to indicate, that the more feeble and refrangible rays of the other colours are either stopped by the way, or at least diverted from their course by accidental deflections.

having the same subtense OE, may be taken to be in the inverse ratio of the lines O*a*, O*b*, O*c*, &c. we shall have $ObE = \frac{1}{2}''$, $OcE = \frac{1}{3}''$, &c.* Now, when the earth is removed to E, we shall have $PEb = EbO = \frac{1}{2}''$, and $PEa - PEb = aEb = \frac{1}{2}''$; that is, the stars *a*, *b*, will appear to be $\frac{1}{2}''$ distant. We also have $PEc = EcO = \frac{1}{3}''$, and $PEa - PEc = aEc = \frac{2}{3}''$; that is, the stars *a*, *c*, will appear to be $\frac{2}{3}''$ distant, when the earth is at E. Now, since we have $bEP = \frac{1}{2}''$, and $cEP = \frac{1}{3}''$, therefore $bEP - cEP = bEc = \frac{1}{2}'' - \frac{1}{3}'' = \frac{1}{6}''$; that is, the stars *b*, *c*, will appear to be only $\frac{1}{6}''$ removed from each other, when the earth is at E.

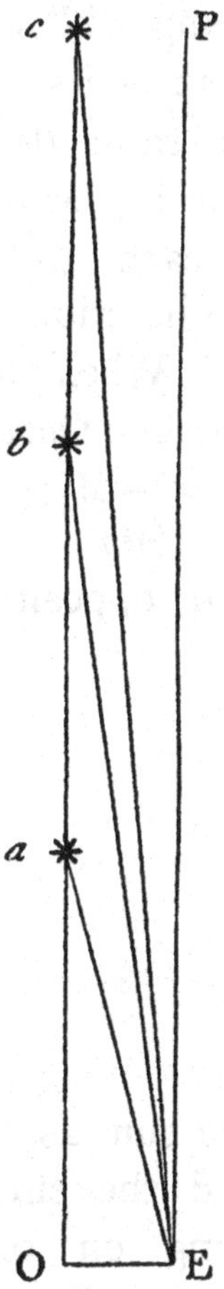

From what has been said, we may gather the following general expression, to denote the parallax that will become visible in the change of distance between the two stars, by the removal of the earth from one extreme of its orbit to the other. Let P express the total parallax of a fixed star of the first magnitude, M the magnitude of the largest of the two stars, *m* the magnitude of the smallest,† and *p* the partial parallax to be observed by the change in the distance of a double star; then will $p = \frac{m - M}{Mm} P$; and *p* being found by observation will give us $P = \frac{pMm}{m - M}$. An example or two will explain this sufficiently. Suppose a star of the first magnitude should have a small star of the twelfth magnitude near it; then will the partial parallax we are to expect to see be $\frac{12 - 1}{12 \times 1} P$; or $\frac{11}{12}$ths of the total parallax of a fixed star of the first magnitude; and if we should, by observation, find the partial parallax between two such stars to amount to $1''$, we shall have the total parallax $P = \frac{1 \times 1 \times 12}{12 - 1} = 1''\cdot 0909$. If the stars are of the third and twenty-fourth magnitude, the partial parallax will be $\frac{24 - 3}{3 \times 24} = \frac{21}{72} P$; and if, by observation,

c P

b

a

O E

FIG. 7.

* This proves what I have before remarked on the parallax of γ Draconis; for that star, (admitting it to be a star of between the second and third magnitude, which ought to be ascertained by experiments, as mentioned in the note above) by the postulata, will have its place assigned somewhere between *b* and *c*, and therefore its parallax will be between $\frac{1}{2}$ and $\frac{1}{3}$ of the parallax of a star of the first magnitude. And if Dr. BRADLEY thought that he should have perceived a parallax in γ Draconis, if at most it had amounted to 2″, it follows, that the angle O*a*E may nearly amount to 4 or 5″ for any thing we can conclude to the contrary from those observations.

† As M and *m* are here taken to express the relative distances of the stars, in measures whereof the distance of the nearest star is taken as unity, those who think the postulata on which these estimations are built cannot be granted, may still use the following formulæ, if instead of the magnitudes M, *m*, they put their own estimations of the relative distances of the stars, according to any other method whatever they may think it most eligible to adopt; for the apparent magnitude of stars is here only proposed as the most probable means we have of forming any conjectures about their relative distances.

p is found to be a tenth of a second, the whole parallax will come out $\frac{\cdot 1 \times 3 \times 24}{24-3} = 0''\cdot 3428$.

It will be necessary to examine some different situations. Suppose the stars, being still in the ecliptic, to appear in one line, when the earth is in any other part of its orbit between O and E; then will the parallax still be expressed by the same algebraic form, and one of the maxima will still lie at O, the other at E; but the whole effect will be divided into two parts, which will be in proportion to each other as radius − sine to radius + sine of the star's distance from the nearest conjunction or opposition.

When the stars are any where out of the ecliptic situated so as to appear in one line O*abc* at rectangles to OE, the maximum of parallax will still be expressed by $\frac{m-M}{Mm}P$; but there will arise another additional parallax in the conjunction and opposition, which will be to that which is found 90° before or after

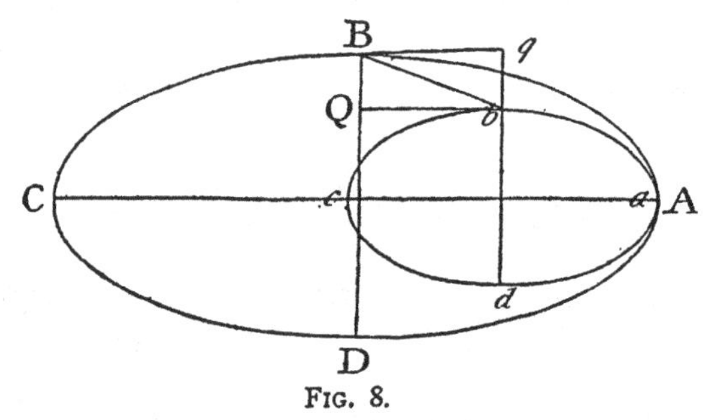

Fig. 8.

the sun, as the sine (S) of the latitude of the stars seen at O is to radius (R); and the effect of this parallax will be divided into two parts; half of it lying on one side of the large star, the other half on the other side of it. This latter parallax, moreover, will be compounded with the former, so that the distance of the stars in the conjunction and opposition will then be represented by the diagonal of a parallelogram, whereof the two semi-parallaxes are the sides; a general expression for which will be $\sqrt{\overline{\frac{m-M}{2Mm}P}\Big|^2 \times \frac{SS}{RR} + 1}$: for the stars will apparently describe two ellipses in the heavens, whose transverse axes will be to each other in the ratio of M to m (fig. 8), and Aa, Bb, Cc, Dd, will be cotemporary situations. Now, if bQ be drawn parallel to AC, and the parallelogram bqBQ compleated, we shall have $bQ = \frac{1}{2}CA - \frac{1}{2}ca = \frac{1}{2}Cc = \frac{1}{2}p$, or semi-parallax 90° before or after the sun, and Bb may be resolved into, or is compounded of, bQ and bq; but $bq = \frac{1}{2}BD - \frac{1}{2}bd =$ the semi-parallax in the conjunction or opposition. We also have R : S :: bQ : $bq = \frac{pS}{2R}$; therefore the distance Bb (or Dd) $= \sqrt{\overline{\frac{p}{2}}\Big|^2 + \overline{\frac{pS}{2R}}\Big|^2}$; and by substituting the value of p into this expression we obtain $\sqrt{\left(\frac{m-M}{2Mm}P\right)^2 \times \frac{SS}{RR}+1}$, as above. When the stars are in the pole of the ecliptic, bq will become equal to bQ, and Bb will be $\cdot 7071\, P\frac{m-M}{Mm}$.

Hitherto we have supposed the stars to be all in one line $Oabc$; let them now be at some distance, suppose 5″ from each other, and let them first be both in the ecliptic. This case is resolvable into the first; for imagine the star a, fig. 9, to stand at x, and in that situation the stars x, b, c, will be in one line, and their parallax expressed by $\frac{m-M}{Mm}P$. But the angle aEx may be taken to be equal to aOx; and as the foregoing form gives us the angles xEb, xEc, we are to add aEx, or 5″ to xEb, and we shall have aEb. In general, let the distance of the stars be d, and let the observed distance at E be D; then will $D = d + p$, and therefore the whole parallax of the annual orbit will be expressed by $\frac{DMm - dMm}{m-M} = P$.

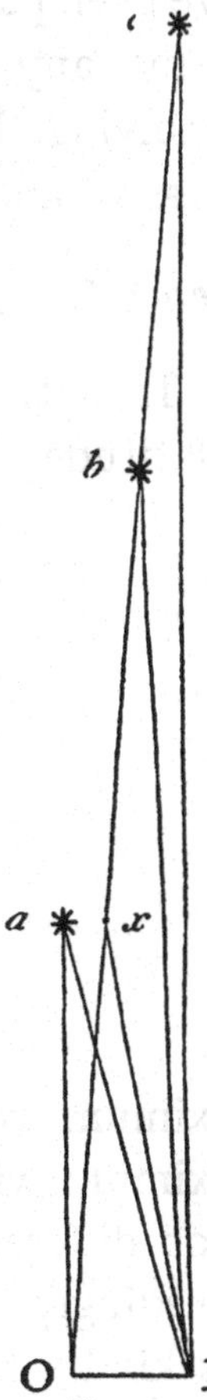

Fig. 9.

Suppose the two stars now to differ only in latitude, one being in the ecliptic, the other, for instance, 5″ north, when seen at O. This case may also be resolved by the former; for imagine the stars b, c, fig. 7, to be elevated at rectangles above the plane of the figure, so that aOb, or aOc, may make an angle of 5″ at O: then, instead of the lines $Oabc$, Ea, Eb, Ec, EP, imagine them all to be planes at rectangles to the figure; and it will appear, that the parallax of the stars in longitude must be the same as if the small star had been without latitude. And since the stars b, c, by the motion of the earth from O to E, will not change their latitude, we shall have the following construction for finding the distance of the stars ab, ac, at E, and from thence the parallax P. Let the triangle $ab\beta$, fig. 10, represent the situation of the stars; ab is the subtense of 5″, that being the angle under which they are supposed to be seen at O. The quantity $b\beta$ by the former theorem is found $\frac{m-M}{Mm}P$, which is the partial parallax that would have been seen by the earth's moving from O to E, had both stars been in the ecliptic; but on account of the difference in latitude it will now be represented by $a\beta$, the hypothenuse of the triangle $ab\beta$: therefore, in general, putting $ab = d$, and $a\beta = D$, we have $\frac{\sqrt{DD - dd} \times Mm}{m-M} = P$. Hence D being taken by observation and d, M, and m, given, we obtain the total parallax.

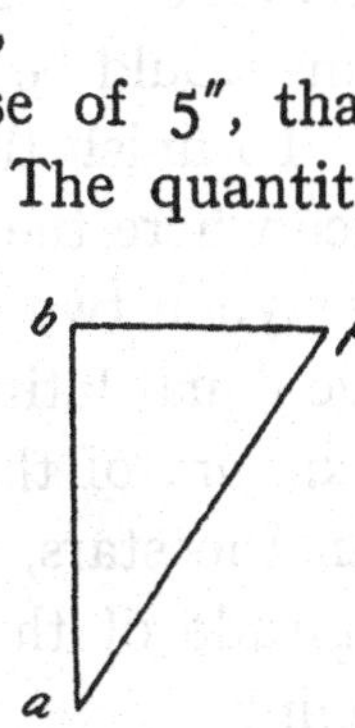

Fig. 10.

If the situation of the stars differs in longitude as well as latitude, we may resolve this case by the following method. Let the triangle $ab\beta$, fig. 11, represent the situation of the stars, $ab = d$ being their distance seen at O, $a\beta = \text{D}$ their distance seen at E. That the change $b\beta$ which is produced by the earth's motion will be truly expressed by $\frac{m-\text{M}}{\text{M}m}\text{P}$, may be proved as before, by supposing the star a to have been placed at α. Now let the angle of position $ba\alpha$ be taken by a micrometer,* or by any other method that may be thought sufficiently exact; then, by solving the triangle $ab\alpha$, we shall have the longitudinal and latitudinal differences $a\alpha$ and $b\alpha$ of the two stars. Put $a\alpha = x$, $b\alpha = y$, and it will be $x + b\beta = aq$, whence $\text{D} = \sqrt{\overline{x + \frac{m-\text{M}}{\text{M}m}\text{P}}\Big|^2 + yy}$; and $\frac{\sqrt{\text{D}^2 - y^2} \times \text{M}^2m^2 - x\text{M}m}{m-\text{M}} = \text{P}$.

If neither of the stars should be in the ecliptic, nor have the same longitude or latitude, the last theorem will still serve to calculate the total parallax whose

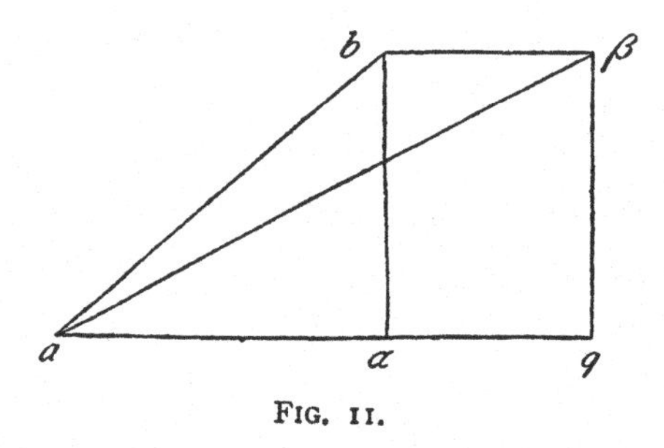

FIG. 11.

maximum will lie in E. There will, moreover, arise another parallax, whose maximum will be in the conjunction and opposition, which will be divided, and lie on different sides of the large star; but as we know the whole parallax to be exceedingly small, it will not be necessary to investigate every particular case of this kind; for, by reason of the division of the parallax, which renders observations taken at any other time, except where it is greatest, very unfavourable, the forms would be of little use.

To finish this theory, I shall only add a general observation on the time and place where the maxima of parallax will happen.

When two unequal stars are both in the ecliptic, or, not being in the ecliptic, have equal latitudes, north or south, and the largest star has most longitude, the maximum of the apparent distance will be when the sun's longitude is 90° more than the stars, or when observed in the morning; and the minimum when the longitude of the sun is 90° less than that of the star, or when observed in the evening.

* The position of a line passing through the two stars, with the parallel of declination of the largest of them, may be had by the micrometer I invented for this purpose in the year 1779, of which a description has been given in a former paper; whence, by spherical trigonometry, we easily deduce their position $ba\alpha$, fig. 11, with regard to the ecliptic.

When the small star has most longitude, the maximum and minimum, as well as the time of observation, will be the reverse of the former.

When the stars differ in latitude, this makes no alteration in the place of the maximum or minimum, nor in the time of observation; that is to say, it is immaterial whether the largest star has the least or the most latitude of the two stars.

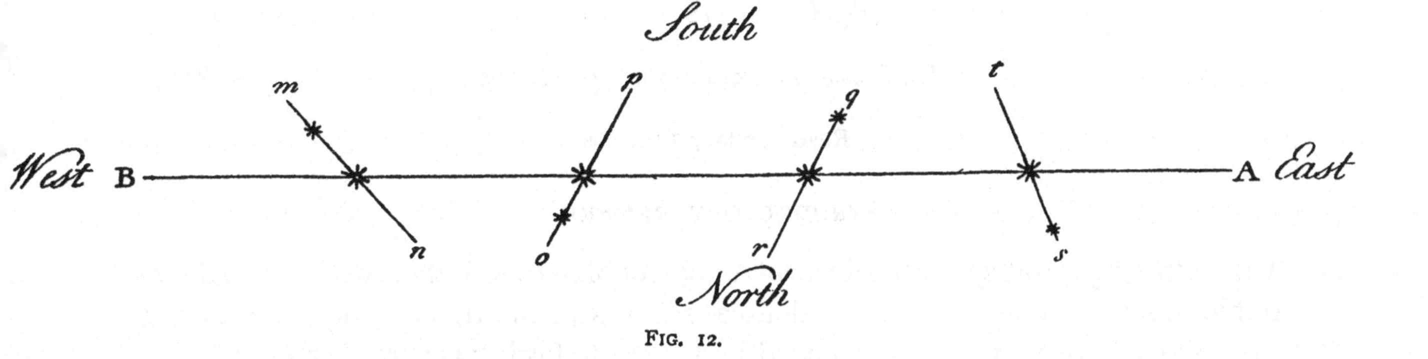

FIG. 12.

VI.

Catalogue of Double Stars. By Mr. HERSCHEL, *F.R.S.; communicated by Dr.* WATSON, *Jun.*

[*Phil. Trans.*, vol. lxxii., 1782, pp. 112–162.]

Read January 10, 1782.

INTRODUCTORY REMARKS.

THE following catalogue contains not only double stars, but also those that are treble, double-double, quadruple, double-treble, and multiple. The particulars I have given of them are comprehended under the following general heads.

I. The names of the stars and number in FLAMSTEED'S Catalogue; or, if not contained therein, such a description of their situation as will be found sufficient to point them out.

II. The comparative size of the stars. On this occasion I have used the terms equal, a little unequal, pretty unequal, considerably unequal, very unequal, extremely unequal, and excessively unequal, as expressing the different gradations to which I have endeavoured to affix always the same meaning.

III. The colours of the stars as they appeared to me when I viewed them. Here I must remark, that different eyes may perhaps differ a little in their estimations. I have, for instance, found, that the little star which is near α Herculis, by some to whom I have shewn it has been called green, and by others blue. Nor will this appear extraordinary when we recollect that there are blues and greens which are very often, particularly by candle-light, mistaken for each other. The situation will also affect the colour a little, making a white star appear pale red when the altitude is not sufficient to clear it of the vapours. It is difficult to find a criterion of the colours of stars, though I might in general observe that Aldebaran appears red, Lyra white, and so on; but when I call the stars garnet, red, pale red, pale rose-colour, white inclining to red, white, white inclining to blue, blueish white, blue, greenish, green, dusky, I wish rather to refer to the double stars themselves to explain what is meant by those terms.

IV. The distances of the stars are given several different ways. Those that are estimated by the diameter can hardly be liable to an error of so much as one quarter of a second; but here must be remembered what I have before remarked on the comparative appearance of the diameters of stars in different instruments.

Those that are measured by the micrometer, I fear, may be liable to an error of almost a whole second; and if not measured with the utmost care, to near 2″. This is, however, to be understood only of single measures; for the distance of many of them that have been measured very often in the course of two years observations can hardly differ so much as half a second from truth, when a proper mean of all the measures is taken. As I always make the wires of my micrometer outward tangents to the apparent diameter of the stars, all the measures must be understood to include both their diameters; so that we are to deduct the two semi-diameters of the stars if we would have the distance of their centers. What I have said concerns only the wire micrometers, for my last new micrometer is of such a construction, that it immediately gives the distance of the centers and its measures (as far as in a few months I have been able to find out) may be relied on to about one-tenth of a second, when a mean of three observations is taken. When I have added *inaccurate*, we may suspect an error of 3 or 4″. *Exactly estimated* may be taken to be true to about one-eighth part of the whole distance; but only *estimated*, or *about*, &c. is in some respect quite undetermined; for it is hardly to be conceived how little we are able to judge of distances when, by constantly changing the powers of the instrument, we are as it were left without any guide at all. I should not forget to add, that the measure of stars, whereof one is extremely small, must claim a greater indulgence than the rest on account of the difficulty of seeing the wires when the field of view cannot be sufficiently enlightened.

V. The angle of position of the stars I have only given with regard to the parallel of declination, to be reduced to that with the ecliptic as occasion may require. The measures always suppose the large star to be the standard, and the situation of the small one is described accordingly. Thus in fig. 12 [p. 57] AB represents the apparent diurnal motion of a star in the direction of the parallel of declination AB; and the small star is said to be south preceding at *mn*, north preceding at *op*, south following at *qr*, and north following at *st*. The measure of these angles, I believe, may be relied upon to 2° or at most 3°, except when mentioned inaccurate, where an error amounting to 5° may possibly take place. In mere estimations of the angle, without any wires at all, an error may amount to at least 10°, when the stars are near each other.

VI. The dates when I first perceived the stars to be double, treble, &c. are marked in the margin of each star.

To shorten the work as much as possible, I have put L. for the large star; S. for the small star; w. for white; r. for red; d. for dusky; n. for north; s. for south; and have likewise occasionally used other abbreviations that will be easily understood.

It may be seen, that this catalogue is yet in a very imperfect state, many of the stars not having even the principal elements of distance and position determined with any degree of accuracy; but having already mentioned the reason why I give it imperfect as it is, I can only add that my endeavours will not be wanting soon to remove those defects. However, since this can only be a work of some time, we may hope, in the mean while, that many lovers of the science will turn their thoughts upon the same subject.

CATALOGUE OF DOUBLE STARS.§

FIRST CLASS.

1. ϵ Bootis. FLAMST. 36. Ad dextrum femur in perizomate.

Sept. 9, 1779. Double. Very unequal. L. reddish; S. blue, or rather a faint lilac. A very beautiful object. The vacancy, or black division between them, with 227 is $\frac{3}{4}$ diameter of S.; with 460, $1\frac{1}{4}$ diameter of L.; with 932, near 2 diameters of L.; with 1159, still farther; with 2010 (extremely distinct) $2\frac{3}{4}$ diameters of L. These quantities are a mean of two years observation. Position 31° 34′ n. preceding.

2. ξ Ursæ majoris. FL. 53. In dextro posteriore pede.

May 2, 1780. Double. A little unequal. Both w. and very bright. The interval with 222 is $\frac{2}{3}$ diameter of L.; with 227, 1 diameter of L.; with 278, near $1\frac{1}{2}$ diameter of L. Position 53° 47′ s. following.

3. σ Coronæ borealis, FL. 17.

Aug. 7, 1780. Treble. The two nearest pretty unequal; the third very faint with powers lower than 460. The two nearest both w.; the third d. Interval of the two nearest with 227, full $1\frac{1}{4}$ diameter of L.; with 460, 2 diameters of L. Position 77° 32′ n. preceding. Distance of the third from L. 24″ by exact estimation. Position 25° n. following by estimation.

4. In constellatione Draconis, FL. 17.

Aug. 8, 1780. Double. It is the star to which a line drawn from ν through μ points, at nearly the same distance from μ as μ from ν. Considerably unequal. L. w.; S. w. inclining to r. With 222, 1 diameter of L.; with 278, $1\frac{1}{2}$ diameter of L. Position 24° 0′ s. following. There is a third star, at some distance, preceding.

5. σ Cassiopeæ, FL. 8. In dextro cubito.

Aug. 31, 1780. Double. It is the star at the vertex of a telescopic isosceles triangle turned to the south. Very unequal. L. w. a little inclining to r.; S. d. With 222, near 1 diameter of L.; with 460, $1\frac{1}{2}$ diameter of L. Position 60° 28′ n. preceding.

§ [Compare the Synopsis of all Sir W. Herschel's measures of Double Stars at the end of Vol. II. of the present edition. For explanation of * and † after the number of an object, see p. 90.—ED.]

I. 6. Quæ infra oculum Lyncis, FL. 12.

Aug. 7, 1780. A curious treble star. Two nearest pretty unequal. L. w.; S. w. inclining to rose colour. With *227*, about ½ diameter; with 460, full ¾ diameter of S. Position 88° 37′ s. preceding. The first and third considerably unequal; second and third pretty unequal. The third pale r. Distance from the first 9″ 23‴; too difficult to be extremely exact. Position with regard to the first 32° 33′ n. preceding.

7. *b* Draconis, FL. 39. Trium in recta, in prima inflectione colli, borea.

Oct. 3, 1780. A minute double star. Extremely unequal, the small star being a fine lucid point. L. w.; S. inclining to r. With *227*, ¾ diameter of L.; with 460, full 1½ diameter of L.; with 932 (extremely fine) full 2 diameters of L. Position 77° 8′ n. following. A third star at some distance; dusky r. Position 63° 55′ n. following.

8. ε Draconis, FL. 63. In quadrilatero inflexionis primæ.

Oct. 3, 1780. A very minute double star. Excessively unequal; the small star can only be seen when the air is perfectly clear. L. w.; S. d. With *227*, less than 1 diameter of L.; with *278*, not a diameter of L. Position 63° 14′ n. preceding. A pretty large third star at about 3 or 4′. Position of this third star with ε 88° 16′ n. following.

9. In cauda Lyncis media, FL. 38.

Nov. 24, 1780. Double. Very unequal. L. w.; S. inclining to r. With *227*, extremely close; with 460, at least ½ diameter of S. A very fine object. Position 25° 51′ s. preceding. A proper motion is suspected in one of the stars.

10. In sinistro anteriore pede Monocerotis, FL. 11.

Dec. 5, 1779. A curious treble star; may appear double at first sight; but with some attention we see that one of them again is double. The first, or single star, is the largest; the other two are both smaller, and almost equal, but the preceding of them is rather larger than the following. They are all w. The two nearest with *227*, 1 diameter of the preceding, or nearly 1¼ of the following; with 460, 1¼ diameter of the preceding. Position of the two nearest 11° 32′ s. following. For an account of the single star, see the second class. As perfect as I have seen this treble star with 460, it is one of the most beautiful sights in the heavens; but requires a very fine evening.

11. In constellatione Cancri, FL. 17.

Mar. 13, 1781. Double. Considerably unequal. Both pale r. With *227*, 1 full diameter of L.; with 460, about 1¾ diameter of L. Position 85° 10′ n. preceding.

12. *d* Serpentis, FL. 59. In Cauda.

July 17, 1781. Double. Very unequal. L. reddish w.; S. fine blue. With *227*, 1 full diameter of L.; with *278*, 1⅓ diameter of L. Position 44° 33′ n. preceding.

I. 13. In constellatione Aquilæ, near FL. 37.

July 25, 1781. A curious treble star. It is the last star of a telescopic trifolium n. following *k*, similar to that in the hand of Aquarius. The two nearest very unequal; the third star excessively small, and not visible with 227. The two nearest with 460, no more than ½ diameter of L.; the farthest about 7 or 8″.

14. In constellatione Aquilæ, FL. 23.

July 30, 1781. Double. In HARRIS's maps it is the star in the elbow of Antinous. Excessively unequal; the small star is but just visible with 227; but with 460 it is pretty strong. L. pale r.; S. d. With 227, 1 full diameter of L.; with 460, 1½ diameter of L. Position 72° 0′ s. following.

15. *i* Bootis, FL. 44.

Aug. 17, 1781. Double. In HARRIS's maps it is marked *i*, but has no letter in FL. Atlas. Considerably unequal. Both w. With 227 they seem almost to touch, or at most ¼ diameter of S. asunder; with 460, ½ or ¾ diameter of S. This is a fine object to try a telescope, and a miniature of α Geminorum. Position 29° 54′ n. following. §

16. η Coronæ borealis, FL. 2.

Sept. 9, 1781. Double. A little unequal. They are whitish stars. They seem in contact with 227, and though I can see them with this power, I should certainly not have discovered them with it; with 460, less than ¼ diameter; with 932, fairly separated, and the interval a little larger than with 460. I saw them also with 2010, but they are so close that this power is too much for them, at least when the altitude of the stars is not very considerable; with 460 they are as fine a miniature of *i* Bootis as that is of α Geminorum. Position 59° 19′ n. following.

17. In constellatione Bootis, near FL. 51.

July 30, 1780. Double. It is a star near μ not marked in FLAMSTEED's Catalogue. Considerably unequal. Both dusky w. inclined to r. The interval with 460 is ¾ diameter of S. The position of the small star is turned towards μ a little following the line which joins L. to μ Bootis. See μ Bootis in the sixth class.

18. In constellatione Coronæ borealis.

Sept. 10, 1781. Double. It is the smallest of two telescopic stars between θ and δ, not contained in FL. Cat. Equal. Both d. With 460, about 1¾ diameters. Position 21° 0′ n. preceding.

19. *h* Draconis, FL. 20.

Sept. 10, 1781. One of the most minute of all the double stars I have hitherto found. It is the small telescopic star near the preceding *h* Draconis. Considerably unequal. Both dusky w. inclining to r. With 460, they seem in

§ [I. 15. A measure of date 1802·246 gave 27° 1′ sp (=242°·98), which agrees with a diagram. But under date March 12, 1803, H. corrected this to nf (*i.e.* 62°·98). Compare Lewis, *Mem. R.A.S.*, lvi. p. 406.—ED.]

contact; I have however had a very good view of a small dark division between them. Position (by exact estimation) 25 or 30° s. preceding. They are too minute for any micrometer I have. It is in vain to look for them if every circumstance is not favourable. The observer as well as the instrument must have been long enough out in the open air to acquire the same temperature. In very cold weather, an hour at least will be required; but in a moderate temperature, half an hour will be sufficient.

I. 20. In dextro humero Orionis, FL. 52.

Oct. 1, 1781. Double. A little unequal. Both w. a little inclining to pale r. With 227, ¼ diameter; with 460, ½ diameter. Position 69° 41′ s. preceding.

21. *c* Trianguli, near FL. 12. and 13.

Oct. 8, 1781. Double. It is the most north of a small telescopic trapezium of unequal stars. Extremely unequal. With 460, ¾ diameter of L. Position (by estimation) 55 or 60° n. preceding.

22. *n* Orionis, FL. 33. Duarum præcedentium 13^{am} (ω) antecedens.

Oct. 22, 1781. Double. Considerably unequal. L. w.; S. w.; inclining to blue. With 227, they seem almost in contact; with 460, ½ diameter of S. Position 61° 23′ n. following. A very pleasing object and easily seen.

23. In posterioribus femoribus Canis minoris.

Nov. 21, 1781. A most minute double star. It is the small telescopic star following Procyon. A little unequal. Both w. With 278, ⅛ of a diameter of S.; with 460, near ¼ of a diameter of S. They are closer than η Coronæ, because their diameters, by which they are estimated, are smaller. Position 27° 21′ s. following. To see this very minute double star well, Procyon should be near its meridian altitude. There is a small telescopic star preceding the double star. Distance 1′ 59″ 39‴ from center to center.

24. ζ Cancri, FL. 16.

Nov. 21, 1781. A most minute treble star. It will at first sight appear as only a double star, but with proper attention, and under favourable circumstances, the preceding of them will be found to consist of two stars, which are considerably unequal. The largest of these is larger than the single star; and the least of the two is less than the single star. The first and second (in the order of magnitude) pretty unequal. The second and third pretty unequal. The two nearest both pale r. or r. With 278, but just separated; with 460, ¼ diameter of S. Position 86° 32′ n. following. For measures relating to the third or single star see ζ Cancri in the third class of double stars.

SECOND CLASS OF DOUBLE STARS.

II. 1. † α Geminorum, FL. 66. In capite præcedentis Π[i].

April 8, 1778. Double. A little unequal. Both w. The vacancy between the two stars, with a power of 146, is 1 diameter of S.; with 222, a little more than 1 diameter of L.; with 227, 1½ diameter of S.; with 460, near 2 diameters of L.; (see [p. 50] fig. 6) with 754, 2 diameters of L.; with 932, full 2 diameters of L.; with 1536 (very fine and distinct) 3 diameters of L.; with 3168, the interval extremely large, and still pretty distinct. Distance by the micrometer 5″·156. Position 32° 47′ n. preceding. These are all a mean of the last two years observations, except the first with 146.

2. † α Herculis, FL. 64. In capite.

Aug. 29, 1779. A beautiful double star. Very unequal. L. r.; S. blue inclining to green; the colours with every power the same. The interval with 222, 1¾ diameter of L.; with 227, above 2 diameters of L.; with 932, above 3 diameters of L. Distance 4″·966. All a mean of two years observations. A single measure with my last new micrometer, from center to center, 4″ 34‴. Position 30° 35′ s. following.

3. * ρ Herculis, FL. 75. Trium in sinistro femore, tertia.

Aug. 29, 1779. Double. Pretty unequal. Both w. With 227, 1¼ diameter of L.; with 460, 2 diameters of S. Distance 2″·969. Position 30° 21′ n. preceding. The distance a mean of two years observations.

4. * p Serpentarii, FL. 70. Tres has sequitur, quasi supra mediam.

Aug. 29, 1779. Double. Considerably unequal. L. w.; S. inclining to r. With 227, 1⅔ diameter of L.; with 460, much above 2 diameters of L. Position 9° 14′ s. following. § Mean of two years observations.

5. et 6. * ε Lyræ, FL. 4. and 5.

Aug. 29, 1779. A very curious double-double star. At first sight it appears double at some considerable distance, and by attending a little we see that each of the stars is a very delicate double star. The first set consists of stars that are considerably unequal. The stars of the second set are equal, or the preceding of them rather larger than the following. The colour of the stars in the first set L. very w.; S. a little inclining to r. In the second set both w. The interval between the stars of the unequal set, with a power of 227, is full 1 diameter of L.; with 460, near 1½ diameter of L.; with 932, full 1½ diameter; with 2010, 2⅓ diameters. The interval between the equal set with a power of 227 is almost 1½ diameter of either; with 460, full 1¾ diameter; with 932, 2 diameters; with 2010, 2½ diameters.

§ [II. 4. That the companion was sf is confirmed by a diagram in the original Journal.—ED.]

These estimations are a mean of two years observations. Position of the unequal set 56° 5′ n. following. Position of the equal set 83° 28′ s. following.§

II. 7. * ζ Aquarii, FL. 55. Trium in manu dextra præcedens.

Sept. 12, 1779. Double. Equal, or the preceding rather the largest. Both w. With 227, 1¼ diameter; with 449, 1⅓ diameter; with 460, 2 diameters; with 910, near 2 diameters; with 932, 2½ diameters; with 2010, pretty distinct, but too tremulous to estimate. With my 20 feet reflector, power 600, full 2 diameters, very distinct. Position, 71° 39′ n. following. Distance 4″·56, mean of two years observation.

8. ζ Coronæ borealis, FL. 7.

Oct. 1, 1779. Double. Considerably unequal. L. fine w. S. w. inclining to r. With 222, almost 3 diameters of L. Distance 5″·468, mean of two years observations. Position 25° 51′ n. preceding.

9. λ Orionis, FL. 39. In capite nebulosa.

Oct. 7, 1779. Quadruple, or rather a double star and two more at a small distance. The double star considerably unequal. L. w.; S. pale rose colour. With 222, 1½ diameter of L.; with 449, above two diameters of L. Distance 5″·833, a mean of all the measures. Position 45° 14′ n. following. As every one of the four stars is perfectly distinct, it is evident, the whole appeared nebulous to FLAMSTEED for no other reason than because his telescope had not sufficient power to distinguish them.

10. and 11. σ Orionis, FL. 48. Ultimam cinguli præcedit ad austrum.

Oct. 7, 1779. A double-treble star, or two sets of treble stars, almost similarly situated. Preceding set. The two nearest equal; the third larger and, compared with either of the former two, pretty unequal. The two nearest with 222, about 2 diameters. Position of the following star of the two nearest with the third 66° 35′ s. preceding. Position of the two nearest, by exact estimation, 2 or 3° n. following or s. preceding the following set. The two nearest very unequal. The largest of the two and the farthest considerably unequal. L. w.; S. blueish. The two nearest with 222, about 2¼ diameters of L.; the two farthest 43″ 12‴. Position of the two nearest 5° 5′ n. following. Position of the two farthest 29° 4′ n. following. A pretty object with 227.

12. α Piscium, FL. ultima. In nodo duorum linorum.

Oct. 19, 1779. Double. Considerably unequal. Both w. With 222, not quite 2 diameters of L.; with 460, about 3 diameters of L. Distance 5″·123 mean measure. Position 67° 23′ n. preceding.

§ [II. 5–6. *Phil. Trans.* has 72° 57′; corrected in MS. "Nov. 29, 1782.: The 20-feet shows me two small stars between ε and *e* Lyræ (Mayer), which I have never before taken notice of in the 7-feet."—ED.]

II. 13. μ Draconis, FL. 21. In lingua.

Oct. 19, 1779. Double. Equal. Both w. With *227*, 1½ diameter; with 460, 2½ diameters. Distance 4″·354 mean measure. Position 37° 38′ s. preceding or n. following.

14. ω Aurigæ, FL. 4.

Oct. 30, 1779. Double. Very unequal. L. w.; S. r. With *227*, almost 2 diameters of L.; with 460, full 3 diameters of L. Position 82° 37′ n. preceding.

15. ψ Cygni, FL. 24. In ala dextra.

Nov. 2, 1779. Double. Extremely unequal; the small star a mere point. L. w.; S. r. With *227*, near 1¼ diameter of L.; with *278*, near 1½ diameter of L.; with 460, 2 diameters of L. Position 89° 32′ s. preceding.§

16. ξ Cephei, FL. 17. In pectore.

Nov. 5, 1779. A fine double star. Considerably unequal. L. w. inclining to r.; S. dusky grey. With *222*, nearly 2 diameters of L. Single measure 5″. Position 20° 18′ n. preceding.

17. * In sinistro anteriore pede Monocerotis, FL. 11.

Dec. 5, 1779. Double. With *222*, about 1½ diameter. Position (taken Oct. 20, 1781) with the farthest of the other two stars 31° 38′ s. following. See the tenth star in the first class.

18. ξ Bootis, FL. 37.

April 9, 1780. Double. Very unequal. L. pale r. or nearly r. S. garnet, or deeper r. than the other. With *222*, 1½ diameter of L.; with 460, full 3 diameters of L. Distance 3″ 23‴ single measure. Position 65° 53′ n. following.

19. ρ Ophiuchi, FL. 5.

May 2, 1780. Double. It is a star in the body of Scorpio, and the double star is at the angular point of the three telescopic *g's* making a rectangle. Pretty unequal. Both w. With *227*, 1½ diameter of L. Position 82° 10′ n. preceding.§§

20. and 21. ξ Libræ, FL. ultima.

May 23, 1710. Double-double. The first set very unequal. L. fine w. With *227*, nearly 2 diameters of L.‡ By the micrometer 6″ 23‴, but too large a measure. Position 1° 23′ n. following. The other set both small and obscure. With *227*, perhaps 5 or 6 of their diameters asunder.

§ [II. 15. *Phil. Trans.* has np, but a diagram shows the companion sp.—ED.]

§§ [II. 19. *Phil. Trans.* has sp, but two distinct diagrams of this date place the small star north of the parallel. Herschel calls the star *g*, following Flamsteed.—ED.]

‡ In a future collection this set will be found as a treble star of the first class, the large white star, with a power of 460 and 932, appearing to be two stars.

II. 22. ε Persei, FL. 45. In sinistro genu.

Aug. 2, 1780. Double. Extremely unequal. L. w.; S. d. With 222, 2½ diameters of L. Position 81° 28′ n. following, a little inaccurate. A third star near at about 1½ or 1¾ min.

23. In constellatione Serpentarii, near FL. 11.

Aug. 7, 1780. Double. It is the smallest and preceding of two in the finder. Pretty unequal. L. pale r.; S. dusky r. With 222, about 1¼ diameter of L.; with 278, about 1½ diameter of L.; with 460, above 2 diameters of L. Position 46° 24′ n. preceding. A little inaccurate.

24. In constellatione Aquarii, FL. 107. In sequenti flexu 4ª.

Aug. 23, 1780. Double. In HARRIS's maps it is marked *i*. Unequal. With 227, 2 diameters; with 460, about 3 diameters.

25. *k* Cygni, FL. 52.

Sept. 8, 1780. Double. Extremely unequal. L. w. inclining to r.; S. d. and extremely faint; with 227, 2½ diameters of L.; with 460, about 4 diameters of L. or more. Position 31° 3′ n. following.§

26. In constellatione Orionis, near FL. 42. In longo ensis.

Oct. 23, 1780. Double. It is the most north of three telescopic stars in a line at the end of a cluster near *c*. Extremely unequal. L. w.; S. d. With 278, 1¾ diameter of L. Position 26° 5′ n. following.

27. δ Geminorum, FL. 55. In inguine sinistro sequentis IIⁱ.

Mar. 13, 1781. Double. Extremely unequal. L. w. inclining to r.; S. r. With 227, about 2½ full diameters of L.; with 460, 4 or 5 diameters. Position 85° 51′ s. preceding.

28. In constellatione Aquilæ, near FL. 54.

July 23, 1781. Double. It is a star following *o*. Excessively unequal. The small star is not visible with 227, nor with 278. It is visible with 460; but not without attention. Distance with 460, about 4 or 5 diameters of L. Position, by very exact estimation, 36° 28′ n. preceding.

29. In constellatione Aquilæ, near FL. 63. In medio capite.

July 31, 1781. Double. It is the star at the vertex of a telescopic isosceles triangle near τ. Extremely unequal. Both r. With 460, 2 diameters of L. Position 75° 48′ n. preceding.

30. ζ Sagittæ, FL. 8. Trium in arundine sequens.

Aug. 23, 1781. Double. Extremely unequal. The small star brighter with 460 than with 227 or with 278; with 460, between 4 or 5 diameters of L.; with 278, 2½ diameters of L. Distance 8″ 50‴ very inaccurate.‡ Position 34° 10′ n. preceding.

§ [II. 25. *Phil. Trans.* has 28° 17′; corrected by C. H. from original observation.—ED.]

‡ [II. 30. According to original observation of Nov. 23, 1781. Not 5″ 27‴ as in *Phil. Trans.*—ED.]

II. 31. In constellatione Draconis.

Sept. 6, 1781. Double. A little unequal. Both w. With 460, near 3 diameters. Distance 5″ 7‴.

32. In constellatione Sagittæ, near FL. 4.

Aug. 23, 1781. Double. It is the star north following ε. L. pale r.; S. d. Distance 27″ 30‴ inaccurate.§

33. β Orionis, FL. 19. In sinistro pede splendida.

Oct. 1, 1781. Double. Extremely unequal. L. w.; S. inclining to r. With 227, 2¼ or 2½ diameters of Rigel. With 460, more than 3 diameters of L. Distance 6″ 27‴. Position 68° 12′ s. preceding. The small star not wanting apparent magnitude is better to be seen with my power of 227 than with 460.

34. ι Trianguli, FL. 6.

Oct. 8, 1781. Double. It is marked *b* in the small triangle of HARRIS's maps. Very unequal. L. pale r. or reddish w.; S. blueish r. With 227, full 1¼ diameter of L.; with 460, full 1½ diameter of L. Position 4° 23′ n. following. A pretty object, somewhat resembling α Herculis, but smaller and not so bright.

35. In constellatione Trianguli, near FL. 6.

Oct. 8, 1781. Double. It is the star following ι. Equal. Both dusky w. With 460, about 2½ diameters.

36. In constellatione Eridani, FL. 32.

Oct. 22, 1781. Double. Considerably unequal. L. reddish w.; S. blue. Distance 4″ 19‴. Position 73° 23′ n. preceding.

37. In capite Monocerotis.

Oct. 22, 1781. Double. It is one of a cluster of six telescopic stars, arranged in pairs.

38. In constellatione Bootis.

Dec. 24, 1781. Double. It is the most north and largest of three in a line, s. following FL. 15. Considerably unequal, L. w.; S. inclining to r. Distance 5″ 10‴. Position 83° 5′ s. preceding.

THIRD CLASS OF DOUBLE STARS.

III. 1. † θ Orionis, FL. 41. Trium contiguarum in longo ensis media.

Nov. 11, 1776. Quadruple. It is the small telescopic Trapezium in the Nebula. Considerably unequal. The most southern star of the following side of the Trapezium is the largest; the star in the opposite corner is the smallest; the remaining two are nearly equal. L. pale r.; the star preceding L. inclined to garnet; following L. inclined to garnet; opposite to L. d. With 460, the

§ [II. 32. Not 5″ 3‴ as in *Phil. Trans.*: error of reduction.—ED.]

stars are all full, round, and well-defined. The two stars in the preceding side distance 9″·06; in the southern side, 12″·81; in the following side 14″·27; in the northern side, 20″·52.

III. 2. ζ Ursæ majoris, FL. 79. Trium in cauda media.

Aug. 17, 1779. Double. Considerably unequal. L. w.; S. w.; inclining to pale rose colour. Distance 14″·5 by two years observation, not a mean but that which I suppose nearest the truth. Position 56° 46′ s. following.

3. η Cassiopeæ, FL. 24. In cingulo.

Aug. 17, 1779. Double. Very unequal. L. fine w.; S. fine garnet, both beautiful colours. Distance 11″·275 mean measure. Position 27° 56′ n. following.

4. In extremitate pedis Cassiopeæ, FL. 55. ι Ptolemæi.

Aug. 17, 1779. Double. Extremely unequal. L. w.; S. blueish r. Distance 7″·5 single measure. Position 10° 37′ s. following.§

5. * γ Andromedæ, FL. 57. Supra pedem sinistrum.

Aug. 25, 1779. Double. Very unequal. L. reddish w.; S. fine light sky-blue, inclining to green. Distance 9″·254 a mean of two years observation. Position 19° 37′ n. following. A most beautiful object.

6. β Cephei, FL. 8. In cingulo ad dextrum latus.

Aug. 31, 1779. Double. Very unequal. L. blueish w.; S. garnet. Distance 13″·125. Position 15° 28′ s. preceding.

7. * β Scorpii, FL. 8. Trium in fronte, lucidarum, borea.

Sept. 19, 1779. Double. Very unequal. L. whitish r.; S. r. Distance 14″·375. Position 64° 51′ n. following.

8. * π Bootis, FL. 29.

Sept. 20, 1779. Double. Pretty unequal. L. w.; S. w. inclining to r. Distance 6″·875. Position 6° 28′ s. following.

9. † γ Arietis, FL. 5. Quæ in cornu duarum præcedens.

Sept. 27, 1779. Double. Equal, or if any difference the following is the largest. Distance 10″·172, a mean of two years observation. L. w. inclining a little to r.; S. w. Position 86° 5′ n. preceding.

10. * γ Delphini, FL. 12. Borea sequentis lateris, quadrilateri.

Sept. 27, 1779. Double. Nearly equal, the following a little larger. Both w. Distance 11″·822, being a mean of the measures taken in Sept. Oct. Nov. and Dec. 1779. As I suspect a motion in one of these stars, I thought it best not to join other observations in that measure. Position 4° 9′ n. preceding.

§ In a future collection this will be found as a treble star of the first class; the large star having a small one preceding, easily seen with 460 and 932.

III. 11. κ Bootis, FL. 17. Trium in sinistro manu præcedens.

Sept. 27, 1779. Double. Very unequal. L. w.; S. d. Distance 12″·503, a mean of the observations in 1779, 80, 81. Position about 30° s. preceding.

12. ι Orionis, FL. 44. Trium contiguarum in ense austrina.

Oct. 7, 1779. Treble. It is the following or largest of the two ι's. One is L.; the other two are extremely small. L. w.; the other two both dusky r. Distance of the nearest 12″·5. Distance of the farthest 48″ 31‴. Position of the nearest 43° 51′ s. following. Position of the farthest 11° 19′ s. following.

13. and 14. ι Orionis, FL. 44. Trium contiguarum in ense austrina.

Oct. 7, 1779. Double-treble. It is the preceding or smallest of the two ι's. The preceding set (forming a triangle) consists of three equal stars. All dusky r. Distance of the two nearest, with 227, about 3 diameters. The following set (forming an arch) consists of three stars of different sizes. The middle star is the largest; that to the south is also pretty large; and the third is very small. L. w.; l. w.; S. pale r. Distance 36″·25.

15. * μ Cygni, FL. 78.

Oct. 19, 1779. Double. Considerably unequal. L. w.; S. blueish. Distance 6″·927 mean measure. Position 20° 15′ s. following.

16. * In constellatione Delphini.

Nov. 15, 1779. Double. It is the star south preceding ε. A little unequal. Both w. Distance 12″·5. Position 9° 42′ s. preceding.

17. In extremitate caudæ Lacertæ, near FL. 1.

Nov. 20, 1779. Double. Considerably unequal. L. w.; S. d. inclining to r. Distance 13″ 43‴ inaccurate. Position 76° 16′ s. preceding.

18. † γ Virginis, FL. 29. De quatuor in ala sinistra, sequens.

Jan. 21, 1780. Double. Equal. Both w. Distance 7″·333 mean measure. Position 40° 44′ s. following.

19. † ζ Cancri, FL. 16.

April 5, 1780. Double. Considerably unequal. L. pale r.; S. pale r. Distance 8″·046 mean measure. Position 88° 16′ s. preceding. See the 24th in the first class.

20. In constellatione Bootis.

June 25, 1780. Double. Draw a line through π and ζ to the small star under the right foot, and erecting a perpendicular towards the left foot of equal length, the end of it will mark out this double star. Pretty unequal. Both r. Distance 7″ 36‴ full measure. Position 59° 32′ n. preceding.

21. ε Equulei, FL. 1.

Aug. 2, 1780. Double. Considerably unequal. L. w.; S. much inclining to r. Distance 9″·375 mean measure. Position 5° 39′ n. following. A third small star follows at some distance.

III. 22. Quæ infra oculum Lyncis, FL. 12.

Aug. 7, 1780. Double. With 222, about 3 diameters of L. Considerably unequal. L. w.; S. pale r. Distance 9″ 23‴, not extremely accurate. Position 32° 33′ n. preceding. See the sixth star in the first class.

23. In constellatione Cassiopeæ, FL. 34.

Aug. 8, 1780. Double. It is one of two telescopic stars, and is marked ϕ in HARRIS's maps. Extremely unequal. L. pale r.; S. d. Distance about 12″ or more.

24. θ Sagittæ, FL. 17.

Aug. 19, 1780. Treble. The two nearest extremely unequal. L. pale r.; S. d. Third star pale r. Distance of the two nearest 11″ 8‴. Distance of the two largest 57″ 49‴.

25. * In constellatione Serpentarii, FL. 39.

Aug. 29, 1780. Double. It is the most south and largest of two in the finder. Very unequal. L. w.; S. inclining to blue. Distance 10″ 2‴, a little inaccurate. Position 87° 14′ n. preceding.

26. * In constellatione Cerberi 1. HEVELII 1ª. FL. Herculis 95.

Sept. 8, 1780. Double. It is the star in the leaf nearest to Hercules's face and hand. Equal. Preceding w. Following blueish w. Distance 6″ 6‴. Position 4° 9′ s. preceding or n. following.

27. In constellatione Navis, near FL. 3.

Feb. 15, 1781. Double. It is a star between η Canis majoris and ξ Navis. Equal. Distance about 15″.

28. In constellatione Navis, near FL. 9.

Feb. 15, 1781. Double. It is one of two telescopic stars under Monoceros. Distance about 8″.

29. In naribus Monocerotis, FL. 8. ::

Feb. 15, 1781. Double. Distance about 12″.

30. * In constellatione Leonis, FL. 54. Duarum supra dorsum sequens.

Feb. 21, 1781. Double. Considerably unequal. L. brilliant w.; S. ash-colour, or greyish w. Distance 7″ 6‴ mean measure. Position 9° 14′ s. following.

31. In constellatione Herculis.§

May 20, 1781. Double. Over ι ::. Equal. Both very small. Distance about 10″.

32. In constellatione Aquilæ, FL. 11.

July 25, 1781. Double. It is the most south of two near ϵ and ζ. Excessively unequal. S. hardly visible with 227, but pretty strong with 460. Distance about 7″.

§ [III. 31. "Over ι Herculis towards γ. Not sure that the letter ι is right." Cannot be identified. —ED.]

III. 33. In constellatione Aquilæ, FL. 5.

July 30, 1781. Double. It is a star preceding the two small stars north of *k* and *l*. Unequal. L. w.; S. blueish w. Distance 11″ 35‴ inaccurate, but not much.

34. In constellatione Aquarii, FL. 94.

Aug. 20, 1781. Double. Between ψ and ω towards δ. Very unequal. Distance 13″ 45‴. L. pale r.; S. d.

35. In constellatione Serpentarii, FL. 54.

Aug. 21, 1781. Double. It is the preceding of two stars in the head. Excessively unequal. L. reddish w.; S. d. Distance about 8″.

36. In constellatione Persei.

Sept. 14, 1781. Double. A little south of γ. Considerably unequal. L. w.; S. w. inclining to r. Distance 11″ 53‴, rather full measure.

37. and 38. In constellatione Persei, near FL. 38.§

Sept. 24, 1781. Double-double. South preceding the first *o*. The equal set with 227, about 4 or 5 diameters. The unequal set about 5 or 6 diameters. Near this last set is also a third star forming an obtuse angle with the stars of this set. Distance about 10″.

39. *o* Persei, FL. 40.

Sept. 24, 1781. Double. It is the second or most northern *o*. Extremely unequal. L. w.; S. d. With 227, S. is hardly visible; with 460, it appears at first sight. Distance 15″ 12‴, inaccurate on account of the obscurity of S.

40. In constellatione Herculis, near FL. 87.

Oct. 10, 1781. Double. Of three stars, forming an obtuse angle, whereof FL. 87. (a star south of μ) is at the angular point, that towards Ramus Cereb. Extremely unequal. L. w.; S. d. Distance 10″ 20‴. Position 47° 19′ s. following.‡

41. * *i* Herculis, FL. 43.

Oct. 10, 1781. Double. Equal. Preceding star w. A little inclined to r. Following w. Distance 11″ 43‴. Position 88° 23′ n. following.

42. In constellatione Trianguli.

Oct. 12, 1781. Double. It is a star north following δ. Unequal. L. reddish. S. blueish. Both d. Distance about 6 or 7″.

43. In sinistro anteriore pede Monocerotis.

Oct. 20, 1781. Double. It is the most south of two telescopic stars preceding the treble star. Extremely unequal. L. w.; S. d. Position 23° 39′ n. preceding.

§ Mr. BRYANT of Bath first observed these stars.

‡ [III. 40. *Phil. Trans.* has 19° 37′ sf. Corrected by MS.—ED.]

III. 44. In ore Monocerotis.§

Oct. 20, 1781. Double. Considerably unequal. L. w.; S. r. Distance 12″ 30‴. Position 60° 14′ n. following.

45. In constellatione Tauri, near FL. 10.

Oct. 22, 1781. Double. It is near the star sub pede et scapula dextra. Extremely unequal. L. pale r.; S. d. Position 35° 33′ s. preceding.‡

46. In constellatione Monocerotis.

Oct. 22, 1781. Double. It is the star following the tip of the ear.

FOURTH CLASS OF DOUBLE STARS.

IV. 1. α Ursæ minoris, FL. 1. Stella Polaris.

Aug. 17, 1779. Double. Extremely unequal. L. w.; S. r. Distance 17″ 15‴. Position 66° 42′ s. preceding.

2. * η Lyræ, FL. 20. Duarum contiguarum ad ortum a testa, borea.

Aug. 29, 1779. Double. Considerably unequal. L. w.; S. r. Distance 25″ 42‴. Position 31° 51′ s. preceding.¶ Three other stars in view.

3. FL. 64. Sagittarii.

Aug. 26, 1780. Double. It is the preceding star of two. Extremely unequal. Distance about 25″.

4. η Persei, I. HEVELII 9. In dextro brachio.

Sept. 20, 1779. Double. Very unequal. L. r.; S. blue. Distance 26″, very inaccurate. Position 20° 5′ n. preceding.

5. In constellatione Arietis, FL. 33. Quatuor inform. sup. dors. præc.

Sept. 27, 1779. Double. It is the first in the head of the fly. L. w.; S. d. Considerably unequal. Distance 25″ 32‴ inaccurate. Position 87° 14′.§§

6. † θ Serpentis, FL. 63. In extremitate Caudæ.

Oct. 17, 1779. Double. Equal. Both w. Distance 19″·375.

7. ψ Draconis, FL. 31. Prima ad ψ.

Oct. 19, 1779. Double. Pretty unequal. L. w.; s. pale r. Distance 28″ 14‴.

§ [III. 44 is = III. 29.—ED.]

‡ [III. 45. " Posit. 35° 33′ sp, but this only exact esti. because the star was invisible with this power. With 227 I could but just see it, with 460 it is a pretty strong star. The angle is too small by 6 or 8 degrees, I believe."—ED.]

¶ [IV. 2. This angle cannot refer to η Lyræ, but possibly to the neighbouring V. 42, as suggested by Sadler.—ED.]

§§ [IV. 5. No quadrant stated and no diagram made. The position angle was probably 87° 14′ nf. No orbital motion.—ED.]

IV. 8. * ζ Piscium, FL. 86. Trium in lino lucidarum sequens.

Oct. 19, 1779. Double. Pretty unequal. L. w.; S. w. inclining to blue. Distance 22″·187, not very accurate. Position 22° 37′ n. following.

9. * Prima ad ψ Piscium, FL. 74. Trium in pinna costarum præcedens.

Oct. 30, 1779. Double. Distance 27″·5. Position about 80° s. following. An obscure star also within 1½ minute.

10. χ Tauri, FL. 59. Australis sequentis lateris quadrilateri, in cervice.

Oct. 30, 1779. Double. Distance 18″·75, very inaccurate.

11. χ Cygni, FL. 17.

Nov. 20, 1779. Double. Very unequal. L. w.; S. dusky r. Distance 24″ 52‴.

12. * ψ Aquarii, FL. 91.

Nov. 26, 1779. Double. It is the first of three ψ's. Unequal. Distance 23″ 5‴, pretty accurate.

13. In constellatione Leonis, FL. 83.

April 6, 1780. Double. It is a small star north preceding τ. A little unequal. Both inclining to r. Distance 29″ 5‴. Position 54° 55′ s. following.

14. In constellatione Aquilæ, FL. 57.

Aug. 2, 1780. Double. It is the preceding of two, near the south end of Antinous's bow. A little unequal. L. w.; S. w. inclining to r. Distance 29″ 28‴, pretty accurate. Position 81° 55′ s. preceding.

15. In dextra aure Camelopardali. I. HEVELII ultima.

Aug. 2, 1780. Double. A little unequal. L. reddish w.; S. reddish w. Distance 20″ 5‴.

16. In constellatione Cassiopeæ, FL. 31.

Aug. 2, 1780. Double. It is marked with the letter A in HARRIS's maps. Distance about 20″ or more.

17. * Cor Caroli, FL. 12. Canum Venaticorum.

Aug. 7, 1780. Double. Very unequal. L. w.; S. inclining to r. Distance 20″ 0‴, inaccurate. Position 41° 47′ s. preceding.

18. * In constellatione Cygni, FL. 61.

Sept. 20, 1780. Double. It is a star preceding τ. Pretty unequal. L. pale r.; S. r.; or L. r.; S. garnet. Distance 16″ 7‴. Position 36° 28′ n. following.

19. In constellatione Aurigæ, FL. 14.

Sept. 24, 1780. Double. It is the preceding star of a cluster of stars that precede φ and χ. Very unequal. L. reddish w.; S. d. Distance 16″ 8‴, a little inaccurate. Position 37° 38′ s. preceding.

IV. 20. *o* Draconis, FL. 47.

Oct. 3, 1780. Double. Very unequal. L. pale r.; S. dusky r. Distance 26″ 39‴. Position 90° n. preceding or following, by exact estimation.

21. ζ Orionis, FL. 50. Trium in cingulo sequens.

Oct. 10, 1780. Double. Very unequal. L. w.; S. d. Distance about 25″. Position 83° 25′ n. following, very inaccurate.

22. *f* Cygni, FL. 59.

Oct. 27, 1780. Double. Extremely unequal. L. fine w.; S. d. Distance 18″ 11‴.

23. In genu dextro Cygni.

Oct. 27, 1780. Double. Considerably unequal. L. reddish w.; S. d. Distance within 30″. Position 7° 23′ n. preceding.

24. 3 ad ω Cygni, FL. 46 adjacens. In genu dextro.

Oct. 27, 1780. Treble. Very unequal, and extremely unequal. L. fine garnet; S. r.; smallest d. All within 30″. Position of the brightest of the two small stars 44° 19′ n. preceding. Position of the faintest —— preceding.

25. In constellatione Ceti, FL. 66.

Dec. 23, 1780. Double. It is a star near the place of the periodical star *o*. Distance 16″·875, a little inaccurate.

26. In constellatione Navis, FL. 19. ::

Feb. 15, 1781. Double. It is a star under the ham of Monoceros's right-foot. Distance about 25″.§

27. In constellatione Comæ Berenices, FL. 24.

Feb. 28, 1781. Double. Considerably unequal. L. whitish r.; S. blueish r. Mean distance 18″ 24‴. Position 3° 28′ n. preceding.

28. In constellatione Geminorum.

Mar. 13, 1781. Double. It is near γ towards ζ Tauri. A little unequal. Both r. Distance 19″ 41‴. Position 57° 0′ s. preceding.

29. *h* Ursæ majoris, FL. 23. Duarum in collo sequens.

Apr. 25, 1781. Double. Extremely unequal. L. reddish w.; S. d. Distance with 460, 19″ 26‴. Position 3° 14′ n. preceding.

30. In constellatione Lyncis, FL. 43[am] præcedens ad boream.

May 26, 1781. Double. It is the eye or nose of Leo minor. Unequal. Distance 24″ 53‴ inaccurate.

31. In constellatione Cephei, near FL. 27.

May 27, 1781. Treble. It is a star near δ. Distance of the nearest about 20″.

§ [IV. 26. "The distance 25″ in my printed catalogue being estimated at random, I find it not right." On Oct. 19, 1782, this object was estimated to belong to Class VI.—ED.]

IV. 32. * In constellatione Serpentarii, FL. 61.

July 15, 1781. Double. It is a star near γ. A little unequal. L. w.; S. grey. Distance 19″ 4‴, inaccurate. Position almost directly following.

33. In constellatione Aquilæ.

July 19, 1781. Treble. It is the first of two stars preceding υ. Distance of the two nearest 21″ 59‴, inaccurate.

34. In constellatione Aquilæ, near FL. 64.

July 25, 1781. Double. It is near a star preceding θ. Equal distance about 30″.

35. β Delphini, FL. 6. Austrina præcedentis lateris quadrilateri.

Aug. 1, 1781. Double. Extremely unequal. Hardly visible with 227; pretty strong with 460. Distance 25″ 54‴, rather narrow measure. Position 78° n. preceding, by exact estimation from a diagram.

36. β Serpentis, FL. 28. In eductione colli.

Aug. 13, 1781. Double. Extremely unequal. L. w.; S. extremely faint. Distance 24″, pretty exactly estimated. Position 3 or 4° s. preceding, too obscure for measuring.

37. δ Equulei, FL. 7. Duarum in ore sequens.

Aug. 13, 1781. Double. Excessively unequal. S. hardly visible with 227; but with 460, visible at first sight. L. w.; S. d. Distance 19″ 32‴. S. too obscure to be very accurate. Position 11° 39′ n. following.

38. In constellatione Aquarii.

Aug. 14, 1781. Double. It is the star in the cheek or hair of the neck. Very unequal. L. w.; S. d. Distance 25″, very inaccurate.

39. In constellatione Cygni.

Oct. 1, 1781. Double. It is a star north following σ. Extremely unequal. L. w.; S. d. Distance 18″ exact estimation. Position 30° 28′ s. following.

40. In constellatione Trianguli.

Oct. 8, 1781. Double. It is the preceding of three telescopic stars. Unequal. Distance 17″ 19‴, pretty accurate.

41. μ Herculis, FL. 86.

Oct. 10, 1781. Double. Excessively unequal. The small star is not visible with 227, nor with 278. I saw it very well with 460. L. inclined to pale r.; S. d. Distance, by pretty exact estimation, 18″. Position, by very exact estimation, 30° s. preceding.

42. In constellatione Herculis.

Oct. 10, 1781. Double. It is a star just by ν. Considerably unequal. L. inclined to r.; S. inclined to blue. Distance 18″ 19‴. Position 4° 58′ n. preceding.

IV. 43. In origine fluvii Eridani.§

Oct. 22, 1781. Double. It is the middle of three telescopic stars. Very unequal. L. w.; S. r.

44. In constellatione Tauri, near FL. 4.

Dec. 22, 1781. Double. It is a small telescopic star south following *s*. Extremely unequal. L. w.; S. d.

FIFTH CLASS OF DOUBLE STARS.

V. 1. δ Herculis, FL. 65. In sinistro humero.

Aug. 9, 1779. Double. Extremely unequal. L. w.; S. inclining to r. Distance 33″·75. Position 72° 28′ s. following.

2. * ζ Lyræ, FL. 6.

Aug. 29, 1779. Double. Pretty unequal. L. w.; S. w. inclining to pale rose colour. Distance 41″ 58‴, perhaps a little inaccurate. Position 62° 18′ s. following, a little inaccurate.

3. * β Lyræ, FL. 10. Duarum in jugimento borea.

Aug. 29, 1779. Quadruple. All w. First and second considerably unequal. First and third very unequal. First and fourth very unequal. The second a little inclining to r. The third and fourth more inclining to r. Distance of the first and second 43″ 57‴. Position 60° 28′ s. following, a little inaccurate.

4. δ Cephei, FL. 27. Sequitur tiaram.

Aug. 31, 1779. Double. Considerably unequal. L. reddish w.; S. blueish w. Distance 38″ 18‴, a bright object.

5. † β Cygni, FL. 6. In ore.

Sept. 12, 1779. Double. Considerably unequal. L. pale r.; S. a beautiful blue. The estimation of the colours the same with 227 and 460. Distance 39″ 42‴, pretty accurate. Position 36° 28′ n. following.

6. * ν Scorpii, FL. 14. Duarum adjacentium boreæ frontis, borea.

Sept. 19, 1779. Double. Very unequal. Both w. Distance 38″ 20‴, pretty accurate. Position 64° 51′ n. preceding.‡

7. μ Sagittarii, FL. 13. In summo arcu, borealis.

Sept. 19, 1779. Treble. Two small stars near on each side. L. w.; S. both r. Distance of the nearest about 30″. Position — preceding, the other — following.

§ [IV. 43 is not λ Eridani, as in *Phil. Trans.*, but a star very near it, = B. 718.—ED.]

‡ [V. 6. 69° 28′ in *Phil. Trans.*: error of reduction.—ED.]

V. 8. κ Herculis, FL. 7. In dextri brachii ancone.

Sept. 20, 1779. Double. A little unequal. L. r.; S. garnet; or L. pale r.; S. r. (when the stars are low the first estimation of the colours will take place). Distance 39″ 59‴. Position 82° 23′ n. following.§ Has also a third star.

9. ι Bootis, FL. 21. Trium in sinistra manu, media.

Sept. 27, 1779. Double. Very unequal. L. w.; S. d. Distance 36″·56. This is not a mean of the measures; for I suspect a motion in one of the stars, which another year or two may shew. Position 52° 51′ n. following.

10. * δ Orionis, FL. 34. Trium in cingulo præcedens.

Oct. 26, 1779. Double. Considerably unequal. L. w.; S. blueish r. Distance 52″·968 full measure. Position 88° 10′ n. preceding.

11. † ν Draconis, FL. 24. and 25. In ore duplex.

Oct. 19, 1779. Double. A little unequal. L. pale r.; S. pale r. Distance 54″ 48‴.‡ Position 44° 19′ n. preceding.

From the right ascension and declination of these stars in FLAMSTEED's catalogue we gather, that in his time their distance was 1′ 11″·418; their position 44° 23′ n. preceding; their magnitude equal or nearly so. The difference in the distance of the two stars is so considerable, that we can hardly account for it otherwise than by admitting a proper motion in either one or the other of the stars, or in our solar system; most probably neither of the three is at rest.

12. * λ Arietis, FL. 9. In vertice.

Oct. 30, 1779. Double. Considerably unequal. L. pale r.; S. dusky garnet. Distance 36″ 44‴, a little inaccurate. Position 42° 0′ n. following.

13. φ Tauri, FL. 52. Borea sequentis lateris quadrilateri in Cervice.

Oct. 30, 1779. Double. Distance 55″·625, inaccurate.

14. In constellatione Monocerotis.

Dec. 5, 1779. Multiple. It is a spot over the right fore-foot; 4 or 5 small stars within one minute.

15. *c* Ursæ majoris, FL. 16.

May 2, 1780. Double. Very unequal. L. whitish r.; S. d. Distance with 460, 48″ 59‴. Position 79° 51′ s. preceding.¶

16. σ Piscium, FL. 76. Duarum in ore piscis sequentis borealior.

Aug. 3, 1780. Double. Extremely unequal. L. pale r.; S. dusky r. Distance 48″·125, pretty accurate. Position 15° 28′ n. preceding.

17. π Andromedæ, FL. 29. In dextro humero.

Aug. 25, 1780. Double. Extremely unequal. L. w.; S. blueish. Distance 34″ 12‴, inaccurate.

§ [V. 8. 79° 37′ in *Phil. Trans.*: error of reduction.—ED.]

‡ [V. 11. Some error in this; the distance is 62″, and there is not any relative motion.—ED.]

¶ [V. 15. 80° 47′ in *Phil. Trans.*: error of reduction.—ED.]

V. 18. α Cassiopeæ, FL. 18. In pectore.

Aug. 31, 1780. Double. Extremely unequal. L. pale r.; S. d. Distance 52″·812. Position 5° 26′ n. preceding.§

19. γ Herculis, FL. 20. In dextro brachio.

Sept. 4, 1780. Double. Extremely unequal. L. reddish w.; S. r. Distance 41″ 49‴, a little inaccurate. Position 19° 30′ s. preceding.

20. *e* Pegasi, FL. 1.

Sept. 8, 1780. Double. Very unequal. L. pale r.; S. d. Distance 37″ 5‴, pretty accurate. Position 38° 19′ n. preceding.

21. τ Aurigæ, FL. 29.

Sept. 26, 1780. Double, about 30″.

22. λ Aurigæ, FL. 15.

Sept. 30, 1780. Has 4 or 5 near. Two are about 20″ or 30″ from each other.

23. In constellatione Orionis.

Oct. 10, 1780. Double. It is a star following *f*. Distance about 40″.

24. In constellatione Ceti, FL. 37.

Oct. 12, 1780. Double. It is a star between η and θ towards the north. Distance 42″·812, inaccurate.

25. τ Orionis, FL. 20. supra talum in tibia.

Oct. 23, 1780. Double. Very unequal. Distance about 30″.

26. *h* Leonis, FL. 6.

Feb. 21, 1781. Double. Very unequal. L. r.; S. d. Distance 36″ 9‴. Position 12° 55′ n. following.

27. In constellatione Libræ, near FL. 31.

May 24, 1781. Double. The most south of three small stars in the finder. Equal, or the preceding rather the largest. Both w. inclining to pale r. Distance 44″ 12‴, a little inaccurate. Position 40° 17′ s. following.

28. In constellatione Cephei, near β Cephei.

May 27, 1781. Double. It is a star near β. Extremely unequal. Distance about 30″.

29. ν Serpentis, FL. 53. Post dextrum femur Serpentarii.

July 16, 1781. Double. Unequal. Distance about 35″.

30. In constellatione Serpentarii, FL. 53.

July 19, 1781. Double. It is a star between α and β one-third of the way from α. Very unequal. L. w.; S. inclining to r. Distance 32″ 21‴, narrow measure.

§ [V. 18. 40° 58′ in *Phil. Trans.*: error of reduction, noticed among Errata.—Ed.]

V. 31. In constellatione Aquilæ.

July 19, 1781. Double. It is the star next but one preceding δ. Very unequal. L. r.; S. d. Distance about 30″.

32. α Andromedæ.

July 21, 1781. Double. Extremely unequal. The small star better with 460 than with 227. L. w.; S. d. Distance 55″ 32‴, rather narrow measure. Position 10° 37′ s. preceding.

33. *h* Aquilæ, FL. 15.

July 25, 1781. Double. Unequal. Both pale r. Distance 33″ 53‴, inaccurate.

34. In constellatione Aquilæ, A. FL. 28.§

July 25, 1781. Double. It is one of two stars near A. Distance about 35″.

35. In constellatione Aquilæ.

July 25, 1781. Double. It is a star near that which follows θ. Very unequal. Distance about 40″.

36. *o* Scuti, FL. 2. in constellatione Aquilæ.

July 30, 1781. Double. Very unequal. L. pale r.; S. d. Distance 42″ 44‴, a little inaccurate.

37. υ Coronæ, FL. 18.

Sept. 21, 1781. Treble. Very unequal. L. w.; S. both r. Distance of the nearest about 50″; the farthest 1½ min.‡

38. In constellatione Herculis, FL. 23.

Sept. 21, 1781. Double. It is the star between ν and ξ Coronæ, the largest of a telescopic triangle. Distance 36″ 27‴, rather narrow measure. L. w.; S. w. inclining to r.

39. α Lyræ, FL. 3. In testa fulgida.

Sept. 24, 1781. Double. Excessively unequal. By moon-light I could not see the small star with 278, and saw it with great difficulty with 460; but in the absence of the moon I have seen it very well with 227. L. fine brilliant w.; S. dusky. Distance 37″ 43‴. Position 26° 46′ s. following.

Oct. 22, 1781. Having often measured the diameters of many of the principal fixed stars, and having always found that they measured less and less the more I magnified, I fixed upon this fine star for taking a measure with the highest power I have yet been able to apply, and upon the largest scale of my new micrometer I could conveniently use. With a power of 6450 (determined by experiments upon a known object at a known distance) I looked at this star for at least a quarter

§ [V. 34. "Near FL. 28" is corrected among Errata to "A. FL. 28." An observation of May 30, 1783, describes V. 34 in the same manner, adding: "Very or considerably unequal, L. r. w., S. p. r." —ED.]

‡ In a future collection the small star at the obtuse angular point will be found as a double star of the second or third class.

of an hour, that the eye might adapt itself to the object; having experimentally found, that the aberration by this means will appear less and less, and, in the telescope I used upon this occasion with powers from 460 to 1500, will often quite vanish, and leave a very well-defined circular disk for the apparent diameter of the stars. The diameter of α Lyræ, by this attention, appeared perfectly round, and occasionally separated from rays that were flashing about it. From the very brilliant appearance of the star with this great power, and a pretty accurate rough calculation founded on its apparent brightness, when observed with the naked eye with 227, with 460, with 6450, I surmise, that it has light enough to bear being magnified at least a hundred thousand times with no more than six inches of aperture, provided we could have such a power, and other considerations would allow us to apply it. When I had as good a view as I expected to have, I took its diameter with my new micrometer upon a scale of eight inches and 4428 ten thousandth to 1″ of a degree, and found it subtended an angle of 0″·3553. I had no person at the clock; but suppose the time of its passing through the field of my telescope (which in this great power is purposely left undefined, and as large as possible) was less than three seconds.

V. 40. ν Lyræ, FL. 8.

Sept. 24, 1781. Treble. Extremely unequal. L. w.; S. both d. One n. preceding, the other s. following. Distance of the following star 56″ 47‴, a little inaccurate. Position of the same 28° 37′ s. following.

41. A Persei, FL. 43.

Sept. 24, 1781. Double. Unequal. L. w. Distance about 50″.

42. In constellatione Lyræ.

Sept. 25, 1781. Double. It is a small star just by η. A little unequal. Both r. Distance 38″ 8‴ Position 26° 18′ n. following.

43. In constellatione Cygni, FL. 76.

Oct. 1, 1781. Double. It is the third star from ρ towards υ. Unequal. Distance 48″ by exact estimation. Position —— preceding.

44. In constellatione Cygni, FL. 69.

Oct. 1, 1781. Treble. Very unequal. L. w.; S. both reddish. Position both —— preceding.

45. In constellatione Cygni.

Oct. 1, 1781. Double. It is the most south of two telescopic stars following τ. Very unequal. L. w.; S. d. Distance 44″ by exact estimation. Position —— following.

46. c Cygni, FL. 16. 1ª ad c.

Oct. 5, 1781. Double. It is the star next following θ. Almost equal. Both pale r. Distance 30″ by pretty exact estimation.

V. 47. *c* Cygni, FL. 26. 2ª ad *c*.

Oct. 8, 1781. Double. Very unequal. L. reddish w.; S. dusky r. Distance 39″ by pretty exact estimation.

48. * In constellatione Piscium.

Oct. 8, 1781. Double. It is a telescopic star just by θ southwards. Both d. Distance about 45″.

49. * In constellatione Arietis, FL. 30.

Oct. 15, 1781. Double. It is a small star over the Ram's back. Nearly equal. Distance 31″ 6‴, inaccurate.

50. γ Leporis, FL. 13. In posterioribus pedibus austrina.

Oct. 22, 1781. Double. Considerably unequal. Distance about 40″.

51. In constellatione Sagittæ.

Nov. 23, 1781. Double. It is a star north following ε. Extremely unequal. Distance 32″ 48‴. L. r.; S. blue.

SIXTH CLASS OF DOUBLE STARS.

VI. 1. *o* Ceti, FL. 68. In pectore nova.

Oct. 20, 1777. Double. Very unequal. L. garnet. S. dusky. §
Dist. { mean of some very accurate measures 1′ 44″·218
mean of other very accurate measures 1′ 53″·032.

As I can hardly doubt the motion of this star, I have given the mean of the most accurate measures separately; and hope in a few years time to be able to give a better account of it.

2. *o* Serpentarii, FL. 67.

Aug. 29, 1779. Double. Distance about 1¼ min.

3. δ Lyræ, FL. 11.

Aug. 29, 1779. Double. Extremely unequal. L. w.; S. d. Distance about 4′, pretty exact estimation.

4. Near α Capricorni, FL. 5.‡

Sept. 19, 1779. Double. Very unequal. L. r.; S. d. Distance about 1¼ min. Position —— s. preceding.

5. In constellatione Arietis, FL. 41. supra dorsum.

Sept. 22, 1779. Double. It is the star in the body of the fly. Distance 2′ 5″ 35‴.

6. ε Capricorni, FL. 39. Duarum in eductione caudæ præced.

Sept. 27, 1779. Double. Unequal. L. pale r. Distance about 1¼ min.

§ [VI. 1. Under date Sept. 9, 1780, H. notes: "Colour of a dark red ink, but darker than I can recollect to have seen any star."—ED.]

‡ [VI. 4. This is not α_1 and α_2 Capricorni, but a faint and unequal pair preceding them.—ED.]

VI. 7. * τ Tauri, FL. 94. In eductione cornu borei.
Oct. 6, 1779. Double. Distance 1′ 1″·25‴, pretty accurate.

8. κ Tauri, FL. 65 and 67.
Oct. 6, 1779. Double. At a considerable distance.

9. * ζ Geminorum, FL. 43. In sinistro genu sequentis Π[1].
Oct. 7, 1779. Double. Very unequal. L. reddish w.; S. dusky r. Distance 1′ 31″ 52‴, rather full measure. Position 81° 14′ n. preceding.

10. o Cygni, FL. 31. Duarum in dextro pede sequens.
Nov. 2, 1779. Double. Considerably unequal. L. pale r. S. blue. It is the following star of the two o's that are close together. Distance 1′ 39″ 57‴. Position 87° 14′ s. following. §

11. * α Leonis, FL. 32. In corde.
Nov. 14, 1779. Double. Very unequal. L. w.; S. d. Distance 2′ 48″ 20‴. Position 30° 5′ n. preceding.

12. * τ Leonis, FL. 84. Quasi in cubito.
April 6, 1780. Double. Considerably unequal. L. r.; S. inclining to blue. Distance 1′ 22″ 42‴. Position 75° 21′ s. following.

13. o Leonis, FL. 95. In extremitate caudæ.
April 6, 1780. Double. Extremely unequal. L. reddish w.; S. d. Distance about 1½ min. Position about 80° n. following.

14. η Serpentis, FL. 58. In cauda.
June 19, 1780. Double. Extremely unequal. L. pale r.; S. d. Distance 1′ 21″ 2‴. Position 9° 7′ s. following.

15. In constellatione Bootis, near FL. 6.
June 25, 1780. Double. It is a telescopic star near that which forms a rectangle with α and η. Distance about 2′.

16. δ Bootis, FL. 49. In dextro humero.
July 23, 1780. Double. Considerably unequal. Distance about 2¼ min. L. reddish w.; S. w. Position 5° 46′ n. following.

17. μ Bootis, FL. 51. In baculo recurvo.
July 30, 1780. Double. Unequal. Distance 2′ 8″, exact estimation. Position 80° 25′ s. following. L. reddish w.; S. pale r. See the 17th star of the first class.

18. ν Coronæ, FL. 21.
July 30, 1780. Double. Very unequal. L. r.; S. garnet. At some considerable distance. Position about 80° n. following.

§ [VI. 10. Not sp, as in *Phil. Trans.* 1781 Oct. 20, the companion is said to be sf.—ED.]

VI. 19. χ Persei.

Aug. 2, 1780. Multiple. An astonishing number of small stars all within the space of a few minutes. I counted not less than 40 within my small field of view.

20. μ Persei, FL. 51. Duarum in dextro poplite sequens.

Aug. 2, 1780. Double. Very unequal. L. w. Distance about 1′½.

21. η Pegasi, FL. 44.

Aug. 23, 1780. Double. Distance about 2¼ min.

22. In constellatione Draconis, I. HEVELII 69.

Aug. 7, 1780. Double. It is the star between α Draconis and the tail of Ursa major. Distance about 3½ min.

23. In naribus Lyncis.

Aug. 7, 1780. Double. Distance about 2′.

24. *d* Cassiopeæ, FL. 4.

Aug. 12, 1780. Treble. Two are large. Distance about 2′. A third is obscure. Distance about 1¾ min. They form almost a rectangle.

25. In constellatione Cassiopeæ, FL. 3.

Aug. 18, 1780. Double. Distance about 2¼ min.

26. ε Sagittæ, FL. 4.

Aug. 19, 1780. Double. Very unequal. L. r.; S. r. inclining to blue. Distance 1′ 31″ 53‴. Position 8° 32′ n. following. §

27. In constellatione Aquilæ.

Aug. 24, 1780. Double. It is a star north of θ. Distance about 1′.

28. β Capricorni, FL. 9. Trium in sequente cornu austrina.

Aug. 26, 1780. Double. Considerably unequal. Distance about 3′. Position —— preceding.

29. ρ Capricorni, FL. 11. Trium in rostro media. ‡

Aug. 26, 1780. Double. Distance about 2½ min.

30. α Aurigæ, FL. 13. In humero sinistro.

Sept. 8, 1780. Double. Extremely unequal. L. w.; S. d. Distance 2′ 49″ 8‴. Position 61° 23′ s. following. With a power of 227, and my common micrometer, the diameter of this star measured 2″·5. ¶ The circumference was remarkably well defined.

31. *d* Tauri, FL. 88. In sinistro cubito.

Sept. 24, 1780. Double. Distance 1′ 10″·625. A little inaccurate.

§ [VI. 26. Pos. is nf by a diagram, not sf as in *Phil. Trans.*—ED.]

‡ [VI. 29. Not π, 10 Capric., as in *Phil. Trans.*—ED.]

¶ [VI. 30. There are two measures of the diameter of Capella on Sept. 8, 1780: 3″·125 full measure, 2″·5 narrow measure.—ED.]

VI. 32. λ Cygni, FL. 54.

Sept. 20, 1780. Double. Extremely unequal. L. blueish w.; S. d. Distance about 1 min Position 12° 42′ s. following.

33. In constellatione Cygni, FL. 32.

Sept. 20, 1780. Double. Distance about 2 min.

34. θ Aurigæ, FL. 37. In dextro carpo.

Sept. 26, 1780. Double. Distance about 2½ min.

35. In constellatione Aurigæ, FL. 9. §

Sept. 26, 1780. Double. It is the star over the goat's head. Distance about 2′.

36. In constellatione Camelopardali, FL. 10.

Sept. 30, 1780. Double. Distance about 1½ min.

37. c Draconis, FL. 46. In flexura colli.

Oct. 3, 1780. Double. Distance 3 or 4′. A rich spot.

38. e Draconis, FL. 65.

Oct. 3, 1780. Double. Distance about 2′.

39. α Orionis, FL. 58. In dextro humero lucida rutilans.

Oct. 10, 1780. Double. Extremely unequal. L. r. but not deep; S. d. Distance 2′ 41″ 46‴. Position 62° 18′ s. following.

40. γ Leporis, FL. 13.

Feb. 21, 1781. Double. Distance about 2½ min.

41. ρ Cancri 5 ad ρ, FL. 67.

Feb. 21, 1781. Double. Very unequal. L. reddish w.; S. d. Distance 1′ 35″ 59‴. Position 50° 33′ n. preceding.

42. β Geminorum, FL. 78. In capite sequentis Πⁱ.

Mar. 13, 1781. Multiple. Extremely unequal. The nearest distance 1′ 56″ 45‴, rather full measure. Position 24° 28′ n. following, not extremely accurate. This is the smallest. The next distance 2′ 40″ 42‴, pretty accurate.‡ Position 15° 56′ n. following. A third I did not measure.

43. θ Virginis, FL. 51. De quatuor ultima et sequens.

May 14, 1781. Double. Extremely unequal. L. w.; S. d. Distance 1′ 3″ 53‴, inaccurate. Position 24° 55′ n. preceding.

44. ι Libræ, FL. 24.

May 24, 1781. Double. Very unequal. L. w.; S. dusky r. Distance 59″ 4‴, not accurate. Position 22° 30′ s. following.

§ [VI. 35. Is not 13 Camelop. as in *Phil. Trans.* "Revised Sept. 12, 1798. My 13 Camelop. does not exist. My D* VI. 35 is 9 Aurigæ."—ED.]

‡ [VI. 42. *Phil. Trans.* has 3′ 17″ 19‴, which has been struck out in MS. and 2′ 40″ 42‴ substituted. The first observation, on Mar. 13, 1781, is merely: "Pollux is followed by three small stars at about 2′ and 3′ distance." Immediately after this follows in the Journal the discovery of Uranus.—ED.]

VI. 45. In constellatione Andromedæ.

July 21, 1781. Double. It is a star near ι towards ο. L. r. Distance about 1½ min.

46. α Aquilæ, FL. 53.

July 23, 1781. Double. Extremely unequal. L. w.; S. d. Distance 2′ 23″ 18‴. Position 64° 44′ n. preceding.

47. In constellatione Aquilæ, near FL. 35.

July 25, 1781. Double. It is one of the preceding stars of a small quartile near *c*, not very near.

48. In constellatione Aquilæ, near FL. 35.

July 25, 1781. Double. It is also one of the preceding stars of a small quartile near *c*, not very near.

49. In constellatione Aquilæ.

July 26, 1781. Double. The following star of a trapezium near l.

50. In constellatione Aquilæ.

July 26, 1781. Double. The following star of a trapezium near l. not near.

51. In monte Mænali Heveliana. §

Aug. 5, 1781. Double. It is a star near the middle. The following of two, not very near.

52. In constellatione Bootis.

Aug. 17, 1781. Double. It is a star between ε and *f*. Distance above 1′. Unequal.

53. In constellatione Bootis.

Aug. 17, 1781. Double. It is a star more south than *i*. Distance above 1′.

54. In constellatione Serpentarii. ‡

Aug. 21, 1781. Double. It is a star more south than *o*. Distance 75″, exact estimation.

55. In constellatione Cassiopeæ, FL. 2.

Sept. 6, 1781. Double. It is a star near *e*. L. r. Distance within 2′½.

56. θ Lyræ, FL. ultima.

Sept. 25, 1781. Double. Very unequal. L. w.; S. inclining to r. Distance about 1½ min. Position —— n. following.

57. In constellatione Cygni, FL. 79.

Oct. 1, 1781. Double. It is the fifth star from ρ to υ. Unequal. L. w.; S. pale r. Distance 1′ 40″ estimation.

§ [VI. 51. "Probably 2 Serpentis" (C. H.). But it is 1 Serpentis, as pointed out by Burnham.—Ed.]

‡ [VI. 54. Cannot be identified. Cannot be = III. 25.—Ed.]

VI. 58. In constellatione Aquarii, FL. 4.

Oct. 5, 1781. Double. It is the most south of two in the arrow of Antinous. Distance above 1′.

59. In constellatione Cygni, near FL. 28.

Oct. 5, 1781. Double. It is a star near *b*. Distance 73″, exact estimation.

60. In constellatione Cygni. §

Oct. 8, 1781. Double. It is a star near the second *c*. Considerably unequal. L. w.; S. d. Distance 88″, exact estimation.

61. In constellatione Piscium, near FL. 7. ‡

Oct. 8, 1781. Treble. It is a star preceding *b*. They form a triangle, each side of which is about 1′.

62. *κ* Piscium, FL. 8. In ventre.

Oct. 8, 1781. Double. Distance near 2′.

63. In constellatione Sagittæ.

Oct. 12, 1781. Double. It is near the star north following *ε*. Extremely unequal. L. w. inclining to r.; S. d. Distance 1′ 30″ 56‴. Position 4° 9′ s. preceding. A third star in the same direction, at a little more than twice the distance. A fourth star in view.

64. In constellatione Eridani.

Oct. 22, 1781. Double. It is the small star near *ν*. Distance about 1¾ min.

65. In capite Monocerotis.

Oct. 22, 1781. Multiple. It is one star with at least 12 around it, all within the field of my telescope.

66. *α* Tauri, FL. 87. Splendida in austrino oculo.

Dec. 19, 1781. Double. Extremely unequal. L. r.; S. d. Distance 1′ 27″ 45‴. Position 52° 58′ n. following. With 460, the apparent diameter of this star, when on the meridian, measured 1″ 46‴, a mean of two very compleat observations, they agreed to 6‴; with 932, it measured 1″ 12‴, also a mean of two excellent observations; they agreed to 8‴. The apparent disk was perfectly well defined with both powers.

POSTSCRIPT TO THE CATALOGUE OF DOUBLE STARS.

Since my having delivered my paper on the Parallax of the Fixed Stars, in which I refer to the above Catalogue of Double Stars, I have received, by the favour of our President Sir JOSEPH BANKS, the fourth volume of the *Acta Academiæ*

§ [VI. 60. Cannot be identified. Cannot be = V. 47, the distance of which is 42″.—ED.]

‡ [VI. 61. Not found by Burnham in 1901.—ED.]

Theodoro Palatinæ, which contains a most excellent Memoir of Mr. MAYER's, "De novis in Cœlo sidereo Phænomenis;" wherein I see that the idea of ascertaining the proper motion of the stars by means of small stars that are situated at no great distance from large ones, has induced that gentleman before me to look out for such small stars. In the course of that undertaking he has discovered a good many double stars, of which he has given us a pretty large list, some of them the same with those in my catalogue. My view being the annual parallax required stars much nearer than those that would do for Mr. MAYER's purpose; therefore I examined the heavens with much higher powers, and looked out chiefly for such as were exceedingly close.*

The above catalogue contains 269 double stars, 227 of which, to my present knowledge, have not been noticed by any person. I hope they will prove no inconsiderable addition to the general stock, especially as in that number there are a great many which are out of the reach of Mr. MAYER's and other mural quadrant or transit instruments. It can hardly be expected, that a power of 70 or 80 would be sufficient to discover those curious stars that are contained in the first class of my catalogue; so that it is not strange they should have intirely escaped Mr. MAYER's notice. We see that it is not for want of his looking at those stars; for we find he has frequently observed ζ Cancri, the star near Procyon, and the star in Monoceros, without perceiving the small stars near them, which I have pointed out. Nor is it only in the first class that his telescope wanted power, light, and distinctness; for the small stars that are near β Orionis, β Serpentis, ζ Orionis, *e* Pegasi, α Lyræ, α Andromedæ, μ Sagittarii, α Aquilæ, η Pegasi, δ Lyræ, ι Libræ, κ Piscium, α Tauri, and many more, have escaped his discovery, though he has given us the places of other more distant small stars not far from them, and therefore must have had them frequently in the field of view of his telescope. In settling the relative situations of very close double stars, neither Mr. MAYER's instruments, nor his method, were adequate to the purpose. It is well known, that whenever we employ time as a measure, the results cannot be very accurate; because a mistake of no more than a tenth part of a second in time will produce an error of a whole second and an half in measure, so that his Æ must be extremely defective. Nor could his micrometer give the declination much better unless the telescope had bore a power of at least 4 or 500. When the angle of position is but small, such as 3, 4, 5, or 6 degrees, and the distance of the stars not above a few seconds, it is evident, that a micrometer must be able to measure tenths of a second at least to give even a tolerable exactness of position. On the contrary; the position being measured with such a micrometer as I have constructed for the purpose, we may from thence deduce the declination, with great confidence, true to a quarter of a tenth of a second for every second of the distance of the stars.

* [Mayer's Catalogue was re-published by Schjellerup, arranged according to constellations, in the journal *Copernicus*, vol. iii. p. 57 *sq.*—ED.]

Mr. MAYER's account of α Geminorum, for instance, gives a difference of 0″·7 of time in Æ, of 3″·8 in declination, and of 1 to 6 in magnitude or degree of light of the stars. These quantities reduced to my notation, and compared with my measures of the same star, give

	Mr. MAYER's	Mine
	Distance 9″·635 from center to center	5″·156 diameters included.
	Position 23° 14′ n. preceding	32° 47′ n. preceding.
	Magnitude extremely unequal	A little unequal.

To account for this difference I ascribe Mr. MAYER's error in distance to his method of measuring by time. The error of position follows always from an observation of the declination taken with the common micrometer, when it is deduced from an erroneous Æ. In my measures the distance and position are independent of each other, which I look upon as no small advantage of my cross-hair micrometer. The error in the magnitudes of the stars I ascribe to the want of power in Mr. MAYER's telescope, which did not separate the stars far enough for him to judge accurately of their size, otherwise he would soon have found, that instead of five there is hardly so much as one single degree of difference in their magnitudes. See fig. 6 [p. 50] for a representation of those stars with my power of 460.

I do not mean to depreciate Mr. MAYER's method, the excellence of which is well known; and with some stars of my third, all those of the fourth, fifth, and sixth classes, as well as with those still farther distant, to which he has applied it with admirable skill, and "magno labore, multisque nocturnis vigiliis" (as he very justly expresses himself) a better can hardly be wished for; but with stars of the second class which generally differ no more than one, two or three-tenths of a second of time in Æ, and can never differ more than four tenths, the insufficiency of measuring by time is obvious. In regard to the declination, it is also no less evident, that it is much more accurate to take an angle, which may be had true to 2 or 3° at most, than to measure its tangent, which in stars of the second class is generally no more than 2, 3, or 4″ of a degree, and can never exceed five. I do not so much as mention the stars of the first class: they must certainly, as to sense, pass the meridian at the same instant of time. Their distance has even eluded the attacks of my smallest silk-thread micrometer armed with an excellent power of 460; but I shall soon apply my last new instrument to them,* not without hopes of success. Now, though I have hitherto not been able to express the distance of the stars of the first class, otherwise than by the proportion it bears to their apparent diameters, I think it a very great point gained, that one of my instruments at least (*viz.* the cross-hair micrometer) has laid hold of them: for their angle of position, I think, is within a very small quantity as well determined

* For a description of which see p. [91].

as it is in those of the second class. This simple but most useful instrument can, by actual measure, discover beyond a doubt a motion in two stars that are very close together, though it should amount to no more than a tenth part of a second of a degree, provided that motion be in such a direction that the effect of it be thrown upon the angle of position; wherein, with some of the stars of the first class, it would occasion an alteration of 10, 20, 30, or more degrees.

I have marked all those stars in my catalogue which have been observed by Mr. MAYER and other astronomers with an asterisk (*) affixed to the number that they may be known; those with the mark of a dagger (†) have been observed by different astronomers before Mr. MAYER. Among the stars which are not marked, will be found several that have been observed by Mr. MAYER; but, on comparing them together, it will be seen, that they are observations of different small stars; for instance, Mr. MAYER (*Act. Acad.* vol. IV. p. 296) observed a small star near Rigel at the distance of 1′ 0″·5 Æ in time, and 2′ 55″·2 in difference of declination north preceding Rigel. In my second class (the 33rd star) we also find Rigel; but the small star I have observed is one which has not been seen by Mr. MAYER, and is at a distance of no more than 6″ 27‴. Position 68° 12′ south preceding; and so on with other stars.

I have used the expression *double-star* in a few instances of the sixth class in rather an extended signification: the example of FLAMSTEED, however, will sufficiently authorize my application of the term. I preferred that expression to any other, such as Comes, Companion, or Satellite; because, in my opinion, it is much too soon to form any theories of small stars revolving round large ones, and therefore I thought it adviseable carefully to avoid any expression that might convey that idea. I am very well persuaded, FLAMSTEED, who first used the word Comes, meant it only in a figurative sense.

I shall not fail to take the first opportunity of looking out for those of Mr. MAYER's double-stars which I have not in my catalogue, amounting to 31; and also for one I find mentioned in *La Connoissance des Temps* for 1783, discovered by Mr. MESSIER.

VII.

Description of a Lamp-Micrometer, and the Method of using it.

By Mr. WILLIAM HERSCHEL, *F.R.S.*

[*Phil. Trans.*, vol. lxxii., 1782, pp. 163–172.]

Read January 31, 1782.

THE great difficulty of measuring very small angles, such as hardly amount to a few seconds, is well known to astronomers. Since I have been engaged in observations on double stars, I have had so much occasion for micrometers that would measure exceeding small distances exactly, that I have continually been endeavouring to improve these instruments.

The natural imperfections of the parallel wire micrometer in taking the distance of very close double stars are the following. When two stars are taken between the parallels, the diameters must be included. I have in vain attempted to find lines sufficiently thin to extend them across the centers of the stars so that their thickness might be neglected. The single threads of the silk-worm, with such lenses as I use, are so much magnified that their diameter is more than that of many of the stars. Besides, if they were much less than they are, the power of deflection of light would make the attempt to measure the distance of the centers this way fruitless: for I have always found the light of the stars to play upon those lines and separate their apparent diameters into two parts. Now since the spurious diameters of the stars thus included, to my certain knowledge, are continually changing according to the state of the air, and the length of time we look at them, we are, in some respect, left at an uncertainty, and our measures taken at different times, and with different degrees of attention, will vary on that account. Nor can we come at the true distance of the centers of any two stars, one from another, unless we could tell what to allow for the semi-diameters of the stars themselves; for different stars have different apparent diameters, which, with a power of 227, may differ from each other (as I have experienced) as far as two seconds.

The next imperfection is that which arises from a deflection of light upon the wires when they approach very near to each other; for if this be owing to a power of repulsion lodged at the surface, it is easy to understand, that such powers must interfere with each other, and give the measures larger in proportion than they would have been if the repulsive power of one wire had not been opposed by a contrary power of the other wire.

Another very considerable imperfection of these micrometers is a continual uncertainty of the real zero. I have found, that the least alteration in the situation and quantity of light will affect the zero, and that a change in the position of the wires, when the light and other circumstances remain unaltered, will also produce a difference. To obviate this difficulty, whenever I took a measure that required the utmost accuracy, my zero was always taken immediately after, while the apparatus remained in the same situation it was in when the measure was taken; but this enhances the difficulty because it introduces an additional observation.

The next imperfection, which is none of the smallest, is that every micrometer that has hitherto been in use requires either a screw or a divided bar and pinion to measure the distance of the wires or divided image. Those who are acquainted with works of this kind are but too sensible how difficult it is to have screws that shall be perfectly equal in every thread or revolution of each thread; or pinions and bars that shall be so evenly divided as perfectly to be depended upon in every leaf and tooth to perhaps the two, three, or four thousandth part of an inch; and yet, on account of the small scale of those micrometers, these quantities are of the greatest consequence; an error of a single thousandth part inducing in most instruments a mistake of several seconds.

The last and greatest imperfection of all is, that these wire micrometers require a pretty strong light in the field of view: and when I had double stars to measure, one of which was very obscure, I was obliged to be content with less light than is necessary to make the wires perfectly distinct; and several stars on this account could not be measured at all, though otherwise not too close for the micrometer.

The instrument I am going to describe, which I call a Lamp-Micrometer, is free from all these defects, and has, moreover, to recommend it, the advantage of a very enlarged scale. The construction of it is as follows.

ABGCFE (fig. 1) is a stand nine feet high, upon which a semi-circular board *qhogp* is moveable upwards or downwards, in the manner of some fire-screens, as occasion may require, and is held in its situation by a peg *p* put into any one of the holes of the upright piece AB. This board is a segment of a circle of fourteen inches radius, and is about three inches broader than a semi-circle, to give room for the handles *r*D, *e*P, to work. The use of this board is to carry an arm L, thirty inches long, which is made to move upon a pivot at the center of the circle, by means of a string, which passes in a groove upon the edge of the semi-circle *pgohq*; the string is fastened to a hook at *o* (not expressed in the figure being at the back of the arm L), and passing along the groove from *oh* to *q* is turned over a pulley at *q*, and goes down to a small barrel *e*, within the plane of the circular board, where a double-jointed handle *e*P commands its motion. By this contrivance we see the arm L may be lifted up to any altitude from the horizontal position to the perpendicular, or be suffered to descend by its own weight below the horizontal to the reverse perpendicular situation. The weight of the handle P is sufficient to keep the arm in any given position; but if the motion should be too easy, a friction spring applied to the barrel will moderate it at pleasure.

In front of the arm L a small slider, about three inches long, is moveable in a rabbet from the end L towards the center backwards and forwards. A string is fastened to the left side of the little slider, and goes towards L, where it passes round a pulley at *m*, and returns under the arm from *m*, *n*, towards the center, where it is led in a groove on the edge of the arm, which is of a circular form, upwards to a barrel (raised above the plane of the circular board) at *r*, to which the handle *r*D is fastened. A second string is fastened to the slider, at the right side, and goes towards the center, where it passes over a pulley *n*, and the weight *w*, which is suspended by the end of this string, returns the slider towards the center when a contrary turn of the handle permits it to act.

a and *b* are two small lamps, two inches high, $1\frac{1}{2}$ in breadth by $1\frac{1}{4}$ in depth. The sides, back, and top, are made so as to permit no light to be seen, and the front consists of a thin brass sliding door. The flame in the lamp *a* is placed three-tenths of an inch from the left side, three-tenths from the front, and half an inch from the bottom. In the lamp *b* it is placed at the same height and distance measuring from the right side. The wick of the flame consists only of a single very thin lamp-cotton thread; for the smallest flame being sufficient it is easier to keep it burning in so confined a place. In the top of each lamp must be a little slit, lengthways, and also a small opening in one side near the upper part, to permit air enough to circulate to feed the flame. To prevent every reflection of light, the side opening of the lamp *a* should be to the right, and that of the lamp *b* to the left. In the sliding door of each lamp is made a small hole with the point of a very fine needle just opposite the place where the wicks are burning, so that when the sliders are shut down, and every thing dark, nothing shall be seen but two fine lucid points of the size of two stars of the third or fourth magnitude. The lamp *a* is placed so that its lucid point may be in the center of the circular board where it remains fixed. The lamp *b* is hung to the little slider which moves in the rabbet of the arm, so that its lucid point, in a horizontal position of the arm, may be on a level with the lucid point in the center. The moveable lamp is suspended upon a piece of brass fastened to the slider by a pin exactly behind the flame upon which it moves as a pivot. The lamp is balanced at the bottom by a leaden weight, so as always to remain upright, when the arm is either lifted above, or depressed below, the horizontal position. The double-jointed handles *r*D, *e*P, consist of light deal rods, ten feet long, and the lowest of them may have divisions, marked upon it near the end P, expressing exactly the distance from the central lucid point in feet, inches, and tenths.

From this construction we see, that a person at a distance of ten feet may govern the two lucid points, so as to bring them into any required position south or north, preceding or following, from 0 to 90° by using the handle P, and also to any distance from six-tenths of an inch to five or six and twenty inches by means of the handle D. If any reflection or appearance of light should be left from the top or sides of the lamps, a temporary screen, consisting of a long piece of paste-board, or a wire frame covered with black cloth, of the length of the whole arm and of any required breadth, with a slit of half an inch broad in the middle, may be affixed to the arm by four bent wires projecting an inch or two before the lamps, situated so that the moveable lucid point may pass along the opening left for that purpose.

Fig. 2 represents part of the arm L, $\frac{3}{8}$ of the real size; S the slider; *m* the pulley, over which the cord *xtyz* is returned towards the center; *v* the other cord going to the pulley *n* of fig. 1, R the brass piece moveable upon the pin *c*, to keep the lamp upright. At R is a wire rivetted to the brass piece, upon which is held the lamp by a nut and screw. Fig. 3, 4, represent the lamps *a*, *b*, with the sliding doors open, to shew the situation of the wicks. W is the leaden weight with a hole *d* in it, through which the wire R of fig. 2 is to be passed when the lamp is to be fastened to the slider S. Fig. 5 represents the lamp *a* with the sliding door shut; *l* the lucid point; and *ik* the openings at the top, and *s* at the sides for the admission of air.

Every ingenious artist will soon perceive that the motions of this micrometer are capable of great improvement by the application of wheels and pinions, and other well known mechanical resources; but, as the principal object is only to be able to adjust the two lucid points to the required position and distance, and to keep them there for a few minutes, while the observer goes to measure their distance, it will not be necessary to say more upon the subject.

I am now to shew the application of this instrument. It is well known to opticians and others, who have been in the habit of using optical instruments, that we can with one eye look into a microscope or telescope, and see an object much magnified, while the naked eye may see a scale upon which the magnified picture is thrown. In this manner I have generally determined the power of my telescopes; and any one who has acquired a facility of taking such observations will very seldom mistake so much as one in fifty in determining the power of an instrument, and that degree of exactness is fully sufficient for the purpose.

The Newtonian form is admirably adapted to the use of this micrometer; for the observer stands always erect, and looks in a horizontal direction, notwithstanding the telescope should be elevated to the zenith. Besides, his face being turned away from the object to which his telescope is directed, this micrometer may be placed very conveniently without causing the least obstruction to the view: therefore, when I use this instrument I put it at ten feet distance from the left eye, in a line perpendicular to the tube of the telescope, and raise the moveable board to such a height that the lucid point of the central lamp may be upon

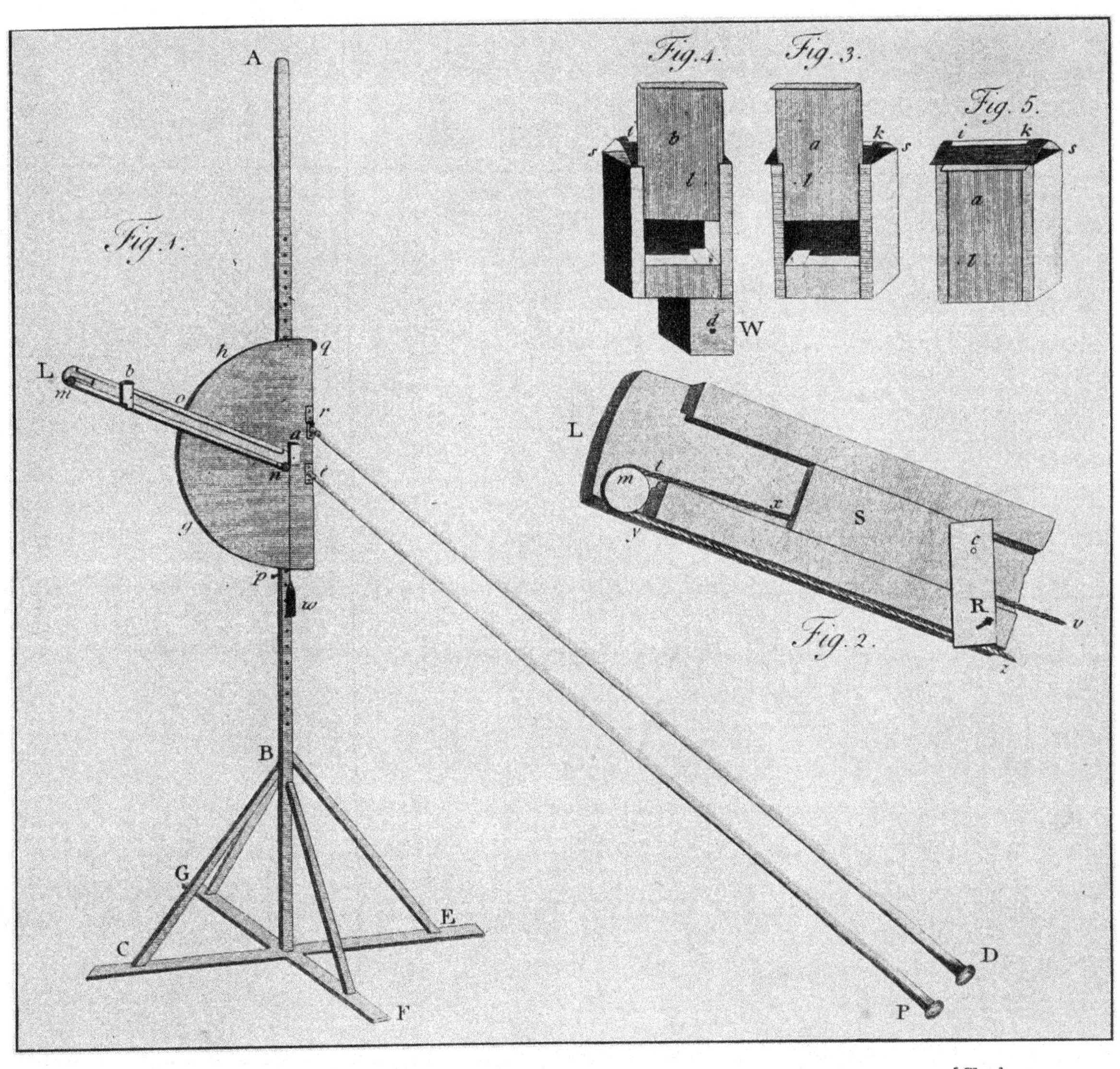

[*To face page* 94.

a level with the eye. The handles, lifted up, are passed through two loops fastened to the tube, just by the observer, so as to be ready for his use. I should observe, that the end of the tube is cut away so as to leave the left eye intirely free to see the whole micrometer.

Having now directed the telescope to a double star, I view it with the right eye, and at the same time with the left see it projected upon the micrometer: then, by the handle P, which commands the position of the arm, I raise or depress it so as to bring the two lucid points to a similar situation with the two stars; and, by the handle D, I approach or remove the moveable lucid point to the same distance of the two stars, so that the two lucid points may be exactly covered by, or coincide with the stars. A little practice in this business soon makes it easy, especially to one who has already been used to look with both eyes open.

What remains to be done is very simple. With a proper rule, divided into inches and fortieth parts, I take the distance of the lucid points, which may be done to the greatest nicety, because, as I observed before, the little holes are made with the point of a very fine needle. The measure thus obtained is the tangent of the magnified angle under which the stars are seen to a radius of ten feet; therefore, the angle being found and divided by the power of the telescope gives the real angular distance of the centers of a double star.

For instance, September 25, 1781, I measured α Herculis with this instrument. Having caused the two lucid points to coincide exactly with the stars center upon center, I found the radius or distance of the central lamp from the eye 10 feet 4·15 inches; the tangent or distance of the two lucid points 50·6 fortieth parts of an inch; this gives the magnified angle 35′, and dividing by the power 460, which I used, we obtain 4″ 34‴ for the distance of the centers of the two stars. The scale of the micrometer at this very convenient distance, with the power of 460 (which my telescope bears so well upon the fixed stars that for near a twelvemonth past I have hardly used any other) is above a quarter of an inch to a second; and by putting on my power of 932, which in very fine evenings is extremely distinct, I obtain a scale of more than half an inch to a second, without increasing the distance of the micrometer; whereas the most perfect of my former micrometers, with the same instrument, had a scale of less than the two thousandth part of an inch to a second.

The measures of this micrometer are not confined to double stars only, but may be applied to any other objects that require the utmost accuracy, such as the diameters of the planets or their satellites, the mountains of the moon, the diameters of the fixed stars, &c.

For instance, October 22, 1781, I measured the apparent diameter of α Lyræ; and judging it of the greatest importance to increase my scale as much as convenient, I placed the micrometer at the greatest convenient distance, and (with some trouble, for want of longer handles, which might easily be added) took the diameter of this star by removing the two lucid points to such a distance as just to inclose the apparent diameter. When I measured my radius it was found to be twenty-two feet six inches. The distance of the two lucid points was *about* three inches; for I will not pretend to *extreme* nicety in this observation, on account of the very great power I used, which was 6450. From these measures we have the magnified angle 38′ 10″: this divided by the power gives 0″·355 for the apparent diameter of α Lyræ. The scale of the micrometer, on this occasion, was no less than 8·443 inches to a second, as will be found by multiplying the natural tangent of a second with the power and radius in inches.

November 28, 1781, I measured the diameter of the new star; but the air was not very favourable, for this singular star was not so distinct with 227 that evening as it generally is with 460: therefore, without laying much stress upon the exactness of the observation, I shall only report it to exemplify the use of the micrometer. My radius was 35 feet 11 inches. The diameter of the star, by the distance of the lucid points, was 2·4 inches, and the power I used 227: hence the magnified angle is found 19′, and the real diameter of the star 5″·022. The scale of this measure ·474 millesimals of an inch, or almost half an inch to a second.

VIII.

A Paper to obviate some Doubts concerning the great Magnifying Powers used. By Mr. HERSCHEL, *F.R.S.*

[*Phil. Trans.*, vol. lxxii., 1782, pp. 173–178.]

TO SIR JOSEPH BANKS, BART. P.R.S.

SIR,—I have the honour of laying before you the result of a set of measures I have taken in order to ascertain once more the powers of my Newtonian seven-feet reflector. The method I have formerly used, and which I still prefer to that which I have now been obliged to practise, requires very fine weather and a strong sun-shiny day; but my impatience to answer the requests of Sir JOSEPH BANKS would not permit me to wait for so precarious an opportunity at this season of the year. The difference in all the powers, as far as 2010, will be found to be in favour of those I have mentioned; and, I believe, a much greater concurrence could not well be expected, where different methods of ascertaining them are used. The variation in the two highest powers is more considerable than I was aware of; but still may easily be shewn to be a necessary consequence of the difference in the methods. However, if upon comparing together the methods it should be thought, that the power 5786 is nearer the truth than 6450, I shall readily join to correct that number. The manner in which I have now determined the powers is as follows: I took one of the eye lenses which magnifies least, and measured its solar focus by the sun's rays as exactly as I could five times, which proved to be 1·01, 1·04, 1·09, 1·01, 1·05, in half-inch measure, a mean of which is 1·04. The sidereal focus of my seven-feet speculum is 170·4 in the same measure. Thence, dividing 170·4 by 1·04 we find that the telescope will magnify 163·8 times when that lens is used. This power being found, I applied the same lens as a single microscope to view with it a certain object, which was a drawn brass wire fastened so as not to turn upon its axis or change its position; for these wires are seldom perfectly round, or of an even size, and it is therefore necessary to use this precaution to prevent errors: then, with a fine pair of compasses, I took four independent measures of the image of the brass wire, which was thrown upon a sheet of paper exactly 8½ inches from the lens, the eye being always as close to the lens as possible. I viewed the same wire, exactly in the same manner, with every one of the lenses, and measured the pictures upon the paper. When I came to the higher powers the wire was exchanged for another 4·37 times thinner than the former, as determined by comparing the proportion of their images 54 to $235\frac{3}{4}$, taken by the same lens.

When the images of these wires are obtained, the power of the telescope, with every one of the lenses, becomes known by one plain analogy: *viz.* as the image of the wire by the first lens ($77\frac{3}{4}$) is to the power it gives to the telescope (163·8), so is the image of the wire by the second lens (119) to the power it will give to the same telescope (250·7). The particulars of all the measures are as follows:

Powers as they have been called in my papers.	Images of a wire thrown upon a paper in hundredths of half inches.	A mean of the four measures.	Powers as they come out by this method.
146	77 78 78 78	$77\frac{3}{4}$	$163·86 = \frac{170·4}{1·04}$
227	119 119 119 119	119	250·7
278	143 143 144 143	$143\frac{1}{4}$	301·8
460 {	236 236 235 236	$235\frac{3}{4}$	} 496·7
	Smaller wire.		
	53 54 55 54	54	
754	83 85 84 85	$84\frac{1}{4}$	775·1
932	107 107 107 108	$107\frac{1}{4}$	986·7
1159	128 128 129 128	$128\frac{1}{4}$	1179·9
1536	An excellent lens, lost	about eight mo	nths ago.
2010	236 236 238 236	$236\frac{1}{2}$	2175·8
3168	281 283 281 280	$281\frac{1}{4}$	·2585·5
6450	635 625 630 626	629	5786·8

I beg leave, Sir, now to give a short description of the method I have formerly used to determine these powers. In the year 1776 I erected a mark of white paper, exactly half an inch in diameter, which I viewed with my telescope at the greatest convenient distance with one of the least magnifiers. An assistant was placed at rectangles in a field, at the same distance from my eye as the object from the great speculum of the telescope. Upon a pole erected there I viewed the magnified image of the half inch, and the assistant marked it by my direction; this being measured gave the power of the instrument at once. The power thus obtained was corrected by theory, to reduce it to what it would be upon infinitely distant objects. The powers of the rest of the lenses I deduced from this by a *Camera-eye-piece*, which I made for that purpose. ABCD (fig. 1) represents a perpendicular section of it. The end A screws into the telescope. Upon the end B may be screwed any of the common single-lens eye-pieces. *lmn* is a small oval plane speculum, adjusted to an angle of 45° by three screws, two whereof appear at *op*. When the observer looks in at B, he may see the object projected upon a sheet of paper on a table placed under the Camera-piece, and measure its picture

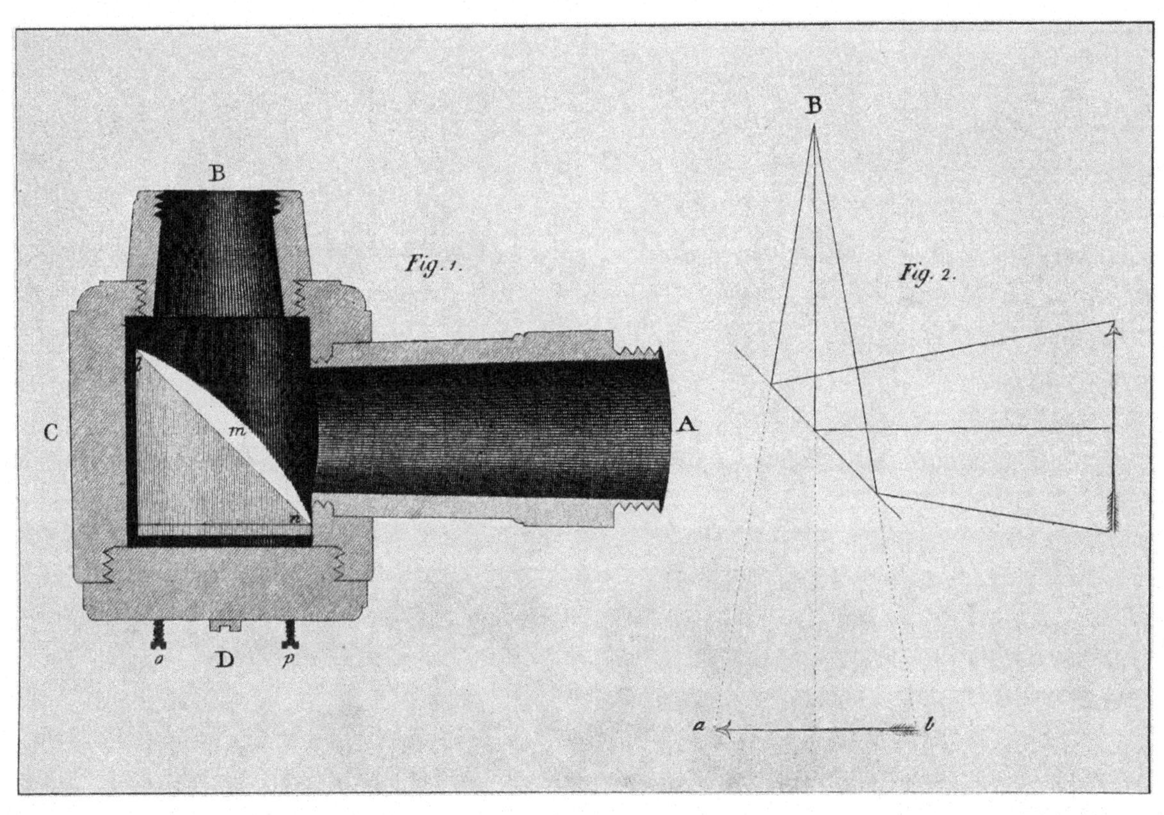

[*To face page* 98.

a, b, as in fig. 2. The power of one lens therefore being known, that of the rest was also found by comparing the measures of the projected images.

It may not be amiss to mention some of the advantages and inconveniences attending each of these methods. When we take the focus of an eye-lens, which the first method requires, we are liable to a pretty considerable uncertainty, and in very small lenses it is not to be done at all. Moreover, in calculating the power by that focus no account is made of the aberration which takes place in all specula and lenses, and increases the image, so that we rather find out how much the telescope *should* magnify than how much it really *does* magnify; but in determining the power by an experiment we avoid these difficulties.

On the other hand, when the power is very great, the latter method becomes inconvenient, both on account of want of light in the object, and a very considerable aberration which takes place, and makes the picture too indistinct to be very accurate in the measure, and of course larger than it ought to be; and this will account for the excess in the measures of my two largest powers. However, when I employed 6450 upon the diameter of α Lyræ, I incline to think the method I had used when I determined that power, ought to be preferred, because my Lamp-micrometer gives the measure of an object as it appears in the telescope, and therefore this aberration is included, and should be taken into consideration.

To prevent any mistakes, I wish to mention again, that I have all along proceeded *experimentally* in the use of my powers, and that I do not mean to say I have used 6450 (or 5786) upon the planets, or even upon double stars; every power I have mentioned is to be understood as having been used just as it is related; but farther inferences ought not as yet to be drawn. For instance, my observations on ε Bootis mention that I have viewed that star with 2010 (or as in the above table with 2175) extremely distinct; but upon several other celestial objects I have found this power of no service. Many plausible suggestions have already occurred to account for these appearances; but I wait till farther experiments shall have furnished me with more materials to reason upon. The use of high powers is a new and untrodden path, and in this attempt variety of new phænomena may be expected, therefore I wish not to be in a haste to make general conclusions. I shall not fail to pursue this subject, and hope soon to be able to attack the celestial bodies with a still stronger armament, which is now preparing.

It remains now only for me to make the most sincere acknowledgement for the favours you have shewn to me, and to say that I shall ever remain, with equal respect and gratitude,

SIR, your most obedient, &c.

P. S. Dr. WATSON junior has done me the favour separately to examine and measure the powers of my telescope; and placing the greatest confidence in his accuracy, I rely on his measures at least as much as my own.

IX.

A Letter from William Herschel, *Esq. F.R.S.*

[*Phil. Trans.*, vol. lxxiii., 1783, pp. 1–3.]

TO SIR JOSEPH BANKS, BART. P.R.S.

Sir,—By the observations of the most eminent Astronomers in Europe it appears, that the new star, which I had the honour of pointing out to them in March, 1781, is a Primary Planet of our Solar System. A body so nearly related to us by its similar condition and situation, in the unbounded expanse of the starry heavens, must often be the subject of the conversation, not only of astronomers, but of every lover of science in general. This consideration then makes it necessary to give it a name, whereby it may be distinguished from the rest of the planets and fixed stars.

In the fabulous ages of ancient times the appellations of Mercury, Venus, Mars, Jupiter, and Saturn, were given to the Planets, as being the names of their principal heroes and divinities.* In the present more philosophical æra, it would hardly be allowable to have recourse to the same method, and call on Juno, Pallas, Apollo, or Minerva, for a name to our new heavenly body. The first consideration in any particular event, or remarkable incident, seems to be its chronology: if in any future age it should be asked, *when* this last-found Planet was discovered? It would be a very satisfactory answer to say, "In the Reign of King George the Third." As a philosopher then, the name of GEORGIUM SIDUS presents itself to me, as an appellation which will conveniently convey the information of the time and country where and when it was brought to view. But as a subject of the best of Kings, who is the liberal protector of every art and science;—as a native of the country from whence this Illustrious Family was called to the British throne;—as a member of that Society, which flourishes by the distinguished liberality of its Royal Patron;—and, last of all, as a person now more immediately under the protection of this excellent Monarch, and owing every thing to His unlimited bounty;—I cannot but wish to take this opportunity of expressing my sense of gratitude, by giving the name Georgium Sidus,

Georgium Sidus
—— *jam nunc assuesce vocari.* VIRG. Georg.

* M. de la Lande's *Ast.* § 639.

to a star, which (with respect to us) first began to shine under His auspicious reign.

By addressing this letter to you, SIR, as President of the Royal Society, I take the most effectual method of communicating that name to the Literati of Europe, which I hope they will receive with pleasure. I have the honour to be, with the greatest respect,

SIR,

Your most humble

and most obedient servant,

W. HERSCHEL.

X.

On the Diameter and Magnitude of the Georgium Sidus ; *with a Description of the dark and lucid Disk and Periphery Micrometers. By* William Herschel, *Esq., F.R.S.*

[*Phil. Trans.*, vol. lxxiii., 1783, pp. 4–14.]

Read November 7, 1782.

It is not only of the greatest consequence to the astronomer, but also gives the highest pleasure to every intelligent person, to have a just idea of the dimensions of the solar system, and the heavenly bodies that belong to it. As far then as they fall within the reach of our instruments, they ought carefully to be examined and measured by all the various methods we can invent. Almost every sort of micrometer is liable to some inconveniences and deceptions : it will, however, often happen, that we may correct the errors of one instrument by the opposite defects of another. The measures of the diameter of the Georgium Sidus, which were delivered in my first paper, differ considerably from each other. However, if we set aside the three first, on a supposition (as I have hinted before) that every minute object, which is much smaller than what we are frequently used to see, will at first sight appear less than it really is ; and take a mean of the remaining observations, we shall have $4'' \, 36\frac{1}{2}'''$ for the diameter of the planet. On comparing the measures then with this mean, we find but two of them that differ somewhat more than half a second from it ; the rest are almost all within a quarter of a second of that measure. This agreement, in the dimensions of any other planet, would appear very considerable ; but not being satisfied, when I thought it possible to obtain much more accurate measures, I employed the lamp-micrometer in preference to the former. The first time I used it upon this occasion I perceived, that if, instead of two lucid points, we could have an intire lucid disk to resemble the Planet, the measures would certainly be still more compleat. The difficulty of dilating and contracting a figure that should always remain a circle, appeared to me very considerable, though nature, with her usual simplicity, holds out to us a pattern in the Iris of the eye, which, simple as it appears, is not one of the least admirable of her inimitable works. However, I recollected, that it was not absolutely requisite to have every insensible degree of magnitude ; since, by changing the distance, I could without much inconvenience make every little intermediate gradation between a set of circles of a proper size, that might be prepared for the purpose. Intending to put this design into practice, I contrived the following apparatus.

A large lanthorn, of the construction of those small ones that are used with my lamp-micrometer,* must have a place for three flames in the middle, which is necessary, in order that we may have the quantity of light required, by lighting one, two, or all of them. The grooves, instead of brass sliding doors, must be wide enough to admit a paste-board, and three or four thicknesses of paper. I prepared a set of circles, cut out in paste-board, increasing by tenths of an inch from two inches to five in diameter, and these were made to fit into the grooves of the lamp. A good number of pieces, some of white, others of light blue paper, of the same size with the paste-boards, were also cut out, and several of them oiled, to render them more transparent. The oiled papers should be well rubbed, that they may not stain the dry papers when placed together. This apparatus being ready, we are to place behind the paste-board circle, next to the light, one, two, or more, either blue or white, dry or oiled, papers; and by means of one or more flames, to obtain an appearance perfectly resembling the disk we would compare it with. It will be found, that more or less altitude of the object, and higher or lower powers of the instrument, require a different assortment of papers and lights, which must by no means be neglected: for if any fallacy can be suspected in the use of this apparatus, it is in the degree of light we must look for it. In a few experiments I tried with these lucid disks, where I placed several of them together, and illuminated them at once, it was found, that but very little more light will make a circle appear of the same size with another, which is one, or even two-tenths of an inch less in diameter. A well known and striking instance of this kind of deception is the moon, just before or after the conjunction, where we may see how much the luminous part of the disk projects above the rest.

The method of using the artificial disks is the same which has been described with the lamp-micrometer, of which this apparatus may be called a branch. We are only to observe, that the Planet we would measure should be caused to go either just under, or just over, the illuminated circle. It may indeed also be suffered to pass across it; but in this case, the lights will be so blended together, that we cannot easily form a proper judgement of their magnitudes. By a good screw to the motions of my telescope I have been able, at any time, to keep the Planet opposite the lucid disk for five minutes together, and to view them both with the most perfect and undisturbed attention. The apparatus I employed being now sufficiently explained, several alterations that were occasionally introduced will be mentioned in the observations and experiments on the Georgium Sidus, as they follow, in the order of time in which they were made.

* *Phil. Trans.* vol. LXXII. p. 166 [above, p. 93].

Observations on the Light, Diameter, and Magnitude, of the Georgium Sidus.

Oct. 22, 1781. The Georgium Sidus was perfectly defined with a power of 227; had a fine, bright, steady light; of the colour of Jupiter, or approaching to the light of the Moon.

Nov. 28, 1781. I measured the diameter of the Georgium Sidus by the lamp-micrometer, and took one measure, which I was assured was too large; and one, which I was certain was too little; then taking the mean of both, I compared it with the diameter of the star, and found it to agree very well.

Hence $\frac{\text{Image}=2{\cdot}4 \text{ inches}}{\text{Distance}=431 \text{ inches}}$ = tang. ·0055684; and $\frac{\text{Angle}=19'\ 8''}{\text{Power}=227}$ = the diameter 5″·06. But the evening was foggy, and the star having much aberration, I was induced to try the above method of extreme and mean diameters, suggested by the method of altitudes, where two equally distant extremes give us a true mean.

Nov. 19, 1781. The diameter measured 32¼ parts of my micrometer, the wires being outward tangents to the disk. On shutting them gradually by the same light, they closed at 24; therefore the difference is 8¼ parts, which, according to my scale, gives 5″ 2‴ for the diameter. This was taken with 227, and the measure seemed large enough. Not perfectly pleased with my light, which was rather too strong, I repeated the measure, and had 33½ parts; then shutting the wires gradually, by *this* light they closed at 25: the difference, which is 8½ parts, gives 5″ 11‴.

Aug. 29, 1782. 15 h. I saw the Georgium Sidus full as well defined with 460, as Jupiter would have been at that altitude with the same power.

Sept. 9, 1782. Circumstances being favourable, I took a measure of the diameter of the Georgium Sidus with the power of 460, and silk-thread micrometer. After a proper allowance for the zero, I found 4″ 11‴.

Oct. 2, 1782. I had prepared an apparatus of lucid disks, and measured the diameter of the Georgium Sidus with it. Having only white oiled papers, I placed two of them together, and used only a single lamp; but could not exactly imitate the light of the Planet. When I first saw the Sidus and luminous circle together, I was struck with the different colours of their lights; which brought to my recollection γ Andromedæ, ϵ Bootis, α Herculis, β Cygni, and other coloured stars. The Planet unexpectedly appeared blueish, while the lucid disk had a strong tincture of red; but neither of the colours were so vivid and sparkling as those of the just mentioned stars. The distance of the luminous circle from the eye (which I always measure with deal rods) was 588·25 inches. The circle measured 2·35 inches. Hence we have the angle 13′ 44″; which, divided by the power 227, gives 3″·63 for the diameter of the Planet. I suspected some little fallacy from the want of a perfect resemblance in the light and colour of the artificial disk to the real appearance of the Planet.

Oct. 4, 1782. I measured the diameter of the Georgium Sidus again, by an improvement in my apparatus, for I now used pale blue papers, both oiled and plain, instead of white; by which means I obtained a resemblance of colours; and by an assortment of one oiled and two dry papers with two lamps burning, I effected the same degree of light which the Planet had, and both figures were equally well defined. By first changing the disk, and, when I had one which came nearest, changing my distance, I came at a perfect equality between the Planet and disk. The measure was several times repeated with great precaution. The result was $\frac{2{\cdot}8}{692{\cdot}6}={\cdot}0040283$; and $\frac{13'\ 53''{\cdot}85}{227}=3''{\cdot}67$. If any thing be wanting to the perfection of this measure, it is perhaps that the Sidus should be in the meridian, in order to have all the advantages of light and distinctness.

Oct. 10, 1782. The measures of the Planet by the lucid disk micrometer appearing to me very small, I resolved to ascertain the power of my telescope again most scrupulously, by an actual experiment, without any deduction from other principles. On a most convenient and level plain I viewed two slips of white paper, and measured their images upon a wall. The distances were measured by deal rods, every repetition whereof was certainly true to half a tenth of an inch; nor did the direction of the measure ever deviate, so much as two inches, from a straight line.

Distance of the object from the eye in inches	7255·5
Distance of the eye from the vertex of the speculum	80·2
Distance of the vertex of the speculum from the object	7335·7
Distance of the eye from the wall	2292·35
Diameter of the largest paper	·99125
Diameter of the smallest	·5075
Image of the largest paper on the wall	73
Image of the smallest on the same	37·8
Angle subtended by the large paper at the vertex of the speculum 27″·87	
Angle subtended by its image on the wall, at the eye 1° 49′ 26″·4	
Power of the telescope deduced from the large paper	235·6
Angle subtended by the small paper at the vertex of the speculum 14″·27	
Angle subtended by its image on the wall, at the eye, 56′ 40″·9	
Power of the telescope deduced from the small paper	238·3
Mean of both experiments, as being equally good	237
Focal length of the speculum upon those objects	86·1625
Upon Capella	85·2
And 237 diminished in the ratio of 85·2 to 86·1625 gives 234·3 for the power of the instrument upon the fixed stars.	

It appears then, from these experiments, that the power of the telescope has not been over-rated; and that, therefore, the measures of the Georgium Sidus cannot be found too small on that account.

There is one cause of inaccuracy or deception in very small measures, long suspected, but never yet sufficiently investigated. That there is a *dispersion* of the rays of light in their passage through the atmosphere, we may admit from various experiments; if then the quantity of this dispersion be, in general, regulated by certain dispositions of the air, and other causes, it will follow, that a *concentration* may also take place: for should the rays of light, at any time, be less dispersed than usual, they might with as much reason be said to be concentrated, as the mercury of a thermometer is said to be contracted by cold, when it falls below the zero.

Oct. 12, 1782. The night was so fine, that I saw the Georgium Sidus very plainly with my naked eye. I took a measure of its diameter by the lucid disk, and found, that I was obliged to come nearer, as the Planet rose higher, and gained more distinct light. At the altitude of 52° it was as follows:

$$\frac{3{\cdot}415}{731{\cdot}3} = {\cdot}0046698;\ \text{and}\ \frac{16'\ 3''{\cdot}2}{227} = 4''{\cdot}24.$$

Oct. 13, 1782. 16 h. I viewed the Georgium Sidus with several powers. With 227 it was beautiful. Still better with 278. With 460, after looking some time, very distinct. I perceived no flattening of the polar regions, to denote a diurnal motion; though, I believe, if it had had as much as Jupiter, I should have seen it. With 625 pretty well defined.

Oct. 19, 1782. The inconvenience arising from the quantity of light contained in the lucid disk, suggested to me the idea of taking only an illuminated periphery, instead of the area of a circle. By this means I hoped to see the circle well defined, and yet have but little light to interfere with the appearance of the Planet. The breadth of my lucid periphery was one-twentieth of an inch. The result of this measure proved

$$\frac{3{\cdot}3}{765{\cdot}45} = {\cdot}0041486;\ \text{and}\ \frac{14'\ 15''{\cdot}69}{227} = 3''{\cdot}77.$$

Oct. 26, 1782. In my last experiment I found the lucid periphery much broader than I could have wished; therefore, I prepared one of no more than one-fortieth part of an inch in breadth, the outer circle measuring very exactly 4 00, and the inner circle 3·95. With this slender ring of light illuminated with only one single lamp, I measured the Georgium Sidus, by removing the telescope to various distances; and found at last the following result:

$$\frac{4}{1033{\cdot}05} = {\cdot}0038720;\ \text{and}\ \frac{13'\ 18''{\cdot}6}{227} = 3''{\cdot}51.$$

Nov. 4, 1782. I was now fully convinced that light, be it in the form of a lucid circle or illuminated periphery, would always occasion the measures to be less than they should be, on account of its vivid impression upon the eye, whereby the magnitude of the object, to which the Planet was compared, would be increased. It occurred to me then, that if a lucid circle encroached upon the surrounding darker parts, a lucid square border, round a dark circle, would in its turn advance upon the artificial disk. In my last measures, where the Planet had been compared to a lucid ring, I had plainly observed that the Sidus, which was but just equal to the illuminated periphery, was considerably larger than the black area contained within the ring. This seemed to point out a method to discover the quantity of the deception arising from the illumination; and consequently, to furnish us with a correction applicable to such measures; which would be *plus*, when taken with a lucid disk or ring; and *minus*, when obtained from a dark ring or circle. Having suspended a row of paste-board circles against an illuminated sheet of oiled paper, I caused the Georgium Sidus to pass by them several times, and selected from their number that to which the Planet bore the greatest resemblance in magnitude. I produced a perfect equality by some small alteration of my distance, and the result was as follows:

$$\frac{3{\cdot}165}{633{\cdot}95}={\cdot}0049925\text{: hence }\frac{17'\,9''{\cdot}8}{227}=4''{\cdot}53.$$

I was desirous of seeing what would be the effect of lessening the light of the illuminated frame, against which the dark disks were suspended, and also waited a short time that the Planet might rise up higher. The measure being then repeated at a different distance, and with a different black disk, I obtained the following particulars:

$$\frac{3{\cdot}59}{803{\cdot}05}={\cdot}0044704\text{; and }\frac{15'\,22''{\cdot}1}{227}=4''{\cdot}06.$$

I intend to pursue these experiments still farther, especially in the time of the Planet's opposition, and am therefore unwilling as yet to draw a final conclusion from the several measures. In a subject of such delicacy we cannot have too many facts to regulate our judgement. Thus much, however, we may in general surmise, that the diameter of the Georgium Sidus cannot well be much less, nor perhaps much larger, than about four seconds. From this, if we will anticipate more exact calculations hereafter to be made, we may gather that the real diameter of that planet must be between four and five times that of the earth: for by the calculations of M. DE LA LANDE, contained in a letter he has favoured me with, the distance of the Georgium Sidus is stated at 18·913, that of the earth being 1. And if we take the latter to be seen, at the sun, under an angle of 17", it would subtend no more than ·"898, when removed to the orbit of the Georgium Sidus. Hence we obtain $\frac{4}{{\cdot}898}=4{\cdot}454$; which number expresses how much the real diameter of the Georgium Sidus exceeds that of the earth.

XI.

On the proper Motion of the Sun and Solar System; with an Account of several Changes that have happened among the fixed Stars since the Time of Mr. FLAMSTEED. *By* WILLIAM HERSCHEL, *Esq. F.R.S.*

[*Phil. Trans.*, vol. lxxiii., 1783, pp. 247–283.]

Read March 6, 1783.

THE new lights that modern observations have thrown upon several interesting parts of astronomy begin to lead us now to a subject that cannot but claim the serious attention of every one who wishes to cultivate this noble science. That several of the fixed stars have a proper motion is now already so well confirmed, that it will admit of no further doubt. From the time this was first suspected by Dr. HALLEY we have had continued observations that shew Arcturus, Sirius, Aldebaran, Procyon, Castor, Rigel, Altair, and many more, to be actually in motion; and considering the shortness of the time we have had observations accurate enough for the purpose, we may rather wonder that we have already been able to find the motions of so many, than that we have not discovered the like alterations in all the rest. Besides, we are well prepared to find numbers of them apparently at rest, as, on account of their immense distance, a change of place cannot be expected to become visible to us till after many ages of careful attention and close observation, though every one of them should have a motion of the same importance with Arcturus. This consideration alone would lead us strongly to suspect, that there is not, in strictness of speaking, one *fixed* star in the heavens; but many other reasons, which I shall presently adduce, will render this so obvious, that there can hardly remain a doubt of the general motion of all the starry systems, and consequently of the solar one among the rest.

I might begin with principles drawn from the theory of attraction, which evidently oppose every idea of absolute rest in any one of the stars, when once it is known that some of them are in motion: for the change that must arise by such motion, in the value of a power which acts inversely as the squares of the distances, must be felt in all the neighbouring stars; and if these be influenced by the motion of the former, they will again affect those that are next to them, and so on till all are in motion. Now as we know several stars, in divers parts of the heavens, do actually change their place, it will follow, that the motion of our solar system is not a mere hypothesis; and what will give additional weight to this consideration is, that we have the greatest reason to suppose most of those very stars, which have been observed to move, to be such as are nearest to us; and, therefore, their influence on our situation would alone prove a powerful argument in favour of the proper motion of the sun, had it actually been originally at rest. But I shall waive every view of this subject which is not chiefly derived from experience.

To begin with my own, I will give a short but general account of the most striking changes I have found to have happened in the heavens since FLAMSTEED's time. I have now almost finished my third review. The first was made with a Newtonian telescope, something less than 7 feet focal length, a power of 222, and an aperture of 4½ inches. It extended only to the stars of the first, second, third, and fourth magnitudes. Of my second review I have already given some account*: it was made with an instrument much superior to the former, of 85·2 inches focus, 6·2 inches aperture, and power 227. It extended to all the stars in HARRIS's maps, and the telescopic ones near them, as far as the eighth magnitude. The catalogue of double stars, which I have had the honour of communicating to the Royal Society, and the discovery of the Georgium Sidus, were the result of that review. My third review was with the same instrument and aperture, but with a very distinct power of 460, which I had already experienced to be much superior to 227, in detecting excessively small stars, and such as are very close to large ones. At the same time I had ready at hand smaller powers to be used occasionally after any particularity had been observed with the higher powers, in order to see the different effects of the several degrees of magnifying such objects. I had also 18 higher magnifiers, which gave me a gradual variety of powers from 460 to upwards of 6000, in order to pursue particular objects to the full extent of my telescope, whenever a favourable interval of remarkably fine weather presented me with a proper opportunity for making use of them. This review extended to all the stars in FLAMSTEED's catalogue, together with every small star about them, as far as the tenth, eleventh, or twelfth magnitudes, and occasionally much farther, to the amount of a great many thousands of stars. To shew the practicability of what I have here advanced, it may be proper to mention, that the convenient apparatus of my telescope is such, that I have many a night, in the course of eleven or twelve hours of observation, carefully and singly examined not less than 400 celestial objects, besides taking measures of angles and positions of some of them with proper micrometers, and sometimes viewing a particular star for half an hour together, with all the various powers of my telescope. The particularities I attended to in this last review were, 1. The existence of the star itself, such as it is given in the British catalogue. 2. To observe well whether it was double or single, well defined or hazy. 3. To view and mark down its particular

* *Phil. Trans.* vol. LXX. LXXI. LXXII.

colour, whenever the altitude and situation of the star would permit it to be done with certainty. 4. To examine all the small stars in the neighbourhood, as far at least as the twelfth magnitude, and note the same particulars concerning them, except the colours, which would have taken up too much time in committing to paper, and be of no very material use. The result of these observations I shall collect under a few general heads in the following articles.

I.

Stars that are lost, or have undergone some capital change, since FLAMSTEED's time.*

In the *British Catalogue* we find two remarkable stars of the fourth magnitude in the constellation of Hercules, *viz.* the 80th and 81st. They are no more to be seen. I looked for them in October, 1781, but could not find them; and have since frequently examined that part of the heavens with no better success, though the small stars κ, z, y, of the sixth magnitude, not far from the place where the former should be, appear very plainly in a fine evening. On referring to my preceding review in August, 1780, I find, that I then also examined the left foot of Hercules, and had these stars been there at that time I must have seen them; for not only ι of the fourth magnitude, but y of the sixth, is in the list of stars examined; and the place of the 80th and 81st is very near directly between them.

In the northern claw of Cancer FLAMSTEED has placed three stars of the sixth magnitude; they are the 53d, 55th, and 56th of his catalogue. The latter of them is vanished. Its place and magnitude are so well pointed out by the other two stars, that there can remain no doubt of this remarkable change. We find a very small telescopic star near the place where the 56th should be: this may possibly be the remains of that vanishing star; but that may be ascertained by those astronomical gentlemen, who, having fixed instruments, can determine the place of this small star, and compare it with the 56th of Cancer, when it will appear how far their places agree. I missed it first in February, 1782, and have since looked often for it in vain.

The 19th Persei, a star of the 6th magnitude, is either lost, or so considerably removed from its place in FLAMSTEED's time, that it is no longer to be known. What gives occasion for a suspicion of its having been removed is, that we find no star of the sixth magnitude in the place where this should be; whereas, about a degree following that situation, there is one of that, or the fifth magnitude, not taken notice of by FLAMSTEED, who could not overlook so considerable a star if it had been there in his time; because, being not far from the parallel of τ and υ,

* [About these stars see Baily's edition of the *British Catalogue*, p. 393. Twenty-two cases of "stars observed but not existing" were left undecided by Baily (p. 646), but they were all cleared up by C. H. F. Peters in a paper published in the *Memoirs of the National Academy of Sciences*, vol. iii. Every single star in the *British Catalogue* has been accounted for.—ED.]

both which he has given us, it must have passed the field of view of his telescope whenever he observed them.

The 108th Piscium, a star of the sixth magnitude, near the head of Aries is lost. The 107th and 109th, both marked as smaller stars in FLAMSTEED's catalogue, and near the place of the 108th, are easily discovered.

Two stars of the sixth magnitude, the 73d and 74th Cancri, in the southern claw, are either both lost, or at least have undergone such a remarkable change of magnitude, and one of them of place, that it is hardly possible to know them any longer. The alteration must evidently appear when we compare them to the 81st and 82d Cancri; the former of which, though only marked of the 7th magnitude, far outshines the brightest of those which may be supposed to be the two stars in question.

The 8th Hydræ is lost. There is a star just by, which I take to be the 31st Monocerotis. If this should be the 8th Hydræ, and a small star near the latter should agree with the place of the 31st Monocerotis, then the magnitudes will be quite contrary to what FLAMSTEED makes them. There must, at all events, have been a very remarkable change.

The 26th Cancri is lost. Near this star is placed the 22d of the same constellation, and, as their distance is not much more than a quarter of a degree, it requires fixed instruments to determine which of the two is the star wanting: from the magnitude, however, I surmise, that the remaining star is the 22d rather than the 26th.

The 62d Orionis is lost; and a star near the 54th and 51st is not taken notice of by FLAMSTEED. Perhaps the 62d has changed place; if this should be the case, it must have a very considerable motion.

The 71st Herculis, a star of the 5th magnitude, is lost. The 70th and 71st are so near each other by FLAMSTEED's catalogue, that it cannot be determined without fixed instruments, which is the star wanting. There is a small telescopic star, within about 30 minutes north following, in a direction towards μ Lyræ; if that should be the 71st, it is wonderfully changed both in place and size. The 40th star in Mr. MAYER's collection of double stars * seems to be the 70th Herculis of FLAMSTEED. Now, as that star is perfectly single in my telescope, with every power I have tried upon it, we may surmise that one of the stars which is now vanished was still visible in the year 1778, when Mr. MAYER observed it, though then already diminished from the 5th to the 8th magnitude.

The 34th Comæ Berenices is lost: FLAMSTEED has marked it as a star of the 5th magnitude.

The 19th of the same constellation is also lost, or moved and changed in magnitude.

* *De novis in Cœlo sidereo phænomenis.*

The 40th and 41st Draconis have undergone so great an alteration of place that we cannot possibly mistake it; for in FLAMSTEED's time they were above three minutes asunder, whereas now their distance is much less than half a minute. A more particular account of these two stars will be given in a second collection of more than 400 new double stars, observed in my third review, which I hope soon to have the honour of presenting to the Royal Society.

There seems to be an alteration in the place of the 65th, 64th, 54th, and 57th Orionis; but without fixed instruments I cannot ascertain in which of the stars it is. Their situation in the heaven does not agree with that which is delineated in FLAMSTEED's *Atlas Cœlestis*, for these two pair of stars are much nearer now than they should be, according to that account.

II.

Stars that have changed their magnitude since FLAMSTEED's time.*

α Draconis is so much less than β, which is set down as a smaller star in FLAMSTEED's catalogue, that the change of magnitude cannot be doubted.

β Ceti marked of the 3d, and α Ceti of the 2d, are evidently the reverse, β being by much the larger star. I have mentioned this circumstance in my observations on the periodical star in Collo Ceti,† and it seems now as if the difference between the magnitudes of these two stars was still increasing.

ζ Serpentis is not near so large as η, and yet we find FLAMSTEED has placed them in the same class: however, we cannot intirely confide in the marks of the magnitudes when two stars are placed in the same class, since every order admits of a considerable variety; but when the marks contradict experience so far as to describe one star, for instance, of the third, and another of the 4th magnitude, when observation shews the latter to be of the 3d and the former of the 4th, I think we can hardly doubt but that there must have been a change.

η Cygni is a brighter star than χ, though marked by FLAMSTEED of a less magnitude.

The 2d Ursæ minoris is marked of the 6th magnitude, but is certainly intitled to the 5th.

η Bootis is much larger than ζ.

ι Delphini is much larger than κ.

β Trianguli is much larger than α.

γ Aquilæ is much larger than β.

σ Sagittarii is larger than δ, γ, and ϵ, though marked of an inferior magnitude.

δ Canis majoris is larger than β, and yet is marked to be less.

η Serpentis is so much larger than ζ, that they certainly should not have been put in the same order of magnitude.

* [Compare Baily, *l.c.*, p. 405.—ED.]

† *Phil. Trans.* vol. LXX. numb. XXI. [above, p. 2.—ED.].

κ Serpentarii is larger than γ and ϵ, though marked to be of a less magnitude than either.

β Equulei is so much less than α that it could hardly deserve to be put in the same class.

δ Delphini is larger than ϵ, though placed in an inferior order.

ϵ Bootis is so much larger than ζ that it should not be put into the same order.

δ Sagittæ is larger than α and β, though placed in a lower order of magnitude.

δ Ursæ majoris is less than either ϵ, ζ, or η, though it is marked of a superior order of magnitude. Besides, it is evidently visible, that δ cannot be intitled to more than the 4th magnitude, or at most to between the 4th and 3d: on the contrary, ϵ, ζ and η, should be of the 2d, or at least between the 2d and 3d; all which is very different from FLAMSTEED's account of those remarkable stars.

α Ursæ majoris is less than any star marked of the same magnitude, and cannot have the least pretension to be called a star of between the 1st and 2d, as FLAMSTEED has marked it, and as I make no doubt it was in his time.

The 1st and 2d Hydræ are noted by FLAMSTEED as being of the 4th magnitude, whereas they now are only of the 8th or 9th. It is remarkable, that the 30th Monocerotis, which is situated between them, has retained the order assigned to it by FLAMSTEED, and being of the 6th serves to point out the change of the other two in a very evident manner.

γ Lyræ is much larger than β.

The change in the magnitudes of the 31st and 34th Draconis is very striking, these two stars being just the contrary of what they are marked in FLAMSTEED's catalogue. The 31st from the 7th is increased to the 4th; and the 34th, from being a star between the 4th and 5th, is reduced to one of the 6th, if not 7th magnitude.

The 44th Cancri is much too small for the 6th magnitude. As ϵ and others are marked of the 6th, this, on being compared to them, can be intitled to no more than the 8th or 9th order.

The 96th Tauri is small enough to be of the 8th magnitude, though marked as one of the 6th.

The 62d Arietis is of the 5th magnitude, though only marked of the 6th.

The magnitudes of the 12th and 14th Lyncis are just the reverse in the heavens to what FLAMSTEED has marked them. This denotes a double change of a star from the 5th to the 7th, and from the 7th to the 5th magnitude.

The 38th Persei, marked of the 6th magnitude, is increased so as to be equal to θ and κ of the 4th. Also, θ is less than τ contrary to FLAMSTEED.

The 8th Monocerotis is less than the 76th Orionis, though the former should be of the 4th, and the latter only of the 6th magnitude.

The 23d Geminorum, though marked of the 5th, is less than the 21st of between the 6th and 7th magnitude.

The 26th Orionis is much too small for the magnitude of which it is marked to be, or rather is lost; for I can hardly take any one of the remaining telescopic stars for it.

ξ Leonis in FLAMSTEED's time was of the 4th; but is now less than a star of the 5th magnitude.

III.

Stars newly come to be visible.

Near Lacerta's tail-end is a star of between the 4th and 5th magnitude, not mentioned in FLAMSTEED's catalogue, though the 1st Lacertæ, not far from that place, is recorded. It is so easy to be seen with the naked eye, and in a spot where but few stars of that magnitude are near, that we can hardly account for its being omitted if it had been visible to FLAMSTEED. Its colour is pale red.

The star of the 5th magnitude following τ Persei, supposed to be υ removed, is most likely new, unless future observations were to favour the supposed motion of this star. It is among the double stars of my 4th class, so that it will be easy to detect its proper motion.

A very considerable star, not marked by FLAMSTEED, will be found near the head of Cepheus. Its right ascension in time, is about 2′ 19″ preceding FLAMSTEED's 10th Cephei, and it is about 2° 20′ 3″ more south than the same star. It is of a very fine deep garnet colour, such as the periodical star ο Ceti was formerly, and a most beautiful object, especially if we look for some time at a white star before we turn our telescope to it, such as α Cephei, which is near at hand.

A considerable star in a direction from the 68th Geminorum towards the 61st is not to be found in FLAMSTEED, its colour is red.

A star of a considerable magnitude preceding the 1st Equulei is not contained in FLAMSTEED's catalogue. It is a double star of the first class, the 61st of my second collection, where measures of it will be found.

A considerable star following the 1st Sextantis, and another following the 7th, are not inserted.

Between β Cancri and δ Hydræ is a very considerable star not marked by FLAMSTEED, though its situation is very remarkable. As the constellation of Cancer contains so rich a collection of very small stars, it is to be wondered how a star of such consequence could be omitted, if it had been visible in FLAMSTEED's time.

Nearly 1½ degree north following δ Herculis, almost in the direction of δ and ν, is a star of the 5th, or between the 4th and 5th magnitude, very visible to the naked eye. We can hardly think FLAMSTEED could have overlooked it, had it been there in his time.

About 3 degrees south preceding γ Bootis, a considerable star not in FLAMSTEED's catalogue of the 6th magnitude; and south preceding λ, another, almost as large.

Here we ought to observe, that it is not easy to prove a star to be newly come; for though it should not be contained in any catalogue whatsoever, yet the argument for its former non-appearance, which is taken from its not having been observed, is only so far to be regarded as it can be made probable, or almost certain, that a star would have been observed had it been visible. For these reasons I will lay no particular stress on the new appearance of the above stars; they are, however, such as do well deserve to have their places settled, while I shall leave it to others to determine how far they may think them to be new visitors to those starry regions that fall within the reach of our sight.

To return to the principal subject of this paper, which is the proper motion of the sun and solar system: does it not seem very natural, that so many changes among the stars,—many increasing their magnitude, while numbers seem gradually to vanish;—several of them strongly suspected to be new-comers, while we are sure that others are lost out of our sight;—the distance of many actually changing, while many more are suspected to have a considerable motion:—I say, does it not seem natural that these observations should cause a strong suspicion that most probably every star in the heavens is more or less in motion? And though we have no reason to think, that the disappearance of some stars, or new appearance of others, nor indeed the frequent changes in the magnitudes of so many of them are owing to their change of distance from us by proper motions, which could not occasion these phenomena without being inconceivably quick; yet we may well suppose, that motion is some way or other concerned in producing these effects. A slow motion, for instance, in an orbit round some large opaque body, where the star, which is lost or diminished in magnitude, might undergo occasional occultations, would account for some of those changes, while others might perhaps be owing to the periodical return of large spots on that side of the surface which is alternately turned towards us by a rotatory motion of the star. The idea also of a body much flattened by a quick rotation, and having a motion similar to the moon's orbit by a change of the place of its nodes, whereby more of the luminous surface would one time be exposed to us than another, tends to the same end; for we cannot help thinking with Mr. DE LA LANDE (*Mem.* 1776), that the same force which gave such rotations, would probably also occasion motions of a different kind by a translation of the center.* Now, if the proper motion of the stars in general be once admitted, who can refuse to allow that our sun, with all its planets and comets, that is, the solar system, is no less liable to such a general agitation as we find to obtain among all the rest of the celestial bodies?†

* Relating to the motion of the fixed stars, the Astronomer Royal has an expression in the second page of the explanation and use of the tables published in his *Astronomical Observations*, which seems to favour this idea, where he mentions the "peculiar but small motions, which many, IF NOT ALL OF THEM, have among themselves, which have been called their *proper motions*, the causes and laws of which are hid for the present in almost equal obscurity."

† See Mr. MICHELL's note, *Phil. Trans.* vol. LVII. p. 252.

Admitting this for granted, the greatest difficulty will be how to discern the proper motion of the sun between so many other (and variously compounded) motions of the stars. This is an arduous task indeed, which we must not hope to see accomplished in a little time; but we are not to be discouraged from the attempt. Let us, at all events, endeavour to lay a good foundation for those who are to come after us. I shall therefore now point out the method of detecting the direction and quantity of the supposed proper motion of the sun by a few geometrical deductions, and at the same time shew by an application of them to some known facts, that we have already some reasons to guess which way the solar system is probably tending its course.

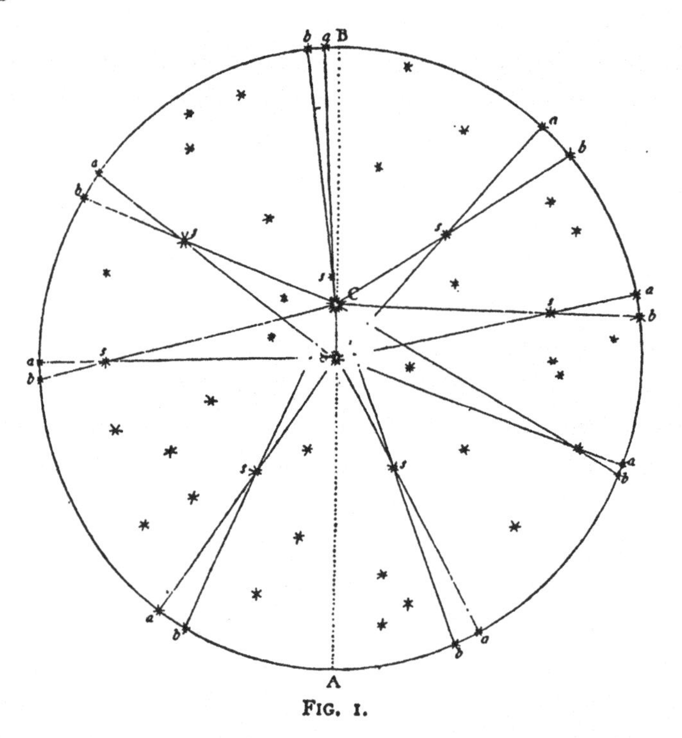

Fig. 1.

Suppose the sun to be at S, fig. 1; the fixed stars to be dispersed in all possible directions and distances around at *s*, *s*, *s*, *s*, &c. Now, setting aside the proper motion of the stars, let us first consider what will be the consequence of a proper motion in the sun; and let it move in a direction from A towards B. Suppose it now arrived at C. Here, by a mere inspection of the figure, it will be evident, that the stars *s*, *s*, *s*, which were before seen at *a*, *a*, *a*, will now, by the motion of the sun from S to C, appear to have gone in a contrary direction, and be seen at *b*, *b*, *b*; that is to say, every star will appear more or less to have receded from the point B, in the order of the letters *ab*, *ab*, *ab*. The converse of this proposition is equally true; for if the stars should all appear to have had a retrograde motion, with respect to the point B, it is plain, on a supposition of their being at rest, the sun must have a direct motion towards the point B, to occasion all these appearances. From a due consideration of what has been said, we may draw the following inferences.

1. The greatest or total systematical parallax of the fixed stars, fig. 2, will fall upon those that are in the line DE, at rectangles to the direction AB of the sun's motion.

2. The partial systematical parallax of every other star, *s*, *s*, *s*, not in the line DE, will be to the total parallax as the sine of the angle BS*a*, being the star's distance from that point towards which the sun moves, to radius.

3. The parallax of stars at different distances will be inversely as those distances; that is, one half at double the distance, one third at three times, and so

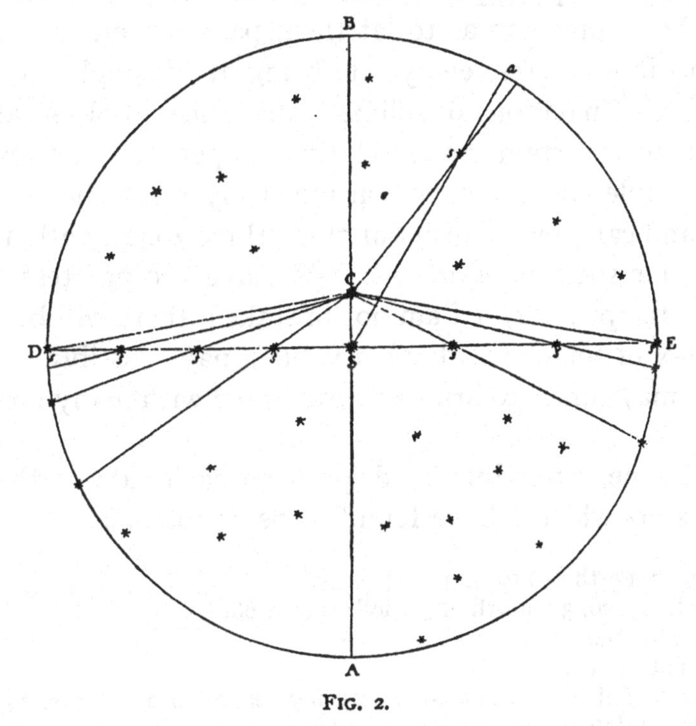

Fig. 2.

on; for the subtense SC remaining the same, and the parallactic angle being very small, we may admit the angle S*s*C, to be inversely as the side S*s*, which is the star's distance.

4. Every star at rest, to a system in motion, will appear to move in a direction contrary to that in which the system is moving.

Corollary. Hence it follows, that if the solar system be carried towards any star situated in the ecliptic: every star, whose angular distance *in antecedentia* (reckoned upon the ecliptic from the star towards which the system moves) is less than 180 degrees, will decrease in longitude. And that, on the contrary, every star, whose distance from the same star (reckoned upon the ecliptic but *in consequentia*) is less than 180 degrees, will increase in longitude, in both cases without alteration of latitude.

From these principles it would be easy to draw general theorems for every possible direction of the motion of the solar system, by which we might find what alteration of longitude or latitude would take place in any given star; but it will be time enough for those investigations when we shall have more immediate occasion for them. What we are now chiefly to endeavour at is, the speedy method of obtaining sufficient facts to proceed upon.

The immense regions of the fixed stars may be considered as an infinitely expanded globe, having the solar system for its center. With this idea it will occur to us, that no method can be so proper for finding out the direction of the motion of the sun as to divide our observations on the systematical parallax of the fixed stars into three principal zones. These, for the convenience of fixed instruments, may be assumed so as to let them pass around the equator and the equinoctial and solstitial colures, every one being at rectangles to the other two, according to the three dimensions of solids. And since no observations can be so conveniently made to ascertain small relative proper motions among the fixed stars as those on double stars, I have continued my researches in that line with great application, and can now furnish out these three zones, with a very complete set of double stars for such observations. We have the greatest reason to hope for success in this attempt; for, if I am not mistaken, there will be found a secular systematical parallax of some considerable value; nay, possibly, so short a space of time as ten years may suffice to bring us acquainted with many hitherto unknown celestial motions.

The equatorial zone, extending 10 degrees on each side of the equator, will contain about 150 stars which I have found to be double, *viz.*

Piscium 38. 51. 77. 86. north of 110. 113.
Ceti south of 13. foll. 25. 26. 37. north of 37. foll. 54. 61. 66.
Eridani 32. near 48. 62. 69.
Tauri near 10. 45. foll. 66. 88.
Orionis foll. 1. near 10. foll. 10. 19. pre. 20. 20. 23. near 26. 28. near 28. pre. 29. near 30. between 30 and 33. 32. 33. 34. foll. 34. pre. 36. 39. 41. near 42. 44. another 44. pre. 46. north of 46. foll. 47. 48. 50. 52. pre. 58. 58. 59.
Monocerotis 8. 11. near 11. In naribus. Sub genam. Inter pedes. near 12. foll. 15. pre. 25. foll. 25. 29. 31. near 31.
Canis minoris near 10. foll. 10. 14.
Cancri 17.
Hydræ pre. 4. pre. 4. foll. 4. 15. 17. pre. 22. 22. foll. 22. 27. 30. 31.
Leonis 3. foll. 3. pre. 43. south of 43. 57. foll. 63. 74. pre. 75. 83. 84.
Virginis 4. 17. 25. 29. pre. 44. 44. 51. 51. 84. pre. 93. 93.
Libræ 17. near 31.
Serpentis 58. 59. 63.
Serpentarii near 11. 53. 61. 67. south of 67. 70.
Aquilæ 2. near 6. near 6. pre. 7 and 8. 15. 23. pre. 30. near 35. near 35. 49. 53. near 54. 57. near 63. 64. near 65. near 65.
Delphini 1.*
Aquarii 4. 4. 22. 24. 51. 55. south of 72.
Equulei 1. pre. 1. south of 2. south of 6. 7.

* [Not Fl. 1, but P. XX. 178 = Σ 2690.—ED.]

Pegasi 3. near 3. 8. near 18.
Piscium near 7. 8. south of 10. 35.
Ceti south of 4.

The zone of the equinoctial colure, extending 10 degrees of a great circle on each side, will contain, as far as it is visible in our hemisphere, about 70 double stars, *viz.*

Ceti foll. 4. near 13.
Aquarii 107.
Piscium 51. 38. 35. 76.
Andromedæ 21. Supra caput. 29. pre. 23. near 27. near 16. near 17.
Cassiopeæ pre. 25. prec. 25. 8. 24. 18. 6. 9. south of 11. 55. 34. 31. 4. 3. 2. near 33. 35. 36. near 3. 44. 47.
Cephei 1. foll. 32. foll. 32. foll. 31.
Ursæ Minoris 1. 18.
Draconis 40 and 41. between 10 and 11. near 77.
Ursæ Majoris 79. between 50 and 38. foll. 42. 65. 57. near 58. south of 69.
Canum venaticorum 12. 2.
Comæ Berenices 24. from 36. to 26. 2. 12.
Leonis 95. 90. 93. pre. 95.
Virginis 29. pre. 44. between 4 and 6. 17. 25. 27. south of 12. 4.
Corvi 7.
Crateris 17. foll. 21.

The zone of the solstitial colure, of the same extent, will include about 120 double stars, *viz.*

Canis majoris north of 13. 17. foll. 5. foll. 2. foll. 5.
Monocerotis 11. 8. 12. near 11. In naribus. foll. 15. Inter pedes. Sub genam.
Leporis 13.
Orionis 48. near 42. 41. 44. 44. 50. foll. 72. 58. In fuste. 59. pre. 70 and 76. pre. 58. pre. 69.
Geminorum 43. 38. pre. 1. between 13 and 18. 21. 15. north of 19. 12. near 34. 27. 37. near 24. near 24. south of 18. foll. 13 and 18.
Tauri north of 123.
Aurigæ 29. 13. 37. 34. 32. south of 34. 56. near 59. pre. 58. pre. Nebulam. 41.
Lyncis 12. In naribus. 5. 13. In pectore. 19.
Camelopardali. In aure.
Ursæ minoris 1. near 18. near 12.
Draconis 39. 63. 20. 21. 56.* 31. 47. 24 and 25. 69. 46. near 31. 40 and 41. 48. near 23. near 77.
Cephei 1. 8.
Herculis 65. 75. 95. over 85. near 87. 100. 86. near 94. near 103.
Lyræ 4. 5. 6. 10. 3. 8. 11. near 3. near 6 and 7. foll. 5. foll. 10. pre. 10. pre. 14 and 15.
Serpentarii 70. 54. 61. 53. 67. near 67. north of 72. north of 72.
Serpentis 59. 63. 58.
Aquilæ 2. 5. near 6. near 6.
Sagittarii 13. 38.

It will not be amiss to add a zone of the ecliptic, which will contain, among others, a great many double stars that may undergo occultations by the moon or planets. This is of the same extent, and includes about 120 double stars.

* [Not Fl. 56, but B. 2421 = Σ 2452.—Ed.]

Arietis south of 4. 5. pre. 6. north of 6. pre. 17. 30. 41. 42. pre. 54. foll. 62. pre. 63. foll. 63. foll. 63. foll. 63.
Tauri near 4. 7. pre. 8. foll. 11. 30. 52. 59. 62. 68. foll. 68. pre. 74. 87. 94. 103. north of 103. 105. 111. foll. 112. 114. foll. 117. 118. north of 123. κ.
Orionis 68. pre. 69. pre. 70 and 76.
Aurigæ 14. 26. pre. Nebulam.
Geminorum pre. 1. 4. 12. foll. 13 and 18. 15. 18. 21. pre. 24. near 24. near 24. 27. 37. 38. 54. foll. 55. pre. 61. 63. north of 63. 66. 78. foll. 78. foll. 81.
Cancri 17. 16. foll. 16. 22. 23. 24. 30. 48. 54. pre. 77. ω. foll. ι.
Leonis 2. 3. foll. 3. 6. 7. 14. 25. 32. 41. south of 43. pre. 43. foll. 44. 57. foll. 63. 74. pre. 75. 83. 84.
Virginis 4. foll. 4. pre. 4. 17. 25. 44. 51.
Hydræ 54.
Libræ pre. 12. 18. 24. 31. 51.
Scorpii 8. 14.
Serpentarii 5. 38. 39.
Sagittarii 13. 38. foll. 43. 64. near 65.
Capricorni north of 1. 5. 7. 9. 11. 12. near 29. 29. 39.
Aquarii 4. 14. 22. foll. 43. 51. 55. 57. 70. 71. 72. 91. 94.
Piscium near 7. south of 8. 35. 38. 51. 77. 86. 110. 113.
Ceti foll. 4. near 13. 26. 54.

An account of each of these double stars, not already in my first catalogue, will be contained in the second collection. It remains now only for me to make an application of this theory to some of the facts we are already acquainted with, relating to the proper motion of the stars. And first let me observe, that the rules of philosophizing direct us to refer all phenomena to as few and simple principles as are sufficient to explain them. Thus, for instance, we see the stars and planets rise and set every day: now, as it is much more simple to admit the earth to turn once in 24 hours, than to suppose every single star to revolve round the earth in that time, we very justly ascribe a diurnal motion to the earth; but yet, since we find that the planets do not every night exactly retain their relative places among the stars, we next admit that such deviations from the law, which all the rest seem to obey, are owing to a proper motion of their own. To apply this to the solar system.—Astronomers have already observed what they call a proper motion in several of the fixed stars, and the same may be supposed of them all. We ought, therefore, to resolve that which is common to all the stars, which are found to have what has been called a proper motion, into a single real motion of the solar system, as far as that will answer the known facts, and only to attribute to the proper motion of each particular star the deviations from the general law the stars seem to follow in those movements.

By Dr. MASKELYNE'S account of the proper motion of some principal stars,* we find that Sirius, Castor, Procyon, Pollux, Regulus, Arcturus, and α Aquilæ, appear to have respectively the following proper motions in right ascension. $-0''\cdot63$; $-0''\cdot28$; $-0''\cdot80$; $-0''\cdot93$; $-0''\cdot41$; $-1''\cdot40$; and $+0''\cdot57$; and two

* *Astronomical Observations made at the Royal Observatory at Greenwich.*

of them, Sirius and Arcturus, in declination, *viz.* 1″·20 and 2″·01, both southward. Let fig. 3 represent an equatorial zone, with the above mentioned stars referred to it, according to their respective right ascensions, having the solar system in its center. Assume the direction AB from a point somewhere not far from the 77th degree of right ascension to its opposite 257th degree, and suppose the sun to move in that direction from S towards B; then will that one motion answer that of all the stars together: for if the supposition be true, Arcturus, Regulus, Pollux, Procyon, Castor, and Sirius, should appear to decrease in right ascension, while α Aquilæ, on the contrary, should appear to increase. Moreover, suppose the sun to ascend at the same time in the same direction towards some point in the northern hemisphere, for instance, towards the constellation of Hercules; then will also the observed change of declination of Sirius and Arcturus be resolved into the single motion of the solar system. I am well aware of the many yet remaining difficulties, such as the correspondence of the exact quantity of each star's observed proper motion with the quantity that will be assigned to it by this hypothesis; but we ought to remember, that the very different and still unknown relative distances of the fixed stars must, for a good while yet, leave us in the dark about the particular and strict application of the theory; and that any deviation from it may easily be accounted for by the still unknown *real* proper motion of the stars: for if the solar system have the motion I ascribe to it, then what astronomers have already observed concerning the change of place of the stars, and have called their proper motion, will become only an *apparent* motion; and it will still be left to future observations to point out, by the deviations from the general law which the stars will follow in those apparent motions, what may be their real proper motions as well as relative distances. But lest I should be censured for admitting so new and capital a motion upon too slight a foundation, I must observe, that the concurrence of those seven principal stars cannot but give some value to an hypothesis that will simplify the celestial motions in general. We know that the sun, at the distance of a fixed star, would appear like one of them; and from analogy we conclude the stars to be suns. Now, since the apparent motions of these seven stars may be accounted for, either by supposing them to move just in the manner they appear to do, or else by supposing the sun alone to have a motion in a direction, somehow not far from that which I have assigned to it, I think we are no more authorised to suppose the sun at rest than we should be to deny the diurnal motion

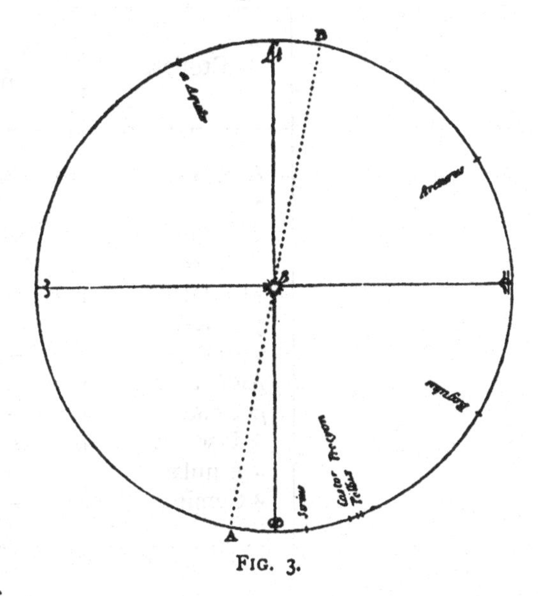

FIG. 3.

of the earth, except in this respect, that the proofs of the latter are very numerous, whereas the former rests only on a few though capital testimonies. But to proceed: I have only mentioned the motions of those seven principal stars, as being the most noticed and best ascertained of all; I will now adduce a farther confirmation of the same from other stars.

M. DE LA LANDE gives us the following table of the proper motion of 12 stars, both in right ascension and declination, in 50 years.*

Etoiles.	Chang. d'asc. droite.	Chang. de déclinaison.
	′ ″	′ ″
Arcturus	−1 11	−1 55
Sirius	− 37	− 52
β Cygni	− 3	+ 49
Procyon	− 33	− 47
ε Cygni	+ 20	+ 34
γ Arietis	− 14	− 29
γ Gemin.	− 8	− 24
Aldébaran	+ 3	− 18
β Gemin.	− 48	− 16
γ Piscium	+ 53	+ 7
α Aquilæ	+ 32	− 4
α Gemin.	− 24	− 1

Fig. 4 represents them projected on the plane of the equator. They are all in the northern hemisphere, except Sirius, which must be supposed to be viewed in the concave part of the opposite half of the globe, while the rest are drawn on the convex surface. Regulus being added to that number, and Castor being double, we have 14 stars. Every star's motion, except Regulus, is assigned in declination as well as in right ascension, so that we have no less than 27 motions given to account for. Now, by assuming a point somewhere near λ Herculis, and supposing the sun to have a proper motion towards that part of the heaven, we shall satisfy 22 of these motions. For β Cygni, α Aquilæ, ε Cygni, γ Piscium, γ Arietis, and Aldebaran, ought, upon the supposed motion of the sun, to have an apparent progression, according to the hour circle XVIII, XIX, XX, &c. or to increase in right ascension, while Arcturus, Regulus, the two stars of α Geminorum, Pollux, Procyon, Sirius, and γ Geminorum, should apparently go back in the order XVI, XV, XIV, &c. of the hour circle, so as to decrease in right ascension; but according to M. DE LA LANDE's table, excepting β Cygni and γ Arietis, all these motions really take place. With regard to the change of declination, we see that every star in the table should go towards the south; and here we find but three exceptions in β and

* *Ast.* par M. DE LA LANDE, tom. IV. p. 685.

ϵ Cygni, and γ Piscium; so that upon the whole we have but five deviations out of 27 known motions which this hypothesis will not account for. And these exceptions must be resolved into the real proper motion of the stars.

There are also some very striking circumstances in the quantities of these motions that deserve our notice. First, Arcturus and Sirius being the largest of the stars, and therefore probably the nearest, ought to have the most apparent motion, both in right ascension and declination, which is agreeable to observation, as we find by the table. Next, in regard to the right ascension only, Arcturus

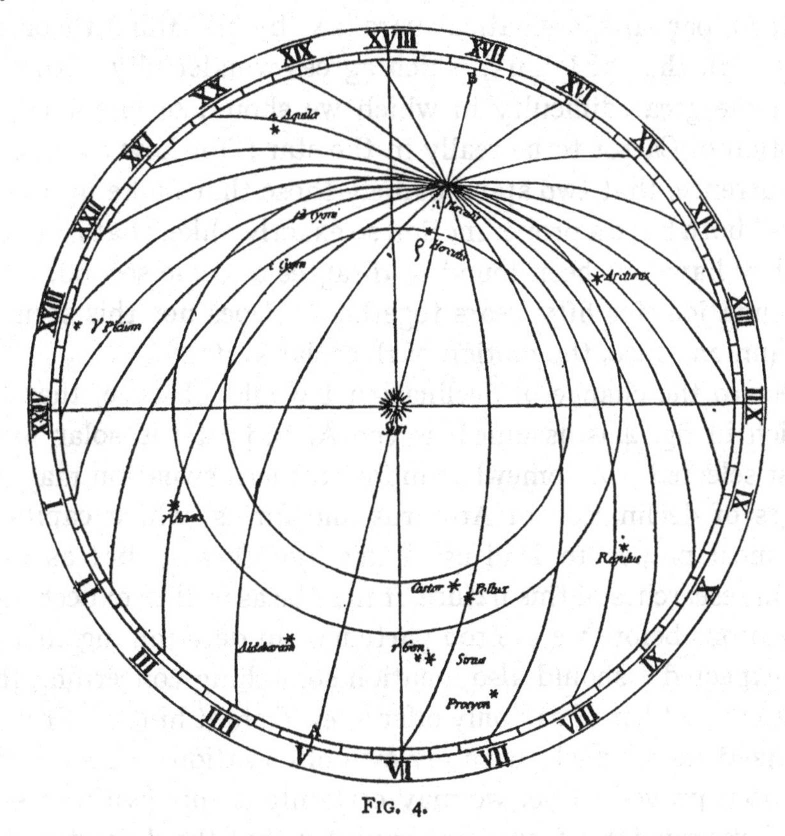

FIG. 4.

being better situated to shew its motion, by theorem 2 [p. 117], ought to have it much larger, which we find it has. Aldebaran, both badly situated and considerably smaller than the two former, by the same theorem ought to shew but little motion. Procyon, better situated than Sirius, though not quite so large, should have almost as much motion; for by the third theorem, on supposing it farther off because it appears smaller, the effect of the sun's motion will be lessened upon it; whereas, on the other hand, by the second theorem, its better situation will partly compensate for its greater distance. This again is conformable to the table. ϵ Cygni very favourably situated, though but a small star, should shew it considerably as well as α Aquilæ; whereas β Cygni should have but little motion: and γ Piscium, best situated of all, should have a great increase of right ascension, and these deductions also agree with the table.

In the last place, a very striking agreement with the hypothesis is displayed in Castor and Pollux. They are both pretty well situated, and we accordingly find that Pollux, for the size of the star, shews as much motion in right ascension as we could expect; but it is remarkable, and seemingly contrary to our hypothesis, that Castor, equally well placed, shews by the table no more than one half of the motion of Pollux. Now, if we recollect that the former is a double star, consisting of two stars not much different in size, we can allow but about half the light to each of them, which affords a strong presumption of their being at a greater distance, and therefore their partial systematical parallax, by the third theorem, ought to be so much less than that of Pollux, which agrees wonderfully with observation.* Not to mention the great difficulty in which we should be involved, were we to suppose the motion of Castor to be really in the star: for how extraordinary must appear the concurrence, that two stars, namely those that make up this apparently single star, should both have a proper motion so exactly alike, that in all our observations hitherto they have not been found to disagree a single second, either in right ascension or declination, for fifty years together! Does not this seem strongly to point out the common cause, the motion of the solar system?

With respect to the change of declination I would observe, that the point of λ Herculis, which in fig. 4 is assumed as the Apex † of the solar motion is not perhaps the best selected. A somewhat more northern situation may agree better with the changes of declination of Arcturus and Sirius, which capital stars may perhaps be the most proper to lead us in this hypothesis; but as we should be guided by facts in researches of this nature, it may be as well to expect the assistance of future observations before we are too particular in determining this point.‡

It may be expected I should also mention something concerning the quantity of the solar motion; but here I can only offer a few distant hints. From the annual parallax of the fixed stars, which, from my own observations, I find much less than it has hitherto been proved to be, we may certainly admit (without entering into a subject which I reserve for a future opportunity) that the diameter of the earth's orbit, at the distance of Sirius or Arcturus, would not nearly subtend an angle of one second; but the apparent motion of Arcturus, if owing to a translation of the solar system, amounts to no less than 2″·7 a year, as will appear if we compound

* If the light of Castor was exactly equal to that of Pollux, and the two stars, which make up the former star, were perfectly of the same size, we might, on that account, suppose the distance of Castor from us to be to that of Pollux as $\sqrt{2} : 1$; but Castor is in fact something less bright; and this consideration, added to the former, will make it probable enough that its distance may perhaps be double that of Pollux.

† I use the term Apex here to denote that point of fig. 4, wherein all great circles, drawn through the supposed direction of the motion of the solar system, intersect, and which, in other stereographic projections, is generally a pole, either of the ecliptic or equator. As this point is in the northern or elevated hemisphere, the sun, by tending to it, may be said to ascend, and the term Apex may perhaps not be an improper one.

‡ From the additional testimony of other capital stars considered in the postscript it now appears, that the point of λ Herculis is probably as well chosen as any we can fix upon in that part of the heavens.

the two motions of 1′ 11″ in right ascension, and 1′ 55″ in declination, into one single motion, and reduce it to an annual quantity. Hence we may in a general way estimate, that the solar motion can certainly not be less than that which the earth has in her annual orbit.

I have now only to add, that it is to be expected future observations will soon throw more light upon this interesting subject, and either fully establish or overturn the hypothesis of the motion of the whole solar system. To this end I have already begun a series of observation upon several zones of double stars; and should the result of them be against these conjectures, I shall be the first to endeavour to point out the fallacy of them.

Datchet near Windsor,
Feb. 1, 1783.

Postscript to the Paper on the Motion of the Solar System.

In my paper on the Motion of the Solar System, I used a table of the proper motion of some fixed stars, which M. DE LA LANDE has given us as an extract from TOB. MAYER's *Opera inedita*. By the favour of my astronomical friend Mr. AUBERT, I am now furnished with the scarce edition of the original. This work contains a catalogue of the place of 80 stars, observed by Mr. MAYER in 1756, and compared with the same stars as given by ROEMER in 1706. From the goodness of the instrument with which the observations to which Mr. MAYER has compared his own were made, he gives it as his opinion, that where the disagreement in the place of a star is but small, it may be attributed to the imperfection of the instrument; but that when it amounts to 10 or 15″, it is a very probable indication of a proper motion of such a star. He adds, that when the disagreement is so much as in some stars which he names, (among which is FOMAHAND, where the difference is 21″ in 50 years) he has not the least doubt of a proper motion.*

By this extensive table I thought it highly necessary immediately to examine the hypothesis of the motion of the solar system, that it might receive an early check from observations, if they should be unfavourable; or that, on the other hand, it might be supported by the additional evidence of more stars, if their apparent proper motions should coincide with the idea I have pointed out in my paper on this subject.

I have followed Mr. MAYER's judgement of his own and ROEMER's observations, and left out of the list all the stars that do not shew a disagreement amounting to 10″ in the places which are given for them in 1706 and 1756. I have also left out

* De motu fixarum proprio Commentatio. *Op. ined.* vol. I. p. 79.

those 13, or rather 14 stars, which have already been examined in my paper, and have been shewn to support the hypothesis I have advanced: the rest are here drawn up in two tables. The first contains the stars that agree with my assigned motion of the solar system; or rather which are thereby resolved into apparent, or partly apparent, and partly proper motions. The second table contains those stars whose motions cannot be accounted for by my hypothesis, and must therefore be ascribed to a real motion in the stars themselves, or to some *still more hidden* cause of a *still remoter* parallax.*

* That I may not be obscure, it will be proper to mention what I allude to, especially as it claims a distant connection with our subject, and may hereafter become of sufficient moment to engage our attention. Mr. MICHELL'S admirable idea of the stars being collected into systems (*Phil. Trans.* vol. LI. p. 249) appears to be extremely well-founded, and is every day more confirmed by observations: though this does not, in my opinion, take away the probability of many stars being still as it were *solitary*, or, if I may use the expression, *intersystematical*. It occurs then naturally, that by the principle of gravitation, which is never at rest, and which we have no reason to doubt extends to all possible distances, one system of stars will act on another as if the stars of each system were all collected into the center of gravity of each respective system. Hence then will arise this evident consequence, that a star, or sun, such as ours, may have a proper motion within its own system of stars, while at the same time the whole starry system to which it belongs may have another proper motion, totally different in quantity and direction. It will require no little abstract consideration to conceive the possibility of what may be thus surmised; therefore an instance or two, to elucidate the matter, may not be improper. If an inhabitant of the 5th satellite of Saturn should have discovered, that his little world revolves at a great distance round a planet, and to his great astonishment should also have found, that this planet again revolves round the sun;—if, farther, our hypothesis of the solar motion should prove to be well-founded (which, in some of the stars, supposing them to be suns surrounded with planets and satellites, must certainly be the case); then a third capital motion will be introduced to this inhabitant of Saturn's satellite; and he will experience, in a narrow compass, what we now surmise may possibly be our case upon a more extended scale, by the motion of the whole system of stars to which our sun may belong. Another view may, perhaps, still better throw a light upon the subject. Let us admit that a very small nebula may be a collection of a thousand stars: and if Mr. MICHELL'S opinion of our system of stars, which he assumes to be about a thousand (*Phil. Trans.* vol. LVII. p. 255) has any foundation, all these stars taken together will only subtend an angle of barely a minute to an eye placed 3438 times as far from the center of the system as the two farthest stars in it are from each other. Now as I have found some of these nebulæ that are so small, that a tolerably good telescope cannot distinguish them from a single star, whole systems of stars, when presented to our imagination under this diminutive shape of nebulæ, will easily, I believe, be admitted among the number of those celestial bodies that may have a proper motion. I ought to carry this hint a little farther, just to shew that it may possibly be applied to the subject of resolving a number of concurrent proper motions of the fixed stars into apparent ones; and thereby, in process of time, to arrive at the knowledge of all the real complicated motions of the planet we inhabit; of the solar system to which it belongs; and even of the sidereal system, of which this sun may possibly be a member. We see then, that while the sun, by a proper motion, is going towards a certain point of the heavens, each of the stars belonging to the sidereal system, of which the sun is one, supposing them to be relatively at rest, with respect to each other, will be affected in the manner I have shewn in the Paper on the proper Motion of the Sun [p. 117] notwithstanding the whole system should have a real motion in absolute space, and change its situation with respect to other systems or intersystematical stars. We see also, that with respect to stars not belonging to our system, no parallax can appear but what is compounded of the proper motion of the sun, and of the whole system to which it belongs. And should there ever be found, in any particular part of the heavens, a concurrence of proper motions of quite a different direction, we shall then, perhaps, begin to form some conjectures besides those already mentioned by Mr. MICHELL (p. 253 of the same volume of *Transactions*) which stars may possibly belong to ours, and which to other systems.

TABLE I.

Names of Stars.	Motion in R.A.	Motion in Decl.	Names of Stars.	Motion in R.A.	Motion in Decl.
β Ceti	+32		ζ Hydræ	−23	
α Arietis	+10		γ Leporis		−10
δ Ceti	+15		ε Ursæ majoris	−33	+10
α Ceti	+16		α Serpentarii	insens.	
α Persei	+16		γ Draconis	+12	
η Pleïadum		−16	α Lyræ	insens.	+14
γ Eridani	+14		γ Aquilæ		−20
ε Tauri		−11	γ Capricorni	+19	
α Aurigæ	+11	−11	ε Pegasi		−28
β Orionis	insens.	insens.	δ Capricorni	+24	−17
β Tauri	−11	−13	α Aquarii	+13	
α Orionis	insens.	−11	ζ Pegasi		−13
μ Geminorum	−16		Fomahand	+21	
ρ Navis	−13	−11	β Pegasi	+12	
β Cancri		−14	α Andromedæ		−21
ι Ursæ majoris	−54		β Cassiopeæ	+34	

TABLE II.

Polaris		+13	ζ Hydræ		+24
γ Ceti	−14		α Hydræ		+13
β Persei	−10		β Herculis	+14	
α Leporis		+11	γ Cygni	−13	
μ Geminorum		+15	ε Pegasi	−14	
ε Canis majoris		+10	ζ Pegasi	−20	

From the first table we gather, that the principal stars, Lucida Lyræ, Capella, α Orionis, Rigel, Fomahand, α Serpentarii, α Aquarii, α Arietis, α Persei, α Andromedæ, β Tauri, β Ceti, and twenty more of the most distinguished of the second and third rank of stars, agree with our proposed solar motion; when, on the contrary, the second table contains but a few stars, and not a single one of the first magnitude amongst them to oppose it. It is also remarkable, that many stars of the first table agree both in right ascension and declination with the supposition of a solar motion, whereas there is not one among those of the second table which opposes it in both directions. This seems to indicate that the solar motion, in some of them at least, has counter-acted, and thereby destroyed the effect of their own proper motion in one direction, so as to render it insensible; otherwise it would appear improbable, that eight stars out of twelve, contained in the latter table, should only have a motion at rectangles, or in opposition to any one given direction. The same may also be said of nineteen stars among those of the former table, that only agree with the solar motion one way, and are as to sense at rest in the other direction ; but these singularities will not be near so remarkable when we have the motion of the sun to compound with their own proper motions. However, I forbear entering too much into refined consideration ; what we are chiefly to determine at present is, an outline or sketch of what many repeated, and farther extended, observations must ripen so far as in time to enable us to apply more particular calculations.

The motions of α Lyræ and ε Ursæ majoris towards the north are placed in the first table ; it will, therefore, be proper to shew the general law by which the apparent declinations of the stars, at present under consideration, are governed. Let an arch of 90 degrees be applied to a sphere representing the fixed stars, so as always to pass through the apex of the solar motion : then, while one end of it is drawn along the equator, the other will describe, on the spherical surface, a curve which will pass through the pole of the equator, and return into itself at the apex. This curve, to borrow a term from natural history, is a *non-descript* as far as I can find at present, and may be called a spherical conchoid from the manner of its generation. The law then is, that all the stars in the northern hemisphere, situated within the nodated part of the conchoid, will seem to go to the north by the motion of the solar system towards its apex ; the rest will appear to go southwards. A similar curve is to be delineated in the southern hemisphere, in the nodated part of which the same appearances will take place. It will require but little attention to see the truth of this construction.

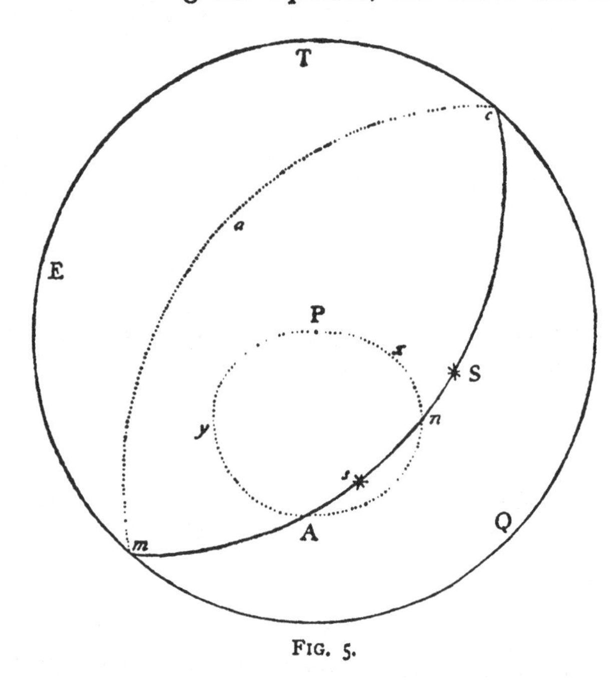

FIG. 5.

Suppose the great circle A*cam*, fig. 5, of which the generating quadrant *mn* is a part, compleated ; then will it intersect the equator EQT in two opposite points *me*. Now, since the apex A, by the hypothesis, is somewhere north of the equator, the great circle will always make some angle A*m*Q with it ; and the point *n*, which is 90 degrees from the intersection *m* with the equator, will be the most northern part of the semicircle *mnc*. From what has been said in the paper on the Motion, &c. [p. 117] it follows, that the apparent motion of any star *s*S will always be in an arch of a great circle A*s*S*c* drawn through the apex A and star *s*S : therefore, if the star be less than 90 degrees of the generating circle distant from its intersection with the equator (having more northern declination than the apex) as at *s*, its northern declination will increase, and it will also fall within the nodated part of the conchoid *AxPy*; but when its distance from the intersection *m* is more than 90 degrees, as at S, the motion will be towards the south, and the star will be situated without the nodated part of the curve. That the star *s* will fall within the nodated part appears because *ms* being less than *mn* by supposition, if *m* be drawn towards E, to describe the conchoid the angle A*m*Q will decrease, and therefore the describing point *n* will be depressed below *s* as it approaches A. For the same reason S will fall without; since, by drawing *m* towards Q, the angle A*m*Q will become greater than S*m*Q, and the describing point *n* will pass above the star S. The application of this theory is very simple; for instance, let it be required to find whether any given star will fall within or without the conchoid. Then, in fig. 6, there will be given P*s*, the polar distance of the star; and QP*c*, the difference of right ascension between the star and the apex of the sun's motion A; also, the polar distance PA, and declination *c*A of the point A. Then, by trigonometry, the sides *s*P, PA, and the included angle being given, we find the side A*s* and angle PA*s*. Again, the side *c*A, and angle *c*A*m*=PA*s* of the right-angled triangle A*cm* being given, we find the hypothenuse A*m*; and if A*m*+A*s* be less than 90 degrees, the star falls within the conchoid, otherwise without.

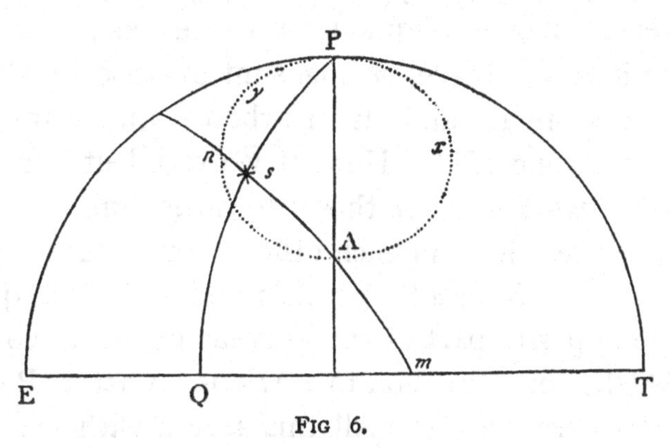

Fig 6.

It will be found, that I have placed the want of sensible motion of α Lyræ and α Orionis in right ascension, and of Rigel both in right ascension and declination to the account of those stars that are in favour. These stars are so bright, that we may reasonably suppose them to be among those that are nearest to us; and if they had any considerable motion, it would most likely have been discovered, since the variations of Sirius, Arcturus, Procyon, Castor, Pollux, &c. have not escaped our notice. Now, from the same principle of the motion of the solar system, by which we have accounted for the apparent *motion* of the latter stars, we may account for the apparent *rest* of the former. Those two bright stars, α Lyræ and α Orionis, are placed so near the direction of the assigned solar motion, that from the application of my second theorem in the paper on the Motion of the Solar System [p. 117] their motion ought to be insensible in right ascension, and not very considerable in declination, all which we find is confirmed by observation. With respect to Rigel and α Serpentarii, admitting them both as stars large enough to have shewn a proper motion, were their situation otherwise than it is, we find that they also should be apparently at rest in right ascension; and Rigel having southern declination, and being a less considerable star than α Orionis, which shews but 11″ motion towards the south in 50 years, its apparent motion in declination may, on that account, be also too small to become visible.

I should not omit to take notice of a very remarkable paragraph of MAYER's, which seems to contain a strong objection against the motion of the solar system, while indeed it may be shewn to be a very good argument in its favour. At the end of his tract, De Motu Fixarum, he says: "Tandem, quum et quæri possit, quæ hujus motus causa sit, hoc unum monere visum, illum explicare non posse per motum totius systematis solaris, etsi nec impossibile sit, solem, ut ejusdem cum fixis naturæ, instar harum quarundam in spatio mundano promoveri. Nam si sol et cum ipso planetæ omnes nostrumque domicilium terra, recta tenderent versus plagam aliquam, universæ fixæ, quæ in ea plaga adparent paullatim a se invicem discedere, et quæ sunt in opposita parte coeli coire viderentur; non secus ac per silvam ambulanti arbores, quæ ante viam sunt, disjungi videntur, quæ a tergo, congredi." Now, if we recollect what has been said of the motion of the stars, we find, that those, towards which I suppose the solar system to move, do really recede from each other: for instance, Arcturus from α Lyræ; α Aquilæ and α Aquarii from α Serpentarii and ε Ursæ majoris; and, on the contrary, those in the opposite part of the heavens do really come nearer to each other; as Sirius to Aldebaran; Procyon to α Arietis; Castor, Pollux, Regulus, &c. to α Ceti, α Persei, α Andromedæ, &c. All this agrees with what MAYER says ought to happen, if the solar system was to have a motion towards a certain part of the heavens; which, by the bye, I find this admirable astronomer mentions as a very possible thing.* However, when he says that *all the stars* in those parts towards which the sun might be supposed to move, should recede from each other, and *vice versâ*; I must add, that this would only take place under the restrictions of my first, second, and third theorems, and therefore it is not to be expected, that we should immediately see the effect of this parallax in any but the stars that are nearest to us. But as we have at present no other method of judging of the relative distance of the fixed stars than from their apparent brightness, those that are most likely, on that account, to be affected by a parallax arising from the motion of the solar system, are the very stars which, by MAYER's own table, I have made use of to point it out to us.†

Datchet, March 13, 1783.

* This paper, De Motu Fixarum, was read at Göttingen in January 1760.

† I have lately been favoured by Dr. WILSON, Professor of Astronomy at Glasgow, with a short tract, called, "Thoughts on general Gravitation, and Views thence arising as to the state of the Universe;" wherein the possibility of a Solar Motion is also shewn. It was printed in 1777. Mr. DE LA LANDE, in the Memoirs for 1776, with his usual felicity of thought, has inferred the probable motion of the system from the sun having a rotation round his axis, when he says, p. 513: "Une force quelconque imprimée à un corps, et capable de le faire tourner autour de son centre, ne peut manquer aussi de déplacer le centre, et l'on ne sauroit concevoir l'un sans l'autre. Il paroit donc très-vraisemblable que le soleil a un mouvement réel dans l'espace absolu," &c.

XII.

On the remarkable Appearances at the Polar Regions of the Planet Mars, *the Inclination of its Axis, the Position of its Poles, and its spheroidical Figure; with a few Hints relating to its real Diameter and Atmosphere.*

[*Phil. Trans.*, vol. lxxiv., 1784, pp. 233–273.]

Read March 11, 1784.

WHAT I have to offer on the subject of the remarkable appearances at the polar regions of Mars, as well as what relates to the inclination of the axis, the position of the poles, and the spheroidical figure of that planet, is founded on a series of observations which I shall deliver in this paper; and after they have been given in the order they were made, it will be easy to shew, by a few deductions from them, that my theory of this planet is supported by facts which will sufficiently authorise the conclusions I have drawn from them. For the sake of better order and perspicuity, however, I shall treat each subject apart, and begin with the remarkable appearances about the polar regions. The observations on them were made with a view to the situation and inclination of the axis of Mars; for to determine these we cannot conveniently use the spots on its surface, in the manner which is practised on the sun. The quantities to be measured are so small, and the observations of the center of Mars so precarious, and attended with such difficulties (since an error of only a few seconds would be fatal) that we must have recourse to other methods.

When I found that the poles of Mars were distinguished with remarkable luminous spots,* it occurred to me, that we might obtain a good theory for settling the inclination and nodes of that planet's axis, by measures taken of the situation of those spots. But, not to proceed upon grounds that wanted confirmation, it became necessary to determine by observation, how far these polar spots might be depended upon as permanent; and in what latitude of the globe of Mars they were situated; for, if they should either be changeable, or not be at the very poles, we might be led into great mistakes by overlooking these circumstances. The following observations will assist us in the investigation of these preliminary points.

1777, April 17. 7 h. 50′. There are two remarkable bright spots on Mars. In fig. 1 Plate VI. they are marked *a* and *b*. The line AB expresses the direction of a parallel of declination. 10 feet reflector, 9 inches aperture, power 211.†

* A bright spot near the southern pole, appearing like a polar zone, has also been observed by M. MARALDI. See Dr. SMITH'S *Optics*, § 1094.

† *Phil. Trans.* vol. LXXI. p. 127 and fig. 17 [above, p. 22].

10 h. 20′. They are both quite gone out of the disk.

1779, This year, in all my observations on Mars, there is no mention of any bright spots, so that I believe there were none remarkable enough to attract my attention. However, as my view was particularly directed to the phænomena of this planet's diurnal rotation, it is possible I might overlook them.

1781, March 13. 17 h. 40′. 20 feet reflector. I saw a very lucid spot on the southern limb of Mars of a considerable extent. See fig. 2.

June 25. 11 h. 36′. 7 feet reflector, power 227. Two luminous spots appeared at *a* and *b*, fig. 3; *a* is larger than *b*.

12 h. 15′. With 460. *a* is thicker than *b*, but *b* is rather longer.

13 h. 12′. *a* is grown thicker, and *b* become thinner.

June 27. 11 h. 20′. The two lucid spots are on Mars.

June 28. 11 h. 15′. They are both visible; *a*, fig. 4, is much thicker than *b*.

12 h. 55′. A line joining *a* and *b* does not go through the center.

June 30. 10 h. 48′. The spot *a* is visible, fig. 5.

11 h. 35′. Both spots are to be seen.

July 3. 10 h. 54′. *a* seems to be larger than I have seen it, fig. 6.

11 h. 24′. *b* is not yet visible, fig. 7.

12 h. 36′. I perceive part of *b*, fig. 8.

July 4. 12 h. 9′. *a* is very full; *b* extremely thin, and barely visible.

12 h. 18′. *a* and *b* are not quite opposite each other.

12 h. 49′. *b* is increased.

July 15. 9 h. 54′. *a* is visible, fig. 9.

11 h. 35′. *b* invisible.

12 h. 12′. *b* not to be seen.

July 16. 11 h. 9′. The bright spot *a* is very large.

July 17. 11 h. 15′. No other bright spot but *a*.

July 19. 13 h. 31′. *a* visible.

July 20. 10 h. 3′. I suppose the bright spot *a* on Mars is, very nearly, the south pole; which therefore must lie in sight. There is no second bright spot *b* visible to night.

10 h. 56′. *b* not visible; the night very fine.

July 22. 11 h. 14′. At *a* and *b*, fig. 10, are bright spots; *a* is larger than *b*. Most probably the south pole is in view, and the north pole just hid from our sight. If the spots are polar, or nearly so, then *a* must, on a supposition of the south pole's being in view, appear larger than *b*; and if *b* extend a little more from the north pole one way than another, it must be subject to some change in its appearance from the revolution of Mars on its axis.

July 30. 9 h. 43′. Both spots visible.

August 8. 10 h. 4′. Only *a* visible, fig. 11.

August 17. 9 h. 21′. Only *a* in sight.

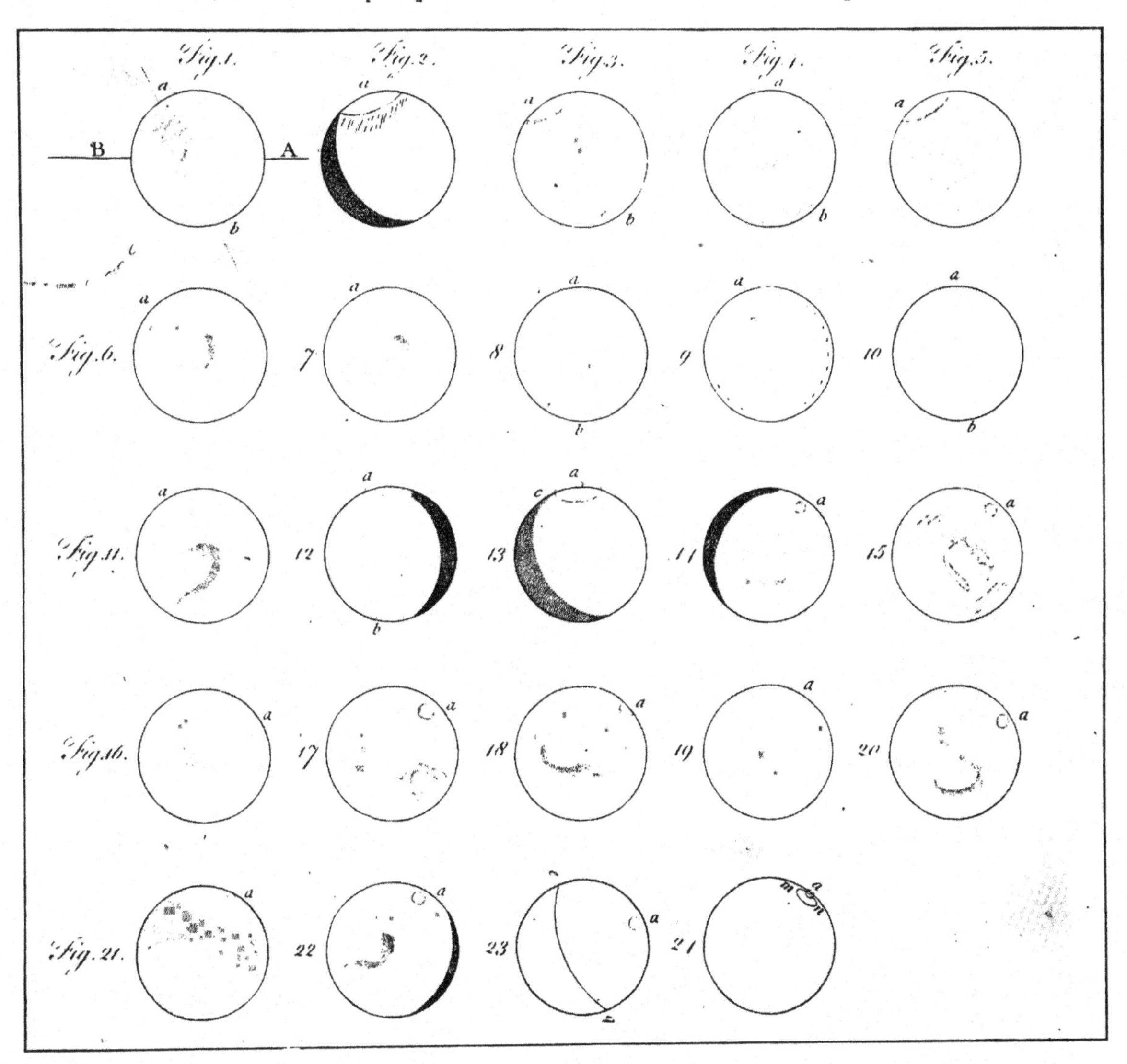

[*To face page* 132.

August 23. 8 h. 44′. *a* as usual, and part of *b* visible, fig. 12.

Sept. 7. The white spot *a* is very large.

1783, May 20. Mars has a singular appearance. At *a*, fig. 13, is the polar spot, which is bright, and seems to project above the disk by its splendour, causing a break at *c*.

July 4. *a* is very bright.

July 23. 14 h. 45′. *a* is very lucid.

August 16. I saw the bright spot with the 20 feet reflector as usual.

Aug. 26. The lucid spot on Mars is its south pole, for it remains in the same place, while the dark equatorial spots perform their constant gyrations: it is nearly circular.

Aug. 29. The south polar spot is in the same situation.

Sept. 9. As usual.

Sept. 22. The south polar spot is of a circular shape, and very brilliant and white. I had a beautiful and distinct view of it when it was about the meridian, and measured its little diameter in the equatorial direction of Mars. With a power of 932 it gave 1″ 41‴, and I saw it very distinctly. The outward edge of the spot came just up to the upper limb; a favourable haziness, taking off every troublesome ray, gave me objects in general exceedingly well defined, especially Mars.

Sept. 23. 9 h. 55′. The polar spot *a*, fig. 14, as usual.

Sept. 24. The same.

Sept. 25. 12 h. 30′. The bright south polar spot *a*, fig. 15, seems to be fixed in its place, and goes nearly up to the margin of the disk; it is perfectly round.

12 h. 55′. The track of the equatorial spots is incurvated, being convex towards the north, see *e*, *q*, fig. 23: this confirms the white spot's being at the south pole. With long attention I can perceive the edge of the disk of Mars beyond the spot, extending about $\frac{1}{4}$ diameter of the spot.

Sept. 26. 12 h. 10′. The spot *a* is in a line with the center and the end of the hook, fig. 16.

Sept. 27, 28, 29. The spot as usual.

Sept. 30. 10 h. 30′. The polar spot as in fig. 17.

Oct. 1. 9 h. 55′. I am inclined to think, that the white spot has some little revolution, and therefore is not with its center exactly at the pole of Mars; it is rather probable, that the real pole, though within the spot, may lie near the circumference of it, or one-third of its diameter from one of the sides. A few days more will shew it, as I shall now fix my particular attention upon it.

10 h. 17′. The bright spot is certainly not so far upon the disk as it used to be formerly, and is either reduced or has a small motion; which of the two is the case will be seen in a few hours.

13 h. 3′. The bright spot has a little motion; for it is now come farther into the disk.

I concluded now, in general, that none of the bright spots on Mars were exactly at the poles, though they could certainly not be far from them: for what has been just related of the 1st, 2d, and 3d of October 1783, shews plainly, that the appearance of the southern spot *a* was a little affected by the diurnal motion of the planet: and the observations of the 3d and 4th of July 1781, shew also that the spot *b* could not be exactly at the north pole; and that, perhaps, the visible branch of the latter extended pretty far towards the equator. However, the south polar spot of the year 1783, being very small and nearly round, afforded a good opportunity for determining its polar distance, by noting the different angles of position it assumed while Mars revolved on its axis; to this end many observations were taken at different hours of the same night, which will be found among the measures of the angles of position in the next division of my subject. And since the different degrees of brilliancy, as well as the proportional apparent magnitude of the spot, would also contribute to the investigation of this point, I continued my remarks on those particulars, as follows.

1783, Oct. 2. 7 h. 59′. The bright spot near the south pole is about half visible.

Oct. 4. 8 h. 0′. The polar spot seems to project above the disk as formerly, and is very small.

Oct. 5. 11 h. 13′. The spot is very small, and seems actually to be in the circumference.

11 h. 30′. The spot is small, and seems to be with its farthest side in the circumference of the disk; or it may, perhaps, be partly beyond it, and therefore not all in sight.

11 h. 50′. I see the spot much clearer than I did before.

13 h. 15′. The white spot is more in sight, and of its usual size, but does not seem much to change its position; however, what change there is shews that it has been beyond the pole, as it appears to have been direct while the equatorial spots were retrograde.

Oct. 9. 11 h. 48′. The white polar spot increases in size. At 10 h. 35′ it was as in fig. 18, but is now larger, and coming round towards that part of its orbit which is nearest to us. See fig. 24.

Oct. 10. 6 h. 20′. I see no white polar spot; but the planet is too low for any observation to be depended on.

6 h. 55′. The white spot beg ns to be visible; at least I see it now, the planet being higher than before, fig. 19.

9 h. 55′. With 460, the white spot is considerably increased, and shews a circular form, fig. 20.

Oct. 11. 7 h. 46′. The bright spot is very visible; the evening fine; with 278.

Oct. 16. 7 h. 7′. The spot is very luminous.

9 h. 55′. It seems rather lengthened; perhaps it may be arrived at the extreme of its parallel of declination.

Oct. 17. 7 h. 47′. The white spot *a*, fig. 21, is very bright.

13 h. 7′. It is less in appearance than it was in the beginning of the evening.

Oct. 23. 6 h. 46′. The bright spot is very large and luminous; I suppose it to be in the nearer parts of its little orbit.

7 h. 11′. It is situated as in fig. 22.

Oct. 24. 7 h. 1′. The white spot is very large.

Oct. 27. 8 h. 45′. It is very large and round.

Nov. 1. 7 h. 47′. The spot is round and bright.

Nov. 11. The deficiency of light which occasions Mars to appear gibbous, reaches over the south polar spot towards the preceding limb, and hides it.

Nov. 14. Mars is gibbous, and the polar spot is thereby rendered invisible.

Nov. 17. 6 h. 0′. The south polar spot is under the falcated defect of light.

6 h. 30′. I do not know whether there be not a faint glimpse of the polar spot left; the weather is too bad to determine it.

I have added fig. 25 to shew the connection of the 15th, 17th, 18th, 19th, 20th, 21st, and 22d figures, which complete the whole equatorial circle of appearances on Mars, as they were observed in immediate succession. The center of the circle marked 17 is placed on the circumference of the inner circle, by making its distance from the center of the circle, marked 15, answer to the interval of time between the two observations, properly calculated and reduced to sidereal measure. The same has been done with regard to the circles marked 18, 19, &c. And it will be found, by placing any one of these connected circles, so as to have its contents in a similar situation with the figures in the single representation which bears the same number, that there is a sufficient resemblance between them; but some allowance must undoubtedly be made for the unavoidable distortions occasioned by this kind of projection.

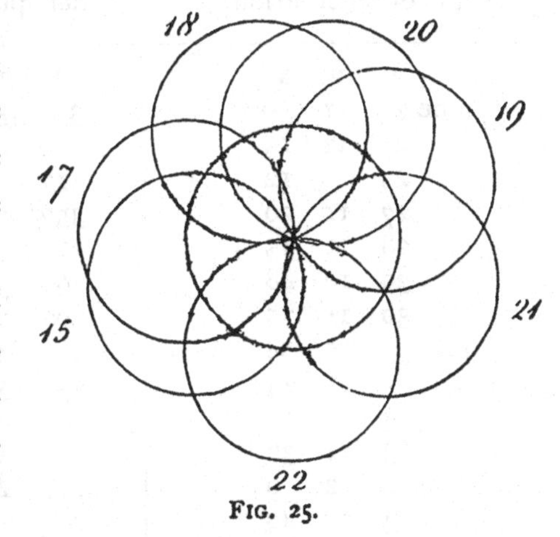

Fig. 25.

In order to bring these observations on the bright spots into one view, I have placed them at the circumference of three circles (see fig. 26, 27, 28 [pp. 138–140]) divided into degrees, representing the parallels of declination in which they revolved about the poles of Mars. The division of the circles marked 360 is where a spot passes that meridian of the planet which is turned towards the earth, and where, consequently, it appears to us in its greatest lustre. The motion of the spot is according to the numbers 30, 60, 90, and so on to 360. In calculating the daily places of the spots I have used the sidereal period of 24 h. 39′ 21″·67 determined in my paper on the rotation of Mars ;* and have also made proper allowances for the alterations of the geocentric longitudes calculated from the situations of that planet given in the *Nautical Almanack* ; by which means the sidereal is reduced to a proper synodical period.

The following three tables contain the result of the calculations, and serve to explain the arrangement of the observations in the circles. In the first column are the times when the observations were made. In the second, the sidereal places of the spot in degrees and minutes. In the third column are the geocentric longitudes of Mars at the time of the observations. In the fourth, the necessary corrections on account of these different longitudes. In the fifth column are the corrected or synodical places of the spots ; and, according to the numbers in that column, they are marked on the circles, where consequently each spot is represented as it must have appeared to be situated at the time of observation.

TABLE I.

Time of observation.	Sider. place.	Geoc. longit.	Correction.	Synod. place.
D. H. M.	D. M.	S. D. M.	D. M.	D. M.
June 25 11 36	350 31	9 24 35	+0 0	350 31
25 12 15	0 0	9 24 35	+0 0	0 0
25 13 12	13 52	9 24 34	−0 1	13 51
27 11 20	357 28	9 24 12	−0 23	357 5
28 11 15	316 40	9 24 1	−0 34	316 6
30 10 48	290 56	9 23 38	−0 57	289 59
30 11 35	302 23	9 23 38	−0 57	301 26
July 3 10 54	263 40	9 22 55	−1 40	262 0
3 11 24	270 58	9 22 55	−1 40	269 18
3 12 10	282 9	9 22 55	−1 40	280 29
3 12 36	288 29	9 22 55	−1 41	286 48
4 12 9	272 20	9 22 40	−1 55	270 25
4 12 49	282 4	9 22 40	−1 55	280 9
15 9 54	134 7	9 19 43	−4 52	129 15
15 11 35	158 42	9 19 42	−4 53	153 49
15 12 12	167 42	9 19 42	−4 53	162 49
16 11 9	142 48	9 19 26	−5 9	137 39
17 11 15	134 40	9 19 9	−5 26	129 14
19 13 31	148 37	9 18 34	−6 1	142 36
20 10 3	88 25	9 18 21	−6 14	82 11
20 10 56	101 19	9 18 20	−6 15	95 4
22 11 14	86 32	9 17 49	−6 46	79 46
30 9 43	347 46	9 16 5	−8 30	339 16

* *Phil. Trans.* vol. LXXI. p. 134 [above, p. 27]

TABLE II.

Time of observation.			Sider. place.		Geoc. longit.			Correction.		Synod. place.	
D.	H.	M.	D.	M.	S.	D.	M.	D.	M.	D.	M.
June 25	11	36	86	51	9	24	35	+1	40	88	31
25	12	15	96	20	9	24	35	+1	40	98	0
25	13	12	110	12	9	24	34	+1	39	111	51
28	11	15	53	0	9	24	1	+1	6	54	6
30	10	48	27	16	9	23	38	+0	43	27	59
30	11	35	38	43	9	23	38	+0	43	39	26
July 3	10	54	0	0	9	22	55	+0	0	0	0
4	12	9	8	40	9	22	40	−0	15	8	25
15	9	54	230	27	9	19	43	−3	12	227	15
15	10	12	234	50	9	19	43	−3	12	231	38
15	11	35	255	2	9	19	42	−3	13	251	49
15	12	12	264	2	9	19	42	−3	13	260	49
16	11	9	339	8	9	19	26	−3	29	235	39
19	13	31	244	57	9	18	34	−4	21	240	36
20	10	3	184	45	9	18	21	−4	34	180	11
20	10	56	197	39	9	18	20	−4	35	193	4
30	9	43	84	6	9	16	5	−6	50	77	16

TABLE III.

Time of observation.			Sider. place.		Geoc. longit.			Correction.		Synod. place.	
D.	H.	M.	D.	M.	S.	D.	M.	D.	M.	D.	M.
Sept. 25	13	30	6	32	0	9	54	+6	44	13	16
Oct. 1	10	17	262	5	0	8	6	+4	56	267	1
1	13	3	302	29	0	8	5	+4	55	307	24
2	7	59	218	55	0	7	50	+4	40	223	25
4	8	0	200	0	0	7	15	+4	5	204	5
4	8	46	211	12	0	7	15	+4	5	215	17
5	11	13	237	23	0	6	55	+3	45	241	8
5	11	30	241	31	0	6	55	+3	45	245	16
5	11	50	246	23	0	6	55	+3	45	250	8
5	13	15	267	4	0	6	54	+3	44	270	48
5	14	0	278	1	0	6	53	+3	43	281	44
7	8	20	176	8	0	6	23	+3	13	179	21
7	10	5	201	41	0	6	22	+3	12	204	53
7	11	50	227	14	0	6	21	+3	11	230	25
9	11	48	207	35	0	5	49	+2	39	210	14
10	6	55	126	42	0	5	37	+2	27	129	9
10	7	50	140	5	0	5	36	+2	26	142	31
10	9	55	170	30	0	5	34	+2	24	172	54
10	12	11	203	36	0	5	33	+2	21	205	57
16	7	7	72	9	0	4	15	+1	5	73	14
16	7	46	81	39	0	4	15	+1	4	82	43
16	9	55	113	2	0	4	14	+1	4	114	6
17	7	47	72	19	0	4	3	+0	53	73	12
17	13	7	150	11	0	4	0	+0	50	151	1
23	6	46	0	0	0	3	10	−0	0	0	0
24	7	1	354	0	0	3	2	−0	8	353	52

From the appearance and disappearance of the bright north polar spot in the year 1781, we collect that the circle of its motion, represented by fig. 26, was at some considerable distance from the pole. By a calculation, made according to the principles hereafter explained, its latitude must have

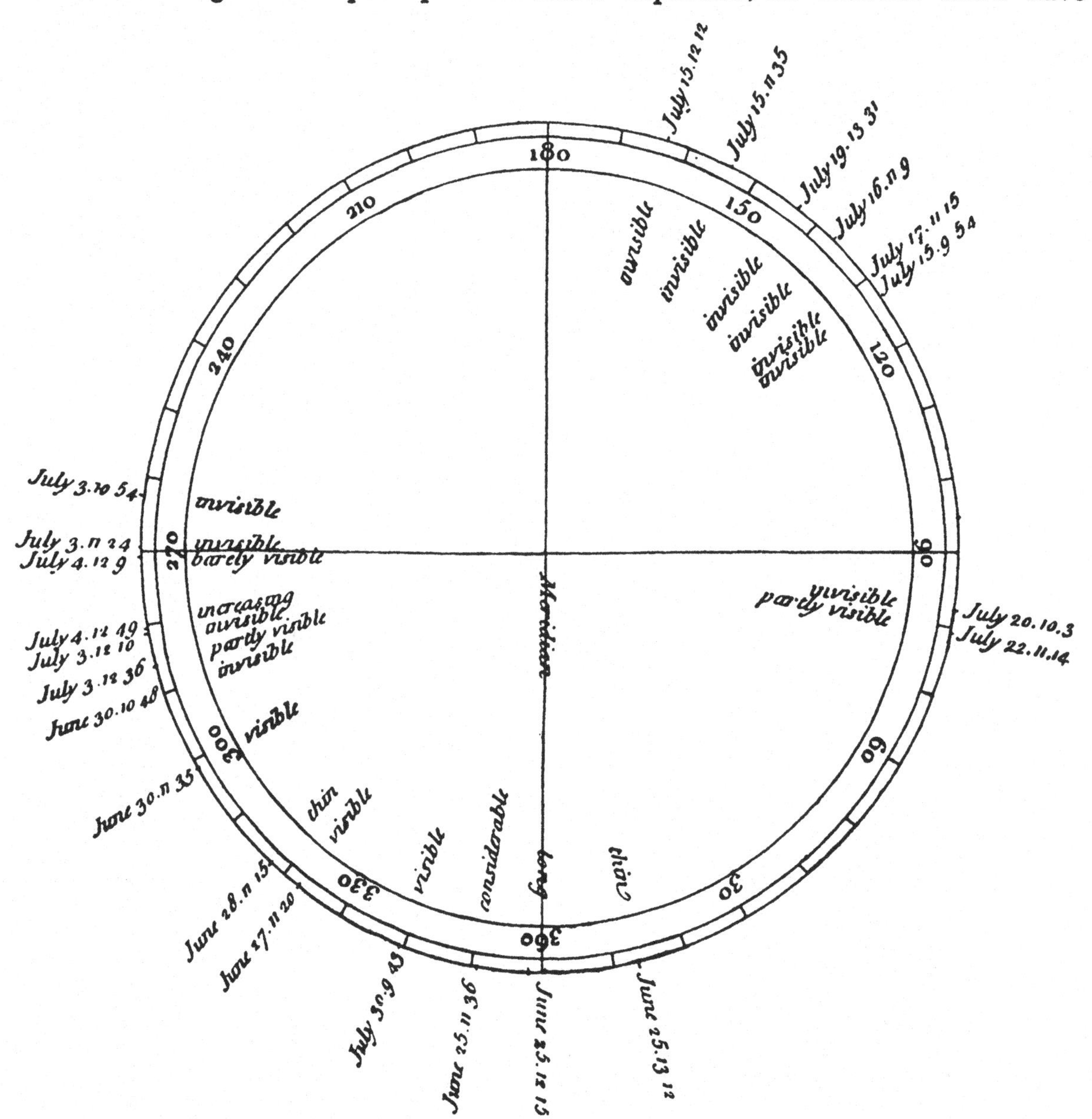

FIG. 26.—Track of north polar spot, June and July 1781.

been about 76° or 77° north; for I find that, to the inhabitants of Mars, the declination of the sun, June 25, 12 h. 15′ of our time, was about 9° 56′ south [p. 148]; and the spot must have been at least so far removed from the north pole as to fall a few degrees within the enlightened part of the disk, to become visible to us.

The south pole of Mars could not be many degrees from the center of the large bright southern spot of the year 1781, whose course is traced in fig. 27; though the spot was of such a magnitude as to cover all the polar regions farther than the 70th or 65th degree, and in that part which was on the meridian July 3, at 10 h. 54′, perhaps a little farther.

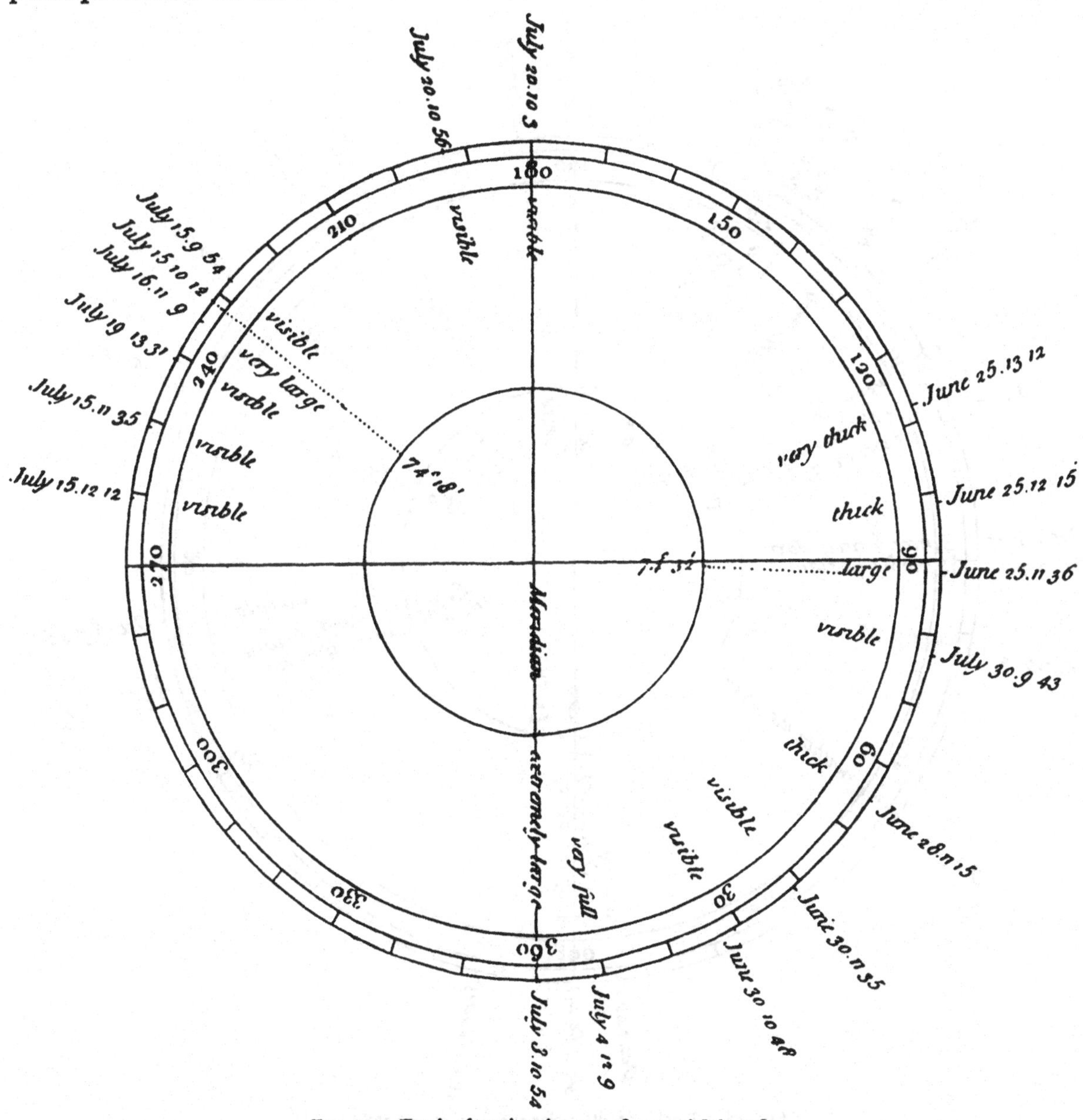

Fig. 27.—Track of south polar spot, June and July 1781.

In the next division of our subject will be shewn, that the inclination and position of the axis of Mars are such, that the whole circle, fig. 28, (which will appear to be in about 81° 52′ of south latitude on the globe of Mars) was in view all the time the observations on the bright south polar spot of the year 1783, which are marked upon it, were made, but in so oblique a situation as to be projected into a very narrow ellipsis. See fig. 24, where *mn* is the little ellipsis in which the spot *a* revolved about the pole. Hence then we may easily account for the observed magnitude and brightness of the spot Oct. 23, 24, and 27, when it was exposed to us in its meridian splendour. Its situations Oct. 16 and 17 on one extreme of the parallel, as well as those of Oct. 5 and Nov. 1 on the other, gave us also a bright

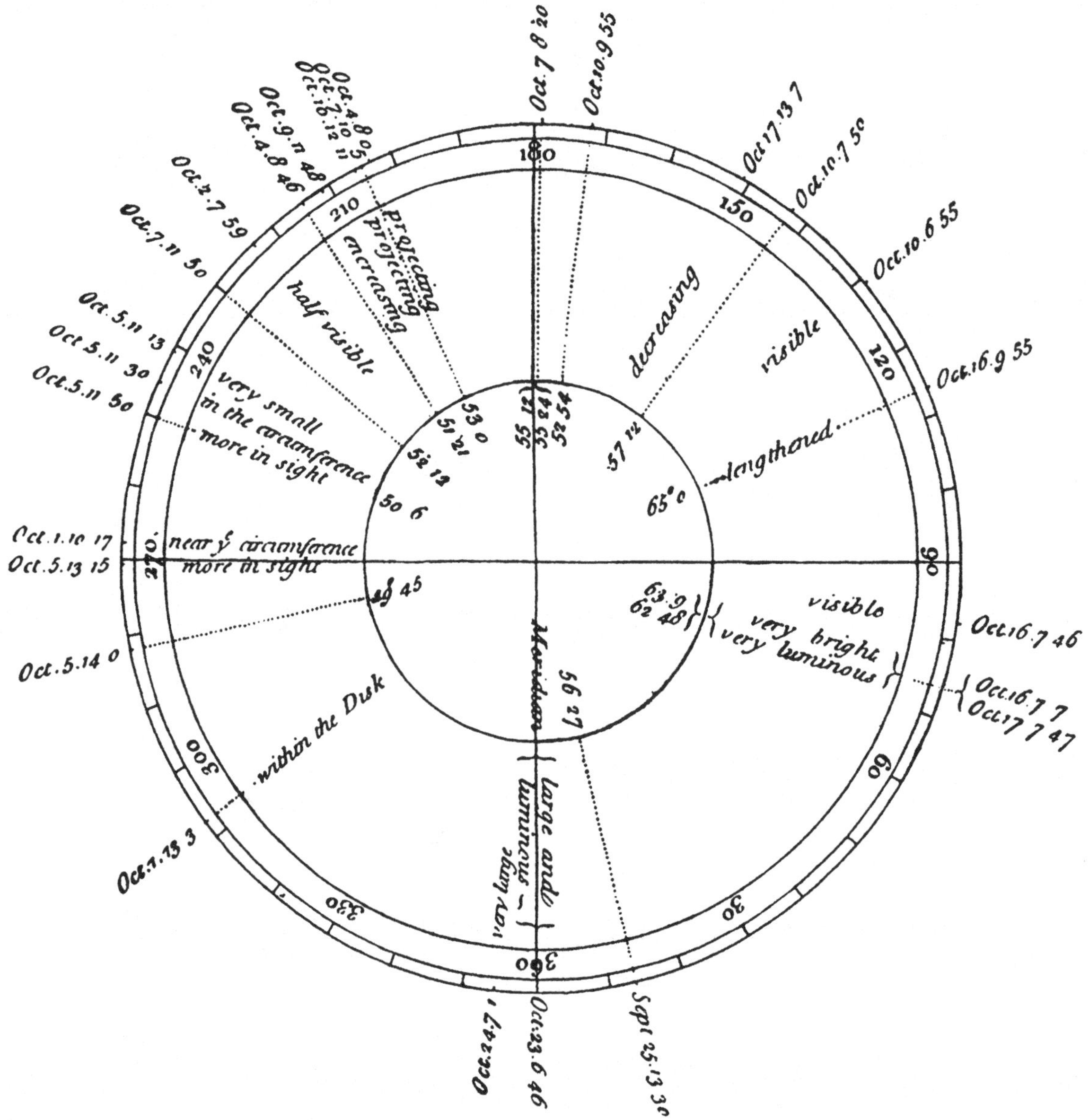

FIG. 28.—Track of south polar spot in October 1783.

view of it: and, when we pass over to that half of the circle which lies beyond the pole, the much greater obliquity into which the spot must there be projected will perfectly account for its being smaller at 13 h. 7′ of Oct. 17 than at 7 h. 47′ of the same evening. It will also explain its smallness Oct. 4 and its increase Oct. 9. We shall have occasion hereafter to recur to the same figure, so that I take no notice at present of the angles of position which are marked upon it.

Of the direction or nodes of the axis of Mars, *its inclination to the ecliptic, and the angle of that planet's equator with its own orbit.*

From the foregoing article we may gather, that the bright polar spots on Mars are the most convenient objects for determining the situation of the axis of this planet; I shall therefore collect, in one view, all the measures I have taken of these spots for that purpose. Before I constructed a micrometer for taking the angle of position, I used to draw a line through the figure delineated of Mars to represent the parallel of declination; in a few of my first observations, therefore, I can only take the situation of the polar spots from such drawings, and of consequence no great accuracy in the angles, as to the exact number of degrees, can be expected.

1777, April 17. 7 h. 50′. A line drawn through the middle of the two bright polar spots *a* and *b*, fig. 1, makes an angle of about 63°, with a parallel of declination AB; the southern spot preceding and the northern following.

My reason for chusing a line drawn through both the spots rather than through one of them and the center is, first, that they were not situated quite opposite each other, and therefore, unless other observations had pointed out which was most polar, I should evidently run the greater risk in fixing on one of them in preference to the other. In the next place, we find by the second observation at 10 h. 20′ that in two hours and a half both spots were intirely gone out of the disk. This plainly denotes, that they were both in the same half of a sphere orthographically projected, and divided by a plane passing through the axis of Mars and the eye, but that neither of them were polar. Now, a line drawn through two points not far from opposite each other, both in the same hemisphere, and both removed from the poles of it, must approach more to a parallelism with the axis, than a line drawn through either of them and the center.

1779, May 9. There being no bright spots by which to judge of the position of the poles, it is estimated from a well-known dark equatorial spot, with a line drawn through the figure to denote a parallel of declination. By very rough estimation it is about 42° south preceding.

May 11. The same figure, being drawn again in another situation, and also with a line giving a parallel of declination, points out, by the same rough estimation, 62° south preceding.

1781, June 25. 11 h. 35′. The position of the spots *a* and *b*, fig. 3, with regard to a parallel of declination, measured with a micrometer 74° 32′. The spot *a* was south preceding, and *b* north following.

July 15. 10 h. 12′. The angle of position, of the center of the spot *a*, fig. 9, through the center of the disk, 74° 18′ south preceding.

1783, August 16. Position of the spot *a*, 64° south following the center; but as the planet is not full, the center becomes dubious, and the measure therefore may not be quite accurate, though taken with a 20 feet reflector; power 200.

Sept. 9. Position of the supposed south pole of Mars 65° 12′ south following; 7 feet reflector; power 460.

Sept. 22. Position of the same 52° 9′ s. following; 460.

Sept. 25. 13 h. 30′. Position of the south polar spot 56° 27′, very accurately taken, by bisecting the disk of Mars through the bright spot, and supposing the planet now near enough the opposition to induce no material error. Hitherto I have taken it through a supposed center by endeavouring to allow a little for what I thought the deficiency in the disk; but not to-night.

Oct. 4. 8 h. 46′. Position of the spot 51° 21′; Mars too low and hazy to depend much on the measure with so high a power as 460.

Oct. 5. The motion of the polar spot being now strongly suspected, or rather already known, I took the following measures, by way of discovering its quantity.

11 h. 50′. Position very exactly taken 50° 6′ s. following.

14 h. 0′. Position of the spot 49° 45′.

Oct. 7. 8 h. 20′. Position 55° 12′. In order to see how far this measure might be trusted to, I set 49° 36′ in the micrometer, which was evidently too small; next I took 51° 36′, which was also too small; after this, I took a new measure, and found 55° 24′, which appeared to me very exact. 10 h. 5′. The position now was 53°. 11 h. 50′. It measured 52° 12′. As there is nothing to distinguish the center, it is extremely difficult to please one's self in bringing the spot into a line with it.

Oct. 10. 7 h. 50′. Position of the polar spot 57° 12′; with 460, very accurate. I tried a few parts less of the micrometer, but found the measure too little. I see pretty distinctly, but the air is tremulous.

9 h. 55′. Position 52° 42′; very distinct.

12 h. 11′. Position 46° 30′; I see not quite so well now as I could wish.

14 h. 1′. Position 44° 12′; but liable to great uncertainty, on account of tremulous air; it becomes more difficult to distinguish the center when the planet is not perfectly defined.

Oct. 16. 7 h. 7′. Position 63° 9′. By way of trial I set 59° 36′, which was too small; also 60° 24′ was too small; again, 61° 24′ was not large enough. Then, taking a fresh measure, I found it 62° 48′, which I thought right.

9 h. 55′. I took three measures, and thought the third, which was 65° 0′, the best of all, for I saw the planet and the spot remarkably well.

Oct. 27. 8 h. 45′. Position of the polar spot 59° 30′. I took three other measures, of which 60° 39′ appeared to me the best; it was taken with long attendance and many changes and trials of the wires in different positions; but the gibbosity of Mars is such, that measures of the situation of the spot are now no longer to be depended on.

These positions, I believe, will be sufficient for the purpose of settling the latitude of the polar spots, and thereby obtaining a correct measure of the situation of the real pole. I have referred those of the south polar spot of the year 1783 to the same circle which contains the observations that were made on the apparent brightness and magnitude of that spot, that they may be compared together. (See fig. 28.) The agreement of the measures, and the phænomena attending the motion of the spot, are sufficient to point out the meridian of the circle; for which, from a due consideration of these circumstances, I have fixed on the place where the spot was Oct. 10, 6 h. 46′.

Of the angles collected in fig. 28, we find 65° 0′ the largest, and 49° 45′ the smallest; but, on account of the different situation of the earth and Mars, the angle measured 7′ less Oct. 16 than it would have done had the planets remained in the places they were in Oct. 5, when the other measure was taken. This being added, we have 65° 7′. The difference between the two positions is 15° 22′. Now, the construction of fig. 28 being admitted, we see that the angles were nearly taken at the opposite extremes of the circle in which the spot moved. However, by the 5th column of Tab. III. Oct. 5, we have the situation of the spot in the circle with respect to the meridian 281° 44′, and Oct. 16, 114° 6′: therefore the south polar distance of the center of the spot is found, by taking half the sum of the sines of these angles to radius, as 7° 41′ (half of 15° 22′) to a fourth number, which is 8° 8′; and the latitude of the circle, in which the spot moved about the pole, therefore is 81° 52′ south. This being determined, we have the following correction for the angles of position: radius is to sine of the angular distance of the spot from the meridian as 8° 8′ to the required quantity. This must be added or subtracted, according as the case requires; and thereby we shall have the position of the true pole from any one of the measures.

I shall now apply the above to determine the situation of the axis of Mars. To this end, we see that, in the first place, the measures must be corrected for the latitude of the spot; next, they must be reduced to a heliocentric observation, which will also correct them from the difference occasioned by the different situation of the planets when they were taken. This being done, we may select two observations at a proper distance; from which, by trigonometry, we shall have the node and inclination of the axis. When these elements are obtained, it will be easy to see how other observations agree with them; which will afford the means of correcting or verifying the former calculations.

Let T, fig. 29 [p. 145], be the earth; ♋Q*q*♑ the ecliptic as seen from T; P the point of the heavens towards which the north pole of the earth is directed; M the place of the orbit of Mars *μm*M, where an observation of the poles of that planet has been made, which is to be reduced to its heliocentric measure. And, first, suppose it to have been made at the time of the opposition of that planet. Then, the place M or Q in the ecliptic being given, we have the sides Q♋, ♋P; whence the angle Q, of the right-angled triangle P♋Q, is found. This being added to, or taken from, the observed angle of position of the axis of Mars, according to circumstances easily to be determined, reduces it to its heliocentric position. But if this observation was not made at the time of an opposition, but at some other place *m*, a second correction is to be applied in the following manner.

Let the angle *q*, of the triangle $P\varpi q$, be found as before, and properly applied to the position of the axis of Mars now at *m*; then make the angle $mS\mu$, at the sun S, equal to the angle SmT, and μ will be the heliocentric place, where the angle of position, when seen from S, will appear to be as it was found at *m*, after the application of the first correction: for $S\mu$ being parallel to Tm, and supposing the axis of Mars to preserve its parallelism while it moves from *m* to μ, appearances of Mars at μ to an eye at S, must be the same as they are at *m* to an eye at T.

The following table contains the result of calculations relating to the angles of fig. 28. In the first column are the times when the observations were made. In the second, the angles as they were taken. In the third column are the quantities of the angles Q, *q*, calculated from the geocentric longitudes contained in the third column of the third table. In the fourth column are the corrections for the situation of the spot in the circle of latitude obtained from the sines of the angles in the fifth column of the third table. In the fifth are the corrections requisite on account of the change of situation of the planets, during the interval between the several days on which the measures were taken; these are obtained from the third column of this table, and I have assumed the 4th of October, as being the observation nearest the opposition, to which I have reduced the other measures. In the sixth column are the angles of the second, corrected by the quantities contained in the fourth and fifth columns, applied according to their signs.

TABLE IV.

Time of observation.	Angles taken.	Angle Q.	First correction.	Second correction.	Angles corrected.
D. H. M.	D. M.	D. M.	D. M.	M.	D. M.
Sept. 25 13 30	56 27	+23 10	−1 52	−8	54 27
Oct. 4 8 46	51 21	+23 18	+4 42	−0	56 3
5 11 50	50 6	+23 19	+7 39	+1	57 46
5 14 0	49 45	+23 19	+7 59	+1	57 45
7 8 20	55 12 55 24	+23 21	−0 7	+2	55 7 55 19
7 10 5	53 0	+23 21	+3 26	+3	56 29
7 11 50	52 12	+23 21	+6 16	+3	58 31
10 7 50	57 12	+23 22	−4 57	+4	52 19
10 9 55	52 42	+23 22	−1 7	+4	51 39
16 7 7	63 9 62 48	+23 25	−7 47	+7	55 29 55 8
16 9 55	65 0	+23 25	−7 23	+7	57 45

As we have no particular reason to select one measure rather than another, a mean of all the 13 will probably be nearest the truth ; so that by these observations, which, as we said before, are reduced to the 4th of October, 1783, we find the position of the axis of Mars that day to have been 55° 41′ south following.

From the appearances of the south polar spot in 1781, represented fig. 27, we may conclude, that its center was nearly polar. We find it continued visible all the time Mars revolved on its axis ; and, to present us generally with a pretty equal share of the luminous appearance, a spot which covered from 45° to 60° of a great circle on the globe of Mars could not have any considerable polar distance : however, a small correction in the angle of position seems to be necessary, which should be taken from the measure of the 15th of July, because that branch of the spot which probably extended farthest towards the equator, was then in the *following*

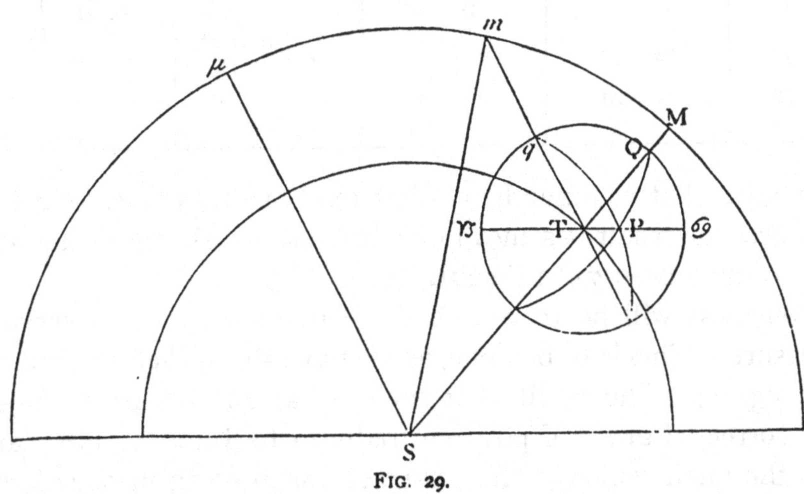

FIG. 29.

quadrant. The measure of both the spots on June the 25th, 1781, is still more to be depended on, as giving us very nearly the position of the true pole ; for it appears evident from the phænomena of the bright north-polar spot in fig. 26 that that spot was in the meridian when the measure was taken, while the southern spot was in the *preceding* quadrant near its greatest limit. Now, since an angle at the circumference of a circle is but half the angle at the center, when the arches which subtend these angles are equal, the correction necessary to be applied to the measure taken through the two spots will be but one half of the correction which would have been requisite had it been taken through the center ; therefore, in order to reduce this to the condition of the former, we may suppose it to have been taken through the center of Mars when the spot was only 30, or 150 degrees from the meridian. It is also necessary to add 1° 54′ to the angle of July 15, which it would have measured more had the planets remained where they were June 25. This done, we may have the polar distance of the center of the spot as before. Half the sum of the sines (of 231° 38′ and 150°) to radius, as 50′ (half the difference between 74° 32′ and 76° 12′) to a fourth number, which is 1° 18′.

I should observe here, that the measures of the angle of position would be too large before the spot came to the meridian, and too small afterwards, the axis of Mars being south preceding; whereas, in fig. 28, they would be too small before, and too large after, the meridian passage, the pole being south following.

These two observations arranged as those in the fourth table, and reduced to the time of the 25th of June, will stand as follows.

TABLE V.

Time of observation.	Angles taken.	Angle Q.	First correction.	Second correction.	Corrected Angle.
D. H. M.	D. M.	D. M.		D. M.	D. M.
June 25 11 36	74 32	−10 14	+ { half of 1° 18′	−0 0	75 11
July 15 10 12	74 18	− 8 20	−1 1	+1 54	75 11

I am to remark, that we have here admitted both measures as equally good; and that, therefore, the result is a mean of them both, and shews the axis of Mars, June 25, 1781, to have been 75° 11′ south preceding.

Our next business will be to reduce these two geocentric observations to a heliocentric measure. This is to be done, as we have shewn before, by a calculation of the angle Q, fig. 29. The result of it shews, that 10° 14′ are to be subtracted from the mean corrected angle of position, reduced to June 25, 1781, and 23° 18′ to be added to the angle which is the corrected mean of 13 measures, reduced to Oct. 4, 1783. Hence we learn, that on those days and hours, when the heliocentric places of Mars were 9^s 24° 35′, and 0^s 7° 15′ (which would happen about July 18, 1781, and Sept. 29, 1783) an observer placed in the sun would have seen, on the former, the axis of Mars inclined to the ecliptic 64° 57′, the north pole being towards the left; and on the latter, he would have seen the same axis inclined to the ecliptic 78° 59′, the north pole being then towards the right.

The first conclusion we may draw from these principles is, that the north pole of Mars must be directed towards some point of the heavens between 9^s 24° 35′ and 0^s 7° 15′; because the change of the situation of the pole from left to right, which happened in the time the planet passed from one place to the other, is a plain indication of its having gone through the node of the axis. Next, we may also conclude, that the node must be considerably nearer the latter point of the ecliptic than the former; for, whatever be the inclination of the axis, it will be seen under equal angles at equal distances from the node.

But, by a trigonometrical process of solving a few triangles, we soon discover both the inclination of the axis, and the place where it intersects the ecliptic at rectangles (which, for want of a better term, I have perhaps improperly called its node). Accordingly I find, by calculation, that the node is in 17° 47′ of Pisces, the north pole of Mars being directed towards that part of the heavens; and that the inclination of the axis to the ecliptic is 59° 42′.

We shall now compare the observations of an earlier date with these principles, to see how far they agree. Some of the particulars and calculations relating to them are as follow.

TABLE VI.

Times of Observation.	Estimations.	Geoc. longit.	Angle Q.	2d correct.
D. H. M.	D.	S. D. M.	D. M.	
1779, May 9 12 0	42	7 22 20	+14 45	+ 0
May 11 12 0	62	7 21 40	+15 11	+26
1777, Apr. 17 7 50	63	6 3 34	+23 26	

May the 9th, 1779, as we have seen, the angle of position was roughly estimated at 42°, and May 11 at 62° The great disagreement of these coarse estimations is undoubtedly owing to the very different situation of the dark spot from which they were taken; however, since we do not mean to use these observations in our calculations, they may suffice in a general way to shew, that the axis of Mars was actually about that time in such a situation as our principles give it: for, reducing the two positions to the 9th of May, that of the 11th, from an allowance of 26′ for the situation of the planets, will become 62° 26′; and a mean of the two, 50° 13′ south preceding; which, reduced to a heliocentric observation, gives 66° 30 the north pole lying towards the left. Now, on calculating from the position of the node and inclination of the axis before determined, we find, that the heliocentric angle was 62° 49′, the north pole pointing towards the left; and a nearer agreement with these principles could hardly be expected from estimations so coarse. If we go to the year 1777, and take the position of the two bright spots observed the 17th of April, we have 63° south preceding; this, reduced to a heliocentric quantity, gives 86° 26′ of inclination, the north pole being to the left. By calculating we find, that that pole was then actually 81° 27′ inclined to the ecliptic, and pointed towards the left as seen from the sun.

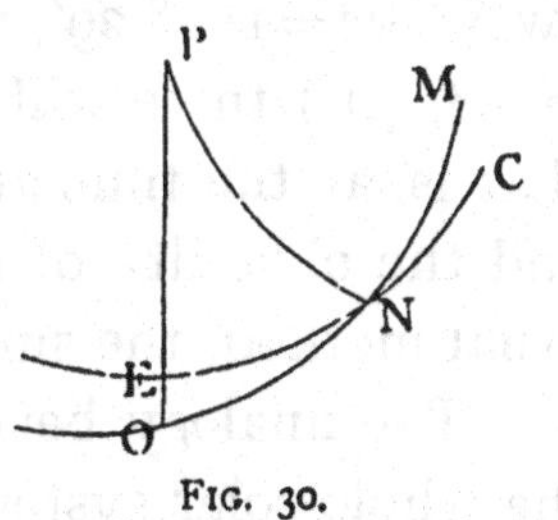

FIG. 30.

The inclination and situation of the node of the axis of Mars with respect to the ecliptic being found may thus be reduced to that planet's own orbit. Let EC, fig. 30, be a part of the ecliptic; OM part of the orbit of Mars; PEO a line drawn from P, the celestial pole of Mars, through E, that point which has been determined to be the place of the node of the axis of Mars in the ecliptic, and continued to O where it intersects the orbit of Mars. Now, if according to Mr. DE

LA LANDE we put the node of the orbit of Mars for 1783, in 1^{s} 17° 58′, we have from the place of the node of the axis (that is, 11^{s} 17° 47′) to the place of the node of the orbit, an arch EN of 60° 11′ ; in the triangle NEO, right-angled at E, there is also given the angle ENO, according to the same author, 1° 51′, which is the inclination of the orbit of Mars to the ecliptic. Hence we find the angle EON 89° 5′, and side ON 60° 12′. Again, when Mars is in the node of its orbit N, we have, by calculation from our principles, the angle PNE = 63° 7′, to which, adding the angle ENO = 1° 51′, we have PNO = 64° 58′ ; from which two angles PON and PNO with the distance ON, we obtain the inclination of the axis of Mars, and place of its node with respect to that planet's own orbit ; the inclination being 61° 18′, and the place of the node of the axis 58° 31′ preceding the intersection of the ecliptic with the orbit of Mars, or in our 19° 28′ of Pisces.

Being thus acquainted with what the inhabitants of Mars will call the obliquity of their ecliptic, and the situation of their equinoctial and solstitial points, we are furnished with the means of calculating the seasons on Mars ; and may account, in a manner which I think highly probable, for the remarkable appearances about its polar regions.

But first it may not be improper to give an instance how to resolve any query concerning the martial seasons. Thus, let it be required to compute the declination of the Sun on Mars, June 25, 1781, at midnight of our time. If ♈ ♉ ♊ ♋, &c. fig. 31, represent the ecliptic of Mars, and ♈♋♎♑ the ecliptic of our planet, A*a*, *b*B, the mutual intersection of the martial and terrestrial ecliptics, then there is given the heliocentric longitude of Mars, ♈*m*$=9^{s}$ 10° 30′ ; then taking away six signs, and ♎*b*, or ♈*a*$=1^{s}$ 17° 58′, there remains *bm*$=1^{s}$ 22° 32′. From this arch, with the given inclination, 1° 51′, of the orbits to each other, we have cosine of inclination to radius, as tangent of *bm* to tangent of BM$=1^{s}$ 22° 33′. And taking away B♈$=1^{s}$ 1° 29′, which is the complement to ♑B (or ♋A, already shewn to be 1^{s} 28° 31′) there will remain ♈M$=0^{s}$ 21° 4′, the place of Mars in its own orbit * ; that is, at the time above mentioned, the sun's longitude on Mars will be 6^{s} 21° 4 , and the obliquity of the martial ecliptic 28° 42′ being also given, we find, by the usual method, the sun's declination 9° 56′ south.

The analogy between Mars and the earth is, perhaps, by far the greatest in the whole solar system. Their diurnal motion is nearly the same ; the obliquity of their respective ecliptics, on which the seasons depend, not very different ; of all the superior planets the distance of Mars from the sun is by far the nearest alike to that of the earth : nor will the length of the martial year appear very different from that which we enjoy, when compared to the surprising duration of the years of Jupiter, Saturn, and the Georgium Sidus. If, then, we find that the

* If no very great accuracy be required, we may add 3^{s} 10° 34′ to any given place of our ecliptic, which will at once reduce it to what it should be called on the orbit of Mars, and will always be true to within a minute.

globe we inhabit has its polar regions frozen and covered with mountains of ice and snow, that only partly melt when alternately exposed to the sun, I may well be permitted to surmise that the same causes may probably have the same effect on the globe of Mars; that the bright polar spots are owing to the vivid reflection of light from frozen regions; and that the reduction of those spots is to be ascribed to their being exposed to the sun. In the year 1781, the south polar spot was

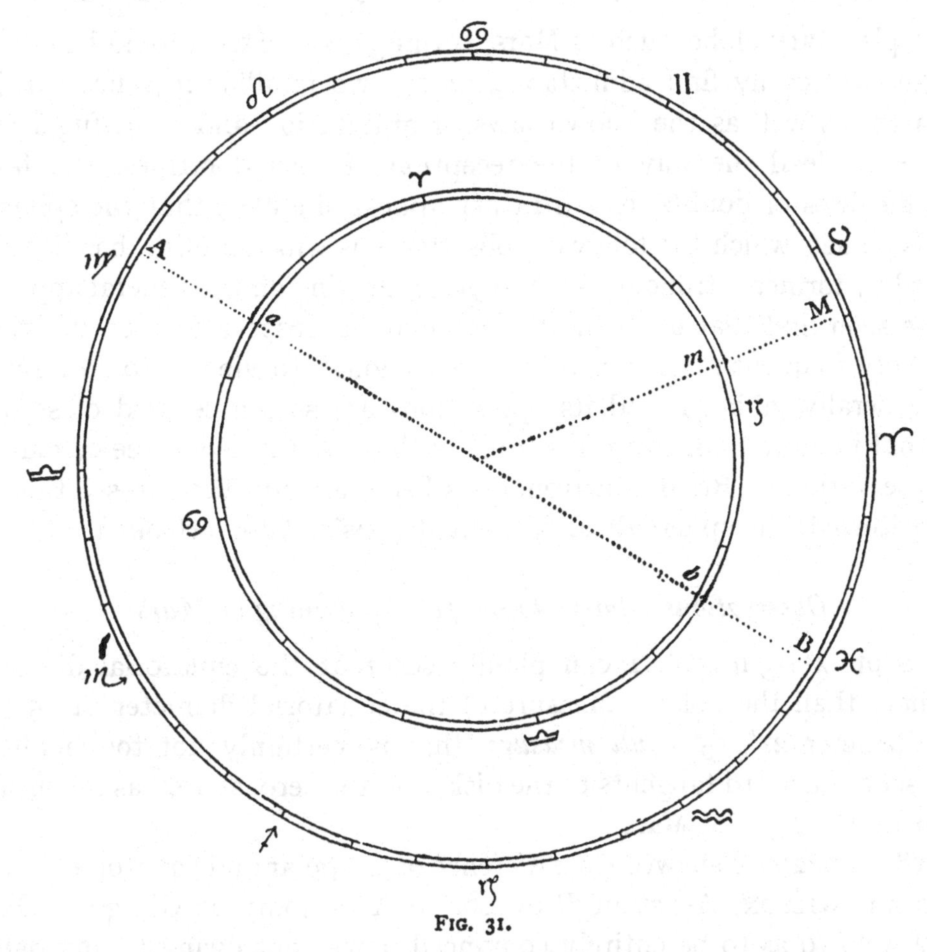

Fig. 31.

extremely large, which we might well expect, since that pole had but lately been involved in a whole twelve-month's darkness and absence of the sun; but in 1783 I found it considerably smaller than before, and it decreased continually from the 20th of May till about the middle of September, when it seemed to be at a stand. During this last period the south pole had already been above eight months enjoying the benefit of summer, and still continued to receive the sun-beams; though, towards the latter end, in such an oblique direction as to be but little benefited by them. On the other hand, in the year 1781, the north polar spot, which had then been its twelve-month in the sun-shine, and was but lately returning to darkness, appeared small, though undoubtedly increasing in size. Its not being visible in the year 1783 is no objection to these phænomena, being owing to the position of the axis, by which it was removed out of sight ; most probably, in the next opposition we shall see it renewed, and of considerable extent and brightness ; as, by the position of the axis of Mars, the sun's southern declination will then be no more than 6° 25′ on that planet.

Of the spheroidical figure of Mars.

That a planetary globe, such as Mars, turning on an axis, should be of a spheroidical form, will easily find admittance, when two familiar instances in Jupiter and the earth, as well as the known laws of gravitation and centrifugal force of rotatory bodies, lead the way to the reception of such doctrines. So far from creating difficulties or doubts, it will rather appear singular, that the spheroidical form of this planet, which the following observations will establish, has not already been noticed by former astronomers ; and yet, reflecting on the general appearances of Mars, we soon find that opportunities for making observations on its real form cannot be very frequent : for, when it is near enough to view it to an advantage, we see it generally gibbous, and its oppositions are so scarce, and of so short a duration, that in more than two years' time we have not above three or four weeks for such observations. Besides, astronomers being already used to see this planet generally distorted, the spheroidical form might easily be overlooked.

Observations relating to the polar flattening of Mars.

1783, Sept. 25. 9 h. 50′. I can plainly see that the equatorial diameter of Mars is longer than the polar. Measure of the equatorial diameter 21″ 53‴ ; of the polar diameter 21″ 15‴ *full measure*, that is, certainly not too small. The wires were set as outward tangents to the disk, and the zero, as well as the measures, were taken by the light of Mars.

Sept. 28. 14 h. 25′. I shewed the difference of the polar and equatorial diameters of Mars to Mr. WILSON, Assistant Professor of Astronomy at Glasgow. He saw it perfectly well, so as to be entirely convinced it was not owing to any defect or distortion occasioned by the eye lens ; and, because I wished him to be satisfied of the reality of the appearance, while he was observing, I reminded him of several well known precautions ; such as causing the planet to pass directly through the center of the field of view, and judging of its figure at the time when it was most distinct and best defined, and so forth.

Sept. 29. I shewed the difference of the polar and equatorial diameters of Mars to Dr. BLAGDEN and Mr. AUBERT. Dr. BLAGDEN not only saw it immediately, but thought the flattening almost as much as that of Jupiter. Mr. AUBERT also saw it very plainly, so as to entertain no manner of doubt about the appearance.

As we cannot take too many opportunities of confirming our own observations by the eyes of other observers, I esteemed it a very fortunate circumstance to have the honour of a visit from these gentlemen at so particular a time, Mars being this day within 37 hours of the opposition, and yesterday when Mr. WILSON saw it, within about two days and a half.

1783, Sept. 30. 10 h. 52′. The difference in the diameters of Mars is very evident and considerable.

Measure of the equatorial diameter	22″ 9‴ with 278.
Second measure	22″ 31‴ full large.
Polar diameter very exact	21″ 26‴.

Oct. 1. 10 h. 50′. I took measures of the diameters of Mars with my 20-feet reflector. The equatorial measured 103 parts of the micrometer; the polar 98. The value of the divisions in seconds and thirds not being well determined, on account of some late change in the focal length of the several 20-feet object metals I use, we have only from these measures the proportion of the diameters as 103 to 98.

13 h. 15′. Every circumstance being favourable, I took the following measures of the diameters of Mars with my 7-feet reflector, and a distinct power of 625.

Equatorial diameter	22″ 12‴ narrow measure.
	22″ 46‴ rather full.
	22″ 35‴ exact.
Polar diameter	21″ 24‴
	21″ 33‴ very exact.

I saw Mars perfectly well all the time I measured, with all its figures upon the disk appearing distinctly; and, I think, these measures may be depended upon better than any I have yet taken.

Oct. 5. 14 h. 0′. The difference of the diameters is very sensible.

Oct. 7. 9 h. 43′. The flattening of the poles is very visible.

13 h. 40′. I turned my Newtonian 7-feet reflector one quarter round, so as to bring the place to look in at to the bottom; and, as well as the uneasy posture would allow, I saw the flattening of the poles the same as when I looked in at the side; power 460.

14 h. 30′. With a 3½ feet achromatic telescope and a single eye lens, I saw the difference of the polar and equatorial diameters very plainly.

Oct. 9. 8 h. 40′. I turned my reflector 90° round, so as now to look in at the upper end, but saw not the least difference in appearances; for, returning it again immediately to its usual position, in both cases the equatorial diameter appeared a little longer than the other; power 278, and the evening fine.

I turned the great speculum one quadrant in its cell, but appearances were not in the least altered; the equatorial diameter still was a little longer than the polar one.

I tried a very fine new object speculum, and found also the equatorial diameter a little longer than the polar one.

Oct. 9. 10 h. 47′. The flattening at the poles very visible.

Oct. 10. 9 h. 55′. A little of the polar flattening is visible, so as to admit of no doubt ; power 460, very distinct.

11 h. 32′. Mars visibly flattened, but not much ; the achromatic shews it also.

11 h. 42′. The disk of Mars is visibly spheroidical.

Oct. 11. 7 h. 37′. Mars is plainly gibbous, therefore measures and estimations of the diameters must for the future be improper.

11 h. 12′. It is rather difficult to say of what shape Mars is now, for it is partly flattened and partly gibbous ; but the gibbous side not being quite in the polar direction of Mars, this produces altogether an odd mixture of shapes : however, upon the whole, the polar diameter is still rather the smallest.

11 h. 13′. The *preceding* side of Mars shews the flattening of the poles, while the *following* is terminated by an elliptical arch.

Oct. 12. 11 h. 12′. The flattening upon the whole is visible.

Oct. 17. 13 h. 7′. The effect of gibbosity is scarcely equal to the flattening ; or, upon the whole, the planet is still rather broader over the equator than over the poles.

Nov. 1. 7 h. 56′. The semi-disk, which is *full*, is evidently part of an oblate spheroid ; but, to an eye not attentively looking for it, and knowing the shape and exact situation of the poles of Mars, this would probably not appear.

Nov. 10. 9 h. 30′. The gibbosity of Mars is now such, that the polar diameter is considerably longer than the equatorial ; but the deficiency not being exactly from pole to pole, makes the disk of a crooked, irregular figure, and renders precision in this estimation impossible ; otherwise the phase of Mars would have made a pretty good micrometer upon the equatorial diameter, and it was with such a view I had directed my attention to this circumstance : appearances, however, are visibly in favour of the polar diameter's being the longest.

We find that the quick alterations in the visible disk of Mars, during the time it is in the best situation for us to observe it, are such, that if we were to use many measures which have been taken of its diameters, we should be obliged to have recourse to a computation of its phases, in order to make proper allowance for them. Now, since these changes are in a longitudinal direction, and the poles of Mars are not perpendicular to the ecliptic, it would bring on a calculation of small quantities, which it is always best not to run into where it can be avoided. For this reason, I shall at once settle the proportion of the equatorial to the polar diameter of this planet, from the measures which were taken on the very day of the opposition. I prefer them also on another account, which is, that they were made in a very fine, clear air, and were repeated with a very high power, and with two different instruments, of whose faithful representation of celestial objects, the many observations on very close double stars I have made with them have given me very evident proofs.

As we are at present only in quest of the proportion of one diameter to the other, the measures of the 20-feet reflector, though not given in angular quantities, will equally suffice for the purpose. By them we have the equatorial diameter to the polar as 103 to 98, or as 1355 to 1289. I have turned the proportion into the latter numbers by way of comparing them the better with the measures of the 7-feet reflector. By that instrument the equator of Mars, Oct. 1, we find, was measured three times; but from the remarks annexed to the different results, I think the third measure should be used. Indeed, on taking the difference of the two first, which is 34‴, and dividing by three, we have the quotient 11⅓‴; then, allotting two-thirds to the first, because the remark says positively "narrow measure," it becomes 22″ 34⅔‴, and taking one-third from the second, which is expressed doubtfully, "rather too full," it becomes 22″ 35⅓‴: this reflection on the two first measures gives additional validity to the third, which is 22″ 35‴, or 1355‴. The polar diameter was measured twice; and as no reason appears against either of the observations, I shall take the mean of both, which is 21″ 29‴, or 1289‴; so that by these measures the equatorial diameter of Mars is to the polar as 1355 to 1289. A less perfect agreement between the proportions of the diameters arising from the measures of the 20-feet reflector and those which we have just now deduced from the 7-feet, would have been sufficient for our purpose, as we might easily have excused one or two thousandths of the whole quantity; however, we have no cause to be displeased with this coincidence, though it should in part be owing to accident, and therefore shall admit the above proportion, and proceed to a farther examination of it.

In the first place, it will be necessary to see whether any correction be required on account of the different heliocentric and geocentric south latitude of Mars; which would apparently compress the polar diameter a little, by the defect of illumination on the north. On computation we find, that a difference arising from that cause would give the longitudinal diameter to the latitudinal as 20000 to 19987; which being much less than one thousandth part of the whole, may therefore be neglected.

But next, a very considerable correction must be admitted, when we take into account the position of the axis of Mars. The declination of the sun on that planet, at the time the measures were taken, was not less than 27° south; so that the poles were not in the circumference of the disk by all that quantity. On a supposition then, that the figure of Mars is an elliptical spheroid, we are now to find the real quantity of the polar diameter from the apparent one. It has been proved, that, in the ellipsis, the excesses of any diameters above the polar one are as the squares of the cosines of the latitudes;* but the diameter at rectangles to

* *Astr.* par M. DE LA LANDE, § 2680.

the equator of Mars, which was exposed to our view in the late opposition, was not the polar one, but such as must take place in a latitude of 63°. Putting therefore m = cosine of 63°, $a = 1355$, $b = 1289$, x = the polar axis, we have $1 : m^2 :: (a-x) : (b-x)$. And $\frac{b-m^2a}{1-m^2} = x$; which gives us 1272 nearly, for the polar diameter. The true proportion, therefore, of the equatorial to the polar diameter will be as 1355 to 1272; which, reduced to smaller but less accurate numbers, is 16 to 15 nearly.

I shall now also mention some of the other measures, but with a view only to shew that they are very consistent with the above determination. From those of the 30th of September, for instance, we collect the proportion of the diameters of Mars as 1340 to 1286; or, reduced to our former numbers, 1355 to 1300. Now, since these measures were taken the night before the opposition, they must on that account be as good as the former; and, had those of the day of opposition not been preferred, because they were oftener repeated, and the superior power of the 7, and great light of the 20-feet reflector, gave them additional weight, I should have taken them into the account; the very small difference, however, cannot but strengthen the results of the former measures.

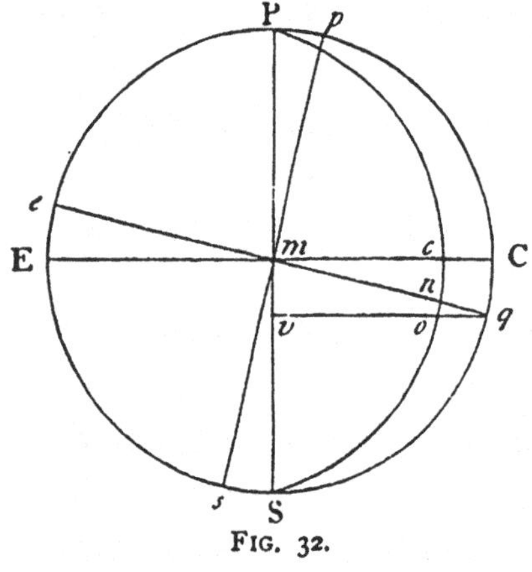

FIG. 32.

From the observations of the 25th of September we have the proportion of the diameters as 1313 to 1275; and if the equatorial measure be increased in the ratio of 20000 to 19953, on account of the different heliocentric and geocentric longitude, Mars not being at the full, it will give the ratio of 1316 to 1275; or, conforming to our former numbers, as 1355 to 1312. I have not been very strict in the application of the correction deduced from the phases of Mars, since no other use was intended to be made of these numbers than merely to shew, that they do not very greatly differ from those we have assigned before.*

It was observed, Oct. 17, 1783, that the equatorial diameter of Mars was still greater than the polar, notwithstanding the depredation of the defect of light upon it. On calculating the phases, we find, that the longitudinal diameter was, that day, to the latitudinal one as 19711 to 20000, which therefore could not be an equal balance to oppose the spheroidical figure so as to render it invisible.

But, Nov. 10, the proportion of the longitudinal diameter to the latitudinal

* If more strictness be required, let EC, fig. 32, be the ecliptic; PS its poles; *ps* the poles of Mars, and *eq* its equator. Then, the angle *pm*C being found, by calculation, we shall have C*m* (radius) to *cm* (cosine of the difference between the heliocentric and geocentric longitude) as *qv* (sine of the angle *qmv* or *pm*C) to *ov*. Then, since with Mars C*c* can never be very great, the small triangle *qno* may be taken for similar to *qvm*; therefore *qm* (radius) is to *qv* (sine of *pm*C) as *qo* ($= qv - vo$) to *qn*; which is the required correction or deficiency of the equatorial diameter *eq* of Mars.

Or, putting $m\mathrm{C} = 1$ and $vq = m$ = cosine of the angle P*mp*; it will be $qn = m^2 \cdot c\mathrm{C}$.

one, from a computation of the phase of Mars, must have been as 18762 to 20000; and accordingly it was by observation found to be more than sufficient to take off all appearance of the polar flattening, and leave a visible excess in the axis above the equator.

To obviate any doubts concerning a fallacy that might arise from the convexity of the eye-glass, or irregular shape of the small speculum, I need only refer, for the latter, to the experiments of the 7th and 9th of October, 1783: for should the short diameter of my small plane speculum have occasioned a compressing of the polar diameter of Mars when exposed to it, half a turn of the telescope must bring the other diameter of that speculum into the same situation, and a contrary effect would have followed. With regard to the former, not only the experiments made with the achromatic, but principally the observation with the 20-feet reflector, where I used a compound eye-piece magnifying only about 300 times, will sufficiently exculpate the eye-glasses. It is also well known, that in a single lens the distortion of the images, if any such there should be, will equally affect the wires of the micrometer, and give a true measure notwithstanding; and the compound eye-piece I used with the 20-feet reflector had likewise the same advantage, for it is constructed on the plan lately proposed by Mr. RAMSDEN in the *Philosophical Transactions*,* which he was so obliging as to communicate to me about a twelve-month ago, and which I immediately adapted to my large micrometers.

On the subject of the figure of Mars I ought to remark also, that perhaps the measures which were taken of its diameters during the last opposition will enable us to ascertain its real size with greater accuracy than has been done before. The micrometer which can distinguish with precision between the equatorial and polar diameters of this small planet, will certainly be admitted as an evidence of considerable consequence; and since the result of these measures is pretty different from what former observations give us, I should not omit mentioning it.

We have seen that the equatorial diameter, on the day of the opposition, measured 22″ 35‴ The distance of Mars from the earth at that time was ·40457, the mean distance of the earth from the sun being 1; therefore, 22″ 35‴ reduced to the same distance will be no more than 9″ 8‴.

I shall conclude this subject with a consideration relating to the atmosphere of Mars. Dr. SMITH † reports an observation of CASSINI's, where "a star in the water of Aquarius, at the distance of six minutes from the disk of Mars, became so faint before its occultation, that it could not be seen by the naked eye, nor with a 3-feet telescope." It is not mentioned what was the magnitude of the star; but, from the circumstance of its becoming invisible to the naked eye, we may conclude, that it must have been of the sixth or seventh magnitude at least. The result of this observation would indicate an atmosphere of such an extraordinary extent, since at the distance of 36 semi-diameters of the planet it should still be dense enough

* Vol. LXXIII. p. 94. † *Optics*, § 1096.

to render so considerable a star invisible, that it will certainly not be amiss to give an observation or two which seem of a very different import.

1783, Oct. 26. There are two small fixed stars preceding Mars, of different sizes; with 460 they appear both dusky red, and are pretty unequal; with 278 they appear considerably unequal. The distance from Mars of the nearest, which is also the largest, with 227 measured 3′ 26″ 20‴. Some time after, the same evening, the distance was 3′ 8″ 55‴, Mars being retrograde. I saw them both very distinctly. I viewed the two stars with a new 20-feet reflector of 18·7 inches aperture, and found them, as I expected, very bright.

Oct. 27. I see the two small stars again. The small one is not quite so bright in proportion to the large one as it was last night, being a good deal nearer to Mars, which is now on the side of the small star; but when I draw the planet aside, or out of view, I see it then as well as I did last night. The distance of the small star measured 2′ 56″ 25‴.*

The largest of the two stars on which the above observations were made cannot exceed the twelfth, and the smallest the thirteenth or fourteenth magnitude; and I have no reason to suppose that they were any otherwise affected by the approach of Mars, than what the brightness of its superior light may account for. From other phænomena it appears, however, that this planet is not without a considerable atmosphere; for, besides the permanent spots on its surface, I have often noticed occasional changes of partial bright belts, as in fig. 1 and 14; and also once a darkish one, in a pretty high latitude, as in fig. 18. And these alterations we can hardly ascribe to any other cause than the variable disposition of clouds and vapours floating in the atmosphere of that planet.

Result of the contents of this paper.

The axis of Mars is inclined to the ecliptic 59° 42′.

The node of the axis is in 17° 47′ of Pisces.

The obliquity of the ecliptic on the globe of Mars is 28° 42′.

The point Aries on the martial ecliptic answers to our 19° 28′ of Sagittarius.

The figure of Mars is that of an oblate spheroid, whose equatorial diameter is to the polar one as 1355 to 1272, or as 16 to 15 nearly.

The equatorial diameter of Mars, reduced to the mean distance of the earth from the sun, is 9″ 8‴.

And that planet has a considerable but moderate atmosphere, so that its inhabitants probably enjoy a situation in many respects similar to ours.

W. HERSCHEL.

Datchet, Dec. 1, 1783.

* The measures were accurate enough for the purpose, though not otherwise to be depended on nearer than, perhaps, six or eight seconds.

XIII.

Account of some Observations tending to investigate the Construction of the Heavens.

[*Phil. Trans.*, vol. lxxiv., 1784, pp. 437-451.]

Read June 17, 1784.

In a former paper I mentioned, that a more powerful instrument was preparing for continuing my reviews of the heavens. The telescope I have lately completed, though far inferior in size to the one I had undertaken to construct when that paper was written, is of the Newtonian form, the object speculum being of 20 feet focal length, and its aperture $18\frac{7}{10}$ inches. The apparatus on which it is mounted is contrived so as at present to confine the instrument to a meridional situation, and by its motions to give the right-ascension and declination of a celestial object in a coarse way ; which, however, is sufficiently accurate to point out the place of the object, so that it may be found again. It will not be necessary to enter into a more particular description of the apparatus, since the account I have now the honour of communicating to the Royal Society regards rather the performance of the telescope than its construction.

It would, perhaps, have been more eligible to have waited longer, in order to complete the discoveries that seem to lie within the reach of this instrument, and are already, in some respects, pointed out to me by it. By taking more time I should undoubtedly be enabled to speak more confidently of the *interior construction* of the heavens, and its various *nebulous and sidereal strata* (to borrow a term from the natural historian) of which this paper can as yet only give a few outlines, or rather hints. As an apology, however, for this prematurity, it may be said, that the end of all discoveries being communication, we can never be too ready in giving facts and observations, whatever we may be in reasoning upon them.

Hitherto the sidereal heavens have, not inadequately for the purpose designed, been represented by the concave surface of a sphere, in the center of which the eye of an observer might be supposed to be placed. It is true, the various magnitudes of the fixed stars even then plainly suggested to us, and would have better suited the idea of an expanded firmament of three dimensions ; but the observations upon which I am now going to enter still farther illustrate and enforce the necessity of considering the heavens in this point of view. In future, therefore, we shall look upon those regions into which we may now penetrate by means of such large telescopes, as a naturalist regards a rich extent of ground or chain of mountains, containing strata variously inclined and directed, as well as consisting of very different materials. A surface of a globe or map, therefore, will but ill delineate the interior parts of the heavens.

It may well be expected, that the great advantage of a large aperture would be most sensibly perceived with all those objects that require much light, such as the very small and immensely distant fixed stars, the very faint nebulæ, the close and compressed clusters of stars, and the remote planets.

On applying the telescope to a part of the *via lactea*, I found that it completely resolved the whole whitish appearance into small stars, which my former telescopes had not light enough to effect. The portion of this extensive tract, which it has hitherto been convenient for me to observe, is that immediately about the hand and club of Orion. The glorious multitude of stars of all possible sizes that presented themselves here to my view was truly astonishing ; but, as the dazzling brightness of glittering stars may easily mislead us so far as to estimate their number greater than it really is, I endeavoured to ascertain this point by counting many fields, and computing, from a mean of them, what a certain given portion of the milky way might contain. Among many trials of this sort I found, last January the 18th, that six fields, promiscuously taken, contained 110, 60, 70, 90, 70, and 74 stars each. I then tried to pick out the most vacant place that was to be found in that neighbourhood, and counted 63 stars. A mean of the first six gives 79 stars for each field. Hence, by allowing 15 minutes of a great circle for the diameter of my field of view, we gather, that a belt of 15 degrees long and two broad, or the quantity which I have often seen pass through the field of my telescope in one hour's time, could not well contain less than fifty thousand stars, that were large enough to be distinctly numbered. But, besides these, I suspected at least twice as many more, which, for want of light, I could only see now and then by faint glittering and interrupted glimpses.

The excellent collection of nebulæ and clusters of stars which has lately been given in the *Connoissance des Temps* for 1783 and 1784, leads me next to a subject which, indeed, must open a new view of the heavens. As soon as the first of these volumes came to my hands, I applied my former 20-feet reflector of 12 inches aperture to them ; and saw, with the greatest pleasure, that most of the nebulæ, which I had an opportunity of examining in proper situations, yielded to the force of my light and power, and were resolved into stars. For instance, the 2d, 5, 9, 10, 12, 13, 14, 15, 16, 19, 22, 24, 28, 30, 31, 37, 51, 52, 53, 55, 56, 62, 65, 66, 67, 71, 72, 74, 92, all which are said to be nebulæ without stars, have either plainly appeared to be nothing but stars, or at least to contain stars, and to shew every other indication of consisting of them entirely. I have examined them with a careful scrutiny of various powers and light, and generally in the meridian. I should mention, that five of the above, *viz.* the 16th, 24, 37, 52, 67, are called clusters of stars containing

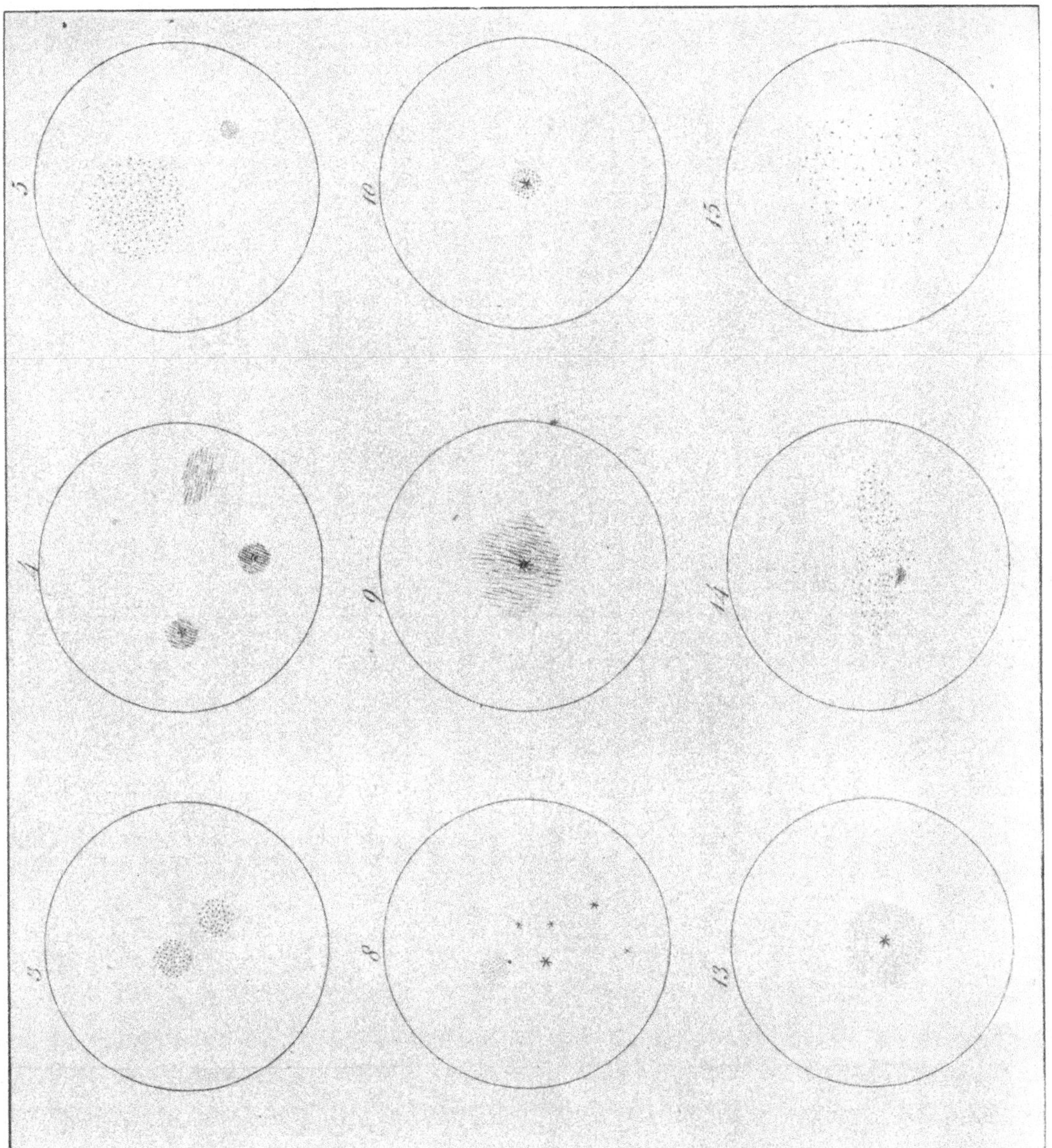

[*To face page* 159.

nebulosity ; but my instrument resolving also that portion of them which is called nebulous into stars of a much smaller size, I have placed them into the above number. To these may be added the 1st, 3d, 27, 33, 57, 79, 81, 82, 101, which in my 7, 10, and 20-feet reflectors shewed a mottled kind of nebulosity, which I shall call resolvable ; so that I expect my present telescope will, perhaps, render the stars visible of which I suppose them to be composed. Here I might point out many precautions necessary to be taken with the very best instruments, in order to succeed in the resolution of the most difficult of them ; but reserving this at present too extensive subject for a future opportunity, I proceed to speak of the effects of my last instrument with regard to nebulæ.

My present pursuits, as I observed before, requiring this telescope to act as a fixed instrument, I found it not convenient to apply it to any other of the nebulæ in the *Connoissance des Temps* but such as came in turn ; nor, indeed, was it necessary to take any particular pains to look for them, it being utterly impossible that any one of them should escape my observation when it passed the field of view of my telescope. The few which I have already had an opportunity of examining, shew plainly that those most excellent French astronomers, Mess. MESSIER and MECHAIN, saw only the more luminous part of their nebulæ ; the feeble shape of the remainder, for want of light, escaping their notice. The difference will appear when we compare my observation of the 98th nebula with that in the *Connoissance des Temps* for 1784, which runs thus : " Nébuleuse sans étoile, d'une lumière extrêmement foible, au dessus de l'aile boréale de la Vierge, sur le parallèle et près de l'étoile N° 6, cinquième grandeur, de la chevelure de Bérénice, suivant FLAMSTEED. M. MECHAIN la vit le 15 Mars, 1781." My observation of the 30th of December, 1783, is thus : A large, extended, fine nebula. Its situation shews it to be M. MESSIER's 98th ; but from the description it appears, that that gentleman has not seen the whole of it, for its feeble branches extend above a quarter of a degree, of which no notice is taken. Near the middle of it are a few stars visible, and more suspected. My field of view will not quite take in the whole nebula. See fig. 1, Plate VII. Again, N° 53, " Nébuleuse sans étoiles, decouverte au-dessous et près de la chevelure de Bérénice, à peu de distance de l'étoile quarante-deuxieme de cette constellation, suivant FLAMSTEED. Cette nébuleuse est ronde et apparente, &c." My observation of the 170th Sweep runs thus : A cluster of very close stars ; one of the most beautiful objects I remember to have seen in the heavens. The cluster appears under the form of a solid ball, consisting of small stars, quite compressed into one blaze of light, with a great number of loose ones surrounding it, and distinctly visible in the general mass. See fig. 2.

When I began my present series of observations, I surmised, that several nebulæ might yet remain undiscovered, for want of sufficient light to detect them ; and was, therefore, in hopes of making a valuable addition to the clusters of stars and nebulæ already collected and given us in the work before referred to, which amount to 103. The event has plainly proved that my expectations were well founded: for I have already found 466 new nebulæ and clusters of stars, none of which, to my present knowledge, have been seen before by any person; most of them, indeed, are not within the reach of the best common telescopes now in use. In all probability many more are still in reserve; and as I am pursuing this track, I shall make them up into separate catalogues, of about two or three hundred at a time, and have the honour of presenting them in that form to the Royal Society.

A very remarkable circumstance attending the nebulæ and clusters of stars is, that they are arranged into strata, which seem to run on to a great length; and some of them I have already been able to pursue, so as to guess pretty well at their form and direction. It is probable enough, that they may surround the whole apparent sphere of the heavens, not unlike the milky way, which undoubtedly is nothing but a stratum of fixed stars. And as this latter immense starry bed is not of equal breadth or lustre in every part, nor runs on in one straight direction, but is curved and even divided into two streams along a very considerable portion of it; we may likewise expect the greatest variety in the strata of the clusters of stars and nebulæ. One of these nebulous beds is so rich, that, in passing through a section of it, in the time of only 36 minutes, I detected no less than 31 nebulæ, all distinctly visible upon a fine blue sky. Their situation and shape, as well as condition, seem to denote the greatest variety imaginable. In another stratum, or perhaps a different branch of the former, I have seen double and treble nebulæ, variously arranged; large ones with small, seeming attendants; narrow but much extended, lucid nebulæ or bright dashes; some of the shape of a fan, resembling an electric brush, issuing from a lucid point; others of the cometic shape, with a seeming nucleus in the center; or like cloudy stars, surrounded with a nebulous atmosphere; a different sort again contain a nebulosity of the milky kind, like that wonderful, inexplicable phænomenon about θ Orionis; while others shine with a fainter, mottled kind of light, which denotes their being resolvable into stars. See fig. 3, &c. But it would be too extensive at present to enter more minutely into such circumstances, therefore I proceed with the subject of nebulous and sidereal strata.

It is very probable, that the great stratum, called the milky way, is that in which the sun is placed, though perhaps not in the very center of its thickness. We gather this from the appearance of the Galaxy, which seems to encompass the whole heavens, as it certainly must do if the sun is within the same. For, suppose a number of stars arranged between two parallel planes, indefinitely extended every way, but at a given considerable distance from each other; and, calling this a sidereal stratum, an eye placed somewhere within it will see all the stars in the direction of the planes of the stratum projected into a great circle, which will appear lucid on account of the accumulation of the stars; while the rest of the heavens, at the sides, will only seem to be scattered over with constellations, more or less

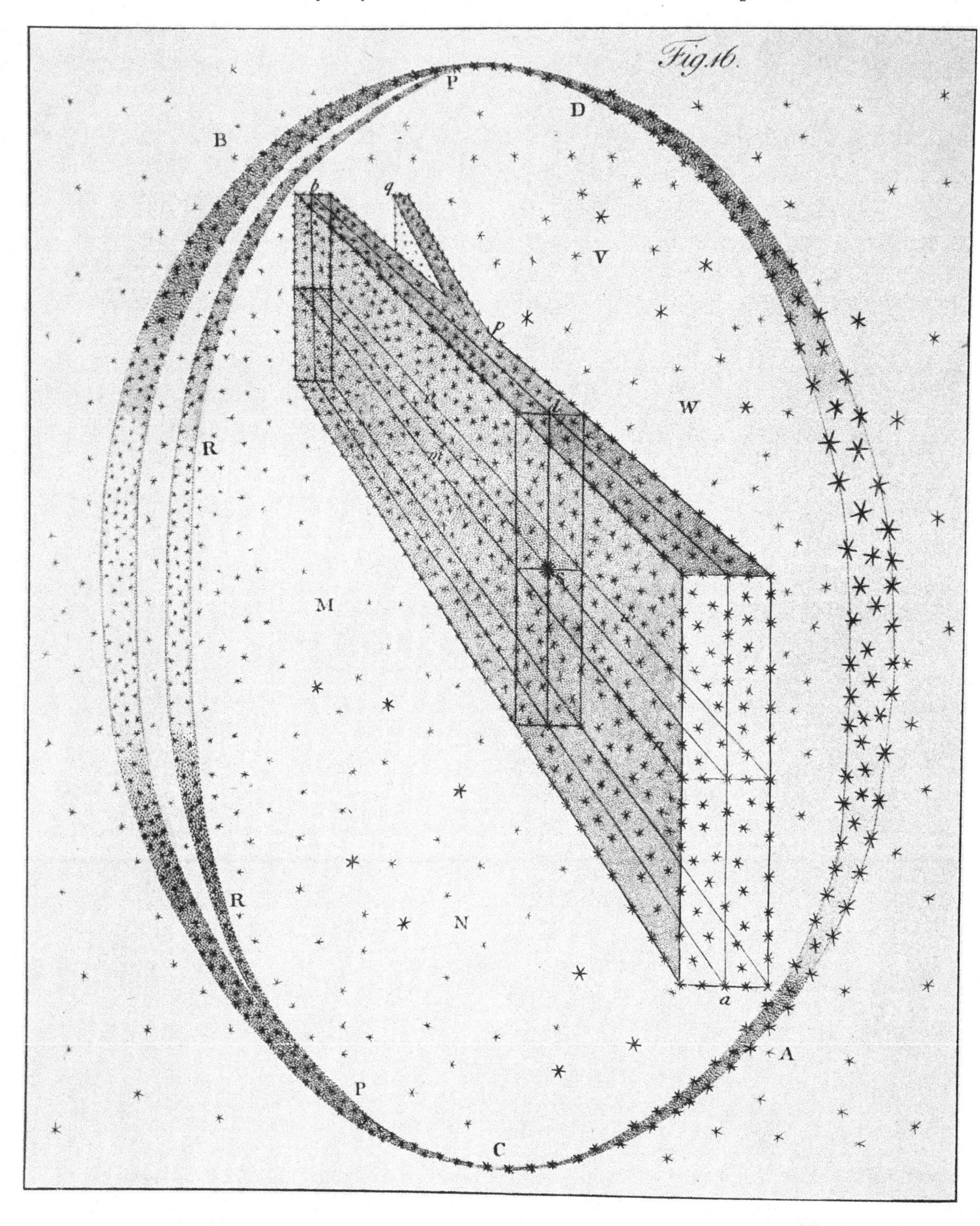

[*To face page* 161.

crowded, according to the distance of the planes or number of stars contained in the thickness or sides of the stratum.

Thus, in fig. 16 (Plate VIII.) an eye at S within the stratum *ab*, will see the stars in the direction of its length *ab*, or height *cd*, with all those in the intermediate situations, projected into the lucid circle ACBD ; while those in the sides *mv*, *nw*, will be seen scattered over the remaining part of the heavens at MVNW.

If the eye were placed somewhere without the stratum, at no very great distance, the appearance of the stars within it would assume the form of one of the less circles of the sphere, which would be more or less contracted to the distance of the eye ; and if this distance were exceedingly increased, the whole stratum might at last be drawn together into a lucid spot of any shape, according to the position, length, and height of the stratum.

Let us now suppose, that a branch, or smaller stratum, should run out from the former, in a certain direction, and let it also be contained between two parallel planes extended indefinitely onwards, but so that the eye may be placed in the great stratum somewhere before the separation, and not far from the place where the strata are still united. Then will this second stratum not be projected into a bright circle like the former, but will be seen as a lucid branch proceeding from the first, and returning to it again at a certain distance less than a semi-circle.

Thus, in the same figure, the stars in the small stratum *pq* will be projected into a bright arch at PRRP, which, after its separation from the circle CBD, unites with it again at P.

What has been instanced in parallel planes may easily be applied to strata irregularly bounded, and running in various directions ; for their projections will of consequence vary according to the quantities of the variations in the strata and the distance of the eye from the same. And thus any kind of curvatures, as well as various different degrees of brightness, may be produced in the projections.

From appearances then, as I observed before, we may infer, that the sun is most likely placed in one of the great strata of the fixed stars, and very probably not far from the place where some smaller stratum branches out from it. Such a supposition will satisfactorily, and with great simplicity, account for all the phænomena of the milky way, which, according to this hypothesis, is no other than the appearance of the projection of the stars contained in this stratum and its secondary branch. As a farther inducement to look on the Galaxy in this point of view, let it be considered, that we can no longer doubt of its whitish appearance arising from the mixed lustre of the numberless stars that compose it. Now, should we imagine it to be an irregular ring of stars, in the center nearly of which we must then suppose the sun to be placed, it will appear not a little extraordinary, that the sun, being a fixed star like those which compose this imagined ring, should just be in the center of such a multitude of celestial bodies, without any apparent reason for this singular distinction ; whereas, on our supposition, every star in this stratum, not very near the termination of its length or height, will be so placed as also to have its own Galaxy, with only such variations in the form and lustre of it, as may arise from the particular situation of each star.

Various methods may be pursued to come to a full knowledge of the sun's place in the sidereal stratum, of which I shall only mention one as the most general and most proper for determining this important point, and which I have already begun to put in practice. I call it *Gaging the Heavens*, or the *Star-Gage*. It consists in repeatedly taking the number of stars in ten fields of view of my reflector very near each other, and by adding their sums, and cutting off one decimal on the right, a mean of the contents of the heavens, in all the parts which are thus gaged, is obtained. By way of example, I have joined a short table, extracted from the gages contained in my journal, by which it appears, that the number of stars increases very fast as we approach the Via Lactea.

N.P.D. 92 to 94°.	
R.A.	Gage.
15ʰ 10′	9·4
15 22	10·6
15 47	10·6
16 8	12·1
16 25	13·6
16 37	18·6

N.P.D. 78 to 80°.	
R.A.	Gage.
11ʰ 16′	3·1
12 31	3·4
12 44	4·6
12 49	3·9
13 5	3·8
14 30	3·6

Thus, in the parallel from 92 to 94 degrees north polar distance, and R. A. 15 h. 10′, the star-gage runs up from 9·4 stars in the field to 18·6 in about an hour and a half; whereas in the parallel from 78° to 80° north polar distance, and R. A. 11, 12, 13, and 14 hours, it very seldom rises above 4. We are, however, to remember, that with different instruments the account of the gages will be very different, especially on our supposition of the situation of the sun in a stratum of stars. For, let *ab*, fig. 17, be the stratum, and suppose the small circle *ghlk* to represent the space into which, by the light and power of a given telescope, we may penetrate; and let GHLK be the extent of another portion, which we are enabled to visit by means of a larger aperture and power; it is evident, that the gages with the latter instrument will differ very much in their account of stars contained at MN, and at KG or LH; when with the former they will hardly be affected by the change from *mn* to *kg* or *lh*. And this accounts for what a celebrated author says concerning the effects of telescopes, by which we must understand the best of those that are in common use.*

* On voit avec les télescopes des étoiles dans toutes les parties du ciel, à peu près comme dans la voie lactée, ou dans les nébuleuses. On ne sauroit douter qu'une partie de l'éclat et de la blancheur de la voie lactée, ne provienne de la lumière des petites étoiles qui s'y trouvent en effet par millions;

It would not be safe to enter into an application of these, and such other gages as I have already taken, till they are sufficiently continued and carried all over the heavens. I shall, therefore, content myself with just mentioning that the situation of the sun will be obtained, from considering in what manner the star-gage agrees with the length of a ray revolving in several directions about an assumed point, and cut off by the bounds of the stratum. Thus, in fig. 18, let S be the place of an observer; Srrr, S*rrr*, lines in the planes rSr, *r*S*r*, drawn from S within the stratum to one of the boundaries, here represented by the plane AB. Then,

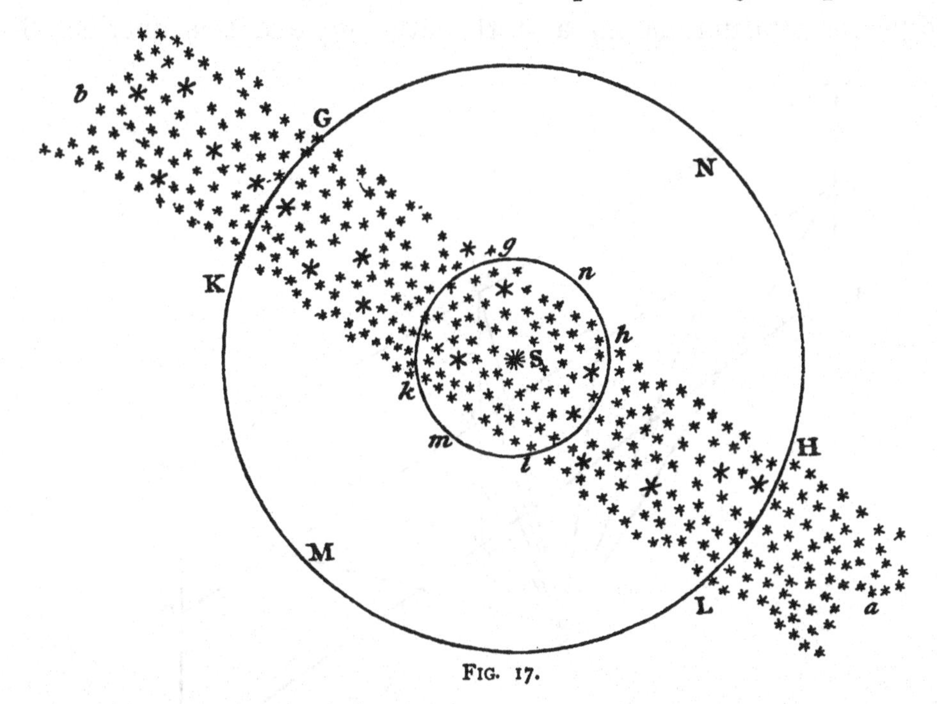

Fig. 17.

since neither the situation of S, nor the form of the limiting surface AB, is given, we are to assume a point, and apply to it lines proportional to the several gages that have been obtained, and at such angles from each other as they may point out; then will the termination of these lines delineate the boundary of the stratum, and consequently manifest the situation of the sun within the same. But to proceed.

If the sun should be placed in the great sidereal stratum of the milky way, and, as we have surmised above, not far from the branching out of a secondary stratum, it will very naturally lead us to guess at the cause of the probable motion of the solar system: for the very bright, great node of the Via Lactis, or union of the two strata about Cepheus and Cassiopeia, and the Scorpion and Sagittarius,

cependant, avec les plus grands télescopes, on n'en distingue pas assez, et elles n'y sont pas assez rapprochées les unes des autres pour qu'on puisse attribuer à celles qu'on distingue la blancheur de la voie lactée, si sensible à la vue simple. L'on ne sauroit donc prononcer que les étoiles soient la seule cause de cette blancheur, quoique nous ne connoissions aucune manière satisfaisante de l'expliquer. *Ast.* M. de la Lande, § 833.

points out a conflux of stars manifestly quite sufficient to occasion a tendency towards that node in any star situated at no very great distance; and the secondary branch of the Galaxy not being much less than a semi-circle seems to indicate such a situation of our solar system in the great undivided stratum as the most probable.

What has been said in a former paper on the subject of the solar motion seems also to support this supposed situation of the sun; for the apex there assigned lies nearly in the direction of a motion of the sun towards the node of the strata. Besides, the joining stratum making a pretty large angle at the junction with the

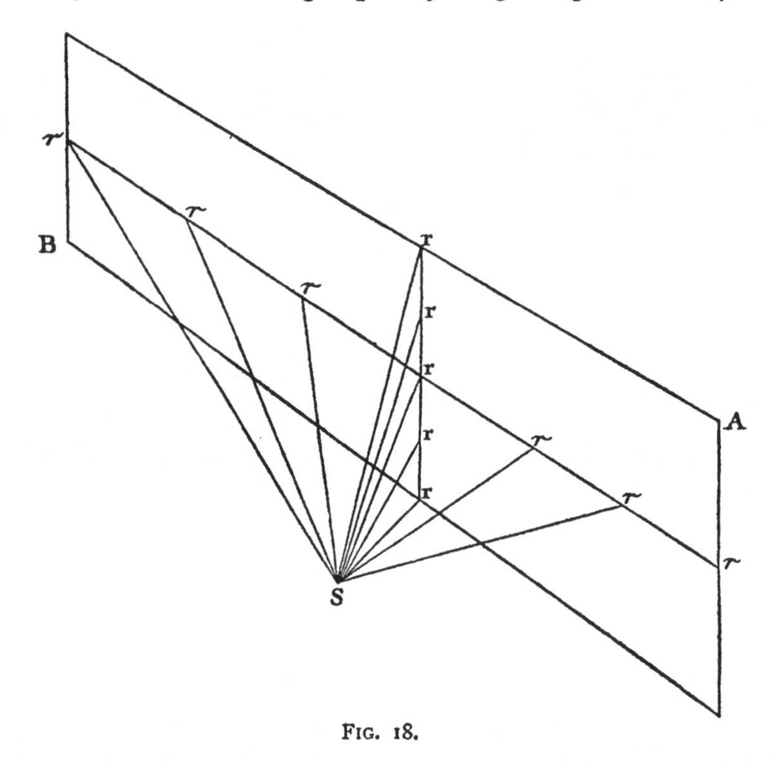

FIG. 18.

primary one, it may easily be admitted, that the motion of a star in the great stratum, especially if situated considerably towards the side farthest from the small stratum, will be turned sufficiently out of the straight direction of the great stratum towards the secondary one. But I find myself insensibly led to say more on this subject than I am as yet authorised to do; I will, therefore, return to those observations which have suggested the idea of celestial strata.

In my late observations on nebulæ I soon found, that I generally detected them in certain directions rather than in others; that the spaces preceding them were generally quite deprived of their stars, so as often to afford many fields without a single star in it; that the nebulæ generally appeared some time after among stars of a certain considerable size, and but seldom among very small stars; that when I came to one nebula, I generally found several more in the neighbourhood; that afterwards a considerable time passed before I came to another parcel; and these events being often repeated in different altitudes of my instrument, and some of them at a considerable distance from each other, it occurred to me, that the intermediate spaces between the sweeps might also contain nebulæ; and finding this to hold good more than once, I ventured to give notice to my assistant at the clock, "to prepare, since I expected in a few minutes to come at a stratum of the nebulæ, finding myself already" (as I then figuratively expressed it) "on nebulous ground." In this I succeeded immediately; so that I now can venture to point out several not far distant places, where I shall soon carry my telescope, in expectation of meeting with many nebulæ. But how far these circumstances of vacant places preceding and following the nebulous strata, and their being as it were contained in a bed of stars, sparingly scattered between them, may hold good in more distant portions of the heavens, and which I have not yet been able to visit in any regular manner, I ought by no means to hazard a conjecture. The subject is new, and we must attend to observations, and be guided by them, before we form general opinions.

Before I conclude, I may, however, venture to add a few particulars about the direction of some of the capital strata or their branches. The well known nebula of Cancer, visible to the naked eye, is probably one belonging to a certain stratum, in which I suppose it to be so placed as to lie nearest to us. This stratum I shall call that of Cancer. It runs from ϵ Cancri towards the south over the 67 nebula of the *Connoissance des Temps*, which is a very beautiful and pretty much compressed cluster of stars, easily to be seen by any good telescope, and in which I have observed above 200 stars at once in the field of view of my great reflector, with a power of 157. This cluster appearing so plainly with any good, common telescope, and being so near to the one which may be seen by the naked eye, denotes it to be probably the next in distance to that within the quartile formed by $\gamma, \delta, \eta, \theta$; from the 67th nebula the stratum of Cancer proceeds towards the head of Hydra; but I have not yet had time to trace it farther than the equator.

Another stratum, which perhaps approaches nearer to the solar system than any of the rest, and whose situation is nearly at rectangles to the great sidereal stratum in which the sun is placed, is that of Coma Berenices, as I shall call it. I suppose the Coma itself to be one of the clusters in it, and that, on account of its nearness, it appears to be so scattered. It has many capital nebulæ very near it; and in all probability this stratum runs on a very considerable way. It may, perhaps, even make the circuit of the heavens, though very likely not in one of the great circles of the sphere: for, unless it should chance to intersect the great sidereal stratum of the milky way before-mentioned, in the very place in which the sun is stationed, such an appearance could hardly be produced. However, if the stratum of Coma Berenices should extend so far as (by taking in the assistance of M. MESSIER'S and M. MECHAIN'S excellent observations of scattered nebulæ, and some detached former observations of my own) I apprehend it may, the direction of it towards the north lies probably, with some windings, through the great Bear onwards to Cassiopeia; thence through the girdle of Andromeda and the northern Fish, proceeding towards Cetus; while towards the south it passes through the Virgin, probably on to the tail of Hydra and the head of Centaurus. But, notwithstanding I have already fully ascertained the existence and direction of this stratum for more than 30 degrees of a great circle, and found it almost every where equally rich in fine nebulæ, it still might be dangerous to proceed in more extensive conjectures, that have as yet no more than a precarious foundation. I shall therefore wait till the observations in which I am at present engaged shall furnish me with proper materials for the disquisition of so new a subject. And though my single endeavours should not succeed in a work that seems to require the joint effort of every astronomer, yet so much we may venture to hope, that, by applying ourselves with all our powers to the improvement of telescopes, which I look upon as yet in their infant state, and turning them with assiduity to the study of the heavens, we shall in time obtain some faint knowledge of, and perhaps be able partly to delineate, *the Interior Construction of the Universe.*

WILLIAM HERSCHEL.

Datchet near Windsor,
April, 1784.

XIV.

Catalogue of Double Stars.

[*Phil. Trans.*, vol. lxxv., 1785, pp. 40–126.]

Read December 9, 1784.

INTRODUCTORY REMARKS.

THE great use of Double Stars having been already pointed out in a former paper, on the Parallax of the Fixed Stars, and in a latter one, on the Motion of the Solar System, I have now drawn up a second collection of 434 more, which I have found out since the first was delivered.

The happy opportunity of giving all my time to the pursuit of astronomy, which it has pleased the Royal Patron of this Society to furnish me with, has put it in my power to make the present collection much more perfect than the former; almost every double star in it having the distance and position of its two stars measured by proper micrometers; and the observations have been much oftener repeated.

The method of classing them is in every respect the same as that which has been used in the first collection; for which reason I refer to the introductory remarks that have been given with that collection* for an explanation of several particulars necessary to be previously known. The numbers of the stars are here also continued, so that the first class ending there at 24 begins here at 25, and the same is done with the other classes.

Most of the double stars in my first collection are among the number of those stars which have their places determined in Mr. FLAMSTEED's extensive catalogue; but of this collection many are not contained in that author's work, I have therefore adopted a method of pointing them out, which it will be proper to describe.

The finder of my reflector is limited, by a proper diaphragm, to a natural field of two degrees of a great circle in diameter. The intersection of the cross wires, in the center of it, points out one degree; and by the eye this degree, or the distance from the center to the circumference, may be divided into $\frac{1}{4}$, $\frac{1}{2}$, $\frac{3}{4}$, $\frac{1}{3}$, and $\frac{2}{3}$. Thus we are furnished with a measure which, though coarse, is however sufficiently accurate for the purpose here intended; and which, if more than two degrees are wanted, may be repeated at pleasure.

* See *Philosophical Transactions*, vol. LXXII. p. 112 [above, p. 58].

In such measures as these I have given the distance of a double star, whose place I wanted to point out, from the nearest star in FLAMSTEED's Catalogue. And since, besides the distance, it is also required to have its position with regard to the star thus referred to, I have used the neighbouring stars for the purpose of pointing it out.

The usefulness of this method is so extensive, that I shall be a little more particular in describing its application. When a star is thus pointed out, as for instance the 32d in the first class, where it is said, "About $\frac{3}{4}$ degree s. preceding the 44th Lyncis, in a line parallel to θ Ursæ majoris and the 39th Lyncis;" we are to apply one eye to the finder, and placing the 44th Lyncis into the center of the field, we are to look at θ Ursæ majoris and the 39th Lyncis in the heavens with the other eye by the side of the finder. The naked eye then will immediately direct us, by means of the two stars just mentioned, towards the place where, in the finder, the armed eye will perceive the double star in question about $\frac{3}{4}$ degree from the 44th Lyncis. I need hardly observe, that we must recollect the inversion of the finder, as those who are in the habit of using telescopes with high powers, always furnished with inverting finders, will of course look for the small star in the upper part of the field, as in fig. 1.

At the 45th star, in the first class, the description says, "About $1\frac{1}{4}$ degree s. preceding μ, towards ι Aurigæ." This double star will accordingly be found by placing μ Aurigæ first into the center of the finder; then, drawing the telescope towards ι, which the naked eye points out, the star we look for will begin to appear in the circumference as soon as μ is about $\frac{1}{4}$ degree removed from the center, as in fig. 2.

It will sometimes happen, that other stars are very near those which are thus pointed out, that might be mistaken for them. In such cases an additional precaution has been used by mentioning some circumstance either of magnitude or situation, to distinguish the intended star from the rest. After all, if any observer should be still at a loss to find these stars without having their right ascension and declination, he may furnish himself with them by means of FLAMSTEED's *Atlas Cœlestis*; for my description will be sufficiently exact for him to make a point in the maps to denote the star's place; then, by means of the graduated margin, he will have its Æ and declination to the time of the Atlas, which he may reduce to any other period by the usual computations.

Before I quit this subject I must remark, that it will be found on trial, that this method of pointing out a double star is not only equal, but indeed superior, to having its right ascension and declination given: for, since it is to be viewed with very high powers, not such as fixed instruments are generally furnished with, the given right ascension and declination would be of no service. We might, indeed, find the star by a fixed or equatorial instrument; and, taking notice of its situation with regard to other neighbouring stars, find, and view it afterwards, by a more powerful telescope; but this will nearly amount to the very same way which here is pursued, with more deliberate accuracy than we are apt to use, while we are employed in seeking out an object to look at.

It will be required, that the observer should be furnished with FLAMSTEED'S *Atlas Cœlestis,* which must have the stars marked from the author's catalogue, by a number easily added to every star with pen and ink, as I have done to mine. The catalogue should also be numbered by an additional column, after that which contains the magnitudes. I hope in some future editions of the Atlas to see this

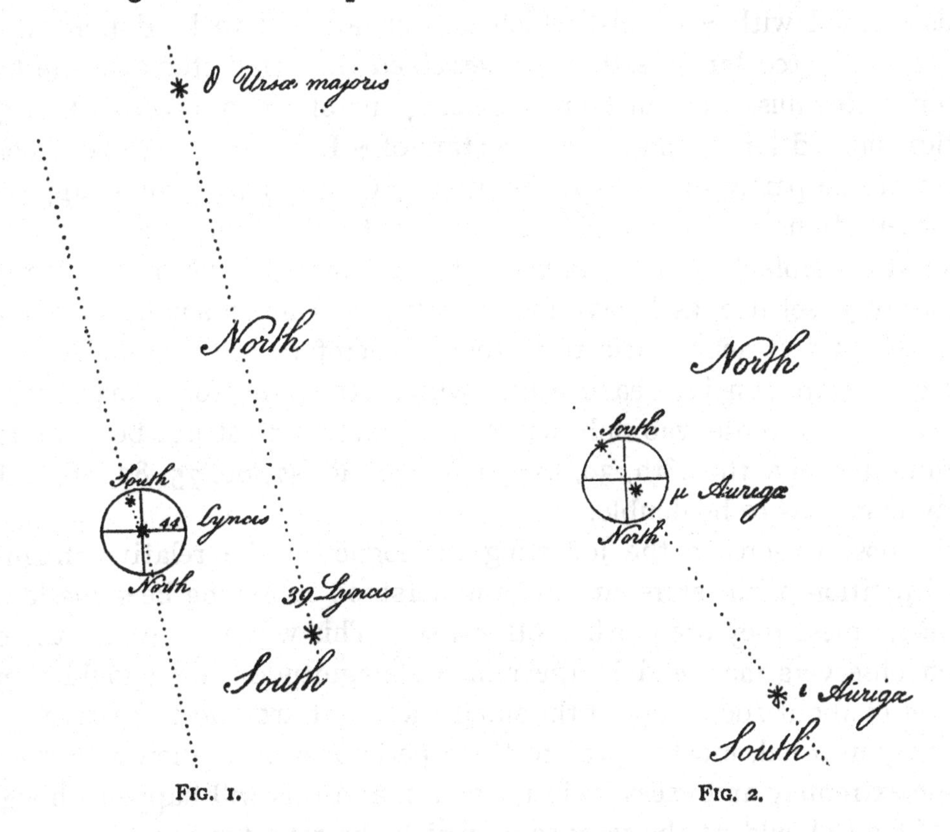

FIG. 1. FIG. 2.

method adopted in print, as the advantage of it is very considerable, both in referring to the catalogue for the place of a star laid down in the Atlas, and in finding a star in the latter whose place is given in the former.

I would recommend a precaution to those who wish to examine the closest of my double stars. It relates to the adjustment of the focus. Supposing the telescope and the observer long enough out in the open air to have acquired a settled temperature, and the night sufficiently clear for the purpose; let the focus of the instrument be re-adjusted with the utmost delicacy upon a star known to be single, of nearly the same altitude, magnitude, and colour, as the star which is to be examined, or upon one star above and another below the same. Let the phænomena of the adjusting star be well attended to; as, whether it be perfectly round and well defined, or affected with little appendages that frequently keep playing about the image of the star, undergoing small alterations while it passes through the field, at other times remaining fixed to it during the whole passage. Such deceptions may be detected by turning or unscrewing the object-glass or speculum a little in its cell, when those appendages will be observed to revolve the same way. Being thus acquainted with the imperfections as well as perfections of the instrument, and going immediately from the adjusting star, which for that reason also should be as near as may be, to the double star which is to be examined, we may hope to be successful. The astronomical Mr. AUBERT, who did me the honour to follow this method with γ Leonis, which he did not find to be double when the telescope was adjusted by γ itself, soon perceived the small star after he had adjusted it upon Regulus. The instrument, being one of Mr. DOLLOND's best $3\frac{1}{2}$ feet achromatics, shewed Mr. AUBERT the two stars of γ Leonis in very close conjunction, or rather one partly hid behind the other. On comparing these appearances with my observations of that double star, we must not be surprised to find that I place them at a visible distance from each other: for the Newtonian reflectors, on the plan of my 7-feet one, as I have found, will give a much smaller image of the stars than the $3\frac{1}{2}$ feet achromatic refractors; wherefore the two stars, which in refractors as it were run into each other, will in the reflector remain separate. For this reason also, those who only use such refractors must not be disappointed if they cannot perceive the 26th, 30, 31, 36, 41, 44, 46, 47, 60, 75, 82, 86, and 87th stars of my first class to be double.

All the observations in the following catalogue on the relative magnitude, colour, and position of the stars, are to be understood as having been made with a power of 460, unless they are marked otherwise. This will account for the difference which observers may find in the relative magnitude; for should they use only a power of about 200, many of the small stars that are said to be very unequal and extremely unequal, must appear to them perhaps a degree lower in the scale, and become extremely and excessively unequal: and this will happen, though the quantity of light should be the very same which the reflector has that served me to settle these particulars. I need not say, that on other accounts, such as a real difference in the light of the telescope, the presence of the moon, twilights, auroræ boreales, or other causes, many of the small stars may be found to be of a different comparative lustre from what is assigned to them in the catalogue. The small star near Rigel, for instance, appears of a beautiful pale red colour, full, round, and well defined, with my 20-feet reflector; the 10-feet instrument shows it also very well in fine evenings; the 7-feet requires more attention, nor is the small star defined, but of a dusky pale red colour. A good $3\frac{1}{2}$ feet achromatic, of a large aperture, when Rigel is on the meridian, may, perhaps, also shew the small star, although I have not been able to see it with a very good instrument of that sort, which shews the small star that accompanies the pole-star; but the evening was not very favourable.

The measures of the distances were all taken with a parallel silk-worm's-thread micrometer, and a power of 227 only. They are not, as in the former catalogue, with the diameters included, but from the center of one star to the center of the other. I have adopted these measures on finding that I could procure threads fine enough to subtend only an angle of about 1″ 13‴, and that by this means there was no longer any great difficulty of judging when the stars were centrally covered by the threads. However, I do not know whether these measures, with stars at a considerable distance, may not be liable to an additional error of perhaps one second, owing to the remaining uncertainty in judging of their exact central position while the measure is taking.

The positions have all been measured (unless marked otherwise) with a power of 460, adapted to an excellent micrometer, executed by Mess. NAIRNE and BLUNT, according to the model given in the *Philosophical Transactions*, vol. LXXI. page 500. fig. iv.;* but with a great and necessary improvement of making the wheel *d, d,* of that figure perform its whole revolution; by which means the two silk-worm's-threads may be adjusted to a greater degree of exactness; for if they are not placed so as perfectly to bisect the circle, the two threads will not coincide exactly after having performed one semi-revolution, which they must be made to do with the utmost rigour. I found the absolute necessity of this precaution when I came critically to examine the positions of the Georgium Sidus, as they are given in table III. *Phil. Trans.* vol. LXXI. p. 497.† The measures were affected with a small and pretty regular error, which I was at a loss to account for; and the distance of this star being then totally unknown, I looked for the cause of the deviation at first in a diurnal parallax of that heavenly body; but soon found it owing to the inconvenience before-mentioned, of not being able experimentally to adjust the moveable thread to that critical nicety which I have now introduced and used in all the angles of the following catalogue.‡

W. HERSCHEL.

Datchet near Windsor, Nov. 1, 1784.

* [Above, p. 37.] † [Above, p. 35.]

‡ The divisions on the moveable circular index (*a*) of this micrometer should be read off by means of a line drawn on a small plate fastened to the side *t*, and projecting with a proper curvature against the plane of the divisions towards *r*, so as to be nearly in contact; a coincidence of lines being by far the best method of ascertaining the situation of the index. A nonius of four sub-divisions may also be used, whereby the 60 divisions, already divided into halves upon the index-plate, will be had in eighths, each of which, on the construction of my present one, will be equal to three minutes of a degree of the circle.

CATALOGUE OF DOUBLE STARS.

FIRST CLASS.

I. 25. A Orionis. FL. 32. Sub humero in consequentia.

Jan. 20, 1782. Double. Considerably unequal. L. fine w.; S. w. inclining to pale rose colour. The distance or black division between the two stars with 278 is about ¼ diameter of L.; with 460, near ½ diameter of L. Position with 278, 52° 10′ s. preceding.

26. ω Leonis. FL. 2. Anteriorem pedem dextrum præcedens.

Feb. 8, 1782. A very minute double star. Considerably unequal. Both r. With 227 there is not the least suspicion of its being double; with 460 it appears oblong, and, when perfectly distinct, we see ¾ of the apparent diameter of a small star as it were emerged from behind a larger star; with 932 they are more clear of each other, but not separated; the focus of every power adjusted upon the 3d and 6th Leonis. November 6th, 1782, I first suspected a separation; and November 13th, fairly saw a division between them. April 4, 1783, with an improved reflector of 20 feet 3 inches focal length and 12 inches aperture, I saw them evidently divided. Position 20° 54′ s. following.‡

27. FL. 90 Leonis. Infra eductionem caudæ.

Feb. 9, 1782. Treble. The two nearest—very unequal. L. w.; S. rw. With 278, 1¼ diameter of L.; with 460, 1½ diameter of L. Position with 278, 61° 9′ s. preceding. The two farthest—very unequal. S. dusky r. Distance from L. 53″ 43‴. Position 35° 12′ s. preceding.

28. γ Leonis. FL. 41. In collo lucida.

Feb. 11, 1782. A beautiful double star. Pretty unequal. L. w.; S. w. inclining a little to pale red. With 227 and 278 distinctly separated; with 460, ⅙ diameter of S.; with 625, ¼ diameter; with 932, full ¼ diameter, or when best ½ diameter of S.; with 1504, ¾ diameter, well-defined, and the difference of colours still visible; with 2176, not quite a diameter of S, pretty well defined, but exceed-

‡ I suspect these stars to recede from each other. It is, however, very possible, that the opening which I observed between them, at the latter end of the year 1782 and beginning of 1783, may be owing to very favourable weather, or to my being better acquainted with the object. Could we increase our power and distinctness at pleasure, we might undoubtedly separate any two stars that are not absolutely in a direct line passing through the eye of the observer, and the centers of both the stars. This will appear when we consider that perhaps 59 thirds out of one second, which the diameter of the star may subtend, are spurious; so that a double star seemingly in contact, or even partly hiding each other in appearance, may still be far enough asunder to admit of a fair and considerable separation by applying an adequate magnifying power. It would have been curious, if a considerable difference in the colours could have led us to discover which of the two stars is before the other! But the far greatest part of their apparent diameters being, as we have observed, spurious, it is probable, that a different coloured light of two stars would join together, where the rays of one extend into those of the other; and so, producing a third colour by the mixture of it, still leave the question undecided.

ingly tremulous; with 2589, less than 1 diameter; with 3168, still pretty distinct, and about $\frac{3}{4}$ diameter of S; with 4294, more than a diameter of S, but attended with the utmost difficulty of managing the motions; with 5489, the interval still somewhat larger, and if the object could be kept in the center of the field, the eye might adapt itself to the focus, and get the better of the violent aberration; but the edges of the glass being of a different focus, the eye is constantly disappointed in its endeavours to define the object; with 6652, I had but a single glimpse of the star quite disfigured; however, I ascribe it chiefly to the foulness of the glass, which, on account of its smallness, is extremely difficult to be cleaned; with a 10-feet reflector, 9 inches aperture, power 626, above $\frac{1}{2}$ diameter of S. very distinct; with a 20-feet reflector, power 350, too bright an object to be quite distinct, though I see it very well. Position 5° 24′ n. following. A third star preceding. Dist. 1′ 51″ 23‴, pretty accurate for so great a distance. Position 31° 0′ n. preceding. A fourth star preceding the third, and somewhat smaller.

I. 29. Parvula juxta FL. 44am Leonis.

Feb. 17, 1782. Double. About 4′ following the 44th Leonis, which being double in the finder, this is the least of the two. Extremely unequal. L. w S. d. With 227, $1\frac{1}{3}$ diameter of L.; with 460, 2 diameters of L. Position 26° 32′ n. following.

30. Secunda ad σ Cancri. FL. 57.

March 5, 1782. Double. Pretty unequal. Both pr. With 227, about $\frac{1}{4}$ diameter; with 278, $\frac{1}{4}$ diameter; with 460, about $\frac{1}{2}$ diameter or less. Position 68° 12′ n. preceding. A beautiful minute object.

31. Inter FL. 41am et 39am Lyncis.

March 5, 1782. Double. Near $1\frac{1}{4}$ degree n. preceding the 41st Lyncis; towards *n* Ursæ majoris. A little unequal. Both w. With 460, $\frac{1}{4}$ or at most $\frac{1}{3}$ diameter. Position 51° 21′ s. preceding.

32. FL. 44a Lyncis australior et præcedens.

April 3, 1782. Double. About $\frac{3}{4}$ degree s. preceding the 44th Lyncis; in a line parallel to θ Ursæ majoris and the 39th Lyncis. Very unequal. L. r.; S. bluish r. With 227, 1 diameter of L. or $1\frac{1}{4}$ when best; with 460, $1\frac{1}{4}$ diameter, or when best, near 2 diameters of L. The diameters are so small that the length of the time, and attention of looking, makes a considerable difference in the estimation of the distance. Position 8° 27′ s. preceding.

33. ξ Libræ. FL. 51. Primam chelam Scorpii attingens.

May 12, 1782. Treble. Without great attention, and a considerable power, it may be mistaken for a double star; but the largest of them consists of two. Very little unequal. Both w. With 460, $\frac{1}{4}$ or at most $\frac{1}{3}$ diameter asunder; with 932, full $\frac{1}{3}$ diameter of L. or near $\frac{1}{2}$ diameter of S. Position, with 278, 82° 2′ n. following. For measures of the third star see the 20th of the second class.

I. 34. FL. 55. Cassiopeiæ. ι Ptolemæi. In pedis extremitate.

June 11, 1782. Treble. The two nearest very unequal. L. w.; S. colour of pale red blotting paper. With 278, ½ diameter of S. Position with 227, 20° 30′ n. preceding. For measures of the third star see the fourth in the third class.

35. FL. 38. Serpentarii. Dextrum infra pedem.

June 11, 1782. Double. Very unequal. L. w.; S. d. With 460, 1¼ diameter of L. As the situation is too low for 460, I tried 227, but it only shewed the star wedge-formed. Position 60° 48′ n. preceding.

36. ζ Herculis. FL. 40. In dextro latere.

July 18, 1782. A fine double star. Very unequal. L. bluish w.; S. ash-colour. With 460, less than ½ diameter of S.; with 932, 1 full diameter of S.‡ Position with 811, 20° 42′ n. following.

37. φ (FL. 11ª.) Herculis borealior et sequens.

July 22, 1782. Double. About ⅓ degree n. following φ; in a line parallel to the 35th and 42d Herculis; the most south of two very small telescopic stars. Considerably unequal. Both reddish. With 227, they can but just be seen as two stars; with 460, near 1 diameter; with 932, not less than 1½ diameter of L. Position 59° 48′ s. following.

38. FL. 18am Persei præcedens ad boream. In capite.

Aug. 20, 1782. Double. About ½ degree n. preceding the 18th; in a line parallel to σ and τ Persei; of two stars that next to the 18th. A little unequal. Both pr. With 278, a most minute and beautiful object; with 460, ½ diameter of either. Position with 278, 8° 24′ n. preceding.

39. β (FL. 11am) Cassiopeiæ præcedens ad austrum.

Aug. 25, 1782. Double. About ¾ degree s. preceding β; in a line parallel to η and α Cassiopeiæ; the following and largest of two very considerable stars. Very unequal. L. pr.; S. r. With 278, ¼ diameter of S.; with 460, ½, or when best, ¾ diameter of S. Position 50° 42′ n. preceding.§

40. FL. 25am Cassiopeiæ præcedens ad boream.

Aug. 28, 1782. Double. About ½ degree n. preceding the 25th; towards α Cassiopeiæ; the first telescopic star in that direction. Very unequal. Both r. With 460, ¾ diameter of S.; difficult to be seen. Position 50° 30′ s. following.

‡ The interval between very unequal stars, estimated in diameters, generally gains more by an increase of magnifying power than the apparent distance of those which are nearer of a size. Instances of the former may be found in the first class, the 1st, 7, 29, 35, 37, 39, 53, 59, 63, 64, 72d stars; of the latter, the 16th, 28, 33, 45, 46, 73, 81st stars. However, this only seems to take place when there is a difficulty of seeing the object well with a low power, which being removed by magnifying more, the distance is, as it were, laid open to the view.

§ [This measure (of Jan. 18, 1783) is marked "a better measure." Just before this is "47° 36′ interrupted by clouds, therefore not exact."—ED.]

I. 41. FL. 31[a] Draconis borealior.

Aug. 29, 1782. A very minute double star. About $\frac{3}{4}$ degree n. of the 31st; in a line parallel to γ and ξ Draconis; the most south and preceding of two. Considerably unequal. Both pr. or r. With 227, they appear only as a lengthened or distorted star; with 460, $\frac{1}{4}$ diameter of S.; or in very fine nights $\frac{1}{3}$ diameter of S.; with a new speculum and 500, near $\frac{1}{2}$ diameter when best; with 932, $\frac{1}{2}$ diameter. Position 84° 21′ n. preceding. Requires every favourable circumstance to be seen double.

42. δ Serpentis. FL. 13. In primo flexu colli.

Sept. 3, 1782. A beautiful double star. Considerably unequal. L. w.; S. greyish. With 227, $\frac{1}{3}$ diameter of S.; with 278, not quite $\frac{1}{2}$ diameter of S.; with 460, near $\frac{3}{4}$ diameter of S.; with 932, near 1 diameter of S.; with 1504, above 1 diameter of S. Position 42° 48′ s. preceding.‡

43. Ad FL. 48[am] Draconis.

Sept. 3, 1782. A very minute double star. The most north of three, forming an arch; or that which is towards o Draconis. Considerably unequal. Both pale pink. In fine nights, with 460, it has the shape of a wedge; with 932, a fine black division just visible; in a very clear dark night a division may be seen with 500, and with 932, it will be about $\frac{1}{8}$ diameter. Position with 500, 88° 24′ n. preceding.

44. FL. 4. Aquarii. Supra vestimentum manus sinistræ.

Sept. 3, 1782. A minute double star. Very unequal. Both pr. With 460, almost in contact, or at most $\frac{1}{6}$ diameter of S. Position 81° 30′ n. preceding. A third star of the sixth class in view, n. preceding.

45. μ Aurigæ (FL. 11[am]) præcedens ad austrum.

Sept. 5, 1782. Double. About $1\frac{1}{4}$ degree s. preceding μ, towards ι Aurigæ; a pretty considerable star in a minute telescopic constellation. A little unequal. Both pr. or r. With 227, $\frac{1}{3}$ diameter of S.; with 278, near $\frac{1}{2}$ diameter of S.; with 460, about $\frac{1}{2}$ diameter, or near $\frac{2}{3}$ diameter of S. Position 47° 33′ s. preceding.

46. ν (FL. 13[am]) Aquarii sequens ad boream.

Sept. 7, 1782. Treble. About $1\frac{3}{4}$ degree n. following ν, in a line parallel to β and α Aquarii; the middle of three that are in the same direction. The two nearest very unequal. L. rw.; S. pr. With 460, about 1 diameter of L. or more. Position 62° 27′ n. preceding. The two farthest very unequal. S. pr. Distance with 227, 1′ 22″ 42‴. Position 35° 51′ n. following.

‡ [Another measure of same date (1782·671) "by the old micrometer" 54° 0′ sp. "I suspect a mistake in reading off one of the measures."—ED.]

I. 47. FL. 29^{am} Capricorni præcedens ad boream.

Sept. 27, 1782. A minute double star. About $\frac{3}{4}$ degree n. preceding the 29th, in a line parallel to γ and α Capricorni. A little unequal. Appears distorted with 227 and 278; nor will 460 shew it separated; with 657, two stars visible; 932 confirms it. Difficult to be seen distinctly on account of its low situation. Position 84° 48′ n. preceding. 20-feet reflector, 200. Both w.

48. FL. 6^{am} Cephei præcedens. In dextro brachio.

Sept. 27, 1782. A very minute and beautiful double star. Near $\frac{3}{4}$ degree preceding the 6th towards η Cephei; a pretty considerable telescopic star. A little unequal. Both pr. Almost in contact with 460; with 625, better divided; with 657 still better. Position 14° 9′ s. preceding.

49. λ Cephei (FL. 22^{am}) sequens ad boream.

Sept. 27, 1782. Double. About $1\frac{1}{4}$ degree n. following λ, in a line from ζ through λ Cephei continued.‡ Extremely unequal. Both dw. Cannot be seen with 278, except with long attention; with 460, $1\frac{1}{2}$ diameter of L. Position 85° 48′ n. following; perhaps a little inaccurate.

50. λ Aquarii (FL. 73^{am}) præcedens.

Sept. 30, 1782. Double. About $2\frac{1}{3}$ degrees preceding, and a little south of λ Aquarii; a considerable star. Very unequal. L. w; S. dw. With 278, less than 1 diameter of L; with 460, $1\frac{1}{4}$ diameter of L. Position with 227, 41° 12′ n. preceding. The measure inaccurate on account of the low power, and probably 3° or 4° too small.

51. Quæ sequitur ι (FL. 32^{am}) Cephei.

Sept. 30, 1782. Double. About $2\frac{1}{4}$ degrees n. following ι, towards γ Cephei; a considerable star. A little unequal. Both pr. A pretty object with 227; with 460, $1\frac{1}{2}$ diameter nearly. Position 3° 36′ s. preceding.

52. Parvula FL. 25^{tæ} Orionis adjecta.

Oct. 2, 1782. Double. A few minutes n. following the 25th Orionis, in a line parallel to *h* Eridani and ϵ Orionis. Very unequal. L. ash w.; S. dw. With 460, 1 diameter of L. Position 52° 48′ n. preceding.

53. Parvula FL. 30^{mæ} Orionis adjecta.

Oct. 2, 1782. Double. About 10′ preceding the 30th, in a line parallel to λ and γ Orionis. Very unequal. L. w.; S. d.; with 460, 1 diameter of L. Position 43° 24′ n. following.

54. τ (FL. 20^{am}) Orionis præcedens. In malleolo sinistri cruris.

Oct. 4, 1782. Double. Near $\frac{3}{4}$ degree preceding τ, in a line from θ through τ Orionis continued. Very unequal. L. r.; S. dr. With 227, about 1 diameter of L.; with 460, about 2 diameters of L. Position 35° 42′ n. preceding; a little inaccurate.

‡ [I. 49 is probably $= \Sigma$ 2880. No double star anywhere near H's place according to Burnham, —ED.]

I. 55. FL. 8am Tauri præcedens ad boream.

Oct. 9, 1782. Double. About 1⅓ degree n. preceding the 8th Tauri, or near 2 degrees s. following the 65th Arietis, in a line parallel to the Pleiades and ε Tauri; a small telescopic star not easily found. A little unequal. L. r.; S.d. With 227, less than 1 diameter of S.; with 460, near two diameters. Position 82° 48′ s. following.

56. FL. 54am Ceti sequens ad austrum.

Oct. 12, 1782. Double. About ⅓ degree s. following the 54th, towards δ Ceti. Nearly equal. Both r. With 227, about 1 diameter; with 460, about 1½ diameter. Position 87° 39′ n. following.

57. FL. 70am et 67am Orionis præiens.

Oct. 12, 1782. Multiple. In a spot which appears nebulous in the finder, and is about 50′ from the 67th, and 45′ from the 70th Orionis. More than 12 stars in view with 460; among them is a double star. The largest of the base of an isosceles triangle, n. preceded by four stars in a line. Considerably unequal. With 460, 1 full diameter of L. Position 19° 48′ s. following.

58. δ Lyræ (FL. 12am) sequens. Inter eductionem cornuum.

Oct. 24, 1782. Double. About ½ degree following the 12th, in a line continued from the 11 through the 12th Lyræ; the last of a small telescopic triangle. Extremely unequal. L. r.; S. d. Not easily seen with 227; with 460, near 2 diameters of L. Position 13° 0′ n. preceding.

59. Ab ι (FL. 18a) Lyræ β versus.

Oct. 24, 1782. Double. The most south of two very small telescopic stars, which are the second pair situated in a line from ι towards β Lyræ. A little unequal. Both d.; the faintest object that can be imagined. With 460, about 1 diameter. Position 75° 0′ s. preceding; the measure is liable to some error from the obscurity.

60. E telescopicis γ et λ Lyræ australioribus et sequentibus.

Oct. 24, 1782. Double. About ¾ degree s. following λ, in a line parallel to α and γ Lyræ; a very small telescopic star. Extremely unequal. Both dr. With 227, 1 full diameter of L.; with 460, near 2 diameters of L. Position 16° 48′ n. preceding.

61. Præiens FL. 1am Equulei.

Oct. 26, 1782. A minute double star. About ¾ degree n. preceding the 1st Equulei, in a line parallel to α Equulei and γ Aquilæ; a large star. Very unequal. Both pr. With 460, ½ diameter of S. Position 18° 24′ n. preceding. A pretty object, but requires fine weather.

I. 62. Sequitur FL. 2am Equulei.

Oct. 29, 1782. Double. About $\frac{3}{4}$ degree s. following the 2d Equulei, in a line parallel to δ Delphini and δ Equulei.‡ Considerably unequal. Both r. With 460, $1\frac{1}{4}$ or $1\frac{1}{2}$ diameter of S. Position 35° 9′ s. preceding.

63. γ Equulei (FL. 5ª) australior.

Oct. 29, 1782. Double. Full $\frac{1}{2}$ degree s. of γ, in a line from the 5th through the 6th Equulei continued. Equal. Both dr. With 227, about $\frac{1}{4}$ diameter scarce visible; with 460, about $\frac{1}{2}$ diameter. Position 5° 57′ s. preceding.

64. π Arietis. FL. 42. In poplite.

Oct. 29, 1782. Treble. Excessively unequal. L. w.; S. both mere points. With 227, neither of the small stars can be seen, except with considerable and long continued attention, when they also appear; the nearest with this power is $\frac{3}{4}$ or $\frac{4}{5}$ diameter of L.; with 460, $1\frac{1}{2}$ or $1\frac{3}{4}$ diameter of L. The third is about 25″ or 26″ distant from L, by exact estimation. Position of both, being all three in a line 19° 9′ s. following; as exact as the obscurity will permit.

65. In Nubecula β Sagittæ adjecta et sequenti.

Nov. 4, 1782. Double. $\frac{1}{3}$ degree n. following β Sagittæ, towards 29th Vulpeculæ; the largest and most south of a cluster of small stars that appear cloudy in the finder. Very unequal. L. rw.; S. pr. With 227, full 1 diameter of L.; with 460, about $1\frac{3}{4}$ or 2 diameters of L. Position 14° 0′ n. preceding. A third star in view, of the 5th or 6th class.

66. β (FL. 23ª) Draconis australior et præcedens.

Nov. 4, 1782. Double. About $1\frac{1}{4}$ degree s. preceding β, in a line from ν continued through β Draconis. Pretty unequal. Both pr. With 460, $1\frac{1}{2}$ or $1\frac{3}{4}$ diameter of L. Position 2° 24′ s. preceding.

67. Nebulam Aurigæ pedem dextrum sequentem, præcedens.

Nov. 4, 1782. Double. About 55′ from the 37th Nebula of M. MESSIER; the largest and most preceding of two stars. Very unequal. Both pr. With 460, near 2 diameters of L. Position 23° 57′ n. following.

68. Parvula FL. 10æ Orionis quam proximè adjecta.

Nov. 5, 1782. Double. The small star not many minutes from the 10th Orionis. A little unequal. Both whitish. With 460, near 1 diameter. Position 84° 54′ s. following; a little inaccurate on account of the difficulty of seeing the stars well.

69. In Lyncis pectore.

Nov. 13, 1782. Double. About 3 degrees s. preceding the 19th Lyncis, in a line drawn from the 19th Lyncis to τ Aurigæ; the 24th and 19th Lyncis also point to it nearly: in a very clear evening it may just be seen with the naked eye.

‡ [I. 62 = β 269 with an error of 1°.—ED.]

A little unequal. Both rw. With 227, $\frac{3}{4}$ diameter; with 460, $1\frac{1}{4}$ or near $1\frac{1}{2}$ diameter. Position 77° 0′ s. following.‡

I. 70. ζ (FL. 123[a]) Tauri borealior et præcedens.

Nov. 13, 1782. A very pretty double star. Near 1 degree n. preceding ζ Tauri towards Capella; the corner of a rhomboid made up of ζ, this, and two more, and opposite to ζ. Considerably unequal. L. pr.; S. a little deeper r. With 227, almost 1 diameter of L.; with 460, $1\frac{3}{4}$ diameter of L. Position 36° 24′ s. preceding.

71. FL. 44[am] Ursæ majoris præcedens ad austrum.

Nov. 19, 1782. Double. Nearly in the intersection of a line from β Ursæ majoris to the 39th Lyncis, crossed by one from ψ to υ Ursæ majoris; the last line should bend a little towards ψ Ursæ majoris. A little unequal. Both whitish. With 460, near 2 diameters of S. Position 2° 6′ n. following.

72. FL. 65. Ursæ majoris.

Nov. 20, 1782. Double. Excessively unequal. L. pr.; S. a point. Not visible with 227, nor hardly to be suspected unless it has been first seen with a higher power; with 460, $1\frac{3}{4}$ diameter of L. or, when long viewed, full 2 diameters of L. Position 53° 45′ n. following. A third star in view. Equal to L. Colour rw. Distance 1′ 0″ 4‴. Position 22° 21′ s. following.

73. β (FL. 6[a]) Arietis borealior et præcedens.

Nov. 22, 1782. Double. About $1\frac{3}{4}$ degree n. preceding β Arietis, towards β Andromedæ; a considerable star. Very unequal. L. r.; S. deeper r. With 227, about $\frac{3}{4}$ diameter of L.; with 460, full $1\frac{1}{4}$ or almost $1\frac{1}{2}$ diameter of L. when best. Position 77° 24′ s. following.

74. FL. 39[a] Arietis borealior et præcedens.

Dec. 22, 1782. Double. About $\frac{5}{8}$ degree n. preceding 39 Arietis, towards γ Trianguli; a pretty large telescopic star. A little unequal. Both pr. With 227, near 1 diameter of L.; with 460, about $1\frac{1}{2}$ diameter of L. Position 20° 36′ n. preceding.

75. FL. 26[am] Orionis præcedens ad austrum.

Jan. 9, 1783. Double. About $\frac{1}{4}$ degree s. preceding the 26th, in a line parallel to δ and β Orionis; the farthest of two; or $\frac{3}{4}$ degree s. preceding the 30th in the same direction. Nearly equal. Both w. or rw. With 460, perhaps a diameter. Position 89° 36′ n. preceding; but not very accurate.

76. In pectore Lyncis.

Jan. 23, 1783. Double. Not easy to be found. A line from the 19th Lyncis to υ Geminorum crossed by one from θ Ursæ majoris to ε Aurigæ, points out a star but just visible in a fine evening; it is perhaps about three degrees from the

‡ [I. 69. In the MS. the Pos. Angle is "77° 24′ sf. very exact," with the note "77° 0′ in printed Cat., J. F. W. H."—ED.]

19th Lyncis; when that star is found, we have the double star about 1 degree n. following the same, in a line parallel to τ Geminorum and the 19th Lyncis. Considerably unequal. Both ash w. With 460, $\frac{1}{4}$ diameter of S. Position 0° 0′ preceding. A third large star in view. Distance 1′ 7″ 46‴. Position 3° 42′ s. preceding.

I. 77. α (FL. 7ª) Crateris borealior.

Jan. 31, 1783. Double. Near $2\frac{3}{4}$ degrees north of α Crateris; a small telescopic star, about $\frac{1}{4}$ degree following the most north of two large ones. Pretty unequal. Both whitish. With 227, less than half diameter of S.; with 460, near 1 diameter; with 625, a little more than 1 diameter. Position 82° 24′ n. following.

78. FL. 11ª Libræ borealior.

Jan. 31, 1783. Double. Near $2\frac{1}{2}$ degrees north of the 11th Libræ, in a line parallel to μ Virginis and the 109th of the same constellation. Equal. Both inclining to r. With 460, full 1 diameter. Position 58° 24′ n. preceding, or s. following.

79. FL. 46 Herculis. In dextro latere.

Feb. 5, 1783. Double. Extremely or almost excessively unequal. L. w.; S. d. With 227, it is hardly visible; with 460, near 1 diameter of L. Position 66° 36′ s. following.

80. FL. 81 Virginis.

Feb. 7, 1783. Double. Equal. Both pr. With 227, near $\frac{1}{2}$ diameter; with 460, $\frac{2}{3}$ diameter. Position 41° 12′ n. following or s. preceding.

81. π Serpentis (FL. 44ᵃᵐ) præcedens ad austrum.

Mar. 7, 1783. Double. About $1\frac{1}{4}$ degree s. preceding π, towards κ; the most north of two. A little unequal. Both r. With 460, $1\frac{1}{2}$ diameter of L. Position 49° 48′ s. preceding. A third large star in view; paler than the other two. Distance from the two taken as one star 56″ 28‴. Position, with L. of the two, 31° 48′ s. preceding.

82. FL. 49 Serpentis.

Mar. 7, 1783. Double. The most north and following of two stars. A little unequal. Both pr. With 227, $\frac{1}{4}$ or $\frac{1}{3}$ diameter, and a very minute and beautiful object; with 460, $\frac{3}{4}$ diameter. Position 21° 33′ n. preceding.

83. λ Ophiuchi. FL. 10. In ancone sinistri brachii.

Mar. 9, 1783. A very beautiful and close double star. L. w.; S. blue; both fine colours. Considerably or almost very unequal. With 460, $\frac{1}{4}$ or $\frac{1}{3}$ diameter of S.; with 932, full $\frac{1}{3}$ diameter of S. Position 14° 30′ n. following.‡

‡ [I. 83. Note in MS., but not in the Journal, where there is no measure given: "The measure was forgot to be wrote down, but it being the last taken, I find my micrometer stands at 14° 30′, which agrees well enough with the figure, but as the instrument has been touched a good deal it may have been altered."—Ed.]

I. 84. FL. 50[a] Aurigæ australior.

Mar. 18, 1783. Double. Near 1 degree s. of the 50th Aurigæ, in a line parallel to β and θ. Very unequal. L. r.; S. dr. With 227, about ¾ diameter of L.; with 460, almost 1¼ diameter of L. Position 14° 0′ n. following.

85. FL. 36[am] Lyncis sequens ad austrum.

Mar. 24, 1783. Double. Near ½ degree s. following the 36th Lyncis, in a line parallel to the 31st Lyncis and *n* Ursæ majoris; of two the nearest to the 31st Lyncis. Considerably unequal. Both w. With 227, 1 diameter of L.; or when long kept in view, 1¼ diameter of L.; with 460, and after long looking, 2 diameters of L.; otherwise not near so much. Position 88° 57′ n. following.

86. FL. 105[a] Herculis borealior.

Mar. 27, 1783. Double. One full degree n. of the 105th Herculis, in a line from the 72d Serpentarii continued through the 105th Herculis; a small telescopic star. Considerably unequal. Both dr. With 460, a little more than 1 diameter of L. Position 79° 24′ n. preceding.

87. *q* Ophiuchi. FL. 73.

April 27, 1783. A very minute double star. Considerably unequal. L. r.; S. r. With 227, not to be suspected unless known to be double, but may be seen wedge-formed, and with long attention I have also perceived a most minute division; with 460, about ¼ or ⅓ diameter of S. Position 2° 48′ s. preceding.

88. τ Ophiuchi. FL. 69. In dextra manu sequens.

April 28, 1783. The closest of all my double stars; can only be suspected with 460; but 932 confirms it to be a double star. Pretty unequal. Both pr. or wr. It is wedge-formed with 460; with 932, one-half of the small star, if not three-quarters seem to be behind the large star. Position of the wedge 61° 36′ n. preceding. ν Ophiuchi, just by, is perfectly free from this wedge-formed appearance.

89. Illas ad FL. 56[am] Andromedæ præcedens ad boream.

July 28, 1783. Double. About ⅜ degree preceding, and a little north of the two stars that are about the place of the 56th Andromedæ, in a line towards μ; a considerable star; and of two in a line parallel to β and γ Trianguli that which is nearest to the 56th Andromedæ. Pretty unequal. L. drw.; S. dpr. With 227, near 1 diameter of L.; with 460, about 1½ diameter of L. Position 75° 30′ s. following.

90. β Aquarii (FL. 22[am]) præcedens ad austrum.

July 31, 1783. Double. About 4½ degrees from β towards μ Aquarii. A little unequal. Both dw. or pr. With 460, 1½ diameter or near 2. Position 77° 36′ s. following.

I. 91. γ Aquilæ (FL. 50^am^) præcedens ad boream.

Aug. 7, 1783. Double. About $\frac{1}{3}$ degree n. preceding γ, in a line parallel to γ and ζ Aquilæ; of two that nearest to γ. Very unequal. L. dpr.; S. d. With 227, hardly visible, and like a star not in focus; with 460, appears nebulous on one side, but is a double star; with 932, about 1½ diameter of L. Position 8° 18′ n. preceding.

92. π Aquilæ. FL. 52. Duarum in sinistro humero sequens.

Aug. 27, 1783. A minute pretty double star. A little unequal. Both pr. With 460, ½ diameter of L. or near ¾ diameter of S. Position 34° 24′ s. following.

93. FL. 62^am^ Aquilæ præcedens ad boream.

Sept. 12, 1783. A minute double star. About ¾ degree n. preceding the 62d, in a line parallel to θ and ζ Aquilæ; a pretty considerable star. Very unequal. Both inclining to pr. With 278, almost in contact; with 460, near ¾ diameter of S.; when in the meridian, and the air fine, near 1 diameter of L. Position 19° 9′ n. preceding.

94. δ Cygni. FL. 18. In ancone alæ dextræ.

Sept. 20, 1783. Double. Very unequal. L. fine w.; S. ash colour inclining to r. With 278, about ½ diameter of L.; with 460, ¾ diameter of L.; with 932, full 1½ diameter of L. in hazy weather, which has taken off the rays of L. and thereby increased the interval. Position 18° 21′ n. following; perhaps a little inaccurate.

95. FL. 33^am^ Cygni sequens ad austrum.

Sept. 22, 1783. Double. Full 1⅓ degree s. following the 33d, towards ξ Cygni; a pretty considerable star. Very unequal. L. w.; S. inclining to r. With 460, at first about ⅔ diameter of L.; but, after looking a considerable time, and in a fine air, near 1½ diameter. Position 72° 15′ n. preceding.

96. η (FL. 21^am^) Cygni sequens ad austrum.

Sept. 23, 1783. Treble. Full 1¾ degree n. following η, in a line parallel to β and λ Cygni. The two nearest considerably unequal. Both pr. With 460, 1 diameter of S. or ¾ diameter of L. Position 89° 18′ s. following. The two farthest considerably unequal; the colour r. Position 56° 3′ n. preceding.

97. FL. 51^am^ Cygni sequens.

Sept. 24, 1783. A minute double star. About 2½ degrees following the 51st, in a line parallel to δ and α Cygni; the largest and most south of an obtuse-angled triangle; a very considerable star. Pretty unequal. Both rw.; but S. a little darker r. With 278, ½ diameter of S. and beautiful; with 460, ¾ diameter of S. Position 46° 24′ n. following.

SECOND CLASS OF DOUBLE STARS.

II. 39. Procyonem juxta.

Feb. 2, 1782. Double. About 2 degrees s. following Procyon, in a line from λ Geminorum continued through Procyon. Excessively unequal. L. pr.; S. not visible with 278; with 460, more than 3 diameters of L. Position, by the assistance of a wall ‡ and micrometer 54° 28′ s. following.

40. * Secunda ad φ Cancri. FL. 23.

Feb. 2, 1782. Double. A little unequal. Both rw. With 227, near 2 diameters; with 460, 2½ diameters of L. Position 56° 42′ n. following.

41. * Prima ad υ Cancri. FL. 24.

Feb. 2, 1782. Double. Considerably unequal. Both pr. With 227, 1½ diameter of L.; with 460, 4 diameters of L. Position 32° 9′ n. following.§

42. E telescopicis *k* Virginis precedentibus.||

Feb. 6, 1782. Double. About 1¼ degree s. preceding *k* Virginis, in a line parallel to ζ and θ; the most south of three forming an arch. Extremely unequal. L. w.; S. hardly visible with 227 (but with a ten-feet reflector S. b.); with 460, above 2 diameters of L. Position 52° 24′ s. following.

43. FL. 43^{am} Leonis præcedens ad austrum. In dextro genu.

Feb. 17, 1782. Double. Near ⅜ degree s. preceding the 43d, in a line parallel to α and the 14th Leonis. Very unequal. L. w.; S. d. With 227, near 2¼ diameters of L. when best. Position 85° 2′ n. following.

44. ο Virginis. FL. 84. Versus finem alæ dextræ.

Feb. 17, 1782. Double. Extremely unequal. L. w. inclining to r.; S. d. Requires attention to be seen with 227; with 460, 2½ diameters of L. Position, with 278, 29° 5′ s. preceding.

45. FL. 54 Virginis.

April 3, 1782. Double. A little unequal. Both w. With 227, 1½ or near 1¾ diameter. Position 57° 0′ n. following.

‡ When the small star is so faint as not to bear the least illumination of the wires, its position may still be measured by the assistance of some wall or other object; for an eye which has been some time in the dark, can see a wall in a star-light evening sufficiently well to note the projection of the stars upon it, in the manner which has been described with the lamp-micrometer, *Phil. Trans.* vol. LXXII. p. 169 and 170 [above, pp. 94–95]. Then, introducing some light, and adapting the fixed wire to the observed direction of the stars on the wall, the moveable wire may be set to the parallel of the large star, which will give the angle of position pretty accurately.

§ [II. 41. There is no doubt an error of 1 rev. =24° in the micrometer-reading, and the Pos. Angle should be 56° 9′, as pointed out by Sir John Herschel.—ED.]

|| See note to IV. 51.

II. 46. FL. 42am Comæ Berenices sequens ad austrum.

April 15, 1782. Double. About $1\frac{3}{4}$ degree from the 42d Comæ towards υ Bootis; the most south of a telescopic equilateral triangle. Excessively unequal. L. pr.; S. d. With 278, $2\frac{1}{2}$ diameters of L.; not so well to be seen with higher powers. Position 6° 42′ s. following. A third star preceding, above 1′.

47. FL. 2 Comæ Berenices.

April 18, 1782. Double. Considerably unequal. L. rw.; S. pr. With 278, 2 diameters of L.; with 460, above 2 diameters of L. Position 27° 42′ s. preceding.

48. Prope FL. 16am Aurigæ.

Aug. 28, 1782. A minute double star. Less than $\frac{1}{4}$ degree s. preceding the 16th, in a line parallel to the 10 and 8 Aurigæ; the preceding star of a small triangle of which the 16th is the largest and following. A little unequal. Both pr. With 227, $1\frac{1}{2}$ or, when best, $1\frac{3}{4}$ diameter of L. Position 15° 48′ n. following.

49. o (FL. 110a) Piscium borealior. In lino boreo.

Sept. 3, 1782. Double. About $\frac{1}{2}$ degree n. of, and a little preceding 110th, towards η Piscium. A little unequal. Both wr. With 460, about 3 diameters of L. Position 59° 6′ n. preceding. A third star in view, about $1\frac{3}{4}$ min.

50. FL. 38. Piscium. In austrino lino.

Sept. 4, 1782. Double. Pretty unequal. Both pr. With 227, full 2 diameters of L.; with 460, about 4 diameters of L. Position 25° 3′ s. preceding.

51. ρ Capricorni. FL. 11. Trium in rostro sequens.

Sept. 5, 1782. Double. Very unequal. Both rw. With 460, $1\frac{1}{2}$ diameter of L. Position 84° 0′ s. following. A third star in view.

52. o (FL. 40am) Persei præcedens ad boream.

Sept. 7, 1782. Double. Almost $\frac{1}{2}$ degree preceding the 40th, in a line parallel to ζ and the 38th Persei. Equal. Both w. With 227, nearly 2 diameters. Position 8° 24′ n. preceding.

53. FL. 12am Camelopardali præcedens.

Sept. 7, 1782. Double. Less than $\frac{1}{4}$ degree preceding the 11th and 12th, in a line from the 1st Lyncis continued through the 12th Camelopardali. Extremely unequal. Both dr. With 227, it appears like a star with a tail; but 932 shews it plainly to be only a double star; with 227, not much above 1 diameter of L.; with 932, about $3\frac{1}{2}$ diameter of L. Position 18° 33′ s. following; a little inaccurate.

54. Quæ præcedit ϵ (FL. 74am, oculum boreum) Tauri.

Sept. 7, 1782. Double. Near $\frac{1}{2}$ degree s. preceding ϵ, in a line parallel to α and γ Tauri; a small star. Extremely unequal. L. rw.; Ṡ. d. With 460, above 3 diameters of L. Position 68° 42′ s. preceding.

II. 55. FL. 4ª Ceti australior et sequens.

Sept. 9, 1782. Double. About 1 degree s. following the 4th and 5th in a line parallel to η and τ Ceti; in the shorter leg of a rectangular triangle. Very unequal. L. r.; S. d. With 278, rather more than 2 diameters. Position 21° 42′ n. preceding.

56. β (FL. 6am) Arietis præcedens ad boream.

Sept. 10, 1782. Double. Almost 1 degree n. preceding β Arietis, towards ζ Andromedæ; a small star. A little unequal. Both reddish. With 227, full 2 diameters of L. Position 23° 12′ n. preceding. A third star 2′ or 3′ preceding, in the same direction with the two stars of the double star.

57. Ad FL. 72am Aquarii.

Sept. 27, 1782. Treble. About 2½ degrees following κ, in a line parallel to α and η Aquarii. The nearest a little unequal. Both r. With 460, 2½ diameters of L. Position 25° 51′ s. preceding. The two farthest a little unequal. Of the 5th class. About 50° or 55° s. following.

58. FL. 56ª Ceti australior et sequens.

Sept. 27, 1782. Double. About ¾ degree s. following the 56th, in a line parallel to η and τ Ceti. Considerably unequal. Both dw. With 278, 1½ diameter of L. Position 25° 12′ n. preceding; too low for accuracy.

59. ρ (FL. 46am) Aquarii sequens ad austrum.

Sept. 30, 1782. Double. About 2 degrees s. following ρ, in a line parallel to β and δ Aquarii; there is a very considerable star between this and ρ, not much out of the line. Pretty unequal. Both dr. With 227, 2½ or 2¾ diameter of L. Position 61° 12′ n. preceding.

60. ξ (FL. 5am) Canis majoris sequens ad boream.

Sept. 30, 1782. Double. About ½ degree n. following the 2d ad ξ, in a line from the 4th continued through the 5th Canis majoris nearly. Very unequal. L. rw.; s. d. With 227, 1¾ diameter. Position 67° 36′ n. preceding.

61. ω (FL. 47am) Orionis sequens ad austrum.

Oct. 2, 1782. Treble. About 1½ degree s. following ω in a line parallel to φ and α Orionis; the smallest and most south of three forming an arch. The two nearest extremely unequal. L. dw.; S. a mere point. With 227, 1½ or 1¾ diameter of L. Position 4° 54′ n. following; too obscure for accuracy. The two farthest extremely unequal. S. a mere point. Of the fourth class. Position about 50° s. following.

62. FL. 3æ Pegasi adjecta.

Oct. 4, 1872. Double. In a line with, and north of, the two stars that are about the place of the third Pegasi. A little unequal. Both dusky r. With 227, about 3 diameters of S. Position 88° 24′ n. preceding; perhaps a little inaccurate.

II. 63. FL. 2am et 4am Navis præcedens.

Oct. 12, 1782. Multiple. Near 2 degrees preceding the 2d and 4th Navis; the middle one of three. One of the multiple is double. Nearly equal. Both w. or ash colour. With 227, about 2½ diameter, and not less than 20 stars more in view; with 460, about 3 diameters. Position 30° 12′ n. preceding.

64. *g* (FL. 81am) Geminorum ad austrum sequitur.

Oct. 13, 1782. Double. About ½ degree s. following *g*, in a line from ζ continued through *g* Geminorum nearly; the nearest and largest of two. Very unequal. L. r.; S. bluish r. With 227, above 3 diameters of L. Position 4° 9′ n. preceding.

65. Pollucem sequens ad boream.

Oct. 13, 1782. Double. Full ¾ degree n. following β, in a line from δ continued through β Geminorum; the star next to the middle one of three, nearly in a line. Excessively unequal. L. rw.; S. d. With 227, above 2½ or near 3 diameters of L. and 5 other stars in view; with 460, above 3 diameters of L. Position 89° 12′ n. following.

66. Juxta γ Delphini.

Oct. 19, 1782. Double. Full ¼ degree s. preceding γ, towards δ Delphini. Considerably unequal. L. pr.; S. r. With 227, 1½ diameter of L. Position 78° 42′ n. preceding.

67. β (FL. 10am) Lyræ præcedens ad boream.

Oct. 19, 1782. Double. The 4th telescopic star about 1½ degree n. preceding β, in a line parallel to γ and α Lyræ. Extremely unequal. L. r.; S. dr. With 227, 1¼ or almost 1½ diameter of L. With 460, above 2 diameters of L. Position 68° 6′ s. following.

68. Proximè ρ Lyræ.

Oct. 24, 1782. Treble. About 2¼ minutes s. following ρ Lyræ. The two nearest, a little unequal. Both dr. With 460, 3 full diameters. Position 8° 24′ n. following. The farthest as large as L. of the two nearest at least. Colour dr. Position with L. 25° 57′ s. preceding. Distance of ρ Lyræ, which is in view, from the two nearest 2′ 17″ 30‴. Position 65° 12′, ρ being n. preceding, or the double star s. following.

69. FL. 4am Cygni sequens ad boream.

Oct. 24, 1782. Double. Near ½ degree n. following the 4th Cygni, in a line from γ Lyræ continued through the 4th Cygni A little unequal. Both w. With 227, about 2 diameters of L. or 2½ when best. Position 29° 12′ n. following.

II. 70. τῶν 8 telescopicarum *z* (FL. 15) Sagittæ sequentium ultima.

Nov. 6, 1782. Double. About 1¼ degree s. following *z* Sagittæ, in a line parallel to γ Sagittæ and γ Delphini. Extremely unequal. Both r.; S. deeper r. With 227, 1½ diameter of L.; with 460, above 2 diameters of L. Position 72° 57′ n. following.

71. FL. 58ª Aurigæ australior.

Nov. 6, 1782. Multiple. About ¾ degree s. of the 58th Aurigæ, in a line parallel to β and θ. A cluster of stars containing a double star of the second, and one of the third class. That of the second very unequal. Both r. With 460, about 2½ diameter of L. Position 44° 36′ n. following; that of the third equal. Both r. With 227, above 20 stars in view. Distance 17″ 41‴. The two double stars are in the following side of a small telescopic trapezium.

72. FL. 13ª Lyncis australior.

Nov. 13, 1782. A pretty double star. About 1¼ degree s. of the 13th Lyncis, towards θ Geminorum; a considerable star. Nearly equal. Both pr. With 227, full 2½ diameters; with 460, almost 4 diameters. Position 11° 0′ s. preceding.

73. FL. 21ª Ursæ majoris.

Nov. 17, 1782. Double. Very unequal. Both rw. With 227, 2¼ diameter of L.; with 460, above 3. Position 36° 45′ n. preceding.

74. ν (FL. 4ª) Crateris borealior.

Nov. 20, 1782. Treble. Near 1 degree n. preceding ν Crateris, towards α Leonis. The two nearest equal. Both dw. With 227, 2½ or 3 diameters. Position 71° 33′ n. following. The farthest larger than either of the two other stars. Of the sixth class. Position about 68 or 69° s. preceding the double star.

75. FL. 118 Tauri.

Dec. 7, 1782. Double. A little unequal. L. w.; S. w. inclining to r. With 278, 2½ diameter of L.; with the same power by the micrometer 4″ 41‴; more exactly with 625, 5″ 2‴. Position 77° 15′ sp. I could just see it with an 18-inch achromatic, made by Mr. NAIRNE; it was as close as possible, and a pretty object.

76. τ (FL. 63ª) Arietis australior et præcedens.

Dec. 23, 1782. Double. About 1 degree s. preceding τ Arietis, towards μ Ceti; the most south of two small telescopic stars. Nearly equal. Both w. With 227, above 3 diameters; by the micrometer 5″ 47‴. Position 15° 24′ s. preceding.

77. * FL. 17 Hydræ.

Dec. 28, 1782. Double. The largest of two. A little unequal. Both w. With 227, 2¼ diameter of L.; with 460, 1¾ diameter. Position 90° 0′ north.

78. χ (FL. 63ªᵐ) Leonis sequens ad austrum.

Jan. 1, 1783. Double. About ⅓ degree s. following χ, towards τ Leonis; the smallest of two. Very or extremely unequal. L. r.; S. d. With 227, 3 full diameters of L. Position 75° 21′ s. following.

II. 79. FL. 39 Bootis.

Jan. 8, 1783. A pretty double star. A little unequal. Both pr. With 227, near 1½ diameter of L.; with 460, near 2 diameters of L. Position 38° 21′ n. following.

80. *d* (FL. 40ᵃᵉ) Eridani adjecta.

Jan. 31, 1783. Double. About 1⅓ min. s. following *d* Eridani. Very unequal. Both dr. With 227, hardly visible; with 460, very obscure. Position 56° 42′ n. preceding. Distance of L. from *d* Eridani, with 227, 1′ 21″ 47‴. Position of L. 17° 33′ s. following *d* Eridani.

81. FL. 49ᵃᵐ Eridani sequens.

Jan. 31, 1783. Double. Near 1 degree following the 49th Eridani, towards δ Orionis. Very unequal. Both dw. With 227, full 1 diameter of L.; with 278, 1½ or 1¾ diameter of L.; with 460, 2½ or 3 diameters of L. Position 51° 36′ n. preceding.

82. FL. 31ᵃᵐ Bootis sequens ad austrum.

Feb. 3, 1783. Double. Near 1 degree s. following the 31st, in a line from υ continued through the 31st Bootis; the most south of two. A little unequal. L. w.; S. dw. With 227, about 1¾ diameter of L; with 460, about 3 diameters of L. Position 1° 0′ s. following. A third star in view, 20° or 30° n. preceding.

83. FL. 22ᵃ Andromedæ borealior.

Feb. 26, 1783. Double. Within ½ degree north of the 22d, in a line parallel to the 19th and 16th Andromedæ; the following and smallest of two. Considerably unequal. L. w.; S. d. With 227, 1¼ or 1½ diameter of L.; with 460, more than 2 diameters of L. Position 5° 48′ n. following.

84. FL. 65 Piscium.

Feb. 27, 1783. Double. Nearly equal. Both pr. With 227, near 1½ diameter of L.; with 460, full 2 diameters. Position 30° 57′ n. preceding.

85. *b* (FL. 36ᵃ) Serpentis borealior et sequens.

Mar. 4, 1783. Double. About 1½ degree n. following *b*, nearly in a line from the 32d continued through the 36th Serpentis. Extremely unequal. L. w.; S. dw. With 227, 1 full diameter of L.; S. hardly to be seen; with 460, full 2 diameters of L. Position 46° 9′ n. preceding.

86. FL. 49ᵃᵐ Serpentis præcedens ad austrum.

Mar. 7, 1783. Double. About 1½ degree s. preceding the 49th, in a line with the 49th and another between this and the 49th Serpentis, each nearly at ¾ degree distance. Very unequal. L. dw.; S. d. With 227, 2 diameters, or 2¼ when best Position 53° 9′ s. following.

II. 87. FL. 29[a] et 30[a] Monocerotis australior.

Mar. 8, 1783. Multiple. It makes nearly an equilateral triangle with the 29th and 30th Monocerotis towards the south. Among many, the fourth from the south end of an irregular long row is double. A little unequal. Both pr. With 227, 1 diameter of L. and 16 more in view. Position 86° 12′ s. following.

88. ω (FL. 51[am]) Serpentis præcedens ad austrum.

Mar. 8, 1783. Double. About ½ degree s. preceding the 51st, towards the 13th Serpentis. Very or extremely unequal. Both r. With 227, 2¼ diameter of L. when best; with 460, near 3 diameters of L. Position 44° 45′ n. preceding.

89. Ad Genam Monocerotis.

Mar. 26, 1783. Double. About 1 degree n. preceding the 12th Monocerotis, in a line parallel to α and λ Orionis; the smallest and most north of two. Considerably unequal. L. r.; S. bluish r. With 227, near 4 diameters of L. when best. Position 50° 51′ n. following.

90. FL. 100[am] Herculis præcedens ad boream.

Mar. 27, 1783. Double. About 1¾ degree n. preceding the 100th, towards μ Herculis; a very small telescopic star; the most towards μ and smallest of three forming an arch. Considerably unequal. Both dw. With 227, about 2 diameters of L. Position 75° 9′ s. following.

91. z (FL. 15[a]) Sagittæ australior.

Apr. 5, 1783. Treble. About twice as far south of z Sagittæ, as z and the star near it are from each other; a small star. The two nearest very unequal. L. pr.; S. r. With 227, 1½ diameter of L. Position 74° 54′ s. preceding. The third with L. extremely unequal. S. d. With 227, about 3 diameters of L. or more. Position about 40° or 50° n. preceding. With more light this would be a fine object.

92. In Camelopardali clune.

Apr. 30, 1783. Double. About four times the distance of the 10th and 12th Camelopardali, north of the 10th, and almost in the same direction with the 10th and 12th, is a star of between the 5th and 6th magnitude not marked in FLAMSTEED; naming that star A, we have the following direction. About ½ degree preceding A Camelopardali, in a line from the 2d Lyncis continued through A; the second from A. Very unequal. L. w.; S. d. With 227, 1½ or 2 diameters of L. Position 22° 42′ s. following. Very inaccurate.

93. ε (FL. 13[a]) Aquilæ australior.

May 25, 1783. Double. Near ¼ degree south of, and a little following ε, towards λ Aquilæ, a very small star. Very unequal. L. dw.; S. dr. With 460, above 2 diameters of L. Position 16° 0′ n. preceding.

II. 94. ι (FL. 17^am^) Andromedæ præcedens ad boream.

Aug. 19, 1783. Double. About 1⅓ degree n. preceding Andromedæ in a line parallel to α and β Cassiopeiæ; in the side of a trapezium of four small stars. Pretty unequal. Both r. With 460, 2½ diameters of L. Position 34° 24′ n. preceding.

95. η (FL. 55^â^) Aquilæ australior.

Sept. 12, 1783. Double. About ⅓ degree south of η, in a line from α continued through η Aquilæ; a small star. A little unequal. Both dusky ash-coloured. With 460, near 3 diameters of L.; with 278, near 2 diameters of L. Position 29° 3′ n. preceding.

96. θ (FL. 65^â^) Aquilæ borealior et sequens.

Sept. 12, 1783. Double. About 1¾ degree n. following θ Aquilæ, towards ε Delphini; more accurate towards 29 Vulpeculæ; a very considerable star. Nearly equal. Both rw. With 278, about 1¼ diameter of L.; with 460, full 2 diameters. Position 56° 12′ s. preceding.

97. ζ (FL. 64^am^) Cygni præcedens.

Sept. 15, 1783. Treble. About 1 degree preceding ζ, towards the 41st Cygni; a large star. The two nearest extremely unequal. L. w.; S. pr. With 460, 2½ diameters of L. Position 45° 15′ n. preceding. The third with L. extremely unequal. Of the 5th or 6th class; about 50° s. preceding.

98. FL. 49 Cygni.

Sept. 15, 1783. Double. Very unequal. L. r.; S. bluish r. With 278, 1½ diameter of L.; with 460, 2½ diameters of L. Position 31° 48′ n. following.

99. β (FL. 6^am^) Cygni sequens ad boream.

Sept. 15, 1783. Double. Near ½ degree n. following β, towards ξ Cygni. Very unequal. Both dw. With 278, 1½ diameter of L.; with 460, about 2 diameters of L. Position 87° 48′ n. following.

100. FL. 51^â^ Cygni borealior et sequens.

Sept. 24, 1783. Double. Near two degrees n. following the 51st Cygni, in a line parallel to ο Cygni and α Cephei; a pretty considerable star. Very unequal. L. w.; S. inclining to blue. With 278, extremely unequal, and 1½ diameters of L. when best; requires attention to be seen well with this power; with 460, full 2 diameters of L. or 2¼ when best, otherwise much less. Position 15° 51′ n. following.

101. FL. 57^am^ :: Camelopardali præcedens ad boream.

Sept. 26, 1783. Double. About 2 degrees n. preceding the 57 :: , towards the 42d Camelopardali; a considerable star near three smaller, forming an arch. About 1 degree from the double star V. 135. Considerably unequal. Both pr. With 278, 1⅔ diameter of L.; with 460, 2½ diameters of L. Position 67° 15′ n. preceding.

II. 102. *e* (FL. 29[a]) Orionis australior et præcedens.

Sept. 27, 1783. Double. About $\frac{1}{2}$ degree s. preceding *e*, in a line parallel to ρ and β Orionis; the largest of several. Very unequal. L. pr.; S. inclining to garnet. With 278, near 2 diameters of L. With 460, $2\frac{1}{2}$ diameters of L. Position 52° 24′ s. following.

THIRD CLASS OF DOUBLE STARS.

III. 47. *e* Pollucis. FL. 38 Geminorum. In calce.

Dec. 27, 1781. Double. Extremely unequal. L. rw.; S. r. Distance, with 460, 7″ 48‴. Position 89° 54′ s. following. Two more in view, the nearest of them perhaps 40″; they form a rectangle nearly.

48. *r* (FL. 61[am]) Geminorum præcedens ad boream.

Dec. 27, 1781. Double. About $\frac{1}{2}$ degree n. preceding *r*, in a line parallel to κ and the 60th Geminorum; near two degrees from δ. A little unequal. Both pr. Distance 6″ 15‴. Position 43° 54′ n. following.

49. δ (FL. 4[am]) Hydræ præcedens ad boream.

Jan. 20, 1782. Double. About $1\frac{1}{4}$ degree n. preceding δ, in a line from η continued through δ Hydræ. Pretty unequal. L. r.; S. garnet. Distance 12″ 30‴ Position 62° 48′ n. following.

50. θ Virginis. FL. 51. De quatuor ultima et sequens.

Feb. 6, 1782. Treble. The two nearest extremely unequal. L. w.; S. d. Distance 7″ 8‴; but inaccurate on account of the obscurity of S. Position 69° 18′ n. preceding. For measures of the two farthest see VI. 43.

51. FL. 88 Leonis. In dextro clune.

Feb. 9, 1782. Double. Extremely unequal. L. rw.; S. r. Distance 14″ 38‴; a little inaccurate. Position 47° 33′ n. preceding.

52. FL. 10[am] Orionis sequens.

Feb. 17, 1782. Double. Above $\frac{3}{4}$ degree n. following the 10th, towards ω Orionis. Considerably unequal. Both pr. Distance with 278, 13″ 40‴. Position 37° 3′ n. following.

53. γ Virginis borealior et sequens.

Feb. 17, 1782. Double. Near $2\frac{1}{2}$ degrees n. following γ, in a line parallel to ϵ and α Virginis; a considerable star; a line from γ to this passes between two of nearly the same magnitude with this star. A little unequal. Both d. Distance 12″ 58‴. Position 79° 0′ n. preceding.

54. Secunda ad σ Ursæ majoris. FL. 13. In fronte.

June 2, 1782. Double. Extremely unequal. L. w.; S. r. Distance 7″ 56‴. Position 13° 0′ n. preceding.

III. 55. υ (FL. 18[am]) Coronæ borealis sequens ad boream.

June 14, 1782. Double. Considerably unequal. L. dr.; S. d. Distance with 227, about 3 or 4 diameters of L. being too obscure for the micrometer. Position 53° 48′ s. preceding. Distance of the largest of the two from υ Coronæ 1′ 18″ 8‴. Position of the same with υ, 64° 24′ n. following.

56. S (FL. 72[a]) Serpentarii borealior.

June 16, 1782. Double. About 2½ degrees n. of the 72d Serpentarii; a considerable star. A little unequal. Both r. Distance 7″ 37‴. Position 9° 42′ s. preceding. A third star about 1′ preceding.

57. In Anseris corpore.

Aug. 11, 1782. A pretty double star. About ¾ degree n. of a cluster of stars formed by the 4th, 5th, 7th, 9th Anseris; in a line parallel to the 6th Vulpeculæ and β Cygni; that of two which is farthest from the cluster. A little unequal. Both r. Distance 7″ 1‴. Position 58° 36′ s. following.

58. θ Persei. FL. 13. In sinistro humero.

Aug. 20, 1782. Double. Extremely unequal. L. w. inclining to r.; S. d. Distance with 932, 13″ 31‴. Position 20° 0′ n. preceding. A third star, very unequal, within 1′; towards the south.

59. Ad FL. 19[am] Persei. In capite.

Aug. 20, 1782. Double. It is perhaps the 19th Persei removed, or more likely a star not marked in FLAMSTEED's Catalogue; the 19th being either vanished, or misplaced by FLAMSTEED.‡ Pretty unequal. L. bw.; S. br. Distance 12″ 2‴. Position 0° 0′ following.

60. Secunda ad p Persei. FL. 20. Illas in larva præcedit.

Aug. 20, 1782. Double. Extremely unequal. L. rw.; S. d. Distance 14″ 2‴. Position 30° 30′ s. following.§

61. Sub finem caudæ Draconis.

Aug. 29, 1782. Double. Of two considerable stars, about half-way between α and Draconis, that which is towards ι. The two stars are parallel to ζ and ϵ Ursæ majoris. Very unequal. L. pr.; S. db. Distance 12″ 30‴; perhaps a little inaccurate. Position 87° 42′ n. preceding.

62. FL. 35 Piscium. In lino austrino.

Sept. 4, 1782. Double. Considerably unequal. L. rw.; S. pr. Distance 12″ 30‴. Position 58° 54° s. following.

‡ [III. 59. It is = P. II. 220.—ED.]

§ [III. 60. "South following" must be a mistake for sp. The object was seen in Sweep 599, Sept. 21, 1786, when it was noted that "the small one precedes the large."—ED.]

III. 63. Prope FL. 65[am] Sagittarii. Ad extremum paludamentum.

Sept. 5, 1782. Double. Near ½ degree s. following the 65th Sagittarii towards ζ Capricorni. Very unequal. Too low for colours; perhaps dw. Distance 14″ 20‴. Position 73° 48′ n. following.

64. FL. 26 Aurigæ. In dextri cruris involucro.

Sept. 5, 1782. Double. Very unequal. L. rw.; S. r. Distance 13′ 25‴ Position 2° 36′ n. preceding.

65. *e* (FL. 58[a]) Persei australior. In dextri pedis talo.

Sept. 7, 1782. Double. About 10′ south of the 58th Persei, in a line parallel to ζ and ι Aurigæ; a small telescopic star. Very unequal. L. r.; S. d. Distance with 625, 11″ 22‴. Position 48° 54′ n. following. Very inaccurate: windy.

66. *e* Tauri. FL. 30. In dextri humeri scapula.

Sept. 7, 1782. Double. Extremely unequal. L. w.; S. r. Distance 11″ 16‴; inaccurate on account of obscurity. Position 17° 15′ n. following.

67. ι Leporis. FL. 3. Borea præcedentis lateris quadrilateri ad aures.

Sept. 7, 1782. Double. Excessively unequal. L. w.; S. d. With 227, there was not a possibility of measuring the distance, though the glass was carefully cleaned; on trying 625, I found the star so strong that it bore a very tolerable good light.‡ Distance with this power 12″ 20‴. Position 89° 21′ n. preceding.

68. η (FL. 17[a]) Arietis australior et præcedens.

Sept. 10, 1782. Double. Full 1 degree south preceding η, in a line parallel to α and γ Arietis. Very unequal. L. pr.; S. d. Distance 8″ 5‴ Position 55° 42′ s. following.

69. Prope FL. 64[am] Aquarii. In dextro femore.

Sept. 27, 1782. Double. Full 1½ degree n. following the 64th ::, in a line parallel to λ and φ Aquarii; the largest of two that follow a very obscure triangle in the finder. Extremely unequal. L. rw.; S. db. Distance 12″ 46‴ Position 20° 3′ s. following.

70. κ Cephei. FL. 1. In dextro crure.

Sept. 27, 1782. A beautiful double star. Extremely unequal. L. fine w.; S. r. Distance 5″ 47‴. Position 32° 30′ s. following.

‡ With regard to small stars, that become visible by an increase of magnifying powêr, we may surmise, that it is partly owing to the greater darkness of the field of view, arising from the increased power, and partly to the real effect of the power; for, though the real diameter of a star, notwithstanding it be magnified a thousand times, should still remain smaller than the minimum visibile, yet since a star of the seventh magnitude may be seen by the naked eye, we may conclude, that the light of a star subtends incomparably a larger angle than its luminous body; and this may be in such a proportion, with very small stars, that the power of the telescope shall be just sufficient to magnify the real diameter so as to bring it within the limits of this proportion, whereby the star will become visible.

III. 71. Tiaram Cephei præcedens.

Sept. 27, 1782. Treble. About 1½ degree preceding the *garnet star*,‡ in a line parallel to ι and ζ Cephei. The two nearest very unequal. L. w.; S. db. Distance 11″ 35‴. Position 35° 24′ s. following. The two farthest considerably unequal. S. db. Distance 18″ 37‴. Position 73° 57′ n. preceding. The place of the *garnet star*, reduced to the time of FLAMSTEED's Catalogue, is about Æ 21 h. 45′. P.D. 32°½.

72. Tiaram Cephei præcedens.

Sept. 27, 1782. Double. Within ¼ degree of the foregoing treble star. Considerably unequal. L. rw.; S. pr. Distance 13″ 7‴. Position 32° 0′ n. following.

73. FL. 25ª Ceti australior et sequens.

Oct. 2, 1782. Double. About ¾ degree s. following the 25th, in a line parallel to θ and τ Ceti. Pretty unequal. Distance with 278, 14″ 50‴. Position 89° 12′ s. preceding; perhaps a little inaccurate.

74. FL. 18ª Pegasi australior. Ad oculum sinistrum.

Oct. 4, 1782. Double. About ¾ degree s. preceding the 18, in a line parallel to η and ϵ Pegasi; the most north and largest of two. A little unequal. Both rw. Distance 14″ 29‴ full measure. Position 31° 33′ n. following.

75. Ad Genam Monocerotis.

Oct. 4, 1782. Double. About 1 degree n. of, and a little preceding the six telescopics in the place of the 12th, in a line parallel to the 12th Monocerotis and μ Geminorum.

76. $\tau\omega\nu$ quatuor telescopicarum, δ Orionis sequentium, penultima.

Oct. 4, 1782. Double. About ¾ degree n. following δ, in a line parallel to τ and ι Orionis. Extremely unequal. L. r.; S. d. Distance with 278, 9″ 12‴. Position 13° 6′ n. preceding.

77. FL. 65am Arietis sequens ad austrum.

Oct. 9, 1782. Double. About ¾ degree s. following the 65th Arietis, in a line parallel to the Pleiades and ϵ Tauri; the preceding of two. Very unequal. L. r.; S. bluish. Distance 8″ 32‴. Position 73° 18′ s. following.

78. FL. 13am Tauri præcedens ad austrum.

Oct. 9, 1782. Double. About 1¾ degree s. preceding the 13th Tauri, in a line parallel to ϵ Tauri and δ Ceti.§ Nearly equal. Both pr. Distance 7″ 10‴. Position 87° 57′ n. preceding.

79. ϵ (FL. 83ª) Ceti borealior.

Oct. 13, 1782. Double. About ⅜ degree n. of ϵ Ceti; the nearest of three forming an arch. Extremely unequal. L. rw.; S. darkish red. Distance with 278, 10″ 48‴. Position 45° 12′ s. preceding.

‡ *Phil. Trans.* vol. LXXIII. p. 257. [Above, p. 114.]

§ [III. 78. The place is 1° in error. It is =Σ 414.—ED.]

III. 80. σ (FL. 76[am]) Ceti præcedens. In sinistro crure.

Oct. 13, 1782. Double. Full 1½ degree preceding σ, towards τ Ceti. Extremely unequal. L. rw.; S. br. Distance 11″ 16‴. Position 22° 24′ n. preceding.

81. Parvula à ζ Lyræ ε versus.

Oct. 19, 1782. Double. Above ½ degree from ζ towards ε Lyræ. Extremely unequal. L. r.; S. dr. Distance 9″ 27‴ full measure. Position 66° 18′ n. following.

82. FL. 41 Aurigæ.

Nov. 6, 1782. A pretty double star. Considerably unequal. L. w.; S. grey inclining to r. Distance 8″ 32‴. Position 80° 0′ n. preceding.

83. FL. 19 Lyncis.

Nov. 13, 1782. Double. A little unequal. L. rw.; S. bw. Distance 14″ 11‴. Position 46° 54′ n. preceding.‡

84. Prope FL. 40 Lyncis. In Ursæ majoris pede.§

Nov. 13, 1782. Double. Very or extremely unequal. L. wr.; S. r. Distance 7″ 11‴. Position 48° 12′ n. preceding.

85. FL. 2 Canum Venaticorum.

Nov. 13, 1782. Double. Very unequal. L. r.; S. bluish. Distance 12″ 12‴. Position 11° 0′ s. preceding.

86. FL. 57 Ursæ majoris.

Nov. 20, 1782. Double. The largest of two stars. Excessively unequal. L. w.; S. a red point without sensible magnitude. With 227, S. is but just visible. Position 75° 36′ n. following.

87. FL. 59[a] Ursæ majoris borealior.

Nov. 20, 1782. A pretty treble star. Near 1½ degree n. of the 59th, in a line parallel to ψ and β Ursæ majoris nearly. The two nearest considerably unequal. L. pr.; S. r. Distance 12″ 30‴. Position 0° 0′ preceding. The two farthest very unequal. S. dr. Distance 32″ 21‴. Position 4° 0′ n. following.

88. FL. 11[a] Tauri borealior et sequens.

Nov. 25, 1782. Double. About ½ degree n. following the 11th Tauri, towards ι Aurigæ. Very unequal. L. w.; S. pr. Distance with 278, 13″ 37‴. Position 89° 51′ n. following.

89. Ad 63[am] Herculis. In linea per δ et ε ducta.

Nov. 26, 1782. Double. About 4 degrees from δ towards ε Herculis, near the 63d. Very unequal. L. r.; S. r. Distance 11″ 53‴. Position 47° 48′ n. following.

‡ [III. 83. Not sp, as in *Phil. Trans.*; diagram in Journal.—ED.]

§ [III. 84. The larger star is not 40 Lyncis, as in *Phil. Trans.*, but is 3′ nf it.—ED.]

III. 90. FL. 103ª Tauri borealior.

Nov. 29, 1782. Double. About three degrees directly n. of the 103 Tauri; the largest of three, forming an obtuse angle. Considerably unequal. L. rw.; S. pr. Distance with 278, 13″ 6‴. Position 64° 0′ n. following.

91. FL. 62ª Arietis borealior et sequens.

Dec. 23, 1782. Double. Near 1 degree n. following the 62d Arietis, towards ε Persei. Nearly equal. Both dw. Distance 11″ 17‴; not very accurate. Position 12° 24′ n. preceding or s. following.

92. ξ (FL. 77am) Cancri præcedens ad boream.

Dec. 28, 1782. Double. About 1 degree n. preceding ξ Cancri, in a line parallel to ε Leonis and the 41st Lyncis; a considerable star. A little unequal. Both rw. Distance 8″ 50‴. Position 65° 12′ s. preceding.

93. In constellatione Tauri.

Dec. 31, 1782. Double. Almost equal. Both rw. Distance 12″ 12‴. Position 52° 27′ s. following.

94. ν (FL. 7am) Leporis præcedens ad boream.

Dec. 31, 1782. Double. About 1⅓ degree n. preceding ν Leporis, in a line parallel to κ and ε Orionis; the second in that line. Equal. Both rw. Distance 11″ 44‴. Position 4° 0′ s. following or n. preceding.

95. ν (FL. 48am) Eridani præcedens ad austrum.

Jan. 2, 1783. Double. Near ⅓ degree s. preceding ν, in a line from the 51st continued through the 48th Eridani. Extremely unequal. L. rw.; S. d. and hardly to be seen with 227. Distance with 278, 15″ 21‴; very inaccurate on account of obscurity. Position 9° 18′ s. preceding.

96. FL. 17 Crateris.

Jan. 10, 1783. Double. Nearly equal. Both rw. Distance 9″ 46‴. Position 64° 27′ s. preceding.‡

97. FL. 54 Hydræ.

Jan. 10, 1783. Double. Very unequal. L. w.; S. bluish r. Distance 11″ 17‴; too low for great accuracy. Position 38° 15′ s. following.

98. Ad Genam Monocerotis.

Jan. 13, 1783. Double. About ⅔ degree s. preceding the most s. of a cluster of six telescopics in the place of the 12th, in a line parallel to the 15th and 12th Monocerotis. Excessively unequal. Position 61° 57′ s. preceding.

99. FL. 55 Eridani.

Jan. 31, 1783. Double. A very little unequal. L. pr.; S. rw. Distance 9″ 9‴. Position 44° 9′ n. preceding.

‡ [III. 96. "A third eF * in the same line p at 3 or 4 times the distance of the other two."—MS.]

III. 100. FL. 55^{am} Eridani præcedens ad austrum.

Jan. 31, 1783. Double. About 2¼ degrees s. preceding the 55th Eridani, in a line parallel to Rigel and γ Eridani. Considerably unequal. L. pr.; S. db. Distance 11″ 53‴. Position 16° 24′ s. preceding.

101. *k* Centauri. FL. 3.

Jan. 31, 1783. Double. Considerably unequal. L. dw.; S. dpr. Distance 11″ 35‴. Position 22° 0′ s. following.

102. *h* (FL. 29^{am}) Herculis præcedens ad austrum.

Feb. 3, 1783. Double. About 1¼ degree s. preceding *h* Herculis towards ε Serpentis; a small star. Very unequal. Both r. Distance 14″ 2‴. Position 67° 12′ n. following.

103. ε (FL. 37^{a}) Serpentis borealior et sequens.

March 4, 1783. Double. Near two degrees s. following ε, in a line parallel to the 13th Serpentis and 10th Serpentarii. Very unequal. L. pr.; S. r.; but a *dry fog*, if I may so call it, probably tinges them too deeply. Distance with 278, 12″ 34‴; with 625, 12″ 23‴. Position 50° 12′ n. preceding.

104. FL. 83^{am} Herculis præcedens.

Mar. 26, 1783. Double. About ⅓ degree preceding the 83rd; the second star towards the 79th Herculis. Very unequal. L. r.; S. darker r. Distance 14″ 20‴. Position 83° 48′ n. following.‡

105. γ (FL. 12^{a}) Sagittæ borealior et præcedens.

April 7, 1783. Double. About 2′ preceding the double star V. 106. Pretty unequal. L. r.; S. d. Distance 14″ 29‴; very inaccurate, on account of obscurity. Position 50° 24′ s. preceding.

106. FL. 5 Serpentis.

May 21, 1783. Double. Excessively unequal. L. rw.; S. db. Too obscure for measures. Of the third class, far. Position about 30° or 40° n. following.

107. Congerie Stellularum Sagittarii borealior.

June 6, 1783. Double. Above 1¼ degree n. of the 20th cluster of stars of the *Connoissance des Temps*, in a line parallel to γ Sagittarii and the cluster: the most south of many. Considerably unequal. Distance with 278, 15″ 10‴. As accurate as the prismatic power of the atmosphere, which lengthens the stars, will permit. Position 54° 48′ s. preceding.§

‡ [III. 104. A diagram in the Journal shows that the companion was nf. So also in the MS. *Phil. Trans.* has np.—ED.]

§ What I call the prismatic power of the atmosphere, of which little notice has been taken by astronomers, is that part of its refractive quality whereby it disperses the rays of light, and gives a lengthened and coloured image of a lucid point. It is very visible in low stars; FOMALHAND, for instance, affords a beautiful prismatic spectrum. That this power ought not to be overlooked in delicate and low observations, is evident from some measures I have taken to ascertain its quantity. Thus

III. 108. FL. 19[am] Aquilæ præcedens ad boream.

July 7, 1783. Double. Above ¾ degree n. preceding the 19th, in a line parallel to β and ζ Aquilæ. Very unequal. L. r.; S. dr. Distance 12″ 58‴. Position 58° 27′ s. following.

109. FL. 19[am] Aquilæ præcedens ad Boream.

July 7, 1783. Double. About 1⅓ degree n. preceding the 19th, in a line parallel to ϵ and δ Aquilæ. Pretty unequal. Both rw. Distance 10″ 13‴. Position 22° 6′ n. preceding.

110. FL. 77[a] Cygni borealior et præcedens.

Sept. 17, 1783. Quadruple. Full ¾ degree n. preceding the 17th, in a line parallel to σ and α Cygni; a small star. The two nearest extremely unequal. L. r.; S. d. Distance with 625, 13″ 54‴. Position 67° 36′ s. following. The two largest a very little unequal. Both r. Distance with 278, 25″ 58‴. Position 40° 33′ n. following. The farthest very unequal. S. d. Position almost in a line with the two largest.

111. ϵ (FL. 46[a]) Orionis borealior et sequens.

Sept. 20, 1783. Treble. About 1¼ degree n. following ϵ, towards α Orionis. The two nearest of the third class.

112. δ (FL. 18[am]) Cygni sequens ad austrum.

Sept. 22, 1783. Double. About 1 degree s. following δ, towards the 47th Cygni; a pretty considerable star. Equal, or perhaps the southern star the smallest. Both pr. Distance with 278, 10″ 8‴. Position 71° 0′ s. following.

113. FL. 27[am] Cygni præcedens ad austrum.

Sept. 23, 1783. Quadruple and Sextuple. About ½ degree s. preceding the treble star I. 96; the middle of three, the most north whereof is the 27th Cygni. In the quadruple or n. preceding set, the two nearest very unequal. Distance with 278, 11″ 16‴. Position 26° 0′ n. preceding; the two largest almost equal. Both r. Distance with 278, 29″ 27‴ Position 57° 12′ n. following. In the sextuple or s. following set, the two largest pretty unequal. Both r. Distance with 278, 19″ 20‴. Position 27° 36′ s. preceding. All

I found, May 4, 1783, that the perpendicular diameter of ϵ, FLAMSTEED'S 20th Sagittarii, measured 16″ 9‴, while the horizontal was 8″ 35‴; which gives 7″ 34‴ for the prismatic effect: the measures were taken with 460, near the meridian, and the air remarkably clear. And though this power, which depends on the obliquity of the incident ray, diminishes very fast in greater altitudes, yet I have found its effects perceivable as high, not only as α or γ Corvi in the meridian, but up to Spica Virginis, and even to Regulus. Experiments on these two latter stars I made November 20, 1782; when Regulus, at the altitude of 49°, shewed the purple rather fuller at the bottom of the field of view than when it was at the upper edge; which shews that the prismatic powers of the edges of the eye lens were assisted in one situation by the power of the atmosphere, but counteracted by it in the other. I turned the eye lens in all situations, to convince myself that it was not in fault. This experiment explains also, why a star is not always best in the center of the field of view; a fact I have often noticed before I knew the cause.

the other stars are as small as the smallest of the quadruple set; and some of them much smaller.

III. 114. FL. 16[am] Monocerotis præcedens ad boream.

Jan. 23, 1784. Double. About 1¼ degree n. preceding the 16th.

FOURTH CLASS OF DOUBLE STARS.

IV. 45. In pectoris crate Orionis.

Dec. 27, 1781. Double. About ⅝ degree following ψ, towards n Orionis. Extremely unequal. L. pr.; S. dr. Distance with 278, 20″ 3‴. Position 62° 24′ s. following.

46. FL. 21 :: Geminorum.‡

Dec. 27, 1781. Double. A little unequal. Both pr. Distance about 25″.

47. FL. 3 Leonis.

Feb. 9, 1782. Double. Excessively unequal. L. r.; S. d.; not visible with 227. Distance estimated with 460, about 24″. Position a little n. following. A third star in view. Distance perhaps 2′. Position about 15° s. following.

48. H (FL. 1[am]) Geminorum præcedens ad boream.

Feb. 6, 1782. Quintuple. In the form of a cross. About ⅝ degree n. preceding H Geminorum, in a line parallel to the 65th Orionis and ζ Tauri; the middle of three. The two nearest or preceding of the five extremely unequal. Distance 20″ 27‴.§ Position 7° 27′ s. preceding. The last of the three, in the short bar of the cross, has an excessively obscure star near it of the third class. Five more in view, differently dispersed about the quintuple.

49. ξ (FL. 4[am]) Virginis sequens ad boream.

Feb. 6, 1782. Double. 1 full degree n. following ξ Virginis, in a line parallel to and β Leonis. A little unequal. L. pr.; S. dr. Distance 27″ 28‴. Position 56° 30′ s. preceding.

50. FL. 17 Virginis. In pectore.

Feb. 6, 1782. Double. Considerably unequal. L. w.; S. bluish. Distance 20″ 9‴. Position 58° 21′ n. preceding.

51. *k* Virginis :: FL. 44 :: .|| In ala austrina.

Feb. 6, 1782. Double. A star south of three forming an arch, and of the same magnitude with the middle one of the arch. Extremely unequal. L. w.; S. db. Distance 22″ 17‴; inaccurate. Position 32° 30′ n. following.

‡ The 21st and 20th Geminorum are not in the heavens as they are marked in FLAMSTEED's Atlas, so that it becomes doubtful whether the N° 21 is right. [It is 20 Geminorum.—ED.]

§ [IV. 48. Not 20″ 57‴ as in *Phil. Trans.*—ED.]

|| Perhaps the 45th; requires fixed instruments to determine. [It is 44 Virginis.—ED.]

IV. 52. * Cancri. FL. 48. In boreali forfice.

Feb. 8, 1782. Double. Considerably unequal. L. rw.; S. d. garnet. Distance 29″ 54‴. Position 39° 54′ n. preceding; a little inaccurate.

53. π Geminorum. FL. 80. Supra capita.

Feb. 9, 1782. Double. Excessively unequal. L. garnet; S. d. Distance with 460, 21″ 30‴. Position s. preceding. Other very small stars in view.

54. δ (FL. 4^am^) Hydræ sequens

Feb. 11, 1782. Double. About ½ degree following δ, towards ζ Hydræ. Pretty unequal. Both pr. S. deeper. Distance 25″ 43‴. Position 59° 24′ n. following.

55. FL. 41^am^ Lyncis sequens. In caudæ fine.

Mar. 5, 1782. Double. About 3½ minutes n. following the 41st Lyncis. Extremely unequal. L. r.; S. dr. Distance 15″ 52‴; a little inaccurate. Position 50° 48′ n. preceding; inaccurate.

56. FL. 18 Libræ.

April 3, 1782. Double. The following of two. Extremely unequal. L. r.; S. b. Distance 17″ 59‴. Position 44° 45′ n. following.

57. FL. 42^am^ Comæ Berenices sequens ad austrum.

April 15, 1782. Double. About 3 degrees s. following the 42d Comæ Berenices towards υ Bootis; the vertex of an isosceles triangle. Extremely unequal. Distance with 625, 16″ 42‴.‡ Position 46° 31′ s. preceding.

58. FL. 36^am^ Comæ Berenices præcedens ad boream.

April 18, 1782. A pretty double star. About 2½ degrees n. preceding the 36th, in a line parallel to the 42d and 15th Comæ Berenices; the following of two unequal stars. A little unequal. Both rw. Distance 15″ 52‴. Position 67° 57′ s. preceding.

59. Prope α Lyræ.

May 12, 1782. Double. About 2 or 3 minutes s. preceding α Lyræ. Very unequal. Both d. Distance with 278, 22″ 20‴. Position 33° 57′ n. preceding. Position of the largest with regard to α Lyræ 59° 12′ s. preceding.

60. FL. 4^am^ Ursæ majoris sequens ad boream.

June 6, 1782. Double. Near 1 degree n. following the 4th, in a line parallel to ο and *h* Ursæ majoris; a pretty large star. Extremely unequal. L. r.; S. d. Distance near 30″; but too obscure for measures.

61. ζ (FL. 7^a^) Coronæ australior et præcedens.

July 18, 1782. Double. Near ½ degree s. preceding ζ, towards η Coronæ bor. Nearly equal. Both pr. Distance 16″ 46‴. Position 4° 57′ n. following.

‡ [IV. 57. "Distance with 227, 17″ 23‴"—MS. It is = Σ 1737, H's place being only roughly determined.—ED.

IV. 62. τ (FL. 22[a]) Herculis australior et sequens.

Aug. 11, 1782. Double. About 2½ degrees s. following τ Herculis, in a line parallel to ι and γ Draconis; a considerable star. Very or extremely unequal. L. w.; S. br. Distance 16″ 51‴ Position 72° 15′ s. preceding.

63. FL. 42 Herculis. Dextrum supra genu.

Aug. 11, 1782. Double. Very unequal. L. r.; S. rw. Distance 21″ 31‴. Position 3° 42′ s. following.

64. Prope *q* (FL. 12[am]) Persei.

Aug. 20, 1782. Double. Within a few minutes of *q* Persei. Pretty unequal. Both pr.; but S. a little darker. Distance 21″ 59‴ Position 57° 57′ s. preceding.

65. Prope FL. 3[am] Cassiopeiæ.

Aug. 25, 1782. Double. Within 10 minutes of the 3d Cassiopeiæ.‡ Very unequal. L. pr.; S. r. Distance 20″ 46‴; very inaccurate. Position 41° 12′ s. preceding.

66. θ (FL. 33[am]) Cassiopeiæ præcedens.

Aug. 28, 1782. Double. About 1¾ degree s. of, and a little preceding θ, in a line from δ continued through θ Cassiopeiæ. Extremely unequal. L. r.; S. db. Distance 24″ 2‴; very inaccurate. Position 13° 12′ n. following; inaccurate.

67. † FL. 40 et 41 Draconis.

Aug. 29, 1782. Double. A little unequal. L. rw.; S. pr. Distance 20″ 39‴ mean measure; very accurate. Position 34° 27′ s. preceding.§ There is a third, much smaller star. Distance 3′ 16″ 33‴. Position about 30° s. following.

68. FL. 77 Piscium. In lini flexu.

Sept. 3, 1782. Double. A little unequal. L. wr.; S. pr. Distance 29″ 36‴. Position 4° 48′ n. following. In both measures the weather too windy for accuracy.

69. FL. 23[am] Andromedæ præcedens.

Sept. 4, 1782. Double. Full 1½ degree preceding the 23d, in a line parallel to ν and ι Andromedæ.|| Of two double stars in the finder the largest of the preceding set. Very unequal. L. r.; S. d. Distance with 278, 21″ 58‴ Position 70° 36′ n. preceding.

‡ [IV. 65. The star of reference is not 3 Cassiopeiæ but Cephei 288 = P. XXIII. 101 by subsequent MS. note. Pos. is sp by a diagram and not sf as in *Phil. Trans.*—ED.]

§ The proper motion of one of these stars at least since the time of FLAMSTEED is evident, as he gives us their difference in Æ 2′, and in PD 3′ 5″. Position s. preceding. Hence we have the hypotenuse or distance above 3′ 40″, instead of 20″ 39‴; and the angle 86° 17′ instead of 34° 27′. [35° 15′ in *Phil. Trans.* is due to an error of reduction.—ED.]

|| [IV. 69. Probably = Σ 3064 rej., which is 1° south. Nothing found by Burnham near H.'s place.—ED.]

V. 70. FL. 51 Piscium. In austrino lino.

Sept. 4, 1782. Double. Very unequal. L. rw.; S. d. Distance with 278, 22″ 29‴. Position 0° 36′ n. following.

71. * *o* Capricorni. FL. 12. Trium in rostro austrina.

Sept. 5, 1782. Double. Pretty unequal. Both rw. Distance 23″ 30‴. Position 30° 45′ s. preceding.

72. FL. 55[a] Persei borealior.

Sept. 7, 1782. Double. About ¼ degree n. of the 55th Persei; of three in a line the most north. Pretty unequal. L. rw.; S. pr. Distance with 278, 16″ 51‴. Position 27° 24′ n. following.

73. In Constellatione Camelopardali.

Sept. 7, 1782. Double. Between FL. 2 and 8 Cam.; the smallest of two that are within ¼ degree of each other. Considerably unequal. Distance 19″ 32‴. Position 85° 0′ s. preceding.

74. δ (FL. 68[am]) Tauri sequens ad boream.

Sept. 7, 1782. Double. Near ½ degree n. following δ, towards ι Tauri. Very unequal. L. pr.; S. r. Distance 16″ 31‴. Position 25° 45′ n. following.

75. γ (FL. 66[am]) Tauri sequens.

Sept. 7, 1782. Double. About 1¼ degree n. following γ, in a line parallel to μ Tauri and the 9th Orionis. Very unequal. L. r.; S. dr. Distance 22″ 35‴. Position 61° 36′ s. following.

76. FL. 13[am] Ceti præcedens ad austrum.

Sept. 9, 1782. Double. About 1 degree s. preceding the 13th, towards the 8th Ceti. Considerably unequal. L. rw.; S. br. Distance with 278, 18″ 35‴. Position 40° 24′ n. following.

77. FL. 37[a] Ceti borealior. In dorso.

Sept. 27, 1782. Double. About ¼ degree n. preceding the 37th, towards the 36th Ceti. Very unequal. L. r.; S. dr. Distance 19″ 36‴. Position 63° 24′ n. preceding.

78. η (FL. 3[am]) Cephei præcedens.

Sept. 27, 1782. Double. About 1¼ degree preceding η, in a line from ε continued through η Cephei. Very unequal. L. r.; S. d. Distance 19″ 32‴. Position 40° 36′ n. following.

79. Prope μ Cephei. FL. 13. Ad coronam.‡

Sept. 27, 1782. Double. A little unequal. L. w.; S. rw. Distance 21″ 13‴. Position 77° 48′ s. preceding.

‡ [IV. 79. Is not μ Cephei, as in *Phil. Trans.*, but B. 2866 sp it.—ED.]

IV. 80. β (FL. 2ª) Canis majoris borealior.

Sept. 30, 1782. Double. About 1¾ degree n. of β Canis majoris towards the 11th Monocerotis; the most n. of two. Considerably unequal. Distance 17″ 59‴; difficult to take, and perhaps a little inaccurate. Position 2° 24′ n. following.

81. ν Canis majoris. FL. 6. In dextro genu.

Sept. 30, 1782. Double. Considerably unequal. L. rw.; S. pr. Dist. 18″ 19‴. Position very near directly preceding.

82. Prope FL. 16ᵃᵐ Cephei. In cingulo.

Sept. 30, 1782. Double. Above ¾ degree following the 16th Cephei, in a line parallel to β and α Cassiopeiæ. Considerably unequal. L. orange. S. r. Distance 28″ 5‴. Position 79° 18′ n. preceding.

83. FL. 26 Ceti. Supra dorsum.

Oct. 2, 1782. Double. Very unequal. L. rw. S. db. Distance 17″ 2‴ mean measure. Position 14° 36′ s. preceding.

84. *m* Orionis. FL. 23. In crate pectoris.

Oct. 2, 1782. Double. Considerably unequal. L. w.; S. pr. Distance with 278, 26″ 9‴. Position 59° 33′ n. following.

85. FL. ultima Lacertæ.

Oct. 4, 1782. Treble. The two nearest extremely unequal. L. rw.; S. d. Distance 20″ 27‴. Position 79° 33′ n. preceding. The next very unequal; S. r. Distance 54″ 57‴; very inaccurate.‡ Position 44° 24′ n. following. A fourth and fifth star in view.

86. FL. 8 Lacertæ. In media cauda.

Oct. 4, 1782. Quadruple. The two largest and nearest a little unequal. Both rw. Distance 17″ 14‴. Position 84° 30′ s. preceding. The two next very unequal, of the fourth class. The two remaining considerably unequal, of the fifth class. They form an arch.

87. *e* (FL. 29ᵃᵐ) Orionis præcedens. In sinistro calcaneo.

Oct. 4, 1782. Double. About 1 degree preceding *e*, in a line parallel to σ Orionis and *h* Eridani nearly.§ Considerably unequal. Both pr. Distance 29″ 18‴. Position 82° 18′ n. following.

88. FL. 7 Tauri. In dorso.

Oct. 9, 1782. Double. Very unequal. L. pr.; S. dr. Distance 19″ 50‴. Position 23° 15′ n. following.

89. E telescopicis caudam Arietis sequentibus.

Oct. 9, 1782. Double. The vertex of an isosceles triangle following τ Arietis; a very small star. Very unequal. L. r.; S. d. Distance with 278, 20″ 3‴. Position 62° 0′ s. following.

‡ [IV. 85. Another measure the same evening 56″ 37‴, somewhat inaccurate. 1783, Apr. 8.—ED.]

§ [IV. 87 *h* Eridani, only mentioned in one obs., should be *b* Eridani.—ED.]

IV. 90. Ad FL. 18^{am} Ursæ minoris. Prope eductionem caudæ.

Oct. 12, 1782. Double. The largest of six or seven stars, and most south of a triangle formed by three of them. A little unequal. L. pr.; S. deeper pr. Distance 26″ 24‴. Position 3° 12′ n. following.

91. FL. 2 Navis.

Oct. 12, 1782. A pretty double star. A little unequal. L. w.; S. w. inclining to r. Distance 17″ 23‴. Position 69° 12′ n. preceding.

92. β inter et ζ Delphini.

Oct. 17, 1782. Treble. Between β and ζ, but nearer to β Delphini. All three nearly equal. All wr. Distance of the two nearest with 278, 21″ 33‴. Position 18° 27′ n. preceding.

93. ϵ (FL. 4^{am}) Lyræ sequens.

Oct. 19, 1782. Double. About 3 degrees following ϵ, in a line parallel to α and θ Lyræ; the largest of two. Extremely unequal. L. w.; S. r. Distance 19″ 50‴. Position 24° 0′ s. preceding.

94. E borealibus telescopicis β Lyræ præcedentibus.

Oct. 19, 1782. Double. Full 2 degrees n. preceding β Lyræ, in a line parallel to the 18th and ϵ; the sixth telescopic star. Considerably unequal. L. rw.; S. pr. Distance 22″ 53‴. Position 5° 24′ n. following.

95. FL. 25^{am} Monocerotis præcedens.

Oct. 19, 1782. Quadruple. About 2½ degrees preceding, and a little n. of the 25th Monocerotis.‡ Two large stars always to be seen, and two more only visible in dark nights. The smallest of the two large ones has an obscure star following; extremely unequal. Distance 20″ 27‴; very inaccurate.

96. FL. 25^{am} Monocerotis sequens. In latere.

Oct. 19, 1782. Double. About 1½ n. following the 25th, in a line parallel to the 21st Monocerotis and Procyon. A little unequal. Both dr. Distance 18″ 19‴ Position 24° 0′ s. preceding.

97. FL. 29 Monocerotis. In femore.

Oct. 19, 1782. Double. Extremely unequal. L. wr.; S. d. Distance 29″ 54‴. Position 15° 12′ s. following. Six more in view.

98. α (FL. 58^{am}) Orionis ad austrum præiens.

Oct. 29, 1782. Double. About ½ degree preceding α, towards ζ Orionis. Equal. Both r. Distance 17″ 59‴; a little inaccurate.

99. Duarum telescopicarum δ Sagittæ ad austrum sequentium borea.

Nov. 6, 1782. Treble. Of a trapezium, consisting of this treble star, δ, ζ, and the 9th Sagittæ, it is the corner opposite to ζ; the nearest to ζ of two. The two nearest very unequal. L. pr.; S. db. Distance 21″ 22‴; inaccurate. Posi-

‡ [IV. 95 is very probably = Σ 1084.—ED.]

tion 0° 0′ following. The two largest a little unequal; of the fifth class. Position 10° 36′ s. preceding.

IV. 100. χ Sagittæ. FL. 13. Infra mediam arundinem.

Nov. 6, 1782. Treble. The largest of three. The two nearest equal. Both r. Distance 23″ 2‴. Position 10° 12′ s. preceding. The third is a large star. Distance above 1 minute. Position about 10° or 15° n. preceding the other two.

101. φ (FL. 24[a]) Aurigæ borealior et præcedens.

Nov. 6, 1782. Double. Near ¾ degree n. preceding φ, in a line parallel to the 21st and 8th Aurigæ. Pretty unequal. L. rw. S. bluish. Distance 25″ 29‴. Position 76° 0′ n. preceding.

102. FL. 59 Aurigæ.

Nov. 6, 1782. Double. The apex of an isosceles triangle. Very or extremely unequal. L. rw.; S. . . . Distance 23″ 30‴. Position 50° 3′ s. preceding.

103. FL. 77[am] Draconis sequitur.

Nov. 13, 1782. Double. Near ¾ degree following the 77th Draconis, in a line parallel to κ Cephei and the 76th Draconis nearly; of a rectangular triangle the leg nearest the 77th. Very unequal. L. r.; S. bluish r. Distance 22″ 35‴. Position 45° 48′ n. following.

104. Inter γ et 55[am] Andromedæ.

Nov. 13, 1782. Double. A little more than 1 degree n. following the 55th Andromedæ, in a line parallel to β Trianguli and Algol. Considerably unequal. L. r.; S. d. Distance with 278, 18″ 57‴. Position 22° 33′ n. following.

105. δ Corvi. FL. 7. Duarum in ala sequente præcedens.

Nov. 13, 1782. Double. Extremely unequal. L. w.; S. r. Distance 23″ 30‴. Position 54° 0′ s. preceding.

106. α (FL. 50[am]) Ursæ majoris sequens ad boream.

Nov. 17, 1782. Double. About 1¾ degree n. following α, in a line parallel to β Ursæ and κ Draconis; the last of three in a row. Extremely unequal. Both r. Distance 18″ 55‴; very inaccurate. Position 44° 33′ s. following. A third small star in view.

107. FL. 79[a] Pegasi australior et præcedens.

Nov. 20, 1782. Double. About ¾ degree s. preceding the 79th, towards τ Pegasi; at the center of a trefoil. Very unequal. L. r.; S. d. Distance with 278, 26″ 12‴. Position 50° 21′ n. following.

108. FL. 69[a] Ursæ majoris australior.

Nov. 20, 1782. Double. Near 2 degrees s. of the 69th, towards the 63d Ursæ majoris. A very little unequal. Both r. Distance 19″ 15‴; very inaccurate. Position 10° 12′ n. following.

IV. 109. FL. 62 Tauri.

Nov. 25, 1782. Double. Considerably unequal. L. w.; S. r. Distance 28″ 5‴. Position 21° 12′ n. preceding.

110. β (FL. 112ª) Tauri borealior et sequens.

Dec. 24, 1782. Double. About 1¼ degree n. following β Tauri, towards θ Aurigæ; the second in that direction. Very unequal. L. r.; S. d. Distance 16″ 1‴. Position 74° 54′ n. preceding.

111. FL. 54 Cancri.

Dec. 28, 1782. Double. A little unequal. Both rw. S. a little darker. Distance 17″ 14‴. Position 29° 0′ s. following.

112. γ (FL. 15ᵃᵐ) Crateris sequens ad boream.

Jan. 1, 1783. Double. About 1 degree n. following γ Crateris, in a line parallel to δ Corvi and Spica. Equal. Both pr. Distance 26″ 15‴; too low for accuracy. Position 58° 42′ n. preceding or s. following.

113. FL. 61ª Cygni borealior et præcedens.

Jan. 6, 1783. Double. About 1¼ degree n. preceding the 61st, in a line parallel to υ and α Cygni. Very or extremely unequal. L. r.; S. db. Distance with 278, 17″ 30‴. Position 28° 24′ n. preceding. A third star in view.

114. *t* (FL. 12ª) Virginis australior.

Jan. 8, 1783. Double. About 1½ degree s. of *t* Virginis. Very unequal. L. pr.; S. d. Distance 23″ 21‴. Position 15° 54′ n. preceding.

115. φ (FL. 11ᵃᵐ) Herculis præcedens ad austrum.

Jan. 10, 1783. Double. About 2½ degrees s. of, and a little preceding φ, in a line parallel to η and ζ Herculis; the largest of three or four. Extremely unequal. L. r.; S. b. Distance 20″ 54‴. Position 43° 48′ n. following.

116*. FL. 83ᵃᵐ Pegasi sequens ad austrum.‡

Jan. 13, 1783. Double. Equal. Both w. Distance 28″ 59‴. Position 68° 21′ s. following. Mr. C. MAYER, in 1777, settled its place Æ 0ʰ. 52′ 53″ in time, and 20° 17′ 53″ in declination N.

117. FL. 42ª Eridani australior.

Jan. 31, 1783. Double. About 1¼ degree s. of the 42d Eridani, in a line parallel to Rigel and μ Leporis; the most south and following of three. Very unequal. L. r.; S. r. Distance 19″ 32‴. Position 31° 48′ s. preceding.

118. ι (FL. 48ᵃᵐ) Cancri sequens.

Feb. 5, 1783. Double. Full ½ degree following the 48th, in a line parallel to δ Cancri and ε Leonis; a very small star, next to two more which are nearer to ι. A little unequal. Distance 24″ 6‴. Position about 25° n. following.

‡ [Not " ad boream " as in *Phil. Trans.*; it is =IV. 9, ψ Piscium.—ED.]

IV. 119. ι (FL. 68am) Virginis præcedens ad austrum.

Feb. 7, 1783. Double. About 1 degree s. preceding the 68th, in a line parallel to the 99th and α Virginis. Extremely unequal. Distance 21″ 49‴. Position 36° 54′ n. preceding.

120. FL. 82am Piscium sequens ad boream.

Feb. 27, 1783. Double. About ¾ degree n. following the 82d Piscium, in a line parallel to α and β Trianguli; the largest of two. Considerably unequal. L. rw.; S. pr. Distance 18″ 19‴. Position 21° 0′ s. preceding. A third star in view.

121. σ Scorpii FL. 20 præcedens trium lucidarum in corpore.

Mar. 1, 1783. Double. Very unequal. L. whitish; S. r. Distance 21″ 40‴. Position 0° 0′ (or perhaps 1°) n. preceding.

122. FL. 32a Ophiuchi borealior et præcedens.

Mar. 7, 1783. Double. Near 1 degree n. of, and a little preceding the 32d Ophiuchi, in a line parallel to α and η Herculis. Very unequal. Distance 21″ 3‴. Position 25° 3′ s. preceding.

123. FL. 19 Ophiuchi.

Mar. 9, 1783. Double. The most south of two. Very unequal. L. pr.; S. d. Distance 20″ 27‴. Position 3° 9′ s. following.

124. ψ (FL. 4am) Ophiuchi præcedens ad austrum.

Mar. 24, 1783. Double. About ¾ degree preceding and a little s. of ψ, in a line parallel to ψ Ophiuchi and ω Scorpii; in the base of a triangle, the nearest to ψ. A little unequal. Both inclining to r. Distance 15″ 24‴. Position 62° 54′ n. following.

125. FL. 29 Camelopardali.

April 2, 1783. Double. Very unequal. L. pr.; S. d. Distance 22″ 26‴; very inaccurate. Position 47° 36′ s. following; a little inaccurate.

126. λ (FL. 22a) Cephei borealior et præcedens.

April 20, 1783. Double. Less than ½ degree n. preceding λ, in a line almost parallel to δ and ζ Cephei; a considerable star. A little unequal. Both dw. Distance 18″ 50‴. Position 45° 39′ n. preceding.

127†. λ (FL. 16am) Aquilæ sequens ad boream.

May 21, 1783. Double. About 2½ degrees n. following the farthest of two which are about 1½ degree from λ, in a line parallel to λ and δ Aquilæ. Very unequal. L. rw.; S. dr. Distance 17″ 14‴; more exact with 932, 15″ 52‴. Position 69° 54′ n. preceding. Mr. PIGOTT, who favoured me with it, gives its place Æ 18h 52′½±, Declination 1° 0′ S.

IV. 128. γ (FL. 57[am]) Andromedæ præcedens ad austrum.

July 28, 1783. Double. About 1⅓ degree s. preceding γ almost towards β Andromedæ; more exact towards σ Piscium; one not in a row of stars which are near that place. Considerably unequal. L. pr.; S. dr. Distance 15″ 42‴ Position 24° 12′ n. following.

129. FL. 59 Andromedæ.

July 28, 1783. Double. A little unequal. L. rw.; S. pr. Distance 15″ 15‴. Position 55° 9′ n. following. A third star in view about 58° or 60° s. preceding.

130. η (FL. 99[a]) Piscium borealior et sequens.

Aug. 2, 1783. Double. About 1½ degree n. of, and a little following η Piscium, in a line parallel to β Arietis and β Trianguli; the last of four in a crooked row. Very unequal. L. r.; S. darker r. Distance with 278, 15″ 49‴. Position 62° 15′ n. following.

131. FL. 100 Piscium.

Aug. 2, 1783. Double. Pretty unequal. L. pr.; S. r. Distance 15″ 52‴. Position 5° 0′ n. following.

132. FL. 46[am] Aquilæ sequens ad boream.

Aug. 6, 1783. Double. About ½ degree n. following 46 Aquilæ, in a line parallel to α and γ Sagittæ. Very unequal. L. r.; S. db. Distance 22″ 44‴. Position 42° 24′ n. preceding.

FIFTH CLASS OF DOUBLE STARS.

V. 52. FL. 15 Geminorum.

Dec. 27, 1781. Double. The second star from ν towards μ Geminorum. Pretty unequal. L. r.; S. b. Distance 35″; inaccurate.

53. *p* Geminorum. FL. 63. In inguine sequentis Π[i].

Dec. 27, 1781. Double. The brightest of two. Extremely unequal. L. pr; S. d. Distance 44″ 15‴.

54. θ Hydræ. FL. 22. In eductione cervicis.

Jan. 20, 1782. Double. Excessively unequal. L. w.; S. a point. Distance near 1 minute, too obscure for measures, and not visible till after having looked a good while at θ. Position about 75° s. following.

55. Ad. FL. 12[am] Geminorum. In pede Π[i] præcedentis sinistro.

Jan. 30, 1782. Treble. A small star near the place of the 12th Geminorum. The two nearest a little unequal. Distance less than 1′.‡

‡ [V. 55. There is a star 7·5 in or near the place indicated, with a companion 10 m in 59°·3, 62″·8 (Burnham 1903·8). But there are many wide pairs in the neighbourhood.—ED.]

V. 56. FL. 15 Geminorum. Dextrum prioris Πⁱ pedem attingens.‡

Jan. 30, 1782. Double. Considerably or very unequal. L. r.; S. d. Distance 32″ 39‴. Position near 60° s. preceding.

57. FL. 9ª Orionis borealior et sequens. In exuviarum summo.

Feb. 4, 1782. Treble. More than 1 degree n. following the 9th Orionis, towards the 113th Tauri; the largest of two. The two nearest considerably unequal. L. rw.; S. rw. Distance with 278, 36″ 26‴. Position 33° 36′ n. preceding. The farthest very unequal. S. r. Distance Vth Class. Position . . . following.

58. FL. 7 Leonis. Supra pedem borealem anteriorem.

Feb. 4, 1782. Double. Very unequal. L. rw.; S. r. Distance 42″ 25‴. Position 8° 36′ n. following.

59. θ Cancri. FL. 31. In quadrilatero circa Nubem.

Feb. 6, 1782. Double. Extremely unequal. L. r.; S. d. Distance 44″ 52‴. Position n. following.

60. o (FL. 95ᵃᵐ) Leonis præcedens; ad caudam.

Feb. 9, 1782. Double. Near ¾ degree s. preceding the 95th, in a line parallel to β and ρ Leonis. Very unequal. L. rw.; S. d. Distance 37″ 15‴. Position 70° 48′ n. following.

61. FL. 81 Leonis. In clune.

Feb. 9, 1782. Double. Extremely unequal. L. rw.; S. r. Distance 57″ 23‴.

62. FL. 57 Leonis. E posteriores pedes præcedentibus.§

Feb. 11, 1782. Double. Very unequal. Distance 33″ 16‴.

63. Prope FL. 25 Leonis. In infimo pectore.||

Feb. 17, 1782. Double. The largest of two. Extremely unequal. L. pr.; S. d. Distance 52″ 46‴.

64. FL. 43ª Leonis australior. Ad sinistrum anteriorem cubitum.

Feb. 17, 1782. Double. Near 1 degree s. of the 43d, in a line parallel to η and α Leonis. Extremely unequal. L. w. inclining to r.; S. db. Distance 59″ 40‴.

65. Secunda ad π Canis majoris. FL. 17. In pectore.

Mar. 3, 1782. Treble. The two nearest very unequal. L. rw.; S. r. Distance 44″ 52‴. Position 64° 12′ s. following. The two farthest very or extremely unequal. S. r. Distance Vth Class. Position about 85° s. preceding. The three stars form a rectangle, the hypotenuse of which contains the largest and smallest.

‡ [V. 56 is identical with V. 52.—ED.]

§ [V. 62 cannot be 57 Leonis, which has not any companion. Perhaps it is S. 617, Burnham 5590.—ED.]

|| [V. 63 is not =25 Leonis, as in *Phil. Trans.* An observation of 1795 Dec. 14 has "a * 1° 25′ s. of 27 Leonis in a line parallel to η and α." It is Leonis 91.—ED.]

V. 66. *p* (FL. 63[a]) Geminorum borealior.

Mar. 3, 1782. Double. About ¾ degree n. of, and a little preceding *p*, in a line parallel to *υ* and *α* Geminorum. Very unequal. L. pr.; S. d. Distance 34″ 39‴. Position 1° or 2° n. preceding.

67. Pollucem prope. In capite sequentis Π[i].

Mar. 3, 1782. Double. Near 1 degree n. following *β*, in a line from *δ* continued through *β* Geminorum nearly; the farthest and smallest of three. Considerably unequal. L. r.; S. dr. Distance 47″ 37‴.‡

68. FL. 75[am] Leonis præcedens ad boream.

Mar. 5, 1782. Treble. One of two n. preceding the 75th, in a line parallel to the 84th and 59th Leonis. The two nearest very unequal. Distance 54″ 37‴. The farthest extremely unequal.

69. FL. 7 Leonis minoris. In extremo anteriore pede.

Mar. 12, 1782. Double. The largest of two. Extremely unequal. L. pr.; S. r. Distance 58″ 18‴.

70. FL. 2[am] Bootis præcedens ad boream.

April 5, 1782. Double. Near 3 degrees n. preceding the 2d Bootis, towards the 43d Comæ Ber.; the preceding of three in a line parallel to *α* and *η* Bootis. A little unequal. L. r.; S. darker r. Distance 56″ 56‴. Position 7° 0′ s. preceding.

71. Prope *γ* (FL. 24[am]) Geminorum.

April 15, 1782. Double. Three or four minutes n. preceding *γ* Geminorum. Of the Vth Class. More in view.

72. † *m* Herculis. FL. 36 et 37. In sinistro Serpentarii brachio.

May 18, 1782. Double. A little unequal. L. bluish w. S. reddish w. Distance 59″ 59‴. Position 36° 57′ s. preceding.§

73. *τ* Ursæ majoris. FL. 14. Duarum in collo præcedens.

June 11, 1782. Double. Extremely unequal. L. w.; S. d. Distance 54″ 46‴. Position about 45° n. following.

74. S (FL. 72[a]) Serpentarii borealior.

June 16, 1782. Double. More than 1 degree n. following the 56th double star of the IIId Class; nearly in a line parallel to the 62d and 72d Serpentarii. Very unequal. L. rw; S. r. Distance 40″ 54‴. Position 39° 15′ s. following; inaccurate.

‡ [V. 67. Either the place or the distance (which is correctly reduced) is wrong, as Burnham only found S. 560 near the place, with distance 89″. H. adds: "Both small, requires a dark night." —ED.]

§ One of these stars, at least, seems to have changed its place since the time of FLAMSTEED, who makes their difference in R.A. 45″, and in P.D. 1′ 35″, Position s. preceding; hence we have the hypotenuse or distance above 1′ 45″, instead of 59″ 59‴, and position 69° 46′ instead of 36° 57′.

V. 75. E telescopicis ε Coronæ borealis sequentibus.

July 18, 1782. Double. About 1 degree s. following ε, in a line parallel to θ and ε Coronæ; the preceding of three forming an arch. Extremely unequal. L. r.; S. darker r. Distance 41″ 12‴. Position 16° 0′ s. following.

76. β Aquarii. FL. 22. In sinistro humero.

July 20, 1782. Double. Excessively unequal. L. w.; S. d. Distance about 33″ 16‴; very inaccurate. Position 55° 48′ n. preceding.

77. *d* (FL. 43[a]) Sagittarii borealior et sequens.

Aug. 4, 1782. Double. A few minutes n. following the 43d, in a line parallel to ο and π Sagittarii; the nearest of two. Extremely unequal. L. w.; S. d. Distance with 278, 36″ 3‴. Position 78° 45′ s. following.

78. ζ Sagittarii. FL. 38. Trium super costis sub axilla.

Aug. 4, 1782. Double. Extremely unequal. L. r.; S. d. Distance Vth Class. Position 28° 6′ n. preceding. A third star. Distance about four times as far as the former. Position also n. preceding.

79. FL. 9 :: Cassiopeiæ.

Aug. 25, 1782. Double. Of two in a line parallel to β and γ, that towards γ Cassiopeiæ.‡ Very unequal. L. w.; S. pr. Distance 52″ 39‴. Position 50° 36′ n. preceding.

80. τ Aquarii. FL. 69. Duarum in dextra tibia borealior.

Aug. 28, 1782. Double. Very unequal. L. rw.; S. d. Distance 36″ 47‴. Position 19° 54′ s. following.

81. FL. 35 :: Cassiopeiæ. In sinistro crure.

Aug. 28, 1782. Double. Considerably unequal. L. rw.; S. br. Distance 42″ 35‴. Position 85° 12′ n. following.

82. ν (FL. 25[am]) Cassiopeiæ præcedens. In sinistra manu.

Aug. 28, 1782. Double. Near ¼ degree n. preceding ν, in a line parallel to α and β Cassiopeiæ. Nearly equal. Both pr. Distance 43″ 26‴. Position 7° 48′ n. following.

83. ψ Cassiopeiæ. FL. 36. Sub pede sinistro.

Aug. 28, 1782. Double. Very unequal. L. pr.; S. r. Distance 33″ 25‴. Position 10° 12′ s. following.

84. FL. 47 :: Cassiopeiæ. Ex obscurioribus infra pedes.

Aug. 29, 1782. Double. The largest of three forming a rectangular triangle on, or near, the place of the 47th Cassiopeiæ.§ A little unequal. L. rw.; S. pr. Distance 50″ 58‴. Position 3° 33′ n. preceding.

‡ [V. 79. If this is 9 Cassiopeiæ, the companion would seem to be Smyth's B in Pos. 330°, Dist. 80″. Observed three times by H., but the description of the place occurs only once, April 7, 1783.—ED.]

§ [V. 84. Is no doubt = Σ C.P. 52 = S. 405, the place being very much in error.—ED.]

V. 85. ρ (FL. 27[a]) borealior et præcedens. In dextro brachio.

Aug. 29, 1782. Double. About $\frac{1}{3}$ degree n. preceding ρ Andromedæ θ versus. Very unequal. L. rw.; S. r. Distance 30″ 57‴. Position 79° 24′ n. following.

86. FL. 12 Ursæ minoris.

Sept. 4, 1782. Treble. Extremely unequal. All three r. The nearest is the smallest. Position some degrees s. following. The farthest also south, but more following.

87. σ Capricorni. FL. 7. Sub oculo dextro.

Sept. 5, 1782. Double. Very, or almost extremely unequal. L. r.; S. d. bluish. Distance 50″ 7‴. Position 85° 12′ s. following.

88. λ (FL. 15[a]) Aurigæ borealior. In sinistra manu.

Sept. 5, 1782. Double. About 3′ or 4′ n. following the 15th Aurigæ. Very unequal. Distance 34″ 15‴, mean measure. Position 54° 6′ s. preceding.

89. θ Aurigæ. FL. 37. In dextro carpo.

Sept. 5, 1782. Double. Excessively unequal. L. fine w.; S. reddish. Distance with 460, 35″ 18‴, narrow measure. Position 16° 0′ n. preceding. A third star in view.

90. ν Aurigæ. FL. 32. In dextri brachii ancone.

Sept. 5, 1782. Double. Excessively unequal. L. orange w.; S. r. Distance 53″ 43‴. Position 61° 48′ s. preceding. S. not visible till after some minutes' attention.

91. β (FL. 34[æ]) Aurigæ adjecta. In dextro humero.

Sept. 5, 1782. Double. Near $\frac{1}{2}$ degree s. following β, in a line from the 27th continued through β Aurigæ; a considerable star. Very or extremely unequal. L. pr.; S. d. Distance 30″ 3‴. Position 45° 6′ n. preceding.

92. FL. 4[a] Arietis australior.

Sept. 10, 1782. Double. Full $\frac{1}{2}$ degree s. following the 4th Arietis, in a line parallel to α Arietis and δ Ceti; the most south of two. Equal. Both reddish. Distance 51″ 16‴. Position 52° 45′ n. preceding or s. following.

93. FL. 103[am] Herculis sequens ad austrum.

Sept. 19, 1782. Double. About $1\frac{1}{4}$ degree s. following the 103d Herculis, in a line parallel to the 1st and 10th Lyræ; the nearest of two. Equal, perhaps the following the smallest. Both r. Distance 47″ 46‴. Position 45° 42′ s. following.

94. FL. 31[a] Cephei borealior.

Sept. 30, 1782. Double. About $\frac{3}{4}$ degree n. of the 31st Cephei, towards α Polaris. Pretty unequal. Both pr. Distance 41″ 40‴. Position 45° 15′ s. following.

V. 95. FL. 51 Aquarii. In dextro cubito.

Oct. 2, 1782. Double. Excessively unequal. L. rw.; S. d. Distance Vth Class. Position n. preceding. Two other stars in view; the nearest of them extremely unequal. Position about 80 or 90° s. preceding. The farthest very unequal. Position about 30° s. following.

96. υ (FL. 59[am]) Aquarii sequens ad austrum.

Oct. 2, 1782. Double. About ½ degree s. following υ, in a line parallel to δ and c Aquarii.‡ Extremely unequal. Distance Vth Class near. Position 15 or 20° s. preceding.

97. FL. 10 Lacertæ.

Oct. 4, 1782. Double. Very unequal. L. w.; S. r. Distance with 278, 52″ 34‴.§ Position 38° 45′ n. following.

98. FL. 3 Pegasi.

Oct. 4, 1782. Double. Pretty unequal. L. wr.; S. dr. Distance 34″ 43‴. Position 82° 48′ n. preceding. Besides II. 62, another star in view. Position . . . following.

99. FL. 33 Pegasi.

Oct. 4, 1782. Double. Considerably unequal. L. pr.; S. r. Distance with 278, 45″ 3‴. Position 89° 12′ n. following.

100. FL. 59 Orionis.

Oct. 4, 1782. Double. The following of two. Extremely unequal. L. w.; S. a point requiring some attention to be seen. Distance 37″ 15‴. Position about 65° s. preceding.

101. υ (FL. 36[am]) Orionis præcedens.

Oct. 4, 1782. Double. About ⅜ degree preceding υ, nearly in a line parallel to κ and β Orionis; the second from υ. Extremely unequal. L. w.; S. r. Distance 44″ 15‴. Position about 15° s. following.

102. FL. 61 Ceti.

Oct. 12, 1782. Double. Extremely unequal. L. rw.; S. dr. Distance with 278, 37″ 53‴. Position 76° 21′ s. preceding. A third star at some distance. A little unequal. Position n. following.

103. Ab ι (FL. 18[a]) Lyræ β versus.

Oct. 24, 1782. Double. Full ⅓ degree s. preceding ι, nearly towards β Lyræ. Extremely unequal. L. w.; S. r. Distance with 278, 45″ 32‴. Position 29° 12′ n. following.

‡ [V. 96 cannot be identified.—ED.]

§ [V. 97. Note to the distance: "Perhaps the measure is a little narrow." It ought to be 61″. Two other measures in April 1783 give 58″; they were afterwards suspected to belong to 12 Lacertæ (VI. 121), but without cause.—ED.]

V. 104. ε (FL. 4ª) Sagittæ australior et præcedens.

Nov. 6, 1782. Double. Full ½ degree s. preceding ε, in a line parallel to γ Sagittæ and γ Aquilæ; the nearest of two. Extremely unequal. L. pr.; S. d. Distance Vth Class. Position 16° 18′ s. following.

105. y (FL. 14ª) Sagittæ australior et sequens.

Nov. 6, 1782. Double. About ⅓ degree s. following y Sagittæ, in a line parallel to Sagitta and Delphinus. Considerably unequal. L. pr.; S. r. Distance 38″ 36‴. Position 74° 15′ s. following.

106. γ (FL. 12ª) Sagittæ borealior et præcedens.

Nov. 6, 1782. Double. About 1¾ degree n. preceding γ Sagittæ, towards the 6th Vulpeculæ; a considerable star. Equal. Both rw. Distance 38″ 54‴. Position 60° 42′ n. preceding or s. following.

107. FL. 56 Aurigæ.

Nov. 6, 1782. Double. Considerably unequal. L. w.; S. pr. Distance 52″ 57‴. Position 72° 36′ n. following.

108. κ (FL. 13ª) Canis majoris borealior.

Nov. 6, 1782. Double. About ¾ degree n. of κ Canis majoris. A little unequal. L. dw.; S. d. Distance 42″ 53‴. Position 23° 18′ n. following.

109. Inter β Cancri et δ Hydræ.

Nov. 6, 1782. Double. A large star not in FLAMSTEED, between β Cancri and δ Hydræ. Excessively unequal. Distance 35″ 24‴. Position 55° 0′ n. preceding.

110. FL. 111 Tauri.

Nov. 13, 1782. Double. Very unequal. L. rw.; S. r. Distance 46″ 42‴. Position 3° 48′ n. preceding.

111. FL. 42ª Ursæ majoris australior et sequens.

Nov. 20, 1782. Double. Full 1 degree s. following the 42d, in a line parallel to the 29th and 48th Ursæ majoris; the middle of three forming an arch. Considerably unequal. L. wr.; S. r. Distance 30″ 40‴. Position 51° 27′ n. following.

112. * Ex obscurioribus μ et ν Geminorum sequentibus.

Dec. 1, 1782. Double. Forms almost an isosceles triangle with μ and ν Geminorum. Nearly equal. The preceding pr. the following wr. Distance Vth Class far.

113. * FL. 9am inter et 11am Orionis.

Dec. 7, 1782. Treble. About 1½ degree s. preceding the 11th Orionis, towards ι Tauri. The two largest considerably unequal. L. w.; S. pr. Distance 37″ 51‴. Position 33° 54′ n. preceding. The third farther off and smaller. S. r. Position n. following.

V. 114. FL. 103 Tauri.

Dec. 7, 1782. Double. Excessively unequal. L. rw.; S. d. Distance with 278 and 625, 30″ 2‴, mean measure. Position 72° 24′.

115. *o* Tauri. FL. 114.

Dec. 7, 1782. Double. Excessively unequal. L. w.; S. a point. Distance 51″ 34‴.‡ Position 77° 54′ s. preceding. Two other small stars following, and a third to the north.

116. FL. 41 Arietis.

Dec. 23, 1782. Treble. The two nearest excessively unequal. L. w.; S. a point. Distance with 278, 39″ 20‴. Position 80° 48′ s. preceding. For the distance of the farthest, see VI. 5.

117. ζ (FL. 58[am]) Arietis præcedens ad boream.

Dec. 23, 1782. Double. About 1¾° n. preceding ζ, towards the 41st Arietis; the following of four forming an arch. Very unequal. Both dr. Distance 34″ 48‴. Position 47° 33′ n. preceding.

118. ε (FL. 46[a]) Orionis borealior et præcedens.

Dec. 28, 1782. Double. The most n. of three preceding ε Orionis, towards μ Tauri. More north is another set of three; care must be taken not to mistake one of them for this. Extremely unequal. L. rw.; S. d. Distance Vth Class. Position 13° 6′ s. preceding. Two more following, excessively unequal; one about 1′, the other about 1½ minute.

119. ε (FL. 46[a]) Orionis australior et præcedens.

Dec. 28, 1782. Double. Full ¾ degree s. preceding ε, in a line parallel to ε Orionis and *h* Eridani; the smallest and most s. of two.§ Very unequal. L. w.; S. r. Distance 30″ 12‴; a little inaccurate. Position 21° 33′ s. preceding. A third star 2 or 3° s. following.

120. FL. 15 Hydræ.

Dec, 28. 1782. Double. Extremely unequal. L. w.; S. r. Distance 43″ 2‴. Position about 70° n. preceding.

121. FL. 12 Comæ Berenices.

Jan. 1, 1783. Double. Considerably unequal. L. rw.; S. pr. Distance 58″ 55‴. Position about 77° s. following.

122. FL. 44[a] Bootis australior et præcedens.

Jan. 8, 1783. Double. Near ⅔ degree s. preceding the 44th, towards the 38th Bootis. Very unequal. L. bw.; S. pr. Distance 34″ 21‴. Position 67° 6′ s. following.

‡ [V. 115. Distance is marked: " Central measure, inaccurate."—ED.]

§ [V. 119. Not *h*, but *b* Eridani, Fl. 62, misnamed *h* on Flamsteed's Atlas. The double star is 30′ p and 30′ s of ε Orionis.—ED.]

V. 123. * In Andromedæ pectore.

Jan. 8, 1783. Double. Equal. Both rw. or pr. Distance 45″ 1‴. Position 32° 24′ s. preceding. Its place, as determined in 1777 by C. MAYER, is Æ 0h 34′ 33″ in time, and 29° 45′ 3″ declination north.

124. *g* (FL. 2am) Centauri sequens ad austrum.

Jan. 31, 1783. Double. About 1½ degree s. following *g* Centauri, in a line parallel to γ Serpentis and θ Centauri; the most s. of two. Considerably unequal. Distance 54″ 1‴; too low for accuracy.

125. FL. 46am Bootis sequens ad boream.

Feb. 3, 1783. Double. Near 2 degrees n. following the 46th, in a line parallel to ζ Bootis and β Coronæ; the third star about that direction. Considerably unequal. L. r.; S. darker r. Distance 33″ 53‴. Position 35° 33′ s. preceding.

126. *r* (FL. 5am) Herculis præcedens ad austrum.

Feb. 3, 1783. Double. Near ½ degree s. preceding *r* Herculis, in a line parallel to γ and δ Serpentis; a small star. A little unequal. Both pr. Distance 37″ 51‴, rather full measure. Position 52° 6′ s. preceding.

127. (FL. 41am) Herculis præcedens ad boream.

Feb. 5, 1783. Double. About ¾ degree n. preceding the 41st Herculis, in a line parallel to κ Serpentarii and β Herculis. Pretty unequal. Both r. Distance 48″ 40‴. Position 19° 45′ n. preceding.

128. ι (FL. 68am) Virginis sequens.

Feb. 7, 1783. Double. About 1½ degree following ι Virginis, in a line parallel to Spica and β Libræ. A little unequal. L. pr.; S. r. Distance 41″ 58‴.

129. *f* (FL. 25am) Virginis sequens ad boream.

Feb. 7, 1783. Double. About 1½ degree n. following *f*, in a line parallel to γ and ε Virginis; a large star. Very unequal. L. r.; S. dark r. Distance 46″ 42‴. Position 6 or 7° s. following. A double star of the Vth Class in view, preceding.

130. FL. 35 Comæ Berenices.

Feb. 26, 1783. Double. Very unequal. L. r.; S. d. Distance 31″ 17‴. Position 36° 51′ s. following.

131. FL. 24am Libræ sequens ad boream.

Mar. 1, 1783. Double. About 1½ degree n. following the 24th Libræ, in a line parallel to π and β Scorpii. Considerably unequal. L. rw.; S. r. Distance 47″ 46‴.

132. FL. 29am inter et 30am Libræ.

Mar. 1, 1783. Double. Of two between the 29th and 30th Libræ that nearest to the 30th. Very unequal. L. w.; S. d. Distance 39″ 59‴; very inaccurate.

V. 133. FL. 60 Herculis.

Mar. 7, 1783. Double. Extremely unequal. L. w.; S. d. Distance 48″ 40‴. Position 37° 0′ n. preceding.

134. ψ (FL. 4[am]) Ophiuchi præcedens ad austrum.

Mar. 24, 1783. Double. About 1 degree preceding and a little s. of ψ, in a line parallel to ψ Ophiuchi and ω Scorpii; the farthest of two in the base of a triangle. Equal. Distance 45″ 47‴.

135. Ad FL. 49[am] Camelopardali.

April 4, 1783. Double. The smallest and most s. of two that are about 20′ asunder. A little unequal. Both r. Distance with 278, 38″ 18‴.‡ Position 85° 0′ s. preceding.

136. θ (FL. 65[a]) Aquilæ borealior.

Sept. 12, 1783. Double. About $\frac{3}{8}$ degree n. of θ, in a line parallel to η and β Aquilæ; a considerable star. Considerably unequal. L. pr.; S. r. Distance with 278, 47″ 5‴. Position 65° 48′ s. preceding.

137. χ (FL. 17[a]) Cygni borealior.

Sept. 22, 1783. Double. About $1\frac{1}{3}$ degree n. of χ, towards δ Cygni; a considerable star. Considerably unequal. L. garnet; S. r. Distance with 278, 35″ 1‴. Position 57° 3′ n. following.

SIXTH CLASS OF DOUBLE STARS.

VI. 67. η Orionis. FL. 28. In extremo ensis manubrio.

Dec. 27, 1781. Double. Excessively unequal. L. w.; S. d. Distance 1′ 50″ 57‴. Position 35° 12′ n. following.

68. η (FL. 28[a]) Orionis australior.

Dec. 27, 1781. Double. About $\frac{1}{2}$ degree s. of, and a little following η, in a line nearly parallel to δ and θ Orionis. Very unequal. L. r.; S. d. Distance 2′ 0″ 11‴. Position 7° 54′ n. preceding.

69. FL. 14 Arietis. Supra caput.

Dec. 27, 1781. Double. Very unequal. L. pr.; S. dr. Distance 1′ 29″ 28‴. Position 11° 12′ n. preceding.

70. FL. 70 Geminorum. Supra caput prioris Π[1].

Dec. 27, 1781. Treble. Or two small stars in view; the nearest a little more than 1 minute; the other not much farther.

‡ [V. 135. There are two measures of the distance. 1783 April 4, equal, 4th class. April 9, 14″ 47‴ perhaps a little inaccurate. Sept. 26, 38″ 18‴ 5th class, with the note appended: "The observation of page 371 [Apr. 9] and this observation do not agree, nor can I surmise what could occasion the mistake of 371." It is, however, the first measure which is correct, the distance being still 15″.—ED.]

VI. 71. τ Hydræ. FL. 31. Trium in flexu colli australissima.

Jan. 20, 1782. Double. Pretty unequal. L. w. inclining to rose colour. S. pr. Distance 1′ 1″ 40‴. Position 88° 36′ n. preceding.

72. Ad FL. 68am Orionis. In fuste.

Jan. 30, 1782. Double. The most n. of two that are 1 degree asunder. Very unequal. L. w.; S. dr. Distance with 278, 1′ 12″ 50‴. Position 41° 0′ s. preceding.

73. ε Geminorum. FL. 27. In boreali genu præcedentis IIi.

Feb. 2, 1782. Double. L. w. Distance 1′ 50″ 30‴.

74. FL. 51 Geminorum.

Feb. 2, 1782. Has two very obscure stars in view. L. r.; S. r. S. r. The nearest about 1½, the next 2 minutes. Position of both about 40 or 50° n. following.

75. ω Cancri. FL. 4. Ad primum borealem forficem.

Feb. 2, 1782. Has a very obscure star in view. L. pr. Distance about 1¼ minute. Position about 30° n. preceding. A third about 2′. Position more north.

76. ο Leonis. FL. 14.

Feb. 2, 1782. Double. Extremely unequal. L. rw.; S. r. Distance 1′ 3″ 29‴. Position 49° 36′ n. following.

77. τ Virginis. FL. 93.

Feb. 4, 1782. Double. Very unequal. L. w.; S. dr. Distance 1′ 8″ 22‴.

78. ζ (FL. 16am) Cancri sequitur.

Feb. 8, 1782. Double. About ½ degree following ζ Cancri, towards η Leonis. Extremely unequal. Distance 1′ 3″ 47‴.

79. φ Leonis. FL. 74.

Feb. 9, 1782. Double. Very unequal. L. w.; S. pr. Distance 1′ 38″ 35‴ Position about 10 or 12° n. preceding.

80. FL. 93 Leonis.

Feb. 9, 1782. Double. Very unequal. L. w.; S. db. Distance 1′ 10″ 13‴.

81. FL. 27 Virginis. In ala dextra.

Feb. 9, 1782. Double. Extremely unequal. L. w. Distance 1′ 28″ 48‴.

82. FL. 31 Monocerotis. In media cauda.

Feb. 9, 1782. Double. Very unequal. L. rw.; S. db. Distance 1′ 10″ 13‴. Position 40° 0′ n. preceding.

83. Prope FL. 1am Orionis.

Feb. 9, 1782. Double. A few minutes s. following the 1st, towards the belt of Orion. Considerably unequal. L. pr; S. r. Distance 1′ 20″ 58‴. Position 88° 15′ n. following.

VI. 84. FL. 14 Canis minoris.

Feb. 9, 1782. Treble. The nearest extremely unequal. L. rw.; S. d. Distance 1′ 5″ 28‴. Position 26° 24′ n. following. The third forms an angle, a little larger than a rectangle, with the other two. Position s. following.

85. FL. 27 Hydræ.

Feb. 9. 1782. Double. Very unequal. L. rw.; S. pr. Distance VIth Class far. Position about 60° s. preceding.

86. Prima ad σ Cancri. FL. 51.

March 5, 1782. Double. Extremely unequal. L. w.; S. d. Position n. following.

87. Tertia ad σ Cancri. FL. 64.

March 5, 1782. Double. Very unequal. L. rw.; S. dr. Distance 1′ 25″ 45‴. Position 25° 12′ n. preceding.

88. β Aurigæ. FL. 34. In dextro humero.

March 5, 1782. Double. Extremely or excessively unequal. L. fine bluish w.; S. d. Distance 2′ 49″ 6‴. Position 54° 12′ n. following. A third farther off. Very unequal. About 40 or 50° n. following.

89. FL. 6æ Bootis adjecta.

Mar. 12, 1782. Double. Just following the 6th Bootis.‡ A little unequal. L. r.; S. deeper r. Distance 1′ 19″ 39‴. Position 58° 6′ s. preceding.

90. FL. 61 Virginis.

Apr. 3, 1782. Double. Very unequal. L. w.; S. d. Distance 1′ 13″ 15‴. Position about 75° n. preceding.

91. Prope γ (FL. 24am) Geminorum.

Apr. 15, 1782. Double. Three or four minutes n. of γ Geminorum. Considerably unequal. Both small; too obscure for measures with 7-feet; my 20-feet shews a third star between them with 12 inches aperture.

92. ξ (FL. 1A) Capricorni borealior.

June 14, 1782. Double. About ⅓ degree n. of ξ Capricorni. Very unequal. Both r. Distance 1′ 2″ 16‴.§ Position 2° 3′ s. preceding.

93. ρ Coronæ borealis. FL. 15. Ad summum.

July 18, 1782. Double. Very unequal. L. w.; S. d. Distance 1′ 27″ 44‴; a little inaccurate. Position 54° 27′ s. following.

94. λ Coronæ borealis. FL. 12.

July 18, 1782. Double. Extremely unequal. L. w.; S. r. Distance 1′ 35″ 14‴. Position 33° 12′ n. following.

‡ [VI. 89 = VI. 15. Nothing else near the place according to Burnham.—ED.]

§ [VI. 92. Note to distance measure: " but very inaccurate on account of obsc." [*sic*].—ED.]

VI. 95. *η* Bootis. FL. 8. Trium in sinistro crure borea.

Aug. 3, 1782. Double. Extremely unequal. L. w. inclining to orange; S. r. Distance about 1½ minute. Position about 25 or 30° s. following.

96. *ζ* Persei. FL. 44. In pede sinistro.

Aug. 25, 1782. Treble. The nearest extremely unequal. L. w.; S. r. Distance 1′ 11″ 26‴. Position 66° 36′ s. preceding. The farthest very unequal. S. r. about 1½ minute. 70 or 75° s. preceding.

97. Secunda ad *τ* Aquarii. FL. 71. In dextro crure.

Aug. 28, 1782. Double. Very unequal. L. r.; S. d. Distance 2′ 3″ 36‴, mean measure. Position 18° 30′ n. preceding.

98. FL. 46am Tauri sequens ad austrum.

Sept. 7, 1782. Double. About 1½ degree s. following the 46th, nearly in a line parallel to the 38th Tauri and the 42d Eridani. A little unequal. L. pr.; S. r. Distance 1′ 2″ 34‴. Position 43° 48′ n. preceding. A double star of the Vth Class in view, following within 3′. Equal. Both small and r. Almost similarly situated with the above, but position more n. preceding.

99. *m* Persei. FL. 57. In dextri pedis talo.

Sept. 7, 1782. Double. Pretty unequal. L. r.; S. rw. Distance 1′ 36″ 27‴. Position 71° 51′ s. preceding.

100. *ι* (FL. 32am) Cephei sequens.

Sept. 30, 1782. Double. About 1¾ degree n. following *ι*, nearly towards *γ* Cephei. A little unequal. Both pr. Distance 1′ 1″ 54‴. Position 8° 9′ n. preceding.

101. *δ* Tauri. FL. 68.

Oct. 31, 1782. Has two stars in view. The nearest excessively unequal. L. w.; S. d. Distance with 278, 1′ 3″ 18‴. Position 35° 24′ s. preceding. The farthest extremely unequal. S. r. About 1½ minute. Position about 50° n. preceding.

102. FL. 5 Lyncis.

Nov. 13, 1782. Double. The largest of a small triangle. Very unequal. L. r.; S. garnet. Distance 1′ 28″ 20‴. Position 2° 0′ n. preceding.

103. *ε* Pegasi. FL. 8.

Nov. 20, 1782. Double. Very unequal. L. pr.; S. dr. Distance 1′ 30″ 56‴. Position 52° 45′ n. preceding.

104. *ζ* Bootis. FL. 30. In dextro calcaneo.

Nov. 29, 1782. Has a very obscure star in view. Extremely unequal. L. w. inclining to r.; S. d. Distance about 1½ minute. Position almost directly preceding.

VI. 105. FL. 105 Tauri.

Dec. 7, 1782. Double. Very unequal. L. pr.; S. r. Distance 1′ 41″ 29‴.* Position 18° 0′ s. preceding.

106. *b* Eridani. FL. 62.

Dec. 7, 1782. Double. Considerably unequal. L. w.; S. pr. Distance 1′ 0″ 26‴. Position 15° 9′ n. following.

107. FL. 31[a] Monocerotis australior et præcedens.

Dec. 21, 1782. Double. About 1¼ degree s. of, and a little preceding the 31st Monocerotis, in a line parallel to ζ Hydræ and the 31st Monocerotis; the most south of two. Considerably unequal. L. r.; S. deeper r. Distance about 1½ minute. Position 50 or 60° s. following.

108. θ (FL. 22[a]) Hydræ borealior et præcedens.

Dec. 28, 1782. Double. About ½ degree n. of, and a little preceding θ, nearly in a line parallel to α and θ Hydræ. Very unequal. L. r.; S. blackish r. VIth Class far. Position 1 or 2° n. preceding. A third star preceding.

109. φ Cancri. FL. 22.

Dec. 29, 1782. Double. One of the two being lost,‡ it does not appear which is the remaining star. Very unequal. L. r.; S. dr.

110. Telescopica ad *o* Ceti.

Jan. 2, 1783. Double. Looking for *o* Ceti, which was invisible to the naked eye, I mistook this for it.§ Pretty unequal. L. rw. of about the eighth magnitude; S. r. Distance 1′ 20″ 52‴. Position 33° 42′ s. following.

111. α Hydræ. FL. 30. Duarum contiguarum lucidior.

Jan. 8, 1783. Has two stars within about 2 minutes; the nearest excessively unequal; the farthest extremely unequal. Both s. following.

112. FL. 13 Bootis.

Jan. 8, 1783. Double. Extremely unequal. L. r.; S. dr. Distance 1′ 17″ 58‴. Position 7° 24′ n. preceding.

113. FL. 4 Virginis.

Jan. 8. 1783. Double. Extremely unequal. L. wr.; S. dr. Distance 2′ 25″ 44‴; too obscure for accuracy.

114. FL. 69[am] Orionis præcedens ad austrum.

Jan. 9, 1783. Double. About ½ degree s. preceding the 69th, nearly towards λ Orionis. Considerably unequal. L. pr.; S. d. Distance 1′ 30″ 38‴. Position 22° 6′ s. following.

‡ See *Phil. Trans.* vol. LXXIII. p. 252. [Above, p. 111.—ED.]

§ [V. 110. Identified by Burnham as S.D. −3°, 340–341.—ED.]

VI. 115. FL. 21^{am} Crateris sequens ad austrum.

Jan. 10, 1783. Double. About 2½ degree s. following the 21st, in a line parallel to the 12th Crateris and 4th Corvi. Very unequal. L. w.; S. r. Position 12° 12′ n. following.

116. FL. 43 Herculis.

Jan. 10, 1783. Double. Very unequal. L. inclining to garnet; S. r. Distance 1′ 14″ 37‴. Position 38° 48′ s. preceding.

117. FL. 12^{a} Libræ borealior et præcedens.

Jan. 10, 1783. Double. About 1¼ degree n. preceding the 12th Libræ, towards Spica. Very unequal. L. rw.; S. r. Position about 40° s. preceding.

118. FL. 30 Monocerotis.

Feb. 12, 1783. Double. Very or extremely unequal. Distance 3′ 30″ 54‴.‡

119. ε (FL. 18^{a}) Piscis austrini australior et præcedens.

July 28, 1783. Double. About 1¼ degree s. of, and a little preceding ε Piscis austrini, in a line from δ Aquarii continued through ε Piscis. Pretty unequal. L. dpr. S. dr. Distance 1′ 26″ 58‴. Position 67° 46′ s. following.§

120. FL. 43^{am} Sagittarii sequens ad austrum.

Aug. 16, 1783. Double. Near 1 degree s. following the 43d, in a line parallel to ξ and o Sagittarii; a considerable star. Very unequal. Both dr. Distance with 278, 1′ 14″ 9‴. Position 37° 0′ n. preceding.‖

121. FL. 12 Lacertæ.

Aug. 18, 1783. Double. Very unequal. L. w.; S. r. Distance with 278, 1′ 0″ 10‴. Position 73° 0′ n. following.

‡ On account of the change in the magnitudes of the 1st and 2d Hydræ, this small star may be of use to ascertain whether the 30th Monocerotis, which is situated between them, has any considerable proper motion. See *Phil. Trans.* vol LXXIII. p. 255 [above, p. 113].

[According to Burnham, 30 Monocerotis has no companion, close or distant. He believes VI. 118 to be 2 Hydræ, 40^{s} f, 5′ s, with a ∗ 10 in 3°·1, 72″·1 (1903·2). Both in the Journal and in the *Phil. Trans.* the distance is 3′ 30″ 54‴, which in J. H.'s Synopsis has been altered to 1′ 30″·9 without any explanation. The Journal has 5R11 +34½ $^{3\ \frac{3}{27}\ \frac{8}{28}}_{\ \ \ \ \ 18}$. One rev. =36″ 38‴.—Ed.]

§ [VI. 119. The following remarks are found in the Journal:—

July 28, 1783. I partly suspect the most south to be double of the 1st class [crossed out by a very thick line].

July 30, 1783. Cannot be verified to be of the 1st class on account of prismatic power of atmosphere. It would be worth while to go to the cape of good hope to view the star there.

The close pair was finally found on the 25th October 1797. It is =N 117 =h 5356, rediscovered by John Herschel at the Cape!—Ed.]

‖ [VI. 120. Error of reduction in the Pos. Angle. It should be 49° 0′ np (J. F. W. H.)].

XV.

On the Construction of the Heavens.

[*Phil. Trans.*, vol. lxxv., 1785, pp. 213–266.]

Read February 3, 1785.

THE subject of the Construction of the Heavens, on which I have so lately ventured to deliver my thoughts to this Society, is of so extensive and important a nature, that we cannot exert too much attention in our endeavours to throw all possible light upon it; I shall, therefore, now attempt to pursue the delineations of which a faint outline was begun in my former paper.

By continuing to observe the heavens with my last constructed, and since that time much improved instrument, I am now enabled to bring more confirmation to several parts that were before but weakly supported, and also to offer a few still further extended hints, such as they present themselves to my present view. But first let me mention that, if we would hope to make any progress in an investigation of this delicate nature, we ought to avoid two opposite extremes, of which I can hardly say which is the most dangerous. If we indulge a fanciful imagination and build worlds of our own, we must not wonder at our going wide from the path of truth and nature; but these will vanish like the Cartesian vortices, that soon gave way when better theories were offered. On the other hand, if we add observation to observation, without attempting to draw not only certain conclusions, but also conjectural views from them, we offend against the very end for which only observations ought to be made. I will endeavour to keep a proper medium; but if I should deviate from that, I could wish not to fall into the latter error.

That the milky way is a most extensive stratum of stars of various sizes admits no longer of the least doubt; and that our sun is actually one of the heavenly bodies belonging to it is as evident. I have now viewed and gaged this shining zone in almost every direction, and find it composed of stars whose number, by the account of these gages, constantly increases and decreases in proportion to its apparent brightness to the naked eye. But in order to develop the ideas of the universe, that have been suggested by my late observations, it will be best to take the subject from a point of view at a considerable distance both of space and of time.

Theoretical view.

Let us then suppose numberless stars of various sizes, scattered over an indefinite portion of space in such a manner as to be almost equally distributed throughout the whole. The laws of attraction, which no doubt extend to the remotest regions of the fixed stars, will operate in such a manner as most probably to produce the following remarkable effects.

Formation of nebulæ.

Form I. In the first place, since we have supposed the stars to be of various sizes, it will frequently happen that a star, being considerably larger than its neighbouring ones, will attract them more than they will be attracted by others that are immediately around them; by which means they will be, in time, as it were, condensed about a center; or, in other words, form themselves into a cluster of stars of almost a globular figure, more or less regularly so, according to the size and original distance of the surrounding stars. The perturbations of these mutual attractions must undoubtedly be very intricate, as we may easily comprehend by considering what Sir ISAAC NEWTON says in the first book of his *Principia*, in the 38th and following problems; but in order to apply this great author's reasoning of bodies moving in ellipses to such as are here, for a while, supposed to have no other motion than what their mutual gravity has imparted to them, we must suppose the conjugate axes of these ellipses indefinitely diminished, whereby the ellipses will become straight lines.

Form II. The next case, which will also happen almost as frequently as the former, is where a few stars, though not superior in size to the rest, may chance to be rather nearer each other than the surrounding ones; for here also will be formed a prevailing attraction in the combined center of gravity of them all, which will occasion the neighbouring stars to draw together; not indeed so as to form a regular or globular figure, but however in such a manner as to be condensed towards the common center of gravity of the whole irregular cluster. And this construction admits of the utmost variety of shapes, according to the number and situation of the stars which first gave rise to the condensation of the rest.

Form III. From the composition and repeated conjunction of both the foregoing forms, a third may be derived, when many large stars, or combined small ones, are situated in long extended, regular, or crooked rows, hooks, or branches; for they will also draw the surrounding ones, so as to produce figures of condensed stars coarsely similar to the former which gave rise to these condensations.

Form IV. We may likewise admit of still more extensive combinations; when, at the same time that a cluster of stars is forming in one part of space, there may be another collecting in a different, but perhaps not far distant quarter, which may occasion a mutual approach towards their common center of gravity.

V. In the last place, as a natural consequence of the former cases, there will be formed great cavities or vacancies by the retreat of the stars towards the various centers which attract them; so that upon the whole there is evidently a field of the greatest variety for the mutual and combined attractions of the heavenly bodies to exert themselves in. I shall, therefore, without extending myself farther upon this subject, proceed to a few considerations, that will naturally occur to every one who may view this subject in the light I have here done.

Objections considered.

At first sight then it will seem as if a system, such as it has been displayed in the foregoing paragraphs, would evidently tend to a general destruction, by the shock of one star's falling upon another. It would here be a sufficient answer to say, that if observation should prove this really to be the system of the universe, there is no doubt but that the great Author of it has amply provided for the preservation of the whole, though it should not appear to us in what manner this is effected. But I shall moreover point out several circumstances that do manifestly tend to a general preservation; as, in the first place, the indefinite extent of the sidereal heavens, which must produce a balance that will effectually secure all the great parts of the whole from approaching to each other. There remains then only to see how the particular stars belonging to separate clusters will be preserved from rushing on to their centers of attraction. And here I must observe, that though I have before, by way of rendering the case more simple, considered the stars as being originally at rest, I intended not to exclude projectile forces; and the admission of them will prove such a barrier against the seeming destructive power of attraction as to secure from it all the stars belonging to a cluster, if not for ever, at least for millions of ages. Besides, we ought perhaps to look upon such clusters, and the destruction of now and then a star, in some thousands of ages, as perhaps the very means by which the whole is preserved and renewed. These clusters may be the *Laboratories* of the universe, if I may so express myself, wherein the most salutary remedies for the decay of the whole are prepared.

Optical appearances.

From this theoretical view of the heavens, which has been taken, as we observed, from a point not less distant in time than in space, we will now retreat to our own retired station, in one of the planets attending a star in its great combination with numberless others; and in order to investigate what will be the appearances from this contracted situation, let us begin with the naked eye. The stars of the first magnitude being in all probability the nearest, will furnish us with a step to begin our scale; setting off, therefore, with the distance of Sirius or Arcturus, for instance, as unity, we will at present suppose, that those of the second magnitude are at double, and those of the third at treble the distance, and so forth. It is not necessary critically to examine what quantity of light or magnitude of a star intitles it to be estimated of such or such a proportional distance, as the common coarse estimation will answer our present purpose as well; taking it then for granted, that a star of the seventh magnitude is about seven times as far as one of the first, it follows, that an observer, who is inclosed in a globular cluster of stars, and not far from the center, will never be able, with the naked eye, to see to the end of it: for, since, according to the above estimations, he can only extend his view to about seven times the distance of Sirius, it cannot be expected that his eyes should reach the borders of a cluster which has perhaps not less than fifty stars in depth every where around him. The whole universe, therefore, to him will be comprised in a set of constellations, richly ornamented with scattered stars of all sizes. Or if the united brightness of a neighbouring cluster of stars should, in a remarkable clear night, reach his sight, it will put on the appearance of a small, faint, whitish, nebulous cloud, not to be perceived without the greatest attention. To pass by other situations, let him be placed in a much extended stratum, or branching cluster of millions of stars, such as may fall under the III[d] form of nebulæ considered in a foregoing paragraph. Here also the heavens will not only be richly scattered over with brilliant constellations, but a shining zone or milky way will be perceived to surround the whole sphere of the heavens, owing to the combined light of those stars which are too small, that is, too remote to be seen. Our observer's sight will be so confined, that he will imagine this single collection of stars, of which he does not even perceive the thousandth part, to be the whole contents of the heavens. Allowing him now the use of a common telescope, he begins to suspect that all the milkiness of the bright path which surrounds the sphere may be owing to stars. He perceives a few clusters of them in various parts of the heavens, and finds also that there are a kind of nebulous patches; but still his views are not extended so far as to reach to the end of the stratum in which he is situated, so that he looks upon these patches as belonging to that system which to him seems to comprehend every celestial object. He now increases his power of vision, and, applying himself to a close observation, finds that the milky way is indeed no other than a collection of very small stars. He perceives that those objects which had been called nebulæ are evidently nothing but clusters of stars. He finds their number increase upon him, and when he resolves one nebula into stars he discovers ten new ones which he cannot resolve. He then forms the idea of immense strata of fixed stars, of clusters of stars and of nebulæ*; till, going on with such interesting observations, he now perceives that all these appearances must naturally arise from the confined situation in which we are placed. *Confined* it may justly be called, though in no less a space

* See a former paper on the Construction of the Heavens.

than what before appeared to be the whole region of the fixed stars; but which now has assumed the shape of a crookedly branching nebula; not, indeed, one of the least, but perhaps very far from being the most considerable of those numberless clusters that enter into the construction of the heavens.

Result of Observations.

I shall now endeavour to shew, that the theoretical view of the system of the universe, which has been exposed in the foregoing part of this paper, is perfectly consistent with facts, and seems to be confirmed and established by a series of observations. It will appear, that many hundreds of nebulæ of the first and second forms are actually to be seen in the heavens, and their places will hereafter be pointed out. Many of the third form will be described, and instances of the fourth related. A few of the cavities mentioned in the fifth will be particularised, though many more have already been observed; so that, upon the whole, I believe, it will be found, that the foregoing theoretical view, with all its consequential appearances, as seen by an eye inclosed in one of the nebulæ, is no other than a drawing from nature, wherein the features of the original have been closely copied; and I hope the resemblance will not be called a bad one, when it shall be considered how very limited must be the pencil of an inhabitant of so small and retired a portion of an indefinite system in attempting the picture of so unbounded an extent.

But to proceed to particulars: I shall begin by giving the following table of gages that have been taken. In the first column is the right ascension, and in the second the north polar distance, both reduced to the time of FLAMSTEED's Catalogue. In the third are the contents of the heavens, being the result of the gages. The fourth shews from how many fields of view the gages were deduced, which have been ten or more where the number of the stars was not very considerable; but, as it would have taken too much time, in high numbers, to count so many fields, the gages are generally single. Where the stars happened to be uncommonly crowded, no more than half a field was counted, and even sometimes only a quadrant; but then it was always done with the precaution of fixing on some row of stars that would point out the division of the field, so as to prevent any considerable mistake. When five, ten, or more fields are gaged, the polar distance in the second column of the table is that of the middle of the sweep, which was generally from 2 to 2½ degrees in breadth; and, in gaging, a regular distribution of the fields, from the bottom of the sweep to the top, was always strictly attended to. The fifth column contains occasional remarks relating to the gages.

I. Table of Star-Gages.

R.A.	P.D.	Stars.	Fields.	Memorandums.
H. M. S.	D. M.			
0 1 41	78 47	9·9	10	
0 4 55	65 36	20·0	10	
0 7 54	74 13	11·3	10	Most of the stars extremely small.
0 8 24	49 7	60	1	
0 9 52	113 17	4·1	10	* The gages marked with an asterisk
0 12 52	113 17	3·2	10	* are those by which fig. 4 [page 251] has been delineated.
0 16 48	67 44	11·9	10	
0 21 52	113 17	3·9	10	*
0 22 21	87 10	5·9	10	
0 28 26	46 54	60	1	
0 31 38	46 54	40	1	
0 33 33	65 32	20·4	10	
0 34 22	56 38	20	1	
0 35 22	55 38	24	1	
0 36 39	76 32	11·3	10	
0 39 56	78 43	8·1	10	
0 40 29	48 43	60	½	
0 44 21	87 10	7·6	10	
0 46 22	69 51	11	10	
0 46 33	65 32	13	10	
0 48 42	58 47	40	1	
0 48 50	58 13	17	1	
0 53 18	67 41	9·8	10	A little hazy.
0 53 40	45 37	73	1	
0 54 10	75 16	13	1	
0 55 10	73 16	14	1	
0 56 4	74 0	15	1	
0 57 52	113 17	3·8	10	*
0 59 10	74 25	14	1	
1 0 16	74 16	11·1	10	
1 1 10	74 5	11·2	10	
1 1 18	111 0	5·2	10	Very clear for this altitude.
1 2 52	52 0	28·1	10	Most of the stars very small.
1 3 52	113 17	2·8	10	*
1 4 15	94 52	7·5	10	
1 4 33	65 32	11·0	10	
1 5 55	78 31	9·2	10	
1 7 27	45 23	58	1	
1 12 0	58 37	20	1	
1 12 48	60 19	13	1	

R.A.	P.D.	Stars.	Fields.	Memorandums.
H. M. S.	D. M.			
1 13 4	94 50	6·3	10	
1 15 51	48 40	30	1	
1 18 21	48 40	58	1	
1 23 21	48 40	44	1	
1 27 30	65 42	12·9	10	
1 31 21	87 7	5·8	10	
1 32 4	94 50	7·3	10	
1 33 10	100 8	6·4	10	
1 33 32	92 35	7·1	10	
1 34 52	60 8	17	1	
1 43 30	65 42	14·4	10	
1 45 24	69 43	7·1	10	
1 48 4	100 12	4·9	10	
1 54 24	76 28	12·1	10	
1 58 55	61 55	15·0	10	
2 4 28	87 5	6·4	10	
2 4 36	78 38	9·3	10	
2 7 12	94 56	7·8	10	
2 8 0	83 3	7·3	10	
2 10 4	100 12	4·3	10	
2 11 30	65 45	14·8	10	
2 16 27	110 54	4·2	10	*
2 19 27	76 24	9·9	10	
2 22 17	45 31	82	1	
2 23 6	60 16	14	1	
2 23 19	113 8	4·2	10	*
2 24 6	58 30	15	1	
2 27 40	115 21	3·0	10	* The situation too low for great accuracy.
2 30 0	94 56	6	10	
2 31 23	76 22	13·8	10	
2 35 14	87 2	5·6	10	
2 38 0	94 56	6·6	10	
2 42 7	61 50	14·8	10	
2 47 32	74 3	11·1	10	Most of the stars exceedingly small.
2 49 22	92 55	9·0	10	
2 49 30	110 55	6·1	10	*
2 50 0	94 56	6·8	10	
2 54 53	76 22	9·2	10	
2 59 56	81 10	6·1	10	
3 1 53	78 37	4·1	10	

R.A.	P.D.	Stars.	Fields.	Memorandums.
H. M. S.	D. M.			
3 1 56	81 10	5·1	10	
3 4 53	78 37	3·5	10	
3 10 20	100 2	6·8	10	
3 11 6	59 29	7·0	5	In a part of the heavens which looks pretty full of stars to
3 13 6	59 29	6·1	10	
3 15 6	59 29	9·4	10	the naked eye.
3 22 57	83 1	10·3	10	
3 23 21	92 49	10·1	10	
3 29 41	46 35	55	1	
3 35 0	62 1	15	1	About 15 stars generally in the field.
3 35 12	100 3	7·4	10	
3 36 1	113 3	4·9	10	*
3 42 49	46 10	54	1	
3 48 16	99 59	8·1	10	
3 55 11	74 2	11·0	10	
4 1 24	92 48	13·8	10	
4 6 18	82 57	13·4	10	
4 8 31	114 55	4·2	10	*
4 12 41	69 33	15·3	10	And many more, extremely small,
4 16 34	112 45	6·2	10	* suspected.
4 26 34	112 45	8·8	10	*
4 27 11	70 41	25	1	
4 28 41	70 1	17	1	
4 29 5	69 24	30	1	
4 30 14	99 50	9·7	10	
4 31 19	67 33	15·6	10	
4 32 29	69 2	36	1	
4 33 31	114 55	8·1	10	*
4 42 14	86 27	19·9	10	
4 53 22	72 59	56	1	
4 57 45	83 22	38	1	
4 58 45	84 36	35	1	
5 1 16	69 23	34	1	
5 3 45	83 29	17·7	6	
5 10 52	69 22	74	1	
5 11 22	96 37	24	1	
5 17 22	96 15	8·9	8	
5 18 0	80 46	30	1	About 30 stars in the field, not
5 21 7	92 52	19·1	10	very exactly gaged.
5 24 12	66 5	36	1	
5 27 3	68 52	58	1	
5 27 48	110 40	17·7	10	*
5 33 4	76 10	65	1	
5 33 12	66 26	86	1	
5 33 17	114 59	13·5	10	*

R.A.	P.D.	Stars.	Fields.	Memorandums.
H. M. S.	D. M.			
5 34 45	70 33	50	1	
5 36 30	62 1	20—30		From 20 to 30 stars in the fields, not very exactly gaged.
5 37 4	74 26	140	½	
5 38 45	70 8	73	1	
5 41 12	66 43	60	1	
5 44 0	116 43	11·5	10	*
5 45 30	83 30	50	1	
5 47 34	112 34	19·3	10	*
5 48 30	62 1	30	1	About 30 stars in the field; not very exactly gaged.
5 48 44	92 51	22·4	5	
5 49 0	80 5	50	1	
5 52 14	93 14	44	1	
5 52 30	83 30	60	1	
5 53 0	80 5	110	1	
5 55 4	92 56	57	1	
5 56 40	70 27	73	1	
5 57 0	80 5	60	1	
5 57 37	110 33	19·6	10	*
5 58 51	88 36	90	1	
5 59 30	83 30	80	1	
6 0 23	86 38	24·1	10	
6 1 0	80 5	70	1	
6 4 0	80 5	90	1	
6 5 4	67 17	120	¼	Very unequally scattered.
6 6 14	96 16	52	1	
6 6 30	83 30	80	1	
6 6 30	80 5	70	1	
6 6 38	91 45	54	1	Like the rest, or many such fields.
6 6 40	68 24	56	1	
6 9 0	80 5	74	1	
6 9 34	113 35	26	1	*
6 11 0	62 1	30—40	1	Between.
6 11 0	80 5	63	1	The least number of stars in the field I could find in this neighbourhood.
6 11 34	112 5	33	1	*
6 11 37	90 15		1	About 60 or 70 generally.
6 14 4	68 11	178	¼	
6 14 38	90 15	77	1	
6 17 45	62 1	50	1	
6 18 14	96 12	38	1	Very unequally scattered.
6 19 14	93 59	72	1	
6 26 17	114 59	15·9	10	
6 27 14	94 36	132	½	*
6 27 32	70 23	50	1	
6 31 48	115 40	40	1	
6 34 44	92 25	94	1	

R.A.	P.D.	Stars.	Fields.	Memorandums.
H. M. S.	D. M.			
6 34 55	79 5	50	1	Generally about 50 stars.
6 36 0	94 56	62	1	Twilight.
6 37 15	75 5	70	1	Generally about 70 stars.
6 39 8	99 7	50	1	*
6 40 0	116 43	31·3	10	
6 43 25	79 5	67	1	
6 44 28	100 30	67	1	*
6 49 5	87 21	120	½	*
6 49 30	77 31	50	1	Many fields like this.
6 49 44	92 33	120	½	
6 51 8	98 33	78	1	*
6 52 0	116 21	48	1	
6 52 25	79 5	60	1	About 60 stars.
6 52 44	92 59	98	1	
6 54 9	111 11	45	1	*
6 57 8	100 1	34	1	*
6 57 38	98 50	83	1	*
6 58 39	112 48	81	1	*
7 0 25	79 5	70	1	
7 4 0	92 3	102	1	*
7 4 38	98 59	70	1	*
7 5 9	111 11	70	1	*
7 8 9	112 15	62	1	*
7 12 8	100 5	118	½	*
7 15 38	98 12	112	½	*
7 19 0	91 51	58	1	*
7 20 0	78 59	48	1	
7 25 9	111 21	168	¼	* One of the richest fields.
7 28 9	112 34	204	¼	* A field like the rest.
7 33 3	115 28	86	1	
7 41 9	113 26	108	½	*
7 53 4	86 39	28·3	10	*
8 1 4	111 15	80	1	*
8 3 4	113 31	66	1	
8 6 38	100 5	40	1	
8 7 38	99 3	45	1	*
8 11 8	99 25	24·2	10	*
8 12 34	112 15	52	1	*
8 22 4	111 30	35	1	*
8 31 4	112 1	33	1	
8 32 24	112 7	30	1	
8 35 4	112 17	24	1	
8 35 14	111 19	20	1	
8 40 4	111 11	22	1	*
8 45 4	113 22	13	1	

R.A.	P.D.	Stars.	Fields.	Memorandums.
H. M. S.	D. M.			
8 46 39	91 26	20·3	10	*
8 48 4	112 23	16·2	10	
8 57 25	66 20	8·3	10	*
9 5 38	91 22	13·8	10	*
9 10 4	115 17	14·0	10	
9 20 4	112 23	15·8	10	
9 20 40	99 12	11·1	10	
9 20 58	88 7	11·5	10	*
9 35 4	112 23	13·0	10	
9 38 4	115 17	10·1	10	
9 38 8	90 23	7·9	10	*
9 42 16	86 16	7·7	10	*
9 45 49	112 21	13·2	10	Strong twilight.
10 0 4	115 17	9·1	10	
10 16 8	88 8	7·2	10	*
10 19 32	91 14	6·5	10	
10 25 8	88 8	4·9	10	*
10 26 0	81 41	5·6	7	*
11 4 4	81 38	5·3	6	*
11 7 36	91 14	5·6	10	
11 10 6	115 23	6·5	10	Twilight.
11 16 52	81 38	3·1	8	*
11 20 37	91 17	4·9	10	
11 53 43	81 39	6·0	5	*
12 5 6	78 57	2·2	13	*
12 30 40	79 3	3·4	11	*
12 46 51	81 40	4·6	13	*
12 48 19	79 4	3·9	13	*
12 53 45	101 45	9·3	10	Twilight.
12 57 8	99 56	8·1	10	Pretty strong day-light.
13 1 19	79 4	3·8	12	*
13 17 27	101 45	8·6	10	Twilight.
13 22 49	100 1	8·4	10	Some day-light.
13 27 57	101 45	11·3	10	
13 31 10	75 55	5—6	1	* Generally about 5 or 6 stars in the [field.
13 38 53	104 27	8·5	10	
13 48 49	100 1	9·2	10	Strong twilight.
13 51 27	101 45	10·0	10	
13 55 44	58 11	7·4	10	* Twilight.
13 57 53	104 27	12·3	10	Most very small.
14 9 49	100 1	11·2	10	Twilight.
14 13 52	113 4	9·7	10	
14 14 57	101 45	8·8	10	
14 24 49	81 53	2·7	6	
14 29 45	100 5	13·3	10	

R.A.			P.D.		Stars.	Fields.	Memorandums.
H.	M.	S.	D.	M.			
14	30	7	66	3	8·8	10	* All sizes.
14	30	8	80	38	3·5	13	
14	33	22	58	7	8·9	10	* Chiefly small.
14	33	52	113	4	10·3	10	
14	39	57	101	45	14·0	10	All sizes.
14	40	36	64	47	6·4	10	
14	44	11	114	54	10·3	10	
14	49	52	113	4	12·8	10	
14	51	14	58	10	9·2	10	* Twilight.
14	52	58	60	41	4·4	10	* Strong Aurora borealis.
14	53	7	66	15	9·0	10	Chiefly large.
14	55	36	64	47	6·6	10	Most very small.
14	59	11	114	54	8·8	10	
15	2	42	62	48	8·3	10	
15	3	7	66	15	9·5	10	
15	4	36	64	47	5·0	10	
15	8	37	113	0	14·1	10	
15	8	45	93	5	9·4	12	Very small.
15	13	42	62	48	8·9	10	
15	15	44	58	17	10·0	10	* Twilight.
15	19	48	60	40	4·9	10	* Strong Aurora borealis, so as to affect the gages.
15	20	0	75	52	9·5	4	
15	21	0	93	5	10·9	12	
15	26	7	81	53	11·0	5	
15	28	48	99	51	13·1	10	
15	29	7	66	15	10·6	10	All sizes.
15	29	44	58	17	8·9	10	* Twilight.
15	32	0	75	51	6	6	
15	33	52	111	32	12·8	10	
15	35	0	75	51	6·5	6	
15	42	2	58	14	13·1	10	* Twilight.
15	42	3	116	56	18·6	10	
15	42	53	113	47	32·5	2	The stars too small for the gage.
15	46	30	93	5	10·8	12	
15	48	37	113	0	17·1	10	
15	48	46	63	4	12·4	10	
15	49	52	111	32	18·1	10	The situation so low that it requires attention to see the stars.
15	50	20	114	55	9·2	10	
15	57	3	116	56	7·2	10	
16	0	2	58	14	12·2	10	* Twilight.
16	0	3	116	56	6·1	10	
16	0	12	114	57	1·6	10	
16	3	12	114	57	2·0	10	
16	4	0	75	43	13	6	All sizes.
16	4	19	113	6	·5	10	Perfectly clear. See p. [253].

R.A.			P.D.		Stars.	Fields.	Memorandums.
H.	M.	S.	D.	M.			
16	4	46	63	4	12·0	10	Most small.
16	4	52	99	57	14·6	10	Moon and twilight.
16	6	28	113	4	·7	10	Perfectly clear.
16	7	12	66	15	13·3	10	
16	8	6	115	1	3·8	6	
16	8	11	93	9	12·2	12	
16	8	16	116	48	11·6	10	
16	9	28	113	4	1·1	10	Perfectly clear. See p. [253].
16	11	28	113	4	1·4	10	The same. [to the naked eye.
16	13	28	113	4	1·8	10	ρ Serpentarii and 19 Scorpii visible
16	13	52	58	24	14·2	10	* Most small.
16	14	42	63	7	15·1	10	Most very small.
16	15	37	80	40	9·7	12	All sizes.
16	17	28	113	4	4·7	10	
16	20	51	81	57	13·8	6	
16	23	0	73	43	24	1	
16	23	28	113	4	13·6	10	
16	24	11	93	9	13·6	12	Require attention to be seen.
16	25	7	80	40	14·6	13	
16	27	32	68	23	21·6	10	Twilight.
16	29	16	116	48	50·4	10	
16	30	37	80	40	34	1	
16	31	12	66	15	18·4	10	Strong twilight.
16	32	28	113	4	20·3	10	Most extremely small.
16	32	52	58	24	15·6	10	* Most small.
16	35	42	63	7	16·5	10	*
16	35	48	93	·15	18·6	12	All sizes.
16	38	12	66	15	20·1	10	Strong twilight.
16	38	45	107	57	19·9	10	Strong twilight.
16	40	51	113	14	41·1	8	
16	45	32	68	23	19·0	4	Hazy.
16	51	45	107	57	29·8	10	
16	52	22	66	26	16·6	10	Day-light pretty strong.
16	55	42	63	7	26·6	10	* Strong twilight.
17	1	34	58	11	18·8	10	* Strong day-light.
17	3	22	66	26	35	1	* Day-light too strong for gaging.
17	6	36	98	38	13·7	10	Most small, and more suspected.
17	9	30	116	55	7·6	10	
17	9	32	68	23	32·3	10	
17	11	10	66	26	38	1	* Day-light pretty strong.
17	13	24	63	21	32·8	10	* Strong day-light.
17	17	36	111	47	15·3	10	Moon and day-light.
17	25	7	108	5	23	10	
17	27	29	116	48	25	1	
17	28	32	68	23	42·2	5	* Twilight.

R.A.	P.D.	Stars.	Fields.	Memorandums.
H. M. S.	D. M.			
17 30 29	116 48	42	1	
17 33 29	116 48	52	1	Day-light very strong.
17 34 36	98 38	18·5	10	Very strong twilight.
17 39 34	120 0	84	1	Most large.
17 40 41	114 52	77	1	Day-light very strong.
17 41 29	116 48	82	1	Day-light very strong.
17 43 45	105 3	80	1	Flying clouds.
17 48 0	61 18	25·6	5	Most large.
17 50 4	56 16	27·2	10	Twilight. [heaven.
17 50 7	108 5	59	1	Like the rest in this part of the
17 52 7	108 5	118	1	Many such fields just by.
17 52 17	98 43	7.6	10	
17 52 30	62 12	40	1	Most large.
17 52 32	68 19	54	1	* Strong day-light.
17 55 7	108 5	232	½	
17 55 15	106 6	112	1	Many such fields.
17 55 38	112 54	112	½	
17 57 30	60 28	38	1	Most large.
17 58 37	103 24	35	1	
17 58 41	118 57	64	1	
17 58 49	122 17	17	1	
17 59 1	108 8	320	½	
17 59 19	104 24	68	1	
18 0 13	122 11	27	1	
18 3 49	120 42	19	1	
18 5 17	98 47	65	1	
18 6 37	90 36	9·4	10	Too soon for gaging, not having
18 7 4	62 14	40	1	Most large. { been long enough
18 7 4	56 16	38·2	5	{ out in the dark.
18 7 37	103 25	88·0	3	
18 10 7	120 58	20	1	
18 10 52	61 8	78	1	Chiefly large.
18 11 49	104 6	170	½	
18 13 37	104 16	238	½	
18 13 52	93 11	2·0	7	
18 14 46	56 20	48	1	
18 15 28	92 42	3·4	7	
18 16 52	92 42	8·9	7	
18 18 40	92 42	13·8	7	
18 19 37	102 34	9·5	2	
18 20 7	103 18	19	1	
18 20 46	92 42	25·8	6	
18 21 1	103 55	22	1	
18 21 12	90 41	8·6	10	
18 21 31	103 36	24	1	

R.A.	P.D.	Stars.	Fields.	Memorandums.
H. M. S.	D. M.			
18 22 4	62 7	48	1	Large and small.
18 22 4	56 16	39·6	5	
18 22 19	104 6	14	1	
18 22 37	103 45	30	1	
18 24 3	115 10	35	1	
18 24 4	109 35	35	1	Twilight.
18 24 7	102 31	30	1	
18 24 10	92 59	88	1	
18 24 43	103 39	25	1	
18 25 37	102 34	39	1	
18 26 17	98 3	111	1	
18 26 25	103 57	60	1	
18 26 47	97 43	250	1	
18 27 1	120 58	30	1	
18 27 55	120 44	32	1	
18 28 7	102 51	13	1	Extremely small.
18 28 8	91 44	39	1	Most small.
18 28 25	103 9	20	1	Extremely small.
18 28 37	122 25	12	1	
18 29 25	103 24	20	1	Extremely small.
18 29 47	97 50	150	1	
18 29 49	121 39	24	1	
18 30 34	57 18	62	1	
18 31 10	92 42	13·7	7	
18 31 10	108 53	74	1	Twilight.
18 31 13	103 19	112	1	All sizes.
18 31 17	97 53	188	½	Many more suspected.
18 31 34	62 34	76	1	* Large and small.
18 31 49	121 39	19·3	10	
18 33 4	108 43	88	1	Twilight.
18 33 7	103 53	146	½	
18 34 5	98 34	130	1	
18 34 47	71 53	78	1	*
18 34 58	60 41	80	1	Large and mall.
18 36 34	110 12	83	1	Twilight.
18 36 34	91 37	176	¼	
18 36 47	72 28	224	½	*
18 37 34	93 29	5	1	
18 38 1	104 14	118	½	
18 39 40	93 52	116	¼	
18 40 28	92 47	10	1	
18 40 47	71 48	236	¼	*
18 41 22	91 37	156	¼	
18 42 49	121 39	15·2	10	Very clear for this altitude.
18 43 17	72 8	368	¼	*

R.A.			P.D.		Stars.	Fields.	Memorandums.
H.	M.	S.	D.	M.			
18	43	33	119	21	21	1	
18	44	34	112	43	53	1	
18	44	34	60	34	84	1	All sizes.
18	47	32	91	14	328	¼	
18	48	4	110	12	83		
18	50	16	60	55	136	½	Many of them small.
18	51	4	57	26	84	1	
18	51	32	108	26	36·8	5	Strong twilight.
18	52	49	115	30	26·2	5	
18	54	4	57	18	93	1	
18	54	8	91	14	328	¼	
18	54	55	104	23	180	½	
18	55	4	108	41	80	1	
18	55	16	62	31	206	½	
18	59	8	91	14	328	¼	
18	59	26	72	37	40	1	Too soon for gaging.
19	1	2	71	40	75	1	
19	1	34	56	47	127	1	Moonlight.
19	2	29	74	53	204	¼	* Twilight.
19	2	37	103	16	160	½	
19	2	49	121	39	14·1	10	
19	3	34	55	56	146	½	☽
19	6	34	61	8	196	½	And many small besides.
19	7	34	56	56	130	½	☽
19	7	52	57	59	116	½	
19	8	38	92	8	120	½	
19	9	37	109	1	60	1	
19	9	40	56	51	130	1	☽
19	12	59	75	21	58	1	*
19	13	50	59	59	256	¼	
19	13	52	59	29	158	½	
19	14	2	72	15	60	1	*
19	14	4	61	21	279	⅓	Too crowded for accuracy.
19	14	55	103	36	64	1	Changeable focus.
19	15	40	55	26	160	1	☽ bright.
19	16	50	60	43	296	¼	
19	16	59	73	23	56	1	*
19	17	44	108	12	50	1	
19	18	23	78	9	196	¼	*
19	18	28	61	21	279	⅓	
19	19	52	57	14	180	½	
19	19	56	108	12	55	1	
19	20	51	60	55	384	¼	
19	21	1	78	47	472	¼	*
19	21	34	55	17	208	½	☽ bright.

R.A.	P.D.	Stars.	Fields.	Memorandums.
H. M. S.	D. M.			
19 22 27	62 29	320	¼	
19 24 36	56 49	224	¼	
19 24 49	104 24	36	1	Changeable focus.
19 24 50	60 43	296	¼	
19 24 53	113 51	18·3	10	
19 25 4	57 9	190	½	☽ bright.
19 25 16	64 18	280	¼	
19 25 22	59 36	340	¼	
19 25 37	103 50	55	1	Changeable focus.
19 27 36	72 34	424	¼	* Too small and too crowded to be [certain of the number.
19 27 44	61 8	240	⅓	
19 28 1	103 30	45	1	Changeable focus.
19 28 6	56 49	288	¼	
19 28 52	59 26	344	¼	
19 28 52	56 47	186	½	☽ very bright.
19 29 46	65 10	34	1	
19 30 36	74 33	588	¼	*
19 30 36	54 53	312	¼	
19 31 33	92 34	62·2	5	*
19 32 9	109 44	23·8	10	
19 32 15	62 35	296	¼	
19 33 4	55 34	212	½	☽
19 33 7	103 12	50	1	Changeable focus.
19 33 14	61 8	240	⅓	
19 33 20	58 59	232	¼	
19 34 51	115 44	14·1	10	
19 35 34	63 19	256	¼	
19 36 6	54 57	384	¼	
19 36 37	102 31	68	1	Changeable focus.
19 36 50	60 35	296	¼	
19 40 33	63 0	296	¼	
19 40 46	59 12	192	¼	
19 40 48	74 33	588	¼	*
19 42 33	73 14	352	¼	*
19 43 30	57 23	130	½	☽
19 43 56	64 27	124	½	Most large.
19 45 36	77 58	140	½	* Faint ☽.
19 45 37	103 3	50	1	
19 46 21	73 14	352	¼	*
19 46 51	115 44	12·8	10	Strong twilight.
19 47 8	60 35	312	¼	
19 47 18	109 46	20·9	10	
19 47 22	57 38	312	¼	Very unequally scattered.
19 49 6	57 13	268	¼	
19 49 48	56 51	120	½	☽

R.A.	P.D.	Stars.	Fields.	Memorandums.
H. M. S.	D. M.			
19 50 5	92 39	39·2	5	* Most small.
19 51 37	62 37	51	1	
19 52 0	57 15	220	½	☽
19 53 1	60 36	80	1	
19 53 28	63 40	52	½	
19 53 40	54 59	306	¼	
19 53 49	121 39	7·7	10	
19 54 0	55 12	160	½	☽
19 54 12	78 3	120	½	* Faint ☽
19 54 22	59 58	136	¼	
19 55 7	62 41	48	1	
19 56 19	60 44	112	½	
19 56 22	57 17	192	¼	
19 57 19	62 34	45	1	
19 57 40	58 29	104	½	
19 59 49	62 37	41	1	
20 0 21	79 3	56	1	* Strong ☽.
20 0 24	55 12	184	½	☽
20 0 25	60 33	80	1	Most of the stars extremely small.
20 0 51	115 44	12·2	10	Twilight.
20 1 39	79 34	68	1	* Strong ☽.
20 5 26	56 34	46	1	☽
20 5 27	72 56	280	¼	
20 6 23	107 27	22·6	10	
20 6 43	62 32	75	1	Many small.
20 8 26	56 27	47·4	5	☽
20 8 27	72 56	280	¼	
20 8 58	103 37	38	1	
20 9 6	109 40	24·2	5	
20 9 52	102 48	31	1	
20 12 22	58 14	76	½	
20 17 20	76 12	184	¼	Some twilight.
20 18 51	115 44	10·6	10	Twilight.
20 20 58	61 27	88	1	
20 21 36	71 28	104	½	Hazy.
20 22 56	56 27	66	1	☽
20 22 58	103 26	20	1	
20 24 51	115 44	9·3	10	Twilight.
20 25 58	103 26	22·8	10	Changeable focus.
20 25 59	67 27	248	¼	
20 26 1	92 44	30·8	5	*
20 26 46	109 37	16·7	10	Not clear.
20 26 49	121 39	7·7	10	A little hazy.
20 27 33	96 7	39	1	Most small.
20 34 51	115 44	9·5	10	☽

R.A.	P.D.	Stars.	Fields.	Memorandums.
H. M. S.	D. M.			
20 35 53	61 20	142	½	
20 37 18	58 28	108	½	
20 37 34	97 6	26·6	10	*
20 38 1	92 44	28·2	5	*
20 39 42	66 37	78	½	
20 40 22	56 21	192	¼	
20 41 11	67 54	108	½	
20 41 56	74 33	116	½	
20 42 59	62 14	112	½	
20 43 1	70 29	76	1	
20 43 30	54 47	260	¼	Most of the stars of the same size.
20 44 59	70 6	80	1	
20 47 13	60 46	120	½	
20 49 1	92 44	27·0	5	*
20 49 10	57 11	248	¼	Most of a size.
20 50 59	103 26	17·2	3	
20 51 23	68 30	70	1	
20 53 29	103 26	17·4	5	
20 54 1	107 47	10·3	10	
20 56 59	103 26	14·9	10	Most extremely small.
20 57 55	61 25	64	1	Twilight.
20 59 1	92 44	21·4	5	*
21 1 6	96 43	40	1	* Most small.
21 3 29	66 39	80	½	
21 3 53	73 9	55	1	
21 6 13	69 23	40	1	A little hazy.
21 6 55	103 32	11·1	10	
21 7 49	109 45	12·8	10	
21 7 59	64 58	110	½	
21 9 25	61 36	75	1	Strong twilight.
21 10 13	60 39	70	1	Strong twilight.
21 11 17	73 18	50	1	
21 11 42	96 13	25	1	*
21 12 1	92 44	16·4	5	
21 15 3	109 56	15·3	10	
21 16 43	59 7	76	½	
21 18 54	57 20	50	1	
21 20 18	96 43	24	1	*
21 21 0	107 49	8·1	10	
21 22 14	76 33	30·0	5	
21 25 31	92 44	8·0	5	
21 29 12	83 11	21·6	5	
21 30 58	78 57	18·9	10	
21 32 10	57 14	25	1	
21 33 1	92 44	15·4	5	Strong twilight.

R.A.			P.D.		Stars.	Fields.	Memorandums.
H.	M.	S.	D.	M.			
21	34	55	97	17	13·6	10	*
21	36	38	65	55	42	1	
21	38	20	65	38	60	1	
21	39	55	96	17	18	1	*
21	41	52	58	42	44	1	
21	43	22	109	55	11·5	10	*
21	45	4	59	39	52	1	
21	48	22	59	30	29	1	
21	51	52	58	56	61	1	
21	51	55	97	17	11·5	10	*
21	54	22	109	55	12·8	10	*
21	57	49	59	37	60	½	
21	58	4	75	7	33	1	
21	58	19	59	6	40	½	
21	58	43	58	34	32·6	5	☽
21	58	49	58	20	34	1	
22	2	25	60	9	42·6	5	
22	2	52	109	55	7·4	10	*
22	3	56	71	48	25·1	10	
22	7	22	109	55	8·9	10	*
22	10	28	75	2	26	1	
22	11	32	97	14	10·7	10	* Twilight.
22	11	35	65	48	26·6	5	
22	18	32	97	14	9·1	10	* Twilight.
22	20	35	109	58	8·3	10	*
22	20	55	78	54	11·7	10	Bright ☽.
22	27	41	95	4	8·1	10	
22	30	35	109	58	5·0	10	*
22	31	28	73	59	17·3	10	
22	33	6	76	52	16·5	10	
22	34	40	61	56	20·1	10	
22	35	35	109	58	7·1	10	*
22	36	49	71	57	18·5	10	
22	39	41	82	5	19	1	
22	40	5	65	48	21·3	10	
22	43	55	60	9	26·7	10	Faint ☽.
22	45	3	80	47	13·2	10	
22	45	30	58	38	17·2	10	☽
22	48	49	71	57	13·4	10	
22	52	9	78	43	8·2	10	☽
22	52	41	95	4	8·9	10	
22	55	40	71	54	11·6	10	
22	56	55	67	53	12·1	10	
22	58	19	78	42	9·2	10	☽
23	0	27	113	12	4·4	10	

R.A.	P.D.	Stars.	Fields.	Memorandums.
H. M. S.	D. M.			
23 0 30	58 38	18·7	10	☽
23 2 59	65 50	21·3	10	
23 5 35	109 58	7·3	10	☽
23 8 52	95 1	7·5	10	Most extremely small.
23 10 4	64 55	26	1	
23 11 40	61 48	21·1	10	
23 12 40	71 54	11·9	10	
23 17 50	81 0	9·7	10	
23 23 58	69 48	12·1	10	
23 25 32	113 12	3·1	10	*
23 32 2	69 51	9·5	10	
23 33 20	79 45	10	1	
23 43 2	69 51	10·9	10	
23 44 47	45 24	50	1	
23 46 52	113 17	4·2	10	*
23 46 55	65 36	15·3	10	
23 59 21	87 10	5·6	10	
23 59 56	95 4	7·8	10	

PROBLEM.

The stars being supposed to be nearly equally scattered, and their number, in a field of view of a known angular diameter, being given, to determine the length of the visual ray.

Here, the arrangement of the stars not being fixed upon, we must endeavour to find which way they may be placed so as to fill a given space most equally. Suppose a rectangular cone cut into frustula by many equidistant planes perpendicular to the axis; then, if one star be placed at the vertex, and another in the axis at the first intersection, six stars may be set around it so as to be equally distant from one another and from the central star. These positions being carried on in the same manner, we shall have every star within the cone surrounded by eight others, at an equal distance from that star taken as a center. Fig. 1 contains four sections of such a cone distinguished by alternate shades, which will be sufficient to explain what sort of arrangement I would point out.

The series of the number of stars contained in the several sections will be 1 . 7 . 19 . 37 . 61 . 91 . &c. which continued to n terms, the sum of it, by the differential method, will be $na + n \cdot \frac{n-1}{2} d' + n \cdot \frac{n-1}{2} \cdot \frac{n-2}{3} d''$, &c.: where a is the first term, d', d'', d''', &c. the 1st, 2d, and 3d differences. Then, since $a = 1$, $d' = 6$, $d'' = 6$, $d''' = 0$, the sum of the series will be n^3. Let S be the given number of stars; 1, the diameter of the base of the field of view; and B, the diameter of the base of the great rectangular cone; and, by trigonometry, we shall have

$B = \frac{\text{Radius}}{\text{Tang. } \frac{1}{2} \text{ field}}$. Now, since the field of view of a telescope is a cone, we shall have its solidity to that of the great cone of stars, formed by the above construction, as the square of the diameter of the base of the field of view, to the square of the diameter of the base of the great cone, the height of both being the same;

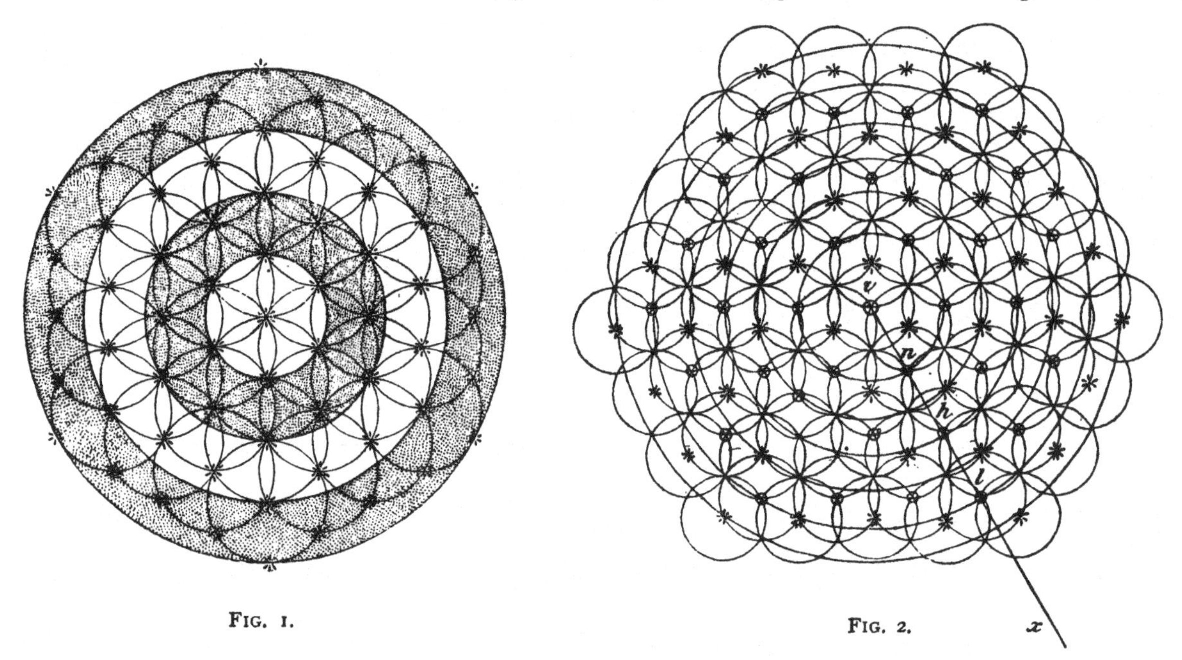

FIG. 1. FIG. 2.

and the stars in each cone being in the ratio of the solidity, as being equally scattered,* we have $n = \sqrt[3]{B^2S}$. And the length of the visual ray $= n - 1$, which was to be determined.

The same otherwise.

If a different arrangement of the stars should be selected, such as that in fig. 2, where one star is at the vertex of a cone; three in the circumference of the first

* We ought to remark, that the periphery and base of the cone of the field of view, in gaging, would in all probability seldom fall exactly on such stars as would produce a perfect equality of situation between the stars contained in the small and the great cone; and that, consequently, the solution of this problem, where we suppose the stars of one cone to be to those of the other in the ratio of the solidity on account of their being equally scattered, will not be strictly true. But it should be remembered, that in small numbers, where the different terminations of the fields would most affect this solution, the stars in view have always been ascertained from gages that were often repeated, and each of which consisted of no less than ten fields successively taken, so that the different deviations at the periphery and base of the cone would certainly compensate each other sufficiently for the purpose of this calculation. And that, on the other hand, in high gages, which could not have the advantage of being so often repeated, these deviations would bear a much smaller proportion to the great number of stars in a field of view; and therefore, on this account, such gages may very justly be admitted in a solution where practical truth rather than mathematical precision is the end we have in view. It is moreover not to be supposed that we imagine the stars to be actually arranged in this regular manner, and, returning therefore to our general hypothesis of their being equally scattered, any one field of view promiscuously taken may, in this general sense, be supposed to contain a due proportion of them; so that the principle on which this solution is founded may therefore be said to be even more rigorously true than we have occasion to insist upon in an argument of this kind.

section, at an equal distance from the vertex and from each other; six in the circumference of the next section, with one in the axis or center; and so on, always placing three stars in a lower section in such a manner as to form an equilateral pyramid with one above them: then we shall have every star, which is sufficiently within the cone, surrounded by twelve others at an equal distance from the central star and from each other. And by the differential method, the sum of the two series equally continued, into which this cone may be resolved, will be $2n^3+1\frac{1}{2}n^2+\frac{1}{2}n$; where n stands for the number of terms in each series. To find the angle which a line vx, passing from the vertex v over the stars v, n, h, l, &c. to x, at the outside of the cone, makes with the axis; we have, by construction, vs in fig. 3, representing the planes of the first and second sections $=2\times\cos. 30°=\phi$, to the radius ps, of the first section $=1$. Hence it will be $\sqrt{\phi^2-1}=vp=\frac{1}{2}vm$; or $vm=2\sqrt{\phi^2-1}$: and, by trigonometry, $\frac{R\phi}{2\sqrt{\phi^2-1}}=T$. Where T is the tangent of the required angle to the radius R *; and putting $t=$ tangent of half the given field of view, it will be $\frac{T}{t}=B$, the base of the cone. And $\frac{\sqrt{\phi^2-1}}{\phi}=d$, will be an expression for vp, in terms of vs, which is the mutual distance of the scattered stars. Then having $\frac{B^2S}{2}=n^3+\frac{3}{4}n^2+\frac{1}{4}n$, we may find n; whence $2dn-d$, the visual ray, will be obtained.

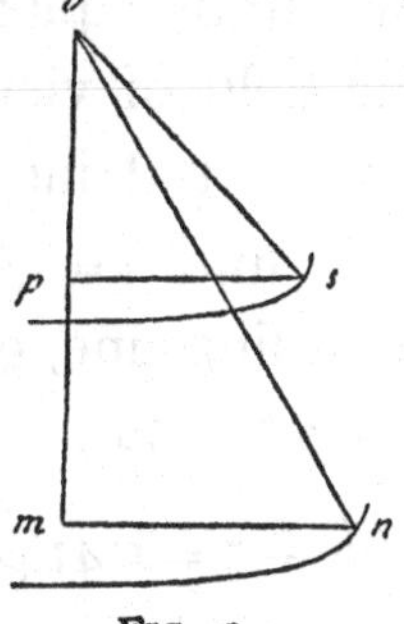

FIG. 3.

The result of this arrangement gives a shorter ray than that of the former; but since the difference is not so considerable as very materially to affect the conclusions, I shall, on account of the greater convenience, make use of the first.

We inhabit the planet of a star belonging to a Compound Nebula of the third form.

I shall now proceed to shew that the stupendous sidereal system we inhabit, this extensive stratum and its secondary branch, consisting of many millions of stars, is, in all probability, *a detached Nebula*. In order to go upon grounds that seem to me to be capable of great certainty, they being no less than an actual survey of the boundaries of our sidereal system, which I have plainly perceived, as far as I have yet gone round it, every where terminated, and in most places very narrowly too, it will be proper to shew the length of my sounding line, if I may so call it, that it may appear whether it was sufficiently long for the purpose.

* In finding this angle we have supposed the cone to be generated by a revolving rectangular triangle of which the line vx, fig. 2, is the hypotenuse; but the stars in the second series will occasion the cone to be contained under a waving surface, wherefore the above supposition of the generation of the cone is not strictly true; but then these waves are so inconsiderable, that, for the present purpose, they may safely be neglected in this calculation.

In the most crowded part of the milky way I have had fields of view that contained no less than 588 stars,* and these were continued for many minutes, so that in one quarter of an hour's time there passed no less than 116000 stars through the field of view of my telescope.† Now, if we compute the length of the visual ray by putting $S = 588$, and the diameter of the field of view fifteen minutes, we shall find $n = \sqrt[3]{B^2S} = 498$; so that it appears the length of what I have called my sounding line, or $n - 1$, was probably not less than 497 times the distance of Sirius from the sun. The same gage calculated by the second arrangement of stars gives $\sqrt{\phi^2 - 1} = 1{\cdot}41421$; $\frac{R\phi}{2\sqrt{\phi^2 - 1}} =$ tangent of 31° 28′ 55″·77; $\frac{T}{t} = B = 280{\cdot}69$; $\frac{\sqrt{\phi^2 - 1}}{\phi} = d = {\cdot}81649$; $\frac{B^2S}{2} = 23163409{\cdot}7 = n^3 + \frac{3}{4}n^2 + \frac{1}{4}n$; where $n = 284{\cdot}8$ nearly; and $2dn - 1 = 464$, the visual ray.

It may seem inaccurate that we should found an argument on the stars being equally scattered, when in all probability there may not be two of them in the heavens, whose mutual distance shall be equal to that of any other two given stars; but it should be considered, that when we take all the stars collectively there will be a mean distance which may be assumed as the general one; and an argument founded on such a supposition will have in its favour the greatest probability of not being far short of truth. What will render the supposition of an equal distribution of the stars, with regard to the gages, still less exposed to objections is, that whenever the stars happened either to be uncommonly crowded or deficient in number, so as very suddenly to pass over from one extreme to the other, the gages were reduced to other forms, such as the border-gage, the distance-gage, &c. which terms, and the use of such gages, I shall hereafter find an opportunity of explaining. And none of those kinds of gages have been admitted in this table, which consists only of such as have been taken in places where the stars apparently seemed to be, in general, pretty evenly scattered; and to increase and decrease in number by a certain gradual progression. Nor has any part of the heavens containing a cluster of stars been put in the gages; and here I must observe, that the difference between a crowded place and a cluster may easily be perceived by the arrangement as well as the size and mutual distance of the stars: for in a cluster they are generally not only resembling each other pretty nearly in size, but a certain uniformity of distance also takes place; they are more and more accumulated towards the center, and put on all the appearances which we should naturally expect from a number

* See the table of Gages, p. [239].

† The breadth of my sweep was 2° 26′, to which must be added 15′ for two semi-diameters of the field. Then, putting $161 = a$, the number of fields in 15 minutes of time; $\cdot 7854 = b$, the proportion of a circle to 1, its circumscribed square; $\phi =$ sine of 74° 22′, the polar distance of the middle of the sweep reduced to the present time; and $588 = S$, the number of stars in a field of view, we have $\frac{a\phi S}{b} = 116076$ stars.

of them collected into a group at a certain distance from us. On the other hand, the rich parts of the milky way, as well as those in the distant broad part of the stratum, consist of a mixture of stars of all possible sizes, that are seemingly placed without any particular apparent order. Perhaps we might recollect, that a greater condensation towards the center of our system than towards the borders of it should be taken into consideration; but, with a nebula of the third form, containing such various and extensive combinations, as I have found to take place in ours, this circumstance, which in one of the first form would be of considerable moment, may, I think, be safely neglected. However, I would not be understood to lay a greater stress on these and the following calculations than the principles on which they are founded will permit; and if hereafter we shall find reason, from experience and observation, to believe that there are parts of our system where the stars are not scattered in the manner here supposed, we ought then to make proper exceptions.

But to return: if some other high gage be selected from the table, such as 472 or 344, the length of the visual ray will be found 461 and 415. And although, in consequence of what has been said, a certain degree of doubt may be left about the arrangement and scattering of the stars, yet when I recollect, that in those parts of the milky way where these high gages were taken, the stars were neither so small, nor so crowded, as they must have been on a supposition of a much farther continuance of them, when certainly a milky or nebulous appearance must have come on, I need not fear to have over-rated the extent of my visual ray. And indeed every thing that can be said to shorten it will only contract the limits of our nebula, as it has in most places been of sufficient length to go far beyond the bounds of it. Thus, in the sides of the stratum opposite to our situation in it, where the gages often run below 5, our nebula cannot extend to 100 times the distance of Sirius; and the same telescope, which could shew 588 stars in a field of view of 15 minutes, must certainly have presented me also with the stars in these situations as well as the former, had they been there. If we should answer this by observing that they might be at too great a distance to be perceived, it will be allowing that there must at least be a vacancy amounting to the length of a visual ray not short of 400 times the distance of Sirius; and this is amply sufficient to make our nebula a detached one. It is true, that it would not be consistent confidently to affirm that we were on an island unless we had actually found ourselves every where bounded by the ocean, and therefore I shall go no farther than the gages will authorise; but considering the little depth of the stratum in all those places which have been actually gaged, to which must be added all the intermediate parts that have been viewed and found to be much like the rest, there is but little room to expect a connection between our nebula and any of the neighbouring ones. I ought also to add, that a telescope with a much larger aperture than my present one, grasping together a greater quantity of light, and thereby enabling us to see farther into space, will be the surest means of compleating and establishing the arguments that have been used: for if our nebula is not absolutely a detached one, I am firmly persuaded, that an instrument may be made large enough to discover the places where the stars continue onwards. A very bright milky nebulosity must there undoubtedly come on, since the stars in a field of view will increase in the ratio of n^3, greater than that of the cube of the visual ray. Thus, if 588 stars in a given field of view are to be seen by a ray of 497 times the distance of Sirius; when this is lengthened to 1000, which is but little more than double the former, the number of stars in the same field of view will be no less than 4774: for when the visual ray r is given, the number S of stars will be $=\frac{n^3}{B^2}$; where $n=r+1$; and a telescope with a three-fold power of extending into space, or with a ray of 1500, which, I think, may easily be constructed, will give us 16096 stars. Now, these would not be so close but that a good power applied to such an instrument might easily distinguish them; for they need not, if arranged in regular squares, approach nearer to each other than $6''\cdot 27$; but what would produce the milky nebulosity which I have mentioned is the numberless stars beyond them, which in one respect the visual ray might also be said to reach. To make this appear we must return to the naked eye, which, as we have before estimated, can only see the stars of the seventh magnitude so as to distinguish them; but it is nevertheless very evident that the united lustre of millions of stars, such as I suppose the nebula in Andromeda to be, will reach our sight in the shape of a very small, faint nebulosity; since the nebula of which I speak may easily be seen in a fine evening. In the same manner my present telescope, as I have argued, has not only a visual ray that will reach the stars at 497 times the distance of Sirius so as to distinguish them (and probably much farther), but also a power of shewing the united lustre of the accumulated stars that compose a milky nebulosity, at a distance far exceeding the former limits; so that from these considerations it appears again highly probable, that my present telescope, not shewing such a nebulosity in the milky way, goes already far beyond its extent: and consequently, much more would an instrument, such as I have mentioned, remove all doubt on the subject, both by shewing the stars in the continuation of the stratum, and by exposing a very strong milky nebulosity beyond them, that could no longer be mistaken for the dark ground of the heavens.

To these arguments, which rest on the firm basis of a series of observations, we may add the following considerations drawn from analogy. Among the great number of nebulæ which I have now already seen, amounting to more than 900, there are many which in all probability are equally extensive with that which we inhabit; and yet they are all separated from each other by very considerable intervals. Some indeed there are that seem to be double and treble; and though with most of these it may be, that they are at a very great distance from each other, yet we allow that some such conjunctions really are to be found ; nor is this what we mean to exclude. But then these compound or double nebulæ, which are those of the third and fourth forms, still make a detached link in the great chain. It is also to be supposed, that there may still be some thinly scattered solitary stars between the large interstices of nebulæ, which, being situated so as to be nearly equally attracted by the several clusters when they were forming, remain unassociated. And though we cannot expect to see these stars, on account of their vast distance, yet we may well presume, that their number cannot be very considerable in comparison to those that are already drawn into systems ; which conjecture is also abundantly confirmed in situations where the nebulæ are near enough to have their stars visible ; for they are all insulated, and generally to be seen upon a very clear and pure ground, without any star near them that might be supposed to belong to them. And though I have often seen them in beds of stars, yet from the size of these latter we may be certain, that they were much nearer to us than those nebulæ, and belonged undoubtedly to our own system.

Use of the gages.

A delineation of our nebula, by an application of the gages in the manner which has been proposed to be done in my former paper, may now be attempted, and the following table is calculated for this purpose. It gives us the length of the visual ray for any number of stars in the field of view contained in the third column of the foregoing table of gages from $\frac{1}{10}$ to 100000. If the number required is not to be found in the first column of this table, a proportional mean may be taken between the two nearest rays in the second column, without any material error, except in the few last numbers. The calculations of resolvable and milky nebulosity, at the end of the table, are founded, the first, on a supposition of the stars being so crowded as to have only a square second of space allowed them ; the next assigning them only half a second square. However, we should consider that in all probability a very different accumulation of stars may take place in different nebulæ ; by which means some of them may assume the milky appearance, though not near so far removed from us ; while clusters of stars also may become resolvable nebulæ from the same cause. The distinctness of the instrument is here also concerned ; and as telescopes with large apertures are not easily brought to a good figure, nebulous appearances of both sorts may probably come on much before the distance annexed to them in the table.

[TABLE

32

TABLE II.

Stars in the field.	Visual ray.	Stars.	Ray.	Stars.	Ray.	Stars.	Ray.	Stars.	Ray.
		31	186	71	245	210	352	700	527
0·1	27	32	188	72	246	220	358	800	551
0·2	34	33	190	73	247	230	363	900	573
0·3	39	34	192	74	249	240	368	1000	593
0·4	43	35	193	75	250	250	374	10000	1280
0·5	46	36	195	76	251	260	378	100000	2758
0·6	49	37	197	77	252	270	383		
0·7	52	38	199	78	253	280	388		
0·8	54	39	201	79	254	290	393		
0·9	56	40	202	80	255	300	397		
1	58	41	204	81	256	310	401		
2	74	42	206	82	257	320	406	636175 or resolvable nebulosity	5112
3	85	43	207	83	258	330	410		
4	93	44	209	84	259	340	414		
5	101	45	210	85	260	350	418		
6	107	46	212	86	261	360	422		
7	113	47	214	87	262	370	426		
8	118	48	215	88	263	380	430		
9	123	49	217	89	264	390	433		
10	127	50	218	90	265	400	437		
11	131	51	219	91	266	410	441		
12	135	52	221	92	267	420	444	2544700 or milky nebulosity	8115
13	139	53	222	93	268	430	448		
14	142	54	224	94	269	440	451		
15	146	55	225	95	270	450	455		
16	149	56	226	96	271	460	458		
17	152	57	228	97	272	470	461		
18	155	58	229	98	273	480	464		
19	158	59	230	99	274	490	468		
20	160	60	232	100	275	500	471		
21	163	61	233	110	284	510	474		
22	166	62	234	120	291	520	477		
23	168	63	236	130	300	530	480		
24	170	64	237	140	308	540	483		
25	173	65	238	150	315	550	486		
26	175	66	239	160	322	560	489		
27	177	67	240	170	328	570	492		
28	180	68	242	180	335	580	495		
29	182	69	243	190	341	590	498		
30	184	70	244	200	347	600	500		

Section of our sidereal system.

By taking out of this table the visual rays which answer to the gages, and applying lines proportional to them around a point, according to their respective right ascensions and north polar distances, we may delineate a solid by means of the ends of these lines, which will give us so many points in its surface; I shall, however, content myself at present with a section only. I have taken one which passes through the poles of our system, and is at rectangles to the conjunction of the branches which I have called its length. The name of poles seemed to me not improperly applied to those points which are 90 degrees distant from a circle passing along the milky way, and the north pole is here assumed to be situated in R.A. 186° and P.D. 58°. The section represented in fig. 4 is one which makes an angle of 35 degrees with our equator, crossing it in 124½ and 304½ degrees. A celestial globe, adjusted to the latitude of 55° north, and having σ Ceti near the meridian, will have the plane of this section pointed out by the horizon, and the gages which have been used in this delineation are those which in table I. are marked by asterisks. When the visual rays answering to them are taken out of the second table, they must be projected on the plane of the horizon of the latitude which has been pointed out; and this may be done accurately enough for the present purpose by a globe adjusted as above directed; for as gages, exactly in the plane of the section, were often wanting, I have used many at some small distance above and below the same, for the sake of obtaining more delineating points; and in the figure the stars at the borders which are larger than the rest are those pointed out by the gages. The intermediate parts are

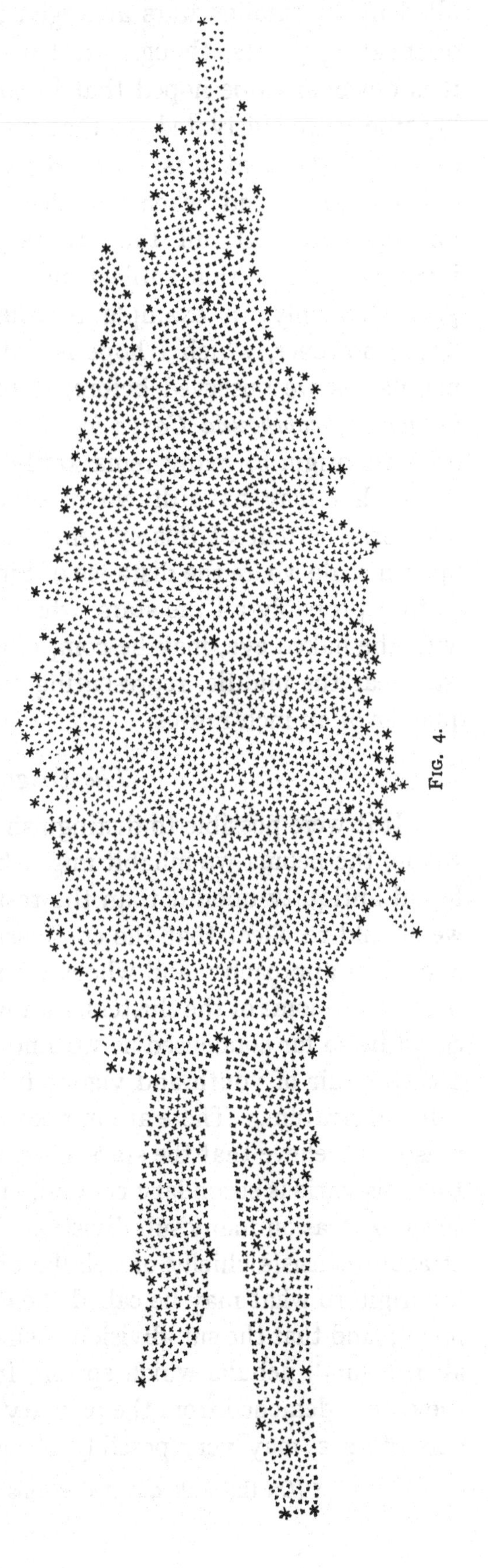

Fig. 4.

filled up by smaller stars arranged in straight lines between the gaged ones. The delineating points, though pretty numerous, are not so close as we might wish; it is however to be hoped that in some future time this branch of astronomy will become more cultivated, so that we may have gages for every quarter of a degree of the heavens at least, and these often repeated in the most favourable circumstances. And whenever that shall be the case, the delineations may then be repeated with all the accuracy that long experience may enable us to introduce; for, this subject being so new, I look upon what is here given partly as only an example to illustrate the spirit of the method. From this figure however, which I hope is not a very inaccurate one, we may see that our nebula, as we observed before, is of the third form; that is: *A very extensive, branching, compound Congeries of many millions of stars*; which most probably owes its origin to many remarkably large as well as pretty closely scattered small stars, that may have drawn together the rest. Now, to have some idea of the wonderful extent of this system, I must observe that this section of it is drawn upon a scale where the distance of Sirius is no more than the 80th part of an inch;* so that probably all the stars, which in the finest nights we are able to distinguish with the naked eye, may be comprehended within a sphere, drawn round the large star near the middle, representing our situation in the nebula, of less than half a quarter of an inch radius.

The Origin of nebulous Strata.

If it were possible to distinguish between the parts of an indefinitely extended whole, the nebula we inhabit might be said to be one that has fewer marks of profound antiquity upon it than the rest. To explain this idea perhaps more clearly, we should recollect that the condensation of clusters of stars has been ascribed to a gradual approach; and whoever reflects on the numbers of ages that must have past before some of the clusters, that will be found in my intended catalogue of them, could be so far condensed as we find them at present, will not wonder if I ascribe a certain air of youth and vigour to many very regularly scattered regions of our sidereal stratum. There are moreover many places in it where there is the greatest reason to believe that the stars, if we may judge from appearances, are now drawing towards various secondary centers, and will in time separate into different clusters, so as to occasion many sub-divisions. Hence we may surmise that when a nebulous stratum consists chiefly of nebulæ of the first and second form, it probably owes its origin to what may be called the decay of a great compound nebula of the third form; and that the sub-divisions, which happened to it in length of time, occasioned all the small nebulæ which sprung from it to lie in a certain range, according as they were detached from the primary one. In like manner our system, after numbers of ages, may very possibly become divided so as to give rise to a stratum of

* [On the scale adopted in this edition, the 100th part of an inch.—Ed.]

two or three hundred nebulæ; for it would not be difficult to point out so many beginning or gathering clusters in it.* This view of the present subject throws a considerable light upon the appearance of that remarkable collection of many hundreds of nebulæ which are to be seen in what I have called the nebulous stratum of Coma Berenices. It appears from the extended and branching figure of our nebula, that there is room for the decomposed small nebulæ of a large, reduced, former great one to approach nearer to us in the sides than in other parts. Nay, possibly, there might originally be another very large joining branch, which in time became separated by the condensation of the stars; and this may be the reason of the little remaining breadth of our system in that very place: for the nebulæ of the stratum of the Coma are brightest and most crowded just opposite our situation, or in the pole of our system. As soon as this idea was suggested, I tried also the opposite pole, where accordingly I have met with a great number of nebulæ, though under a much more scattered form.

An Opening in the heavens.

Some parts of our system indeed seem already to have sustained greater ravages of time than others, if this way of expressing myself may be allowed; for instance, in the body of the Scorpion is an opening, or hole, which is probably owing to this cause. I found it while I was gaging in the parallel from 112 to 114 degrees of north polar distance. As I approached the milky way, the gages had been gradually running up from 9·7 to 17·1; when, all of a sudden, they fell down to nothing, a very few pretty large stars excepted, which made them shew 0·5, 0·7, 1·1, 1·4, 1·8; after which they again rose to 4·7, 13·5, 20·3, and soon after to 41·1. This opening is at least 4 degrees broad, but its height I have not yet ascertained. It is remarkable, that the 80th *Nébuleuse sans étoiles* of the *Connoissance des Temps*, which is one of the richest and most compressed clusters of small stars I remember to have seen, is situated just on the western border of it, and would almost authorise a suspicion that the stars, of which it is composed, were collected from that place, and had left the vacancy. What adds not a little to this surmise is, that the same phænomenon is once more repeated with the fourth cluster of stars of the *Connoissance des Temps*; which is also on the western border of another vacancy, and has moreover a small, miniature cluster, or easily resolvable nebula of about 2½ minutes in diameter, north following it, at no very great distance.

Phænomena at the Poles of our Nebula.

I ought to observe, that there is a remarkable purity or clearness in the heavens when we look out of our stratum at the sides; that is, towards Leo, Virgo, and Coma

* Mr. MICHELL has also considered the stars as gathered together into groups (*Phil. Trans.* vol. LVII. p. 249); which idea agrees with the sub-division of our great system here pointed out. He founds an elegant proof of this on the computation of probabilities, and mentions the Pleiades, the Præsepe Cancri, and the nebula (or cluster of stars) in the hilt of Perseus's sword, as instances.

Berenices, on one hand, and towards Cetus on the other; whereas the ground of the heavens becomes troubled as we approach towards the length or height of it. It was a good while before I could trace the cause of these phænomena; but since I have been acquainted with the shape of our system, it is plain that these troubled appearances, when we approach to the sides, are easily to be explained by ascribing them to some of the distant, straggling stars, that yield hardly light enough to be distinguished. And I have, indeed, often experienced this to be actually the cause, by examining these troubled spots for a long while together, when, at last, I generally perceived the stars which occasioned them. But when we look towards the poles of our system, where the visual ray does not graze along the side, the straggling stars of course will be very few in number; and therefore the ground of the heavens will assume that purity which I have always observed to take place in those regions.

Enumeration of very compound Nebulæ or Milky-Ways.

As we are used to call the appearance of the heavens, where it is surrounded with a bright zone, the Milky-Way, it may not be amiss to point out some other very remarkable Nebulæ which cannot well be less, but are probably much larger than our own system; and, being also extended, the inhabitants of the planets that attend the stars which compose them must likewise perceive the same phænomena. For which reason they may also be called milky-ways by way of distinction.

My opinion of their size is grounded on the following observations. There are many round nebulæ, of the first form, of about five or six minutes in diameter, the stars of which I can see very distinctly; and on comparing them with the visual ray calculated from some of my long gages, I suppose, by the appearance of the small stars in those gages, that the centers of these round nebulæ may be 600 times the distance of Sirius from us.

In estimating the distance of such clusters I consulted rather the comparatively apparent size of the stars than their mutual distance; for the condensation in these clusters being probably much greater than in our own system, if we were to overlook this circumstance and calculate by their apparent compression, where, in about six minutes diameter, there are perhaps ten or more stars in the line of measures, we should find, that on the supposition of an equal scattering of the stars throughout all nebulæ, the distance of the center of such a cluster from us could not be less than 6000 times the distance of Sirius. And, perhaps, in putting it, by the apparent size of the stars, at 600 only, I may have considerably underrated it; but my argument, if that should be the case, will be so much the stronger. Now to proceed.

Some of these round nebulæ have others near them, perfectly similar in form, colour, and the distribution of stars, but of only half the diameter: and the stars in them seem to be doubly crowded, and only at about half the distance from each other: they are indeed so small as not to be visible without the utmost attention. I suppose these miniature nebulæ to be at double the distance of the first. An instance, equally remarkable and instructive, is a case where, in the neighbourhood of two such nebulæ as have been mentioned, I met with a third, similar, resolvable, but much smaller and fainter nebula. The stars of it are no longer to be perceived; but a resemblance of colour with the former two, and its diminished size and light, may well permit us to place it at full twice the distance of the second, or about four or five times that of the first. And yet the nebulosity is not of the milky kind; nor is it so much as difficultly resolvable, or colourless. Now, in a few of the extended nebulæ, the light changes gradually so as from the resolvable to approach to the milky kind; which appears to me an indication that the milky light of nebulæ is owing to their much greater distance. A nebula, therefore, whose light is perfectly milky, cannot well be supposed to be at less than six or eight thousand times the distance of Sirius; and though the numbers here assumed are not to be taken otherwise than as very coarse estimates, yet an extended nebula, which in an oblique situation, where it is possibly fore-shortened by one-half, two-thirds, or three-fourths of its length, subtends a degree or more in diameter, cannot be otherwise than of a wonderful magnitude, and may well outvie our milky-way in grandeur.

The first I shall mention is a milky Ray of more than a degree in length. It takes *k* (FL. 52) Cygni into its extent, to the north of which it is crookedly bent so as to be convex towards the following side; and the light of it is pretty intense. To the south of *k* it is more diffused, less bright, and loses itself with some extension in two branches, I believe; but for want of light I could not determine this circumstance. The northern half is near two minutes broad, but the southern is not sufficiently defined to ascertain its breadth.

The next is an extremely faint milky Ray, above $\frac{3}{4}$ degree long, and 8 or 10′ broad; extended from north preceding to south following. It makes an angle of about 30 or 40 degrees with the meridian, and contains three or four places that are brighter than the rest. The stars of the Galaxy are scattered over it in the same manner as over the rest of the heavens. It follows ε Cygni 11·5 minutes in time, and is 2° 19′ more south.*

The third is a branching Nebulosity of about a degree and a half in right ascension, and about 48′ extent in polar distance. The following part of it is divided into several streams and windings, which, after separating, meet each other again towards the south. It precedes ζ Cygni 16′ in time, and is 1° 16′ more north. I suppose this to be joined to the preceding one; but having observed them in different sweeps, there was no opportunity of tracing their connection.†

The fourth is a faint, extended milky Ray of about 17 in length, and 12′ in breadth. It is brightest and broadest in the middle, and the ends lose themselves.

* [V. 15 = N.G.C. 6960.—ED.]

† [They are identical, = V. 14 = N.G.C. 6992.—ED.]

It has a small, round, very faint nebula just north of it; and also, in another place, a spot, brighter than the rest, almost detached enough to form a different nebula, but probably belonging to the great one. The Ray precedes α Trianguli 18′·8 in time, and is 55′ more north. Another observation of the same, in a finer evening, mentions its extending much farther towards the south, and that the breadth of it probably is not less than half a degree; but being shaded away by imperceptible gradations, it is difficult exactly to assign its limits.*

The fifth is a Streak of light about 27′ long, and in the brightest part 3 or 4′ broad. The extent is nearly in the meridian, or a little from south preceding to north following. It follows β Ceti 5′·9 in time, and is 2° 43′ more south. The situation is so low, that it would probably appear of a much greater extent in a higher altitude.†

The sixth is an extensive milky Nebulosity divided into two parts; the most north being the strongest. Its extent exceeds 15′; the southern part is followed by a parcel of stars which I suppose to be the 8th of the *Connoissance des Temps.*‡

The seventh is a wonderful, extensive Nebulosity of the milky kind. There are several stars visible in it, but they can have no connection with that nebulosity, and are, doubtless, belonging to our own system scattered before it. It is the 17th of the *Connoissance des Temps.* §

In the list of these must also be reckoned the beautiful Nebula of Orion. Its extent is much above one degree; the eastern branch passes between two very small stars, and runs on till it meets a very bright one. Close to the four small stars, which can have no connection with the nebula, is a total blackness; and within the open part, towards the north-east, is a distinct, small, faint nebula, of an extended shape, at a distance from the border of the great one, to which it runs in a parallel direction, resembling the shoals that are seen near the coasts of some islands.

The ninth is that in the girdle of Andromeda, which is undoubtedly the nearest of all the great nebulæ; its extent is above a degree and a half in length, and, in even one of the narrowest places, not less than 16′ in breadth. The brightest part of it approaches to the resolvable nebulosity, and begins to shew a faint red colour; which, from many observations on the colour and magnitude of nebulæ, I believe to be an indication that its distance in this coloured part does not exceed 2000 times the distance of Sirius. There is a very considerable, broad, pretty faint, small nebula near it; my Sister discovered it August 27, 1783, with a Newtonian 2-feet sweeper. It shews the same faint colour with the great one, and is, no doubt, in the neighbourhood of it. It is not the 32d of the *Connoissance des Temps*; which is a pretty large round nebula, much condensed in the middle, and south following the great one; but this is about two-thirds of a degree north preceding it, in a line parallel to β and ν Andromedæ.

* [V. 17 = N.G.C. 598.—Ed.]

† [V. 20 = N.G.C. 247.—Ed.]

‡ [V. 13 = N.G.C. 6533.—Ed.]

§ [M. 17 = N.G.C. 6618.—Ed.]

To these may be added the nebula in Vulpecula:* for, though its appearance is not large, it is probably a double stratum of stars of a very great extent, one end whereof is turned towards us. That it is thus situated may be surmised from its containing, in different parts, nearly all the three nebulosities; *viz.* the resolvable, the coloured but irresolvable, and a tincture of the milky kind. Now, what great length must be required to produce these effects may easily be conceived when, in all probability, our whole system, of about 800 stars in diameter, if it were seen at such a distance that one end of it might assume the resolvable nebulosity, would not, at the other end, present us with the irresolvable, much less with the colourless and milky sort of nebulosities.

A Perforated Nebula, or Ring of Stars.

Among the curiosities of the heavens should be placed a nebula, that has a regular, concentric, dark spot in the middle, and is probably a Ring of stars. It is of an oval shape, the shorter axis being to the longer as about 83 to 100; so that, if the stars form a circle, its inclination to a line drawn from the sun to the center of this nebula must be about 56 degrees. The light is of the resolvable kind, and in the northern side three very faint stars may be seen, as also one or two in the southern part. The vertices of the longer axis seem less bright and not so well defined as the rest. There are several small stars very near, but none that seem to belong to it. It is the 57th of the *Connoissance des Temps*. Fig. 5 is a representation of it.

FIG. 5.

Planetary Nebulæ.

I shall conclude this paper with an account of a few heavenly bodies, that from their singular appearance leave me almost in doubt where to class them.

The first precedes ν Aquarii 5′·4 in time, and is 1′ more north.† Its place, with regard to a small star Sept. 7, 1782, was, Distance 8′ 13″ 51‴; but on account of the low situation, and other unfavourable circumstances, the measure cannot be very exact. August 25, 1783, Distance 7′ 5″ 11‴, very exact, and to my satisfaction; the light being thrown in by an opaque-microscopic-illumination.‡ Sept. 20, 1783, Position 41° 24′ south preceding the same star; very exact, and by the same kind of illumination. Oct. 17, 1783, Distance 6′ 55″ 7‴; a second measure 6′ 56″ 11‴, as exact as possible. Oct. 23, 1783, Position 42° 57′; a second measure 42° 45′; single lens; power 71; opaque-microscopic-illumination. Nov. 14, 1783,

* [M. 27 = N.G.C. 6853.—ED.] † [IV. 1 = N.G.C. 7009.—ED.]

‡ It may be of use to explain this kind of illumination for which the Newtonian reflector is admirably constructed. On the side opposite the eye-piece an opening is to be made in the tube, through which the light may be thrown in, so as to fall on some reflecting body, or concave perforated mirror, within the eye-piece, that may throw it back upon the wires. By this means none of the direct rays can reach the eye, and those few which are reflected again from the wires do not interfere sensibly with the faintest objects, which may thus be seen undisturbed.

Distance 7′ 4″ 35‴. Nov. 12, 1784, Distance 7′ 22″ 35‴; Position 38° 39′. Its diameter is about 10 or 15″. I have examined it with the powers of 71, 227, 278, 460, and 932; and it follows the laws of magnifying, so that its body is no illusion of light. It is a little oval, and in the 7-feet reflector pretty well defined, but not sharp on the edges. In the 20-feet, of 18·7 inch aperture, it is much better defined, and has much of a planetary appearance, being all over of an uniform brightness, in which it differs from nebulæ: its light seems however to be of the starry nature, which suffers not nearly so much as the planetary disks are known to do, when much magnified.

The second of these bodies precedes the 13th of FLAMSTEED'S Andromeda about 1′·6 in time, and is 22′ more south.* It has a round, bright, pretty well defined planetary disk of about 12″ diameter, and is a little elliptical. When it is viewed with a 7-feet reflector, or other inferior instruments, it is not nearly so well defined as with the 20-feet. Its situation with regard to a pretty considerable star is, Distance (with a compound glass of a low power) 7′ 51″ 34‴. Position 12° 0′ s. preceding. Diameter taken with 278, 14″ 42‴.

The third follows B (FL. 44) Ophiuchi 4′·1 in time, and is 23′ more north. It is round, tolerably well defined, and pretty bright; its diameter is about 30″.†

The fourth follows η Sagittæ 17′·1 in time, and is 2′ more north. It is perfectly round, pretty bright, and pretty well defined; about ¾ min. in diameter.‡

The fifth follows the 21st Vulpeculæ 2′·1 in time, and is 1° 46′ more north. It is exactly round, of an equal light throughout, but pretty faint, and about 1′ in diameter.

The sixth precedes *h* (FL. 39) Cygni 8′·1 in time, and is 1° 26′ more south. It is perfectly round, and of an equal light, but pretty faint; its diameter is near 1′, and the edges are pretty well defined.§

The planetary appearance of the two first is so remarkable, that we can hardly suppose them to be nebulæ; their light is so uniform, as well as vivid, the diameters so small and well defined, as to make it almost improbable they should belong to that species of bodies. On the other hand, the effect of different powers seems to be much against their light's being of a planetary nature, since it preserves its brightness nearly in the same manner as the stars do in similar trials. If we would suppose them to be single stars with large diameters we shall find it difficult to account for their not being brighter; unless we should admit that the intrinsic light of some stars may be very much inferior to that of the generality, which however can hardly be imagined to extend to such a degree. We might suspect them to be comets about their aphelion, if the brightness as well as magnitude of the diameters did not oppose this idea; so that after all, we can hardly find any hypothesis so probable as that of their being Nebulæ; but then they must consist of stars that

* [IV. 18 = N.G.C. 7662.—ED.] † [IV. 11 = N.G.C. 6369.] ‡ [IV. 16 = N.G.C. 6905.]
§ [The fifth and sixth are identical, = IV. 13 = N.G.C. 6894. See below, p. 293.—ED.]

are compressed and accumulated in the highest degree. If it were not perhaps too hazardous to pursue a former surmise of a renewal in what I figuratively called the Laboratories of the universe, the stars forming these extraordinary nebulæ, by some decay or waste of nature, being no longer fit for their former purposes, and having their projectile forces, if any such they had, retarded in each others' atmosphere, may rush at last together, and either in succession, or by one general tremendous shock, unite into a new body. Perhaps the extraordinary and sudden blaze of a new star in Cassiopea's chair, in 1572, might possibly be of such a nature. But lest I should be led too far from the path of observation, to which I am resolved to limit myself, I shall only point out a considerable use that may be made of these curious bodies. If a little attention to them should prove that, having no annual parallax, they belong most probably to the class of nebulæ, they may then be expected to keep their situation better than any one of the stars belonging to our system, on account of their being probably at a very great distance. Now to have a fixed point somewhere in the heavens, to which the motions of the rest may be referred, is certainly of considerable consequence in Astronomy; and both these bodies are bright and small enough to answer that end.*

W. HERSCHEL.

Datchet near Windsor,
January 1, 1785.

* Having found two more of these curious objects, I add the place of them here, in hopes that those who have fixed instruments may be induced to take an early opportunity of observing them carefully.

Feb. 1, 1785. A very bright, planetary nebula, about half a minute in diameter, but the edges are not very well defined. It is perfectly round, or perhaps a very little elliptical, and all over of an uniform brightness: with higher powers it becomes proportionally magnified. It follows γ Eridani 16′ 16″ in time, and is 49′ more north than that star. [IV. 26 = N.G.C. 1535.]

Feb. 7, 1785. A beautiful, very brilliant globe of light; a little hazy on the edges, but the haziness goes off very suddenly, so as not to exceed the 20th part of the diameter, which I suppose to be from 30 to 40″. It is round, or perhaps a very little elliptical, and all over of an uniform brightness: I suppose the intensity of its light to be equal to that of a star of the ninth magnitude. It precedes the third *b* (Fl. 6) Crateris 28′ 36″ in time, and is 1° 25′ more north than that star. [IV. 27 = N.G.C. 3242.]

XVI.

Catalogue of One Thousand new Nebulæ and Clusters of Stars.

[*Phil. Trans.*, vol. lxxvi., 1786, pp. 457–499.]

Read April 27, 1786.

THE following Catalogue, which contains one thousand new Nebulæ and Clusters of stars, is extracted from a series of observations (or Sweeps of the heavens), which was begun in the year 1783, and which I am still continuing till the whole be completed. As I may, perhaps, find an opportunity hereafter to publish these observations at full length, I shall now only mention such circumstances, relating to the instrument and apparatus with which they were made, as will be necessary to shew what degree of accuracy may be expected in the determination of the places of these Nebulæ and Clusters of stars; and also to serve any astronomer, who wishes to review them, to form a judgment what instrument will suffice for this purpose.

The telescope I have used, as has been observed on a former occasion,* is a Newtonian reflector of 20-feet focal length, and $18\frac{7}{10}$ inches aperture. The sweeping power has been 157, except where another is expressly mentioned. The field of view 15′ 4″.

My eye-glass is mounted on that side of an octagon tube, which, in the horizontal position of the instrument, makes an angle of 45° with the vertical; having found, by experience, that this position, resembling the situation of a reading desk, is preferable to the perpendicular one commonly used in the Newtonian construction.

In the present improved state of the apparatus this telescope will, in general, give the relative place of an object by a single observation true to within $1\frac{1}{2}$ or 2 minutes of polar distance, and 4 or 6 seconds of time in right ascension. But when there is an opportunity of repeating the observation, it will hardly differ a single minute in the former, and seldom so much as 3 or 4″ in the latter. My apparatus, however, has not been equally perfect from the beginning; for, being from time to time adapted to the different views I had in sweeping, it could only arrive to its present degree of perfection by many experiments and gradual improvements.

* *Philosophical Transactions*, vol. LXXIV. p. 437 [above, p. 157].

To begin a short history of this 20-feet telescope. In the month of October of the already mentioned year I began to use it, being then mounted on its present stand, but with a lateral motion under the point of support of the great speculum, by which its direction could be changed about 15 degrees. It had also a kind of moveable gallery in front, about nine feet long, which permitted me to follow a celestial object near 15 degrees more; by which means I obtained a range of 30 degrees without moving the stand. The Newtonian form has the capital advantage of rendering observations equally commodious in all altitudes; I had therefore placed the instrument in the meridian, that I might view the stars in their most favourable situation.

When I had seen most of the objects I wished to examine, I proceeded to the work of a general review of the heavens. The first method that occurred was, to suffer the telescope to hang freely in the center; then, walking backwards and forwards on the moveable gallery, I drew the instrument from that position by a handle fastened to a place near the eye-glass, so as to make it follow me, and perform a kind of very slow oscillations of 12 or 14 degrees in breadth, each taking up generally from 4 to 5 minutes of time. At the end of each oscillation I made a short memorandum of the objects I chanced to see; and when a new nebula or cluster of stars came in my way, I made a delineation of the stars in the field of view, both of the finder and of the telescope, that it might serve me to find them again. This being done, the instrument was, by means of a fine motion under my hands, either lowered or raised about 8 or 10 minutes, and another oscillation was then performed like the first. Thus I continued generally for about 10, 20, or 30 oscillations, according as circumstances would permit; and the whole of it was then called a *Sweep*, and as such numbered and registered in my journal.

When I had completed 41 Sweeps, the disadvantages of this method were too evident to proceed any longer. By going into the light so often as was necessary to write down my observations, the eye could never return soon enough to that full dilatation of the iris which is absolutely required for delicate observations. The difficulty also of keeping a proper memorandum of the parts of the heavens which had been examined in so irregular a manner, intermixed with many short and long stops while I was writing, as well as the fatigue attending the motion, upon a not very convenient gallery, with a telescope in my hands of no little weight, especially at the extremes of the oscillations, where it made a considerable arch upwards, were sufficient motives to induce me to look out for another method of sweeping. And it is evident, that the places of nebulæ hitherto determined, which was till the 13th of December, 1783, must be liable to great inaccuracy. I therefore began now to sweep with a vertical motion; and as this increased the labour of continually elevating and depressing the telescope by hand, I called in the assistance of a workman to do that part of the business, by which means I could observe very commodiously, and for a much longer time than before.

Soon after I removed also the only then remaining obstacle to seeing well, by having recourse to an assistant, whose care it was to write down, and at the same time loudly to repeat after me, every thing I required to be written down. In this manner all the descriptions of nebulæ and other observations were recorded; by which I obtained the singular advantage that the descriptions were actually writing and repeating to me while I had the object before my eye, and could at pleasure correct them, whenever they disagreed with the picture before me without looking from it.

In about half a dozen sweeps, done according to this new way, I found that the stars of FLAMSTEED'S Catalogue entered nearly at the time when they were expected; this suggested the possibility of converting my telescope into a transit instrument. By way of trial, Dec. 18, 1783, I began to use a watch, and noted the times of the transits of stars and nebulæ to the nearest minute; and, this succeeding, Dec. 24, a sidereal time-piece was introduced.

I found also that, by the turns of the handle which gave motion to the telescope, it was practicable, in a coarse way, to ascertain the difference of altitude between any two objects that passed the field of view; on which account, Dec. 30, I began to use an index-board, divided into inches, and marked with numbers, which, being placed behind the rope that moved the telescope, would point out at what altitude a certain index, affixed to the rope, was situated. My tackle of ropes and pullies was such that, while the telescope traversed an arch of two degrees, the mark on the rope passed over about 24 inches of the index-board: but the exact measure was always to be determined experimentally, as it varied according to the situation of the instrument. I perceived immediately that the quantity of rope used in the motion of the telescope would be much better observed by the assistant, if the index were brought within doors near the writing desk: to effect this, I used a small cord, which, being led off from the great one, was carried over a pulley into the observatory, so as to pass over a set of numbers, which I now divided into such parts as, in an equatorial situation of the instrument, would give nearly each equal to one minute.

It would exceed the limits of this Paper to enumerate the various trials I made to bring the right ascension to greater perfection; such as causing the tube sometimes to hang inclining or rubbing against a perpendicular plane; at others, drawing it against the same by a small weight, fastened to a cord, passing over a side pulley, &c. I shall also pass over the several changes in the form of the machine shewing the polar distance, which, for convenience sake, was soon brought to an index moving over a dial, in the manner of a clock.

By way of directing the person who gives motion to the telescope, a small machinery was added, which strikes a bell at each extreme of the breadth of the sweep, and is adjustable to any required number of turns of the handle.

In June, 1784, I introduced a small quadrant of altitude, the use of which became soon after of the greatest consequence in determining the value of the numbers of the polar distance piece. Hitherto I had settled this value by causing a star to pass vertically through the field of the finder, which was very accurately limited to two degrees; but now I found, by many comparisons between the degree determined by the quadrant and by the finder, that I had generally under-rated the value of the numbers. Fortunately so many stars of FLAMSTEED's Catalogue had been taken, that the numbers between their different polar distances were sufficient to recover the value of the degree; but this occasioned a laborious recalculation of the places of all objects taken in near 300 sweeps. The quadrant being once introduced, I carried the refinements of the determination, in high sweeps where the ropes acted very unequally, so far as to ascertain by it separately the value of every 20 or 30 minutes throughout the whole breadth of a sweep of two degrees, and the numbers were then accordingly cast up by so many different tables calculated on purpose.

Being still disappointed in many instances, when, on a review of a nebula whose place I had before determined, I perceived a difference of 4 or 5 minutes in polar distance, I began at last intirely to new model the machinery of the polar distance piece, and on Sept. 24, 1785, completed one with the following capital improvements. My former piece shewed a set of numbers whose value differed in every situation of the telescope, and therefore required different and very extensive tables to cast them up in degrees and minutes. This shews at once both the degree and minute of the polar distance of every celestial object, without requiring any tables to cast up numbers. In the next place, the considerable inaccuracy arising from the unequal tension of the great ropes, and their expansion or contraction by moisture or dryness, is intirely taken away; for now my index cord is contrived so as to go off from the front of the telescope itself, in the direction of a tangent to the arch it describes when moving; by which means this cord will even serve as an hygrometer to shew the variations of the ropes that suspend the telescope. If a shower of rain, for instance, should shorten them so as to elevate the telescope 2, 4, or 6 minutes, which has happened sometimes, notwithstanding they have all been well saturated with oil, the index cord will immediately make the polar-distance-clock shew this effect of the rain, by pointing out an equal change on the dial. As to the variations of the cord itself, they are in the first place very trifling, since it consists merely of a few threads of hemp, very loosely twisted, well oiled, and always equally stretched; but especially these variations are of no consequence, as they are so easily to be discovered by the check of the quadrant of altitude affixed to the telescope, or the successive transits of known stars, and may either be immediately corrected by the adjustable hand of the polar distance dial, or be left to be accounted for afterwards.

The improvement of the right ascension has not been less attended to; and the Royal Society having kindly intrusted me with an excellent time-piece, I succeeded at last by means of the addition of the following apparatus. Against the side of the tube is fixed a vertical iron plate, and the point of suspension of the telescope is disposed so as to permit this plate to be just in contact with a roller which remains fixed during the time of a sweep. There is also a considerable spring applied on the opposite side, in such a manner as, by always exerting a pressure nearly uniform, to cause the iron plate to rub against the fixed roller as the telescope sweeps up and down. By this means I have frequently, in very stormy weather, observed many hours without finding my time materially affected, and the corrections will seldom, in accurate observations, exceed a few seconds.

To those who are accustomed to the accuracy of transit instruments in regular observatories, this telescope, notwithstanding the above-mentioned improvements, may perhaps appear far from being brought to perfection; but they should recollect the size of the instrument as well as its extensive use, since I can not only follow any object for near a quarter of an hour, without disturbing the situation of the apparatus, but can at pleasure, in a few minutes, turn it to any part of the heavens, and view a celestial object wheresoever it may chance to be situated, even the zenith not excepted.

From this account it will be understood, that the places of a few of the nebulæ and clusters of stars, determined before the 13th of December, 1783, may be faulty in right ascension as far as 1′ of time, and in polar distance to 8 or 10′ of space. Afterwards the errors will be found to become gradually less considerable till the latter end of the year 1784, when, I suppose, they will seldom exceed half that quantity. From that period to Sept. 24, 1785, they will diminish, and probably not often amount to so much as 3 or 4′ in polar distance, and 10 or 12″ in right ascension. And now I flatter myself that all places, determined since the last mentioned time, will generally be true to a very small quantity; such as 4 or 6″ in right ascension, and 1½ or 2′ in polar distance, and often much nearer.

Some of the nebulæ in that part of the heavens which, in a former Paper, I have called the stratum of Coma Berenices, are indeed so crowded that there was no possibility of taking them all in the center of the field of view, and a somewhat less degree of accuracy may therefore be expected; but having used myself by very frequent estimations of the parts of the field of view to judge of their value in time as well as in space, I corrected this defect at the moment of observation by affixing to the transits of these excentric nebulæ such proper marks of *plus* or *minus* in right ascension and polar distance as I judged would bring them to a central observation. A similar method, well known to good astronomers in estimating their tenths of seconds by the proportional space over which the stars move in their meridian passage, makes it unnecessary to expatiate on the degree of accuracy that long practice enables us herein to obtain.

If, however, I had been willing to delay giving this catalogue till, by a repeated review of the heavens, the places had been more accurately determined, the work would undoubtedly have been more perfect; but whoever considers that it requires years to go through such observations will perhaps think with me, that it is the best way to give them in their present state, if it were but to announce the existence of such objects by way of inducing other astronomers also to look out for them. Another motive for not delaying this communication is to shew that my late endeavours to delineate the construction of the heavens have been guided by a careful inspection of them; and, probably, a catalogue which points out no less than one thousand instances of such systems as those are into which I have shewn the heavens to be divided, will considerably support what has been said on this subject in my two last Papers.

When the diurnal motion of the earth was first maintained, it could not but greatly add to the reception of this opinion when the telescope exposed to our view Jupiter, Mars, and Venus, revolving on their axes; * and if these instances of the similar condition of other planets support the doctrine of the diurnal motion, the view of so many sidereal systems, some of which we may discern to be of a most surprising extent and grandeur, will in like manner add credit to what I have proposed with regard to the condition of our situation within a system of stars: for, to the inhabitants of the nebulæ of the present catalogue, our sidereal system must appear either as a small nebulous patch; an extended streak of milky light; a large resolvable nebula; a very compressed cluster of minute stars hardly discernible; or as an immense collection of large scattered stars of various sizes. And either of these appearances will take place with them according as their own situation is more or less remote from ours.

In the distribution of the nebulæ and clusters of stars into classes, I have partly considered the convenience of other observers: thus, in the first class, the degree of brightness of the nebulæ has been the leading feature, as most likely to point out those which their several instruments may give them expectation to reach. The first class, therefore, contains the brightest of them; the second, those that shine but with a feeble light; and in the third are placed all the very faint ones. Besides this general division, I have added a fourth and a fifth class, which contain nebulæ that, on different accounts, seemed to deserve a more particular description than I had allotted to the three former divisions.

The clusters of stars are sorted by their apparent compression, in the manner of my former Catalogues of double, treble, and multiple stars; so that the closest and richest clusters take up the first class; the brightest, largest, and pretty much compressed ones, the second; and those, which consist only of scattered and less collected large stars, are put into the last.

In every class the order of time when the nebulæ and clusters of stars were discovered, or first observed with my 20-feet telescope, has been followed; and

* To these may now also be added Saturn, on whose body I have, in the year 1780, seen several belts, with spots that changed their situation in the course of a few nights.

that I might describe all these objects in as small a compass as could well be done, I have used single letters to express whole words, an explanation of which, with an example of the manner of reading those letters, is given. It should be observed, that all estimations of brightness and size must be referred to the instrument with which the nebulæ and clusters of stars were seen ; the clearness and transparency of the atmosphere, the degree of attention, and many more particular circumstances, should also be taken into consideration ; so that probably some of the nebulæ which I have called very bright, and very large, may only be just perceivable, as very small faint patches, in many of our best common telescopes.

The Identity of each nebula in this catalogue has been well ascertained by a projection on a proper map, made on purpose, which pointed out all other nebulæ near its place, and thus afforded the means of a rigorous examination. When, therefore, several nebulæ are found within the limits of the accuracy with which my telescope can discriminate them, in different nights, it may be concluded, that they were seen either at once in the same field of view, or otherwise in immediate succession during the same sweep.

In the same manner these nebulæ have been compared with those that are contained in the two volumes of the *Connoissance des Temps*, for the years 1783 and 1784, of which none have been inserted in this catalogue. It was indeed easy enough to distinguish the nebulæ of that excellent collection from those of mine which in several places are very near them : The quantity of good light in my telescope having enabled me, even in bright moon-light nights, to see occasionally some of the most feeble of the former, when the latter could not by any means be perceived.

Perhaps it will not be displeasing to those who may look out for some of the objects contained in this catalogue, to know that the pictures which were given in a former Paper, representing the various shapes and appearances of several nebulæ, have been actually taken from nature, by Drawings made of them while I had them in view ; I have therefore added a reference to these figures, as the descriptions of the originals which they represent occur in their order in the catalogue.

Arrangement of the columns, and explanations of the abbreviations.

The first column contains the class and number of the nebulæ.

In the second are the dates when the nebulæ were first observed.

The third column contains the star, or other object, by which the place has been determined.

In the fourth column the letter p or f shows that the nebula is either preceding or following the star.

In the fifth is the time, in sidereal minutes and seconds, by how much it precedes or follows the same star.

The letter n or s, contained in the sixth column, denotes that the nebula is north or south of the determining star.

In the seventh is the quantity, in degrees and minutes, by how much the nebula is more north or more south than the same star.

The eighth column contains the number of observations that have been made of each nebula;* and it is to be noted, that the determination of the place is generally taken from the last observation, on account of the more perfect state of the telescope.

The ninth column contains the description of the nebulæ, by means of single letters, or now and then a few words added to them.†

The abbreviations are to be understood as follows.

B. Bright.	v. very.
F. Faint.	c. considerably.
L. Large.	p. pretty.
S. Small.	e. extremely.

Of these letters I have composed vB. cB. pB. pF. vF. eF. vL. pL. pS. vS. eS.; all which require no farther explanation.

R. Round.	l. a little.
E. Extended.	i. irregularly.
M. in the middle.	g. gradually.
b. brighter.	s. suddenly.
m. much.	

When these are joined we have iR. mE. lE. bM. gbM. sbM. mbM. lbM. glbM. gmbM. smbM., and by taking in some of the former letters BM. vBM. cBM.; where no other remark will be necessary than that writing for instance bM, or brighter in the middle, it is intended to express, that a nebula, which is faint at the borders, is less so towards the middle. And these degrees of brightness happening sometimes to be so well united from the most imperceptible border to a very luminous center, I have, on such occasions, used the expression vgmbM, or very gradually much brighter in the middle.

r. resolvable. m. milky.
er. (joined) easily resolvable.
iF. (joined) of an irregular figure.
C. Cometic, or resembling a telescopic comet.
N. having a Nucleus, or bright compressed spot.
l, b, or d. (joined to minutes) long, broad, or diameter.
st. a star. stars.

* [That is, up to the time when this paper was sent to the Royal Society.—Ed.]

† [The last column has been added in the present edition. It gives the numbers of the objects in the "New General Catalogue" (*Mem. R.A.S.*, vol. 49). I.C. refers to the "Index Catalogues." An asterisk at the number refers to the Notes added in this edition at the end of the Catalogue.—Ed.]

n. north. north of.
s. south. south of.
p. preceding. np. north preceding. sp. south preceding.
f. following. nf. north following. sf. south following.
betw. between. ver. 240. verified by a power of 240.
bran. branches.
che. chevelure.
mer. in the direction of the meridian.
par. in the direction of the parallel of declination.
np sf. in a direction from north preceding to south following.
sp nf. in a direction from south preceding to north following.

Example. I. 13. 22. 69 Leon. p. 7. 57. n. 0. 2. 3. vB. mE. mer. smbM. 7 or 8′ l.
13th nebula of the 1st class. Feb. 22, 1784. It precedes the 69th Leonis of FLAMSTEED's Catalogue 7′ 57″ in time, and is 0° 2′ more north than that star. 3 observations. Very bright, much extended in the direction of the meridian of the nebula, suddenly much brighter in the middle, 7 or 8′ in length.

I. 32. . . . . p. 5. 11. n. 0. 28. 3. cB. S. BN. and 2vF. bran.
32d nebula of the first class. April 13, 1784. It precedes the 31st (or 1st d) Virginis of FL. Cat. 5′ 11″ in time, and is 0° 28′ more north than that star. 3 observations. Considerably bright, small, having a bright nucleus, and two very faint branches.

First class. Bright nebulæ.

I.	1783	Stars.		M. S.		D. M.	Ob.	Description.	N.G.C.
1	Dec. 19	82 (δ) Ceti	f	2 17	n	0 8	7	cB. cL. iF. bM.	1055
2	—	3 Leonis	p	18 7	s	1 12	5	cB. cL. vgbM. N. R.	2775
3	—	34 Sextant	p	28 55	s	0 13	4	cB. pL. C. mbM.	3166
4	—	- -	p	28 27	s	0 10	4	cB. pL. C. mbM.	3169
5	30	81 Leonis	p	2 42	n	0 7	2	B. pS. iR. bM. r.	3655
	1784								
6	Jan. 19	64 Virginis	f	33 56	s	0 1	3	vB. pL. gmbM.	5363
7	23	49 Leonis	f	126 45	s	0 40	1	vB. L. R. The place inac.	4472*
8	—	32 (2*d*) Virg	f	2 50	n	0 48	5	cB. pL. iR. mbM. r.	4698
9	24	10 (*r*) Virg	f	3 12	s	0 35	4	cB. E. np sf. N and 2 bran. 3′ l.	4179
10	—	- -	f	33 37	n	0 4	4	vB. pL. lE. gmbM. 2′ l. 1½′ b.	4643
11	Feb. 15	5 Comæ Be.	p	1 30	s	2 11	1	B. pL. lE. bM. m.	4153
12	19	6 Comæ	f	9 12	s	0 9	2	B. pS. R. BM. r.	4377
13	22	69 Leonis	p	7 57	n	0 2	3	vB. mE. mer. smbM. 7 or 8′ l. Fig. 11. Note.	3521
14	—	29 (γ) Virg	f	0 43	n	1 23	2	cB. cL. mE. near par. 3 or 4′ l.	4632
15	—	- -	f	3 23	n	0 58	2	cB. mE. sp nf. sbM. 4 or 5′ l.	4666
16	—	- -	f	10 34	n	0 13	2	cB. vL. iF. vgmbM.	4753
17	} Mar. 11	46 (*i*) Leo	f	15 50	s	1 32	5	{ The 2 p of 3. Both vB. cL.	3379
18	}		f	16 18	s	1 29	5	{ mbM. C. II. 41. Fig. 4. Note.	3384

I.	1784		Stars.		M. S.		D. M.	Ob.	Description.	N.G.C.
19	Mar.	14	11 Comæ	p	10 30	n	0 46	1	vB. pL. gbM.	4147
20		15	73 (*n*) Leonis	f	8 52	s	1 57	2	vB. mE. nearly par.	3666
21		—	– –	f	25 31	s	1 49	3	vB. cL. R. gmbM.	3810
22		—	34 Virginis	p	22 24	s	0 17	2	cB. pS.	4371
23		—	– –	p	18 24	s	0 19	2	B. S. mE.	4452
24		—	30 (ρ) Virg	p	1 42	s	0 5	2	vB. pL. r. near 2 Bst.	4596
25		—	34 Virginis	f	4 45	s	0 40	1	B. S. in a line with 2 st.	4754
26		19	52 (K) Leonis	p	3 45	s	2 9	1	cB. pL. not R. mbM.	3345*
27	Apr.	8	46 (*i*) Leonis	f	18 47	s	0 43	3	vB. BNM. and 2 F. bran. np sf.	3412
28		—	34 Virginis	p	19 36	n	1 8	2	{ One of two, at 4 or 5′ dist. B. cL. Note.	4435 4438
29		12	73 (*n*) Leonis	p	1 9	s	0 30	3	vB. cL. E. par. mbM.	3593
30		13	31 (1 *d*) Virg	p	17 41	n	0 32	2	vB. cL. lE. iF.	4365
31		—	31 (1 *d*) Virg	p	8 0	n	0 37	1	vB. E. mbM. r. betw. 2 Bst.	4526
32		—	– –	p	5 11	n	0 28	4	cB. S. BN and 2 vF. bran.	4570
33		15	9 (*o*) Virgin	f	3 12	n	1 39	1	B. L. mE. mbM. r.	4124
34		—	59 (*e*) Virgin	f	20 42	s	0 34	2	vB. cL. E. np sf. SBN.	5248
35		17	34 Virginis	p	31 42	n	1 5	1	B. vmE. vBM. 9 or 10′ l.	4216*
36 } 37 }		—	– –	p	11 24	n	0 20	1	{ Two. Both B. S. lE.	4550 4551
38		18	32 (2 *d*) Virg	p	11 36	n	0 0	1	B. vL. mE. mbM.	4526
39		24	51 (θ) Virg	p	21 36	s	0 14	1	vB. vL. smbM. rN.	4697
40		—	– –	p	5 48	n	0 2	1	cL. vBSNM.	4941
41		25	28 Virginis	f	9 24	n	1 2	1	B. L. iR. lbM.	4731
42		—	26 (χ) Virgin	f	30 27	n	0 8	2	cB. L. iR. vgbM.	4995
43	May	9	49 (*g*) Virgin	p	28 6	s	0 51	1	E. vBM. 5 or 6′ l.	4594*
44		21	51 (*e*) Ophiu	f	7 18	n	0 0	2	cB. pL. N.	6401
45		24	43 Ophiuchi	p	6 36	n	0 4	2	B. R. vgmbM.	6316
46		—	– –	f	0 54	n	1 46	1	pB. cL. R. BM. r.	6355
47	June	16	1 (*m*) Aquilæ	f	17 48	s	0 33	1	B. vL. iF. er. st visible.	6712
48		17	43 (*d*) Sagit^{i}	p	114 6	n	1 44	1	B. L. R. gbM. er.	6356
49		24	10 (γ) Sagit^{i}	p	2 18	n	0 23	1	B. pL. bM. r.	6522
50		—	19 (δ) Sagit^{i}	f	3 0	s	0 33	1	cL. R. vBM. m.	6624
51	July	12	22 (λ) Sagit^{i}	f	3 12	s	0 13	1	cL. R. vBM. er.	6638
52	Aug.	21	17 Delphini	f	6 0	n	2 24	4	vB. S. R. gmbM. r.	7006
53	Sept.	5	66 (υ) Cygni	f	78 6	s	0 51	2	vB. cL. mE. mbM. r.	7331
54	Oct.	5	35 (ν) Andr	f	18 36	s	1 26	1	pB. S. R. vgbM.	393*
55		19	66 Pegasi	p	17 59	n	0 2	3	cB. mE. mer. gbM. 4′ l. 2′ b.	7479
56 } 57 }	Nov.	16	4 (λ) Leonis	f	0 46	s	1 29	1	{ Two, at 1′ distance. Both cB. cL. appear like one mE.	2903 2905
58		17	19 Eridani	f	5 9	s	1 22	2	B. S. lE. mbM.	1395
59		20	15 (ι) Navis	f	64 18	n	0 21	1	S. cBM. lE. m.	2784
60	Dec.	9	19 Eridani	p	6 51	n	0 16	1	vB. S. lE. mbM.	1332
	1785									
61	Jan.	6	6 Sextantis	p	8 42	n	0 31	2	vB. S. iF. 1′ nfcBst.	2974
62		10	55 (ζ) Ceti	p	0 25	n	0 37	2	cB. pL. E. bM.	701
63		—	80 Ceti	f	5 12	s	0 25	1	B. R. mbM. 1′ d.	1052
64		—	8 (1 ρ) Erid	p	15 9	n	0 2	2	vB. pL. lE. mbM.	1084
65	Feb.	7	31 Crateris	f	23 30	n	0 52	1	vB. pL. iR. bM. like 2 N.	4361
66		8	12 Hydræ	f	25 2	s	1 7	1	B. vS. iF. mbM.	2781

I.	1785	Stars.		M. S.		D. M.	Ob.	Description.	N.G.C.
67	Feb. 8	8 (η) Corvi	p	37 17	n	2 10	3	cB. pL. iF. mbM. 2 or 3′ d.	3962
68	—	53 Virginis	p	12 40	n	1 11	1	cB. iR. mbM.	4856
69	—	– –	p	11 4	n	1 41	1	cB. pL. iR.	4902
70	Mar. 5	106 Virginis	f	1 2	n	0 54	1	vB. cL. iF. vgbM.	5634
71	—	19 (δ) Libræ	p	0 3	n	1 4	2	cB. vS. b towards f side.	5812
72	13	23 Leonis min	f	13 7	n	0 13	1	cB. cL. E. mbM.	3254
73	—	13 Can. vena	p	50 17	s	0 22	1	vB. S.	4150
74	—	– –	p	43 5	s	1 11	1	cB. R. mbM.	4245
75	—	– –	p	40 35	s	1 9	1	vB.	4274
76	—	– –	p	38 3	s	0 52	1	cB. L. E.	4314
77	—	– –	p	34 15	n	0 23	1	vB. L. broadly E. bM.	4414
78	April 3	27 Ursæ	f	7 46	n	0 4	1	vB. cL. vsmbM.	2985
79	—	– –	f	33 52	n	1 17	1	cB. pL. R. vgmbM.	3147*
80	—	– –	f	67 10	n	0 46	1	cB. S. i. elliptical.	3348
81	6	41 Leonis min	p	0 6	n	1 40	2	cB. cL. m. just p. 2 st.	3344
82	—	14 (b) Comæ	p	37 40	s	0 14	2	cB. pL. lE. mer. vgbM.	3900
83	—	21 (g) Comæ	f	0 10	n	1 12	1	cB. pL. iR. mbM.	4494
84	—	– –	f	19 34	n	0 55	1	cB. iR. sBM. m. 7 or 8′ d.	4725
85	10	40 Comæ	f	5 9	n	0 18	1	cB. pL.	5012
86	11	39 Leonis min	p	13 14	n	0 59	1	cB. pL. mbM. brightness lE.	3245
87	—	44 Leonis min	f	10 30	n	1 1	1	vB. vL. gbM.	3486
88	—	– –	f	13 30	n	0 1	1	cB. cL. iR. mbM.	3504
89	—	14 (b) Comæ	p	8 18	n	0 55	1	vB. S. lE.	4251
90	—	– –	p	6 30	n	1 57	1	The np of 2 cB. pL. R. II. 377.	4278
91	—	15 (c) Comæ	f	1 10	n	0 19	1	vB. E. par. pBLN. and 2 bran.	4448
92	—	– –	f	9 8	s	0 19	1	vB. vL. mE. np sf. 10 or 12′ l. 4 st. in it.	4559
93	—	31 Comæ	f	2 56	n	1 24	1	cB. pL.	4793

Second class. Faint nebulæ.

II.	1783	Stars.		M. S.		D. M.	Ob.	Description.	N.G.C.
1	Oct. 28	41 Aquarii	p	11 45	n	0 17	2	F. cL. mE. bM. er.	7184
2	30	24 (α) Pis. aust	f	14 40	n	1 2	3	pB. S. iF. mbM.	7507
3	Dec. 13	17 (1 φ) Ceti	p	9 ::	n	2 ::	2	F. L. mE. between 2 cBst.	157*
4	—	41 Ceti	f	15 13	n	0 37	6	pB. pS. R. mbM. C.	596
5	18	82 (δ) Ceti	p	0 5	n	0 46	8	pB. S. lE. bM.	1032
6	—	– –			n	½ ::	1	S. C, between 2 L and 1 S st. Note.	1055*
7	19	45 Eridani	p	1 13	n	0 54	3	F. pL. iR. vlbM.	1589
8 }	—	44 Eridani	f	2 11	s	0 41	4	Two. The first. F. S. r.	1587
9 }			f	2 18	s	0 42	4	The second, F. vS. r.	1588
10	24	88 (γ) Pegasi	p	13 38	s	0 23	2	F. pL. E. sp nf. bM. r.	7800
11	30	6 Comæ	f	1 13	n	0 30	2	F. pL. nearly R. r.	4237
12	—	27 Comæ	p	3 15	s	0 9	2	pB. cL. lE. mbM. r.	4651

II.	1784	Stars.		M. S.		D. M.	Ob.	Description.	N.G.C.
13	Jan. 18	78 (ι) Leonis	f	6 18	s	1 10	2	pB. pL. mbM. r.	3705
14	—	59 (e) Virginis	p	69 0	n	0 12	1	lE. not C.	4119*
15	—	20 Virginis	f	4 12	s	0 42	2	F. pL. pR.	4578
16	23	56 Leonis	p	0 32	n	1 32	4	F. vS. nearly R.	3462
17	—	9 (o) Virginis	f	11 54	s	1 33	3	F. pL. E. followed by III. 91.	4235
18	—	31 (1 d) Virg	p	12 28	n	1 1	2	F. S.	4470*
19	—	I. Class 7 Neb	p	0 0	s	0 12	1	F. pS. R.	4470*
20	—	31 Bootis	p	118 45	s	0 38	2	vS.	4612*
21	—	32 (2 d) Virg	f	9 28	n	0 22	4	pB. pL. b towards the p side.	4795
22	—	31 Bootis	p	80 15			1	F. vS.	5106*
23	24	75 Leonis	f	70 0	n	0 26	2	F. mE.	4420
24	—	– –	f	97 0	n	0 7	2	F. pL.	4772*
25	—	78 Virginis	p	7 50	s	1 34	4	pB. cL. nearly R. mbM.	5147
26	28	11 (s) Virg	f	18 0	n	0 45	1	pB. cL. b towards the f side.	4453*
27	30	31 Bootis	p	9 6	s	0 2	3	F. pL. R. lb not M.	5665
28 }	Feb. 15	41 (γ) Leonis	f	3 45	n	0 18	1	Two, about 2′ asunder. Both	3226
29 }								F. cL. R. Fig. 3.	3227
30	—	68 (δ) Leonis	f	6 30	s	2 23	1	pB. r.	3626*
31	22	29 (γ) Virg	p	2 31	n	0 55	2	pB. cL. lE. par. r.	4592
32	23	84 (τ) Leonis	p	6 30	n	0 7	1	pB. vS. bM.	3645
33	—	– –	p	6 0	n	0 20	1	pB. pL. R. bM.	3640
34	—	60 (σ) Virgin	p	51 23	s	1 27	4	F. S.	4412
35	—	16 (c) Virgin	f	8 45	n	0 15	2	pB. mbM.	4457
36	—	60 (σ) Virgin	p	46 5	s	1 29	3	F. vL. iR. bM. 6′ l. 4′ b.	4496
37	—	16 (c) Virgin	f	13 15	s	0 40	2	pB. E. np sf. mbM.	4527
38	—	35 Virginis	p	5 30	s	0 53	2	pB. pL. iF. r.	4636
39	—	– –	p	3 0	s	0 20	1	pB. contains 2 stM.	4665*
40	Mar. 11	6 (h) Leonis	f	1 43	n	0 27	3	F. pL. lbM.	2911
41	—	46 (i) Leonis	f	16 30	s	1 35	4	The f of 3. F. E.	3389
42	—	78 (ι) Leonis	p	14 0	n	0 12	1	F. S.	3547
43	12	36 (ζ) Leonis	p	3 10	s	0 40	2	pB. cL. iF.	3162
44 }	—	20 Leonis	f	28 15	n	0 48	1	Two. Both F. E. lbM. r.	3190
45 }									3193
46	—	68 (δ) Leonis	p	37 30	n	1 29	1	pB. S. r.	3301*
47	—	54 Leonis	p	2 26	s	1 49	2	pB. pL. lE. r. 3 or 4 st in it.	3437
48	14	85 (l) Gemin	f	54 45	s	0 42	1	pB. pL. lbM. contains 1 st.	2677*
49	—	86 Leonis	p	15 0	s	0 19	1	vgbM. r.	3599
50 }	—	– –	p	13 30	s	0 22	1	Of three that M. pB. cL. R. bM.	3607
51 }								That to the n. S. R. bM. III. 27.	3608
52	—	– –	p	10 0	s	0 5	1	pB. S. lE. bM.	3626
53	—	81 Leonis	p	1 36	n	1 19	2	F. E. r.	3659
54	—	85 Leonis	p	1 24	n	1 29	2	F. S. R.	3691
55	—	11 Comæ	f	4 45	n	0 24	1	The f. of 2. r. See Note.	4394
56	—	25 Comæ	p	8 30	n	0 1	2	pL. iR. bM. 2 or 3′ d.	4450*
57 }	15	5 (ξ) Leonis	p	7 15	n	0 18	1	Two, distant 1′ np sf. The p.	2872*
58 }								pS. lbM. r. The f. pL. lbM. r.	2874*
59	—	– –	f	26 30	s	0 53	1	vS. C. in a row with 2 F and 1 Bst.	3070
60	—	12 (t) Virg	p	5 15	n	0 10	1	F. S.	4124
61 }	—	34 Virginis	p	26 12	s	0 31	2	Two, nearly par. The first F.	4294
62 }								pL. E. The second F. pL. R.	4299

II.	1784	Stars.		M. S.		D. M.	Ob.	Description.	N.G.C.
63	Mar. 15	34 Virginis	p	24 42	s	0 9	2	F. pL. mE.	4313
64	—	12 (*t*) Virg	f	11 45	n	0 57	1	F. vS.	4352*
65	—	– –	f	14 45	n	0 50	1	pB. not vS.	4429
66	—	30 (ρ) Virg	p	9 30	n	0 56	1	pB.	4503
67	—	– –	p	8 30	n	1 6	1	pB. vS.	4528*
68	—	– –	p	5 30	n	1 12	2	pB.	4564*
69	—	– –	p	0 48	s	0 4	2	pB. pL. R. mbM. r.	4608
70	—	30 (ρ) Virg	f	0 45	n	1 7	1	A nebula.	4638
71	—	– –	f	2 0	n	1 0	1	S.	4660
72	—	34 Virginis	f	..	s	0 54	1	S. lE.	4694*
73	—	– –	f	4 0	s	1 3	1	F. not vS.	4733
74 }	—	– –	f	5 30	s	0 42	1	Two, nearly par. The p. pB. nearly R. The f. pB. vmE. 8 or 10′ distance.	4754
75 }									4762
76	—	20 (χ) Serp^{s}	p	3 12	s	0 42	2	pB. pL. lE. gbM. r.	5970
77	19	52 (K) Leonis	p	4 42	s	0 27	2	pB. pL. E. bM. r. f. pBst.	3338
78	—	– –	p	0 12	s	0 27	2	pB. pL. r.	3367
79	—	15 Bootis	f	18 30	s	0 15	1	F. L. R. lbM. r. 4 or 5′ dia^{r}	5669
80	21	47 (δ) Cancri	f	4 45	n	0 55	1	pB. pL. E. r. 2 or 3 st in it.	2672*
81	—	51 (*m*) Leon	f	1 15	s	1 36	1	pB. pL. not R. r.	3370
82	—	– –	f	8 15	s	1 35	1	F. S. lE. r. s. pBst.	3455
83	—	3 Comæ	p	0 15	s	0 43	1	F. pL. r.	4152
84	—	– –	f	12 30	s	0 59	1	F. S. R. r. Note.	4328
85 }	—	25 Comæ	p	13 0	s	0 21	1	Two. The p. pB. S.	4340
86 }								The f. F. S.	4350
87	—	– –	p	11 15	s	1 33	1	S. bM. r.	4379
88	—	– –	p	10 45	s	0 53	1	S. bM. r.	4405
89	—	6 Comæ	f	11 12	n	0 31	2	S. bM. r. near Bst.	4421
90	—	25 Comæ	p	7 30	s	0 3	1	pB. bM. r. near Bst.	4450*
91	—	– –	p	6 0	s	0 18	1	vS.	4489
92	—	– –	p	5 30	s	0 24	1	S.	4502
93	—	– –	p	4 0	s	0 51	1	F. vS.	4515
94	—	– –	p	2 0	s	1 35	1	F. S.	4540
95	—	27 Comæ	f	2 45	s	1 23	1	pB. vmE. nearly mer.	4710
96	—	15 Serpentis	f	1 0	s	1 5	1	pB. pL. not R. bM. r.	5962
97	—	26 Serpentis	f	2 30	n	0 35	1	pF. S. r. p. 2 pBst.	5996
98	23	8 Leonis	f	15 45	n	0 20	2	F. cL. iR. mbM. 4 or 5′ dia^{r}	3041*
99	April 8	52 (K) Leo	f	0 42	s	0 12	1	pB. S.	3377
100	—	– –	f	13 30	n	0 42	1	F. pS. r.	3485
101	—	– –	f	14 0	s	0 20	1	pB. S. mbM.	3489
102	—	70 (θ) Leonis	f	0 48	s	0 38	1	F. pS. R. lbM. r.	3596
103	—	94 (β) Leonis	p	9 12	n	0 47	1	F. S. E. r. 2 or 3 st visible in it.	3800
104	—	– –	p	3 18	s	0 48	2	pB. S. R. r. pLrN.	3872
105	—	34 Virginis	p	35 6	n	1 17	2	pB. pL. R. vgmbM. r.	4168
106	—	– –	p	33 18	n	1 28	2	F. pL.	4189
107	—	6 Comæ	p	0 24	s	1 0	1	pL.	4212*
108	—	– –	p	0 6	s	1 2	1	mE. r.	4212*
109	—	– –	f	0 24	s	1 54	1	r.	4222*
110	—	– –	f	3 12	s	0 3	1	S. r.	4262

II.	1784	Stars.		M. S.		D. M.	Ob.	Description.	N.G.C.
111	April 8	6 Comæ	f	5 18	s	0 18	1	Two, about 2′ distant. The first R. r. The 2d, E. r.	4298
112									4302
113	—	– –	f	10 54	n	0 9	1	E. r.	4419
114	—	34 Virginis	p	16 54	n	1 28	2	F. r.	4473
115	—	6 Comæ	f	14 0	s	1 21	1	Two.	4477
116								Both r.	4479
117	—	– –	f	14 24	s	0 54	1	r.	4474
118	—	– –	f	15 36	s	0 28	1	S. Note.	(4501)*
119	—	– –	f	18 54	n	0 35	1	pL. r.	4540
120	—	– –	f	19 6	s	0 25	1	L. r.	4548
121	12	34 Virginis	p	17 54	n	1 21	1	Two. Both pF. S. bM.	4458
122									4461
123	—	– –	p	17 0	n	0 23	2	The two p. of 3. Both F. S. bM. Note.	4476*
124									4478*
125	—	– –	p	4 18	n	1 25	1	not vF. S. r.	4639
126	—	– –	p	3 0	n	1 11	2	pB. L. E. r.	4654
127	—	– –	p	2 30	n	1 39	1	F. vS. R. lbM. r.	4659
128	—	– –	f	0 42	n	1 52	1	L. R. bM. r.	4689*
129	—	41 Virginis	f	19 0	n	0 11	1	F. pL. lbM. R. r.	5020
130	—	20 (χ) Serp^{s}	p	12 30	n	0 12	1	F. not S. iF. r.	5936
131	13	56 Leonis	p	3 48	s	0 26	1	pB. vL. nearly R. lbM.	3423*
132	—	8 (π) Virgin	p	4 48	n	0 7	1	pL. E. pBM. r.	3976
133	—	9 (o) Virginis	f	7 46	s	1 41	2	not vF. S. E. mer.	4180
134	—	11 (s) Virgin	f	4 24	s	0 0	1	F. mE.	4197
135	—	– –	f	5 54	n	0 38	1	S. E. pBM.	4215
136	—	– –	f	6 30	n	1 39	1	F. S. iF. r.	4224
137	—	9 (o) Virginis	f	12 32	s	2 2	2	F. pL. r.	4241*
138	—	11 (s) Virgin	f	9 6	n	0 20	3	F.	4260
139	—	– –	f	9 30	n	0 0	3	Two. The 1st is the largest. The 2d vF.	4261
140									4264
141	—	– –	f	13 18	n	0 18	3	Three nebulæ. The last is the largest.	4326
142									4333
143									4339
144	—	31 (1 d) Virg	p	17 9	n	0 42	2	F. pL. the largest of 2.	4370
145	—	60 (σ) Virg	p	50 28	n	0 30	2	vF. S. E.	4423
146	—	31 (1 d) Virg	p	14 6	s	0 34	1	F. pL.	4430
147	—	– –	p	7 36	s	0 20	1	pB. pL. mE. r.	4532
148	—	– –	p	0 18	n	0 31	1	not F. R. vgbM.	4612
149	—	– –	f	0 9	n	0 53	4	F. pL. iF. r.	4623
150	—	24 (α) Serp^{s}	p	73 42	n	1 10	1	F. pL. nearly R. er.	5645
151	—	12 Herculis	f	4 18	s	0 26	1	not vF. pL. iR. bM. r.	6106
152	15	78 (ι) Leonis	f	4 24	s	1 1	1	F. mE. r.	3692
153	—	2 (1ξ) Virgin	p	3 0	n	2 2	1	Two, about 5′ distant. Both F. pS. C.	3822
154									3825
155	—	20 Virginis	p	6 6	s	0 40	1	F. pL. lE. lb towards p side.	4417
156	—	– –	p	4 36	s	0 26	1	F. pL. lE. r.	4442
157	—	20 Virginis	p	3 36	s	1 29	1	F. pL. mE. bM. r.	4469
158	—	31 (1 d) Virg	p	8 38	n	1 51	3	pF. pL. nearly R. r.	4519
159	17	81 Leonis	f	0 36	n	0 24	1	pB. S. bM. almost stellar.	3681
160	—	– –	f	1 0	n	0 45	1	cL. R. vgbM.	3686*

II.	1784	Stars.		M. S.		D. M.	Ob.	Description.	N.G.C.
161	April 17	90 Leonis	f	5 36	n	0 53	1	F. not S. R. bM.	3801
162	—	34 Virginis	p	51 54	n	0 0	2	not vF. pL. iR. lb towards f side.	3968
163	—	– –	p	33 6	n	1 13	1	pS.	4193
164	—	– –	p	32 48	n	0 13	1	pS. vmE.	4200
165	—	– –	p	32 30	n	1 13	1	F. vmE.	4206*
166	—	– –	p	27 36	n	0 53	1	pB. vS.	4267
167	—	– –	p	21 30	n	0 49	1	Two nebulæ.	4387
168								The most s. E.	4388
169	—	– –	p	20 30	n	0 40	1	S.	4413
170	—	– –	p	19 42	n	0 49	1	F.	4425
171	—	– –	p	19 6	n	0 20	1	Three nebulæ.	4431
172								The two first vS.	4436
173								The third S.	4440
174	—	– –	p	17 48	n	1 16	1	F.	4461
175	—	– –	p	12 36	n	1 9	1	pF. L.	4531
176	—	– –	p	3 48	s	0 37	1	F.	4638
177	—	20 Bootis	f	3 30	s	1 42	1	pF. not S. lbM. r.	5600*
178	—	28 (β) Serp[s]	p	12 6	s	0 7	2	Two, very close. Both S. stellar. The s. is largest.	5953*
179									5954*
180	22	15 (η) Virg	f	8 59	s	1 18	3	pB. L. iR. er.	4454
181	—	29 (γ) Virg	f	5 18	s	1 15	1	pF. pL. E. r.	4684
182	—	– –	f	6 24	s	1 54	1	pF. pL. E. r.	4691
183	24	51 (θ) Virg	p	30 36	n	0 14	1	pB. cL. E. vsmbM.	4593
184	—	– –	p	28 30	n	0 26	1	not F. L. lE. lbM. r.	4602
185	—	– –	p	1 0	n	0 10	1	F. S. iF. near pBst.	4989
186	25	28 Virginis	f	12 6	n	0 51	1	pF. cL. R. r.	4775
187	—	– –	f	12 42	n	0 37	1	pF. pL. r.	4786
188	—	– –	f	22 54	n	0 57	1	F. cL. E. r.	4951
189	—	72 (1 l) Virg	p	21 54	s	0 18	1	pB. R. vsmbM. near Bst.	4981
190		26 (χ) Virg	f	23 44	s	0 6	2	F. pL. iR. lbM. r.	4928*
191	May 9	49 (g) Virg	p	4 6	s	0 46	1	pF. pS. R. r. near some Sst.	4933
192	—	18 Libræ	f	10 36	s	0 16	2	pF. pL. lE. mer. nearly.	5861
193	11	100 (λ) Virg	p	59 30	n	0 48	2	The most n of 3. pB. vS. bM.	5077
194	19	12 (d) Bootis	f	7 42	s	0 2	2	F. pL. R. mbM.	5548
195	21	39 Ophiuchi	p	12 54	n	1 42	2	pB. cL. iR. lbM. r.	6287
196	22	54 Hydræ	p	6 42	s	1 2	1	pB. S. nearly R. bM. r.	5694
197	—	51 (e) Ophiu	f	35 36	s	1 13	1	pB. pL. iR. r.	6544
198	24	3 (φ) Sagitt[1]	f	18 42	s	0 4	1	pF. not L. crookedly E. er.	6540
199	June 16	64 (ν) Ophiu	f	2 48	n	0 48	1	pB. pL. R. gbM. r.	6517
200	24	10 (γ) Sagitt[1]	p	1 6	n	0 22	1	F. pS. r. unequally B.	6528
201	July 13	18 Sagittarii	p	7 54	s	0 55	1	F. pL. lbM. r.	6569
202	17	12 (φ) Cygni	f	17 36	s	0 53	1	A resolvable nebulous patch of st.	6847*
203	—	65 (ζ) Cygni	p	9 30	s	0 16	2	pB. pL. iE. bM.	7013
204	Aug. 7	24 Sagittarii	p	9 18	n	0 50	1	pB. S. stellar, not verified.	6629*
205	—	– –	p	1 42	n	0 33	1	pB. cL. iE. bM.	6642
206	Sept. 7	52 (k) Cygni	f	5 36	n	1 22	1	F. S. crookedly E. r.	6979*
207	—	44 (η) Pegasi	p	34 27	n	1 15	1	cL. R. gmbM. er.	7217
208	10	84 (ψ) Pegasi	p	13 48	n	1 0	1	F. cL. R. vgbM. sf. st.	7741
209	—	34 (ζ) Andr	p	5 57	n	1 12	2	F. pL. iR. equally B. r.	214

II.	1784	Stars.		M. S.		D. M.	Ob.	Description.	N.G.C.
210	Sept. 11	31 (δ) Andr	f	18 12	s	0 26	1	F. pL. unequally B. near pBst.	315
211	—	13 Triang	f	5 24	s	0 35	1	F. pL. lE. bM. n. 2 st.	972
212	12	63 Pegasi	p	19 42	s	0 15	1	pB. pL. lE. mbM. r. s. 2 Fst.	7457
213	—	79 Pegasi	p	2 36	n	0 42	1	F. pL. vlE. lbM.	7753
214	—	40 Androm	p	7 18	s	0 15	1	F. E. p. Bst.	296
215 }	—	– –	f	4 30	n	0 41	1	Three. mer. Nearly equal in	379
216 }								size. All F. vS. R. proportion	380
217 }								of dist. s to n. 2 to 1.	383
218	—	– –	f	5 30	n	1 22	1	F.	392
219 }	—	– –	f	7 36	n	1 22	1	Two. The p. F. vS.	407
220 }								The f. pL.	410
221	—	3 (ε) Triang	p	6 12	s	0 15	1	F. pL. mE. r. 1½′ l.	740
222	—	– –	p	5 12	s	0 10	1	Just like the former.	750
223	—	– –	p	2 12	s	1 52	1	pB. pS. R.	777
224	13	43 (β) Andr	p	0 18	n	0 5	1	pB*. cL. R. bM. {* Though β And. in the field.	404
225	—	9 (γ) Trian	f	4 18	s	0 39	1	F. vS. R.	890
226	15	71 (γ) Pegasi	p	4 54	s	0 5	1	F. pL. bM. elliptical.	7678
227	—	89 (χ) Pegasi	p	10 18	n	0 32	2	F. cL. mE. r.	7817
228 }	—	6 (β) Arietis	p	5 12	n	1 7	1	Two. Both F. pS. iR.	678
229 }									680
230	18	81 (φ) Pegasi	p	1 27	n	1 4	1	F. pL. R. bM. r.	7769
231	—	– –	p	1 3	n	0 59	1	F. pL. E. par. contains a stell. neb. or st.	7771
232	—	– –	f	6 45	n	1 35	1	F. S. R. or large stellar.	7798
233 }	19	47 (λ) Pegasi	p	9 3	n	0 12	3	Two. The p. pB. lE. nearly mer.	7332
234 }			p	8 33	n	0 14	3	The f. F. E. nearly par. 1′½ l.	7339
235	20	11 Piscium	p	13 43	s	0 36	2	F. pL. broadly E.	7556
236	—	90 (φ) Aqua	f	3 53	n	1 22	4	pB. pL. iR. mbM.	7585
237	—	79 Ceti	p	4 48	n	0 36	1	F. E. mer. 2′ l.	958
238	Oct. 6	26 (β) Persei	p	28 34	s	0 10	2	pB. mE. near par. mbM. 4′ l. 1′ b.	1003
239	7	27 (κ) Persei	p	8 27	n	0 2	1	The 1st of 2. pB. pS. r.	1161*
240	8	39 Piscium	p	2 24	n	1 0	1	pF. pL. iR. er.	7814
241	—	– –	p	14 24	s	0 11	1	pS. C.	57
242	11	87 Pegasi	p	39 50	s	0 54	2	F. S. iR. near and p. 2 or 3 st.	7681
243	—	– –	f	6 27	s	0 54	2	F. S. iR.	57
244	14	54 (α) Pegasi	f	30 48	n	0 6	2	F. S. lE.	7711
245	—	58 Piscium	p	3 36	n	2 16	4	pB. pL. R. lbM.	234
246	—	19 Arietis	f	4 54	s	0 49	1	F. pL. E. 4 or 5′ s is a c st.	877
247	15	13 Pegasi	f	10 0	n	0 28	1	pB. R. bM. 1′ d.	7177
248	—	54 (α) Pegasi	p	38 0	n	0 59	2	F. pS. a quartile with 3 Sst.	7280
249	—	– –	p	3 36	n	1 11	2	F. pS. E. f. pBst.	7454
250	—	47 Piscium	p	67 12	s	0 37	1	F. lE. p. vBst.	7625
251	16	54 (α) Pegasi	p	4 36	n	0 44	1	pB. cL. E. r.	7448
252	—	102 (π) Pisc	p	12 48	n	0 45	1	F. pL. oval. lbM. p. pBst.	514
253	—	– –	f	5 54	n	1 30	1	pB. pL. E. bM. r.	660
254	—	38 Arietis	f	8 48	n	0 34	1	F. S. iR. r.	1134
255	18	82 Pegasi	p	8 21	s	0 11	2	pB. pS. R. gbM. r.	7742
256	—	77 Pegasi	f	1 0	s	0 25	1	F. R. gbM.	7743
257	—	34 Piscium	f	12 6	s	0 39	2	F. pL. iR. mbM.	95

II.	1784	Stars.		M. S.		D. M.	Ob.	Description.	N.G.C.
258	Oct. 20	15 Eridani	p	8 54	n	1 54	3	F. vL. lbM. R. 7 or 8′ d.	1232
259	Nov. 16	43 (γ) Cancri	p	20 58	n	1 2	1	F. S. iF.	2577
260	—	4 (λ) Leonis	f	3 22	s	1 16	1	F. pS. lE.	2916
261	17	12 Pegasi	f	2 8	s	0 46	1	F. iR. less than 1′ d.	7137
262	—	27 Eridani	p	11 51	s	1 40	1	F. l and iE. above 1′ d.	1371
263	—	- -	p	9 28	s	1 15	1	not vF. bM. 1′½ d.	1385
264	—	25 (δ) Canis	p	67 42	n	2 20	1	F. S. I. C.	2154*
265	20	4 (ξ 1) Canis	p	19 20	n	1 28	1	pF. pS. iF. lE. bM.	2196
266	—	15 (ι) Nav	f	25 33	n	1 25	1	F. E. bM. r. 1′½ d.	2613
267	Dec. 9	27 Eridani	p	6 1	n	0 40	1	F. vS. R. lbM.	1415
268	—	8 (ι) Crateris	p	63 16	s	0 16	1	F. S. R. SB point M. C.	3078
269	—	10 Crateris	f	4 26	n	1 22	1	pB. pL. lE. mbM.	3585
270	13	106 (ν) Pisc	f	11 56	s	1 11	1	pB. S. iR. mbM.	718
271 }	—	- -	f	14 54	n	0 11	3	Two, very close, not far from par.	741*
272 }								The f smallest and most n.	742*
273	—	86 (γ) Ceti	p	0 14	n	1 44	1	F. S. iR.	1070
274	—	92 (α) Ceti	p	3 54	s	0 47	1	F. vS. iE. er.	1153
275	20	32 (2 τ) Hyd	f	9 55	n	1 32	2	pB. cL. iR.	2967
276	—	10 (r) Virgin	p	6 58	n	0 5	3	F. pL. R. lbM.	4045
277	—	- -	p	5 14	s	0 1	3	F. S.	4073
	1785								
278	Jan. 6	75 Ceti	p	1 38	s	0 5	1	pB. S. E.	955
279	—	35 Eridani	f	2 55	s	0 38	2	F. mE. vlbM. about 4′ l.	1507
280	—	14 Hydræ	f	5 2	n	0 21	1	F. vS. lE. ver. 240.	2695
281	—	28 (A) Hydr	p	29 27	n	1 40	2	F. vS. E.	2708
282	10	41 Ceti	f	17 28	n	0 20	2	pB. cL. lE. mbM.	615
283	—	- -	f	21 26	n	0 10	2	pB. S. mbM.	636
284	—	80 Ceti	f	3 34	s	0 19	1	F. mE. about 3′ l and ¾′ b.	1035
285	—	55 (ζ) Ceti	f	74 50	n	1 2	2	pB. E. sp nf. about 1′½ l.	1208
286	—	13 (ζ) Eridani	p	4 34	s	0 9	1	F. pL. R. lbM. s. Sst.	1241
287	27	17 Eridani	p	10 24	s	1 12	2	F. vS. lE. er. unequally B.	1299
288	28	21 Eridani	p	1 55	n	0 35	3	F. pL. iR. r.	1376
289	31	7 (ν) Lepor	f	2 32	n	0 51	1	F. pL. i triangular. F. r.	1888
290	Feb. 1	89 (π) Ceti	f	49 17	n	0 21	3	F. pL. R. lbM. s. pLst.	1357
291	—	26 (π) Erid	p	3 39	s	1 25	1	pF. mE. mer. 3 or 4′ l and 1′ b.	1421
292	4	5 (μ) Lepor	p	0 50	n	0 29	1	pB. iR. mbM. sp. pcst.	1832
293	7	6 (3 b) Crater	p	52 51	n	0 23	1	pB. S. iR. bM.	3091
294	—	31 Crateris	p	6 45	n	0 6	1	F. S. E. r.	3957
295	—	- -	p	1 48	n	1 18	1	F. vS. iF. bM.	4024*
296	—	- -	p	0 12	n	0 24	1	pB. pL.	4027*
297	—	89 Virginis	p	11 47	n	0 18	1	pF. L. mbM.	5247
298	8	8 (η) Corvi	f	18 44	n	1 51	1	F. pL. lbM. ½′ p. is a S suspected stellar.	4727
299	—	53 Virginis	p	12 30	n	0 55	1	pB. pL. mbM.	4877*
300	—	- -	p	11 0	n	2 8	2	pF. cL.	4899*
301	—	- -	p	3 8	n	0 41	1	pB. pL. iR. mbM.	4984
302	28	2 (1 ω) Cancri	p	3 5	s	1 40	1	pF. vS. bM. er.	2481*
303	—	19 (λ) Cancri	p	2 22	s	0 35	1	F. S. mbM. r.	2554
304	Mar. 4	11 Monoc	f	30 53	s	0 37	3	Some Sst with pB nebulosity.	2316
305	5		p	7 14	n	0 49	1	F. S. lE. er.	3110*

II.	1785	Stars.		M. S.		D. M.	Ob.	Description.	N.G.C.
306	Mar. 5	88 Virginis	f	0 52	s	0 24	1	F. vS. iF. r.	5306
307	—	– –	f	3 58	n	0 43	1	F. cL. iF. bM.	5324
308	—	82 (*m*) Virg	f	12 28	n	1 6	2	F. S. iR. lbM.	5343
309 }	—	99 (*ι*) Virg	p	12 31	s	0 1	1	Two, nearly mer. dist. 4′ Sst. betw.	
310 }								che. touch. { n. pB. cL. mbM.	5427
								{ s. F. S.	5426
311	10	6 (3 *b*) Crater	p	68 34	s	1 18	2	cB. S. mbM.	2986
312	—	45 (*ψ*) Hydr	f	9 41	n	2 0	1	F. L. iR. vgbM.	5068
313	—	– –	f	10 53	n	1 16	1	pB. lE. par. b towards f side.	5084
314	—	– –	f	15 57	n	1 55	1	F. S. iF. bM.	5134
315	11	23 (2 *φ*) Can	f	0 29	s	1 0	2	F. S. R. bM. C. N.	2592
316 }	12	64 (1 *b*) Gem	p	4 16	n	1 17	1	Two. sp nf. dist. 1′ che. mix.	2371
317 }								Both F. S. equal. N.	2372
318	—	22 (1 *φ*) Can	f	8 38	n	0 36	1	F. pL. lE. mbM. r.	2608
319	—	48 (1 *ι*) Canc	p	9 10	s	0 5	1	F. S. bM. r.	2619
320	13	23 Leonis min	p	12 38	n	1 50	1	F. pS. R. lbM.	3106
321	—	13 Can. ven.	p	51 31	s	0 50	1	pB. L. gbM.	4136
322 }	—	– –	p	40 19	s	1 28	1	The two first of 3 in a line. of	4278
323 }								unequal size and brightness.	4283
324	—	– –	p	38 3	n	0 17	1	F. S.	4317*
325	—	– –	p	26 51	s	0 30	1	F. pL. E. bM.	4525
326	—	– –	p	14 11	s	0 4	1	F. mE. mer.	4676
327	—	– –	f	19 43	s	0 35	1	F. pS.	5089
328	—	– –	f	23 43	n	0 38	1	pB. pS. nearly R. mbM.	5127
329	—	49 (*δ*) Bootis	p	48 50	n	0 5	3	pF. S. R. r. n. 2 pBst.	5623
330	—	– –	p	45 45	s	2 2	1	pB. pL. R. bM.	5653
331	16	11 Ursæ min	p	60 36	s	0 2	1	F. pS. er.	5607
332	—	– –	p	20 10	s	0 2	1	pB. cL. b towards p side.	5832
333 }	Apr. 3	27 Ursæ	f	20 14	s	0 2	1	{ Two. Nearly mer. Most n. pB. }	3063
334 }								{ pS. bM. Most s. F. S. bM. }	3065
335	—	– –	f	73 0	n	1 41	1	pF. cL. iE.	3403
336	—	– –	f	88 16	n	0 30	1	pB. vS. iR.	3516
337	—	– –	f	94 42	n	0 48	1	pF. pS. bM.	3562
338	6	44 Leonis min	f	31 8	s	0 59	2	F. cL. iR. gvlbM.	3629
339	—	53 Leonis min	f	19 26	n	1 0	1	pF. pS. iF.	3689
340	—	72 Leonis	f	25 8	n	1 35	2	F. vS. stellar. short ray p side.	3798
341	—	4 Comæ	p	29 46	n	0 35	1	F. stellar.	3826
342	—	– –	p	22 2	n	0 37	1	F. pL.	3912
343	—	21 (*g*) Comæ	f	4 36	n	1 56	1	not L.	4555
344	—	– –	f	20 56	n	1 11	1	F. pL. lE.	4747
345	—	– –	f	23 34	n	2 28	1	just s. pBst.	4789
346	—	31 Comæ	f	4 54	s	0 34	2	F. pL. iF.	4819
347	10	36 (*ζ*) Leonis	f	11 14	s	0 33	1	pB. S. bM. r.	3248
348	—	41 Leonis min	p	3 34	n	0 51	1	F. S. lE.	3327
349	—	72 Leonis	f	14 12	n	1 1	1	F. pL. i triangular F.	3701
350	—	– –	f	16 2	s	0 17	1	F. S.	3710
351	—	– –	f	18 2	n	1 22	1	F. S.	3728
352	—	92 Leonis	p	3 9	n	1 16	2	F. pS.	3772
353	—	7 (*h*) Comæ	p	4 37	n	0 11	1	pB. cL. iF. bM.	4162
354	—	– –	p	0 43	n	0 2	1	F. vS.	4213

II.	1785	Stars.		M. S.		D. M.	Ob.	Description.	N.G.C.
355	Apr. 10	22 Comæ	p	4 32	s	1 28	1	pF. L. broadly E.	4455
356	—	40 Comæ	f	5 38	n	1 25	1	pB. S.	5016
357	—	12 (*d*) Bootis	f	18 46	s	1 53	1	F. S. iF. lbM.	5637
358	11	39 Leonis mi	p	8 4	n	0 8	1	F. pL.	3274
359	—	- -	p	7 38	n	1 1	1	pB. pS. nearly R. bM.	3277
360	—	44 Leonis mi	p	1 46	n	0 39	1	F. pL. iF.	3380
361	—	- -	f	0 50	n	0 30	1	F.	3400
362	—	- -	f	1 18	n	0 1	1	pB. pL.	3414
363	—	- -	f	1 32	n	0 8	1	F. S.	3418
364	—	- -	f	4 35	s	0 44	1	pF. pL. lE. b towards sf side.	3451
365	—	- -	f	14 2	n	0 56	1	F. mE. 1′½ l. but v. narrow.	3510
366	—	- -	f	14 24	n	0 4	1	pF. pL.	3512
367	—	- -	f	42 12	n	0 14	1	F. vS.	3713
368	—	14 (*b*) Comæ	p	28 42	n	0 59	1	pF. bM.	4008
369	—	- -	p	27 58	n	0 12	1	F. pL. E. b towards f side.	4017
370	—	- -	p	20 16	n	0 55	1	pB. cL. mb towards nf side.	4104
371	—	- -	p	17 40	n	1 55	1	One of three. F. iF.	4134
372	—	- -	p	14 24	n	1 55	1	One of 4. The most n. of the	4173
								p. side of a quartile. F. S.	
373	—	- -	p	13 28	n	1 16	1	F. L. bM.	4281
374	—	- -	p	12 22	n	1 12	1	F. S.	4196
375	—	- -	p	11 4	n	1 14	1	F. pS.	4209*
376	—	- -	p	6 38	n	0 22	1	pF. S. almost R. bM.	4275
377	—	- -	p	6 30	n	1 57	1	About 6′ sf I. 90. pB. S. the	4283
								place is that of the np.	
378	—	- -	p	4 10	n	1 57	1	F. cL. lE.	4310
379	—	- -	p	1 36	n	1 18	1	F. S.	4375
380	—	15 (*c*) Comæ	f	9 8	s	1 22	1	F. pL.	4556
381	—	31 Comæ	p	3 46	s	0 20	1	F. S.	4692
382	—	- -	f	3 16	s	0 9	1	F. pS.	4798
383	—	- -	f	4 26	n	0 12	1	F. pL.	4816
384	—	- -	f	5 2	s	0 23	1	F. pL.	4827
385	—	- -	f	5 40	n	0 4	1	F. pL.	?4840*
386	—	- -	f	5 54	s	0 6	1	F. pL.	?4839*
387	—	- -	f	5 48	n	0 55	1	F. pL.	4841
388 }	—	41 Comæ	p	7 46	n	0 22	1	Two. The time taken between	4869
389 }								them.	4872
390	—	- -	p	7 10	s	0 43	1	F.	4892
391	—	- -	p	7 18	n	0 23	1	F.	4889
392 }	—	- -	p	5 46	n	0 14	1	Three. The 2 f. p near each	4911
393 }								other. The sp about 8′ dist.	4921
394 }								The time is that of the 2.	4923
395	—	- -	p	3 26	n	0 33	1	F. S.	4944
396	—	- -	p	2 16	n	1 29	1	F. S.	4952
397	—	- -	p	2 2	s	0 4	1	F. S.	4957
398	—	- -	p	1 30	n	0 8	1	F. S.	4961
399	—	3 (β) Coron	f	6 54	s	0 27	1	pF. pL. iR. bM. r.	5958
400	13	26 Bootis	f	47 12	s	1 33	1	F. pL. er.	5910
401	14	11 Serpentis	p	2 14	s	1 35	1	pF. pL. vlbM. r. p 3 Sst.	5937
402	—	12 Ophiuchi	p	14 32	n	0 4	1	F. cL. E. sp nf. r. 3′ l 2′ b.	6118

Third class. Very faint nebulæ.

III.	1783	Stars.		M. S.		D. M.	Ob.	Description.	N.G.C.
1	Nov. 3	36 (υ) Orion	f	3 39	n	1 57	2	vF. S. mE. In the L. neb.	1982*
2	Dec. 21	60 Ceti	f	13 ::	n		1	eF. vS. R. lbM.	867*
3	30	95 (o) Leonis	f	4 15	n	0 36	2	vF. vS. lE. r.	4028*
	1784								
4	Jan. 18	6 (h) Leonis	f	6 4	s	0 9	3	eF. vS. iE. sp. a triangle of Bst.	2939
5	—	47 (ρ) Leonis	f	10 0	n	0 9	1	eF. eS. viewed also with 240.	3342*
6	—	59 (e) Virgin	p	28 ::			1	vS.	4698*
7	23	3 (β) Can. mi	f	36 30	n	0 19	1	Stellar. 240 left some doubt.	2508*
8	—	3 Leonis	f	1 6	s	0 28	3	E. er. 3 of the st visible.	2894
9 }	—	32 (2 d) Virg	f	46 54	s	0 25	2	Two. Both vF. and vS.	5208
10 }									5209
11	—	31 Bootis	p	38 15	s	0 1	1	vF. stellar.	5417*
12	—	– –	p	21 15	s	0 34	1	vF. forming an arch with 3 st.	5570
13	28	11 (s) Virg	f	27 30			1	eF. not verified.	4577*
14	30	31 Bootis	p	12 30	s	0 9	1	eF. vL. not verified.	5621*
15 }	Feb. 15	68 (δ) Leonis	f	7 30	s	0 24	1	Two. The p. vF. L. 5 or 6′	3646
16 }								dia. The f. eF. S. Fig. 5.	3649
17	23	16 (c) Virgin	f	6 0	s	0 47	1	vF. pS. r.	4409*
18	—	– –	f	11 45	n	0 38	1	vF. cL. r.	4505
19	Mar. 11	2 (ε) Can. mi	f	5 16	n	0 28	1	2 vS and close st. with nebulosity left doubtful.	2402
20	—	53 (l) Leonis	f	1 45	s	0 26	1	vF. r.	3433*
21	—	73 (n) Leonis	p	15 36	s	1 11	2	vF. S. C. ver. 240.	3491
22	—	53 (l) Leonis	f	14 0	n	0 31	1	vF. vS. with 240 cL.	3506
23	—	73 (n) Leonis	p	9 6	s	1 56	2	vF. vS. lE. ver. 240.	3524
24	12	20 Leonis	f	11 30	n	1 19	1	vS. 240 left some doubt.	3088
25	—	– –	f	26 15	n	0 0	1	vF. S.	3177*
26	—	20 Comæ	f	4 30	s	0 37	1	eF. L. left doubtful.	4529*
27	14	86 Leonis	p	13 30	s	0 22	1	The most s. of 3. vF. vS. II. 50. 51.	3605
28	—	– –	p	2 30	s	1 10	1	vF. L. r.	3686
29	—	– –	f	7 15	s	0 34	1	vF. eS. stellar. ver. 240.	3768*
30	—	– –	f	10 0	s	0 40	1	vF. pS. f. 2 vBst.	3802
31	—	11 Comæ	f	3 15	s	0 10	1	eF. forms a triangle with 2 Sst.	4344
32	—	8 (η) Bootis	f	16 0	s	0 56	1	vS. or nebulous double st. ver. 240.	5490
33	—	5 (r) Herculis	f	3 30	n	1 30	1	eF. pL. partly ver. 240	6046*
34	15	5 (ξ) Leonis	f	13 0	s	0 10	1	eF. vS. completely ver. 240.	2984
35 }	—	78 (ι) Leonis	f	20 30	s	0 15	1	Two on par. 3 or 4′ dist.	3848*
36 }								Both eF. vS.	3852*
37	—	12 (t) Virgin	p	8 15	n	0 40	1	eF. vS. with 240. cL.	4067*
38	—	– –	f	11 15	n	0 20	1	vF. vS.	4368*
39	—	– –	f	12 45	n	0 15	1	vF. near some Bst.	4390
40	—	30 (ρ) Virg	p	11 15	n	0 31	1	eF. pL. easily overlooked.	4482*
41	—	– –	p	10 45	n	1 14	1	vF.	4491
42	—	– –	p	10 15	n	1 23	1	vF.	4497
43	—	34 Virginis	p	6 6	s	0 4	2	vF. pL. lE. contains two st.	4606
44	—	– –	p	3 30	s	0 25	2	The p. of 2. vF. S. Note.	4647
45 }	—	71 Virginis	f	0 37	n	0 12	1	Two. mistaken for one; but	5174
46 }								240 shewed them both. cL. vF.	5175

III.	1784	Stars.		M. S.		D. M.	Ob.	Description.	N.G.C.
47	Mar. 15	32 Bootis	p	25 0	s	0 47	1	vF. r. 2 or 3 st. in it.	5532
48	—	- -	f	3 30	n	0 27	1	eF.	5747*
49	19	62 (1 o) Can.	p	14 33	s	1 5	2	F. cS. lE. np sf. like 2 joined.	2648
50	—	45 (1A) Can	f	3 15	s	0 4	1	eF. ver. 240. and cL. R.	2661*
51 }	—	27 (ν) Leonis	p	7 0	n	0 21	1	Two. about 20° np sf. 6 or 7′	3020*
52 }								dist. Both eF. p is the largest.	3024*
53	—	34 Leonis	f	1 0	s	0 41	2	eF. S. lE. r. 3 or 4 st in it.	3153
54	—	52 (K) Leon	p	10 45	s	1 27	1	eF. cL. R. r. no N.	3299*
55	—	46 (i) Leonis	f	4 18	n	0 3	2	vF. vS. iR. r. some st in it.	3300
56	—	15 Bootis	p	13 0	s	0 40	1	eF. vS. E. r.	5416
57	—	- -	p	10 30	s	0 28	1	eF. S. ver. 240.	5446*
58	—	- -	p	8 30	s	0 43	1	eF. S. ver. 240 and lE.	5463
59	—	- -	p	6 15	s	1 10	1	eF. S. ver. 240.	5482
60	21	47 (δ) Cancri	f	20 0	n	0 23	1	vF. S. with 240 near Sst.	2744
61	—	- -	f	26 30	s	0 18	1	eF. 240 shewed 5 Sst with nebulos. ?	(2774)*
62 }	—	- -	f	31 30	n	0 50	1	Two. nearly mer. Both vF. pS.	2802*
63 }								R. lbM. r. with 240 cL.	2803*
64	—	- -	f	36 0	n	0 52	1	eF. 240 shewed some Sst with susp. neb.	2843*
65	—	51 (m) Leon	p	38 15	s	0 33	1	vS. E. r. better with 240.	3129
66	—	- -	p	9 15	s	0 44	1	vF. S. E. r. the same with 240	3303
67	—	- -	f	11 45	s	1 45	1	vF. nebul. betw. 2 st. 2′ l. ver. 240	3473
68	—	3 Comæ	p	1 45	s	0 40	1	2 vSst with susp. neb. 240 doubtf.	4126
69	—	25 Comæ	p	5 0	s	0 18	1	vF. S.	4498
70	—	27 Comæ	f	6 0	s	0 42	1	vF. not S.	4758
71	—	42 Comæ	f	19 30	s	0 41	1	3 Sst with suspect. nebul. 240 left some doubt.	5180
72	—	4 (τ) Bootis	p	10 15	s	1 26	1	eF. vS. ver. 240 and cL.	5249
73	—	5 (r) Herc	p	4 0	s	1 50	1	eF. vS. easily ver. 240.	6021
74	—	48 Serpentis	p	1 15	n	0 5	1	vF. S. ver. 240.	6073
75	Apr. 8	70 (θ) Leonis	p	12 24	s	1 7	1	eF. not S. doubtful.	3498*
76	—	- -	f	4 0	s	0 41	1	eF. pL. easily ver. 240.	3616*
77	—	94 (β) Leonis	f	12 12	s	1 12	1	eF. pL. R. r.	4037
78	—	6 Comæ	f	17 18	s	0 19	1	vF. r. by moon-light.	4516
79	12	73 (n) Leonis	p	5 6	s	1 25	1	eF. not L. lE. r.	3559
80	—	- -	f	18 36	s	0 48	1	vF. vS. R. bM. stellar. ver. 240	3731
81	—	- -	f	22 35	s	1 11	2	vF. vS. R. stellar.	3773
82	—	41 Virginis	p	1 42	n	1 7	1	vF. S. E. r.	4752*
83	—	- -	f	6 18	n	0 0	1	vF. S. iF. r.	4880
84	—	70 Virginis	p	3 42	s	0 4	1	eF. vS. stellar. ver. 240.	5136
85 }			f	6 12	n	0 1		Three. The two p. vF. S. R.	5222
86 }	—	- -	..	.. ..	..	.. ..	1	The last vF. pL. R. Place	5221
87 }			f	6 48	s	0 9		of the 2d not taken.	5230
88	13	56 Leonis	p	5 42	s	0 23	1	eF. no time to ver.	3401*
89	—	63 (χ) Leon	f	6 24	s	1 29	1	eF. a little doubtful.	3567
90	—	3 (ν) Virginis	f	4 54	n	0 1	1	vF. vS. vlbM.	3914
91	—	11 (s) Virg	f	7 48	n	1 19	1	The f. of 2. eF. II. 17.	4246
92 }	—	9 (o) Virginis	f	16 15	s	2 4	2	Two. One vF. vS. The other	4296*
93 }								just by. eF. eS. left doubtful.	4297*

III.	1784	Stars.		M. S.		D. M.	Ob.	Description.	N.G.C.
94 } 95 } 96 }	Apr. 13	9 (ο) Virginis	f	18 22	s	1 46	2	Three. All. eF. vS. R. In the 2d observation two of them were overlooked.	4343* 4341* 4342*
97	—	31 (1 d) Virg	p	17 9	n	0 42	1	The smallest of 2. eF. The other is II. 144.	4366*
98	—	– –	p	3 6	n	0 0::	1	eF. eS. The place not accurate.	4588*
99	—	32 (2 d) Virg	f	47 36	s	0 33	1	eF. S.	5210
100	—	– –	f	50 42	s	1 8	1	eF. E.	5235
101	—	– –	f	51 0	s	0 23	1	eF. pL. R. er. The st almost visible.	5239
102	15	2 (1 ξ) Virgin	p	1 48	n	1 54	1	eF. pL.	3833
103	—	– –	f	0 24	n	0 58	1	vF. r.	3876
104	—	4 (2 ξ) Virgin	p	2 12	n	0 19	1	vF. vS. left doubtful. Twilight.	3874*
105	—	31 (1 d) Virg	p	1 52	n	1 35	2	eF. vL. lbM.	4598
106	—	33 Virginis	f	7 30	n	0 8	1	vF. pL. vlbM. r.	4779
107	17	48 Leonis	f	6 54	s	0 8	1	Suspected. eF. pL. but too much daylight.	3356*
108	—	63 (χ) Leonis	p	13 18	n	1 7	1	eF. eS. r.	3425*
109	—	90 Leonis	f	5 18	n	0 49	1	8 or 10′ sp. II. 161. vS. stellar. not ver.	3790
110	—	20 Bootis	f	1 54	s	2 29	1	vF. vS. lE. ver. 240.	5587
111	18	{ 58 (d) Leon { 84 (τ) –	f ..	8 36	.. n	.. .. 1 43	1	vF. vS. r. ver. 240.	3535*
112	24	74 (φ) Leonis	f	10 6	s	1 52	1	eF. cL. R. r. near vBst. ☽ light.	3679*
113	—	– –	f	34 18	s	1 3	1	eF. eS. with 240. 2 vSst and nebu.	3915*
114	25	28 Virginis	p	14 18	n	1 37	1	2 vSst with nebulosity. 240 rather confirmed it. ☾ light.	4422*
115	May 9	67 (α) Virg	f	1 12	s	1 10	1	vS. vF. stellar. ver. 240.	5146
116	—	31 (ε) Libræ	p	8 48	n	0 15	1	vF. cL. nearly R. rather m.	5885
117 } 118 }	11	100 (λ) Virg	p	59 30	n	0 48	1	The two most s. of 3. That M. vF. vS. The most s. eF. eS. ver. 240. II. 193.	5076 5079
119	—	– –	p	55 42	n	0 29	1	eF. vS. stellar. ver. 240.	5111
120	—	– –	f	6 24	n	0 9	1	eF. pL. iR. lb towards f side.	5605
121 } 122 }	14	9 (α) Libræ	p	27 0	s	0 36	1	Two np sf. The f. eF. 1′ d. nearly R. The p. vF. vS. R. dist. 5′.	5597 5595
123	15	18 Herculis	f	40 30	s	0 47	1	vF. pL. R. lbM.	6267
124	—	– –	f	43 30	s	0 47	1	vF. stellar. ver. 240.	6278
125	16	25 (ρ) Bootis	p	33 12	s	1 10	1	vF. S. iR. lbM. almost stellar.	5396*
126	—	– –	p	2 18	s	0 24	1	2 Sst. with suspected nebul. almost ver. 240.	5642
127 } 128 }	—	28 (σ) Bootis	f	3 48	n	0 45	1	Two. 3′ dist. par. The f. vF. vS. iR. The p. eF. vS. ver. 240.	5706* 5709*
129 } 130 }	—	– –	f	17 48	n	0 3	1	Two. about 6′ dist. Both eF. vS. R. ver. 240.	5771 5773
131	—	– –	f	22 54	n	0 11	1	vF. E. close to a st. contains 2 st.	5798
132	17	36 (ε) Bootis	p	16 54	n	0 26	1	eF. S. lE. the same with 240.	5635
133	—	– –	p	2 36	n	1 35	1	cF. cL. iR. lbM.	5735
134	19	12 (d) Bootis	f	4 28	n	0 12	2	vF. pL. E. par. r.	5523

III.	1784	Stars.		M. S.		D. M.	Ob.	Description.	N.G.C.
135	May 19	12 (*d*) Bootis	f	12 30	n	1 5	1	eF. vS. stellar. ver. 240.	5594*
136	—	– –	f	14 8	s	0 30	2	vF. S. E. nearly par. with 240 like two stel.	5610
137	—	76 (λ) Hercu	p	2 54	n	0 22	1	vF. not S. iE.	6372
138	21	20 (1 γ) Libr	f	13 36	n	1 9	3	Two. nearly par. 7′ dist. Both	5898
139								vF. not vS. R.	5903
140	June 11	27 (β) Hercu	p	23 30	s	0 51	1	vF. vS. r. ver. 240. np. pBst.	6064*
141	July 12	16 (ψ) Capri	p	20 42	n	0 33	1	vF. cL. lE. lbM. 240. same.	6907
142	21	70 Aquilæ	p	3 39	n	0 31	2	vF. E. about 2′ l.	6926
143	Aug. 7	35 (2 ν) Sagit^{i}	p	0 0	s	0 3	1	3 vSst with suspected nebulosity.	6717
144	Sept. 5	39 (*h*) Cygni	p	21 18	n	1 20	1	Some eSst. with neb. iE. ver. 240.	6857*
145	10	10 (κ) Pegasi	p	25 48	n	0 53	2	vF. lE. stellar.	7052
146	—	69 Pegasi	f	11 24	n	1 53	1	vF. E. some Sst. with nebulosity.	7720*
147	—	85 Pegasi	f	7 54	s	1 13	1	2 or 3 st. with seeming nebulosity.	23
148	11	28 Androm	p	4 12	s	0 32	1	vF. pL. lbM.	108
149	—	31 (δ) Andr	f	4 24	s	0 15	1	eF. vS. R.	233
150	—	2 (α) Triang	p	18 48	n	1 4	2	Near V. 18. vF. SR. bM.	604
151	—	6 (ι) Triang	p	7 0	s	1 18	1	vF. vS. stellar. betw. vL. and Sst.	807
152	—	39 Arietis	p	8 12	n	0 49	1	vF. pS. of equal light.	1012
153	12	40 Andr	p	13 0	n	0 29	1	vF. pL. lE. vlb. towards f side.	266
154	—	– –	f	9 18	n	0 20	1	Two. Both eF. vS.	421*
155								The f is the largest.	420*
156	—	43 (β) Andr	f	13 6	s	2 8	2	Three forming a rect. triangle.	495
157								In the legs eF. vS. at the	496
158								rectangle vF. pL.	499
159	—	40 Androm	f	20 6	n	1 30	1	Two. Both eF. S. but un-	507
160								equal.	508
161	—	17 (*r*) Persei	p	14 30	s	1 47	2	vF. S. iE. r.	987
162	—	21 Persei	p	13 42	n	0 30	1	Two. Both vF. pS. R. lbM.	1060
163			p	13 18	n	0 32	1		1066
164	—	– –	f	15 36	s	1 19	1	eF. vS. 240 left a doubt.	1240*
165	13	66 (υ) Cygni	f	43 0	n	0 4	1	5 or 6 st. forming a parallelogr. with mixed nebul. ver. 240.	7186
166	—	– –	f	78 18	s	0 47	1	eF. vS. E. nf. and 4 or 5′ dist. from I. 53.	7335
167	—	43 (β) Andr	f	15 30	s	2 12	1	Two. Both stellar.	515*
168									517*
169	—	– –	f	15 12	s	1 46	1	stellar.	513*
170	—	– –	f	16 30	s	1 31	1	stellar.	537*
171	—	– –	f	17 30	s	0 56	1	stellar.	536*
172	—	– –	f	18 0	s	2 8	1	Two. Both vS. stellar. a little	552*
173								doubtful.	553*
174	—	3 (ε) Triang	p	25 24	n	0 22	1	stellar. ver. 240.	614*
175	—	– –	p	12 48	n	2 29	1	stellar.	679
176	—	– –	p	6 6	n	1 0	1	eF. stellar. 240 left some doubt.	735*
177	—	9 (γ) Triang	f	9 36	s	0 17	1	vF. cL. iR. r. 2 or 3′ d.	925
178	—	17 (*r*) Persei	f	10 0	n	0 13	1	vF. pL. R. SB place M.	1167
179	15	6 (β) Arietis	p	3 0	n	1 30	1	vF. pL. lE.	697
180	18	40 Pegasi	p	3 0	n	0 47	1	eF. vS. R. n. st. 9 m.	7316
181	—	65 Pegasi	p	6 48	s	1 49	1	vF. vS. R. ver. 240.	7515*

III.	1784	Stars.		M. S.		D. M.	Ob.	Description.	N.G.C.
182	Sept. 18	40 Pegasi	f	38 24	s	0 51	2	4 or 5 Sst. with nebul. 240 doubt.	7578
183	—	89 (χ) Pegasi	f	0 30	s	1 38	1	eF. S. iE.	52
184	20	11 Piscium	p	17 44	s	0 22	2	eF. vS. stellar. ver. 240.	7506
185	—	– –	p	12 50	s	0 32	2	vF. E. er. 3 Fst. visible in it.	7566
186	—	20 Piscium	p	29 15	s	1 41	1	eF. vS.	7592
187	—	– –	p	14 39	n	0 1	1	eF. stellar. ver. 240 and cL.	7694
188	—	– –	p	13 33	s	0 8	1	eF. stellar. just like 187.	7701
189	—	– –	p	8 15	s	1 52	1	eF.	7725
190	—	29 Piscium	f	4 54	s	0 40	1	vF. vS.	7832
191	—	34 Ceti	p	9 12	s	1 53	2	vF. mE.	352
192	—	62 Ceti	p	17 24	s	1 43	1	eF. S. ver. 240. with difficulty.	702
193	—	– –	p	12 12	s	2 6	1	eF. ver. 240. with difficulty. nr. a S. star.	748
194	—	81 Ceti	f	38 6	n	0 55	1	eF. eS.	1266*
195	—	– –	f	41 6	n	0 49	1	eF. eS. ver. 240.	1287
196 }	—	– –	f	47 0	n	0 36	1	Two. Both eF. ver. 240 but	1321
197 }								just suspected with 157.	1320
198	Oct. 6	12 (q) Persei	p	3 3	n	0 40	2	cB. mE. vgmbM. near 4′ l.	1003*
199	7	27 (κ) Persei	p	8 27	n	0 2	2	The n. of 2. vF. iF. pS. II. 239.	1160*
200	14	53 Piscium	f	4 24	n	1 13	2	2 Sst with nebulosity ver. 240.	213
201	—	19 Arietis	f	4 6	s	0 47	1	vF. vS. E. s. pcst.	871
202	15	47 Piscium	p	83 54	s	1 15	1	eF. vS. stellar. ver. 240.	7468
203	—	– –	p	78 18	n	0 18	1	vF. cL. E. 2′ l.	7497
204	—	59 Piscium	f	0 42	n	0 2	1	vF. S. sp. 2 vSst.	251
205	—	92 Piscium	p	5 30	s	0 10	1	eF. ver. 240. discovered in gaging.	459*
206	—	– –	p	3 30	s	1 20	1	eF. S.	473
207	—	8 (ι) Arietis	f	5 12	n	0 32	1	eF. vS. stellar. plainly. ver. 240.	794
208	—	– –	f	6 30	s	1 49	1	eF. vS. iR. just f. pBst.	803
209	16	17 Delphini	f	18 6	s	0 11	1	vF. S. R.	7042
210 }	—	54 (α) Pegasi	p	2 48	n	0 46	1	Two. The p. vF. S. lE.	7463
211 }								The f. vF. vS. stellar.	7465
212	—	– –	f	21 6	s	0 59	1	eF. eS. ver. 240. completely though with difficulty.	7659
213	—	– –	f	27 36	n	0 40	1	eF. cL. ver. 240. betw. 2 Bst.	7691
214	—	31 Arietis	p	36 48	n	1 24	1	vF. stellar. ver. 240.	774
215	—	– –	p	36 6	n	0 6	1	eF. stellar. discovered by 240.	781
216 }	18	46 (ξ) Pegasi	f	3 15	s	0 37	3	Two. The p. vF. pS. R. vlbM.	7385
217 }			f	3 25	s	0 32	3	The f. vF. pS. R. vlbM.	7386
218	—	58 (n) Pegasi	f	13 51	n	0 4	1	eF. pS. lE.	7648*
219	19	15 Delphini	p	5 24	n	0 2	1	eF. vS. stellar. ver. 240. with dif.	6956
220	—	66 Pegasi	p	10 10	n	0 23	4	F. R. bM. $1'\frac{1}{2}$ d.	7515
221	—	– –	p	7 10	n	1 0	2	vF. S.	7559
222	—	– –	p	7 7	n	0 54	2	eF. S. R.	7563*
223	20	7 (h) Ceti	f	23 12	s	1 1	1	vF. lE. or oval. 1′ d. np. 2 pBst.	175*
224	—	1 (1 τ) Erid	p	21 42	s	2 11	2	vF. S. iR.	907
225	—	15 (δ) Lepor	f	6 24	n	0 49	1	eF. E. r. near 1′ l. ver. 240.	2124
226	21	70 (q) Pegasi	p	1 50	s	0 18	2	eF. S. stellar. ver. 240.	7671*
227	Nov. 7	64 Ceti	p	2 24	s	0 37	1	2 or 3 Sst. with neb. rather confirmed 240.	827*

III.	1784	Stars.		M. S.		D. M.	Ob.	Description.	N.G.C.
228	Nov. 7	73 (2ξ) Ceti	f	12 54	n	0 17	1	Two about 1′ dist. The p. eF.	1044
229								vS. ver. 240. The f. eF. eS. 240. doubtf.	1046
230	12	55 (*l*) Pegasi	p	3 36	s	0 29	1	eF. eS. 240 left some doubt.	7469
231	—	31 (1 *c*) Pisc	p	9 0	s	1 0	1	Two. Both vF. stellar.	7778
232									7779
233	—	– –	p	8 27	s	1 0	2	eF. pL. glbM.	7782
234	16	43 (γ) Canc	p	11 24	n	1 6	1	vF. stellar.	2599
235	—	– –	p	3 20	n	2 4	1	eF. S. ver. 240.	2628
236	—	4 (λ) Leonis	p	23 22	s	1 37	1	eF. lE. betw. 2 pBst. ver. 240.	2764
237	17	33 Pegasi	f	12 54	n	0 46	1	eF. vS.	7321
238	—	66 Pegasi	p	6 6	n	1 10	1	eF. eS. ver. 240. with difficulty.	7570
239	—	4 Eridani	p	32 26	s	1 1	1	vF. S. 1′ dia. or more.	922
240	20	12 Leporis	p	7 55	s	0 59	1	vF. vS. stellar.	1979
241	—	– –	f	3 39	n	0 23	1	eF. vS. lE. par.	2073
242	—	15 (ι) Nav	f	68 16	n	0 53	1	vF. lE. S. 1′ d.	2815
243	Dec. 2	56 Pegasi	p	9 16	n	0 42	1	vF. S. er.	7436
244	9	48 Ceti	p	48 34	n	0 27	1	eF. vS. E.	216
245	—	15 Eridani	p	15 49	s	0 27	1	vF. cL. iE. r. unequally B.	1187
246	—	19 Eridani	p	1 38	n	0 50	2	vF. E. equally B.	1353
247	—	– –	f	6 5	s	1 4	1	eF. vS.	1401
248	—	27 Eridani	p	4 23	n	1 7	1	vF. vS. lE.	1426
249	—	– –	p	2 19	n	1 18	1	vF. vS.	1439
250	13	89 (*f*) Pisci	f	2 25	s	0 14	1	Two. nearly par. 4 or 5′ dist.	470*
251								Both vF. vS. R.	474*
252	—	– –	f	3 42	n	1 38	1	vF. pL. iR. lbM.	488
253	—	– –	f	6 48	n	0 11	1	eF. cL. E.	520
254	—	15 Sextantis	p	14 34	n	1 52	2	vF. E. np sf. 5′ l. ½′ b.	3044
255	—	7 Sextantis	f	20 27	n	0 42	1	vF. vS. p. triangle of Bst.	3156
256	20	13 (ζ) Can. mi	f	26 5	s	0 58	1	vF. vS. ver. 240.	2555
257	—	– –	f	44 59	s	0 55	1	eF. pL. iF.	2618*
258	—	10 (*r*) Virgin	p	5 2	s	0 7	2	vF. S. E.	4077
	1785								
259	Jan. 6	70 Ceti	p	10 34	s	0 38	1	eF. eS. iF.	850
260	—	– –	p	7 10	n	0 4	1	eF. vS. stellar.	863*
261	—	75 Ceti	p	3 46	s	0 6	1	vF. cL.	941
262	—	94 Ceti	p	1 16	s	1 15	1	eF.stellar.ver.240 with difficulty.	1239*
263	—	24 Eridani	p	3 22	s	0 11	1	eF. stellar. or lE. almost ver. 240.	1409
264	—	28 (A) Hydræ	p	26 48	n	1 19	2	vF. vS. R. ver. 240.	2722*
265	10	45 (θ) Ceti	f	32 28	s	0 46	1	eF. stellar. ver. 240.	755
266	—	– –	f	31 6	s	0 43	1	vF. lE. ver. 240.	731
267	Feb. 4	14 (ζ) Lepor	f	0 1	s	1 56	1	vF. pS. iE. bM.	2076
268	6	11 (α) Lepor	p	24 51	s	0 31	1	eF. vS. stellar. ver. 240. easily.	1794*
269	—	19 Leporis	p	32 23	n	1 11	1	eF. vS. stellar. ver. 240. easily.	1993
270	—	– –	p	20 0	n	1 28	1	vF. eS. stellar. ver. 240 difficulty.	2089
271	—	8 (3 ν) Can^s	f	8 0	n	0 4	1	3 or 4 Sst with neb. vF. ver. 240.	2283
272	7	6 (3 *b*) Crater	p	58 39	n	1 21	1	vF. pS. iF. vlbM.	3052
273	—	– –	p	55 43	n	0 39	1	eF. vS. iF.	3072
274	—	31 Crateris	p	4 40	s	0 14	1	vF. pL. iF.	3981
275	8	12 Hydræ	f	20 30	s	1 49	1	vF. vS. bM. ½′ s. Sst.	2763

III.	1785	Stars.		M. S.		D. M.	Ob.	Description.	N.G.C.
276	Feb. 8	38 (κ) Hyd	p	9 20	s	0 26	1	vF. vS. stellar. 240. the same.	2902
277 278	—	39 (1 υ) Hyd	p	5 0	n	0 30	1	Two. 3 or 4 dist. The most n. vF. S. The s. vF. vS. Both stell.	2992* 2993*
279	—	8 () C orvi	p	31 26	n	0 16	1	eF. pL. better with 157 than 240.	4035
280	—	– –	f	18 44	n	1 51	1	½′ p. II. 298. eF. eS. stell. 240. doubtful.	4724
281	—	– –	f	20 38	n	0 46	1	vF. pS. r.	4756
282	—	53 Virginis	f	7 12	n	1 19	1	vF. mE. sf np. v. narrow.	5073
283	17	41 (ω) Bootis	p	27 54	n	0 27	1	vF. vS.	5677
284	Mar. 5	25 (f) Virg.	p	54 12	s	0 19	1	vF. S. iE. lbM.	3818*
285	—	88 Virginis	f	8 45	n	1 17	1	eF. vS.	5369
286	—	99 (ι) Virg	p	9 22	n	0 31	1	vF. L. b towards n.	5468
287	—	– –	p	7 58	s	0 7	1	vF. pS. iF.	5476
288	6	15 (ι) Navis	f	11 16	s	1 7	1	vF. cL. er. some of the st. vis.	2566
289	10	6 (3 b) Crat	p	69 14	s	0 31	2	F. vS. large stellar. lbM.	2983
290	—	2 (ε) Corvi	p	16 1	n	2 3	1	eF. pL. broadly E. nearly par.	3956
291	11	75 Cancri	p	2 53	s	1 13	1	vF. pL. R. bM.	2750
292	12	46 Cancri	p	11 46	s	1 14	2	vF. pL. R. lbM. r.	2604
293	—	23 Leonis min	p	17 46	s	0 22	1	eF. eS. ver. 240. doubtful.	3068*
294	13	57 (2 ι) Canc	p	2 44	n	0 15	1	vF. vS. R. bM. large stellar.	2679
295	—	72 (τ) Canc	f	5 47	n	0 24	1	vF. vS. R. nf. 2 pBst.	2783
296	—	– –	f	8 42	n	1 17	1	vF. S. R. lbM.	2796
297	—	15 (f) Leon	p	13 8	s	0 34	1	eF. eS. 240 left a doubt.	2893
298	—	18 Leonis min	p	20 56	s	0 44	2	vF. vS. iR. lbM.	2918
299	—	13 Can. ven.	p	40 51	s	0 27	1	eF.	4272
300	—	– –	p	40 19	s	1 28	1	The most f of 3. vF. II. 322. 323.	4286
301	—	– –	p	28 58	s	1 41	1	vF. vS. R.	4495
302	—	– –	p	27 40	s	1 2	1	eF. vS.	4514
303	—	– –	f	5 43	s	1 43	1	eF. vS. ver. 240.	4962*
304	—	– –	f	6 26	s	1 47	1	eF. vS. ver. 240. just nf a star 8 or 9 m.	4966*
305	—	– –	f	11 0	s	1 9	1	vF. vS. lE.	5004
306 307	—	– –	f	16 12	n	0 6	1	Two. The p. vF. vS. The f. 7 or 8′ nf the first. vF. vS.	5056 5057
308	—	– –	f	17 29	n	0 13	1	vF. S.	5065
309	—	– –	f	18 31	n	0 34	1	eF. vS.	5074
310	—	49 (δ) Bootis	p	43 12	s	1 32	1	vF. vS. iF.	5672
311	16	11 Ursæ min	p	24 18	n	1 18	1	vF. S. iR. betw. 2 pSst. 6′ apart.	5808*
312	—	– –	p	19 9	n	2 6	1	eF. vS. lE. 2 vSst in it.	5836*
313	—	13 (γ) Ur. mi	f	27 8	n	0 12	1	vF. vS. lE.	6011
314	—	– –	f	49 18	n	0 24	1	eF. vS. lE. er.	6094
315	Apr. 3	27 Ursæ	f	3 42	n	0 46	1	eF. vS. ver. 240.	2963
316	—	– –	f	51 42	n	1 43	1	eF. pS. mE. r.	3252
317	—	– –	f	65 18	n	1 19	1	vF. vS.	3343
318	—	– –	f	69 10	n	0 20	1	vF. pL. r.	3364
319	—	7 (β) Urs. mi	p	32 2	s	2 26	1	eF. not verified.	5620
320	6	72 Leonis	f	26 8	n	1 44	2	vF. vS. stellar.	3812
321	—	4 Comæ	p	22 54	n	0 15	1	vF. pS.	3902

III.	1785	Stars.		M. S.		D. M.	Ob.	Description.	N.G.C.
322	Apr. 6	4 Comæ	p	19 8	n	0 18	1	vF. stellar.	3944
323 }	—	– –	p	14 43	s	0 40	1	vF. lE. Suspected another nf.	3987*
324 }				.. ..		.. ..	1	eF. 5 or 6′ dist. pretty sure.	3993*
325	—	– –	p	13 46	s	0 45	1	eF. vS.	4005*
326	—	– –	p	5 47	s	0 17	1	eF. vS. ver. 240. discovered in gaging.	4101
327	—	– –	p	1 45	n	0 33	1	vF. pS.	4146
328	—	31 Comæ	p	6 28	s	0 25	2	F. S.	4670
329	—	21 (*g*) Comæ	f	14 45	n	2 26	2	vF. S.	4673
330	10	36 (ζ) Leonis	f	4 54	n	0 29	1	vF. pS. vlbM. iR.	3216
331	—	41 Leonis mi	p	12 16	n	1 36	1	vF. vS. vlbM.	3270
332	—	54 Leonis	f	2 46	s	0 34	1	vF. 1′ n. Sst.	3475
333	—	72 Leonis	f	2 48	n	0 18	1	vF. vS. ver. 240.	3615
334	—	– –	f	3 17	n	0 23	1	vF. S.	3618
335 }	—	– –	f	7 12	n	1 12	1	Two. 2 or 3′ distant. Both vF.	3651
336 }								vS. the most s. faintest.	3653
337	—	– –	f	9 34	n	0 52	1	vF. S.	3670
338	—	– –	f	25 56	s	0 38	1	vF. vS. 240. the same.	3808
339	—	– –	f	26 30	n	1 44	1	vF. vS. 240. the same.	3815
340	—	– –	f	28 36	s	0 19	1	vF. vS. pL. two stellar suspected near it.	3832
341	—	7 (*h*) Comæ	p	26 41	n	0 56	1	vF. vS. ver. 240. easily.	3911
342	—	– –	p	22 55	s	0 32	1	vF. vS. lE.	3951
343	—	– –	p	20 7	s	0 4	1	vF. vS. 240. the same.	3983
344 }	—	– –	p	18 31	s	0 43	1	Two. 5 or 6′ distant. Both eF.	4002
345 }								vS. ver. 240.	4003
346	—	40 Comæ	f	1 38	n	2 8	1	eF. pL. lE. ver. 240.	4979*
347	—	12 (*d*) Bootis	f	7 40	s	0 17	1	vF. lE. S.	5559
348	11	23 Leonis min	f	3 12	s	1 38	1	eF. lE. a little doubtful.	3196
349	—	39 Leonis min	p	9 28	n	1 18	1	eF. 240 shewed a few Sst. with neb. but doubtf.	3265
350	—	44 Leonis min	f	17 36	n	0 35	1	vF. S.	3527
351 }	—	– –	f	20 58	n	0 51	1	Two. Both vF. vS. the most s.	3550*
352 }								is the faintest.	3552
353	—	– –	f	43 4	n	0 26	1	eF. 240 left it doubtful.	3714
354	—	14 (*b*) Comæ	p	28 29	n	0 43	1	vF. vS. discovered in gaging.	4004
355	—	– –	p	21 49	s	0 16	1	vF. S. pmE.	4080
356 }	—	– –	p	17 40	n	1 55	1	Two of 3. the place is that of	4131
357 }								II. 371. Both vF. mE. A 4th suspected.	4132
358 }	—	– –	p	14 24	n	1 55	1	Three of a quartile. The place	4169
359 }								is that of II. 372. All vF.	4174
360 }								vS. and all within 3′.	4175
361	—	– –	p	0 40	n	0 18	1	vF. vL.	4393
362	—	15 (*c*) Comæ	f	3 2	s	1 3	1	eF. cL. 4 or 5′ l. 2′ b.	4475
363	—	41 Comæ	p	6 16	n	0 23	1	vF. I. C.	4051*
364	—	– –	p	5 24	n	0 25	1	vF.	4927
365	—	– –	f	1 8	n	0 41	1	vF.	4983
366	—	– –	f	2 26	n	1 18	1	vF. pS.	5000
367	—	43 Comæ	f	1 24	s	0 2	1	vF. pL.	5032

III.	1785	Stars.		M. S.		D. M.	Ob.	Description.	N.G.C.
368	Apr. 11	43 Comæ	f	11 2	s	0 53	1	vF. mE. 1½′ l. r. discov. gaging.	5116
369	—	– –	f	25 41	s	0 29	1	eF. vS. 240 left a little doubt.	5251
370	—	– –	f	28 8	n	0 31	1	vF. S. mE. nearly mer.	5263
371	—	14 (ι) Coron	p	13 52	s	1 8	1	vF. S. R. ver. 240 easily.	6001
372	13	93 Leonis	p	1 25	n	0 25	1	vF. cL. moon-light.	3883
373	14	11 Libræ	f	1 18	s	0 12	1	vF. just n. Sst.	5768
374	—	11 Serpentis	p	12 8	s	1 18	1	eF. pL. r.	5913
375	25	93 Leonis	p	7 28	n	0 7	2	vF. vS. r.	3805
376	26	– –	p	5 57	n	0 5	2	eF. vS.	3821

Fourth class. Planetary nebulæ.

Stars with burs, with milky chevelure, with short rays, remarkable shapes, &c.

IV.	1782	Stars.		M. S.		D. M.	Ob.	Description.	N.G.C.
1	Sept. 7	13 (ν) Aquarii	p	5 24	n	0 2	11	vB. nearly R. planetary not well defined disk.	7009
	1783								
2	Dec. 26	13 Monocer	f	6 4	n	1 27	4	cB. fan-shaped. about 2′ l. from the center. Fig. 7.	2261
	1784								
3	Jan. 16	15 Monocer	p	8 18	n	0 15	4	pB. m. like a st. with an electrical brush. Fig. 8.	2245
4	Feb. 22	69 Leonis	f	10 3	s	1 3	2	eF. S. like an st. with a vF. brush sp. 240 shews the st.	3662
5	—	29 (γ) Virg	p	9 0	n	1 33	3	A pBst. with a milky ray s. par. 15 or 20′ l. Fig. 6.	4517
6	23	59 (c) Leonis	p	9 0	s	0 18	1	F. L. C. A central B. point with eF. m. chev.	3423*
7	Mar. 14	51 (m) Leonis	f	17 0	s	0 39	2	F. pL. m. between 2 Bst. like an electrical brush to the most n. but is not connected. R.	3507
8 } 9 }	15	34 Virginis	p	10 12	s	0 51	2	A double Nebula. The che. run into each other. close. not vF	4567 4568
10	21	51 (m) Leonis	p	21 15	s	1 48	1	A pcst. with a vF. brush nf. with 240 2 vSst. visible in it, but not connected.	3239
11	May 21	51 (e) Ophiu	p	1 42	n	0 14	2	pB. R. p. well defined planetary disk. 30 or 40″ d.	6369
12	22	3 (p) Sagitt[i]	f	22 0	n	1 47	1	F. L. iR. inclining to m. 3 or 4′ d. like a brush to a np. st. but probably unconnected.	6553
13	July 17	39 (h) Cygni 21 Vulpecu	p f	8 6 2 6	s n	1 35 1 51	2	pF. exactly R. of equal light. the edges p. well def. 1′ d. See note.	6894

IV.	1784	Stars.		M. S.		D. M.	Ob.	Description.	N.G.C.
14	July 21	27 (*d*) Aquilæ	p	6 6	s	1 45	2	vF. of equal light. r. 1′ d. in the midst of numberless st. of the milky way.	6772
15	Sept. 8	21 (*a*) Andr	f	2 6	s	1 21	1	A Fst. with S. chev. and 2 burs.	16*
16	16	16 (*η*) Sagittæ	f	17 12	n	0 1	2	pB. perfectly R. pretty well defined. ¾′ d. r.	6905
17	20	81 Ceti	f	36 30	n	0 36	1	A Sst. with a vF. nebulous brush 1½ or 2′ l. discovered with 240.	1253*
18	Oct. 6	14 Androm	p	6 11	n	3 16	4	B. R. a planetary p. well defined disk. 15″ dia^{r} with a 7 feet reflector.	7662
19	16	5 Monoc.	p	7 6	s	0 10	1	A st. of the 9 magnitude, with m. chev. i. elliptical.	2170
20	—	– –	p	3 42	n	0 3	1	A st. of the 11 or 12 mag. affected like the foregoing, but vF.	2185
21	Nov. 20	12 Leporis	p	8 48	n	0 24	1	vS. stellar. vBN. and vF. chev. not quite central.	1964
22	Dec. 9	7 (*ξ*) Navis	f	3 10	s	1 28	2	L. pB. R. er. 6 or 7′ d. a faint red colour visible. A st. 8 mag. not far from the center, but not connected. 2d ob. 9 or 10′ d.	2467
23	1785 Jan. 6	75 Ceti	p	4 40	s	0 6	1	cB. a vBN. with a chev. of 3 or 4′ d.	936
24	—	50 (*ζ*) Orio	f	0 57	s	0 17	1	A Bst. with m. chev. 5′ l. 4′ b.	2023
25	31	19 Navis	p	67 0	n	1 15	1	A pcst. with vF. and vS. m. chev. iF.	2327
26	Feb. 1	34 (*γ*) Erid	f	16 16	n	0 49	2	vB. perfectly R. or vl. elliptical. planetary but ill defined disk. 2d obs. r. on the borders, and is probably a very compressed cluster of stars at an immense distance.	1535
27	7	6 (3*b*) Crater	p	28 39	n	1 25	2	Beautiful, brilliant, planetary disk ill defined, but uniformly B. the light of the colour of Jupiter. 40″ d. 2d obs. near 1′ d. by estimation.	3242
28	—	31 Crateris	f	1 0	n	0 47	1	pB. L. opening with a branch, or two nebulæ very faintly joined. The s. is smallest.	4038 4039
29	8	4 (*ν*) Crateris	f	3 46	n	0 16	1	A Sst. with an eF. brush p. perceived in gaging, otherwise I should certainly have overlooked it. ver. 240.	3456*

Fifth class. Very large nebulæ.

V.	1783	Stars.		M. S.		D. M.	Ob.	Description.	N.G.C.
1	Oct. 30	18 (*s*) Pis. aust.	f	128 17	n	1 39	6	cB. mE. sp nf. mbM. Above 50′ l. and 7 or 8′ b. C. H. See note.	253
	1784								
2	Jan. 24	10 (*r*) Virgin	f	24 46	n	0 17	4	cB. mE. np sf. mbM. er. 9 or 10′ l. with a branch towards the np.	4536
3	—	75 Leonis	f	104 0	s	0 24	1	eF. vL. er. R. 7 or 8′ d.	4910*
4	Feb. 23	7 (*b*) Virgin	f	8 15	s	0 45	2	vF. R. 5 or 6′ d.	4123
5	Mar. 14	11 Comæ	f	0 45	n	0 32	1	L. E. r. 6 or 7′ l.	4293
6	21	4 (τ) Bootis	p	0 45	s	1 6	1	vL. eF. r.	5293
7	Apr. 8	52 (K) Leonis	p	3 0	n	0 41	1	vL. F. r. almost R.	3346
8	—	73 (*n*) Leonis	f	4 34	n	0 18	3	B. E. almost par. but l. np sf. near 15′ l.	3628
9	May 22	51 (*e*) Ophiu	f	32 48	s	0 40	1	L. E. broad. m. F I.C.	1271*
10, 11, 12	July 12	5 (*i*) Sagitt[i]	f	2 42	n	0 49	1	Three nebulæ, faintly joined, form a triangle. In the middle is a double st. vF. and of great extent.	6514
13	—	– –	f	4 54	s	0 38::	1	Extensive m. neb. divided into 2 parts. the most n. above 15′. The most s. followed by stars.	6533*
14	Sept. 5	52 (*k*) Cygni	f	11 24	n	0 44	2	Branching nebulosity, extending in R.A. near 1½ deg. and in P.D. 52′. The f. part divides into several streams uniting again towards the s.	6992
15	7	– –	f	0 0	n	0 0	3	Extended; passes thro' *k* Cygni. By the Newtonian view above 1 degree l. By the *Front-view* near 2 deg. l. See note.	6960
16	11	28 Androm.	p	11 12	n	0 17	1	eF. 5 or 6′ d.	68
17	—	2 (α) Triang.	p	18 48	n	0 55	2	m. nebulosity not less than ½ deg. broad. perhaps ¾ degree long, but not determined.	598
18	Oct. 5	35 (ν) Andr	p	9 11	n	0 37	4	vB. mE. 30′ l. 12′ b. C. H.	205
19	6	26 (β) Persei	p	45 11	n	1 16	3	cB. mE. above 15′ l. 3′ b. a black division 3 or 4′ l. M.	891
20	20	7 (*h*) Ceti	f	33 9	s	1 48	1	A streak of light, nearly mer. 26′ l. 3 or 4′ b. pB.	247
	1785								
21	Jan. 31	18 (μ) Canis	f	22 18	n	1 2	2	A broad E nebulosity. forms a parallelogram with a ray southwards; the parallelogram 8′ l. 6′ b. vF.	2359
22	Feb. 7	61 Virginis	f	10 59	n	0 17	1	mE. sf. np. 5 or 6′ l. pF.	5170

V.	1785	Stars.		M. S.		D. M.	Ob.	Description.	N.G.C.
23	Apr. 3	27 Ursæ	f	13 18	n	0 0	1	L. F. lE. r. 6 or 7′ l. 5 or 6′ b.	3027
24	6	21 (g) Comæ	f	5 20	n	1 25	1	A lucid ray 20′ l. or more. 3 or 4′ b. np sf. vBM. a beautiful appearance.	4565

Sixth class. Very compressed and rich clusters of stars.

Additional abbreviations: Cl. Cluster. sc. scattered. com. compressed. co. coarsely.

VI.	1783	Stars.		M. S.		D. M.	Ob.	Description.	N.G.C.
1	Nov. 19	63 (*p*) Gemi	f	11 0	n	0 12	3	A beautiful Cl. of many L. and com. S. st. about 12′ d.	2420
2	Dec. 30	18 (ν) Gemi	f	27 10	s	2 9	3	A v. com. Cl. of eSst. iF. 5 or 6′ d.	2304
	1784								
3	Jan. 24	12 Monocer	f	11 30	s	0 18	1	A Cl. of v. com. and eS. st. E.	2269
4	—	4 Sextantis	f	5 30	s	0 5	1	A Cl. of v. com. S. st.	3055
5	Feb. 11	31 (2 ξ) Gem	p	31 0	s	0 15	1	A Cl. of v. com. S. 7 or 8′ d.	2194
6	Mar. 8	67 Gemin	p	18 0	s	1 57	1	A Cl. of st. of various sizes pm. com. M. p. rich.	2355*
7	14	42 Comæ	f	8 30	n	0 8	1	An eF. Cl. of eS. st. with r. neb. 8 or 10′ d. ver. 240. beyond doubt.	5053*
8	Apr. 25	..	..	..	..	..	1	A v. com. Cl. of st. 8 or 9′ d. e rich. iR. or lE.	.. *
9	May 17	11 Bootis	f	4 18	n	1 7	1	A Cl. of eS. and com. st. 6 or 7′ d. many of the st. visible, the rest so S. as to appear nebulous.	5466
10	22	21 (α) Scorpii	p	1 48	n	0 24	1	A v. com. and cL. Cl. of the smallest stars imaginable. all of a dusky red colour. the next step to an er. neb.	6144
11	—	39 Ophiuchi	p	13 24	s	0 26	1	A fine miniature of the 19 nebula of the *Connois. des Temps* (which is a Cl. of v. com. st. much accumulated M. 4 or 5′ d. all the st. red.) 2 or 2′½ d. the st. F. red.	6284
12	24	43 Ophiuchi	p	12 42	n	1 36	1	Another miniature Cl. like the preceding, but rather coarser.	6293
13	June 24	10 (γ) Sagitt[i]	p	14 48	n	0 18	1	A Cl. of S. and p. com. st. of several mag. 5 or 6′ d. not v. rich.	6451
14	July 11	9 Vulpec	p	4 0	n	0 33	1	A Cl. of eS. and v. com. st. a parallelogram of 4′ l. 2′ b. mer.	6802

VI.	1784	Stars.		M. S.		D. M.	Ob.	Description.	N.G.C.
15	July 12	34 (σ) Sagitt[i]	p	6 54	n	0 27	1	A suspected Cl. of vFst. of considerable extent. not ver.	6678
16	Aug. 18	12 (γ) Sagittæ	p	4 18	s	1 32	1	A vS. Cl. of com. st.	6839
17	Nov. 16	42 (1 ω) Gem	p	54 53	s	0 29	2	A v. rich Cl. of v. com. and eSst. 4 or 5′ d. A miniature of the 35 Cl. of the Conn. des T. which it precedes 1′ 18″ and is 2′ n.	2158
	1785								
18	Mar. 4	11 Monocer	f	27 15	s	0 2	4	A v. com. and rich Cl. of vSst. iF. 8 or 9′ d.	2309
19	10	24 (1st ι) Libr	f	5 0	s	1 16	1	A beautiful L. Cl. of the most minute and most com. st. of different sizes. 6 or 7′ d. iR. F. red colour.	5897

Seventh class. Pretty much compressed clusters of large or small stars.

VII.	1784	Stars.		M. S.		D. M.	Ob.	Description.	N.G.C.
1	Jan. 18	90 (1 c) Tauri	f	11 0	s	1 30	2	A Cl. of L. scat. st. 10 or 12′ in extent, with a vacancy M.	1662
2	24	8 Monocer	f	8 17	n	0 23	3	A beautiful Cl. of sc. st. chiefly of 2 sorts. the first L. the second arranged in winding lines. contains the 12th Monoc.	2244
3	Feb. 8	3 Leporis	p	72 30	s	0 30	1	A S. Cl. of com. st. some pL.	1498*
4	19	15 Orionis	f	3 6	n	1 10	2	A Cl. of pL. and p. com. st. c. rich. 20 or 25′ d. iR.	1817
5	23	13 Monocer	p	3 15	s	0 28	1	A Cl. of com. st. of various mag. p. rich in Sst. not R.	2236
6	Mar. 16	51 Gemino	f	3 55	s	2 9	1	A p. rich and com. Cl. of st.	2356
7	May 24	3 (p) Sagitt[i]	f	15 54	s	0 8	1	A c. rich, but p. co. sc. Cl. of st. l. more com. M.	6520
8	July 17	41 (i) Cygni	f	5 42	s	2 1	5	A v. rich Cl. of pS. sc. st. most of the same size. 20′ d.	6940
9	19	12 Vulpecu	p	0 5	n	0 30	2	A L. Cl. of p. com. st. most of one size.	6830
10	Nov. 20	7 (ξ) Navis	f	5 56	n	0 40	3	A vL. Cl. of sc. st. c. rich and com. more than 15′ d.	2482
	1785								
11	Jan. 31	19 Navis	p	0 40	n	0 5	1	A c. rich Cl. of co. sc. st. above 20′ d.	2539
12	Feb. 4	6 Navis	p	31 59	n	1 25	4	A beautiful Cl. of p. com. st. near ½ deg. d. C. H.	2360
13	6	2 (β) Canis	p	7 10	s	0 44	1	A Cl. of sc. Sst. not v. rich above 15′ d.	2204
14	8	18 (μ) Canis	f	3 17	n	0 20	1	A Cl. of co. sc. st. 20′ d.	2318

VII.	1785	Stars.		M. S.		D. M	Ob.	Description.	N.G.C.
15	Mar. 6	26 Canis	f	1 22	n	1 52	1	A S. Cl. of p. com. st. not v. rich	2352
16	—	- -	f	1 56	n	0 16	1	A Cl. of sc. st. c. rich. 20′ d.	2354
17	—	- -	f	6 26	n	1 1	2	A v. beautiful Cl. of pL. st. v. rich. contains the 30 Canis.	2362

Eighth class. Coarsely scattered clusters of stars.

VIII.	1783	Stars.		M. S.		D. M.	Ob.	Description.	N.G.C.
1	Dec. 3	14 Navis	p	4 0	n	0 40	2	A Cl. of co. sc. st. The place is that of the most com. part which is not M.	2509
1B	18	18 Monoc	p	11 0	..	..		A Cl. of vS. st. not rich.	2319*
2	26	58 (α) Orion	p	8 28	n	1 16		A S. Cl. of vS. sc. st.	2063
3	—	13 Monocer	f	1 30	n	1 2	2	An E. Cl. of L. sc. st.	2251
	1784								
4	Jan. 16	112 (β) Tauri	p	0 51	n	0 38	3	A Cl. of co. and i. sc. pLst.	1896
5	18	15 Monocer	p	0 0	n	0 0	3	Double and attended by more than 30 cLst.	2264
6	24	8 Monocer	p	14 20	n	0 4	2	A Cl. of co. sc. st. not rich.	2180
7	Feb. 10	4 Orionis	p	4 0	s	1 7	1	A Cl. of L. and S. sc. st. not rich.	1663
8	15	97 (i) Tauri	p	5 28	n	0 13	2	A Cl. of cL. v. co. sc. st. perhaps a projecting point of the m. way.	1647
9	19	24 (γ) Gemi	p	8 15	n	0 15	1	A Cl. of vm. sc. st. of various magnit. near ½ deg. not rich.	2234
10	Mar. 15	50 (2 A) Canc	f	3 0	s	0 44	1	A Cl. of v. co. sc. st. not rich.	2678
11	16	51 Gemi	f	15 55	s	2 19	1	A Cl. of sc. st.	2395
12	June 16	1 (m) Aquilæ	f	1 42	n	0 2	1	A Cl. of v. co. sc. st.	6664
13	—	20 Aquilæ	p	12 48	s	0 56	1	A Cl. of co. sc. st. not rich.	6728
14	18	43 (d) Sagitt[i]	p	44 48	n	1 54	1	A Cl. of sc. pLst.	6647
15	July 15	63 Sagittarii	p	103 36	n	2 1	1	A Cl. of co. sc. st.	6604
16	17	12 (φ) Cygni	f	13 6	s	0 44	1	A Cl. of not v. com. st. closest M. It may be called (if the expression be allowed) a forming Cl. or one that seems to be gathering.	6834
17	18	33 Vulpec	p	24 18	n	0 4	1	A Cl. of many L. sc. st.	6938
18	Sept. 4	61 (φ) Aquilæ	p	2 54	n	0 18	1	A S. forming Cl. of st.	6837
19	—	- -	p	0 42	n	0 40	1	A Cl. of co. sc. L. st. not rich.	6840
20	9	18 Vulpec	f	1 0	s	0 27	1	A Cl. of co. sc. st. not rich.	6885
21	10	6 Vulpec	p	2 27	n	0 29	2	A Cl. of cL. co. sc. st.	6800
22	—	18 Vulpec	f	1 12	s	0 12	1	A Cl. of co. sc. st.	6882
23	Oct. 15	12 (γ) Delph	p	5 18	n	0 33	1	A Cl. of co. sc. st.	6950
24	—	67 (ν) Orion	f	1 0	s	0 46	1	A SCl. of pL. white st.	2169
25	16	10 Monocer	f	0 0	s	0 0	1	The 10 Monoc. surrounded by many Bst.	2232

VIII.	1784	Stars.		M. S.		D. M.	Ob.	Description.	N.G.C.
26	Nov. 16	1 (H) Gem	p	2 16	n	0 3	1	A Cl. of st. of various magnit. not v. rich. 6 or 7′ d.	2129
27	20	11 (*e*) Navis	p	36 41	n	0 46	1	A S. Cl. of sc. st. not rich, nor v. com.	2367
28	Dec. 5	54 (1st χ) Orion	p	11 53	s	0 15	1	A Cl. of pL. sc. st. not rich.	2026
29	9	101 (4 *b*) Aqu^{i}	f	32 30	n	0 11	1	A Cl. of a few co. sc. L. st.	7826
30	—	25 (δ) Canis	f	57 10	s	1 15	1	A vL. Cl. of many co. sc. L. st.	2527
	1785								
31	Jan. 6	19 Monocer	p	15 36	n	1 3	1	A L. Cl. of sc. st. not v. rich.	2286
32	10	26 Monocer	p	34 32	s	0 41	1	A Cl. of co. sc. st. of many magn. p. rich. above 15′ d.	2335
33	—	— —	p	32 50	s	1 15	1	A Cl. of sc. L. st.	2343
34	—	— —	p	26 36	s	0 52	1	An extensive Cl. of sc. st.	2353
35	31	2 Navis	p	21 23	n	1 21	3	A Cl. of pL. sc. st. p. rich. about 20′ l. crooked fig.	2374
36	—	19 Navis	p	43 20	n	1 0	1	A forming Cl. of co. sc. st. 20 or 30′ dia.	2396
37	Feb. 4	6 Navis	p	16 47	n	1 43	2	A S. Cl. of p. com. st. of various sizes. not v. rich.	2414
38	—	2 Navis	p	8 55	n	0 10	2	A Cl. of p. com. L. and S. st. R. above 15′ d.	2422
39	Mar. 4	11 Monocer	f	23 36	n	0 3	3	An extensive Cl. of sc. st. of various sizes.	2302
40	11	47 Geminor	p	4 2	n	0 18	1	Clustering L. sc. st. many of equal size.	2331

Notes to some Nebulæ and Clusters of Stars. [By the Author.]

I. 7. This remarkable appearance being no longer in the place it has been observed, we must look upon it as a very considerable telescopic comet. It was visible in the finder and resembled one of the bright nebulæ of the *Connoissance des Temps* so much, that I took it for one of them till I came to settle its place; but this not being done till a month or two after the observation, the opportunity of pursuing and investigating its track was lost. [See Notes, infra.]

I. 13. The figures referred to, in the description of this and some other nebulæ, may be found in the *Philosophical Transactions*, vol. LXXIV. tab. XVII. p. 450. [Above, Plate VII. p. 159.]

I. 18. The numbers annexed to some of the nebulæ refer to the class and number of the preceding Catalogue: thus, II. 41 denotes that the 41st in the second class is the third nebula, following the two here described.

I. 28. Near the 84 and 86 neb. of the *Connoissance des Temps.*

II. 6. This has probably been a telescopic comet, as I have not been able to find it again, notwithstanding the assistance of a drawing which represents the telescopic stars in its neighbourhood.

II. 55. The preceding is the 85 of the *Connoissance des Temps.*

II. 84. 6 or 8′ following the 100 of the *Connoissance des Temps.*

II. 118. Just following the 88 of the *Connoissance des Temps.*

II. 123. 124. The third is the 87th of the *Connoissance des Temps.*

III. 44. The following is the 60th of the *Connoissance des Temps.*

IV. 13. Before the value of the degree was more strictly ascertained, the two observations were thus:

21 Vulpeculæ	f	2′ 6″	n	1° 51′
39 (*h*) Cygni	p	8 6	s	1 35

which, if there be no error in the place of the stars in FLAMSTEED's Catalogue, differ about 14′ in polar distance, for which reason in the second Paper on the Construction of the Heavens this nebula was put down twice, whereas it now appears, that both observations belong to the same.

V. 1. This nebula was discovered Sept. 23, 1783, by my sister CAROLINE HERSCHEL, with an excellent small Newtonian *Sweeper* of 27 inches focal length, and a power of 30. I have therefore marked it with the initial letters, C. H. of her name. See also V. 19, discovered Aug. 27, 1783, and VII. 13, discovered Feb. 26, 1783.

V. 15. The *Front-view* is a method of using the reflecting telescope different from the Newtonian, Gregorian, and Cassegrain forms. It consists in looking with the eye glass, placed a little out of the axis, directly in at the front, without the interposition of a small speculum; and has the capital advantage of giving us almost double the light of the former constructions. In the year 1776 I tried it for the first time with a 10 feet reflector, and in 1784 again with a 20 feet one; but the success not immediately answering my expectations, it was too hastily laid aside. By a more careful repetition of the same experiment I find now, that several other considerable advantages, added to the brilliant light before mentioned, make it so valuable a construction that a judicious observer may avail himself of it at least in all cases where light is more particularly wanted; and from the experience of 30 sweeps, which I have already made with it, I may venture to announce it to be a very convenient and pleasant, as well as useful, way of observing. With regard to the position of objects, it differs from other constructions, by inverting the north and south, but not the preceding and following.

[*Notes to Sir W. Herschel's First Catalogue of Nebulæ and Clusters.*

I. 7. Sweep 105, Jan. 23, 1784: "A beautiful nebula, not cometic. On comparing its place with Messier's Nebulæ I find it to be his 61st. It is visible in the finder and vB in the telescope. It precedes 31 Bootis $2^h\ 0\frac{3}{4}^m$ or follows 49 Leonis $2^h\ 6^m\ 45^s$. These two agree to 12^s, therefore a mean correction of both gives $+51^s$, and in present time the R.A. comes out $12^h\ 30^m\ 21^s$" [in pencil is added "P.D. 81°±"]. Feb. 23, 1784: "The beautiful nebula of the 105 Sweep is not Messier's 61st, as was hastily surmised."

On October 27, 1801, Herschel wrote to Bode as follows*: "After my return from North Wales where I have spent some time I received the favour of your last letter, by Mr Thoelden. On looking over the sweep in which the nebula I. 7 was seen, I find that there is a considerable uncertainty in the polar distance of five successive sweeps, owing to a change in the altitude of the telescope and want of well ascertained stars. There was no mistake in writing south for north, but the identity of 49

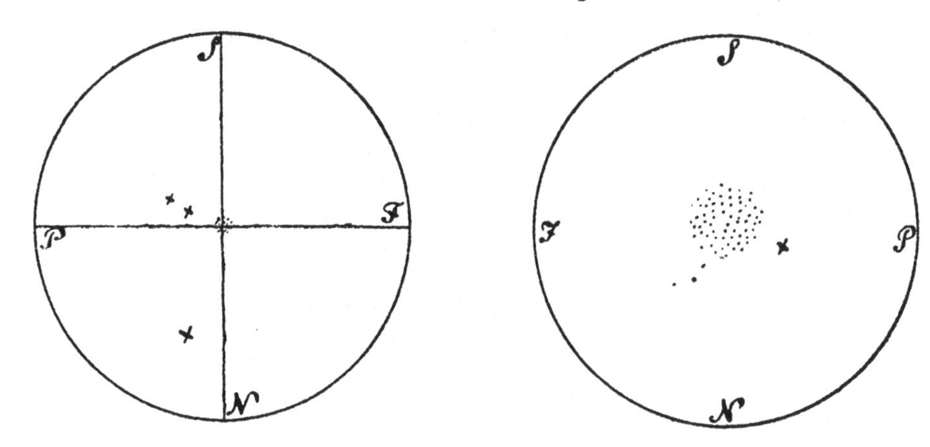

Leonis is doubtful. My apparatus in 1784 was not so complete as it is now, and to solve the doubt about the place it will be necessary to sweep the zone in which the nebula or rather comet was taken over again. There are objects enough in the sweep to ascertain its place with considerable accuracy, and I will repeat it as soon as the constellation comes to the meridian. Meanwhile I send you two

* Printed in translation by Bode, *Astron. Jahrbuch*, 1805, p. 211 (without the sketches). The letter is here given from the draft in Herschel's Letter Book. In the rough sketch in the original journal the letters P, F, S, N do not occur; neither do the three little dots below and to the left of the neb. (H. always denoted stars by small crosses).

drawings, one representing the appearance of the comet in the finder, the other as it was seen in the field of view of the telescope. From this you see that its situation was within half a degree north following two small stars and about one degree south following a pretty considerable one. The field of view of the finder is three degrees. The comet, by the field of the telescope which is 15 minutes, must have been about 4 or 5 minutes in diameter. It had no nucleus, and there was a small star which could not be seen in the finder very near and a little north preceding it."

The letters P, F, S, N in the above sketches are not in the original sketches in Herschel's journal, and everything indicates that he, in January 1784, only used the Newtonian construction. In the sweep (105) this nebula is exactly 2^m preceding a nebula which is undoubtedly II. 20 = II. 148. The R.A. must therefore have been $12^h\ 32^m{\cdot}5$ (for 1860). The fine nebula M. 49 is exactly 10^m p., and this nebula has a ∗ 12 mag. 4^s f. on the parallel, so that it seems fairly certain that I. 7 = M. 49. In the journal H. wrote next the two sketches, " Messier's 61, I believe," but "61" is struck out and "49" substituted. (See below, under II. 19 and 20.)

I. 26. Some error in the observed position, as there is no nebula near the place. According to the sweep (177), it was $\frac{3}{4}^m$ f. and 1° 41′ south of a nebula which turns out to be II. 77. It was probably M. 95, which is not mentioned, and which is $1^m\ 52^s$ f., 2° 2′ s. of II. 77.

I. 35. See below, under II. 165.

I. 43. Second obs., Sw. 819, Mar. 11, 1788, mE. from about 20° sp. to nf., BN., 4 or 5′ long, 49 Virg. p. $27^m\ 45^s$ s. 0° 51′. In 1784 " the B. place in the middle is pL., but breaks off abruptly."

I. 54. Given thus among errata in *Phil. Trans.*, 1786. The nebula originally entered as I. 54 turned out to be M. 32.

I. 79. R.A. $1\frac{1}{2}^m$ too small. (See below, note to III. 312.)

II. 3. Observed twice, Dec. 13 and 24, 1783. Place very rough, deduced from four diagrams of fields of stars between it and 17 Ceti. The correct place is $1^m\ 8^s$ f., 22′ n. of the place found thus.

II. 6. Very rough sketch shows it in line with 2 st. p. and one f. Afterwards assumed = I. 1.

II. 14. *Phil. Trans.* has: " f. ν Virginis $2^m\ 20^s$, 1° 22′ n." This was deduced from four diagrams of stars. Subsequent note: " It precedes 59 Virginis $1^h\ 9^m$, 12′ n.," but this is crossed out in pencil, with the addition: " Was not ν Virginis; is now calculated right, R.A. $11^h\ 57^m\ 48^s$ P.D. 79° . . . for 1800."—According to Frost, there is no nebula here. It is the only object in Sweep 85 (" $11^h\ 41^m$ a nebula N. 22 "), and in Sweep 86 the only one is " $12^h\ 0^m$ a nebula 22." The latter is = II. 15, also observed 15 April 1784, the place of which is nearly correct. It was probably I. 33 = II. 60, 50′ north of the place and $29^m\ 25^s$ p. II. 15. After the transit of 59 Virginis is written: " Hazy, this sweep must be done again some other time, I saw no S."

II. 18. H. afterwards assumed that this was the nebula which he observed 28 Dec. 1785 together with M. 49 (viz. II. 498). But this certainly cannot be the case, as he did not mention M. 49, only 9^s f., 10′·6 n. of it. The entry in the Register (Sweep 105) is: " Preceding M. Messier's 61st Nebula [that is, I. 7] and not far from its parallel is a nebulous star or S. neb., its P.D. must be about 80° 23′. The time of its entrance not being taken leaves it uncertain to some minutes, but probably it is about $12^h\ 15^m$."

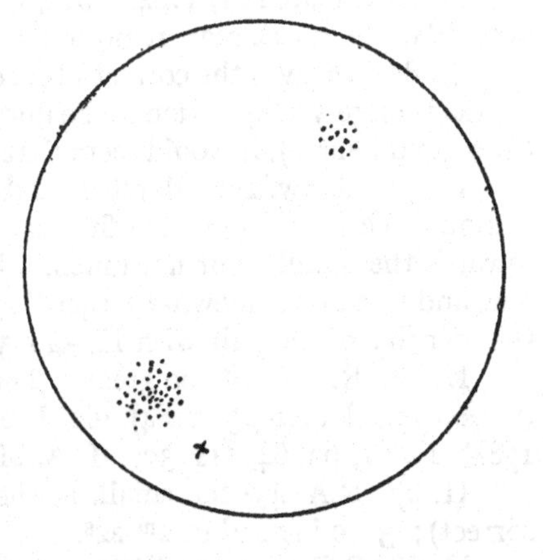

II. 19. Sweep 105, Jan. 23, 1784. " South of M. Messier's 61 Nebula [*i.e.* I. 7], at rectangles to Great Nebula and the small star near it. My field [15′ diam.] takes them both in together. It is incomparably more faint than M. Messier's, its A.R. is about $12^h\ 26^m$. It precedes 31 Bootis $2^h\ 0\frac{3}{4}^m$." Feb. 23, 1784, memorandum: " This neb. of the 105 sweep and that of the same sweep [II. 18] are not near Messier's 61st but near that large new one which I had mistaken for his 61st."

There is no nebula in the place of G.C. 3147 (*Ann. Harvard Coll. Obs.*, xiii. p. 81, and Max Wolf, List II.). M. 49, which, as shown above, is most probably = I. 7, has a smaller nebula (II. 498) 9^s p., $10\frac{1}{2}$′ s., and the agreement of the sketch with the appearance of M. 49, the star following it, and II. 498 in a Newtonian field is perfect.

II. 20. This is quite certainly the same as II. 148. There is a sketch of it under date Jan. 23, 1784 (Sw. 105), which agrees perfectly with the description of II. 148 in Sw. 498, Dec. 28, 1785: " preceding a row of pretty considerable stars and near the most south of them, making a rectangle" [*i.e.* a right angle]. Also in Sw. 560, May 1, 1786, II. 20 is said to be "preceding the most s. of a row of stars, p. 31 Virginis 0^m 27^s n. 0° 25′."

II. 22. Sw. 108, Jan. 23, 1784. "A vS. and F. neb. sp. 59 Virginis. Its A.R. is about 13^h $6\frac{1}{4}^m$. While I looked into the finder to determine its situation I lost it, but shall endeavour to find it another night." The transit is given as 13^h 10^m, that of 31 Bootis (Sweep 111) as 14^h $30\frac{1}{4}^m$. It is probably =Marth 255 (N.G.C. 5100), 35^s p., 30′ n. of the assumed place of II. 22; no neb. in H.'s place.

II. 24. R.A. 40^s too great. A better obs. in Sw. 558. April 30, 1786: 37 Virg. f. 1^m 51^s s. 0° 54′.

II. 26. There is no nebula in this place (Bigourdan and M. Wolf), and h. did not see it either, as he only gives h. 1283, eF. 44^s f., 3′ s. of the place of II. 26. The observations in Sw. 131 are very unsatisfactory; that of σ Leonis differs enormously from ν and 11 Virginis (which agree well *inter se*), and there is a note stating that "σ Leonis was not well taken, the telescope not very settled, ν and *s* were better." 9^m 6^s before the nebula and 22′ south there is a "very large star," which cannot be identified, unless it is + 6°·2588 (7·5 mag.). If there were an error of 20′ in P.D., II. 26 might be =II. 146.

II. 30. Sw. 146, Feb. 15, 1784. "A pB. neb., it seems to contain stars; it is of some extent." Not seen by d'Arrest (5 times), is no doubt =II. 52 only 24^s p. and 12′ n. The latter was observed 14 March 1784," a neb. like II. 51 but a little longish."

II. 39. Sw. 158, Feb. 23, 1784. "A pB. neb., it contains two stars in the centre and is preceded by a small star at the distance of ½ or ¾′." Neither H. nor h. nor d'Arrest saw more than one nebula here, it is therefore =I. 142 (which has a star 10 mag. 4^s·8 p., very little south) with an error of 10′ in P.D.

II. 46. P.D. 8′ too small and the neb. =h. 728. The next object in the same sweep (166, Mar. 12, 1784) is II. 47, the P.D. of which, as given by C. H. (2° 29′ n. of δ Leonis), is 6′ smaller than that resulting from 54 Leonis (10 Apr. 1785), which is right.

II. 48 is quite certainly =II. 80, only one seen at a time. II. 48, Sw. 169, March 14, 1784, r., pL., lbM., contains one star f. the brightness and v. nr. to it. Sw. 698, Feb. 13, 1787, I looked for this neb. but could not find it; it, was, however, not so well looked for as it ought to be, in order to ascertain its absence (II. 80 observed this night, the description agreeing with that of II. 48 in 1784). There was some doubt about the contraction of the rope in Sw. 169, hence the error in the P.D. of II. 48.

II. 56 and II. 90. There is only one nebula here, but H. appears to have seen two on one occasion. The original records are:—

Sw. 170, March 14, 1784. A r. neb. of an irreg. shape of about 2 or 3′ diam. It is near a pB. star, and follows 11 Comæ about $7\frac{1}{4}^m$, 39′ south. [In the sweep it precedes 25 Comæ 10^m, 3′ n. The night was "very windy."]

Sw. 182, March 21, 1784. pL., r., bM., it precedes 25 Comæ $8\frac{1}{2}^m$, and is 1′ more north. A pB., r. neb., bM., near a B. star, it precedes 25 Comæ $7\frac{1}{2}^m$ and is 3′ more south.

$\Delta\alpha$ 8^m 30^s gives the correct place of h. 1282, which has a star 8 mag. 15^s p., 2′ s. Unless the second nebula is merely a repetition of the first, one minute past the meridian (and the pB. * is said to be near the second nebula), it would seem that a faint comet must have been seen on Mar. 21, 1^m following II. 56.

II. 57, 58. Sw. 172, March 15, 1784. "Two neb. about ¾′ or a little more from each other; of the r. kind. The position of the first is about 15° or 20° np. the second; they are pS. and rather brighter towards the middle, but not much. The neby. of the f. one is rather more diluted than that of the p. one, and it is also somewhat larger." R.A. is exactly one minute too small and P.D. 12′ too small, but the identity of the pair with II. 546–547 is certain, both relative positions and descriptions agreeing.

II. 64. R.A. is 1^m too great. The same is the case with several other nebulæ observed this night (Sweep 174, March 15, 1784), viz. I. 21, I. 22 (the place of which in *P.T.* was determined on Apr. 17, 1784), II. 61, 63, 64, III. 37. R.A. of II. 65 is 40^s too great.

II. 67. R.A. 43^s too small, in the sweep (174, 15 Mar. 1784) it precedes II. 68 (obs. of which is correct); 3^m·0 instead of 2^m 22^s.

II. 68. *P.T.* gives 34 Virg. p. 10^m 24^s s. 0° 36′, from Sweep 199, Apr. 17, 1784, which is very in accurate.

II. 72. Minute of transit omitted in sweep, though the fraction ½ is given. *P.T.* gives $\Delta\alpha = 0^m\ 15^s$ which makes the R.A. 45^s too small. It should probably be 1^m 15^s.

II. 80. See II. 48.

II. 90. See II. 56.

II. 98. Two later and better observations are:

Sweep 539, Mar. 18, 1786. 30 Leonis p. 14^m 21^s s. 0° 8′.
Sweep 690, Jan. 14, 1787. 30 Leonis p. 14^m 20^s s. 0° 7′.

II. 107, 108. There is only one nebula here (N.G.C. 4212). The record in Sweep 187 is:

12^h 0^m·2	57	1° 21′	a pretty large	[II. 107]
0·6	17	24′	6 Comæ	
0·5	58	1° 23′	a much extended	[II. 108]
1	92	2° 11′	a neb.	[II. 109]

As II. 108 was recorded *after* 6 Comæ, though it is supposed to precede it by 6^s, we may assume that H., after observing the star, again moved the telescope 1° south and took the nebula a second time without noticing that it was the same object.

II. 109. P.D. is 20′ too great. The observation must have been hurried.

II. 118. Same sweep. M. 88 "with a S. one after it, moon light so strong that I had nearly overlooked the latter." M. 88 was observed again on Jan. 14, 1787, when there is no mention of the S. neb. It is only two F. stars, involved in the neb. according to Tempel, not involved on the Birr drawing (1880, Pl. V. fig. 11).

II. 123–124. *P.T.* has "p. 16^m 24^s n. 0° 29′," which is from Sw. 199, Apr. 17, 1784, and is very incorrect.

II. 128. Second and better obs. Sw. 691, Jan. 14, 1787: pB., vL., vgbM., but the brightness taking up a large place, 29 Comæ p. 1^m 8^s s. 0° 22′.

II. 131 is the same as IV. 6. Five obs. by H., who never saw more than one nebula, but the R.A. of the first obs. (II. 131 in *P.T.*) is 1^m too great.

II. 137. Obs. in *P.T.* is from Sw. 498, Dec. 28, 1785; it gives for 1860 12^h 10^m 37^s, 82° 31′. The first obs., in Sw. 191, Apr. 13, 1784, has "f. 11 Virginis 7^m 18^s, 0° 55′ n." This makes the R.A. 24^s less and the neb. is =h. 1165, while h. 1152 (rough place only) is to be struck out.

II. 160. Sweep 198, observed between II. 159 and II. 54. The transit must have been recorded 1^m late. It is =III. 28.

II. 165. According to Sweep 199, it is 0^m·8 p., 7′ n. of I. 35. In reality it is 37^s·5 p., 7′ 21″ south. By interchanging the P.D.'s of the two, both are made to agree with an obs. by Dreyer.

II. 177. R.A. 28^s too small (Sw. 200). There is a good obs. in Sw. 720, Mar. 19, 1787, 30 Bootis p. 17^m 20^s n. 0° 58′.

II. 178–179. Much better observations in Sw. 571 and 720. The result of the latter, 9 Scorpii f. 8^m 44^s s. 0° 16′ is very accurate.

II. 190. Description and place are from Sw. 536, 3 March 1786. It was also observed in Sw. 916, March 23, 1789 (f. 23^m 39^s s. 0° 9′). A nebula numbered II. 190 had been found in Sw. 208, 25 Apr. 1784, not vF., cL., E., r., but it was afterwards rejected, as there was no determining star owing to clouds (compare VI. 8).

II. 202. Sw. 239, July 17, 1784. "A resolvable nebulous patch; there are great numbers of them in this neighbourhood like forming nebulæ, but this is the strongest of them; they are evidently congeries of small stars." According to Bigourdan, there is no second class nebula here; a region rich in stars, but if there is any nebulosity about, it is very diffused (4 obs.).

II. 204. Time of transit in Sw. 245 was corrected by -1^m·5 : : , which was too much, as it made the R.A. 1^m 13^s too small.

II. 206. Birr Castle, three times not found, once eF. nebulosity seen. Bigourdan, stellar, neb?

II. 239 and III. 199. Three obs. by H. agree perfectly with those at Birr Castle, III. 199 being 3′ 20″ n. of II. 239. In Oct. 1786 H. made it about 3^s f., 3′ n.; in Dec. 1786 3′ due north. There must be an error in the transit of Oct. 24, 1786, which J. H. preferred in the G.C., and whereby he made R.A. of II. 239 2^m too great. The third obs. is right (Dec. 11, 1786, 27 κ Persei p. 8^m 2^s, n. 3′ and 0′).

II. 264. Note in sweep: "The A.R. cannot be above 10 or 15^s out; the roller went off the apparatus which occasions the uncertainty." Auwers' place is wrong owing to a misprint in *P.T.* It is the only nebula in this sweep; 4^m 18^s f. and 2° 36′ s. is a star 6 mag. which is P.V. 327. This gives P.D. =113° 41′ (for 1860), and the nebula is =I.C. 2154 (Swift, XI. 90), 23^s f., 8′·5 n. of the place of G.C., in which nothing has been found by Howe and Bigourdan.

II. 271–272. Observed three times by H., but he only once saw it double. The following one is very slightly *south* of the p. one.

II. 295. R.A. 31^s too great according to Howe. Correction to centre of field probably forgotten.

II. 296. H − h = $+67^s$. There is probably an error of 1^m in the transit, for $48^m\ 57^s$ read $47^m\ 57^s$.

II. 299. The P.D. is correct (compare G.C., p. 32), but according to Tempel the R.A. is 1^m too small. In the sweep the transit is given as $12^h\ 47^m$, and that of I. 68 as $12^h\ 46^m\ 50^s$; obviously the R.A. of II. 299 was not accurately determined.

II. 300. A second obs. (agreeing closely with the first) in Sw. 548: 68 Virginis, p. $25^m\ 43^s$ s. 1° 12′.

II. 302. R.A. $\frac{1}{2}^m$ too great (transit given to whole minute only). Second obs. in Sw. 697, Feb. 10, 1787, F., pL., iE., 9 Cancri p. $9^m\ 14^s$, 1° 8′ n., which is correct.

II. 305. Looked for but not found in 1787. It was the only object compared with "20 Sextantis," but the star was in reality B. 1414. This gives for 1860 $9^h\ 57^m\ 4^s$ 95° 49′, in perfect agreement with N.G.C. 3110 (Stephan XIII.).

II. 324. Not found by Bigourdan.

II. 375. Not noticed by h., who has four obs. of II. 374, nor by d'A., who, however, has only one obs. between clouds. Not found by Bigourdan. In the sweep it is $1^m\ 18^s$ f., 2′ n. of II. 374, but the transit of the latter is only given to the nearest minute.

II. 385–386. There are no nebulæ exactly in the relative positions of II. 385–386. II. 383 in the same sweep (396) is very nearly right, so are II. 384 and II. 346, immediately before II. 385. This is an extremely crowded place, close to the Pole of the Milky Way.

III. 1. is an appendage to the north of M. 43.

III. 2. Sweep 61, Dec. 21, 1783. "An almost invisible F. neb., it is R. and about 8 or 10″ diameter, being brighter in the centre than outwards. It can only be seen when the glass is perfectly clean and the attention confined to the object." By two diagrams it is about $1\frac{1}{2}$° nf. a star which was taken to be 69 Ceti, but obs. was interrupted by clouds. Not found by Bigourdan twice.

III. 3. Sw. 72, Dec. 30, 1783. "About 1° nf. 95 Leonis, vF., not of the cometic kind though almost R. It forms an isosceles triangle with two small stars [by a diagram these are about 6′ sp.]. The nebula is probably of the resolvable kind but eF.; it may be a very distant, compressed cluster of stars, but would require a great quantity of light to resolve. It follows θ Leonis 46^m, 54′ n." Sw. 182, March 21, 1784. "A vF., vS., r. neb. lE. It follows 95 Leonis $4\frac{1}{4}^m$, 36′ n." Not found by Bigourdan twice.

III. 5. Sw. 83, 1784. "The faintest and smallest neb. imaginable. I viewed it a long while and with a higher power than the sweeper. Having no person at the clock, I went in to write down the time and found it impossible to recover the neb. It appeared like a vS. neb. *, and is probably of the cometic sort; there was another vS. * sf. (I think, or rather, am pretty sure), and it preceded a pB. * [neb. is sp. of * by a diagram, about 6′]. It should have been secured before I went into the light. Its place must be about $2\frac{1}{2}$° f. ρ Leonis and about 10′ more n. than that star."

Nothing seen by Spitaler on a very clear night with the 28-inch Vienna refractor, but "a vF. double star (?)." Not seen by Bigourdan.

III. 6. The place agrees sufficiently with that of I. 8, and a sketch made also agrees with one of I. 8. The identity seems certain, and was assumed to be so by J. H. The reason why the nebula was on one occasion classed as III., although the register sheet expressly calls it "a nebula of the first class," was probably a footnote added in pencil after this: "that is, vS. at that time" [*i.e.* like a D. * of Class I.]. In the Journal and in the sweep-book (Jan. 18, 1784) it is simply "a nebula."

III. 7. Place rather rough, d'A. − H. = $-1^m\ 44^s$, −7′.

III. 11. R.A. only approximate. Sw. 109, Jan. 23, 1784: "A neb. star, extremely obscure or faint. P.D. about 80° 40′ and R.A. $13^h\ 48\frac{1}{4}^m$, but the latter was not taken at the moment. It p. 31 Boötis : : $38\frac{1}{4}^m$, and is 0° 1′ more south."

III. 13. Sw. 132. Jan. 28, 1784. "A minute before [the transit of 24 Virginis] I suspected a S. neb., but while I took out another piece to examine it I lost it again." P.D. not taken, clouds. Not seen with certainty by Bigourdan.

III. 14. Sw. 134, Jan. 30, 1784. "I suspect an almost imperceptible cl. of stars or nebulosity. It p. 31 Boötis $12\frac{1}{2}^m$, 9′ south." The Journal has: "or L. F. nebula or nebulosity." h. 1807 is 82^s p. III. 14 was not found by Bigourdan twice, though he suspected eF. neby. 3′ in diam. in h.'s place.

III. 17. Not found by Bigourdan.

III. 20. R.A. is 1^m too small. The error must be in the transit of the neb., as the stars agree *inter se.*

III. 25. R.A. 40^s too small.

III. 26. Sw. 167, March 12, 1784. "Suspected a L., eF. neb., but tho' I looked at it a good while I could not verify the suspicion, nor could I convince myself that it was a deception." [P.D. apparently only approximate.] Sweep 944, March 16, 1790, eF., 26 Comæ Ber. p. $5^m\ 5^s$ s. 0° 32′. Observed by Bigourdan, place correct.

III. 29. R.A. 26^s too great. In Sw. 170 there is this note to 86 Leonis: "Extremely windy, therefore the time a little uncertain."

III. 33. Sw. 171, March 14, 1784. "A neb. suspected by 157 and the suspicion strengthened by 240, but the latter power does not remove all doubt. It follows 3 pB. stars making an arch [concave towards np. or nnp. direction by a diagram], south of which arch there is a still brighter star." Not found twice by Bigourdan.

III. 35. Observed by Bigourdan, place correct.

III. 36. R.A. possibly 1^m too great (see under II. 64). Not found by Bigourdan.

III. 37. =h 1069. R.A. 1^m too great (see under II. 64).

III. 38. R.A. possibly 1^m too great (see under II. 64). Not found by Bigourdan.

III. 40. R.A. half a minute too great (Bigourdan), same sweep as the three last, 174.

III. 48. Sw. 175, March 15, 1784. "An eF. neb., it is S., and required some time to look at it before it could be well seen." Not found twice by Bigourdan.

III. 50. R.A. is 28^s too great (*Ann. Harv. Coll.*, xiii., and Bigourdan).

III. 51–52. R.A. 1^m too great. Transits both of neb. and star only given to whole minute.

III. 54. R.A. 44^s too small. Minute of transit has been corrected, and is doubtful.

III. 57. Not seen by Bigourdan.

III. 61. Sw. 181, March 21, 1784. "Suspected a neb. with power 157; 240 shewed 5 S. st. with a little seeming nebulosity, of which however I still have some doubts; most probably a higher power would have shown them free from it." Observed at Birr Castle as a S. irr. cl., 56′ south of h. 565, which is a different object.

III. 62–63. R.A. $\frac{1}{2}^m$ too small. Sweep 181, immediately after III. 61.

III. 64. Sw. 181, March 21, 1784. "A suspected neb., but 240 shewed some S. st. with suspected nebulosity, probably a deception from want of light and power." Transit only given to nearest minute. Not seen by Bigourdan.

III. 75. Sw. 187, Apr. 8, 1784. "eF., not S. I had some doubt and put on 240, but there being no stars very near it, I could not adjust the focus, and therefore could not verify it." Not seen by d'A.; looked for repeatedly when observing II. 101 sp. it.

III. 76. Same sweep. "Some doubts were removed by putting on 240." Not found on plate by M. Wolf (1907, List VII.), not found by Bigourdan. In the sweep it is $3^m{\cdot}2$ f., 3′ s. of II. 102, the place of which is correct.

III. 82. In the sweep (189, 12 Apr. 1784) it is $4^m{\cdot}3$ f. and 15′ south of II. 128. Not found by Bigourdan.

III. 88. Only seen in Sw. 191, Apr. 13, 1784; place in N.G.C. is that of Auwers from 56 Leonis. In the sweep it is $1^m{\cdot}9$ p., 3′ n. of II. 131.

III. 92–93. A second obs., Sw. 498, Dec. 28, 1785, "left doubtful for want of time, it follows *o* (9) Virg. $16^m\ 15^s$ and is 2° 4′ more south." Places from the two obs. agree well *inter se* and with d'A.'s place of III. 92. III. 93 not seen by d'A. and is not in Wolf's second list. In 1785 H. evidently only saw or suspected one nebula.

III. 94, 95, 96. All three observed by Kobold.

III. 97. Only seen once; not seen Dec. 28, 1785, when II. 144 was reobserved. Never seen by h, d'A., nor at Birr. In G.C. and N.G.C. III. 97 ought to have been given the same place as II. 144.

III. 98. Sw. 191, Apr. 13, 1784. Place very uncertain, especially the P.D., which was not noted. It may be =I.C. 3591 or 3617. No object on Wolf's plate in the place of N.G.C. 4588.

III. 104. Not seen by Bigourdan, at least not with certainty.

III. 107. *P.T.* (Sweep 196) gives a place $2^m\ 30^s$ p. h. 744 on the parallel. In Sw. 1098, 12 Apr. 1801, 59 Leonis p. $16^m\ 30^s$ n. 0° 40′. This gives exactly the place of h. 744. III. 107 was looked for in Sw. 497, Dec. 28, 1785, but not found; a note says: "I am pretty sure it does not exist, as the evening is very favourable." Sweep 196 was the first one on Apr. 17, 1784, and was only 19^m long, when the twilight was barely over.

III. 108. R.A. is 34^s too great (Tempel). Transit only given to nearest minute.

III. 111. The only neb. in Sw. 201, the first one on Apr. 18, 1784. The reason why two comparison stars are given is, that H., after observing 58 Leonis, "lowered the telescope, but without measure." No transit taken of τ Leonis, only P.D., but $11^m \cdot 9$ f. and 4′ n. of neb. is "L," which means a large star. It can only be P. XI. 41, which gives for 1860 $11^h\ 1^m\ 52^s$ 84° 26′ or 34^s f., 1′ south of the correct place. $\Delta\delta$ of this star and τ Leonis should be 2° 1′, but the sweep gives 1° 47′, so that the telescope must again have been disturbed. Hence the erroneous P.D. of Auwers.

III. 112. Sw. 205, Apr. 23, 1784. "A cL., eF., R., r. neb., just p. and v. near a vB. star; the nebulosity touches the star. There is so much moonlight that I do not see it satisfactorily (and am not even without some doubts as to its reality, but must defer the verification till a darker night). It follows ϕ Leonis $10^m \cdot 1$, and is 1° 52′ more south. But ϕ being the first star the determination is probably affected with considerable inaccuracy."

Sw. 673, Dec. 29, 1786. "Looked for this neb., and tho' the night is apparently not a bad one, I could not find it. I examined a great part of the heavens in the neighbourhood but saw nothing of it."

Sw. 912, Mar. 20, 1789. "Looked for III. 112 but could not perceive it nor the B. star. In looking back for it I saw another making a trapezium with 3 S. stars, vF., E., vS. (between $11^h\ 9\frac{1}{2}^m$ and $19\frac{1}{2}^m$, P.D. 94° 32′ + perhaps 4 to 8′)."

In her Zone Cat., C. H. used instead of ϕ Leonis Mayer 510, p. $61^m\ 48^s$, 1° 20′ s., giving for 1800 $1^h\ 15^m\ 49^s$, 94° 50′; while Auwers' place from ϕ Leonis for 1800 was $11^h\ 16^m\ 36^s$, 94° 25′. The nebula in Sweep 205 had probably no real existence; that of Sw. 912 is probably Spitaler's neb. (*A.N.*, 3168, p. 388), for 1800 $11^h\ 15^m\ 59^s$, 94° 29′·1. lEns., perhaps D. or bi-N., between 2 st. 13, * 9·5 6^s p., 3′ n. and another np.

III. 113. Better star (adopted by C. H.), Mayer 510, p. $37^m\ 36^s$ s. 0° 31′. This agrees within 18^s and 1′ with Peters' place.

III. 114. Two later observations are more accurate:—

Sw. 706. Feb. 22, 1787, vF., vS., stellar, 25 Virg. p. $9^m\ 38^s$ s. 0° 1′.
Sw. 913. Mar. 20, 1789. F., S., bM., iF., 25 Virg. p. $9^m\ 41^s$ s. 0° 0′.

III. 125. h. 1711 is $2^m\ 8^s$ p., 4′ n., three obs.; twice h. adds that there is no other neb. on the parallel for some distance following. In the sweep there is before III. 125 a star 8 mag. 2° 49′ more north than III. 125, no transit, and no other object since the remark "Clear again," 10^m before the transit of III. 125. It is therefore possible that the latter was 2^m late. Not found by Bigourdan.

III. 127–128. G.C. and N.G.C. (following C. H.) place these two nebulæ 1° too far south (Auwers' places correct). They are identical with Stephan's nebulæ N.G.C. 5706 and 5709.

III. 135. Compare J. H.'s remarks, G.C. p. 34. In this particular instance C. H. was not justified in altering Δ P.D. from 1° 5′ to 1° 16′, as shown by a comparison with III. 134 and II. 194 just preceding it in the sweep. Not found by Bigourdan once; perhaps he did not search 10′ south.

III. 140. Not found by Bigourdan. Is no doubt =N.G.C. 6052 (m. 302) $1^m\ 38^s$ p. Auwers' place. H. did not observe the neb. in the centre of the field but applied a correction of $-0^m \cdot 7$, which appears to have been too small.

III. 144. Sw. 258. A better comparison star is 17 χ Cygni, f. $15^m\ 30^s$ s. 0° 17′, which agrees well with modern observations.

III. 146. Sw. 264. R.A. 40^s too great. A reduction to the meridian of $-1^m \cdot 3$ has been applied, evidently not great enough.

III. 154–155. Nothing said in the sweep about their distance apart. h, d'A. (only once, in moonlight), an observer at Birr Castle, and Bigourdan have only seen one neb., no doubt the following one.

III. 164. Sw. 268, Sept. 12, 1784. "Suspected, 240 left a doubt; eF. and vS., most probably two close stars; between two stars." No nebula found by Bigourdan, twice.

III. 167, 168, 169, 170, 171. Auwers (p. 201) remarks that it looks as if these, which depend on β Andromedæ, were 1^m out, though II. 224 (in the field with β) is right. In the sweep (271, 13 Sept. 1784) the star agrees well with the next one (ϵ Trianguli), but probably owing to the quick succession of five nebulæ the transits of all but one (III. 169) are only given to the nearest minute. The R.A. of 167, 168, 169 are about 40^s and that of 171 1^m too great; that of III. 170 is correct.

III. 172–173. Observed immediately after III. 171. Not found by Bigourdan, twice. The Birr observation may be of some other object.

III. 174. Sweep 271. R.A. $1^m\ 24^s$ too great. Some error in recording the obs., unless it is a different object.

III. 176. Sw. 271, Sept. 13, 1784. "Stellar, the faintest imaginable, even 240 left some little doubt." Not found by Bigourdan.

III. 181. R.A. is 34^s too great. In Sw. 277 it precedes III. 182 $1^m{\cdot}7$. 51 Pegasi gives the same R.A.

III. 194. P.D. is 11′ too great. At the beginning of the sweep (280, Sept. 20, 1784) is the note: "The rope being broken the P.D. is coarsely marked in revolutions of the axel." Of the other eight nebulæ in this sweep only two are in error, III. 191 (corr. = +8′) and IV. 17 (corr. = −7′).

III. 198. The obs. in *P.T.* (and here) is from Sw. 618, Oct. 18, 1786, and belongs to II. 238. But the original III. 198 was observed in Sw. 283, Oct. 6, 1784: "Suspected, but the haziness will not permit to verify it. 57 γ Andromedæ f. $26^m{\cdot}4$, n. 1′." This might be N.G.C. 937, and there is in Sweep 283 another object, F., lE., r., 35 ν Andromedæ f. 1^h $47^m{\cdot}3$, 1′ s, which is = II 238. As there were flying clouds and haziness, and the telescope was "in the east" (how far from the meridian not stated), these observations should be rejected.

III. 199. See above, under II. 239.

III. 205. Sw. 291, Oct. 15, 1784. "eF., 240 left a doubt, tho' it rather confirmed it. I perceived it in counting a field, otherwise I should never have suspected it." Probably this is the same as a stellar object, perhaps slightly nebulous, found by Bigourdan 29^s p. 4′ s. of H.'s place.

III. 218. P.D. 14′ too small. Reductions correct, but probably an error was made in reading the scale. If 81 be corrected to 91, the nebula would be $10\frac{1}{2}$′ south of 58 Pegasi instead of 4′ n.

III. 222. A second obs. of Nov. 23, 1785, calls it vF., S. Suspected of variability by h.

III. 223. Place correct (error of 1° by C. H.), is =h. 2334.

III. 226. Second obs., Nov. 23, 1785. "vF., vS., stellar, 240 confirmed it." Suspected of variability by h.

III. 227. Place perfectly correct, no other nebula near, but it is not a nebulous cluster but a neb., vF., S., lbM., dif.

III. 250–251. Observed again in 1785 and 1787 (Sw. 462, 788); Dec. 3, 1787, they followed 1^m 54^s 14′ s. and 2^m 18^s 13′ s. of 89 Piscium.

III. 257. R.A. 48^s too great (Bigourdan). There was no star in Sweep 346, so that H. had to use 13 Canis from the previous one.

III. 260. R.A. 30^s too great. In Sweep 351, 70 and 75 Ceti agree within 4^s; the neb. follows 3^m 24^s (should be 3^m 18^s) after III. 269.

III. 262. Same sweep. R.A. 30^s too great. No error to be found.

III. 264. R.A. resulting from the observation given in *P.T.* (Sweep 520, Feb. 1, 1786) is correct, but that of G.C. is 44^s too great. It is founded on Sweep 353, Jan. 6, 1785, f. 14 Hydræ 10^m 2^s, 15′ s., and in this sweep a reduction to the meridian of -34^s has been applied to two preceding nebulæ (II. 280–281) and ought probably also to have been applied to III. 264.

III. 268. In *P.T.* $\Delta\alpha$ from α Leporis is given as 27^m 51^s instead of 24^m 51^s; it was overlooked that the clock had been put on 3^m in the interval between III. 268 and α Leporis. This was also overlooked by C. H. in her Zone Catalogue, where she entered 27^m 51^s, though all the same she gives the correct R.A. from both stars. Sir J. H.'s remarks, G.C. p. 19, are therefore to be cancelled, and the first star was 58 Eridani beyond a doubt. The correctly reduced R.A. agrees closely with that of Howe for N.G.C. 1794, viz. 5^h 1^m 45^s for 1860.

III. 277–278. H. −h. = $+46^s$. Nothing found wrong in Sweep 371.

III. 284. R.A. 33^s too great, the star P. XI. 148 gives the same R.A. Second obs., Sw. 673, Dec. 29, 1786. 74 Leonis f. 25^m 12^s : : , 2° 29′ s., which is correct.

III. 293. "Suspected, eF., eS., stellar, 240 left it doubtful but shew'd the same suspicious nebulous appearance which other stars of equal size were free from." Not found by Bigourdan.

III. 303–304. The second one is certainly =h. 1531, as H. says it is "nf. a vB. star about 8 or 9 m.," although H.'s R.A. is 31^s too great and his P.D. 4′ too great. If H. was right as to their relative positions (III. 303 p. 43^s, 4′ n. of 304), then 303 cannot be =M. Wolf IX. 58, which is 45^s p., 3′ south of h. 1531, nor can it be W. IX. 68, 18^s p., 2′ south. The identity of III. 303 is therefore very doubtful, particularly on account of the error in the place of III. 304. Auwers' place of III. 303 is very close to that of W. IX. 68, H. −W. = $+6^s$, −2′. There is not any possibility of an error of 1°, as suggested by Wolf, as this would throw the object outside the limits of the sweep.

III. 311. d'A. and Bigourdan have only seen a neb. (5819) $+1^m$ −7′ from H.'s place, but it is not in the middle between two stars 6′ apart. B. has an object in $+48^s$ −3′ from H.'s place, but it is probably only a vF. *.

III. 312. Bigourdan's R.A. is $2^m\ 16^s$ greater than that of H. (Auwers). Transits in sweep give $\Delta\alpha = 17^m\ 10^s$, which would agree with B. But under the transit of the neb. is the reduction to the meridian $\left\{\begin{matrix} -\text{sf.} \\ \frac{31''}{\cdot 26} = 1'\ 59'' \end{matrix}\right\}$, whereby $\Delta\alpha$ became $19^m\ 9^s$. Just after the neb. the star 9 Ursæ min. was observed, with the correction $\left\{\begin{matrix} -\text{sf.} \\ = 1'\ 40'' \end{matrix}\right.$ which was justified, since it makes $\Delta\alpha$ from 11 Ursæ min. $= 17^m\ 10^s$, while it was in 1785 $= 17^m\ 17^s$. But in the case of the nebula the correction would seem to have been unnecessary or too great.

At the end of Sw. 392, 3 Apr. 1785, is the note: " It appears that the time of stars so near the pole cannot be had with any degree of precision by my instrument, unless perhaps by more carefully placing it in the meridian."

III. 323–324. Auwers' place of III. 323 is $11^h\ 50^m\ 5^s$ 64° 1′, in excellent accordance with θ on the Birr diagram (by Dreyer), while III. 324 is η, 294″ nf. θ.

III. 325 is α on the Birr diagram. N.P.D. is given in G.C. (2648) and N.G.C. (4007) as 66° instead of 64°.

III. 346. The place of Auwers is $13^h\ 1^m\ 10^s$ 64° 30′ =N.G.C. 4979,
while Javelle 1235 is $13^h\ 0^m\ 56^s$ 64° 26·6 F., cS., R., distinct from 4979,
=M. Wolf IX. 105 $13^h\ 0^m\ 57^s$ 64° 26·5 pF., pS., E., bM.
Wolf has only this one near the place, and his description agrees with that of H.

III. 351. As the brightness of this nebula has been suspected to be variable, it may be well to give H.'s observations in full:—

April 11, 1785. " Two, both vF., vS.; the most south [III. 352] is the faintest, and but for the other could not have been observed."

March 16, 1790. " vF., 53 ξ Ursæ p. $7^m\ 40^s$ s. 2° 50′ [III. 351].
eF., — — p. 7 35 s. 2 53" [III. 352].

d'A. calls III. 351 F., Kobold pB. The only serious discrepancy is that h. on March 30, 1827, described it as B., and on April 19, 1827, as eF.

III. 363. In the G.C. this was put =h. 1510, which is 17^s f., 2′ s. of Auwers' place. In the sweep it follows 62^s on the parallel after II. 391, which shows that it was not h. 1510. The place of I.C. 4051 (Big. 308, Kobold, W. III. 449) agrees closely with that of III. 363.

IV. 6. See above, under II. 131.

IV. 15. is =h. 4, $1^m\ 20^s$ p. H.'s place. Some error in recording the transit, probably simply of 1^m; reductions correct.

IV. 17. See above, under III. 194.

IV. 29. H. −h. = -43^s. Correction to centre of field -96^s, perhaps too much. In Sweep 503, 31 Dec. 1785, " Looked for the neb. of 371 Sweep but could not possibly find it."

V. 3. " The place of this neb. is not determined with accuracy." No modern observations known.

V. 9. N.P.D. in G.C. and N.G.C. is 1° too small; the error dates from C. H.'s Zone Catalogue. The neb. is just following M. 8, and is =I.C. 1271 (Swift, VIII.).

V. 13. Note to N.P.D.: " The number a little uncertain, because the index board stood on the ground, but was supposed would have shewn 153." In the sweep the index reading is 148 ::, ι Sagittarii is 107, 2° =138. The northern part is =V. 9. About the nebulosities f. M. 8, see Barnard, *A.N.*, 3111 and 4239.

VI. 6. H.'s R.A. is $1^m\ 40^s$ too small, but on the " register sheet " it is stated that the star was not well taken. Transits both of cluster and star only given to whole minutes. The Journal adds after the star: " Windy, and liable to some small uncertainty on that account, especially the P.D."

VI. 7. Only observed once, in Sweep 170. R.A. is more than 2^m greater than that of h. 1569. The last object before this in the sweep was M. 53, $\Delta\alpha = 4\frac{1}{2}^m$, so that the entry may be 2^m wrong. H. adds: " I have suspected many such in this neighbourhood."

VI. 8. Sw. 209, April 25, 1784. " A very close, compressed cluster of stars 8 or 9′ in diameter, extremely rich, of an irregular round figure, a little extended. The stars are so small as hardly to be visible, and so accumulated in the middle as to look nebulous."

Three sweeps were observed this night, often interrupted by clouds. The Journal says: " The moon is very bright, but in pursuit of the nebulous stratum I look in hopes of seeing some of the brighter nebulæ in it." In Sweep 207 in the original sweep-book (which ends at $12^h\ 17^m{\cdot}7$ with 74 Virginis) the clock was $1^h\ 3^m$ slow, as shown both by stars and nebulæ observed; the Journal (which is a copy) gives different figures, ending with $13^h\ 15^m\!\cdot\!7$ for 74 Virginis. In Sweep 208 ($12^h\ 20^m$ to 27^m by clock, or in the Journal $12^h\ 32^m - 39^m$) there is only one nebula, numbered II. 190, but rejected, as the place could not be fixed. Sweep 209 is as follows:—

$13^h\ 57^m$		..	..	flying clouds and hazy.
14	1·1 / –·5	62	1° 19′	7·8 m.
	10·7	89	1 45	7 m.
	12·5	0	19	a very close, compressed cl., etc.
	25·2	59	1 16	star.
	25·4	–16	4	6·7 m.
	31	..	..	cloudy.

The Journal gives the same hours and minutes, but they had first been written down 12 minutes greater. The Journal has this note at the top: "In the same situation I believe, but was forgot to be marked." This was crossed over.

Unless the clock was put on one hour between Sweeps 207 and 209 (there is not any note to that effect) the cluster ought to be in R.A. $15^h\ 16^m$ (for 1784). In the sweep the first three stars are marked in pencil thus:—

— supposed to be 665 Bode Virginis.
— Mayer's 577, Bode's 8 Libræ.
— is 37 Libræ Bode.

There is really no indication as to the Polar Distance. At the beginning of Sw. 208 is the note: "Zero 96° 11′, lower than the former sweep." The Journal has: "To follow the stratum I lowered and changed the position (the clock unknown)." The $\Delta\alpha$ and Δ P.D. assigned to this cluster in *P.T.* 1786 really belong to the nebula found in 1786, to which the number II. 190 (*q.v.*) was given. The description of the object is very like the bright globular cluster I. 70, but the stars cannot be identified.

VII. 3. There is no very pronounced cluster near the place.

VIII. 1B. Not given in *Phil. Tr.* but inserted by C. H. in her Zone Catalogue. Sweep 48, Dec. 18, 1783: "began at 8 and 5 Orionis upwards 1¼ degree." The cluster followed 11^m after 18 Monocerotis and there is no P.D. For 1860 its R.A. was therefore about $6^h\ 52^m$ and its P.D. between 87° 48′ and 86° 33′. It is the only cluster or nebula in the sweep. "That part of the milky way thro' which I swept was all compleatly resolved without the least seeming nebulosity."]

XVII.

Investigation of the Cause of that Indistinctness of Vision which has been ascribed to the smallness of the Optic Pencil.

[*Phil. Trans.*, vol. lxxvi., 1786, pp. 500–507.]

Read June 22, 1786.

SOON after my first essays of using high powers with the Newtonian telescope, I began to doubt whether an opinion which has been entertained by several eminent authors, "that vision will grow indistinct, when the optic pencils are less than the 40th or 50th part of an inch," would hold good in all cases. To judge according to so rigid a criterion, I perceived that I was not intitled to see distinctly with a power much more than about 320, in a 7-feet telescope which bore an aperture of 6·4 inches; whereas in many experiments on double stars I found myself very well pleased with magnifiers that far exceeded such narrow limits. This induced me, as it were, by way of apology to myself, for seeing well where I ought to have seen less distinctly, to make a few experiments on the subject of the diameter of optic pencils. It occurred to me, that an opinion which limits them to any given size cannot be supported by theory, which does not determine on subjects of this nature, but must be decided, like many other physical questions relating to matters of fact, by careful experiments made upon the subject. The way, therefore, to come at truth, in a case which seemed to me of considerable importance, lay still open to me, as it had done to former observers; and I thought myself authorised, according to a Cartesian maxim (*Dubia etiam pro falsis habenda*), to suppose, for a while, the size of optic pencils, requisite for distinct vision, intirely undecided.

The first opportunity I had of making the proposed experiments was in the year 1778, and the result of them proved so decisive that I have never since resumed the subject; and had it not been for a late conversation with some of my highly esteemed and learned friends, I might probably have left the papers, on which these experiments were recorded, among the rest of those that are laid aside when they have afforded me the information I want. But a doubt seeming still to be entertained on the subject of the smallness of the optic pencils, it may now be proper for me to communicate these experiments, that it may appear how far the conclusions I have drawn from them are warranted by the facts on which I suppose them to rest.

Experiments with the naked eye.

Exp. 1. Through a very thin plate of brass I made a minute hole with the fine point of a needle; its magnified diameter, very accurately measured under a double microscope, I found to be ·465 of an inch, while under the same apparatus a line of ·05 in length gave a magnified image of 3·545 inches. Hence I concluded, that the real diameter of the perforation was about the 152d part of an inch. Through this small opening, held close to the eye, I could very distinctly read any printed letters on which I made the trial. Proper allowance must be made for the very inconvenient situation of the eye, which by the unusual closeness to the paper cannot be expected to see with its common facility. Besides, the continual motion of the letters, which is required on account of the smallness of the field of view, must needs take up a considerable time.

Exp. 2. In some other pieces of brass I made smaller holes; and among many, that were measured with the same accuracy as in the former experiment, I found one whose magnified diameter was ·29: hence the real diameter could not exceed the 244th part of an inch. Through this opening I could also read the same letters; but the difficulty of managing so as not to intercept all the incident light, as well as the very uneasy situation of the eye, were sufficient reasons for not carrying the intended experiments any further under this form. Besides, I should hardly have allowed them to be fair, if, on a further contraction of the hole in the brass plate, an indistinctness had come on; as we might well have suspected at least two other causes, besides the smallness of the pencils, to contribute to such an imperfection; *viz.* want of light, and a deflection of it on the contracted edges of the hole.

Microscopic Experiments.

Exp. 3. I had now recourse to a double microscope, consisting, for simplicity's sake, of only two lenses. The focal length of the eye-glass, carefully ascertained by an object half a mile off, being ·9; the distance of the object-glass from the eye-glass 9·36; and the aperture of the object-glass ·0405. Hence we compute that the diameter of the optic pencil, when it entered the eye, could not exceed the 232d part of an inch; yet with this construction I saw very distinctly every object I placed under the microscope.

Exp. 4. I reduced the aperture of the object-glass to ·013; hence the pencil was found to be the 724th part of an inch; and yet I saw with this construction very distinctly every object that was placed under the magnifier.

Exp. 5. I made a second reduction of the aperture of the object-glass, so that now it was no more than ·0052; and therefore the optic pencil less than the 1800th part of an inch; and yet I could very well count the bristles on the edge of the wing of a fly, and distinguish their length and thickness.

Exp. 6. Changing the construction of the microscope, I now reduced the pencils by an increase of power. Solar focus of the eye-glass ·52; distance between the object-glass and eye-glass 7·6; aperture the same as in the third experiment. This gave me a pencil of the 336th part of an inch, with which I saw very distinctly.

Exp. 7. Applying now the reduced aperture of the fourth experiment, I had a pencil of the 1139th part of an inch, with which I saw very well.

Exp. 8. I changed the eye lens for another of ·171 focal length; the object-glass and distance between the two lenses remaining as in the two last experiments; aperture ·02. This gave a pencil of the 2173d part of an inch, with which I could count, or rather successively see, the bristles before-mentioned very well; the field, on account of the great power, not taking in more than two large and a small one at a time.

Exp. 9. I was now convinced, that we may see distinctly with pencils incomparably less than the 40th or 50th part of an inch; and indeed so far from expecting any obstruction to distinct vision from the smallness of the pencils, it appeared to me now as if their size might in future be intirely left out of the account. With a view, however, of seeing what other cause might bring on that indistinctness which had been ascribed to the smallness of the optic pencils, I continued these experiments with a variation in the apparatus, and used now an object lens of a different focal length; the aperture and other particulars being as in the 4th experiment. By this construction, which gave me a pencil of the 724th part of an inch, I could see objects very well; but though they appeared distinctly, they were not so sharp on the edges as one would wish to see them. This being compared with the 4th experiment, it appeared that, with equal pencils, unequal degrees of distinctness may take place; and a pretty striking circumstance, which served to lead me in the following experiments, was, that the smallest power gave me the least distinct image; notwithstanding, from former trials, the goodness of the lenses I employed could not be doubted.

Exp. 10. On an examination of circumstances it occurred to me, as indeed I had already before surmised, that a certain proportion of aperture might be necessary to a given focal length of an object-lens or speculum; and that a failure in this point might probably bring on that indistinctness which had been ascribed to the smallness of the pencils. In order, therefore, to put this to a trial, I used now an object-lens of 1·25 focal length, with an aperture confined to ·01; the rest of the apparatus being as in the 3d, 4th, and 5th experiments. The pencil in this case was about the 1000dth part of an inch; and though by a different construction I had already seen very well with a pencil of not half that diameter, I found this to give me, as now I had reason to expect, a very indistinct picture, so much so indeed, that it could hardly be called a representation of the object.

Exp. 11. Increasing the aperture of the object-lens to ·0124, I had a pencil of the 758th part of an inch, but could see no better with it.

Exp. 12. Proceeding in the track now pointed out to me, I admitted an aperture of ·017, which gave a pencil of the 550th part of an inch, but could see not much better with it than before.

Exp. 13. On a farther increase of the aperture to ·0231, and a pencil of the 406th part of an inch, I saw a little better; but still had not distinctness enough even to see the bristles before-mentioned at all. Hence we may conclude, that, in such constructions as the present one, the aperture of the object-glass must bear a considerable proportion to its focal length; since the 54th part (for ·0231 : 1·25 :: 1 : 54) is here not nearly sufficient.

Exp. 14. To the same apparatus I applied a higher power, by an exchange of the eye-glass; but the indistinctness remained as before.

Exp. 15. Returning again to the former construction, I admitted an aperture of about ·037; and having now a pencil of nearly the 250th part of an inch, I could but just perceive some of the large bristles, which shews that even the 34th part (for ·037 : 1·25 :: 1 : 34) of the focal length is not a sufficient aperture for object-lenses that act under such circumstances as the present.

So far I have only related experiments that were made in the year 1778; and my opinion that the smallness of the optic pencils could be no objection to seeing well being thus supported by evident facts, I hesitated not, in a Paper on the Parallax of the Fixed Stars (*Phil. Trans.* vol. LXXII. p. 96),* to affirm, that we might see distinctly with pencils much smaller than the 40th or 50th part of an inch. It did not appear to be necessary, nor would the subject of that Paper permit me to enter into a detail of experiments; but having, in the course of my reading about that time, met with an account of some very small globules made for microscopic uses, I contented myself with an instance of small pencils taken from them. I shall, however, now proceed just to hint at a few inferences that may be drawn from these related experiments; as, upon a mature consideration, we may find reason to believe they point out a cause of indistinctness of vision hitherto never noticed by optical writers; and which, when properly investigated, cannot but influence, and in some respects contribute to the improvement of, our theories in optics. For, admitting that every object-glass or speculum, whose aperture bears less than a certain ratio to its focal length, will begin to give an indistinct picture, it will follow, that while former opticians have been endeavouring to diminish the aberrations arising from the spherical figure, and the different refrangibility of rays, by increasing the focal length, they have been unaware of exposing themselves to the consequences of the cause of indistinctness here pointed out. And till its influence shall be well ascertained and brought to a proper theory, we must suspect that such tables as those which are given in our best authors of optics, pointing out an aperture of less than 6 inches for a glass of 120 feet focal length (or a ratio of 1 to 240) must be far from having that degree of perfection

* [Above, p. 47.]

which may yet be obtained. No wonder that telescopes, made according to theories or tables, where one of the causes of indistinctness is unsuspected, and therefore left out of the account, can bear no smaller pencil than the 40th or 50th part of an inch! If then, on one hand, by increasing our apertures we certainly run into great imperfections, we ought nevertheless also to consider what dangers, on the other, we may incur by lessening them too much.

As soon as convenient, I intend experimentally to pursue this subject, in order to obtain proper *data* for submitting this cause of optical imperfection to theory; at present my engagement with the work of a 40-feet reflector will hardly permit so much leisure; and till I shall have repeated, extended, and varied these experimental investigations, I would wish them to be looked upon as mere hints that may afford matter for future disquisitions to the theoretical optician.

An Account of a new Comet. In a Letter from Miss Caroline Herschel *to* Charles Blagden, *M.D. Sec. R.S.*

[*Phil. Trans.*, vol. lxxvii., 1787, pp. 1–3.]
Read Nov. 9, 1786.

Sir,—In consequence of the friendship which I know to exist between you and my Brother, I venture to trouble you in his absence with the following imperfect account of a comet.

The employment of writing down the observations, when my Brother uses the 20-feet reflector, does not often allow me time to look at the heavens; but as he is now on a visit to Germany, I have taken the opportunity of his absence to *sweep* in the neighbourhood of the sun, in search of comets;

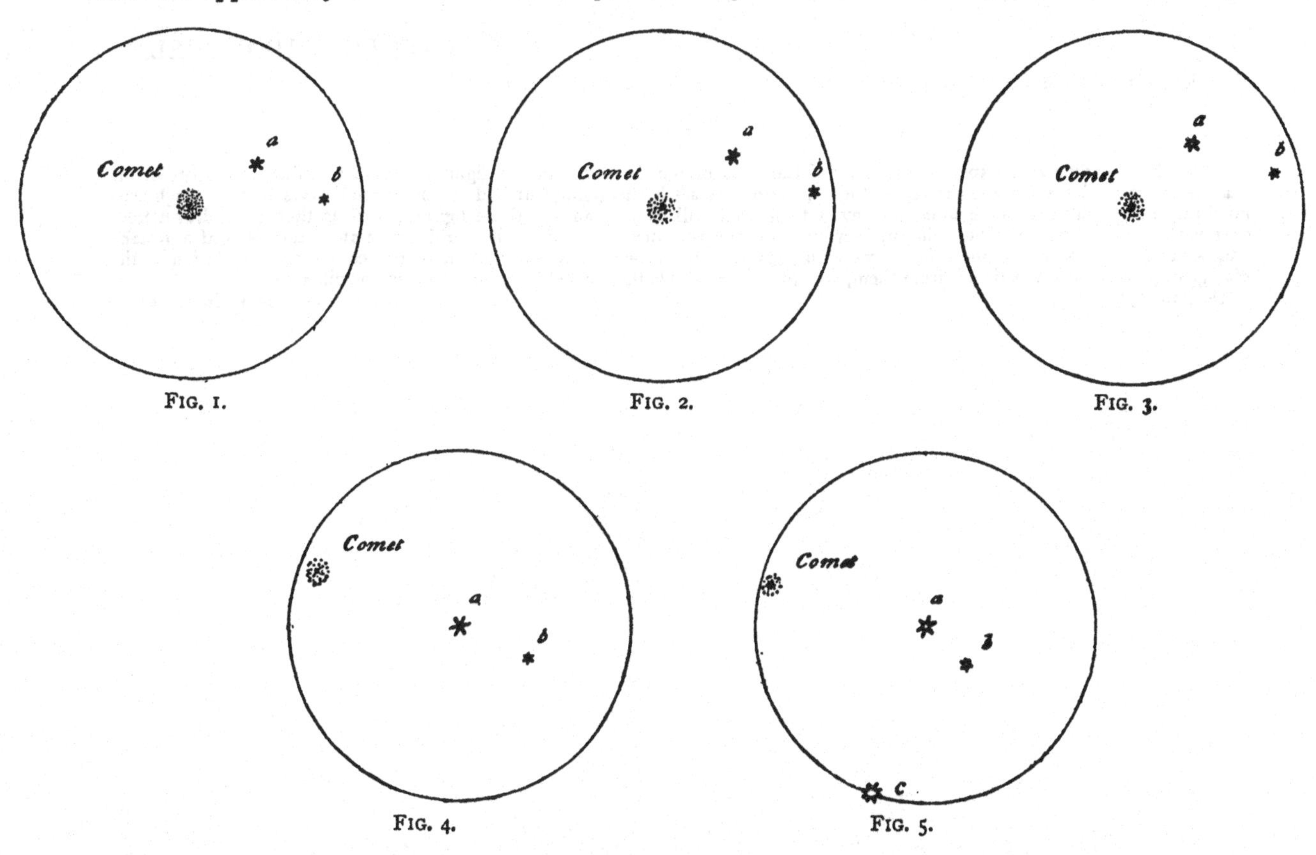

Fig. 1. Fig. 2. Fig. 3. Fig. 4. Fig. 5.

and last night, the 1st of August, about 10 o'clock, I found an object very much resembling in colour and brightness the 27th nebula of the *Connoissance des Temps*, with the difference however of being round. I suspected it to be a comet; but a haziness coming on, it was not possible intirely to satisfy myself as to its motion till this evening. I made several drawings of the stars in the field of view with it, and have inclosed a copy of them, with my observations annexed, that you may compare them together.

August 1, 1786, 9 h. 50′, the object in the center is like a star out of focus, while the rest are perfectly distinct, and I suspect it to be a comet. Fig. 1.

10 h. 33′, fig. 2, the suspected comet makes now a perfect isosceles triangle with the two stars *a* and *b*.

11 h. 8′, I think the situation of the comet is now as in fig. 3; but it is so hazy that I cannot sufficiently see the small star *b* to be assured of the motion.

By the naked eye the comet is between the 54th and 53d Ursæ majoris, and the 14th, 15th, and 16th Comæ Berenices, and makes an obtuse triangle with them, the vertex of which is turned towards the south.

August 2, 10 h. 9′, the comet is now, with respect to the stars *a* and *b*,* situated as in fig. 4, therefore the motion since last night is evident.

10 h. 30′, another considerable star *c* may be taken into the field with it, by placing *a* in the center; when the comet and the other star will both appear in the circumference, as in fig. 5.

These observations were made with a Newtonian sweeper of 27 inches focal length, and a power of about 20, the field of view is 2° 12′. I cannot find the stars *a* and *c* in any catalogue; but suppose they may easily be traced in the heavens; whence the situation of the comet, as it was last night at 10 h. 33′, may be pretty nearly ascertained.

You will do me the favour of communicating these observations to my brother's astronomical friends.

I have the honour to be, &c.

CAROLINE HERSCHEL.

Slough, near Windsor,
Aug. 2, 1786.

* A doubt having arisen about the identity of the stars marked *a* and *b* in the figures, I have examined that part of the heavens in which the comet was the 1st of August, in order to settle this point, but find so many small stars in that neighbourhood that I have not been able to fix on any of them that will exactly answer these figures; and as they were drawn from observations made by moonlight, twilight, hazy weather, and very near the horizon, it would not be at all surprising if a mistake had been made: however, as these figures were only given with a view to shew the motion of the comet, the conclusion of the change of place, which was drawn from them, was equally good whether these stars were the same or different.
Dec. 14, 1786. WILLIAM HERSCHEL.

XVIII.

Remarks on the new Comet. In a Letter from William Herschel, *LL.D. F.R.S. to* Charles Blagden, *M.D. Sec. R.S.*

[*Phil. Trans.*, vol. lxxvii., 1787, pp. 4-5.]

Read Nov. 16, 1786.

Dear Sir,—As my Sister's letter of the 2d of August, relative to the comet discovered by her, has had the honour of being communicated to the Royal Society, I beg leave to add the following remarks upon it.

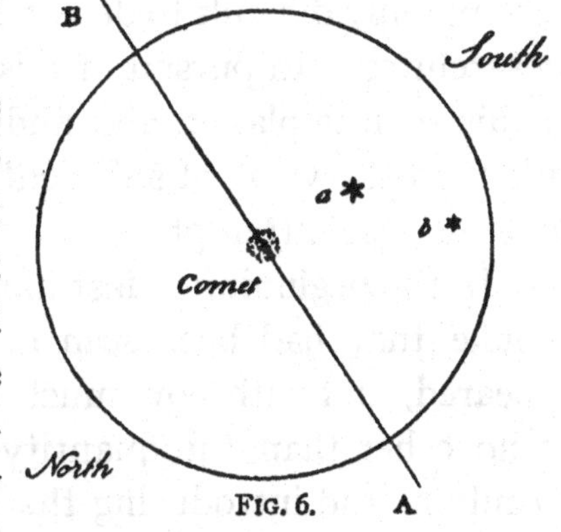

Fig. 6.

The track of the parallel not being taken at the time of her observations, I have endeavoured to recover it by means of directing the same instrument which was used on this occasion towards that part of the heavens where it was placed the 1st and 2d of August. Hence, from the annexed figure (see fig. 6) in which AB represents a parallel of declination, we may conclude, that the comet was nearly in the same meridian with the star *a*; but more north than it by an interval equal to the distance of the small star *b* from *a*. This will consequently give us a pretty good opportunity to ascertain the comet's place with some accuracy.

I have the honour to be, &c.

WILLIAM HERSCHEL.

Slough, near Windsor,
Nov. 15, 1786.

P.S. The first view I had of the comet, after my return from Germany, was the 19th of August, when with a 10-feet reflector it appeared not much unlike the third nebula of the *Connoissance des Temps*, with which it might be very conveniently compared on account of its proximity. It was, however, considerably brighter, and seemed to have a very imperfect and confused kind of gathered light about the middle, which could hardly deserve the name of a nucleus. It had also, besides a diffused coma, a very faint, scattered light towards the north following part, extending to about three or four minutes, and losing itself insensibly.

XIX.

An Account of the Discovery of Two Satellites revolving round the Georgian *Planet.*

[*Phil. Trans.*, vol. lxxvii., 1787, pp. 125–129.]

Read Feb. 15. 1787.

THE great distance of the Georgian planet, and its present situation in a part of the zodiac which is scattered over with a multitude of small stars, has rendered it uncommonly difficult to determine whether, like Jupiter and Saturn, it be attended by satellites. In pursuit of this inquiry, having frequently directed large telescopes to this remote planet, and finding myself continually disappointed, I ascribed my failure to the want of sufficient light in the instruments I used; and, for a while, gave over the attempt.

In the beginning of last month, however, I was often surprised when I reviewed nebulæ that had been seen in former *sweeps*, to find how much brighter they appeared, and with how much greater facility I saw them. The cause of it could be no other than the quantity of light that was gained by laying aside the small speculum, and introducing the *Front-view*; an account of which has been inserted, by way of note, to the Catalogue of Nebulæ contained in the *Philosophical Transactions*, vol. LXXVI. p. 499 [above, p. 294].

It would not have been pardonable to neglect such an advantage, when there was a particular object in view, where an accession of light was of the utmost consequence; and I wondered why it had not struck me sooner. The 11th of January, therefore, in the course of my general review of the heavens, I selected a *sweep* which led to the Georgian planet; and, while it passed the meridian, I perceived near its disk, and within a few of its diameters, some very faint stars whose places I noted down with great care.

The next day, when the planet returned to the meridian, I looked with a most scrutinizing eye for my small stars, and perceived that two of them were missing. Had I been less acquainted with optical deceptions, I should immediately have announced the existence of one or more satellites to our new planet; but it was necessary that I should have no doubts. The least haziness, otherwise imperceptible, may often obscure small stars; and I judged, therefore, that nothing less than a series of observations ought to satisfy me, in a case of this importance. To this end I noticed all the small stars that were near the planet the 14th, 17th,

18th, and 24th of January, and the 4th and 5th of February; and though, at the end of this time, I had no longer any doubt of the existence of at least one satellite, I thought it right to defer this communication till I could have an opportunity of seeing it actually in motion. Accordingly I began to pursue this satellite on Feb. the 7th, about six o'clock in the evening, and kept it in view till three in the morning on Feb. the 8th; at which time, on account of the situation of my house, which intercepts a view of part of the ecliptic, I was obliged to give over the chace: and during those nine hours I saw this satellite faithfully attend its primary planet, and at the same time keep on, in its own course, by describing a considerable arch of its proper orbit.

While I was chiefly attending to the motion of this satellite, I did not forget to follow another small star, which I was pretty well assured was also a satellite, especially as I had, on the night of the 14th of January, observed two small stars which were wanting the 17th, and again missed other two the 24th which had been noticed the 18th; but, whether owing to my great attention to the former satellite, or to the closeness of this latter, which was nearly hidden in the rays of the planet, I could not be well assured of its motion. Indeed, towards morning, when a change of place, in so considerable an interval as nine hours, would have been most conspicuous, the moon interfered with the faint light of this satellite, so that I could no longer perceive it.

The first moment that offered for continuing these observations was on February the 9th, when I saw my first discovered satellite nearly in the place where I expected to find it. I perceived also, that the next supposed satellite was not in the situation where I had left it on the 7th, and could now distinguish very plainly that it had advanced in its orbit, since that day, in the same direction with the other satellite, but at a quicker rate. Hence it is evident, that it moves in a more contracted orbit; and I shall therefore call it in future the first satellite, though last discovered, or rather last ascertained; since I do not doubt but that I saw them both, for the first time, on the same day, which was January the 11th, 1787.

I now directed all my attention to the first satellite, and had an opportunity to see it for about three hours and a quarter; during which time, as far as one might judge, it preserved its course. The interval which the cloudy weather had afforded was, however, rather too short for seeing its motion sufficiently, so that I deferred a final judgment till the 10th; and, in order to put my theory of these two satellites to a trial, I made a sketch on paper, to point out before-hand their situation with respect to the planet, and its parallel of declination.

The long expected evening came on, and, notwithstanding the most unfavourable appearance of dark weather, it cleared up at last. And the heavens now displayed the original of my drawing, by shewing, in the situation I had delineated them, *The Georgian Planet attended by two Satellites.*

I confess that this scene appeared to me with additional beauty, as the little secondary planets seemed to give a dignity to the primary one, which raises it into a more conspicuous situation among the great bodies of our solar system.

For upwards of five hours I saw them go on together, each pursuing its own track; and I left them situated, about two o'clock in the morning on February the 11th, as they are represented in the figure, fig. 1. The letters S, N, P, F, denote the south, north, preceding, and following parts of the heavens, as they are seen, by the *front-view*, in my telescope. The south preceding satellite is the second, or that whose motion was first ascertained; the other is that which moves in a smaller orbit, or what I have called the first satellite; and the direction of their motion is according to the order P, S, F, N, of the letters.

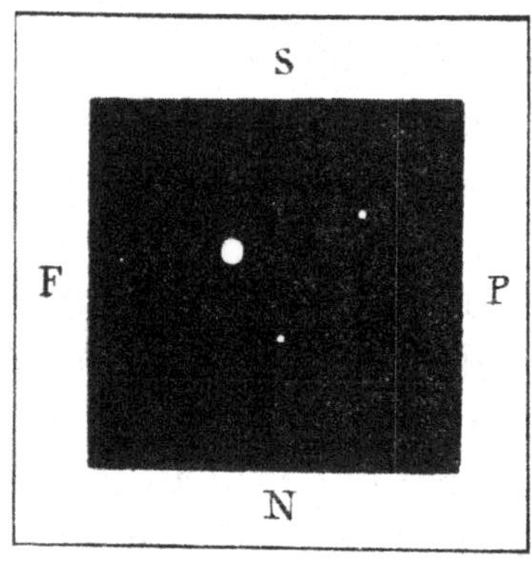

FIG. 1.

I have not seen them long enough to assign their periodical times with great accuracy; but suppose that the first performs a synodical revolution in about eight days and three-quarters, and the second in nearly thirteen days and an half.

Their orbits make a considerable angle with the ecliptic; but to assign the real quantity of this inclination, with many other particulars, will require a great deal of attention, and much contrivance: for, as estimations by the eye cannot but be extremely fallacious, I do not expect to give a good account of their orbits till I can bring some of my micrometers to bear upon them; which, these last nights, I have in vain attempted, their light being so feeble as not to suffer the least illumination, and that of the planet not being strong enough to render the small silk-worm's threads of my delicate micrometers visible. I have, nevertheless, several resources in view, and do not despair of succeeding pretty well in the end.

W. HERSCHEL.

Slough, near Windsor,
February 11, 1787.

XX.

An Account of Three Volcanos in the Moon.

[*Phil. Trans.*, vol. lxxvii., 1787, pp. 229–232.]

Read April 26, 1787.

It will be necessary to say a few words by way of introduction to the account I have to give of some appearances upon the moon, which I perceived the 19th and 20th of this month. The phænomena of nature, especially those that fall under the inspection of the astronomer, are to be viewed, not only with the usual attention to facts as they occur, but with the eye of reason and experience. In this we are however not allowed to depart from plain appearances; though their origin and signification should be indicated by the most characterising features. Thus, when we see, on the surface of the moon, a great number of elevations, from half a mile to a mile and an half in height, we are strictly intitled to call them mountains; but, when we attend to their particular shape, in which many of them resemble the craters of our volcanos, and thence argue, that they owe their origin to the same cause which has modelled many of these, we may be said to see by analogy, or with the eye of reason. Now, in this latter case, though it may be convenient, in speaking of phænomena, to use expressions that can only be justified by reasoning upon the facts themselves, it will certainly be the safest way not to neglect a full description of them, that it may appear to others how far we have been authorized to use the mental eye. This being premised, I may safely proceed to give my observations.

April 19, 1787, 10 h. 36′ sidereal time.

I perceive three volcanos in different places of the dark part of the new moon. Two of them are either already nearly extinct, or otherwise in a state of going to break out; which perhaps may be decided next lunation. The third shews an actual eruption of fire, or luminous matter. I measured the distance of the crater from the northern limb of the moon, and found it 3′ 57″·3. Its light is much brighter than the nucleus of the comet which M. Méchain discovered at Paris the 10th of this month.

April 20, 1787, 10 h. 2′ sidereal time.

The volcano burns with greater violence than last night. I believe its diameter cannot be less than 3″, by comparing it with that of the Georgian planet; as Jupiter was near at hand, I turned the telescope to his third satellite, and estimated the diameter of the burning part of the volcano to be equal to at least twice that of the satellite. Hence we may compute that the shining or burning matter must be above three miles in diameter. It is of an irregular round figure, and very sharply defined on the edges. The other two volcanos are much farther towards the center of the moon, and resemble large, pretty faint nebulæ, that are gradually much brighter in the middle; but no well defined luminous spot can be discerned in them. These three spots are plainly to be distinguished from the rest of the marks upon the moon; for the reflection of the sun's rays from the earth is, in its present situation, sufficiently bright, with a ten-feet reflector, to shew the moon's spots, even the darkest of them: nor did I perceive any similar phænomena last lunation, though I then viewed the same places with the same instrument.

The appearance of what I have called the actual fire or eruption of a volcano, exactly resembled a small piece of burning charcoal, when it is covered by a very thin coat of white ashes, which frequently adhere to it when it has been some time ignited; and it had a degree of brightness, about as strong as that with which such a coal would be seen to glow in faint daylight.

All the adjacent parts of the volcanic mountain seemed to be faintly illuminated by the eruption, and were gradually more obscure as they lay at a greater distance from the crater.

This eruption resembled much that which I saw on the 4th of May, in the year 1783; an account of which, with many remarkable particulars relating to volcanic mountains in the moon, I shall take an early opportunity of communicating to this Society. It differed, however, considerably in magnitude and brightness; for the volcano of the year 1783, though much brighter than that which is now burning, was not nearly so large in the dimensions of its eruption: The former seen in the telescope resembled a star of the fourth magnitude as it appears to the natural eye; this, on the contrary, shews a visible disk of luminous matter, very different from the sparkling brightness of star-light.

WILLIAM HERSCHEL.

Slough near Windsor,
April 21, 1787.

P.S. M. MÉCHAIN having favoured me with an account of the discovery of his comet, I looked for it among the Pleiades, supposing its track since the 10th of this month to lie that way; and saw it April 19th, at 10 h. 10′ sidereal time, when it preceded FL. *d* Pleiadum about 54″ in time, with nearly the same declination as that star; but no great accuracy was attempted in the determination of its place. As I have mentioned the comet in a foregoing paragraph of this Paper, I thought it proper here to add my observation of it. "The comet is nearly round, with a small tail towards the north following part; the chevelure extends to about four or five minutes; and it has a central, very small, ill-defined nucleus, of no great brightness."

XXI.

On the Georgian *Planet and its Satellites.*

[*Phil. Trans.*, vol. lxxviii., 1788, pp. 364–378.]

Read May 22, 1788.

In a Paper, containing an account of the discovery of two satellites revolving round the Georgian planet, I have given the periodical times of these satellites in a general way, and added that their orbits made a considerable angle with the ecliptic. It is hardly necessary to mention, that it requires a much longer series of observations, to settle the mean motions of secondary planets with accuracy, than I can hitherto have had an opportunity of making; but since it will be some satisfaction to astronomers to be acquainted with several of the most interesting particulars, as far as they can as yet be ascertained, I shall communicate the result of my past observations; and believe that, considering the difficulty of measuring objects which require the utmost attention even to be at all perceived, the elements here delivered will be found to be full as accurate as we can at this time expect to have them settled.

The most convenient way of determining the revolution of a satellite round its primary planet, which is that of observing its eclipses, cannot now be used with the Georgian satellites, as will be shewn when I come to give the position of their orbits; and as to taking their situations in many successive oppositions of the planet, which is likewise another very eligible method, that must of course remain to be done at proper opportunities. The only way then left, was to take the situations of these satellites, in any place where I could ascertain them with some degree of precision, and to reduce them afterwards by computation to such other situations as were required for my purpose.

In January, February, and March, 1787, the positions were determined by causing the planet to pass along a wire, and estimating the angle a satellite made with this wire, by a high magnifying power; but then I could only use such of these situations where the satellite happened to be either directly in the parallel of declination, or in the meridian of the planet; or where, at least, it did not deviate above a few degrees from either of them; as it would not have been safe to trust to more distant estimations. In October I had improved my apparatus so far as to measure the positions by the same angular micrometer with which I have formerly determined the relative positions of double stars.*

* For a description of this instrument, see *Phil. Trans.* vol. LXXI. p. 500 and vol. LXXV. p. 46. [*Supra*, pp. 37 and 171.]

In computing the periods of the satellites I have contented myself with synodical appearances, as the position of their orbits, at the time when the situations were taken from which these periods are deduced, was not sufficiently known to attempt a very accurate sidereal calculation. By six combinations of positions at a distance of 7, 8, and 9 months of time, it appears that the first satellite performs a synodical revolution round its primary planet in 8 days 17 hours 1 minute and 19·3 seconds. The period of the second satellite deduced likewise from four such combinations, at the same distance of time, is 13 days 11 hours 5 minutes and 1·5 seconds. The combinations of which the above quantities are a mean do not differ much among themselves; it may therefore be expected that these periods will come very near the truth; and, indeed, I have for many months past been used to calculate the places of the satellites by them, and have hitherto always found them in the situations where these computations gave me reason to expect to see them.

The epochæ, from which astronomers may calculate the positions of these satellites, are October 19, 1787; for the first 19 h. 11′ 28″; and for the second 17 h. 22′ 40″. They were at those times 76° 43′ north following the planet; which, as will be shewn in the sequel, is the place of the greatest elongation of the second satellite; where, consequently, its real angular situation is the same as the apparent one. And I have brought the first satellite to the same place, as hitherto there has not been time to discriminate the situation of its orbit from that of the second.

The next thing to be determined in the elements of these satellites is their distance from the planet; and as we know that, when the periodical times are given, it is sufficient to have the distance of one satellite in order to find that of any other, I confined my attention to the discovery of the distance of the second. As soon as I attempted measures, it appeared, that the orbit of this satellite was seemingly elliptical; it became therefore necessary, in order to ascertain its greatest elongation, to repeat these measures in all convenient situations; the result of which was, that on the 18th of March, at 8 h. 2′ 50″, I found the satellite at the distance of 46″·46; this being the largest of all the measures I have had an opportunity of taking. Hence, by computation, it appears, that the satellite's greatest visible elongation from its planet, at the mean distance of the Georgium sidus from the earth, will be 44″·23.

It ought to be mentioned, that in the reduction of this measure I have used MAYER's tables for the sun, and the tables published in the *Connoissance des Temps* of the year 1787, reduced to the time of Greenwich, for the Georgian planet.

Very possibly this distance might not be taken exactly at the time when the satellite happened to be at the vertex of the transverse axis of its apparently elliptical orbit; but, from other measurements, we have reason to conclude, that it could not be far from that point. For instance, the 9th of November, at 15 h. 56′ 15″, by a mean of four good measures, the satellite was 44″·89 from the planet;

which, by calculation, reduced to the same distance of the Georgian planet from the earth as the former, gives 42″·88. And likewise, the 19th of March, at 7 h. 47′ 59″, the distance measured 44″·24; which, computed as before, gives 42″·15. Now, we find, when the places are calculated in which the satellite happened to be at the times when these two measures were taken, that they fall on different sides of the former measure, and also on opposite parts of the satellite's orbit; but that nevertheless they agree sufficiently well with the position of the transverse axis which we have adopted in the sequel.

Admitting, therefore, at present, that the satellite moves in a circular orbit about its planet, we cannot be much out in taking the calculated quantity of 44″·23 for the true measure of its distance. And, having ascertained this point, we calculate, by the law of KEPLER, and the assigned period of the first satellite, that its distance from the planet must be 33″·09. I ought however to remark, that, in this computation, a true sidereal period should have been used; but, as that cannot as yet be had, the trifling inaccuracy thence arising may well be excused, till, at some future opportunity, we may be permitted to repeat these calculations in a more rigorous manner.

As we are now upon the subject of such parts of the theory of planets as may be determined by calculation, it will not be amiss to see how the quantity of matter and density of our new planet will stand, when compared with the tables that have been given of the same in the other planets; and in order to do this, let us admit the following *data* as a foundation for our computation.

The parallax of the sun 8″·63.

The parallax of the moon 57′ 11″.

Its sidereal revolution round the earth 27 d. 7 h. 43′ 11″·6.

The mean distance of the Georgian planet from the sun 19·0818.

The mean distance of its second satellite from the planet 44″·23.

The periodical time of this satellite 13 d. 11 h. 5′ 1″·5.

Hence we find, that a spectator, removed to the mean distance of the Georgian planet from the earth, would see the radius of the moon's orbit under an angle of 27″·1866; and if 1, d, t, represent the quantity of matter in the earth, the distance of the moon, and its periodical time; M, D, T, be made to stand for the same things in our new planet and its second satellite, we obtain, by known principles, $M = \frac{t^2D^3}{T^2d^3}$. And, consequently, the quantity of matter in the Georgian planet is to that contained in the earth as 17·740612 to 1.

In order to calculate the density, I compare the mean of the four bright measures of the planet's diameter 3″·7975 to the mean of the two dark ones 4″·295; as they are given in my Paper on the diameter and magnitude of the Georgium Sidus, printed in Volume LXXIII. of the *Philosophical Transactions*, p. 9, 11, 12, 13.*

* [Above, pp. 104 *sq.*]

Whence we obtain another mean diameter 4″·04625; which is probably the most accurate of any that we have hitherto ascertained. And let us suppose this measure to belong to the situations of the earth and of the new planet as they were at 10 o'clock, the 25th of October, 1782; which is about the middle of the several times when those measures from which this is deduced were taken. Then, by the tables already referred to, we compute the distance of the two planets from the sun and the angle of commutation; whence, by trigonometry, we find the distance of our new planet from the earth for the supposed 25th of October; and thence deduce its mean diameter, which is 3″·90554. This, when brought to what it would appear if it were seen from the sun at the earth's mean distance, gives 1′ 14″·5246; which, compared with 17″·26, the earth's mean diameter, is as 4·31769 to 1. The Georgium sidus, therefore, in bulk, is 80·49256 times as large as the earth; and consequently its density less than that of the latter in the ratio of ·220401 to 1.

To these particulars, though many of them may be of no other use than merely to satisfy our curiosity, we may also add, that the force of gravity, on this planet's surface, is such as will cause an heavy body to fall through 15 feet $3\frac{1}{3}$ inches in one second of time.

It remains now only, in order to complete our general idea of the Georgian planet, to investigate the situation of the orbits of its satellites. I have before remarked, that when I came to examine the distance of the second, I perceived immediately that its orbit appeared considerably elliptical. This induced me to attempt as many measures as possible, that I might be enabled to come at the proportion of the axes of the apparent ellipsis; and thence argue its situation. But here I met with difficulties that were indeed almost insurmountable. The uncommon faintness of the satellites; the smallness of the angles to be measured with micrometers which required light enough to see the wires; the unwieldy size of the instrument, which, though very manageable, still demanded assistant hands for its movements, and consequently took away a great share of my own directing power, a thing so necessary in delicate observations; the high magnifiers I was obliged to use by way of rendering the spaces and angles to be measured more conspicuous; in short, every circumstance seemed to conspire to make the case a desperate one. Add to this, that no measure could possibly succeed which had not the most beautiful sky in its favour; and we may easily judge how scarce the opportunities of taking such measures must be in the variable climate of this island. As far then as a small number of select measures will permit, which, out of about twenty-one that were taken, amounts only to five, I shall enter into our present subject of the position of the second satellite's orbit.

The following table contains in the first column the correct mean time when the measures were taken. The second gives the quantity of these measures. In the third column are the same measures reduced to the mean distance of the Georgian planet from the earth. The fourth contains the calculated positions of the satellite as it would have appeared to be situated if it had moved in a circular orbit at rectangles to the visual ray; and the degrees are numbered from the first observation supposed to have been at zero, and are carried round the circle from right to left.

	D.	H.	′	″	″	″	°	′
March	18	8	2	50	46·46	44·23	0	0
	19	7	47	59	44·24	42·15	26	28
	20	7	44	8	40·23	38·37	53	8
April	11	9	18	27	35·32	34·35	283	13
Nov.	9	15	56	15	44·89	42·88	199	59

In the use of this table I shall partly content myself with the construction of a figure, and only apply calculation to the most material circumstances. By the third column we see that 44″·23 is the greatest, and 34″·35 the least, distance of the satellite. Let therefore an ellipsis be drawn (fig. 1) having the transverse

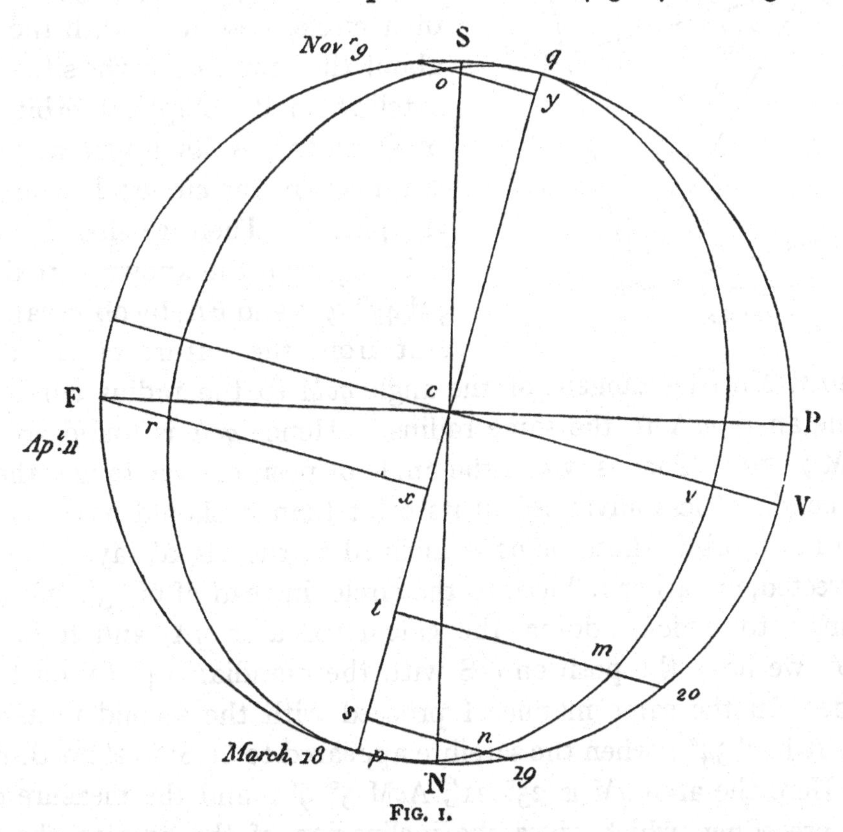

FIG. 1.

and conjugate diameters *cp* and *cv*, in the proportion of the above-mentioned measures. About the center *c*, with the radius *cp*, describe the circle PSFN; and set off the points March 18, 19, 20, April 11 and Nov. 9, according to the tabular order of degrees, beginning at *p*, the supposed zero. From these points to the transverse draw the ordinates March 19 *s*, 20 *t*, April 11 *x*, Nov. 9 *y*. Then, if the satellite moved in a circular orbit at rectangles to the visual ray, we should have seen it at the time given in the table, as the points are placed in the circumference of our circle; but, supposing the plane of the orbit inclined to the visual ray, these points will be projected in the direction of the ordinates; and, falling on the places *p n m r o*, will form the ellipsis we have delineated. Now, on comparing the tabular measures of the third column with the distances of *p n m r* and *o* from the center *c*, we find, that they agree full as well as we could expect; and thus, as far as a few observations can do, these measures establish the truth of the above hypothesis.

That we may have a point in our ellipsis from which to depart, I shall have recourse to two measures of positions. The first was taken October 14 d. 16 h. 28′ 42″, when the satellite was 66° 3′ south-following the planet. In fig. 2 let *qmv* be a portion of an ellipsis, constructed on the semi-transverse *cq*, and semi-conjugate *cv*, taken as 44·23 to 34·35; *q*MF an arch of a circle, described with the radius *cq* about the center *c*; *m* the situation of the satellite in its elliptical orbit, the 14th of October; A its apparent, and M its real place in the circle; F*c* the parallel of the planet. Then we shall have, by calculating from the known period, the arch *q*M 45° 17′; and FA, by observation, 66° 3′. But from the nature of the ellipsis, as

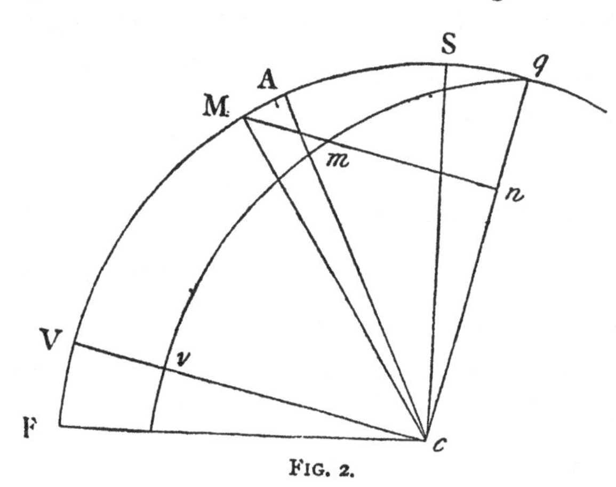

Fig. 2.

V*c* is to *vc* so is M*n* (the tangent of the angle *qc*M to the radius *cn*) to *mn* (the tangent of the angle *qc*A to the same radius). Hence *qc*A is found 38° 7′; and therefore A*c*M 7° 10′. That is, when the angle of position was taken, the satellite appeared to be 7° 10′ less advanced in its orbit than it should have done, owing to its motion in an orbit whose plane is inclined to the visual ray. The measure therefore corrected, or rather reduced to the circle, instead of 66° 3′, will be 58° 53′ south-following; to which, adding the calculated arch *q*A, and from the sum deducting 90°, we have the position *qc*S with the meridian 14° 10′ on the south-preceding side. In the same manner I proceed with the second measure taken October 20 d. 16 h. 7′ 34″; when the satellite appeared to be 82° 12′ north-preceding the planet. Here the arch *q*M is 25° 21′, A*c*M 5° 9′; and the measure corrected 77° 3′ north-preceding, which gives the inclination of the axis to the meridian 12° 24′ on the north-following side. I have no reason to prefer either of the measures, and therefore take a mean of both, which is 13° 17′ from south-preceding to north-following the meridian, as probably nearest the truth; and this position of the axis we may suppose to belong to a time which is about the mean of those from which it has been deduced; or October, 17 d. 16 h.

We are, in the next place, to find the angle which the plane of the meridian made at that time with the plane of the orbit of the Georgian planet. To this end we calculate its longitude and latitude for the given time.* Then, in fig. 3, where ♋ NE is part of the solstitial colure, *nc* a portion of the orbit, and ♋ *s* of the ecliptic; there is given the arch NE, 23° 28′; E*c*, 89° 27′ 56″·1 the complement of the planet's latitude; and the angle ♋ E*s*, 30° 4′ 35″·8, or planet's distance from Cancer. By these we find the angle E*c*N between the circle of latitude N*c* and the meridian E*c* 12° 18′ 59″·5. Now, let *c*, in fig. 4, be the place of the Georgian planet, and G☊*c* a part of its orbit; *e*☊*s* part of the ecliptic; N*c* the meridian; ☊ the place of the planet's ascending node; *pcq* the position of the axis of the apparently elliptical orbit of the second satellite; E*c*N the angle of position of the

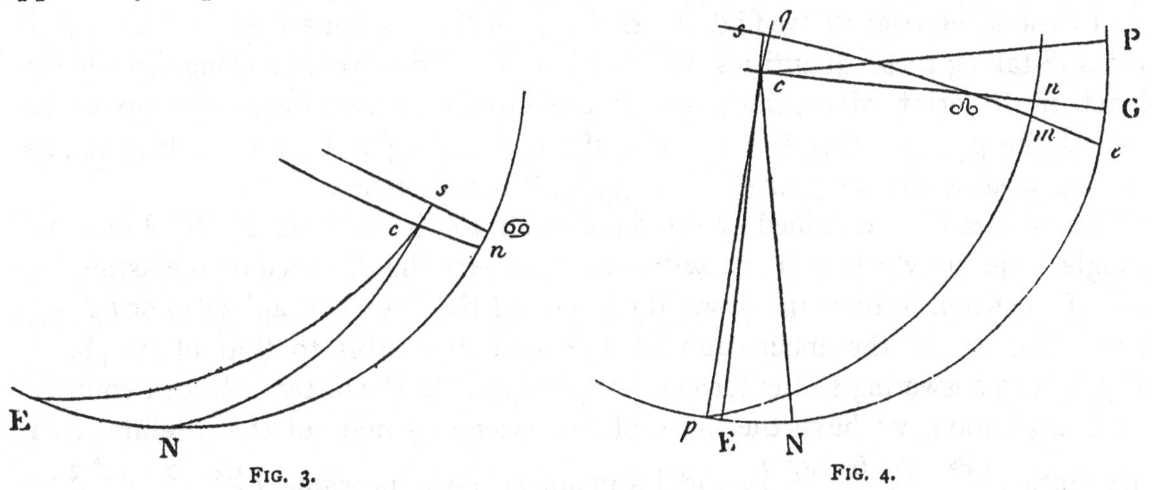

FIG. 3. FIG. 4.

Georgium sidus. Then, by calculation, for the above-mentioned day we have ☊*c*, the planet's distance from the node on its orbit 44° 8′ 17″; *c*☊*s* the inclination of its orbit to the ecliptic 46′ 13″; and ☊*sc* a right-angle; whence E*c*☊ 90° 33′ 10″·1 the supplement of the angle *sc*☊ is found; from which, taking the angle of position N*c*E, before obtained, we have the remaining angle, N*c*G, 78° 14′ 10″·6; or inclination of the planet's orbit to the plane of the meridian, which was required.

From the proportion of the transverse *cp*, fig. 1, to the conjugate *cv*, we calculate the angle *vpc*, which may be either acute or obtuse. For here I must take notice, that observations cannot immediately determine whether the satellite, in passing from *p* through *n m v* to *q*, be in the farthest or nearest part of its orbit; as we shall presently shew that this orbit is not in a situation to permit the satellite to suffer either eclipses or occultations for some time to come. The angle *vpc*, therefore, if the arch *pvq* be turned towards us will be 129° 2′ 46″·5; but, if directed the contrary way, 50° 57′ 13″·5. There is one circumstance which will bring on a discovery of this particular, without waiting for eclipses; for if the apparent ellipsis of the satellite's orbit should contract in a year or two, we may conclude

* [The numerical data and results are here given as subsequently corrected by the Author.—ED.]

this arch to lie towards the sun; if, on the contrary, it opens, we shall know that the satellite has passed through one of its nodes about eight or nine years ago; and that, therefore, we must not expect to see it eclipsed for more than thirty years to come.

Now, having already determined the position of the axis *pc* with respect to the meridian, by adding the angles N*cp* and N*c*G, fig. 4, we obtain *pcn*, 91° 31′ 10″·6; and having also now calculated the ambiguous angle *npc*, we may resolve the quadrantal triangle *pcn*, in which the angle *cnp* gives the inclination of the orbit of the satellite to the orbit of its planet, which will be 90° 57′ 25″·9, if the satellite be approaching to its ascending node; but 89° 2′ 34″·1, if it be lately past the descending one.

In the same triangle we find the side *nc*, which is either 50° 57′ 49″ or 129° 2′ 11″; and taking these quantities, increased by six signs, from the longitude of the planet in its orbit, gives the place of the satellite's ascending node upon the orbit of the planet, either 8ˢ 9° 6′ 55″·0, if the preceding arch of the orbit *pmn*, be concave towards the sun; or 5ˢ 21° 2′ 33″·0, if it be convex.

These elements obtained, we reduce them to the ecliptic by resolving the triangle ☊*nm*, in which we have *m*☊*n*, 46′ 13″; *n*☊ the distance of the ascending node of the satellite from the descending node of the planet 6° 49′ 32″·0 or 84° 53′ 54″·0; and ☊*nm*, the inclination of the satellite's orbit to that of its planet 90° 57′ 25″·9 answering to the former, or 89° 2′ 34″·1 to the latter. In consequence of this resolution, we have the place of the ascending node of the satellite upon the ecliptic, $\left\{\begin{matrix} 8^s & 9^\circ & 6' & 59'' \\ 5 & 21 & 1 & 48 \end{matrix}\right\}$, and its inclination to the same $\left\{\begin{matrix} 91^\circ & 1' & 32''\cdot 2 \\ 89 & 48 & 27\cdot 5 \end{matrix}\right\}$. The orbit being situated so, that when the planet will be in the ascending node of this satellite, which will happen about the year $\left\{\begin{matrix} 1799 \\ 1818 \end{matrix}\right\}$, the northern half of it will be turned towards the $\left\{\begin{matrix} \text{East} \\ \text{West} \end{matrix}\right\}$ of the pole of the ecliptic at the time of its meridian passage.

In justice to the foregoing calculations I should add, that the result of them must be considerably affected by any small alteration in the measures upon which they are founded; the general theory, however, will certainly stand good, and a greater perfection in particulars could not have been obtained, unless I had waited some years, at least, in order to multiply good observations. But with objects that are out of the reach of common telescopes, and which therefore cannot be much attended to, even by our most assiduous astronomers, a general theory will perhaps nearly answer all the ends that may be required of it.

The measures of the distances were taken by a good parallel-wire micrometer, contrived so that one of the wires, which is moveable, can pass over the other; by which means central measures may be obtained with more accuracy than by allowing for the thickness of the wires, the ascertaining of which is liable to some difficulties in other constructions; but here, as we can note the divisions on the first appearance of light at either side of the fixed wire, when the moveable one passes over it backwards and forwards, we may very conveniently determine that part of the scale to which the zero ought to answer in central measures. The value of the scale was ascertained by the transit of stars over the two wires opened to a certain number of divisions, and a chronometer beating five times in two seconds of mean time; and in a number of several sets of experiments, the mean of each seldom differed so much as the 500dth part of a second of space for each division, and these are large enough to be sub-divided and read off, with good exactness to tenths; and yet the space answering to each part amounts only to 282 millesimals of a second. The measures of the distances also were as often repeated as the opportunities would permit, and a mean of them has been used.

The light of the satellites of the Georgian planet is, as we may well expect, on account of their great distance, uncommonly faint. The second is the brightest of the two, but the difference is not considerable; besides, we must allow for the effect of the light of the planet, which is pretty strong within the small distances at which they are revolving. I have seen small fixed stars, as near the planets as the satellites, and with no greater light, which, on removal of the planet, shone with a considerable lustre, such as I had by no means expected of them. A satellite of Jupiter, removed to the distance of the Georgian planet, would shine with less than the 180th part of its present light; and may we not conclude, that our new satellites would be of a very considerable brightness if they were brought so near as the orbit of Jupiter, and thus appeared 180 times brighter than at present? Nay, this is only when we take both the planets at their mean distance; for, in their oppositions, a satellite brought from the superior planet to the orbit of the inferior one, would reflect nearly 250 times the former light; from all which it is evident, that the Georgian satellites must be of a considerable magnitude.

If we draw together the results of the foregoing calculations into a small compass, they will stand as follows:

The first satellite revolves round the Georgian planet in 8 days 17 hours 1 minute and 19 seconds.

Its distance is 33″.

And on the 19th of October, 1787, at 19 h. 11′ 28″, its position was 76° 43′ north-following the planet.

The second satellite revolves round its primary planet in 13 days 11 hours 5 minutes and 1·5 seconds.

Its greatest distance is 44″·23.

And on the 19th of October, 1787, its position at 17 h. 22′ 40″, was 76° 43′ north-following the planet.

Last year its least distance was 34″·35 ; but the orbit is so inclined, that this measure will change very considerably in a few years, and by that alteration we shall know which of the double quantities put down for the inclination and node of its orbit are to be used.

The orbit of the second satellite is inclined to the ecliptic $\left\{\begin{matrix} 91^{\circ} & 1' & 32''\cdot 2 \\ 89 & 48 & 27\ \cdot 5 \end{matrix}\right\}$.

Its ascending node is in $\left\{\begin{matrix} \text{21 degrees of Virgo} \\ \text{9 degrees of Sagittarius} \end{matrix}\right\}$.

When the planet passes the meridian, being in the node of this satellite, the northern part of its orbit will be turned towards the $\left\{\begin{matrix} \text{East} \\ \text{West} \end{matrix}\right\}$ of the pole of the ecliptic.

The situation of the orbit of the first satellite does not seem to differ materially from that of the second.

We shall have eclipses of these satellites about the year $\left\{\begin{matrix} 1799 \\ 1818 \end{matrix}\right\}$, when they will appear to ascend through the shadow of the planet almost in a perpendicular direction to the ecliptic.

The satellites of the Georgian planet are probably not less than those of Jupiter.

The diameter of the new planet is 34217 miles.

The same diameter seen from the earth, at its mean distance, is 3″·90554.

From the sun, at the mean distance of the earth, 1′ 14″·5246.

Compared to that of the earth as 4·31769 to 1.

This planet in bulk is 80·49256 times as large as the earth.

Its density as ·220401 to 1.

Its quantity of matter 17·740612 to 1.

And heavy bodies fall on its surface 15 feet 3⅓ inches in one second of time.

W. HERSCHEL.

Slough, March 1, 1788.

XXII.

Observations on a Comet. In a letter from WILLIAM HERSCHEL, *LL.D. F.R.S. to Sir* JOSEPH BANKS, *Bart. P.R.S.*

[*Phil. Trans.*, vol. lxxix., 1789, pp. 151–153.]

Read April 2, 1789.

Slough, March 3, 1789.

SIR,—The last time I was in town, you expressed a wish to see my observations on the comet which my sister, CAROLINE HERSCHEL, discovered in the evening of the 21st of last December, not far from β Lyræ.

As she immediately acquainted the Rev. Dr. MASKELYNE, and several other gentlemen, with her discovery, the comet was observed by many of them. The Astronomer Royal, in particular, having, I find, obtained a very good set of valuable observations on its path, it will be sufficient if I communicate only those particulars which relate to its first appearance, and a few other circumstances that may perhaps deserve to be noticed.

December 21, 1788, about 8 o'clock, I viewed the comet which my sister had a little while before pointed out to me with her small Newtonian *sweeper*. In my instrument, which was a ten-feet reflector, it had the appearance of a considerably bright nebula; of an irregular, round form; very gradually brighter in the middle; and about five or six minutes in diameter. The situation was low, and not very proper for instruments with high powers.

December 22, about half after five o'clock in the morning, I viewed it again, and perceived that it had moved apparently in a direction towards δ Lyræ, or thereabout. I had been engaged all night with the twenty-feet instrument, so that there had been no leisure to prepare my apparatus for taking the place of the comet; but in the evening of the same day, I took its situation three times, as follows:

	h.			
Dec. 22. at	23	42	19	sidereal time, the comet passed the wire,
at	23	49	24	β Lyræ passed the same,
Difference		7	5	very accurate.

	h.	′	″	
Dec. 22. at	23	52	52	the comet passed,
at	23	59	58	β Lyræ passed,
Difference		7	6	accurate.
at	0	6	35	the comet passed,
at	0	13	40	β Lyræ passed,
Difference		7	5	very accurate.

I found in every observation the small star which accompanies β Lyræ,* exactly in the parallel of the comet.

These transits were taken with a ten-feet reflector ; and the difference in right ascension, I should suppose, may be depended upon to within a second of time. The determination also of the parallel can hardly err so much as fifteen seconds of a degree.

This, and several evenings afterwards, I viewed the comet again with such powers as its diluted light would permit, but could not perceive any sort of nucleus, which, had it been a single second in diameter, I think, could not well have escaped me. This circumstance seems to be of some consequence to those who turn their thoughts on the investigation of the nature of comets ; especially as I have also formerly made the same remark on one of the comets discovered by M. MÉCHAIN in 1787, a former one of my sister's in 1786, and one of Mr. PIGOTT's in 1783 ; in neither of which any defined, solid nucleus could be perceived.

I have the honour to remain, &c.

WILLIAM HERSCHEL.

* For this small star see my Catalogue of Double Stars, in the *Philosophical Transactions* for the year 1782, Part I. Class V. Star 3, where its distance and position are given, and consequently its parallel may be found.

XXIII.

Catalogue of a second Thousand of new Nebulæ and Clusters of Stars; with a few introductory Remarks on the Construction of the Heavens.

[*Phil. Trans.*, vol. lxxix., 1789, pp. 212–255.]

Read June 11, 1789.

By the continuation of a review of the heavens with my twenty-feet reflector, I am now furnished with a second thousand of new Nebulæ.

These curious objects, not only on account of their number, but also in consideration of their great consequence, as being no less than whole sidereal systems, we may hope, will in future engage the attention of Astronomers. With a view to induce them to undertake the necessary observations, I offer them the following catalogue, which, like my former one, of which it is a continuation, contains a short description of each nebula or cluster of stars, as well as its situation with respect to some known object.

The form of this work, it will be seen, is exactly that of the former part, the classes and numbers being continued, and the same letters used to express, in the shortest way, as many essential features of the objects as could possibly be crowded into so small a compass as that to which I thought it expedient to limit myself.

The method I have taken of *analyzing* the heavens, if I may so express myself, is perhaps the only one by which we can arrive at a knowledge of their construction. In the prosecution of so extensive an undertaking, it may well be supposed that many things must have been suggested, by the great variety in the order, the size, and the compression of the stars, as they presented themselves to my view, which it will not be improper to communicate.

To begin our investigation according to some order, let us depart from the objects immediately around us to the most remote that our telescopes, of the greatest *power to penetrate into space*, can reach. We shall touch but slightly on things that have already been remarked.

From the earth, considered as a planet, and the moon as its satellite, we pass through the region of the rest of the planets, and their satellites. The similarity between all these bodies is sufficiently striking to allow us to comprehend them under one general definition, of bodies not luminous in themselves, revolving round the sun. The great diminution of light, when reflected from such bodies, especially when they are also at a great distance from the light which illuminates them, precludes all possibility of following them a great way into space. But if we did not know that light diminishes as the squares of the distances encrease, and that moreover in every reflection a very considerable part is intirely lost, the motion of comets, whereby the space through which they run is measured out to us, while on their return from the sun we see them gradually disappear as they advance towards their aphelia, would be sufficient to convince us that bodies shining only with borrowed light can never be seen at any very great distance. This consideration brings us back to the sun, as a refulgent fountain of light, whilst it establishes at the same time beyond a doubt that every star must likewise be a sun, shining by its own native brightness. Here then we come to the more capital parts of the great construction.

These suns, every one of which is probably of as much consequence to a system of planets, satellites, and comets, as our own sun, are now to be considered, in their turn, as the minute parts of a proportionally greater whole. I need not repeat that by my analysis it appears, that the heavens consist of regions where suns are gathered into separate systems, and that the catalogues I have given comprehend a list of such systems; but may we not hope that our knowledge will not stop short at the bare enumeration of phænomena capable of giving us so much instruction? Why should we be less inquisitive than the natural philosopher, who sometimes, even from an inconsiderable number of specimens of a plant, or an animal, is enabled to present us with the history of its rise, progress, and decay? Let us then compare together, and class some of these numerous sidereal groups, that we may trace the operations of natural causes as far as we can perceive their agency. The most simple form, in which we can view a sidereal system, is that of being globular. This also, very favourably to our design, is that which has presented itself most frequently, and of which I have given the greatest collection.

But, first of all, it will be necessary to explain what is our idea of a cluster of stars, and by what means we have obtained it. For an instance, I shall take the phænomenon which presents itself in many clusters: It is that of a number of lucid spots, of equal lustre, scattered over a circular space, in such a manner as to appear gradually more compressed towards the middle; and which compression, in the clusters to which I allude, is generally carried so far, as, by imperceptible degrees, to end in a luminous center, of a resolvable blaze of light. To solve this appearance, it may be conjectured, that stars of any given, very unequal magnitudes, may easily be so arranged, in scattered, much extended, irregular rows, as to produce the above described picture; or, that stars, scattered about almost promiscuously within the frustum of a given cone, may be assigned of such properly diversified magnitudes as also to form the same picture. But who, that is acquainted with the doctrine of chances, can seriously maintain such improbable conjectures? To consider this only in a very coarse way, let us suppose a cluster to consist of 5000 stars, and that each of them may be put into one of 5000 given places, and have one of 5000 assigned magnitudes. Then, without extending our calculation any further, we have five and twenty millions of chances, out of which only one will answer the above improbable conjecture, while all the rest are against it. When we now remark that this relates only to the given places within the frustum of a supposed cone, whereas these stars might have been scattered all over the visible space of the heavens; that they might have been scattered, even within the supposed cone, in a million of places different from the assumed ones, the chance of this apparent cluster's not being a real one, will be rendered so highly improbable that it ought to be intirely rejected.

Mr. MICHELL computes, with respect to the six brightest stars of the Pleiades only, that the odds are near 500000 to 1 that no six stars, out of the number of those which are equal in splendour to the faintest of them, scattered at random in the whole heavens, would be within so small a distance from each other as the Pleiades are.*

Taking it then for granted that the stars which appear to be gathered together in a group are in reality thus accumulated, I proceed to prove also that they are nearly of an equal magnitude.

The cluster itself, on account of the small angle it subtends to the eye, we must suppose to be very far removed from us. For, were the stars which compose it at the same distance from one another as Sirius is from the sun; and supposing the cluster to be seen under an angle of 10 minutes, and to contain 50 stars in one of its diameters, we should have the mean distance of such stars twelve seconds; and therefore the distance of the cluster from us about seventeen thousand times greater than the distance of Sirius. Now, since the apparent magnitude of these stars is equal, and their distance from us is also equal,—because we may safely neglect the diameter of the cluster, which, if the center be seventeen thousand times the distance of Sirius from us, will give us seventeen thousand and twenty-five for the farthest, and seventeen thousand wanting twenty-five for the nearest star of the cluster;—it follows that we must either give up the idea of a cluster, and recur to the above refuted supposition, or admit the equality of the stars that compose these clusters. It is to be remarked that we do not mean intirely to exclude all variety of size; for the very great distance, and the consequent smallness of the component clustering stars, will not permit us to be extremely precise in the estimation of their magnitudes; though we have certainly seen enough of them to know that they are contained within pretty narrow limits; and do not, perhaps, exceed each other in magnitude more than in some such proportion as one full grown plant of a certain species may exceed another full-grown plant of the same species.

If we have drawn proper conclusions relating to the size of stars, we may with still greater safety speak of their relative situations, and affirm that in the same

* *Phil. Trans.* vol. LVII. p. 246.

distances from the center an equal scattering takes place. If this were not the case, the appearance of a cluster could not be uniformly encreasing in brightness towards the middle, but would appear nebulous in those parts which were more crowded with stars; but, as far as we can distinguish, in the clusters of which we speak, every concentric circle maintains an equal degree of compression, as long as the stars are visible; and when they become too crowded to be distinguished, an equal brightness takes place, at equal distances from the center, which is the most luminous part.

The next step in my argument will be to shew that these clusters are of a globular form. This again we rest on the sound doctrine of chances. Here, by way of strength to our argument, we may be allowed to take in all round nebulæ, though the reasons we have for believing that they consist of stars have not as yet been entered into. For, what I have to say concerning their spherical figure will equally hold good whether they be groups of stars or not. In my catalogues we have, I suppose, not less than one thousand of these round objects. Now, whatever may be the shape of a group of stars, or of a Nebula, which we would introduce instead of the spherical one, such as a cone, an ellipsis, a spheroid, a circle or a cylinder, it will be evident that out of a thousand situations, which the axes of such forms may have, there is but one that can answer the phænomenon for which we want to account; and that is, when those axes are exactly in a line drawn from the object to the place of the observer. Here again we have a million of chances of which all but one are against any other hypothesis than that which we maintain, and which, for this reason, ought to be admitted.

The last thing to be inferred from the above related appearances is, that these clusters of stars are more condensed towards the center than at the surface. If there should be a group of stars in a spherical form, consisting of such as were equally scattered over all the assigned space, it would not appear to be very gradually more compressed and brighter in the middle; much less would it seem to have a bright nucleus in the center. A spherical cluster of an equal compression within, —for that such there are will be seen hereafter,—may be distinguished by the degrees of brightness which take place in going from the center to the circumference. Thus, when a is the brightness in the center, it will be $\sqrt{a^2-x^2}$ at any other distance x from the center. Or, putting $a=1$, and $x=$any decimal fraction; then, in a table of natural sines, where x is the sine, the brightness at x will be expressed by the cosine. Now, as a gradual encrease of brightness does not agree with the degrees calculated from a supposition of an equal scattering, and as the cluster has been proved to be spherical, it must needs be admitted that there is indeed a greater accumulation towards the center. And thus, from the above-mentioned appearances, we come to know that there are globular clusters of stars nearly equal in size, which are scattered evenly at equal distances from the middle, but with an encreasing accumulation towards the center.

We may now venture to raise a superstructure upon the arguments that have been drawn from the appearance of clusters of stars and nebulæ of the form I have been examining, which is that of which I have made mention in my "*Theoretical view—Formation of Nebulæ—Form I.*"* It is to be remarked that when I wrote the paragraph I refer to, I delineated nature as well as I do now; but, as I there gave only a general sketch, without referring to particular cases, what I then delivered may have been looked upon as little better than hypothetical reasoning, whereas in the present instance this objection is intirely removed, since actual and particular facts are brought to vouch for the truth of every inference.

Having then established that the clusters of stars of the 1st Form, and round nebulæ, are of a spherical figure, I think myself plainly authorized to conclude that they are thus formed by the action of central powers. To manifest the validity of this inference, the figure of the earth may be given as an instance; whose rotundity, setting aside small deviations, the causes of which are well known, is without hesitation allowed to be a phænomenon decisively establishing a centripetal force. Nor do we stand in need of the revolving satellites of Jupiter, Saturn, and the Georgium Sidus, to assure us that the same powers are likewise lodged in the masses of these planets. Their globular figure alone must be admitted as a sufficient argument to render this point uncontrovertible. We also apply this inference with equal propriety to the body of the sun, as well as to that of Mercury, Venus, Mars, and the Moon; as owing their spherical shape to the same cause. And how can we avoid inferring, that the construction of the clusters of stars, and nebulæ likewise, of which we have been speaking, is as evidently owing to central powers?

Besides, the step that I here make in my inference is in fact a very easy one, and such as ought freely to be granted. Have I not already shewn that these clusters cannot have come to their present formation by any random scattering of stars? The doctrine of chance, by exposing the very great odds against such hypotheses, may be said to demonstrate that the stars are thus assembled by some power or other. Then, what do I attempt more than merely to lead the mind to the conditions under which this power is seen to act?

In a case of such consequence I may be permitted to be a little more diffuse, and draw additional arguments from the internal construction of spherical clusters and nebulæ. If we find that there is not only a general form, which, as has been proved, is a sufficient manifestation of a centripetal force, what shall we say when the accumulated condensation, which every where follows a direction towards a center, is even visible to the very eye? Were we not already acquainted with attraction, this gradual condensation would point out a central power, by the remarkable disposition of the stars tending towards a center. In consequence of

* *Phil. Trans.* vol. LXXV. p. 214 [above, p. 224].

this visible accumulation, whether it may be owing to attraction only, or whether other powers may assist in the formation, we ought not to hesitate to ascribe the effect to such as are *central*; no phænomena being more decisive in that particular, than those of which I am treating.

I am fully aware of the consequences I shall draw upon myself in but mentioning other powers that might contribute to the formation of clusters. A mere hint of this kind, it will be expected, ought not to be given without sufficient foundation; but let it suffice at present to remark that my arguments cannot be affected by my terms: whether I am right to use the plural number,—central powers,—or whether I ought only to say,—the known central force of gravity,—my conclusions will be equally valid. I will however add, that the idea of other central powers being concerned in the construction of the sidereal heavens, is not one that has only lately occurred to me. Long ago I have entertained a certain theory of diversified central powers of attractions and repulsions; an exposition of which I have even delivered in the years 1780, and 1781, to the Philosophical Society then existing at Bath, in several mathematical papers upon that subject. I shall, however, set aside an explanation of this theory, which would not only exceed the intended limits of this paper, but is moreover not required for what remains at present to be added, and therefore may be given some other time, when I can enter more fully into the subject of the interior construction of sidereal systems.

To return, then, to the case immediately under our present consideration, it will be sufficient that I have abundantly proved that the formation of round clusters of stars and nebulæ is either owing to central powers, or at least to one such force as refers to a center.

I shall now extend the weight of my argument, by taking in likewise every cluster of stars or nebula that shews a gradual condensation, or encreasing brightness, towards a center or certain point; whether the outward shape of such clusters or nebulæ be round, extended, or of any other given form. What has been said with regard to the doctrine of chance, will of course apply to every cluster, and more especially to the extended and irregular shaped ones, on account of their greater size: It is among these that we find the largest assemblages of stars, and most diffusive nebulosities; and therefore the odds against such assemblages happening without some particular power to gather them, encrease exceedingly with the number of the stars that are taken together. But if the gradual accumulation either of stars or encreasing brightness has before been admitted as a direction to the seat of power, the same effect will equally point out the same cause in the cases now under consideration. There are besides some additional circumstances in the appearance of extended clusters and nebulæ, that very much favour the idea of a power lodged in the brightest part. Although the form of them be not globular, it is plainly to be seen that there is a tendence towards sphericity, by the swell of the dimensions the nearer we draw towards the most luminous place, denoting as it were a course, or tide of stars, setting towards a center. And —if allegoral expressions may be allowed—it should seem as if the stars thus flocking towards the seat of power were stemmed by the crowd of those already assembled, and that while some of them are successful in forcing their predecessors sideways out of their places, others are themselves obliged to take up with lateral situations, while all of them seem equally to strive for a place in the central swelling, and generating spherical figure.

Since then almost all the nebulæ and clusters of stars I have seen, the number of which is not less than three and twenty hundred, are more condensed and brighter in the middle; and since, from every form, it is now equally apparent that the central accumulation or brightness must be the result of central powers, we may venture to affirm that this theory is no longer an unfounded hypothesis, but is fully established on grounds which cannot be overturned.

Let us endeavour to make some use of this important view of the constructing cause, which can thus model sidereal systems. Perhaps, by placing before us the very extensive and varied collection of clusters, and nebulæ furnished by my catalogues, we may be able to trace the progress of its operation, in the great laboratory of the Universe.

If these clusters and nebulæ were all of the same shape, and had the same gradual condensation, we should make but little progress in this inquiry; but, as we find so great a variety in their appearances, we shall be much sooner at a loss how to account for such various phænomena, than be in want of materials upon which to exercise our inquisitive endeavours.

Some of these round clusters consist of stars of a certain magnitude, and given degree of compression, while the whole cluster itself takes up a space of perhaps 10 minutes; others appear to be made up of stars that are much smaller, and much more compressed, when at the same time the cluster itself subtends a much smaller angle, such as 5 minutes. This diminution of the apparent size, and compression of stars, as well as diameter of the cluster to 4, 3, 2 minutes, may very consistently be ascribed to the different distances of these clusters from the place in which we observe them; in all which cases we may admit a general equality of the sizes, and compression of the stars that compose them, to take place. It is also highly probable that a continuation of such decreasing magnitudes, and encreasing compression, will justly account for the appearance of round, easily resolvable, nebulæ; where there is almost a certainty of their being clusters of stars. And no Astronomer can hesitate to go still farther, and extend his surmises by imperceptible steps to other nebulæ, that still preserve the same characteristics, with the only variations of vanishing brightness, and reduction of size.

Other clusters there are that, when they come to be compared with some of the former, seem to contain stars of an equal magnitude, while their compression appears to be considerably different. Here the supposition of their being at different distances will either not explain the apparently greater compression, or, if admitted to do this, will convey to us a very instructive consequence: which is, that the stars which are thus supposed not to be more compressed than those in the former cluster, but only to appear so on account of their greater distance, must needs be proportionally larger, since they do not appear of less magnitude than the former. As therefore, one or other of these hypotheses must be true, it is not at all improbable but that, in some instances, the stars may be more compressed; and in others, of a greater magnitude. This variety of size, in different spherical clusters, I am however inclined to believe, may not go farther than the difference in size, found among the individuals belonging to the same species of plants, or animals, in their different states of age, or vegetation, after they are come to a certain degree of growth. A farther inquiry into the circumstance of the extent, both of condensation and variety of size, that may take place with the stars of different clusters, we shall postpone till other things have been previously discussed.

Let us then continue to turn our view to the power which is moulding the different assortments of stars into spherical clusters. Any force, that acts uninterruptedly, must produce effects proportional to the time of its action. Now, as it has been shewn that the spherical figure of a cluster of stars is owing to central powers, it follows that those clusters which, *ceteris paribus*, are the most compleat in this figure, must have been the longest exposed to the action of these causes. This will admit of various points of view. Suppose for instance that 5000 stars had been once in a certain scattered situation, and that other 5000 equal stars had been in the same situation, then that of the two clusters which had been longest exposed to the action of the modelling power, we suppose, would be most condensed, and more advanced to the maturity of its figure. An obvious consequence that may be drawn from this consideration is, that we are enabled to judge of the relative age, maturity, or climax of a sidereal system, from the disposition of its component parts; and, making the degrees of brightness in nebulæ stand for the different accumulation of stars in clusters, the same conclusions will extend equally to them all. But we are not to conclude from what has been said that every spherical cluster is of an equal standing in regard to absolute duration, since one that is composed of a thousand stars only, must certainly arrive to the perfection of its form sooner than another, which takes in a range of a million. Youth and age are comparative expressions; and an oak of a certain age may be called very young, while a cotemporary shrub is already on the verge of its decay. The method of judging with some assurance of the condition of any sidereal system may perhaps not improperly be drawn from the standard laid down [page 332]; so that, for instance, a cluster or nebula which is very gradually more compressed and bright towards the middle, may be in the perfection of its growth, when another which approaches to the condition pointed out by a more equal compression, such as the nebulæ I have called *Planetary* seem to present us with, may be looked upon as very aged, and drawing on towards a period of change, or dissolution. This has been before surmised, when, in a former paper, I considered the uncommon degree of compression that must prevail in a nebula to give it a planetary aspect; but the argument, which is now drawn from the powers that have collected the formerly scattered stars to the form we find they have assumed, must greatly corroborate that sentiment.

This method of viewing the heavens seems to throw them into a new kind of light. They now are seen to resemble a luxuriant garden, which contains the greatest variety of productions, in different flourishing beds; and one advantage we may at least reap from it is, that we can, as it were, extend the range of our experience to an immense duration. For, to continue the simile I have borrowed from the vegetable kingdom, is it not almost the same thing, whether we live successively to witness the germination, blooming, foliage, fecundity, fading, withering, and corruption of a plant, or whether a vast number of specimens, selected from every stage through which the plant passes in the course of its existence, be brought at once to our view?

WILLIAM HERSCHEL.

Slough near Windsor, May 1, 1789.

First Class. Bright Nebulæ.

I.	1785	Stars.	M. S.			D. M.	Ob.	Description.	N.G.C.
94	April 28	61 Ursæ	f	0 6	n	2 17	2	cB. pL. E. sp nf. vgmbM. 3½′ l. 2′ b.	3813*
95	—	— —	f	35 0	n	2 7	2	cB. cL. E. np sf. bM. 4′ l. 3′ b.	4214
96	May 1	14 Canum	f	5 30	n	1 12	2	vB. cL. mE. sp nf. smbM. 6′ l. 1′½ b.	5005
97	—	— —	f	7 58	n	0 47	1	vB. pL. E. nearly mer. gmbM.	5033
98	—	— —	f	36 50	s	0 12	1	cB. pL. R. vgmbM.	5273
99	—	27 (γ) Bootis	p	13 46	s	1 46	2	vB. S. R. vsmbM.	5557
100	Sept. 10	41 Ceti	f	13 43	n	0 48	1	cB. pS. R. mbM. See III. 431.	584
101	—	67 —	p	17 19	n	0 25	2	cB. pL. E. near mer. mbM. 5′ l.	779
102	—	— —	f	21 37	s	0 13	2	cB. pL. R. mbM.	1022
103	24	14 Delphini	p	16 10	s	0 3	1	vB. L. gmbM. er. beautif. object.	6934*
104	28	93 (Ψ) Aqua.	f	1 8	n	0 42	1	cB. cL. E. near. mer. gmbM. F. rays.	7606
105	Oct. 3	47 Ceti	f	26 24	s	0 37	1	cB. pL. iR. mbM.	720
106	—	89 (π) —	f	38 10	s	1 24	2	cB. cL. iR. bM. 3′ dia.	1309
107	6	20 Eridani	f	4 3	s	1 4	2	vB. R. BNM. 1′½ dia.	1407
108	8	111 (ξ) Pisc^{m}	p	34 22	s	0 1	1	cB. vL. iR. p. vBst.	467
109	26	12 Eridani	p	8 17	n	2 54	3	cB. pS. lE. mer. mbM. r. 1′½ l.	1201*
110	Nov. 27	9 Ceti	p	44 0	s	0 47	2	cB. cL. lE. gmbM. iF.	7723
111	—	— —	p	43 3	s	0 6	2	cB. cL. iR. gmbM.	7727
112	29	5 (γ) Arietis	f	5 48	s	0 17	1	vB. L. R. mbM. not er. 4′ dia.	772

I.	1785	Stars.		M. S.		D. M.	Ob.	Description.	N.G.C.
113	Dec. 7	66 (4th σ) Can.	f	18 22	n	1 34	2	cB. cL. lE. iF. mb foll. side.	2830*
114	—	18 Leo. min.	p	13 39	s	0 35	1	cB. cL. iF. mbM.	2964
115	—	— — —	p	5 47	n	1 10	2	cB. pL. lE. iF. mbM.	3021
116 } 117 }	—	37 — —	f	11 5	n	1 1	1	Two; the 1st, cB. cL. iE.; the 2d, pB. pL. iE. Dist. 1' at the vertex at the north ends.	3395 3396
118	—	46 Ursæ	p	3 41	s	1 32	1	cB. cL. iR. mbM.	3430
119	28	31 (1st d) Vir.	p	6 0	n	0 55	1	vB. pS.	4560*
120	31	30 (η) Crateris	p	9 0	n	0 17	1	cB. L. iR. bM. 5' l. 4' b.	3887
	1786								
121	Jan. 1	13 (n) Virgin.	p	18 15	s	0 19	1	vB. cL. lE. mbM. 3' l. 2'½ b. bet. 2 pBst.	4030
122	Feb. 1	57 (μ) Eridani	p	4 0	n	0 22	1	cB. vL. iR. bM. er. 5 or 6' dia.	1637
123	2	60 (σ) Virg.	p	52 27	s	0 30	2	cB. S.	4378
124	—	— — —	p	39 57	s	0 3	2	cB. cL. R.	4580
125	—	— — —	p	39 12	s	1 6	2	cB. cL. E. mbM.	4586
126	24	108 —	p	0 35	n	1 15	1	eB. mE. par. BN. 8 or 9' l.	5746
127	—	110 —	p	1 47	s	0 23	1	cB. pS. mbM.	5813
128	—	— —	f	3 37	s	0 30	1	vB. pL. bM.	5846
129	Mar. 3	26 (χ) —	f	9 46	s	0 41	1	v brilliant. iR. vgmbM.	4699
130	—	— — —	f	26 35	s	0 3	2	vB. lE. mer. BN. and F. br. 2' l.	4958
131	4	14 (ε) Crate.	f	0 29	n	1 3	1	cB. E. gbM. 5' l. 4' b.	3672
132	19	26 Hydræ	f	1 44	n	0 4	2	cB. pL. lE. vgbM. 1'½ diam.	2855
133	25	49 (g) Virgin.	p	16 4	n	0 18	1	cB. vS. BN.	4742
134	—	— — —	p	13 27	n	0 13	1	cB. 7 or 8' l. 3' b.	4781
135 } 136 }	27	68 (ι) —	p	32 2	n	0 11	2	Two; both cB. cS. R. mbM. Dist. 1' near. mer. chev. mixed.	4782 4783
137	28	40 Lyncis.	f	3 13	n	0 8	1	vB. R. vsmbM. chev. 3' dia.	2859
138	—	[1]1102 (e) Hy.	f	33 45	n	1 27	1	cB. R. psmbM. [1] See note.	5061
139	Apr. 17	11 (s) Virgin.	f	12 1	s	1 21	2	eB. vBN. r. 6 or 7' dia.	4303
140	—	— — —	f	39 55	s	0 31	2	cB. pL. mbM.	4713
141	—	— — —	f	45 50	s	1 32	1	vB. cL. E. np sf.	4808
142	30	37 —	p	6 35	n	0 0	1	cB. pL. iR. gmbM.	4665
143	—	43 (δ) Virgin.	f	4 55	s	0 54	1	cB. np. pBst. and close to it.	4900
144	—	109 —	p	25 58	n	2 7	1	cB. cL. R. gmbM.	5566
145 } 146 }	—	— —	p	25 14	n	1 27	1	Two; the p. pB. pL. E. Dist. 3 or 4' sp nf. The f cB. R. pL. Place of 2d.	5574 5576
147	—	43 Ophiuchi	p	8 54	s	1 17	1	vB. R. gmbM. 2'½ dia.	6304
148	May 1	24 (α) Serpen.	p	22 26	s	1 16	1	cB. cL. iR. bM.	5921
149	28	40 (ξ) Ophiu.	f	0 14	n	1 32	1	cB. pS. lE. er.	6342
150	—	— — —	f	27 53	n	0 36	1	cB. R. vgmbM. about 1'½ dia.	6440
151	Sept. 4	71 (ε) Piscium	f	21 41	n	1 41	1	cB. cL. R. C. vgmbM. N.	524
152	—	24 (ξ) Arietis	p	16 23	n	0 20	2	pB. vS. R. bM. 1' sf. cst.	821*
153	20	59 (2d υ) Ceti	f	23 16	s	0 6	1	cB. vL. E. sp nf. above 15' l.	908
154	21	14 Triang.	p	1 23	n	0 59	2	cB. pL. E. np sf. vgmbM. 3' l. 2' b.	949
155	30	32 Eridani	p	7 49	s	1 1	2	cB. S. gmbM.	1453
156	Oct. 18	12 (q) Persei	p	1 41	s	1 10	2	cB. mE. 12° sp nf. vBN. near 10' l.	1023
157	26	90 (υ) Piscium	f	28 9	n	0 13	1	cB. cL. E. par. mbM. 7' l. 3' b.	672
158	Nov. 26	48 (ν) Eridani	p	4 32	s	1 46	2	cB. pL. iR. vgmbM.	1600

I.	1786	Stars.		M. S.		D. M.	Ob.	Description.	N.G.C.
159	Dec. 11	20 (π) Cassiop	f	8 30	n	0 33	3	vB. R. vgbM. 1'½ dia.	278
160	29	29 (γ) Virgin.	p	6 17	s	2 19	2	vB. cL. E. sp nf. vgBN. F. bran.	4546
	1787								
161	Jan. 14	6 Comæ	f	12 58	s	0 55	1	vB. pL. iR.	4459
162	—	29 ——	f	10 35	n	0 2	1	vB. E. sp nf. Sst in it ½' p. N.	4866
163	Feb. 22	20 Sextantis	p	8 29	s	0 22	1	eB. cL. mE. 45° sp nf. N. 2' l. F. br. 5' l.	3115
164	Mar. 17	38 Leo. min.	p	2 54	s	0 36	3	cB. E. 30° np sf. mbM. er. 4' l. 2' b.	3294
165	—	6 Canum	p	15 42	n	0 25	2	vB. BN. not M. or 2 joined the n. N.	4156
166	—	— ——	p	1 20	n	0 23	2	vB. S. R. mbM.	4369
167	18	10 (n) Ursæ	f	13 43	s	1 40	1	cB. R. BN. 1'½ dia.	2782
168	—	34 (μ) ——	p	4 9	s	0 6	3	cB. R. vgbM. 8' dia. cst. in it, unconnected.	3184
169	—	6 Canum	p	16 16	n	0 53	1	cB. cL.	4145
170	—	20 ——	f	28 12	n	1 6	2	cB. E. near par. SNM. 2' l.	5290
171	—	53 (2d ν) Boot	p	49 57	n	1 10	2	cB. S. R. r. mbM.	5739
172	19	31 Leo. min.	f	25 2	s	0 3	1	cB. E. sp nf. few st. in p. 1 in n. unconnected.	3432
173	—	— —— ——	f	86 19	n	0 23	1	vB. R. vgNM. 2'½ dia.	3941
174	20	53 (ξ) Ursæ	f	46 14	n	0 24	1	cB. E. 5' l. 1'½ b.	4062
175	—	13 Canum	p	46 3	n	2 28	1	vB. S. R. mbM.	4203
176 177	—	— ——	p	16 33	n	1 26	1	Two. The s. cB. E. mbM. The n. pB. E. sp nf. Both join and form the letter S.	4656 4657
178 179	Apr. 9	8 ——	f	7 36	s	0 12	1	Two. The n. vB. vmbM. The s. pB. Their nebul. run together.	4618
180	—	20 ——	f	29 9	n	3 15	1	cB. mE. 60° np sf. vsBM.	5297
181	—	— ——	f	40 13	n	1 11	1	cB. cL. mbM.	5383
182	11	1 Serpentis	p	17 22	s	0 2	2	cB. pL. iR. mbM.	5713
183	—	— ——	p	11 19	n	0 1	2	cB. pL. iR. or lE.	5750
184	May 7	8 Libræ	p	8 21	s	1 15	1	cB. pL. E. sp nf. mbM.	5728*
185	11	19 (λ) Bootis	f	11 6	n	0 1	2	c or pB. S. R. psmbM.	5633
186	12	— — ——	p	47 14	n	1 20	2	cB. pL. R or lE. vgbM. 3' np. the 51st of the *Conn. des Temps*.	5195
187	—	— — ——	p	20 15	n	1 14	1	cB. E. 30° sp nf. BN. vgF. bran.	5377
188	—	38 (2d h) ——	p	13 24	n	2 44	2	cB. lE. par. mbM. F. bran. 1'½ l.	5689
189	15	24 (g) ——	f	3 57	s	0 23	1	cB. cL. E. sp nf. broad.	5676
190 191	16	[1] Canum 6m.	f	11 32	s	1 11	1	Two. The s. cB. cL. The n. pB. S. dist. 1'½. [1] See note.	5395 5394
192	Oct. 14	3 Lacertæ	p	80 46	n	2 32	3	cB. iF. 3' l. 2'½ b. Nebulosity.	7008
193	Nov. 12	54 (φ) Andro.	p	1 26	n	0 54	1	Two close together. Both vB. dist. 2' sp nf. One is 76 of the *Conn.*	651
	1788								
194	Jan. 14	56 Ursæ	f	3 19	n	0 5	2	vB. cL. mE. mer. BN. 6' l. 2' b. chev.	3675
195	—	67 ——	f	4 49	n	0 2	2	E. vBN. and F. branches.	4111
196	—	— ——	f	7 17	n	0 38	2	cB. cL. iF. vgbM. sf. st.	4138

I.	1788	Stars.		M. S.		D. M.	Ob.	Description.	N.G.C.
197	Jan. 14	8 Canum	p	3 32	n	0 19	1	Two. The s. vB. vL. iE. The	4485
198								n. B. pS. iF. dist. 1′½.	4490
199	15	15 Leo. min.	f	32 1	s	0 24	2	cB. mE. sp nf. vgbM. 5′ l. 2 or 3′ b.	3198
200	Feb. 5	59 (2d σ) Can	p	4 29	n	0 29	1	v brilliant. mE. sp nf. 8′ l. 3′ b. beautiful.	2683
201	—	63 (χ) Ursæ	f	0 5	s	0 17	2	cB. mE. sp nf. near. mer. 5′ l. 1′ b.	3877
202	—	— — —	f	7 47	n	0 4	2	cB. S. lE.	3949
203	6	59 —	f	14 42	n	0 31	1	cB. cL. R. pBNM.	3938
204	Mar. 9	9 (ι) —	p	16 27	n	2 7	1	cB. vS. lE. m.	2639
205	—	— — —	f	22 18	n	3 1	1	vB. lbM. chev. bran. m. neb. 6′ l. 4′ b.	2841
206	—	3 Canum	p	14 39	n	1 35	3	cB. E. 45° np sf. 6′ l. 4′ b. almost equally B.	4088
207	—	— —	p	14 0	s	1 32	3	cB. mE. 70° sp nf. 6 or 7′ l. 2′ b.	4096
208	—	— —	p	9 9	n	1 32	3	cB. mE. sp nf. SBNM. 5′ l. 1′ b.	4157
209	—	— —	p	3 33	s	1 6	2	cB. cL. E. mbM.	4220
210	Apr. 1	60 Ursæ	f	46 0	n	0 9	2	vB. S. lE. near. par. BN. eF. bran.	4346
211	—	11 Canum	f	5 47	s	1 58	3	cB. S. R. bM. f. vSst.	4800
212	10	60 Ursæ	f	50 50	s	1 58	1	cB. pL. E.	4460
213	27	19 (λ) Bootis	p	110 25	s	1 48	1	v brilliant. cL. E. sp nf. with difficulty r. has 3 or 4 BN.	4449
214	May 1	17 (κ) —	p	8 26	n	1 56	1	cB. cL. n. ends abruptly. s. vg.	5474
215	5	Neb. II. 757	p	3 27	s	1 14	1	vB. cL. E. f. 2 st.	5866

Second class. Faint nebulæ.

II.	1785	Stars.		M. S.		D. M.	Ob.	Description.	N.G.C.
403	Apr. 26	1 Comæ	p	8 50	s	1 21	3	F. cL. iF. lbM.	3947*
404	27	5 —	p	11 40	s	0 29	1	pB. pL. R. C. mbM.	4032
405	—	— —	p	1 0	s	0 24	2	pB. pL. iF. lE. bM. p. pcst.	4158
406	—	20 —	p	6 8	s	1 27	1	pF. pL. mbM. S neb. joined to it. or lb. in the n.	4336
407	—	— —	f	6 44	s	1 35	1	pB. pS. lE.	4561
408	28	61 Ursæ	f	7 54	n	0 46	2	F. S. R. gbM. near ½′ dia.	3897
409	May 1	— —	f	33 54	n	2 25	2	pB. pL. vgbM. r.	4190
410	—	14 Canum	p	32 8	s	0 14	2	pB. cL. R. smbM. r.	4534
411	—	— —	p	24 25	s	0 43	2	pB. pL. R. lbM. 2′ np. pBst.	4619
412	—	— —	p	17 8	s	0 28	2	F. S. lE. glbM. er.	4711
413	—	— —	p	0 50	s	0 36	2	pB. S. R. bM. and vsF. on the edges.	4956
414	—	— —	f	5 58	n	0 27	1	F. S. lE.	5014
415	—	— —	f	48 34	n	0 15	1	F. S. iF.	5352
416	—	— —	f	58 10	s	1 8	2	pB. pL. iE. mbM.	5440
417	—	— —	f	58 18	s	0 47	1	pB. pL. iE. bM.	5444
418	—	51 (μ) Bootis	p	69 38	s	1 48	1	pB. iR. mbM.	5533

II.	1785	Stars.		M. S.		D. M.	Ob.	Description.	N.G.C.
419	May 1	51 (μ) Bootis	p	68 31	s	0 37	1	F. pL.	5544
420	—	— — —	p	61 32	s	2 17	1	pB. vS. R. vgmbM.	5614
421	—	— — —	p	55 14	s	1 53	1	F. pL. iF.	5656
422	—	— — —	p	52 36	s	0 52	1	F. cL. iF. unequally B.	5675
423	—	— — —	p	47 57	s	0 37	1	pF. pS. iF. bM.	5695
424	2	49 (δ) —	p	83 12	n	0 31	1	F. pL. lbM.	5347
425	5	34 (ω) Serpen	p	4 0	n	0 15	3	F. cS. iR. stellar.	5990
426 }	Aug. 12	1 Aquarii	f	7 50	s	0 12	1	Two. The p. F. S. iR. mbM.	6962
427 }								The f. vF. vS. lbM. 3 or 4′ dist. Place of 1st.	6964
428	30	35 Pegasi	f	6 22	n	0 47	2	pB. S. iR. lbM. r.	7311
429 }	—	6 (γ) Piscium	p	2 16	n	1 14	1	Two. The f. pB. mE. par. mbM. 4′ l. 1′ b.	7541
430 }								The p. vF. cS. 3 or 4′ dist. Place of largest.	7537
431	Sept. 10	92 (χ) Aqua.	f	2 0	n	0 9	2	pB. S. lE. par. vgFNM. 1′ l.	7600
432	—	— — —	f	22 5	n	1 9	4	pB. cL. E. 75° sp nf. 3′ l.	7721
433	—	41 Ceti	p	18 0	s	0 4	1	pB. pL. bM. i. parallelogram. mer.	337
434	—	— —	p	14 23	n	1 18	1	F. S. iF. bM. r.	357
435	—	67 —	p	15 52	s	0 27	1	F. S. iR. bM.	788
436	—	— —	f	1 45	s	0 14	1	F. pS. lE. s. 2 or 3 uneq. st.	881
437	—	— —	f	2 7	s	0 24	1	F. pS. lE.	883
438	—	— —	f	4 33	n	0 54	2	pB. vL. iF. mbM. r.	895
439	26	59 (ϕ) Pegasi	f	8 34	s	0 30	1	pB. pS. mbM.	7619
440	—	— — —	f	9 1	s	0 30	1	pB. pS. bM.	7626
441	—	— — —	f	10 1	n	0 10	1	F. S.	7634
442	Oct. 1	62 (η) Aqua.	f	9 4	s	0 5	3	F. S. r. lbM. or f. M.	7364
443	—	— — —	f	15 19	s	1 29	2	F. S. iR. lbM. 1′½ s. S. st.	7391
444	—	20 Ceti	p	10 20	s	0 24	1	F. pL. lbM.	227
445	—	— —	p	6 50	s	0 35	1	F. iF. er. 1′ b.	245
446	—	— —	p	2 16	s	0 45	2	pB. S. R. mbM. m.	271
447	—	34 —	f	1 3	n	2 0	2	F. S. Two more near it. See III. 592. 593.	430
448 }	—	43 —	f	3 28	s	0 53	1	Two. Both stellar. within 1′ dist.	545
449 }								Nebulosities run together.	547
450 }	3	71 (1st τ) Aqu	f	11 10	n	0 45	2	Two. Both F. S. lE. different	7443*
451 }								directions. er. 2 or 3′ from each other.	7444*
452	—	18 Ceti	p	5 33	s	0 59	1	pB. pS. mbM. r. st. 1′½ dist.	210*
453	5	63 (κ) Aqua.	f	13 50	s	1 19	1	F. pL. E. par. r.	7393
454	—	90 (ϕ) Aqua.	f	3 11	n	1 17	1	F. S. almost stellar.	7576
455 }	—	17 Eridani	f	11 19	n	0 26	2	Two. The p. pB. cL. E. lbM.	1417
456 }			f	11 46	n	0 25	2	The f. eF. vS. E.	1418
457	—	61 (ω) —	p	4 31	s	0 2	2	F. cL. lbM.	1665
458	6	20 —	f	8 52	s	0 46	1	pB. R. bM.	1440
459	—	— —	f	9 14	s	1 4	1	F. R. lbM.	1452
460	—	— —	f	12 7	n	1 6	1	pB. S. lE. mbM. N.	1461
461	8	111 (ξ) Pisciu	p	28 48	s	1 32	3	F. pL. iR. vgbM. 1′½ dia.	521
462	—	— — —	p	27 52	s	1 32	2	pB. R. vgbM. 1′¼ dia.	533
463	—	— — —	p	26 40	s	1 15	3	F. S. ilE. par. mbM.	550
464	—	44 Eridani	p	9 2	n	0 0	1	F. vS. r.	1550*

II.	1785	Stars.		M. S.		D. M.	Ob.	Description.	N.G.C.
465	Oct. 9	82 (δ) Ceti	f	7 12	s	0 34	3	F. pL. iR. lbM.	1090
466	—	— — —	f	7 4	s	0 49	3	pB. cL. iR. mbM.	1087
467	25	7 (b) Piscium	p	4 23	n	1 22	1	pB. pL. iF.	7562
468	—	26 ——	f	0 11	s	1 10	1	F. pL. iF. r.	7785
469	26	49 Aquarii	f	5 14	s	0 4	1	F. pS. lE. er. some of the st. visible.	7284
470	Nov. 22	67 Ceti	f	37 51	s	3 27	2	pB. S. stellar.	1140
471	23	34 Piscium	f	20 53	s	0 55	1	F. iF. lbM.	137
472	27	18 Ceti	f	2 18	n	1 24	1	F. pS.	255
473	—	47 ——	f	6 3	n	0 54	1	F. S. iF. er. some of the st. visible.	599
474	—	72 (ρ) ——	p	9 28	n	0 56	2	pB. pL. lE. lbM.	873
475	—	83 (ε) ——	f	24 23	s	0 3	1	pF. pL. iF. bM.	1200
476	28	58 Aquarii	f	2 43	n	0 31	1	F. pL. iR. lbM.	7319
477	—	70 ——	p	2 28	s	0 27	1	pB. pL. iR. lbM.	7371
478	—	17 Ceti	p	10 10	n	0 53	1	pB. L. lE. lbM.	151
479	—	— ——	p	5 13	n	1 35	1	pB. mE. mer. 2′ l.	191
480	—	— ——	p	2 34	n	0 34	1	F. pL. lE. lbM.	217
481	—	53 (χ) ——	p	0 24	n	0 23	1	pB. cL. R. 1′½ f. Sst.	681*
482 } 483 }	—	55 (1st ζ) ——	f	17 54	n	0 15	1	Four. The p. 2, both F. E. S. within 1′ dist. par.	833 835
484 } 485 }	—	— — ——	f	17 56	n	0 11	1	The f. two, both pF. pS. E. about 2′ dist. and nearly mer.	838 839
486	—	— — ——	f	20 13	n	1 5	1	F. S. E.	853
487	—	— — ——	f	37 18	s	0 7	1	F. cL. iF. lbM.	945
488	—	— — ——	f	49 13	s	0 50	1	F. S. iF. bM.	1045
489	29	23 (2d θ) Arie	f	8 36	n	0 42	1	F. S. lE. contains 3 st. uncon.	932
490	Dec. 7	66 (4 σ) Canc.	f	8 10	n	0 54	1	pF. mE. r. 3′ l. 1′½ b.	2770
491	—	18 Leo. min.	p	13 13	s	0 30	1	pB. pL. iF. lbM.	2968
492	—	— —— ——	f	1 47	n	0 0	1	pB. pL. lE. near par.	3067
493	—	37 —— ——	f	13 7	n	0 49	1	F. S.	3413
494	—	46 Ursæ	p	3 47	s	0 36	1	pB. pL. iR.	3424
495	28	3 Leonis	f	3 34	n	0 16	1	F. pL. E. iF.	2906
496	—	9 (o) Virgin.	f	11 52	s	1 5	1	F.	4233
497	—	31 (1st d) ——	p	14 27	n	1 25	1	pF. vS.	4434
498	—	— —— ——	p	12 30	n	1 3	1	F. pL.	4470
499	—	— —— ——	p	10 55	n	1 18	1	F.	4492
500	—	— —— ——	p	7 43	n	1 24	1	vL. er. some st. visible.	4535
501	30	52 (ι) Ceti	f	4 36	n	1 1	1	F. S. R. vSpBN.	682
502	—	76 (σ) ——	f	29 37	n	0 30	1	F. eS. stellar. p. pBst.	1172
503	—	— — ——	f	31 37	s	0 15	1	pB. S. iF. mbM.	1199
504	—	20 Eridani	p	30 24	n	1 44	1	pB. S. lE. mbM.	1209
505	31	9 Hydræ	f	34 16	s	0 15	1	pB. S. lE. sp nf. smbM.	2811
506	—	— ——	f	49 32	s	0 37	1	pB. S. lE. lb sfM.	2907
507	—	4 (ν) Crater	f	13 25	s	0 3	1	F. S. E.	3508
508	—	30 (η) ——	f	4 26	s	0 41	1	pB. S. lE. bM.	4033
509	—	— — ——	f	6 52	n	0 46	1	F. cL. iR. lbM.	4050
510	—	53 Virginis	f	2 58	s	0 25	1	F. lE. 1′½ l.	5037
511	—	— ——	f	3 21	s	0 12	2	pB. pL. R. bM.	5044
512	—	— ——	f	3 55	s	0 12	2	F. S.	5049
513	—	— ——	f	4 53	s	0 27	2	pB. pL. iF. mbM.	5054

II.	1786	Stars.		M. S.		D. M.	Ob.	Description.	N.G.C.
514	Jan. 1	49 Eridani	p	0 34	s	1 9	1	F. pL. E. sp nf. 2′ l. 1′ b.	1620
515	—	— ——	f	2 57	s	1 33	1	F. or pB. S. bM.	1635
516	—	— ——	f	21 45	s	1 16	1	F. S. iR. lbM.	1713
517	—	29 (γ) Virgin.	f	19 8	n	1 22	2	pB. pL. R. bM.	4904
518	} 2	13 Canum	p	44 34	n	2 49	2	Two. The p. F. S. E. The f. F.	4227
519		— ——	p	44 31	n	2 51	2	S. E. in a different direction.	4229
520	27	7 (η) Hydræ	f	24 25	n	0 7	2	F. S. lE. par. er.	2765
521	—	77 (σ) Leonis	p	3 42	s	1 28	3	F. vS. iF. smbM. er.	3611
522	30	47 Eridani	f	6 29	s	0 21	1	F. pS. iE. r. 1′ sp. Sst.	1636
523	—	— ——	f	10 15	s	0 17	1	F. vS. iR. bM. almost stellar.	1646
524	Feb. 1	57 (μ) ——	p	9 24	n	0 3	1	F. S. iF. lbM. p. 2 Sst.	1618
525	—	— — ——	p	4 5	n	1 27	1	F. pL. lE.	1638
526	—	— — ——	f	0 16	n	0 51	1	F. cS. R. lbM.	1653
527	—	— — ——	f	7 30	n	0 12	1	F. S. lbM.	1682*
528	—	— — ——	f	7 40	n	0 12	2	pB. S.	1684*
529	—	28 (A) Hydr.	p	26 37	n	0 8	1	F. S.	2721
530	2	60 (σ) Virg.	p	52 32	n	0 19	1	F. S.	4376
531	—	— — ——	p	47 19	s	1 12	2	pB. pL. E. b. s. M. 3′ l.	4480
532	—	— — ——	p	35 12	s	1 28	2	F. pL. lbM.	4630
533	—	64 — ——	f	26 8	s	1 17	2	F. pL. vlbM. 6 or 7′ l. 4′ b.	5300
534	—	— ——	f	34 2	s	0 15	2	pB. vL. glbM.	5364
535	24	10 (τ) ——	f	43 43	s	0 39	1	F. mE. np sf. 2′ l. ¾′ b.	4771
536	—	— — ——	f	48 21	s	0 1	1	pB. mE. mbM. 2′½ l. 1′ b.	4845
537	—	92 ——	p	46 53	n	0 3	1	F. pL. iR. er.	4999
538	—	108 ——	p	1 8	n	0 9	1	pB. cL. iR.	5740
539	—	110 ——	p	2 58	s	0 1	1	pB. cL. lE. gbM.	5806
540	—	— ——	f	1 11	s	0 3	1	pB. S. mbM.	5831
541	—	— ——	f	2 31	s	0 8	1	F.	5839
542	—	— ——	f	2 31	n	0 0	1	pB.	5838
543	—	— ——	f	4 14	s	0 4	1	F.	5850
544	—	— ——	f	4 52	n	0 7	2	pB. vS.	5854
545	—	— ——	f	6 51	s	1 39	4	pB. S. iE. lbM.	5869*
546	} Mar. 3	6 (h) Leonis	p	6 16	n	1 42	1	Two. Both F. S. The place in-	2872*
547								accurate in R.A.	2874*
548	—	14 Virginis	p	10 27	s	0 8	1	F. pL. mE. np sf. but near. par.	4129
549	—	26 (χ) ——	f	17 34	s	0 33	1	pB. vL. iF. lbM.	4818
550	} 4	14 (ε) Crateris	p	4 13	n	0 35	2	Two. Both F. S. lbM. cBst. be-	3636
551			p	4 0	n	0 36	2	tween, but 1′½ s. of them.	3637
552	—	21 (θ) ——	p	2 24	s	0 2	1	F. pS. iR. f. vSst.	3732
553	—	— — ——	f	11 21	s	1 9	2	pB. pL. iF. gbM. sp. is Sst.	3892
554	18	1 Cancri	f	4 36	s	0 4	2	pB. pL. er. vgmbM.	2507
555	19	26 Hydræ	f	7 26	n	0 21	2	pB. pL. iR. b. s. M.	2889
556	20	6 (3d *b*) Crat	p	76 10	s	1 11	3	pB. cL. iR. vgmbM.	2935
557	24	16 (ζ) Hydræ	f	3 21	n	0 22	1	F. mE. unequally B. 3′ l. 1′ b.	2718
558	25	21 (*q*) Virgin.	f	10 43	s	0 38	1	F. E. mer. 3′ l. f. cBst.	4658
559	—	49 (*g*) ——	p	14 43	n	1 33	1	F. S.	4759
560	—	— — ——	p	13 0	n	0 31	1	pF. pS. iR.	4790
561	—	— — ——	p	3 39	n	0 24	1	pB. pL. R. vgmbM.	4939
562	27	16 (κ) Crater	f	4 56	s	1 54	2	F. S. iR. bM. r.	3715*
563	—	68 (ι) Virgin.	p	29 28	s	0 55	1	pB. iF. bM.	4825

II.	1786	Stars.		M. S.		D. M.	Ob.	Description.	N.G.C.
564	Mar. 28	19 Ursæ	p	3 1	n	0 23	1	pB. S. R. mbM.	2778
565	—	46 Leo. min.	p	5 3	n	0 28	1	pB. cL. iF. lbM.	3381
566	—	[1]1102 (e) Hy.	f	35 28	n	0 53	1	F. pS. E. [1] See note.	5078
567	—	——— — —	f	37 17	n	0 51	1	pB. pL. iF. gbM.	5101
568	Apr. 17	11 (s) Virgin.	f	10 14	n	0 34	1	Four nebulæ. They are scattered about. The place is that of the last.	4270*
569									4273*
570									4277*
571									4281*
572	—	— — ——	f	11 34	s	0 26	1	A nebula.	4300
573	23	— — ——	f	10 18	s	0 26	1	A nebula, but cloudy.	4281*
574	29	3 Serpentis	p	40 48	s	0 20	1	F. S. lE. r. p. 2 vcst.	5668
575	—	— ——	p	36 3	n	0 33	1	pB. cL. iR. mbM.	5701
576	—	— ——	p	21 26	s	0 54	1	F. S. lE. like 2 stellar. joined closely.	5770*
577	30	37 Virginis	p	11 22	n	0 4	1	F. S. making a triangle with 2 Bst	4600
578	—	— ——	p	2 29	n	0 20	1	F. S.	4701
579	—	109 ——	p	26 11	n	2 10	1	pB. cL. E.	5560
580	—	— ——	p	16 35	n	1 24	1	Two. The s. pB. pL. R. gbM. The n. eF. cL. dist. 2′. The place is of s.	5638
581									5636
582	—	— ——	p	8 33	n	0 25	1	F. mE. r. 2′ l. ½′ b. f. st. 6 m. 16″ in time.	5690
583	May 3	14 (1st A) Ser	f	17 48	s	1 2	2	pB. S. E. nearly par. bM.	6010
584	26	5 (g) Ophiuc	f	27 48	n	1 8	1	pB. cL. gbM. er. undoubtedly st.	6235
585	27	3 Serpentis	p	5 43	s	1 52	1	F. S. iE. r.	5864
586	28	40 (ξ) Ophiuc	f	28 13	n	0 57	1	pB. S. iF.	6445
587	June 3	61 Ophiuchi	f	0 23	n	0 36	1	F. cL. iF.	6426
588	Sept. 4	24 (ξ) Ariet.	p	39 40	s	0 17	2	F. S. lE. r. bM.	665
589	—	— — ——	p	36 21	n	0 50	2	F. pL. E. b. f. M. 2′ sp. cBst.	673
590	18	2 Piscium	f	2 2	n	0 48	1	F. S. bM.	7458
591	—	88 (γ) Pegasi	p	4 29	n	0 38	1	F. pL. iF. unequally B.	14
592	—	85 Ceti	p	3 19	n	0 5	1	pB. S. E. bM.	1024
593	20	54 Eridani	p	61 14	n	0 43	1	pB. pS. R. resembling I. 107. but less.	1400
594	—	— ——	p	55 40	n	0 10	1	pB. vS. R. bM.	1442*
595	23	66 Aquarii	p	41 2	s	0 1	2	F. cL. l and iE. nearly par. lbM.	7183
596	30	51 Ceti	f	10 14	n	0 51	1	F. S. bM. 1′ s. Sst.	706
597	—	32 Eridani	p	8 30	s	1 10	2	F. S. E. iF. in a row with some st	1441
598	Oct. 13	59 (υ) Aqua.	f	13 11	s	1 39	1	pB. pL. iR. vgmbM.	7377
599	17	77 Cygni	f	20 15	s	0 6	1	F. pS. E. er.	7197
600	—	10 Andromedæ	f	2 5	s	1 14	2	pB. mE. np sf. but near. mer. lbM. r. 5′ l. 1′½ b. also ob. 1784	7640
601	—	26 (β) Persei	p	15 16	n	1 14	1	F. S. iF. r.	1123
602	—	— — ——	p	13 38	n	0 34	1	F. pS. iR. lbM.	1129
603	—	— — ——	f	11 27	n	0 35	1	pB. stellar. or pcst. with S. vF. chev.	1278
604	18	59 Andromedæ	p	2 10	s	0 17	1	pB. cL. lE. mbM.	818
605	—	— ——	p	0 54	n	0 9	1	pB. S. iF.	828
606	24	6 Lacertæ	p	17 44	n	2 18	3	F. S. er. or rather a patch of st.	7231
607	—	30 Persei	p	12 50	s	1 44	1	F. cL. E.	1175

II.	1786	Stars.		M. S.		D. M.	Ob.	Description.	N.G.C.
608	Oct. 24	30 Persei	p	11 45	n	0 19	1	F. cL. er. some st. visible.	1193
609	26	65 (ι) Piscium	p	1 55	s	0 6	1	pB. S. iR. gbM.	252
610	—	90 (υ) ——	f	24 26	n	1 31	1	F. S. bM. r.	661
611	—	— — ——	f	27 38	n	0 41	1	F. S. lE.	670
612	—	10 (α) Triang	p	28 30	s	1 8	1	pB. pL. lE. nearly par. mbM.	684
613	—	— — ——	p	4 46	s	0 47	1	F. S. lE. par. bM.	855
614 }	—	34 (θ) Gemin	p	5 37	s	0 25	1	Two. The s. F. S. R. bM. The	2274
615 }								n. F. cS. R. bM.	2275
616	—	66 (α) ——	f	9 32	s	0 11	1	F. S. lbM.	2435
617	Nov. 13	6 (β) Arietis	p	3 55	n	0 56	1	F. cL. vglbM.	691
618	—	— — ——	p	3 23	n	1 45	1	vS. stellar.	695
619	—	52 ——	p	5 39	s	0 3	1	pB. cL. pmE. mer. r. 1′ s. st.	1156
620	Dec. 11	27 (κ) Persei	p	5 48	n	1 31	2	F. S. iR. bM. L. stellar.	1169
621	13	34 Ceti	p	23 45	s	0 34	1	F. E. np sf. lbM. 1′½ l.	259
622	20	26 ——	f	9 8	s	0 22	1	F. R. bM. er.	428
623	21	2 (ε) Corvi	p	16 4	s	0 33	2	F. S. E. mer. or few deg. np sf. lb. s. M.	3955
624	29	1 Sextantis	f	8 54	s	1 8	1	F. lE. nearly par. 1′½ l.	2990
625	—	29 (γ) Virgin	p	17 56	s	1 58	2	pB. mE. 20° sp nf 2′ l.	4348
626	30	77 (σ) Leonis	p	4 44	s	1 30	1	pB. S. lE. mbM.	3604*
	1787								
627	Jan. 11	55 (δ) Gemin	f	54 51	s	0 26	3	F. S. iF. lE. sp nf.	2545*
628	14	6 Comæ	f	6 36	n	0 38	1	pB. cL. E.	4312
629	—	— ——	f	13 46	s	0 49	1	F.	4474
630	—	— ——	f	13 20	s	0 56	1	cL.	4468
631	—	— ——	f	16 3	s	1 31	1	F.	4506
632	—	29 ——	p	8 57	n	1 12	1	F. pL. R. vgbM.	4595
633	17	16 (1st p) Pers	p	7 2	s	1 1	1	F. cL. lbM. 4′ dia.	1058
634	Feb. 13	33 (η) Cancri	p	12 7	n	0 34	1	F. S. bM.	2563
635	22	21 (θ) Crater	p	13 5	n	1 9	1	F. pS. iR. vgbM.	3660
636	—	65 Virgin	p	43 8	s	0 49	1	F. vL. bM.	4597
637	Mar. 11	44 (k) ——	f	12 41	s	0 36	1	F. cL. iR. lbM. time inaccurate.	5015*
638	15	[1]1139 (r) Cent	f	22 49	s	0 12	1	pB. S. lE. sp nf. [1] See note.	5253
639	17	32 Leo. min.	p	16 31	s	0 11	1	pB. cS. r.	3158
640	—	—— ——	p	16 11	s	0 18	1	F. vS. r. with 300 the same.	3163
641	—	38 —— ——	f	2 41	s	0 36	2	F. vS.	3334
642	—	6 Canum	p	15 18	n	0 30	2	pB. S. E.	4156
643	—	10 ——	p	0 37	s	2 11	1	F. pL. gbM. r.	4662
644	—	— ——	f	2 55	s	1 1	1	pB. S. R. mbM. among scattered st.	4868
645	—	— ——	f	4 33	s	1 2	1	pB. S. R. mbM.	4914
646	—	17 ——	f	12 21	n	0 12	1	pB. L. iF. uneq. B. 3 or 4′ dia.	5112
647	—	12 (λ) Coronæ	f	33 4	n	1 27	1	F. S. iF. The time inaccurate.	6158*
648	18	53 (2d ν) Boot	p	55 31	n	1 11	2	pB. pL. lbM.	5696*
649	—	— —— ——	p	54 11	s	0 13	2	F. S. E. nearly mer. r.	5704
650	—	— —— ——	p	16 19	n	1 13	3	pB. E. BNM. and F. br. 2′ l. ¼′ b.	5899
651	—	— —— ——	p	5 42	n	0 51	2	pB. pL. iE. er.	5930
652	—	30 (g) Hercul	p	0 57	s	0 57	1	F. pL. r.	6160
653	19	70 Virginis	p	4 21	n	0 11	1	pB. vS. mbM. just p. pcst.	5129
654	—	9 Serpentis	f	7 56	s	0 28	1	F. E. np sf. 1′½ l.	5951

II.	1787	Stars.		M. S.		D. M.	Ob.	Description.	N.G.C.
655	Mar. 19	9 Serpentis	f	15 44	n	0 16	1	F. E. mer. 1'½ l.	5980
656	—	— ——	f	16 59	s	1 17	1	pB. E. np sf. bM. 1'½ l.	5984
657	—	28 (β) ——	f	8 2	s	0 52	1	F. iF. bM. 1'¼ dia. between 2 Bst.	6012
658	20	43 Lyncis	p	47 39	s	0 23	1	pF. vS. mbM.	2691
659	—	13 Canum	p	18 44	n	1 47	1	F. S. R. just np. V. 42.	4627*
660	Apr. 9	8 ——	f	7 58	s	0 5	1	pB. pL. R. mbM.	4625
661	—	— ——	f	9 42	s	0 20	1	pB. vS. stellar. just p. Sst.	4655
662	—	— ——	f	15 2	n	0 36	1	F. S. R. bM.	4704
663	—	19 ——	p	9 58	n	0 56	1	pB. vS. stellar. near and n. Sst.	4963
664	—	— ——	p	3 47	n	3 13	2	pB. mE. sp nf. near. mer. 5' l. ¾' b.	5023
665	—	20 ——	f	2 52	n	2 31	1	pB. cS. E. 300 showed it like a st. with burrs.	5103
666	—	— ——	f	5 24	n	2 30	1	pB. S. iR. mbM.	5123
667	—	— ——	f	7 35	n	2 42	1	pB. vS. lE. bM.	5145
668	—	— ——	f	27 51	n	0 51	1	F. E. par. miniature of I. 170.	5289
669	—	— ——	f	33 20	n	0 41	1	pB. pL. vgmbM.	5320
670	—	— ——	f	35 10	n	2 37	1	pB. pL.	5336
671	—	— ——	f	37 51	n	0 35	1	pB. pL. E.	5362
672	—	— ——	f	43 59	n	0 17	1	pF. pS. bM.	5410
673	—	— ——	f	66 36	n	1 0	1	F. pL. E. vlbM.	5608
674	—	— ——	f	71 16	n	0 27	1	pB. E. nearly par. 1'½ l. ½' b.	5630
675	—	— ——	f	80 7	n	0 51	1	F. vS.	5697*
676	—	— ——	f	98 12	n	1 42	1	pB. vS. stellar.	5784
677	—	— ——	f	99 9	n	1 39	1	F. pS. lbM.	5787
678	—	— ——	f	117 42	n	1 1	1	F. S. r. in a row with 3 st.	5893
679	11	79 (ζ) Virgin	p	4 17	s	1 1	2	Two. The p. F. pS. iF.	5183*
680			p	4 7	s	1 4	2	The f. pB. pL. iF. bM.	5184*
681	—	1 Serpentis	p	19 44	s	0 7	2	pB. pL. iF.	5691
682	—	— ——	p	16 35	s	0 4	2	pB. cS. lE.	5719
683	—	— ——	f	0 49	s	0 55	1	pB. pL. R. mbM. sf. cst.	5792
684	—	4 ——	p	6 6	n	0 7	1	Two. The 2d pB. S. iE. for the 1st see II. 545.	5868*
685	15	90 (p) Virgin	p	2 37	s	0 44	2	F. pL. iR. f. and par. with 2 Fst.	5327
686	—	— — ——	p	0 37	n	0 4	2	pB. S. mbM.	5345
687	—	102 (1st υ) ——	p	6 18	s	0 57	2	pB. cL. mE. 20° sp nf.	5506
688	May 11	19 (λ) Bootis	p	30 37	n	0 7	2	F. mE. 15° sp nf. lbM. 4' l. ¾' b.	5301
689	12	— — ——	p	47 20	n	0 46	3	pB. pL. R. mbM.	5198
690	—	22 (τ) Hercu	f	7 2	n	2 3	2	F. pL. iF. gbM.	6155
691	15	85 (η) Ursæ	f	15 34	s	0 12	1	pB. pL. E. nearly par. mbM.	5448
692	—	— — ——	f	19 36	n	1 20	1	Two. The p. F. pS. R. vgbM.	5480
693								The f. F. vS. stellar. smbM. dist. 2'½.	5481
694	—	24 (g) Bootis	p	6 31	n	0 43	1	pF. pS. lE. mbM.	5602
695	—	— — ——	f	1 7	s	0 12	1	pB. cL. iR. vgmbM.	5660
696	—	— — ——	f	3 40	n	0 3	1	pB. S. E.	5673*
697	16	[1] C Canu. 6m	f	6 23	s	0 39	1	F. E. par. bM. 1'½ l. 1' b. [1] See note.	5351
698	—	— —— —	f	10 0	s	0 58	1	F. S. R. vsmbM.	5380
699	—	— —— —	f	13 19	n	0 20	3	F. pL. R. lbM. 1'½ dia.	5406
700	—	27 (γ) Bootis	f	5 15	n	0 9	1	pF. S. iE.	5698

II.	1787	Stars.		M. S.		D. M.	Ob.	Description.	N.G.C.
701	May 16	25 Herculis	f	17 43	s	0 40	1	pB. pS. E. sp nf. vgmbM.	6207
702	Sept. 11	68 (2d g) Aqu	f	4 23	s	1 1	1	pF. pL. E. np sf. but near. par. mbM. 1′½ l.	7392
703	—	13 Ceti	f	12 53	n	0 49	1	F. cL. E.	259
704	16	47 Cassiop	f	61 37	n	3 48	2	F. pL. mE. np sf. mbM.	1184
705	Nov. 3	25 Cephei	f	21 6	s	1 35	1	pB. S. iR. er. almost equally B.	7354
706	—	1 (e) Cassiop	f	6 26	n	2 5	2	pBM. 2cst. involved in nebulosity. 2′ l. 1′½ b.	7538
707	30	19 (ξ) ——	p	2 50	s	2 12	1	pB. vL. iR. vgmbM. r. 5 or 6′ dia.	185
	1788								
708	Jan. 14	36 Lyncis	f	3 50	s	1 15	1	pB. S. stellar.	2798
709	—	56 Ursæ	p	6 51	s	1 57	2	pB. S. lE. mer. bM.	3600
710	—	27 (γ) Bootis	p	43 42	n	1 51	1	F. S.	5311
711	—	— — ——	p	42 47	n	1 50	1	pB. cL. iF.	5313
712	—	— — ——	p	41 48	n	1 25	1	F. S. R. bM.	5326
713	—	— — ——	p	39 13	n	2 12	2	pB. pL.	5350
714 }	—	— — ——	p	39 5	n	2 9	2	Two. Both pB. S. R. 2′ dist. in	5353
715 }								the same mer.	5354
716	—	— — ——	p	36 48	n	2 19	2	pB. L. iR. FN. mbM. 4 or 5′ dia.	5371
717	15	15 Leo. min.	f	0 58	s	1 58	1	F. pL. iF. lbM.	2998
718	—	45 (ω) Ursæ	p	2 24	n	0 32	2	pB. S. lE. the np. corner of a S. trapezium.	3415
719	Feb. 3	32 Lyncis	p	20 34	s	0 16	1	F. pL. iR. bM.	2543*
720	—	34 (μ) Ursæ	p	2 13	n	1 29	1	F. vS.	3202
721	—	— — ——	p	1 57	n	1 26	1	F. vS. stellar.	3205
722	—	— — ——	p	1 43	n	1 27	1	F. vS. stellar.	3207
723	—	13 Canum	p	73 0	s	0 22	1	pB. S. lE.	3891
724	—	— ——	p	65 22	s	0 44	1	F. vS.	3971
725	—	— ——	p	61 59	s	0 19	1	pB. E. sp nf. but nearer mer. mbM. 2′ l.	4020
726	5	80 (π) Gemi	f	22 56	n	0 38	1	pF. pL. iR. lbM. r. s. 2 st. par.	2532
727	—	59 (2 σ) Canc	p	13 13	n	1 47	1	pF. pL. iR. r.	2649
728	—	60 Ursæ	p	25 2	n	1 32	2	pB. pL. vgmbM.	3583*
729	—	— ——	p	20 38	s	1 4	2	F. cL. lE. par. lbM.	3614
730	—	— ——	p	5 27	n	0 14	1	pB. bM. r. 4′ l. 3′ b.	3726
731	—	— ——	p	0 54	n	1 5	2	pB. S. E. sp nf.	3769
732	—	— ——	f	0 47	s	0 19	1	F. S. almost betw. 2 sp. st. chev. touches them.	3782
733	6	59 ——	f	20 28	n	0 21	1	pB. mE. near mer. pBSN. and vF. br. 4′ l. ¾′ b.	4013
734	9	20 Lyncis	p	14 20	n	0 28	1	F. pL. iF. mbM. sf. a triangle of S. st.	2326
735	—	— ——	p	12 44	s	1 37	1	F. stellar.	2329
736	—	— ——	p	11 9	s	0 3	1	pF. vS. lbM. r.	2340
737	—	63 (χ) Ursæ	p	4 55	s	0 4	2	pF. pS. iR. lbM.	3811
738	—	— — ——	f	2 30	n	0 57	1	pB. pL. R. mbM.	3893
739	—	— — ——	f	2 50	n	0 56	1	F. vS.	3896
740	—	— — ——	f	5 48	n	0 56	1	pF. pS. stellar.	3928
741	—	— — ——	f	17 7	n	0 53	1	pF. S. R. gbM.	4047
742	—	3 Canum	p	1 50	s	1 35	2	F. S. E.	4248

II.	1788	Stars.		M. S.		D. M.	Ob.	Description.	N.G.C.
743	Feb. 9	3 Canum	f	5 22	s	0 10	1	F. S.	4381*
744	—	— —	f	21 50	n	1 27	2	pF. S. er.	4617
745	Apr. 1	Neb. II. 728.	p	35 22	s	0 57	2	pF. pS. E. s. and lp. a star among st. not con.	3320
746	8	54 Virginis	p	0 24	s	0 43	1	pB. S. pBN.	5018
747	10	60 Ursæ	f	31 58	s	0 24	1	pB. E. 15 or 20° np sf. 3′ l.	4144
748	—	— —	f	38 3	n	0 16	1	pB. pL. E. sp. and in a line with 2 st.	4217
749	—	— —	f	47 57	s	1 10	2	pB. pL. iF.	4389
750	27	19 (λ) Bootis	p	109 46	s	1 1	1	pF. pL. E. sp nf.	4460
751 }	—	37 (ξ) —	f	16 12	n	0 25	2	Two. The p. cF. cS. The f. pF.	5857
752 }			f	16 20	n	0 24	2	pL. Both lE. np sf. but nearer par.	5859
753	28	27 (β) Hercul.	f	2 50	s	1 42	1	pF. pS. vlE. mbM.	6181*
754	29	27 (γ) Bootis	p	11 15	n	1 27	1	pB. pL. R. FN.	5582
755	May 1	23 (θ) —	f	44 59	n	0 31	1	pB. pL. lE.	5875
756	5	Neb. II. 757.	p	11 47	s	3 7	2	pB. pL. iF. r.	5820*
757	—	12 (ι) Draco	p	16 38	s	1 56	3	pB. S. iR. or lE. mbM.	5879*
758	—	Neb. II. 757.	f	5 28	s	1 31	1	pF. pS. iR.	5905*
759	—	— — —	f	6 6	s	0 42	1	pB. FNM. 8 or 10′ l. 2′ b.	5907*
760	—	— — —	f	6 29	s	1 37	1	pF. pS. R.	5908*
761	—	— — —	f	24 8	s	0 33	1	pF. pS. iF.	5963*
762	—	— — —	f	24 37	s	0 25	1	pF. pL. E.	5965*
763	25	12 (ι) Draco	p	13 7	n	0 54	2	pB. mE. nearly mer. 2′ l. ½′ b.	5894
764	—	— — —	f	13 58	n	0 20	1	pB. S. iR. one p. suspected vF. lE.	5982
765	—	— — —	f	14 36	s	0 58	1	pF. cS.	5987
766	—	— — —	f	15 0	n	0 18	1	pB. cL. iE. r.	5985
767	June 6	31 (1st Ψ)—	p	31 23	n	0 15	1	pB. pL. R. vgmbM.	6340
768	Nov. 4	14 Camelop	p	42 52	n	1 57	1	pB. S. lE. BN. just s. pB. st.	1569

Third class. Very faint nebulæ.

III.	1785	Stars.		M. S.		D. M.	Ob.	Description.	N.G.C.
377 }	Apr. 26	92 Leonis	f	3 6	s	1 24	2	Two. The n. F. S. lbM. The s.	3837*
378 }								vF. vS. dist. 5′ sp. the place of n.	3842*
379	—	1 Comæ	p	10 26	s	0 6	3	vF. vS. lE. er. or S. patch of st.	3926
380	—	— —	p	8 56	s	1 7	2	F. S.	3940
381	—	— —	p	7 56	s	1 12	1	vF. R.	3954
382 }	—	2 —	f	1 58	s	0 49	2	Three. The place is of the last	4093*
383 }								which is vF. S. The other two	4095*
384 }								are sp. eF. vS.	4098*
385	27	93 Leonis	p	2 54	s	0 35	1	vF. vS. r.	3862
386	—	— —	p	3 10	s	0 27	1	vF. vS. r.	3860
387	—	— —	p	2 10	s	0 27	1	vF. vS. r.	3875
388	—	— —	p	1 48	n	0 11	2	vF. pL. iR. lbM. r. 7′ nf. cBst.	3884

III.	1785	Stars.		M. S.		D. M.	Ob.	Description.	N.G.C.
389	Apr. 27	93 Leonis	f	4 49	n	0 25	1	vF. vS.	3937
390	—	5 Comæ	p	8 56	s	1 47	1	Suspected.	4049
391	—	— —	p	7 54	s	0 9	2	Six nebulæ. The places belong	4070
392			p	7 56	s	0 13	2	to the three first which are vF.	4069
393			p	7 54	s	0 15	2	vS. The other three are 10	4074
394								or 12′ more south, but there	4061
395								was not time to take their	4065
396								places, more suspected.	4076
397	—	— —	f	3 8	n	0 5	2	vF. vL. iR. bM. 6′ l. 5′ b.	4204*
398	—	26 —	f	8 12	s	1 33	1	vF. vS. r.	4685
399	28	61 Ursæ	f	31 26	n	1 57	2	vF. pL. lE. r.	4163
400	May 1	— —	f	25 34	n	2 38	4	vF. vS. stellar. 2′½ n. Sst.	4097
401	—	14 Canum	f	2 36	s	0 36	1	vF. stellar. with 300 the same.	4986
402	—	— —	f	19 23	n	0 35	1	Two. Both vF. cS. The place	5141
403								is that of the p. The 2d, 3′ nf.	5142
404	—	— —	f	20 44	n	0 7	1	Two. Both vF. pS. The place is	5149
405								of the p. The 2d, 5 or 6′ nf.	5154
406	—	— —	f	25 14	s	0 59	1	vF. vS. lE.	5199
407	—	49 (δ) Bootis	p	102 40	n	1 37	2	Two. Both vF. vS. A star be-	5223*
408			p	102 22	n	1 39	2	tween them about half way.	5228*
409	—	14 Canum	f	30 38	s	0 17	1	vF. pL. R. lbM.	5240
410	—	— —	f	35 6	n	1 1	1	vF. S. lE. er.	5265
411	—	— —	f	54 30	s	1 8	1	eF. vS.	5399
412	—	— —	f	54 45	n	0 25	1	vF. vS.	5401
413	—	— —	f	58 30	s	0 53	1	vF.	5445
414	—	51 (μ) Bootis	p	70 6	s	0 55	1	vF. mE.	5529
415	—	— — —	p	65 4	s	1 57	1	eF. pL.	5579
416	—	— — —	p	64 2	s	1 57	1	Two. Both vF. S. dist. 6 or 7′.	5589
417								The place is that of the sf.	5590
418	—	— — —	p	62 52	s	0 6	1	eF. stellar.	5533
419	—	— — —	p	61 4	s	0 42	1	vF. vS. E. er.	5616
420	—	— — —	p	54 54	s	0 50	1	vF. S.	5654
421	—	— — —	p	49 26	s	0 40	1	vF. vS.	5684
422	2	49 (δ) —	p	86 2	n	0 36	1	Two. Both eF. stellar. dist. 4 or	5312
423								5′. nearly mer. The n. faintest.	5318
424	3	— — —	p	147 32	n	0 11	1	vF. stellar. or little larger.	4719
425	—	— — —	p	101 48	n	1 39	1	vF. vS. in the field with III. 407. 408.	5233*
426	Aug. 30	17 (ι) Piscium	p	8 48	s	1 42	1	eF. pL. iR. requires great attention.	7685
427	—	19 —	f	0 14	n	0 19	1	vF. S. lE. nearly mer.	7750
428	Sept. 10	30 —	f	14 30	s	0 19	2	vF. S. iF. lbM.	61
429	—	41 Ceti	p	26 42	n	0 35	1	vF. pS. E.	274
430	—	— —	p	26 54	n	0 44	1	vF. vS.	273
431	—	Neb. I. 100	f	0 22	n	0 0	1	The 2d of two. eF. S. 5 or 6′ dist. from I. 100.	586
432	—	41 Ceti	f	15 36	n	0 22	1	eF.	600
433	—	67 —	p	15 39	n	0 59	1	vF. vS.	790
434	—	— —	f	18 40	s	0 42	1	vF. cL. iF. lbM. 4 or 5′ l. 2 or 3′ b.	991
435	26	59 (*p*) Pegasi	f	8 42	s	0 20	1	vF. vS.	7623

III.	1785	Stars.		M. S.		D. M.	Ob.	Description.	N.G.C.
436	Sept. 26	32 (2d *c*) Pisc	f	1 20	s	1 1	1	vF. pL. lbM.	7816
437	27	26 ——	p	7 39	s	0 13	1	eF. vS. er. confirmed by 240.	7750*
438	28	93 (2 Ψ) Aqua	f	9 22	s	0 15	1	eF. S. stellar. p. 1′½. pBst.	7665
439	Oct. 1	20 Ceti	p	0 42	s	1 3	2	vF. S. iE.	279
440	—	38 ——	f	1 5	n	0 7½	1	vF. vL. requires great attention.	450*
441	—	43 ——	f	5 8	s	1 29	1	vF. vS. iE.	560
442	—	— ——	f	5 23	s	1 26	1	vF. vS. iE.	564
443	5	17 Eridani	p	17 51	s	0 13	1	vF. vS. confirmed by 240.	1248
444	—	— ——	p	9 23	n	0 25	1	eF. vS.	1304
445	—	— ——	p	5 37	s	0 41	1	vF. pS. E.	1324
446	—	— ——	f	3 4	s	0 3	1	vF. S. between some Sst.	1358
447	—	20 (τ) Orion	f	10 23	n	1 32	2	vF. cL. iR. near a hook of vSst.	1924
448	—	— — ——	f	34 45	s	0 26	3	vF. S. R. r. lbM.	2110
449	6	1 (1st τ) Erid	f	4 8	n	1 34	1	vF. pL. broadly E. lbM.	1114
450	—	— — ——	f	6 30	n	1 56	2	vF. S. lE.	1125
451	—	20 Eridani	f	2 30	s	0 59	1	vF. S. R.	1393
452	8	52 (π) Aqua	p	30 46	n	1 39	1	vF. pL. R. r.	7156
453	—	10 Orionis	f	5 7	s	0 4	1	vF. vS. confirmed 240.	1762
454	9	60 Ceti	p	27 18	n	0 27	1	eF. pL. 240. left doubtful.	622
455	—	82 (δ) ——	f	4 11	n	1 2	2	vF. vL. lbM. er. 6 or 7′ dia.	1073
456	25	28 (ω) Piscium	f	13 6	s	0 28	1	vF. pS. iF.	36*
457	—	78 (ν) Ceti	p	20 29	n	0 20	1	vF. cL. vlbM. m. p. Bst. and joining.	864
458	26	49 Aquarii	p	2 52	n	0 6	1	vF. S. er. time inaccurate.	7252
459	—	56 (1st υ) Ceti	p	7 44	s	1 17	1	vF. vS. er.	686
460	—	— — ——	p	2 55	s	1 16	1	vF. vS.	723
461	27	18 (ε) Pis. Au	f	90 20	n	1 56	1	vF. cL. lE. glbM. 4 or 5′ l.	24
462	Nov. 7	82 (δ) Ceti	f	8 1	s	0 36	1	vF. S.	1094
463	22	25 ——	p	12 56	s	0 23	2	vF. pL. iR. r.	268
464	—	67 ——	p	20 11	n	0 59	1	eF. S. found in gaging.	762
465	23	46 (ξ) Pegasi	f	11 21	n	0 54	1	eF. S. iF. 240 the same.	7432
466	—	82 ——	f	5 54	s	0 15	1	vF. S. R. lbM.	7794
467	27	18 Ceti	p	11 15	n	0 12	1	eF. vS. 240 left some doubt.	154
468	—	72 (ρ) ——	p	27 13	n	0 43	1	vF. E. nearly mer. lbM. 1′½ l. 1′ b.	773
469	—	83 (ε) ——	f	19 27	s	0 30	1	vF. stellar. 240 left some doubt.	1162
470	28	91 Aquarii	p	1 53	s	0 7	1	eF. vS. 240 left doubtful.	7526
471	—	53 (χ) Ceti	p	13 54	n	0 40	1	A few Sst. mixed with nebulosity	624
472	—	55 (1st ζ) ——	f	41 48	s	0 18	1	vF. pL. vlbM. near scattered st.	977
473	29	87 (*u*) Pegasi	p	44 53	s	1 26	1	eF. cL. some doubt. p. a row of st.	7647
474	—	23 (2d θ) Arie	f	7 29	n	0 50	1	eF. vS. iR. confir. 240.	924
475	—	34 (μ) ——	p	1 44	s	0 44	1	vF. S. confir. 240.	1036
476	Dec. 5	34 (ζ) Andro	p	11 14	s	0 23	1	vF. vS. stellar. sp. pBst.	160
477	—	36 ——	p	2 25	n	0 44	1	vF. S. R. just p. vFst.	280
478	7	20 Leo. min.	f	1 20	n	0 47	1	eF. S. left doubtful.	3099
479	26	2 (ε) Can. min	f	26 18	n	0 25	1	suspected. eF. vS. lE.	2459*
480	28	9 (*o*) Virgin	f	12 46	s	2 5	1	vF. L. seen by looking at II. 137. I.C.	3115*
481	—	31 (1st *d*) ——	p	17 49	n	1 44	1	vF.	4356
482	—	— — ——	p	15 22	n	1 39	1	eF.	4415
483	—	— — ——	p	12 49	n	1 24	1	vF.	4464

III.	1785	Stars.		M. S.		D. M.	Ob.	Description.	N.G.C.
484	Dec. 28	31 (1st *d*) Virgin	p	11 8	n	1 34	1	vF.	4488
485	30	46 Ceti	p	40 9	s	1 4	2	vF. S. iF. r.	244
486	—	76 (σ) ——	p	12 32	s	0 52	1	vF. vS. iF. better with 240.	887
487	—	20 Eridani	p	3 52	n	2 14	1	vF. S. E.	1354
488	31	9 Hydræ	f	38 13	s	0 26	1	vF. cL. gvlbM. 3′ l. 2′ b. p. pBst.	2848
489	—	53 Virginis	p	18 36	s	0 47	1	vF. S. lbM.	4763
	1786								
490	Jan. 1	45 Eridani	p	11 41	s	0 42	1	vF. vS. lE. better with 240.	1552
491	—	13 (*n*) Virgin	p	16 10	n	0 35	2	vF. S. R. bM.	4044
492	—	15 (η) ——	f	7 0	s	0 15	2	vF. cL. mE. r.	4418*
493	—	29 (γ) ——	p	6 35	n	1 12	2	eF. S. iF.	4541
494	—	— — ——	f	1 24	n	0 48	2	vF. pS. E.	4642
495	2	61 Ursæ	f	58 0	s	0 46	1	eF. S. iF. r.	4583
496	—	— ——	f	70 52	s	0 3	1	eF. vS. pmE.	4737
497	27	36 Sextantis	f	6 47	n	1 20	2	cF. S. R. vlbM.	3434
498	—	58 (*d*) Leonis	f	0 43	n	0 1	1	vF. mE.	3495
499	30	39 (A) Eridan	p	6 26	n	1 25	1	vF. S. E. er.	1516
500	—	69 (λ) ——	p	3 50	s	0 24	1	cF. S. iF. bM.	1779
501	Feb. 1	57 (μ) ——	f	4 13	n	0 30	1	vF. vS.	1670
502	—	— — ——	f	6 2	n	0 39	1	vF. S.	1678
503	—	— — ——	f	14 49	s	0 1	1	vF. pL. sp. 2pBst. equil. triang.	1729
504	2	60 (σ) Virg	p	38 27	n	0 34	2	vF. pS.	4591
505	—	64 ——	f	16 1	s	0 39	2	vF. vS. R.	5252
506	—	— ——	f	32 47	n	0 7	1	vF. E. 2′ l.	5356
507	4	82 ——	p	9 23	s	0 4	1	vF. vS. er. 240 rather confir.	5203
508	—	19 Libræ	p	18 52	s	0 27	1	vF. cL. iE. nearly mer.	5729
509	22	5 (β) Virgin	f	49 54	s	0 35	1	vF. vS.	4599
510	24	55 Orionis	f	1 13	n	0 7	1	eF. E. er. probably a patch of st.	2110
511	—	110 Virginis	f	3 11	s	0 25	1	vF. R. precedes I. 128. 7′ ½. and is 5′ n.	5845
512	Mar. 3	17 (β) Cancri	p	14 · 9	n	0 9	1	vF. S. R. mbM. 240 ditto.	2513
513	—	6 (*h*) Leonis	f	2 1	n	0 25	1	eF. vS. stellar. 240 verif.	2914
514	—	26 (χ) Virgin	f	10 4	s	1 8	2	eF. vS. E.	4703*
515	—	— — ——	f	12 19	s	0 26	1	vF. S. E.	4739
516	—	— — ——	f	14 18	s	0 41	1	vF. S.	4773
517	—	— — ——	f	14 43	s	0 48	1	vF. S.	4777
518	19	41 (λ) Hydræ	p	0 28	s	0 5	1	vF. S. R. in the field with λ.	3145
519	24	1 Sextantis	f	1 47	n	0 7	1	vF. pL. vgvlbM. betw. 2 groups of st. np sf.	2948
520	25	27 Hydræ	f	3 9	s	0 51	1	vF. S. E.	2863
521	—	— ——	f	22 39	s	0 45	1	cF. pS. lE.	2979
522	—	14 (ε) Crater	p	34 1	s	2 2	2	cF. pL. iR. lb. near M.	3411
523	—	21 (*q*) Virgin	f	13 23	s	0 38	1	vF. E. sp nf. 4′ l. 3′ b.	4682
524	—	— ——	f	15 14	s	1 59	2	cF. mE. r. 4′ l. ¾′ b.	4700
525	—	49 (*g*) ——	p	14 19	n	1 14	1	vF. vS.	4770
526	—	— ——	p	13 15	n	0 8	1	eF. eS. some little doubt.	4784
527	27	8 Sextantis	p	10 33	s	0 31	3	vF. S. iR. vgbM.	2969
528	—	— ——	p	9 10	s	1 32	1	vF. S. E. nearly mer.	2980
529	—	16 (κ) Crater	p	13 0	s	1 46	1	eF. S.	3591
530	—	— — ——	p	3 32	s	1 30	1	vF. stellar.	3661

III.	1786	Stars.		M. S.		D. M.	Ob.	Description.	N.G.C.
531	Mar. 27	16 (κ) Crater	p	2 47	s	1 32	1	cF. stellar. vlbM.	3667
532	—	— — —	f	1 7	s	0 51	1	vF. lE. vlb. about M.	3693
533	—	24 (ι) —	f	28 31	s	0 59	1	vF. S. iF. time a little inacc.	4114
534	—	— — —	f	33 51	s	0 48	1	vF. pL. of uneq. light.	4177
535	—	— — —	f	40 50	n	0 58	2	vF. pS. iF.	4265*
536	—	68 (ι) Virgin	p	36 17	s	0 33	1	cF. stellar.	4714
537	—	— — —	p	34 23	s	0 39	1	vF. vS. iF.	4748
538	—	— — —	p	31 24	n	0 8	2	eF. S. er.	4794
539	—	— — —	p	5 57	s	1 21	1	vF. vS.	5094
540	28	19 Ursæ	p	15 27	n	1 3	2	vF. S. E. 20° np sf. contains 2 vFst.	2719
541	—	8 Leo. min.	f	9 41	n	0 50	3	cF. S. iR. gbM. r. 1′½ dia.	2955
542	—	21 — —	p	7 55	n	0 8	3	cF. vL. iF. 5′ l. 4′ b. sp. a double st.	3074
543	Apr. 17	11 (s) Virgin	f	37 39	s	1 31	1	eF. pL.	4688
544	—	— — —	f	43 12	s	1 23	1	vF. vS.	4765
545	—	— — —	f	62 44	s	1 9	1	eF. eS. er.	5019
546 }	29	64 —	f	36 17	n	1 4	1	Two. Both vF. vS. r. the place	5382
547 }								betw. them. sp nf. but near. mer.	5386
548	30	43 (δ) —	p	0 31	s	0 30	1	vF. cS. with 240 lE. near vSst.	4799
549	—	84 (o) —	f	9 1	s	1 13	1	eF. vS. stellar. confir. 240.	5329
550	—	109 —	p	5 32	n	1 35	1	vF. S. p. and in a line with 2Bst.	5718
551 }	May 1	31 Bootis	p	23 9	s	0 42	1	Two. Both eF. vS. The place	5546*
552 }								is that of the f. The first p.	5549*
								the last 3 or 4′.	
553	3	50 (σ) Serpen.	p	12 7	s	0 15	1	cF. iF. r. 5′ l. 3′ b.	6070
554	27	3 —	p	21 20	s	1 19	1	vF. S. E. np sf. but nearly mer.	5775
555	June 22	101 Hercul	p	2 55	s	1 28	1	cF. S. lE. iF. r.	6550
556	Sept. 4	71 (ε) Piscium	f	22 10	n	1 24	1	vF. mE. 75° sp nf. 1′½ l.	532
557	18	85 Ceti	p	6 18	n	0 52	1	vF. vS. lE. r. 240 the same.	990
558	20	97 Aquarii	p	14 9	s	0 33	1	eF. cL. iR. 5 or 6′ dia.	7492
559	—	54 Eridani	p	65 18	s	0 59	1	3 vSst. in a line with vF. nebulosity.	1370
560	21	45 Androm	f	16 14	s	0 32	1	vF. S. E. among st.	551
561	—	58 —	p	17 32	s	1 34	1	vF. stellar.	687
562 }	—	— —	p	15 22	s	1 48	1	Four. stellar. unequal. Three in	703
563 }								a row, and the fourth making	704
564 }								a rectangle with them. That at	705
565 }								the angle is much larger.	706
566	—	— —	p	5 4	n	0 14	2	vF. pL. iR.	797
567	—	— —	f	2 29	s	0 12	1	vF. S. lE.	834
568	30	17 Eridani	p	8 17	n	2 17	1	eF. S. iF. among 3 or 4 st.	1308
569	—	— —	f	9 13	n	0 27	1	eF. lE. er.	1397
570	Oct. 17	26 (β) Persei	p	43 39	n	0 51	1	eF. vS. lE.	898*
571	—	— — —	p	42 9	n	0 44	1	eF. stellar. not verified.	910*
572 }	—	— — —	p	32 26	s	0 11	1	Two. Both vF. vS. er. dist. 4′.	980
573 }								the place between them.	982
574 }	—	— — —	f	13 6	n	0 27	1	Two. Both vF. stellar. vlbM.	1293
575 }								but the s. is the brightest and	1294
								largest.	
576	18	12 Androm	p	24 27	s	1 47	1	vF. S. iR. stellar.	7426

III.	1786	Stars.		M. S.		D. M.	Ob.	Description.	N.G.C.
577	Oct. 18	53 (τ) Androm	p	18 55	s	0 8	1	vF. pL. lE. lbM.	477
578	—	28 (ω) Persei	p	2 50	s	1 16	1	vF. vS.	1207
579	24	17 (ι) Andro	p	3 21	n	1 3	1	vF. vS. just f. pBst.	7707
580	—	30 Persei	p	20 43	s	1 3	1	suspected. r. some st. visible.	1138
581	25	40 Arietis	p	8 24	s	0 18	1	vF. E. iF. time inaccurate.	1030
582	—	— ——	p	1 17	s	2 7	1	vF. S. iF.	1088
583	26	10 (α) Triang	p	18 21	s	0 29	1	vF. vS. E. or 3Fst. with vF. Nebul.	780
584	—	35 Arietis	p	0 41	n	0 50	1	vF. S. bM.	1056
585	Nov. 26	48 (ν) Eridani	p	3 33	s	1 2	1	suspected; hazy weather.	1609
586	—	— — ——	p	3 6	s	0 56	2	eF. S. E. nearly par. another suspec. 3' sf. stellar.	1611*
587	28	42 (ξ) ——	f	2 34	n	0 9	1	vF. S. bM. betw. 2 st.	1576
588	—	— — ——	f	7 35	s	1 57	1	vF. S.	1643*
589	—	— — ——	f	10 16	s	1 22	1	vF. cL. iE. nearly par. bM.	1659
590	Dec. 14	8 Leporis	f	9 18	s	0 6	1	suspected eF. stellar. not very doubtful.	1954
591	15	13 (ζ) Eridani	p	4 35	s	0 6	1	eF. stellar. about 1' nf. II. 286.	1242
592 }	20	Neb. II. 447	p	0 6	s	0 2	1	Two. The p. vF. vS. The next	426
593 }			p	0 0	s	0 5	1	eF. eS. and left doubtful.	429
594	—	26 Ceti	f	18 21	s	0 23	1	vF. mE. bM. 3'½ l. 1'½ b.	493
595	21	29 ——	p	28 42	n	1 17	1	vF. S. some st. in it.	196
596	—	44 Hydræ	p	34 21	n	0 50	2	vF. S. lbM. sf. a trapezium of S. st.	3081
597	24	— ——	p	59 13	n	2 44	1	vF. S. R. vglbM.	2921
598	30	59 (c) Leonis	f	2 40	s	1 19	1	eF. S. lE. I could not verify it.	3509*
	1787								
599	Jan. 11	55 (δ) Gemin	f	68 4	s	0 12	1	eF. pL. r.	2595
600	14	30 (η) Leonis	p	11 47	s	0 20	1	vF. S. iR.	3053
601	—	— — ——	p	11 4	n	0 3	1	vF. cS. lE. er.	3060
602	—	29 Comæ	p	12 7	n	0 6	1	vF. cL. vgbM. s. cBst.	4571
603	—	— ——	p	6 16	n	0 11	1	vF. E. np sf. 2'½ l.	4634
604	17	58 Androm	f	2 45	s	0 21	1	vF. stellar. confir. 240.	845
605	Feb. 10	9 (1st μ) Canc	p	3 15	n	0 46	1	vF. S. iF.	2512
606	13	10 (2d μ) ——	f	11 31	s	1 1	2	vF. S. stellar.	2558
607	—	33 (η) ——	p	12 23	n	0 38	1	vF. vS.	2562
608	22	69 (ν) ——	f	2 5	n	0 33	1	eF. S. R. vlbM.	2743
609	—	21 (θ) Crater	f	2 28	n	0 28	1	vF. vS. R. with 240 gbM.	3791*
610	—	65 Virginis	p	33 51	s	0 12	1	cF. pL. E.	4705
611	—	— ——	p	32 29	n	0 50	1	vF. S. no time to verify.	4720*
612	Mar. 11	87 (e) Leonis	f	23 23	s	0 57	1	vF. cS. E.	3952
613	—	44 (k) Virgin	p	1 42	n	0 8	1	vF. E. er.	4843
614	—	— — ——	f	0 57	s	0 49	1	cF. S. iR.	4890
615	17	38 Leo. min.	p	1 27	s	0 27	2	cF. S. er.	3304
616	—	6 Canum	p	34 40	s	1 1	2	vF. cL. iF. 4' dia. 5' s. st. 6 m.	3930*
617	—	— ——	p	27 3	s	1 13	2	eF. pL. iR. 1' dia. or more.	4025
618	—	12 ——	p	3 8	s	1 31	1	eF. vS.	4774
619	—	17 ——	f	11 29	n	0 2	1	vF. S. E. nearly mer.	5107
620	—	— ——	f	26 36	s	0 14	1	cF. E. nearly par. r. ¾' l.	5243
621	—	— ——	f	38 29	s	0 46	1	vF. S. iR. conf. 300.	5305

III.	1787	Stars.		M. S.		D. M.	Ob.	Description.	N.G.C.
622	Mar. 17	12 (λ) Coronæ	f	7 17	s	0 37	1	vF. S. R. discov. in gaging.	6038
623	—	— — —	f	24 7	s	0 19	1	vF. vS. n. 2 st. 300 confir.	6120
624	—	— — —	f	27 24	s	0 9	1	vF. S. bM. discov. with 300.	6137
625	18	10 (n) Ursæ	p	2 41	s	2 27	1	vF. vS. 300. I.C.	2424?*
626	—	— — —	f	7 14	s	0 4	2	vF. S. iF. lbM. r.	2755
627	—	42 Lyncis	p	17 50	s	1 2	2	vF. vS. stellar. 300.	2838
628	—	— —	p	16 48	s	0 9	1	cF. cS.	2844
629 }	—	— —	p	15 24	s	0 8	1	Two. Both vF. vS. dist. 3′. nearly mer. 300.	2852
630 }									2853
631	—	34 (μ) Ursæ	f	3 39	s	1 55	2	vF. S. R. 300.	3237
632	—	47 —	p	2 8	n	0 32	2	cF. vS. lE. mer. gmbM.	3468
633	—	20 Canum	f	1 58	s	0 13	1	vF. S. lbM.	5093
634	—	54 (φ) Bootis	p	1 24	s	0 36	1	vF. vS. conf. 300 sp. 2 vBst.	5966
635 }	—	— — —	f	6 42	n	0 46	1	Two. The nf. vF. vS. verif. 300. The sp. discov. with 300 eF. S. iF.	5992
636 }									5993
637	—	30 (g) Hercul	p	24 18	s	1 5	1	vF. eS. 300. shewed 2 vSst. with nebulosity.	6058
638	—	— — —	p	3 33	s	0 59	1	vF. vS.	6146
639	—	— — —	p	2 53	s	1 24	1	eF. eS.	6150
640	—	— — —	f	0 54	s	1 5	1	vF. vS.	6173
641	—	— — —	f	1 10	s	1 16	1	vF. vS.	6175
642	19	70 Virginis	f	1 37	s	0 26	1	vF. S. iF. time l. inaccurate.	5185
643	—	— —	f	3 43	n	0 5	1	vF. S. lE. just sf. st.	5207
644	—	5 (υ) Bootis	f	25 41	s	0 44	1	vF. vS. E. confir. 300.	5522*
645	—	30 (ζ) —	p	10 16	n	0 17	1	eF. vS. lbM. betw. 2 vFst. 300.	5649
646	—	28 (β) Serpen	f	11 15	n	0 24	1	vF. S. lE.	6018
647	20	43 Lyncis	p	33 11	s	2 16	1	vF. vS. verif. 300.	2759*
648	—	13 Canum	p	36 41	n	0 46	1	vF. E. par. 1′ l.	4359
649	—	— —	f	12 29	n	1 0	1	vF. S. lE.	5025
650	—	— —	f	19 54	n	2 18	1	eF. vS.	5096
651	—	— —	f	27 11	n	1 11	1	vF. S.	5157
652	—	— —	f	29 47	n	0 17	1	eF. vS.	5187
653	—	— —	f	62 59	n	1 34	1	vF. pS. E. mer. 300.	5433
654	April 9	19 —	p	7 36	n	0 51	1	vF. vS. lbM.	4985
655	—	— —	p	* 26	n	2 57	1	vF. pS. lbM. * forgot, but is 5, 6, or 7′.	5003*
656	—	20 —	f	15 21	n	1 15	1	vF. vS. lbM.	5214
657 }	—	— —	f	83 33	n	1 57	1	Two. Both vF. vS. E. in differ. directions. 2 or 3′ dist. par. each. s. Sst.	5730
658 }									5731
659	—	— —	f	117 16	n	0 18	1	vF.vS. r.	5888
660	—	— —	f	119 18	n	1 16	1	eF. cS.	5900
661	—	— —	f	125 25	n	0 44	1	eF. S.	5922
662	11	29 (γ) Virgin	f	2 8	n	0 54	1	vF. pL.	4653
663	—	— — —	f	3 33	n	0 55	1	vF. S. iF.	4668
664	—	— — —	f	6 6	s	0 14	1	vF. S.	4690
665	15	90 (p) —	p	1 48	n	0 23	2	cF. cL. R. vlbM. r. 5′ dia.	5334
666	—	102 (1st υ) —	p	19 18	s	0 55	1	eF. vS.	5392*
667	—	— — —	p	18 53	s	0 33	2	eF. vS. verif. 300.	5400*
668	—	105 (φ) —	p	3 29	s	0 58	1	cF. S. r.	5604

III.	1787	Stars.		M. S.		D. M.	Ob.	Description.	N.G.C.
669	May 7	61 Virgin	p	5 42	n	1 30	1	vF.	5017
670	—	— ——	p	2 45	n	1 46	1	vF.	5047
671	—	8 Libræ	p	9 38	s	1 28	1	cF. S. R. sp and joining 2 Sst.	5716
672	12	19 (λ) Bootis	p	48 58	n	0 41	3	cF. vS. stellar. 300.	5173
673	—	— — ——	p	38 30	n	2 21	1	cF. S. R. or lE.	5256
674	—	— — ——	p	5 52	n	2 31	1	cF. cS. iR.	5500
675	—	38 (2d *h*) ——	p	11 18	n	0 35	1	vF. pS. iF. sp. 2 S. unequal st.	5714
676	15	24 (*g*) ——	p	16 34	n	0 35	2	cF. cS. lE. nearly par.	5520
677	—	— — ——	p	2 56	s	1 15	1	vF. pS. lE.	5622
678 }	—	[1]A Bootis 7 m	f	0 3	n	0 3	1	Two. The p. vF. vS. The f.	5797
679 }			f	0 48	n	0 1	1	eF. eS. [1] See note.	5804
680	—	42 Herculis	p	13 17	n	1 0	2	vF. S. R. lbM. er. near some Sst.	6154
681	16	[2] C Can[m] 6 m.	f	0 44	s	0 15	1	cF. vS. lE. [2] See note.	5303
682	—	— — ——	f	7 35	s	0 7	1	eF. cS. E. sp. Sst.	5361
683	—	— — ——	f	12 35	s	0 25	1	cF. pL. iF.	5403
684	—	— — ——	f	13 49	n	0 34	1	vF. vS. R.	5407
685	—	27 (γ) Bootis	p	19 48	n	1 6	2	vF. cS. R. sbM.	5515
686	—	— — ——	f	8 42	n	0 18	1	eF. cS. lbM.	5732
687	—	— — ——	f	13 27	n	0 23	1	cF. pS. another suspec. 2′ n. 300.	5754
688	—	16 (τ) Coronæ	f	7 34	s	0 48	1	vF. cS. iR.	6104
689	—	67 (π) Hercu	p	20 32	s	0 11	1	eF. cL. iE. nearly par.	6255
690	19	10 Libræ	p	4 6	s	0 43	1	vF. cS. iF. lbM.	5757
691	—	— ——	f	6 51	s	0 59	1	cF. smbM. stellar.	5791
692	Aug. 12	33 (ι) Aquarii	p	5 23	n	0 38	1	eF. E. np sf. 2′ l. 1′ b.	7171
693	Sept. 11	41 ——	p	11 30	n	0 36	1	eF. vS. 360 confirmed it.	7180*
694	Oct. 11	50 (*f*) Cassio	f	90 22	n	0 30	1	vF. vS. iR. bM.	1343
695	Nov. 3	10 Camelop	p	155 0	n	0 53	1	eF. pL. iF. Polar Dist. inaccurate	896*
696	5	17 (ξ) Cephei	p	16 35	s	0 47	2	vF. S. R. lbM. r. 1′ dia.	7139
	1788								
697	Jan. 14	67 Ursæ	f	11 9	n	0 40	3	vF. E. np sf. 5′ l. 1′ b.	4183
698	—	27 (γ) Bootis	p	39 53	n	1 29	2	vF. S.	5337
699	—	— — ——	p	38 20	n	2 8	2	vF. S. iF.	5355
700	Feb. 3	45 (ω) Ursæ	p	14 53	s	1 33	1	cF. L. iE. mb. s. of M. 4′ l. 2′½ b.	3319
701	—	— — ——	p	6 6	s	0 1	1	vF. vS. iF.	3374
702	—	13 Canum	p	42 13	s	0 55	1	vF. vS.	4253
703	5	71 (*o*) Gemin	p	10 3	s	0 46	1	vF. vS. perhaps a patch of st.	2389
704	—	60 Ursæ	p	81 11	s	0 22	1	eF. vS. perhaps a patch of Sst.	3192*
705	—	— ——	p	39 57	s	0 43	1	vF.	3478
706	—	— ——	p	23 49	n	0 38	2	vF. vS. lE. s. of a cBst.	3595*
707	—	63 (χ) ——	f	11 2	n	0 34	2	vF. vS. another susp. sf. eF. eS.	3985
708	6	59 ——	f	30 14	s	0 31	1	vF. vS. in a line with 2 st. nf sp.	4117
709	Mar. 9	21 Lyncis	f	34 50	n	1 41	1	vF. R. vgbM. 2′½ dia.	2500*
710	—	9 (ι) Ursæ	p	45 51	n	0 49	1	vF. iF. 2′½ l. 1′¾ b.	2541
711	—	— — ——	p	41 10	n	1 49	1	eF. E. sp nf. 3′½ l. 2′½ b.	2552
712	—	— — ——	p	4 49	n	1 6	1	eF. cS. r. p. some Fst.	2684
713	—	— — ——	f	25 7	n	1 15	1	cF. cS. lE.	2856
714	—	— — ——	f	24 58	n	1 11	1	cF. cS. lE.	2854
715	—	63 (χ) ——	f	3 26	n	0 39	1	eF. pS.	3906
716	—	— — ——	f	5 2	n	2 26	1	vF. vS.	3922*
717	—	3 Canum	p	14 1	n	0 37	1	cF. mE. nearly mer. 5′ l.	4100

III.	1788	Stars.		M. S.		D. M.	Ob.	Description.	N.G.C.
718	Mar. 9	3 Canum	p	4 6	s	0 51	1	vF. vS.	4218
719 }	—	— ——	p	2 47	s	1 31	1	Two. Both vF. vS. dist. 1′ in	4231
720 }								the same meridian.	4232
721	—	— ——	f	32 1	s	1 21	1	vF. S.	4741
722	11	49 (g) Virgin	p	18 9	s	0 21	1	eF. S.	4708
723	Apr. 1	Neb. II. 728	p	0 25	s	0 2	1	eF. vS.	3577*
724	8	61 Virginis	f	1 43	s	2 23	1	cF. vS. iF.	5087
725	10	60 Ursæ	f	39 40	s	1 14	2	eF. cL. iR. lbM. 3′ dia.	4242
726	—	— ——	f	42 45	s	0 34	2	vF. pS. R.	4288
727	12	35 (σ) Hercul	f	16 11	n	0 14	1	cF. S. E. par.	6239
728	13	42 ——	f	20 46	n	0 54	1	vF. cS. iR.	6283
729	27	19 (λ) Bootis	p	113 28	s	0 3	1	vF. S.	4392
730	28	27 (β) Hercul	f	4 6	n	0 2	1	eF. vS. E.	6186
731	29	27 (γ) Bootis	p	15 47	n	1 16	1	vF. vS.	5536
732	—	— — ——	p	15 33	n	1 22	1	vF. vS. lE.	5541
733	—	— — ——	p	9 25	n	2 4	1	vF. vS.	5598
734	—	— — ——	p	8 52	n	2 8	1	cF. pS.	5603
735	—	22 (τ) Hercul	f	30 17	s	1 2	1	eF. pS. with 300 iF.	6241
736	30	21 (1st ν) Libr	f	7 7	n	1 59	1	vF. pL. E. mer. lbM. 300.	5878
737	May 1	23 (θ) Bootis	f	49 59	s	1 46	1	vF. vS. stellar.	5902
738	25	12 (ι) Draco	f	17 8	n	0 44	1	vF. vS.	5989
739	June 2	14 (η) ——	p	32 47	n	0 57	1	vF. R. vgbM. er. 3′ dia.	6015
740	3	15 (A) ——	p	10 14	s	3 21	1	cF. pL. iE.	6140*
741	6	31 (1st ψ)——	p	5 13	s	0 5	1	eF. stellar. with 300 lE. par.	6434
742	July 8	[1]B Draco 7 m	f	4 25	s	0 27	1	vF. stellar. verif. 300. [1]See Note.	6742
743	30	19 Aquilæ	f	9 24	n	0 26	1	cF. iR. r. 3 or 4′ dia.	6781
744	Aug. 2	51 ——	p	8 8	n	0 29	1	vF. pL. R. vgmbM.	6814
745	Nov. 1	27 (δ) Cephei	f	26 10	s	1 24	1	vF. pL. iF. er.	7423*
746	—	36 Camelop	f	64 5	s	0 38	1	vF. S. R. lbM.	2347
747	Dec. 3	[2]22 Cam Hev	p	37 1	s	0 8	1	cF. pL. iF. mbM. er. some st. visible. [2]See note.	I.C. 2133*

Fourth class. Planetary nebulæ.

Stars with burs, with milky chevelure, with short rays, remarkable shapes, etc.

IV.	1785	Stars.		M. S.		D. M.	Ob.	Description.	N.G.C.
30	May 1	14 Canum	p	6 48	s	0 55	2	Two st. dist. 3′ connected with a vF. narrow nebulosity.	4861
31	Oct. 3	50 Aquarii	f	7 55	s	0 37	1	F. S. stellar. with pL. chev.	7302
32	5	62 (b) Eridani	f	0 35	n	0 21	2	vB. vS. mbM. like a st. affected with irregular burs.	1700
33	—	49 (d) Orion	p	2 33	n	0 28	4	A st. with m. chev. or vBN. with m. nebulosity.	1999

IV.	1785	Stars.		M. S.		D. M.	Ob.	Description.	N.G.C.
34	Dec. 28	40 (2d ϕ) Orio	f	5 41	s	0 12	2	cB. S. nearly R. like a st. with L. dia. with 240 like an ill defined planetary neb.	2022
35	31	9 Hydræ	p	8 19	s	0 14	1	A S st. with a brush sp. FS. it resembles fig. 7. *Phil. Trans.* Vol. LXXIV. Tab. 17. [Plate VII.]	2610
	1786								
36	Jan. 1	60 Orionis	p	11 38	s	0 20	3	A st. affected with vF. extensive m. chev. The st. not quite central.	2071
37	Feb. 15	28 (ω) Draco	f	20 33	s	2 12	1	A planetary neb. vB. has a disk of about 35″ dia. but very ill defined edge. With long attention a vB. well defined R. center becomes visible.	6543
38	24	55 Orionis	f	18 3	n	1 17	2	A cst. affected with vF. m. chev.	2182
39	Mar. 19	2 Navis	p	3 32	s	0 5	1	pB. R. r. within the 46th of the *Connoiss. des Temps* almost of an equal light throughout, 2′ dia. no connection with the cluster, which is free from nebulosity.	2438
40	27	68 (ι) Virgin	p	30 45	s	0 18	1	Suspected. A pBst. with a seeming brush to it np. may be a vS. neb. close to it. No time to verify it.	4804*
41	May 26	14 Sagittarii	p	11 58	s	1 15	1	A double st. with extensive nebulosity of different intensity. About the double st. is a black opening resembling the neb. in Orion in miniature.	6514
42	Sept. 30	51 Ceti	f	7 26	n	0 27	1	A st. about 8 or 9 m. with vF. bran. mer. each branch 1′ l.	676
43	Oct. 17	26 (β) Persei	p	2 48	n	1 54	2	A pBst. with 2 F. branches.	1186*
44	Nov. 28	5 Monocero	p	7 16	s	0 2	1	A st. involved in m. chev.	(2167)*
	1787								
45	Jan. 17	55 (δ) Gemin	f	9 6	s	1 1	2	A st. 9 m. with a pB. m. nebulosity. equally dispersed all around. A very remarkable phænomenon.	2392
46	Feb. 22	99 (ι) Virgin	p	4 38	n	0 57	1	pB. almost cB. vS. stellar. like a star with burs.	5493
47	Mar. 11	44 (k) ——	f	1 48	s	0 46	1	pB. stellar. resembles a st. with a bur all around.	4915
48	18	19 Leo. min.	f	6 32	s	0 17	1	A vF st affected with vF. nebulosity. E. sp nf. 1′ l. 300.	3104
49	Apr. 15	102 (1st υ) Vir	p	6 9	s	0 52	2	pB. stellar. like a st. with a S. bur all around.	5507
50	May 12	77 (κ) Hercul	p	40 13	s	0 28	1	vB. R. 4′ dia. almost equally B. with a F. r. margin.	6229

IV.	1787	Stars.		M. S.		D. M.	Ob.	Description.	N.G.C.
51	Aug. 8	61 (*g*) Sagitt	p	13 56	n	1 23	2	A cB. S. beautiful planetary nebula; but c. hazy on the edges, of a uniform light; 10 or 15″ dia. perfectly R. I shewed it to M. DE LA LANDE.	6818
52	Nov. 3	4 (*d*) Cassio	p	4 0	s	1 6	2	A st. 9 m. with vF. nebulosity of S. extent about it.	7635
53	—	10 Camelop	p	55 42	n	0 11	2	A pB. planetary nebula. near 1′ dia. R. of uniform light and pretty well defined, with 360 magnified in proportion; but still the borders pretty abruptly defined, and a little elliptical.	1501
	1788								
54	Jan. 14	67 Ursæ	f	7 32	s	0 30	1	cB. S. N. with F. chev.	4143
55	Feb. 6	34 Lyncis	p	28 4	n	0 2	2	pB. R. almost of an even light throughout, approaching to planetary, but ill defined and a little fainter on the edges, ¾ or 1′ dia. p. 1′ pc st.	2537
56	—	59 Ursæ	f	25 11	n	0 56	1	cB. iR. cBNM. with extensive chev. 5′ dia.	4051
57	June 11	35 (σ) Hercul	f	34 27	s	0 18	2	A vS. F. st. involved in eF. nebulosity.	6301
58	Nov. 25	24 Cephei	f	116 28	n	0 2	1	A st. 9 m. surrounded with vF. m. nebulosity. The st. is either double, or not R. Less than 1′ dia.	40

Fifth class. Very large nebulæ.

V.	1785	Stars.		M. S.		D. M.	Ob.	Description.	N.G.C.
25	Nov. 27	18 Ceti	f	1 30	n	1 2	1	Four or five pL. st. forming a trapezium of about 5′ dia. The inclosed space is filled up with faintly terminated m. nebulosity. The st. seem to have no connexion with the nebulosity.	246
26	Dec. 7	18 Leo. min.	p	8 7	n	1 1	2	cB. mE. par. 8′ l. 3′ b.	3003
27	26	15 Monocero	p	0 12	s	0 6	2	Some pBst. 7 or 8′ sp. 15th Monoc. are involved in eF. m. nebulosity which loses itself imperceptibly.	2264

V.	1786	Stars.		M. S.		D. M.	Ob.	Description.	N.G.C.
28	Jan. 1	48 (σ) Orion	f	2 46	n	0 44	2	Wonderful black space included in remarkable m. nebulosity, divided in 3 or 4 large patches; cannot take up less than ½ degree, but I suppose it to be much more extensive.	2024
29	2	61 Ursæ	f	45 38	s	0 40	1	eF. vL. vlbM. r. 10′ l. 8 or 9′ b.	4395 4401
30	18	42 } c Orionis 45 }	p	0 0	n	0 0	2	The 1st and 2d c Orionis, and the stars about them, are involved in eF. unequally B. m. nebulosity.	1977
31	31	44 (ι) ——	p	0 0	n	0 0	2	ι Orionis with its neighbouring st. are involved in eF. m. nebulosity to a great extent.	1980
32	Feb. 1	28 (η) ——	p	17 26	s	1 4	2	cB. vL. m. diffused and vanishing. near and sf. Bst.	1788
33	—	— — ——	f	1 26	s	0 7	1	Diffused eF. m. nebulosity. The means of verifying this phænomenon are difficult.	1908
34	—	46 (ε) Orionis	p	0 0	n	0 0	1	I am pretty certain ε Orionis is involved in unequally diffused nebulosity.	1990
35	—	36 (υ) —— 56 ——	f p	3 39 2 16	s n	0 40 0 28	4	Diffused m. nebulosity, extending over no less than 10 degrees of PD. and many degrees of RA. It is of very different brightness, and in general extremely F. and difficult to be perceived. Most probably the nebulosities of the 28th, 30, 31, 33, 34, and 38th of this class are connected together, and form an immense stratum of far distant stars, to which must also belong the nebula in Orion.	.. *
36	Oct. 17	35 (ν) Andro	p	9 8	s	0 20	2	vF. vL. E. nearly mer. or a little from np sf. about 20′ l.	206
37	24	57 Cygni	f	5 1	s	1 1	1	vL. diffused nebulosity. bM. 7 or 8′ l. 6′ b. and losing itself vg. and imperceptibly.	7000
38	Dec. 20	19 (β) Orion	f f	11 9 11 35	n s	1 19 0 52	1 1	Strongly suspected nebulosity of v. great extent. Not less than 2° 11′ of PD. and 26″ of RA. in time.	1909
39	21	11 (β) Crater	p	8 15	s	0 17	2	vF. mE. near the par. or about 10° sp nf. vgbM. 8′ l. 3′ b.	3511*
40	—	— — ——	p	7 49	s	0 26	2	vF. mE. 15° sp nf. vlbM. about 7′ l. 4′ b.	3513

V.	1787	Stars.		M. S.		D. M.	Ob.	Description.	N.G.C.
41	Mar. 17	6 Canum	p	8 27	s	1 12	1	vB. E. 60° sp nf. 20′ l. 2′ b.	4244
42	20	13 ——	p	18 39	n	1 48	1	vB. mE. sp nf. but near the par. mbM. 16′ l.	4631
43	1788 Mar. 9	3 ——	p	0 38	s	1 41	3	v. brilliant. BN. with Fm. bran. np sf. 15′ l. and to the sf. running into vF. nebulosity extending a great way. the N. is not R.	4258
44	Nov. 1	36 Camelop	f	84 33	n	0 23	2	cB. R. vgbM. BN. 6 or 7′ dia. with a F. branch extending a great way to the np side; not less than ½ degree. and to the n. or nf. the nebulosity diffused over a space, I am pretty sure, not less than a whole degree.	2403

Sixth class. Very compressed and rich clusters of stars.

Additional abbreviations. Cl. Cluster. sc. scattered. com. compressed. co. coarsely.

VI.	1785	Stars.		M. S.		D. M.	Ob.	Description.	N.G.C.
20	Oct. 27	18 (ε) Pis. Aust	f	133 24	n	0 23	2	cB. iR. 8 or 9′ dia. a great many of the st. visible, so that there can remain no doubt but that it is a Cl. of vS. stars.	288
21	Dec. 7	25 Gemino	f	2 15	s	1 15	1	A v. rich and v. com. Cl. st. of about 5′ dia. some of the largest st. are in a row.	2266
22	1786 Feb. 1	31 Monocero	p	30 4	n	1 20	4	A beautiful Cl. of much com. st. consid. rich. 10 or 12′ dia. C. H. discovered it in 1783.	2548
23	June 27	46 (υ) Sagitt	p	49 15	s	0 42	1	A beautiful Cl. of vS. st. of various sizes. 15′ dia. very rich.	6645
24	Oct. 17	58 (ν) Cygni	f	15 56	n	1 18	2	A v. com. and v. rich Cl. of eSst. about 6′ l. 4′ b. nearly par.	7044
25	Dec. 11	27 (κ) Persei	f	5 55	n	2 25	2	A beautiful com. and rich Cl. of S. and L. st. 7 or 8′ dia. the L. st. arranged in lines like interwoven letters.	1245
26	—	53 (d) ——	f	13 34	s	1 13	1	A vF. and v. com. Cl. of eS. st. near 4′ dia.	1605
27	27	22 Monocero	p	20 9	n	0 51	1	A v. beautiful Cl. of much com. S. and L. st. above 20′ dia.	2301

VI.	1787	Stars.		M. S.		D. M.	Ob.	Description.	N.G.C.
28	Jan. 11	75 (*l*) Orionis	f	21 25	n	1 2	1	A Cl. of e. com. and eS. st. c. rich. iF. the f. and most com. part R.	2259
29	Oct. 14	3 Lacertæ	p	7 52	n	2 7	1	A com. Cl. of eS. st.	7245
30	18	7 (ρ) Cassiop	f	3 10	s	0 46	3	A beautiful Cl. of v. com. Sst. v. rich. C. H. discovered it 1783.	7789
31	Nov. 3	37 (δ) ——	f	19 48	n	1 2	1	A beautiful Cl. of pL. st. near 15′ dia. cons. rich.	663
32	1788 Sept. 21	80 (1st π) Cyg	p	11 26	n	0 28	1	A beautiful Cl. of p. com. st. 8 or 9′ dia. nearly R. c. rich.	7086
33	Nov. 1	7 (χ) Persei	f	1 7	s	0 22	1	A v. beautiful and brilliant Cl. of L. st. v. rich. the M. contains a vacancy.	869
34	—	— — ——	f	4 0	s	0 23	1	A v. beautiful, brilliant Cl. of L. st. iR. v. rich. near ½ degree in dia.	884
35	26	15 (κ) Cassiop	p	1 22	s	1 26	1	A S. Cl. of vF. and e. com. st. about 1′ dia. The next step to an er. neb.	136

Seventh class. Pretty much compressed clusters of large or small stars.

VII.	1785	Stars.		M. S.		D. M.	Ob.	Description.	N.G.C.
18	July 17	12 Vulpeculæ	p	7 56	n	0 44	1	An E. Cl. of i. sc. st. of various sizes. c. rich.	6823
19	30	21 Aquilæ	p	5 49	n	1 55	1	A p. com. Cl. of p. sc. st. of var. sizes, magnitudes, and colours. iF. and unequally com. 12 or 15′ dia.	6755
20	Nov. 1	7 Monocero	f	1 3	n	0 35	3	A beautiful Cl. of p. com. and equally sc. st. 10 or 12′ dia.	2215
21	Dec. 26	109 (*n*) Tauri	p	14 59	n	1 37	1	A Cl. of p. com. st. with many eS. st. mixed with them.	1758
22	28	13 Monocero	f	2 48	n	0 21	1	A S. Cl. of p. com. vS. st.	2254
23	30	31 (η) Canis	f	32 6	s	0 39	1	A com. Cl. of pL. st. c. rich.	2489
24	1786 Jan. 1	60 Orionis	p	5 9	s	0 9	2	A Cl. of p. com. pS. sc. st. with many eS. suspec. betw. them 7 or 8′ dia.	2112
25	27	8 Monocero	p	11 46	n	0 49	1	A Cl. of p. com. st. of several sizes. 4 or 5′ dia. with extensively straggling ones.	2186
26	30	6 ——	f	8 59	n	1 7	1	A Cl. of eS. and pm. com. st. with a few L. but not rich. in the shape of a hook.	2225

VII.	1786	Stars.		M. S.		D. M.	Ob.	Description.	N.G.C.
27	Feb. 24	11 Monocero	f	42 13	s	1 21	2	An i. Cl. of eS. st. c. com. 9 or 10′ l. 4 or 5′ b. with an extending bran. towards sp. C. H. discov. 1783.	2349
28	Mar. 19	2 Navis	p	8 23	n	0 47	1	A Cl. of pS. st. p. rich. 15′ dia.	2423
29	Apr. 30	5 (ρ) Scorpii	p	7 14	n	0 38	1	A Cl. of vS. st. p. rich 6′ l. 4′ b. in the form of a parallelogram.	5998
30	May 26	14 Sagittarii	p	1 35	n	0 9	1	A Cl. of pS. sc. st. above 15′ dia.	6568
31	—	— ——	f	1 29	s	0 25	1	A Cl. of vS. and p. com. st. c. rich. 2 or 3′ dia.	6583
32	Sept. 21	58 Androm	p	10 49	s	0 8	4	A vL. co. sc. Cl. of vL. st. iR. v. rich. takes up ½ degree, like a nebulous st. to the naked eye.	752
33	Oct. 18	11 (μ) Aurigæ	f	6 32	n	0 54	1	A Cl. of p. com. pS. st. c. rich. contains one L. the rest are all of a size.	1857
34	Dec. 11	13 (α) ——	f	9 7	n	0 32	1	A Cl. of vF. and vSst. p. com. but not rich. iF. 3′ dia.	1883
35	24	70 (ξ) Orionis	f	15 53	s	1 29	1	A Cl. of S. pm. com. st. with suspected m. nebulosity.	2224
36	26	18 Monocero	p	3 48	n	1 0	1	A Cl. of v. sc. st. c. rich. and of great extent.	2270
37	27	77 Orionis	f	12 24	n	0 55	1	A Cl. of v. com. eSst. c. rich. 3 or 4′ dia. most com. M.	2262
38	—	22 Monocero	p	7 39	n	1 31	2	A beautiful Cl. of vSst. of several sizes. c. com. and rich M. 10 or 12′ dia.	2324
	1787								
39	Jan. 17	21 (σ) Aurigæ	f	3 25	s	2 6	1	A p. com. Cl. of Sst. 4′ dia.	1907
40	Oct. 14	3 Lacertæ	p	38 31	n	1 35	1	A Cl. of Sst. of several sizes. 3 or 4 dia. p. rich. like a forming one.	7128
41	—	— ——	f	5 8	n	0 2	2	A S. Cl. of st. p. com. e. rich in vS. st. The com. part 4 or 5′ dia.	7296
42	18	24 (η) Cassio	f	29 41	n	0 26	2	A brilliant Cl. of L. and vS. st. c. rich.	457
43	Nov. 3	1 ——	p	11 41	n	1 25	1	A S. Cl. of vSst. c. com. and p. rich.	7419
44	—	— — ——	f	4 34	n	1 8	2	A Cl. of p. com. pLst. c. rich. The st. arranged chiefly in lines from sp. nf.	7510
45	—	37 (δ) ——	p	9 29	s	1 28	2	A S. p. com. Cl. of st. not rich. iF. like a forming one.	436
46	—	— — ——	f	17 23	n	1 44	2	A S. Cl. of pL. st. c. rich.	654
47	—	10 Camelop	p	55 40	n	1 37	2	A Cl. of st. p. rich and c. com. lE. 3 or 4′ dia. iF.	1502
48	9	32 Cassiop	f	17 1	s	1 40	1	A com. Cl. of some pL. and many vS. st. iR. 6 or 7′ dia.	559

VII.	1787	Stars.		M. S.		D. M.	Ob.	Description.	N.G.C.
49	Nov. 9	45 (ϵ) Cassiop	p	11 8	n	0 20	1	A Cl. of some cL. st. and many eS. so as hardly to be seen. The Lst. arranged in circular order 3 or 4′ dia.	637
	1788								
50	Sept. 27	81 (2d π) Cyg	p	22 13	s	1 14	1	A few Sst. with suspected nebulosity. with 300 many vS. st. intermixed with the former, so as to make a Cl.	7067
51	Oct. 19	71 (*g*) ——	p	5 49	s	0 9	1	A p. com. Cl. of pS. st. c. rich iR. 5 or 6′ dia.	7062
52	—	— — ——	p	0 42	n	0 34	1	An extensive Cl. of Lst. c. rich above 20′ dia.	7082
53	—	73 (ρ) ——	f	30 41	n	0 48	2	A L. Cl. of p. com. cLst. above 15′ dia. c. rich.	7209
54	Nov. 1	36 Camelop	f	29 1	n	0 16	1	A vF. patch. or S. Cl. of eSst.	2253*
55	23	32 (ι) Cephei	f	57 34	n	1 47	3	A Cl. of cS. st. iF. p. rich and com. contains a vacancy M.	7762

Eighth class. Coarsely scattered clusters of stars.

VIII.	1785	Stars.		M. S.		D. M.	Ob.	Description.	N.G.C.
41	Dec. 7	98 (*k*) Tauri	f	12 11	s	0 54	1	A co. Cl. of st. or projecting point of the m. way.	1802
42	—	125 ——	p	1 22	s	0 4	2	A Cl. of co. sc. st. above 15′ dia. The st. nearly of a size and equally sc.	1996
43	26	109 (*n*) ——	p	15 30	n	1 29	1	A Cl. of v. co. sc. Lst. join. to VII. 21.	1750
44	28	4 (γ) Can. min	f	0 38	s	1 54	1	A Cl. of v. co. sc. Lst. form a cross. not rich.	2394
45	31	6 Navis	p	32 48	s	0 1	1	A co. sc. Cl. of st. not rich.	2358
46	—	— ——	p	10 18	n	0 49	1	A vL. but co. sc. Cl. of st.	2430
47	—	— ——	p	10 27	n	0 39	1	A Cl. of sc. st. or the m. way crowded with st. of equal size and colour.	2428
	1786								
48	Jan. 1	78 Orionis	f	10 59	s	1 9	1	A Cl. of v. sc. st. of various sizes. above ½ degree of extent.	2260
49	3	[1]B Gemi. 6 m	p	33 23	n	0 35	1	A Cl. of co. sc. Lst. not rich. [1] See note.	2240
50	27	8 Monocero	f	10 58	n	0 49	2	A Cl. of st. arranged in a broad row. 25′ l. 6 or 8′ b. not v. com. but p. rich.	2252
51	Feb. 23	11 ——	f	25 25	s	0 1	1	A Cl. of v. sc. st.	2306
52	Mar. 19	2 Navis	p	12 16	n	1 32	1	A Cl. of vL. co. sc. st. not rich.	2413

VIII.	1786	Stars.		M. S.		D. M.	Ob.	Description.	N.G.C.
53	June 27	46 (υ) Sagitta	p	82 10	s	1 4	1	A Cl. of sc. Sst. 8′ dia. not v. rich	6507
54	—	— — —	p	71 19	s	0 25	1	A co. sc. Cl. of cLst. The place is that of a S. triangle.	6561
55	—	— — —	p	64 17	s	0 23	1	A co. sc. Cl. of Lst.	6596
56	Oct. 17	37 (γ) Cygni	f	0 53	n	0 32	1	A S. Cl. of co. sc. st. of various sizes. E. like a forming one.	6910
57	—	58 (ν) —	f	8 47	n	0 20	1	A Cl. of co. sc. pS. st. of several sizes. not rich.	7024
58	24	57 —	f	3 19	n	0 16	2	A Cl. of pL. sc. st. not v. rich.	6997
59	—	59 Persei	f	7 59	n	0 21	1	A Cl. of co. sc. pL. st. not v. rich	1664
60	Nov. 26	19 Monocero	p	5 3	s	0 23	1	A Cl. of pL. sc. st. not v. rich. may be a projecting point of the m. way.	2311
	1787								
61	Jan. 17	21 (σ) Aurigæ	p	16 38	s	0 30	1	A Cl. of co. sc. Lst. iF. not rich. like a forming one.	1778
62	Sept. 19	35 (γ) Cephei	p	4 43	s	4 50	2	A Cl. of co. sc. Lst. not rich. but the st. are brilliant. one 7 m.	7708
63	Oct. 16	21 (ζ) —	f	1 21	s	0 56	1	A S. Cl. of pL. st.	7234
64	Nov. 3	27 (γ) Cassiop	f	11 12	n	0 53	2	A forming cluster of p. com. st. C. H. disc. 1783.	381
65	—	37 (δ) —	f	17 56	n	0 29	2	A S. Cl. of Sst. not v. rich. C. H. 1783.	659
66	—	45 (ε) —	f	47 9	s	1 58	2	A Cl. of co. sc. cLst. 8 or 10′ dia. one 7 m, near M.	1027
67	9	17 (ξ) Cephei	p	10 0	s	2 0	1	A Cl. of co. sc. L. and S. st. 7′ dia. like a forming one.	7160
68	12	41 Aurigæ	p	8 57	n	1 9	1	A S. Cl. of sc. st. not rich. one 7 m. towards the n. but this does not seem connected with the Cl.	2126
69	Dec. 3	18 Androm	p	8 59	s	1 20	1	A Cl. of co. sc. pL. st. one 8 m. in the sf. part.	7686
	1788								
70	Feb. 3	41 (υ) Persei	f	46 17	n	1 28	1	A Cl. of co. sc. Lst. p. rich above 20′ dia.	1582
71	Mar. 4	58 Aurigæ	p	1 22	s	0 44	1	A Cl. of co. sc. pL. st. p. rich. the place is that of a double st. of the 3d class.	2281*
72	July 30	62 Serpentis	p	27 26	n	0 6	3	A Cl. of co. sc. Lst. C. H. 1783.	6633
73	—	59 (ξ) Aquilæ	p	4 2	s	0 34	1	A Cl. of co. sc. st. with one pBst. M.	6828
74	Sept. 21	80 (1st π) Cyg	p	34 12	s	0 12	1	A Cl. of co. sc. Lst. not rich 6′ dia	7031
75	26	3 Lacertæ	p	7 29	s	2 21	2	A Cl. of co. sc. Lst. lE. sp nf. 16′ l.	7243
76	27	59 (1st f) Cyg	p	4 1	s	0 7	1	A st. 6 m. surrounded by many cst. forming a brilliant sc. Cl. the Lst. not M. but f.	6991
77	Nov. 1	27 (δ) Cephei	f	17 23	s	0 22	2	A Cl. of co. sc. st. 8′ dia. C. H. 1787.	7380
78	26	15 (κ) Cassio	f	10 56	s	1 8	2	A Cl. of v. co. sc. Lst. take up 15 or 20′. C. H. disc. 1784.	225

Notes to some nebulæ and clusters of stars. [By the Author.]

I. 138. The number refers to DE LA CAILLE's southern catalogue in the *Cœlum Australe Stelliferum.*

I. 190. A star of the sixth magnitude, not contained in any catalogue. I have called it C Canum Venaticorum. It follows FL. 17, Can. Ven. 37′ 34″ in time, and is 0° 2′ more south than that star.

II. 566. See the note to I. 138.

II. 638. See the note to I. 138.

II. 697. See the note to I. 190.

III. 678. A star of the 7th magnitude, not contained in any catalogue. I have called it A Bootis. It follows FL. 39 Bootis 6′ 56″ in time, and is 0° 55′ more north.

III. 681. See the note to I. 190.

III. 742. A star of the 7th magnitude, not contained in any catalogue. I have called it B Draconis. Its place very coarsely is R.A. 18 h. 47′. P.D. 41°¾.

III. 747. See Mr. WOLLASTON's general catalogue. Zone 20°.

VIII. 49. A star of the 6th magnitude, not contained in any catalogue. I have called it B Geminorum. Not having settled its place, I can only give it in a coarse way, R.A. about 6 h 52′ 4″. P.D. about 55° 17′.

P.S. The planet Saturn has a *sixth satellite* revolving round it in about 32 hours, 48 minutes. Its orbit lies exactly in the plane of the Ring, and within that of the first satellite. An account of its discovery with the forty-feet reflector, and a more accurate determination of its revolution and distance from the planet will be presented to the Royal Society at their next Meetings.

WILLIAM HERSCHEL.

[Notes to the Second Catalogue of Nebulæ and Clusters.

I. 94. April 28, 1785: mE. nearly in the meridian. March 19, 1787: E. from sp. to nf. According to h. it was E. in the parallel, and in G.C. he says: "Surely it does not rotate?" "Nearly in the meridian" is no doubt merely a slip for "nearly in the parallel." Pos. of E. 83° (Dreyer, 1891).

I. 103. This position is to be corrected by $+34^s +20'$. It is the only neb. observed that night, the first occasion when H. used "a new P.D. machine contrived so as to shew the P.D. of the tube in every situation." But the setting part was not finished, and the two first and three last stars differ 10′ in zero of P.D.; ξ Aquilæ would give P.D. of neb. 11′ greater.

I. 109. In *P.T.* the $\Delta\alpha$ is given as $7^m\ 17^s$. This was from Sw. 593, Sept. 19, 1786, and made the R.A. 52^s too great. The error was in the obs. of 12 Eridani, which differs 1^m from ϵ Pisc. Austr. Two other obs. in Sw. 466 and 686 give the correct R.A.

I. 113. A second and better obs. in Sw. 549, Mar. 28, 1786, 40 Lyncis, p. $1^m\ 11^s$, s. 39.

I. 119. In Sweep 498, Dec. 28, 1785, this precedes I. 32 by 49^s, and is 27′ n. of it. h. has one obs., B., L., R., gbM., the R.A. is marked $\pm$. d'A. at Leipzig (4½-inch refractor) "seen, but eF." Not mentioned by d'A. (at Copenhagen) and Schultz, who both observed I. 32; not found by Winnecke and Bigourdan; not on M. Wolf's plate (List II.). Has it disappeared?

I. 152. A second obs. (Sw. 591, Sept. 18, 1786) describes it as vB., vS., lE., vBN. But the neb. is in reality only pB., second class.

I. 184. Suspected of variability by h. H. has only this one obs.

II. 403. R.A. 20^s too small. A correction to centre of field of -34^s was applied. Two later obs. give the correct place.

II. 450–451. A better obs. in Sw. 478, Nov. 27, 1785, 74 Aquarii f. $6^m\ 37^s$, s. 1° 13′.

II. 452. In Sw. 451 18 Ceti differs about 40^s from several other stars, and the R.A. of II. 452 is therefore 37^s too small. The neb. was 8^m 54^s p., 19′ s. of a star 6 mag. =P. $0^h{\cdot}198$, which gives the correct R.A.

II. 464. The neb. is exactly 1° north of the place given. In the Sweep (462) the P.D. of the neb. and of 44 Eridani are both given as 89° 11′ $-1\frac{1}{2}'$.

II. 481. P.D. 7′ too small. Nothing wrong in sweep, stars agree.

II. 527–528. Relative positions correct, but R.A. 40^s too great. Only the second one seen on Nov. 28, 1786 (=h. 334).

II. 545. See II. 684.

II. 546–547. Are identical with II. 57–58. See above, p. 296.

II. 562. R.A. 33^s too great, probably a correction to the meridian was forgotten. Two later obs. give exactly the correct α.

II. 568–571. There are six nebulæ between 12^h $12^m{\cdot}2$ 83° 45′ and 12^h $13^m{\cdot}2$ 83° 52′ (1860), while there are none at or near 82° 50′. There can therefore be no doubt that the quadrant was read 1° wrong, and that the four nebulæ are 26′ south of 11 Virginis and not 34′ north of it.

II. 573. The place agrees with that of II. 571 after applying the above-named correction.

II. 576. R.A. 36^s too great. Several neb. in Sweep 557 have a correction to centre of field of -35^s; perhaps this one should also have had one.

II. 594. There is no nebula in this N.P.D., so that II. 594 is doubtless identical with II. 458 in exactly the same R.A. and 1° north.

II. 626. Not found by Bigourdan. Should probably be rejected, together with III. 88 and III. 598, the only other neb. this night, as there was fog " which indeed was so strong as to make everything swim about me."

II. 627. R.A. by Sw. 683 29^s too great, P.D. $3\frac{1}{2}'$ too great. But in the interval between the star and the neb. H. had discovered two satellites of Uranus, whereby the telescope may have been slightly disturbed. Two obs. of Feb. 1787 give 10 Cancri f. 6^m 32^s s. 0° 13′ and f. 6^m 28^s s. 0° 11′.

II. 637. In the Sweep (709) a correction of -105^s has been applied to the transit to reduce it to the centre of the field, but it was apparently overestimated, as the resulting R.A. is 69^s less than that of Harvard No. 372, same P.D., eF., eS., cE. 55°. They are no doubt identical.

II. 647. A much better obs. in Sw. 1015, May 30, 1791, 25 Herculis f. 2^m 14^s, n. 2° 0′.

II. 648 and II. 675. These were observed together in Sw. 725, Apr. 9, 1787, II. 675 p. 36^s, 10′ s., which does not agree too well with Bigourdan's obs. But both have large reductions to the meridian of $+20^s$ and -40^s. The place of II. 648 in *P.T.* is from Sw. 718, March 18, 1787.

II. 659. It *is* " np. the large, following one" [V. 42], as stated in the sweep, but the Polar Distances 56° 24′ and 56° 23′ contradict this. The figure 4 has been written over another, possibly a 3. $\Delta\delta =$ $1'{\cdot}8$. No change, as d'A. half suspected.

II. 675. See II. 648.

II. 679–680. The place is from Sw. 726, Apr. 11, 1787, where both have the same P.D., but with a correction of $-3'$ to the p. one, while in reality the f. one is 3′ n. of the p. one ($\Delta\alpha = 5^s$). Sw. 709 has: " Two, the place is that of the most s., cL., pB., bM., the smallest about 6^s p. and 4′ s., F., S., iF. 90 Virg. p. 24^m 28^s s. 0° 5′."

II. 684. In Sweep 727, Apr. 11, 1787, H. saw two nebulæ, both described as pB., S., iE., 4 Serpentis p. 6^m 4^s n. 0° 11′, and p. 6^m 6^s n. 0° 7′, noting that one was =II. 545. This agrees with d'A. and Bigourdan, but the north one is not pB. but eF. Perhaps the description of the south one was by a mistake given twice. N.G.C. 5865 =5868. Only one seen at Birr Castle and by h.

II. 696. Sw. 736. The transit must have been entered 1^m too late, as the R.A. is about 1^m too great and $\Delta\alpha$ from I. 189 following is 0^m 17^s instead of 1^m 20^s.

II. 719. Second obs., Sw. 937, Mar 10, 1790. " In a line with a np. star, pB., cL., iR., vgbM., 70 Gemin. f. 34^m 29^s n. 1° 25′." This gives the same R.A. as Sw. 803, but 6′ less in P.D. R.A. in N.G.C. is 1^m greater (h., one obs.). Not found by Bigourdan twice.

II. 728. Also observed in Sw. 822, Apr. 1, 1788, as f. 3^m 39^s, 32′ n. of a * 8 m., which is D.M. $+48^\circ{\cdot}1919$. This agrees well with the first obs.

II. 743. Not found by Bigourdan in this place, but 1^m 6^s preceding, in same P.D. In the sweep the transit is 12^h 2^m 53^s $+20^s$, in which there is probably an error of 1^m.

II. 753. There must be an error of 1^m in the transit. A second observation of May 24, 1791 (Sw. 1007), 20 γ Herculis f. 10^m 31^s n. 0° 37′ agrees perfectly with Schultz, Schönfeld, Engelhardt.

II. 756–762. In Sw. 842, the only one on May 5, 1788, no Flamsteed stars were observed, and II. 757, determined in Sw. 843 (May 25), was used instead of a comparison star. Adopting its place from modern observations, all the others (II. 756, I. 215, II. 758–762) agree well. There were, however, two stars observed thus :—

$15^h\ 26^m\ 13^s$	32° 41′	II. 762
27 54	34 45	6 m. not in Fl. [=G. 2257]
45 0	33 37	6 m. not in Fl. [=G. 2288]

To the observation of II. 757 in Sweep 843 a reduction to the centre of the field of -57^s was applied, which was evidently too great and caused the error of -42^s, by which it differs from modern observations.

III. 377–378. "Two, the place is that of the most n., which is the largest and brightest and is vF., pS., the most south eF., vS., but twilight is too strong." April 27, 1785, "93 Leonis p. $4^m\ 4^s$, 16′ s. F., S., lbM. I suspected a still fainter and smaller about 5′ sp." Places agree perfectly, and there can therefore be no doubt that III. 378 = h. 962 and III. 377 = h. 961 (and not = h. 966 and 962).

III. 382–384. A second obs. is: Sw. 671, Dec. 27, 1786: two, the place is that of the most north which is the largest, both vF.; 20 Comæ p. $23^m\ 54^s$ s. 0° 17′. This gives for 1860 $12^h\ 58^m\ 48^s$ 68° 37′, while Auwers found from the first $59^m\ 4^s$, 35′. As there are many nebulæ close to this place, it is difficult to say which were seen by H. In 1785 he probably saw N.G.C. 4093, 95, 98, and in 1786, 4093 and 4095. 4099 is probably superfluous.

III. 397. *P.T.* gives $\Delta\alpha = 3^m\ 36^s$ (Sw. 403), but this makes the R.A. 29^s greater than that of h. Probably a correction to centre of field was forgotten. Sw. 699, 13 Feb. 1787, has the same star, f. $3^m\ 8^s$ n. 7′, which is correct.

III. 407–408. Results in *P.T.* are from Sw. 407, May 3, 1785, but a better star in this sweep is P. XIII. 51, "very accurate," f. $16^m\ 12^s$, 31′ n. and f. $16^m\ 30^s$, 33′ n. But III. 408 is not on the parallel of III. 425 but 7′ more north. Sw. 405 gives only the place of III. 407, 25 Canum p. $3^m\ 10^s$ s. 1° 35′.

III. 425. A better star is P. XIII. 51, f. $17^m\ 4^s$, 33′ n.

III. 437. R.A. is 30^s too great (d'A. and Bigourdan). It is the only nebula observed this night, between 17 and 26 Piscium, the $\Delta\alpha$ of which is correct.

III. 440. R.A. 22^s too great. 42 Ceti gives the same place.

III. 456. R.A. 1^m too great. Stars in sweep agree well. Error of observation.

III. 479. Only a small cluster of vF. stars.

III. 480 was only seen in Sweep 498 after II. 137, "L., vF., would not have been seen if it had not been for the preceding." The resulting place is absolutely coincident with that of Sn. 8, vF., pL., E., = I. C. 3115.

III. 492. Sw. 726, April 11, 1787, looked for the nebula of the 507th sweep, but did not see it. Auwers' place agrees exactly with that of h. 1261, but the latter is called F., S., R., * nr. No later obs.

III. 514. Sw. 546, Mar. 25, 1786, S., mE., 21 *q* Virginis f. $15^m\ 30^s$ n. 0° 19′. Both obs. agree well with Bigourdan's place.

III. 535. Sw. 825, April 8, 1788 [$31\frac{1}{2}^m$ after the transit of II. 553], looked for the neb. of 548 Sweep. It is not stated whether it was found, but on the register sheet there is added in pencil: "but could not find it." It is beyond a doubt = N.G.C. 4265 (Swift III.). Howe saw only one, but gives no place.

III. 551–552. The distance was underestimated, if they were h. 1770 and 1772, $\Delta\alpha$ of which is 30^s, the second being 11′ south. A second obs. of May 12, 1793 (Sw. 1042), has only one, pB., S., NM., Bootis 19 Hev. p. $5^m\ 29^s$ n. 1° 33′. This agrees perfectly with the correct place of h. 1772 and differs only 2′ from the first obs. It is, however, very probable that H. in 1786 observed N.G.C. 5542 and 5546 (h. 1770), 4′ pf., but in that case his R.A. was 30^s too great and P.D. 12′ too great. He adds: "A little inaccurate. I would not stay to verify it properly, so that there remains some little doubt."

III. 570. R.A. is 40^s too great (Bigourdan). The neb. before (V. 19) and the next four have corrections to centre of field of -42^s to -45^s applied to their transits. Perhaps this should also have been applied to III. 570. III. 571 observed in the same sweep immediately after; place correct.

III. 586. Two observations in excellent agreement, but the suspected object 3′ sf. was only seen on the second night (Nov. 28, 1786). Nobody has seen anything 3′ sf.; probably it was 3′ nf., and the object was N.G.C. 1613.

III. 588. Suspected of variability. h. eF., d'A. F. or vF., Dreyer 1877 F., Roberts 1903 B., pL. (*M.N.*, 63, p. 301).

III. 598. Not found by Bigourdan. See above, under II. 626.

III. 609. R.A. is $\frac{1}{2}^m$ too small, as the neb. is 6′ due north of D.M. −8°·3211.

III. 611. Verified in Sw. 709, March 11, 1787, "vF., vS., lE., er., may be only a few vF stars, 38 Virg. p. 2^m 35^s s. 0° 36′."

III. 616. The P.M. of the star G. 1830 has now carried it to a position sf. the neb.

III. 625. Not found by Bigourdan. Is probably = his No. 271 (I.C. 2424), 1^m 2^s p. and 2′ n. of H.'s place. In the sweep the transit is recorded thus $\begin{matrix}37^m\ 50^s \\ -40^s\end{matrix}$ and the minute is probably wrong.

III. 644. "vF., vS., E., 300 confirmed it, but shewed two small round patches united, which seemed to be like vF. aberrations of two stars without the stars. I viewed them with many different adjustments of the focus."

III. 647. A nearer star in Sw. 721 is P. VIII. 245, f. 2^m 19^s, 49′ s. Both this star and a second obs. in Sw. 938, Mar. 10, 1790, 38 Lyncis p. 10^m 9^s, 46′ n., agree much better with h. in R.A.

III. 655. In the sweep the transit is $\begin{matrix}12^h\ 5\,..^m\ 28^s \\ -28^s\end{matrix}$, that of 19 Canum being 12^h 59^m 28^s. The transits before and after III. 655 are 12^h 49^m 50^s and $\begin{matrix}12^h\ 53^m\ 49^s \\ -10^s\end{matrix}$. Not found by Bigourdan.

III. 666. H. −h. = +52^s. In Sw. 730 the transit of the neb. is $\begin{matrix}13^h\ 44^m\ 58^s \\ -30^s\end{matrix}$. Probably 44 should be 43; otherwise III. 667 (the next object) would only be 25^s later.

III. 667. A second obs., Sw. 914, March 20, 1789, vF., S., 90 Virg. f. 5^m 56^s s. 1° 22′.

III. 693. R.A. 32^s too great. In Sw. 754 a corr. to the meridian of -32^s has been applied, apparently too little.

III. 695. In the Sweep (774) there is the following: "Mem. The P.D. must be reckoned inaccurate, the string having been touched since the last cluster [VI. 31] was taken." No modern observations known.

III. 704. Not found by Bigourdan. Perhaps =h. 691, F., S., R., bM., one obs., which is 8′ due south of H.'s place.

III. 706. Second obs., Apr. 1, 1788 (Sw. 822), "south of a cB. star, stellar or a little E." There is no Flamsteed star in this sweep, but a ∗ 8 m. which is D.M. +48°·1919 is 5^m 2^s p., 22′ n. of the neb. The cB. ∗ is +48°·1925, 7·6 mag.

III. 709. R.A. (from Sw. 815) 26^s too small. Sw. 945, March 17, 1790: "cB., iR., vgbM., about 3 or 4′ diam., 27 Lyncis p. 6^m 32^s s. 0° 49′."

III. 716. This occurs also in Sw. 946, March 17, 1790, where it is called II. 825, and where a much better star is G. 1807, f. 13^m 46^s s. 0° 25′, which agrees very well with the place of III. 716.

III. 723. Not in Sweep 808, where II. 728 was first seen. Not found by Bigourdan.

III. 740. R.A. is 1^m 35^s too small. A correction to centre of field of -94^s (exactly equal to the error) has been applied. The transit of 15 Draconis is correct, as shown by $\Delta\alpha = 51^m\ 3^s$ between it and +69°·806. The only nebula in this sweep.

III. 745. Unless there is an error in the transit in Sweep 876, this is not the same as h. 2191, but is about 1^m f. In the sweep a star 6 mag. = +56°·2923 is 3^m 41^s f. 10′ s.

III. 747. C. H. has used the place of the comparison star in Wollaston's Catalogue for 1790, which is very erroneous. Auwers assumed it to be B.A.C. 1985, hence his very erroneous P.D. But it is =G. 1100, agreeing with two other stars, 42 and 43 Camelop. The place of the neb. found from this coincides with that of I.C. 2133 =Bigourdan 385.

IV. 40. Not found by Bigourdan.

IV. 43. Description is from Sw. 621, Oct. 24, 1786. When first seen in Sw. 614 it was described as "a pB. star with a vF. nebulosity to the nf. side of very little extent."

IV. 44. Occurs only in Sw. 640, 2^m 0^s p., 4′ n. of IV. 38: "Situated between two stars with a third star at rectangles to the former." This cannot be h. 378 (as hitherto assumed), nor does the description quite fit IV. 19, which does not occur in this sweep, though this star has a star 11 m. ssp. and a vF. star north and a third farther off npp. But 70^s f. h. 378 on the same parallel there is a star 11 mag. between two others sp. and nf. with a third star p., forming a striking rectangular triangle. If this is H.'s object, his R.A. is 33^s too small.

V. 35. Diffused nebulosity in Orion. The following are the original notes; the places are for 1786.

Sw. 518, Feb. 1, 1786. I am pretty sure the places of which the following are the boundaries are all full of diffused nebulosity; but notwithstanding I used every means of ascertaining it by motion of the telescope, my range was neither far enough nor sufficiently quick to put it beyond doubt.

R.A.	P.D.
$5^h\ 30^m\ 50^s$	. . .
5 31 40	93° 47′
5 31 55	91 27
5 32 18	93 47
5 32 37	91 27
5 32 58	93 47
5 33 25	91 43
5 34 0	91 54

Sw. 526, Feb. 22, 1786. I am pretty sure the following space is affected with milky nebulosity:—

From R.A. $5^h\ 37^m\ 54^s$ to $5^h\ 39^m\ 4^s$
and P.D. 87° 45′ to 90° 7′.

Sw. 528, Feb. 23, 1786. From the transit of the four stars θ Orionis $5^h\ 28^m\ 12^s$ to this last $5^h\ 31^m\ 7^s$, the whole breadth of the sweep [2° 15′] contains very evidently milky nebulosity of different degrees of brightness in different parts.

From R.A. $5^h\ 24^m\ 46^s$ to $5^h\ 27^m\ 41^s$
and P.D. 95° 52′ to 98° 5′

Sw. 529, Feb. 24, 1786. Affected with milky nebulosity all the breadth of the sweep [2° 16′], some time before v Orionis was taken.

	From	to
	$5^h\ 21^m\ 31^s$	$5^h\ 27^m\ 19^s$
	95° 54′	98° 10′
Still certainly affected	$5^h\ 27^m\ 19^s$	$5^h\ 28^m\ 4^s$
	95° 52′	98° 8′
It becomes doubtful now	$5^h\ 28^m\ 39^s$	$5^h\ 29^m\ 33^s$
	95° 52′	98° 8′
Still a suspicion	$5^h\ 30^m\ 19^s$	$5^h\ 30^m\ 40^s$
	95° 52′	98° 8′
Pretty pure	$5^h\ 31^m\ 47^s$	$5^h\ 32^m\ 10^s$
	95° 51′	98° 7′
Still a pretty strong suspicion . .	$5^h\ 33^m\ 45^s$	$5^h\ 35^m\ 29^s$
	95° 51′	98° 7′

V. 39. Also observed in Sw. 664, Dec. 24, 1786. "The neb. of 660 Sw. as described there." Places agree. V. 40 observed in the same two sweeps, and both seen by Bigourdan, yet Innes with a 7-inch refractor at the Cape could only see V. 40 (*M.N.*, 59, p. 339).

VII. 54. No cluster seen by Bigourdan. In the sweep there is only one star (36 Camelop.), but the positions of the two other neb., III. 746 and V. 44, are correct.]

VIII. 71. The double star is H. II. 71.

XXIV.

Account of the Discovery of a Sixth and Seventh Satellite of the Planet Saturn; with Remarks on the Construction of its Ring, its Atmosphere, its Rotation on an Axis, and its spheroidical Figure.

[*Phil. Trans.*, vol. lxxx., 1790, pp. 1–20.]

Read November 12, 1789.

IN a short Postscript, added to my last Paper on Nebulæ, I announced the discovery of a *sixth satellite* of Saturn, and mentioned, that I intended to communicate the particulars of its orbit and situation to the Members of the Royal Society, at their next meeting. I have now the honour to present them, at the same time, with an account of two satellites instead of one; and have called them the *sixth* and *seventh*,* though their situation in the Saturnian system intitles them, very probably, to the first and second place. This I have done to the end that in future we may not be liable to mistake, in referring to former observations or tables, where the five known satellites have been named according to the order they have hitherto been supposed to hold in the range of distance from the planet.

It may appear remarkable, that these satellites should have remained so long unknown to us, when, for a century and an half past, the planet to which they belong has been the object of almost every astronomer's curiosity, on account of the singular phænomena of its ring. But it will be seen presently, from the situation and size of the satellites, that we could hardly expect to discover them till a telescope of the dimensions and aperture of my forty-feet reflector should be constructed; and I need not observe how much we Members of this Society must feel ourselves obliged to our Royal Patron, for his encouragement of the sciences, when we perceive that the discovery of these satellites is intirely owing to the liberal support whereby our most benevolent King has enabled his humble astronomer to complete the arduous undertaking of constructing this instrument.

The planet Saturn is, perhaps, one of the most engaging objects that astronomy offers to our view. As such it drew my attention so early as the year 1774; when, on the 17th of March, with a 5½-feet reflector, I saw its ring reduced to a very minute line, as represented in fig. 1 [Plate IX.]. On the 3d of April, in the same year, I found the planet as it were stripped of its noble ornament, and dressed in

* [Enceladus and Mimas.—ED.]

the plain simplicity of Mars. See fig. 2 [Plate IX.]. I pass over the following year, in which, with a 7-feet reflector, I saw the ring gradually open, till it came to the appearance expressed in fig. 3 [Plate X.], the original of which was delineated from nature, on the 20th of June, 1778, by means of a very good 10-feet reflector.

It should be noticed, that the black disk, or belt, upon the ring of Saturn is not in the middle of its breadth; nor is the ring subdivided by many such lines, as has been represented in divers treatises of astronomy; but that there is one single, dark, considerably broad line, belt, or zone, upon the ring, which I have always permanently found in the place where my figure represents it. I give this, however, only as a view of the northern plane of the ring, as the situation of the planet has hitherto not afforded me any other. The southern one, which is lately come to be exposed to the sun, will shortly be opened sufficiently to enable me to give also the situation of its belts, if it should have any.

From my observations it appears, that the zone on the northern plane of the ring is not, like the belts of Jupiter or those of Saturn, subject to variations of colour and figure; but is most probably owing to some permanent construction of the surface of the ring itself. That however, for instance, this black belt cannot be the shadow of a chain of mountains, may be gathered from its being visible all round on the ring; for at the ends of the ansæ there could be no shades visible, on account of the direction of the sun's illumination, which would be in the line of the chain; and the same argument will hold good against supposed caverns or concavities. It is moreover pretty evident, that this dark zone is contained between two concentric circles, as all the phænomena answer to the projection of such a zone. Thus, in fig. 4 [Plate X.], which was taken the 11th of May, 1780, we may see, that the zone is continued all round the ring, with a gradual decrease of breadth towards the middle, answering to the appearance of a narrow circular plane, projected into an ellipsis.

As to the surmise, which might occur to us, of a division of the ring, or rather of two rings, one about the other, with a distance of open space between them, it does not appear eligible to venture on so artificial a construction, by way of explaining a phænomenon that does not absolutely demand it. If one ring, of a breadth so considerable as that of Saturn, is justly to be esteemed the most wonderful arch that, by the laws of gravity, can be held together, how improbable must it appear to suppose it subdivided into narrow slips of rings, which by this separation will be deprived of a sufficient depth, and thus lose the only dimension which can keep them from falling upon the planet? It is however true, that as yet we do not know of the rotation of the ring, which may be of such a proper velocity as greatly to assist its strength; and that, in the subdivisions, of course the different velocities for each division may be equally supposed to keep them up. If the southern plane should prove to be very differently marked, it will at once remove every surmise of such a division; but if it should offer us the same appearance of a dark zone, in the same situation, and of an equal breadth with the one I have observed on the northern side, I would still remark, that, since a most effectual way to verify the duplicity of the ring is within our reach, it will be the best way to suspend our judgement till that can be put to the trial. The method I allude to is an occultation of some considerable star by Saturn, when, if the ring be divided, it will be seen between the openings, as well as between the ring and Saturn.

With regard to the nature of the ring, we may certainly affirm, that it is no less solid and substantial than the planet itself. The same reasons which prove to us the solidity of the one will be full as valid when applied to the other. Thus we see, in fig. 3 and 4, the shadow of the body of Saturn upon the ring, which, in fig. 3, is eclipsed towards the north, on the following side, and in fig. 4, about the middle, according to the opposite situation of the sun. In the same manner we see the shadow of the ring cast on the planet, where in fig. 1 and 2, we find it on the equatorial part; and May 28, 1780, I saw it towards the south. If we deduce the quantity of matter, contained in the body, from the power whereby the satellites are kept in their orbits, and the time of their revolution, it must be remembered, that the ring is included in the result. It is also in a very particular manner evident, that the ring exerts a considerable force upon these revolving bodies, since we find them strongly affected with many irregularities in their motions, which we cannot properly ascribe to any other cause than the quantity of matter contained in the ring; at least we ought to allow it a proper share in the effect, as we do not deny but that the considerable equatorial elevation of Saturn, which I shall establish hereafter, must also join in it.

The light of the ring of Saturn is generally brighter than that of the planet: for instance, April 19, 1777, I saw the southern part of the ring, which passed before the body, very plainly brighter than the disk of Saturn, on which it was projected; and on the 27th of the same month, I found, that with a power of 410, my seven-feet reflector had hardly light enough for Saturn, when the ring was notwithstanding sufficiently bright. Again, the 11th of March, 1780, I tried the powers of 222, 332, and 449, successively, and found the light of Saturn less intense than that of the ring; the colour of the body with the high powers turning to a kind of yellow, while that of the ring still remained white. The same result happened on June 25, 1781, with the power 460.

I come now to one of the most remarkable properties in the construction of the ring, which is its extreme thinness. The situation of Saturn, for some months past, has been particularly favourable for an investigation of this circumstance; and my experiments have been so complete, that there can remain no doubt on this head.

When we were nearly in the plane of the ring, I have repeatedly seen the first, the second, and the third satellites,* nay even the sixth and seventh, pass before and behind the ring in such a manner that they served as excellent micro-

* [Respectively Tethys, Dione, Rhea.—Ed.]

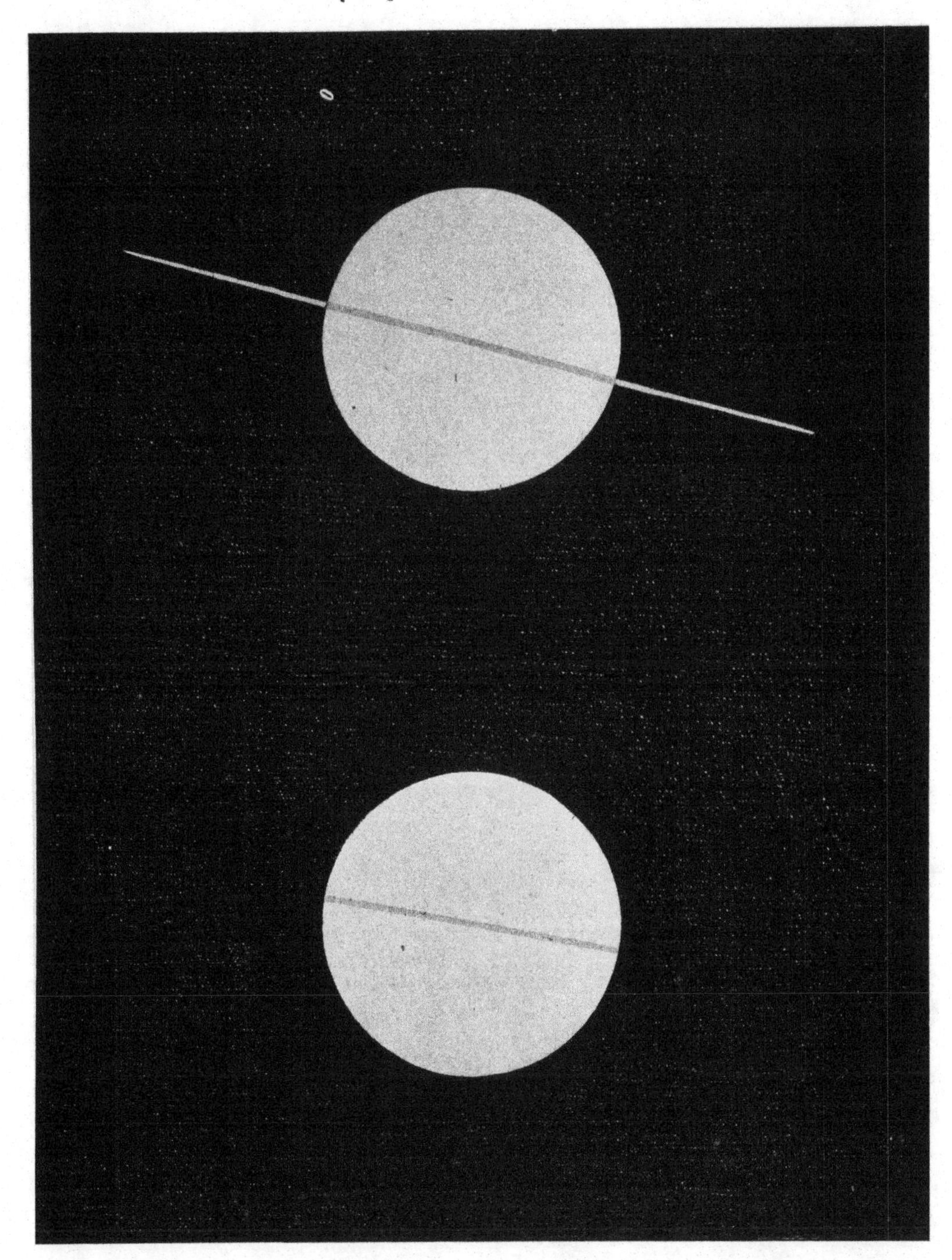

FIG. I.
March 17, 1774.

FIG. 2.
April 3, 1774.

[*To face page* 372.

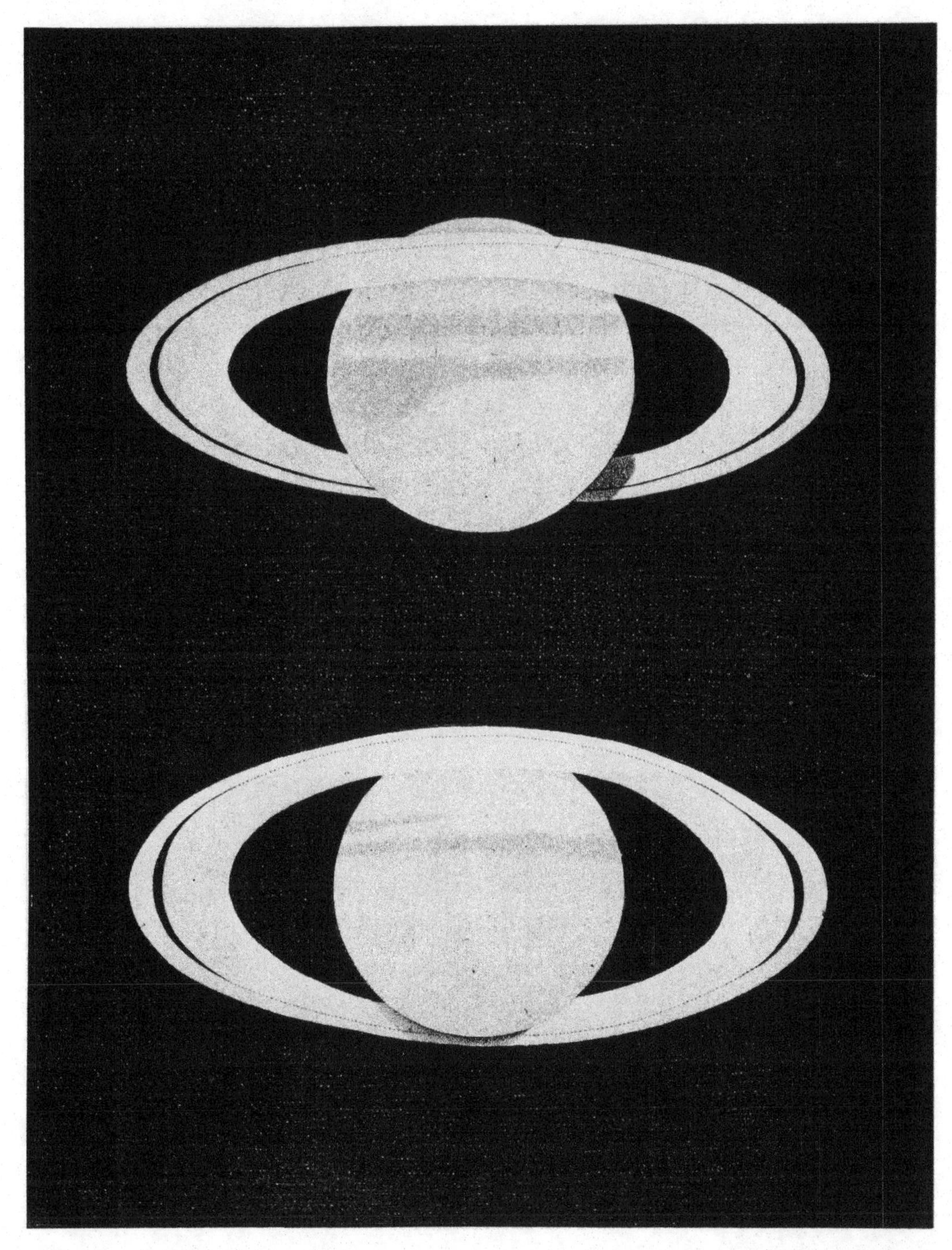

FIG. 3.
June 20, 1778.

FIG. 4.
May 11, 1780.

[*To face page* 372

meters to estimate its thickness by. It may be proper to mention a few instances, especially as they will serve to solve some phænomena that have been remarked by other astronomers, without having been accounted for in any manner that could be admitted, consistently with other known facts. July 18, 1789, at 19 h. 41′ 9″, sidereal time, the first satellite seemed to hang upon the following arm, declining a little towards the north, and I saw it gradually advance upon it towards the body of Saturn; but the ring was not so thick as the lucid point. July 23, at 19 h. 41′ 8″, the second satellite was a very little preceding the ring; but the ring appeared to be less than half the thickness of the satellite. July 27, at 20 h. 15′ 12″, the second satellite was about the middle, upon the following arm of the ring, and towards the south; and the sixth satellite on the farther end, towards the north; but the arm was thinner than either of them. August 29, at 22 h. 12′ 25″, the third satellite was upon the ring, near the end of the preceding arm; and my remark at the time when I saw it was, that the arm seemed not to be the fourth, at least not the third, part of the diameter of the satellite, which, in the situation it was, I took to be less than one single second in diameter. At the same time I also saw the seventh satellite, at a little distance following the third, in the shape of a bead upon a thread, projecting on both sides of the same arm: hence we are sure, that the arm also appeared thinner than the seventh satellite, which is considerably smaller than the sixth, which again is a little less than the first satellite. August 31, at 20 h. 48′ 26″, the preceding arm was loaded about the middle by the third satellite. October 15, at 0 h. 43′ 44″, I saw the sixth satellite, without obstruction, about the middle of the preceding arm, though the ring was but barely visible with my forty-feet reflector, even while the planet was in the meridian; however, we were then a little inclined to the plane of the ring, and the third satellite, when it came near its conjunction with the first, was so situated that it must have partly covered the first a few minutes after the time I lost it behind my house. In all these observations the ring did not in the least interfere with my view of the satellites. October 16, I followed the sixth and seventh satellites up to the very disk of the planet; and the ring, which was extremely faint, opposed no manner of obstruction to my seeing them gradually approach the disk, where the seventh vanished at 21 h. 46′ 44″, and the sixth at 22 h. 36′ 44″.*

I might bring many other instances, if the above were not quite sufficient for the purpose. There is, however, some considerable suspicion, that, by a refraction through some very rare atmosphere on the two planes of the ring, the satellites might be lifted up and depressed, so as to become visible on both sides of the ring, even though the ring should be equal in thickness to the diameter of the smallest satellite, which may amount to a thousand miles. As for the argument of its incredible thinness, which some astronomers have brought from the short time of its being invisible, when the earth passes through its plane, we

* [These observations were made with the 20-feet reflector, not with the 40-feet.—ED.]

cannot set much value upon them ; for they must have supposed the edge of the ring, as they have also represented it in their figures, to be square ; but there is the greatest reason to suppose it either spherical or spheroidical, in which case evidently the ring cannot disappear for any long time. Nay, I may venture to say, that the ring cannot possibly disappear on account of its thinness ; since, either from the edge or the sides, even if it were square on the corners, it must always expose to our sight some part which is illuminated by the rays of the sun : and that this is plainly the case, we may conclude from its being visible in my telescopes during the time when others of less light had lost it, and when evidently we were turned towards the unenlightened side, so that we must either see the rounding part of the enlightened edge, or else the reflection of the light of Saturn upon the side of the darkened ring, as we see the reflected light of the earth on the dark part of the new moon. I will, however, not decide which of the two may be the case ; especially as there are other very strong reasons to induce us to think, that the edge of the ring is of such a nature as not to reflect much light.

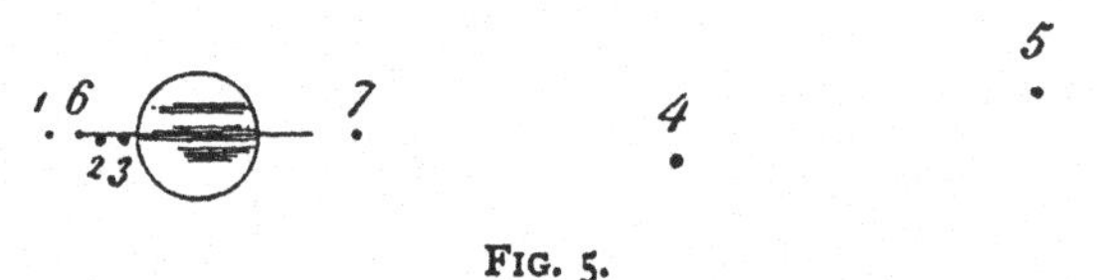

FIG. 5.

I cannot leave this subject without mentioning both my own former surmises, and those of several other astronomers, of a supposed roughness in the surface of the ring, or inequality in the planes and inclinations of its flat sides. They arose from seeing luminous parts on its extent, which were supposed to be projecting points, like the moon's mountains ; or from seeing one arm brighter or longer than another ; or even from seeing one arm when the other was invisible. I was, in the beginning of this season, inclined to the same opinion, till one of these supposed luminous points was kind enough to venture off the edge of the ring, and appeared in the shape of a satellite. Now, as I had collected every inequality of this sort, it was easy enough for me afterwards to calculate all such surmises by the known periodical time of the first, second, third, sixth, and seventh satellites ; and I have always found that such appearances were owing to some of these satellites which were either before or behind the ring. The 20th of October, for instance, at 22 h. 35′ 46″, I saw four of Saturn's satellites all in one row, and at almost an equal distance from each other, on the following side ; and yet the first satellite, which was the farthest of them all, was only about half-way towards its greatest elongation from the body of Saturn, as may be seen in fig. 5. How easily, with an inferior telescope, this might have been taken for one of the arms of Saturn, I leave those to guess who know what a degree of accuracy it must require to distinguish objects that are so minute, and at the same time so faint, on account of their nearness to the disk of the planet. Upon the whole, therefore, I cannot say, that I had any one instance that could induce me to believe the ring was not of an uniform thickness; that is, equally thick at equal distances from the center, and of an equal diameter throughout the whole of its construction. The idea of protuberant points upon the ring of Saturn, indeed, is of itself sufficient to render the opinion of their existence inadmissible, when we consider the enormous size such points ought to be of, for us to see them at the distance we are from the planet.

From these supposed luminous points I am, by imperceptible steps, brought to the discovery of two satellites of Saturn, which had escaped unnoticed, on account of their little distance from the planet, and faintness; which latter is partly to be ascribed to their smallness, and partly to being so near the light of the ring and disk of Saturn. Strong suspicions of the existence of a sixth satellite I have long entertained; and, if I had been more at leisure two years ago, when the discovery of the two Georgian satellites took me as it were off the scent, I should certainly have been able to announce its existence as early as the 19th of August, 1787, when, at 22 h. 18′ 56″, I saw, and marked it down as being probably a sixth satellite, which was then about 12 degrees past its greatest preceding elongation. But, as I observed before, not having time to give my thoughts to the subject, I reserved a full investigation of the number of satellites, and the nature of the ring of Saturn for a future opportunity. Besides, not having any tables of the satellites, I could not confidently say, whether the fifth satellite was not one of the five which I perceived in motion that night, though afterwards I found, that the real fifth had also been in view, and was marked down as a star, by the letter *b*, in a figure I delineated of Saturn and its satellites that evening.

In the year 1788 very little could be done towards a discovery, as my twenty-feet speculum was so much tarnished by *zenith sweeps*, in which it had been more than usually exposed to falling dews, that I could hardly see the Georgian satellites. In hopes of great success with my forty-feet speculum, I deferred the attack upon Saturn till that should be finished; and having taken an early opportunity of directing it to Saturn, the very first moment I saw the planet, which was the 28th of last August, I was presented with a view of six of its satellites, in such a situation, and so bright, as rendered it impossible to mistake them, or not to see them. The retrograde motion of Saturn amounted to nearly 4½ minutes *per* day, which made it very easy to ascertain whether the stars I took to be satellites really were so; and, in about two hours and an half, I had the pleasure of finding, that the planet had visibly carried them all away from their places. I continued my observations constantly, whenever the weather would permit; and the great light of the forty-feet speculum was now of so much use, that I also, on the 17th of September, detected the seventh satellite, when it was at its greatest preceding elongation.

As soon as I had observations enough to make tables of the motion of these new satellites,* I calculated their places backwards, and soon found that many

* [All these observations were made with the 20-feet telescope.—Ed.]

suspicions of these satellites, in the shape of protuberant points on the arms, were confirmed, and served to correct the tables, so as to render them more perfect. Fig. 6 represents the seven satellites of Saturn, as they were situated October 18, at 21 h. 22′ 45″. The small star *s* served to shew the motion of the planet in a striking manner; as, in about 3¾ hours after the above-mentioned time, the whole Saturnian system was completely moved away, so as to leave the star *s* as much following the second and first satellites, which then were in conjunction, as it now was before the second.

By comparing together many observations of the sixth satellite, I find, that it completes a sidereal revolution about Saturn in one day, 8 hours, 53′ 9″. And if we suppose, with M. DE LA LANDE,* that the fourth is at the mean distance of 3′ from the center of Saturn, and performs one revolution in 15 d. 22 h. 34′ 38″, we find the distance of the sixth, by KEPLER's law, to be 35″·058. Its light is considerably strong, but not equal to that of the first satellite; for, on the 20th of October, at 19 h. 56′ 46″, when these two satellites were placed as in fig. 7, the

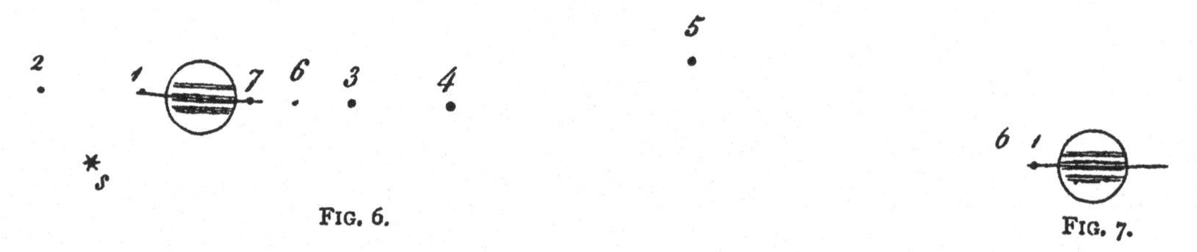

FIG. 6. FIG. 7.

first, notwithstanding it was nearer the planet than the sixth, was still visibly brighter than the latter. It would, however, be worth while to try whether a good achromatic telescope, of a large aperture, might not possibly shew it at the time of its greatest distance from the planet, and when no other satellite is near; that is, provided it will shew the other five satellites with great ease, as otherwise there will be no reason to expect it should shew the sixth.

In the period of this satellite I have employed the observation of the 19th of August, 1787, as, from other calculations, it seems the revolution is determined near enough to reach back so far.

The most distant observations of the seventh satellite, being compared together, shew, that it makes one sidereal revolution in 22 hours, 40 minutes, and 46 seconds: and, by the same *data* which served to ascertain the dimension of the orbit of the sixth, we have the distance of the seventh, from the center of Saturn, no more than 27″·366. It is incomparably smaller than the sixth; and, even in my forty-feet reflector, appears no bigger than a very small lucid point. I see it, however, also very well in the twenty-feet reflector; to which the exquisite figure of the speculum not a little contributes. It must nevertheless be remembered, that a satellite once discovered is much easier to be seen than it was before we were acquainted with its place.

* *Astr.* § 2996, 2997

The revolution of this satellite is not nearly so well ascertained as that of the former. The difficulty of having a number of observations is uncommonly great; for, on account of the smallness of its orbit, the satellite lies generally before and behind the planet and its ring, or at least so near them that, except in very fine weather, it cannot easily be seen well enough to take its place with accuracy. On the other hand, the greatest elongations allow so much latitude for mistaking its true situation, that it will require a considerable time to divide the errors that must arise from imperfect estimations.

The orbits of these two satellites, as appears from many observations of them, are exactly in the plane of the ring, or at least deviate so little from it, that the difference cannot be perceived. It is true, there is a possibility that the line of their nodes may be in, or near, the present greatest elongation, in which case the orbits may have some small inclination; but as I have repeatedly seen them run along the very minute arms of the ring, even then the deviation cannot amount to more than perhaps one or two degrees; if, on the contrary, the nodes should be situated near the conjunction, this quantity would be so considerable that it could not have escaped my observation.

From the ring and satellites of Saturn we now turn our thoughts to the planet, its belts, and its figure.

April 9, 1775. I observed a northern belt on Saturn, which was a little inclined to the line of the ring.

May 1, 1776. There was another belt, inclined about 15 degrees to the same line, but it was more to the south, and on the following side came up to the place in which the ring crosses the body.

July 13, 1776. The belt was again depressed towards the north, almost touching the line where the ring passed behind the body.

April 8, 1777. There were two fine belts, both a little inclined to the ring.

June 20, 1778. There were two belts parallel to the ring; but the northern one had some faint, cloudy appearance, towards the preceding, or western side.

May 11, 1779. Two equatorial belts.
— 13, —— A bright belt over a dark one.
— 22, —— One dark, and one very faint white belt.
— 23, —— A dark belt, and a pretty bright white one.
Jan. 21, 1780. Two belts; the most north clouded.
— 22, —— Faint belts.
May 17, —— A dark, equatorial belt.
— 23, —— A strong, equatorial belt.

June 19, at 10 h. 15′, With a new, excellent seven-feet speculum, I see two belts, and a cloudy appearance, which is not come up to the middle; but, as it is a large figure, some part of it is already past the center (this is, provided Saturn turns upon its axis the same way as Jupiter does). See fig. 8, where the ring is omitted.

June 20, 1780. 10 h. 10′, The same figure is on the disk, but seems to be more central than it was yesterday.

June 21, 9 h. 25′, The same two belts; a strong, dark spot, near the margin of the disk; see fig. 9; ring not expressed. 10 h. 1′, The spot not so remarkable as it was at 9 h. 25′.

June 26, Small twenty-feet telescope; an equatorial belt, and another less marked.

June 29, Two dark equatorial belts.

August 19, Two belts.

—— 23, Two belts, a little declining from the equatorial position.

—— 26, A broad belt much inclined; with 200, 250, 300, 400, faint appearances of a second and of a third belt.

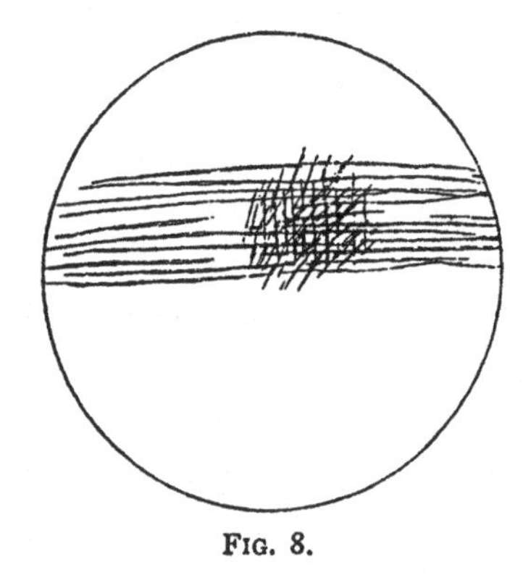

Fig. 8.

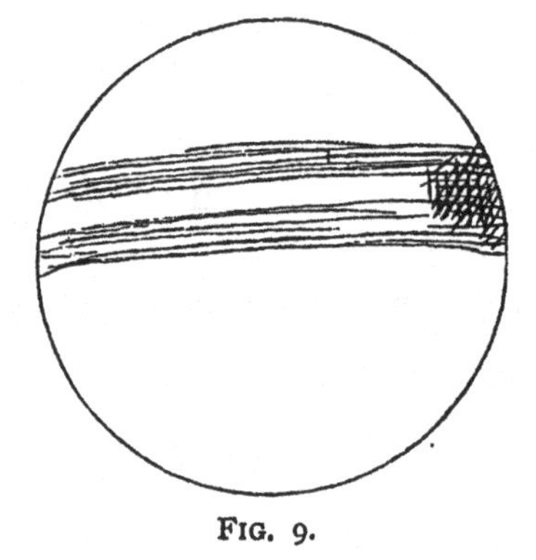

Fig. 9.

August 27, The belts less inclined.

Sept. 2, A darkish belt, but very little inclined; and a fine white belt, close to the ring.

Sept. 5, The belt a little inclined.

— 6, The belt not inclined.

— 8, The bright belt close to the ring, and two dark equatorial belts.

It will not be necessary to continue the account of these belts up to the present time; but I have constantly observed them, and found them generally in equatorial situations, though now and then they were otherwise.

We may draw two conclusions from what has been reported. The first, which relates to the changes in the appearance of the belts, is, that Saturn has probably a very considerable atmosphere, in which these changes take place; just as the alterations in the belts of Jupiter have been shewn, with great probability, to be in his atmosphere. This has also been confirmed by other observations: thus, in occultations of Saturn's satellites, I have found them to hang to the disk a long while before they would vanish. And though we ought to make some allowance for the encroachment of light, whereby a satellite is seen to reach up to the disk sooner than it actually does, yet, without a considerable refraction, it could hardly be kept so long in view after the apparent contact. The time of hanging upon the disk, in the seventh satellite, has actually amounted to 20 minutes. Now, as its quick motion during that interval carries it through an arch of near six degrees, we find, that this would denote a refraction of about two seconds, provided the encroaching of light had no share in the effect. By an observation of the sixth satellite, the refraction of Saturn's atmosphere amounts to nearly the same quantity; for this satellite remained about 14 or 15 minutes longer in view than it should have done; and as it moves about 2¾ degrees in that time, and its orbit is larger than that of the seventh, the difference is inconsiderable. It is not my present intention to enter into a consideration of the amount of these refractions, otherwise we might perhaps find data enough to subject them to some calculation. But what has been said will suffice to shew, that very probably Saturn has an atmosphere of a considerable density.

The next inference we may draw from the appearance of the belts on Saturn is, that this planet turns upon an axis which is perpendicular to the ring. The arrangement of the belts, during the course of fourteen years that I have observed them, has always followed the direction of the ring, which is what I have called being equatorial. Thus, as the ring opened, the belts began to advance towards the south; and to shew an incurvature answering to the projection of an equatorial line, or to a parallel of the same. When the ring closed up, they returned towards the north; and are now, while the ring passes over the center, exactly ranging with the shadow of it on the body; generally one on each side, with a white belt close to it. When I say, that the belts have always been equatorial, I pass over trifling exceptions, which certainly were owing to local causes. The step from equatorial belts to a rotation on an axis is so easy, and, in the case of Jupiter, so well ascertained, that I shall not hesitate to take the same consequence for granted here. But, if there could remain a doubt, the observations of June 19, 20, and 21, 1780, where the same spot was seen in three different situations, would remove it completely.

There is another argument, of equal validity with the former, which now I shall bring on. It is founded upon the following observations, and will shew that Saturn, like Jupiter, Mars, and the Earth, is flattened at the Poles; and therefore ought to be supposed to turn on its axis.

July 22, 1776. I thought Saturn was not exactly round.

May 31, 1781. It appears as if the body of Saturn was at least as much flattened as that of Jupiter; but as the ring interferes, this may be better ascertained eight years hence.

August 18, 1787. The body of Saturn is of unequal diameters, the equatorial one being the longest.

Sept. 14, 1789, 23 h. 36′ 32″. Having reserved the examination of the two diameters of Saturn to the present as the most favourable time, I measured them with my twenty-feet reflector, and a good parallel-wire micrometer.

	″
Equatorial diameter, 1st measure,	21·94
2d . . .	23·11
3d . . .	21·73
4th . . .	22·85
Mean	22·81
Polar diameter, 1st measure,	20·57
2d . . .	20·10
3d . . .	21·16
Mean	20·61

By this it appears that Saturn is considerably flattened at the poles. And as the greatest measures were taken in the line of the ring and of the belts, we are assured that the axis of the planet is perpendicular to the plane of the ring; and that the equatorial diameter is to the polar one nearly as 11 to 10.

We may also infer the real diameter of Saturn from these measures, which are perhaps more to be depended upon than any that have hitherto been given. But as in my journal I have measures that were repeatedly taken these ten years past, not only of the diameter of Saturn, but of the ring, and its opening, whereby its inclination may be known; as well as of the distance of the fourth, and fifth, and other satellites, which will be of great use in ascertaining the quantity of matter contained in the planet, I reserve a full investigation of these things for another opportunity; since, from the date of this Paper, it will be sufficiently evident, that there can be no time for me to enter properly into the subject.

One beautiful observation of the transit of the shadow of the fourth satellite over the disk of Saturn, I must add, to conclude this Paper.

Last night, November 2, 1789, at 23 h. 13′ sidereal time, being always in quest of any appearance that may afford the means of ascertaining the rotation of Saturn on an axis, I discovered a black spot on the following margin of the disk of that planet.

At 23 h. 21′, I perceived a protuberance on the south preceding edge of the disk, which I supposed to be the fourth satellite going to emerge.

At 23′, I found that the black spot had advanced a little towards the preceding side.

At 30′, with a power of 300, I found it still advancing, and saw that the spot was a little to the north of the equatorial belt, but so that a small part of it was upon the belt.

At 35′, the black spot was a little more than one-eighth of the diameter of Saturn advanced from the following edge towards the center.

At 39′, the satellite was detached.

At 49′, the spot was advanced so as to be about one-third of its way towards the center; and the fourth satellite near half its own apparent diameter clear of the edge.

In this situation of the planet I took an eye-draught of it (see fig. 10), as it appeared with the black spot on the belt; the lately emerged fourth satellite; two parallel dark belts, the intermediate space between them and the equatorial one being a little brighter than the rest of the disk; the sixth, third, and second satellites on the preceding side; the ring projecting like two very slender lines on each side of the disk, and containing the first satellite upon the following arm, with the fifth at a considerable distance following.

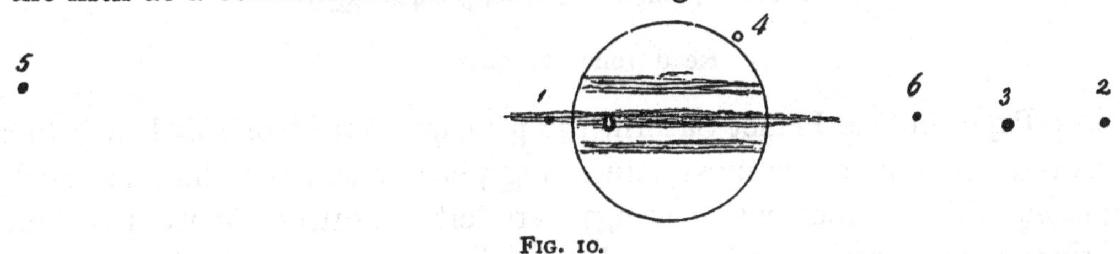

FIG. 10.

At 0 h. 5′, the black spot was got a little more than half way towards the center. It was much darker than the belt, and more upon it than before.

At 1 h. 2′, by advancing gradually towards the south, it was now almost intirely drawn upon the equatorial belt.

At 1 h. 13′, the black spot approached towards a central situation.

At 1 h. 21′ 51″, it was perfectly central, and at the same time upon the middle of the equatorial belt.

I followed the shadow of the satellite with great attention up to the center, in order to secure a valuable epocha, which may serve to improve our tables of the mean motion of this satellite.

WILLIAM HERSCHEL.

Slough, near Windsor,
November 3, 1789.

XXV.

On the Satellites of the Planet Saturn, and the Rotation of its Ring on an Axis.

[*Phil. Trans.*, vol. lxxx., 1790, pp. 427–495.]

Read June 17, 1790.

IN my last Paper on the Planet Saturn, the principal object of which was to give an immediate account of the most interesting phænomena that had occurred till the beginning of November, many things were left unnoticed for want of time to treat of them with sufficient accuracy; but having now before me the whole series of observations from the 18th of July till the 25th of December, 1789, I can enter into a proper examination, assisted by such necessary calculations as then could not conveniently be made.

One of the principal motives which have induced me to hasten this inquiry, is the frequent appearance of protuberant and lucid points on the arms of the ring of Saturn. I have mentioned before that such phænomena had been resolved by the situation of satellites that put on these appearances; but as my observations were continued near two months afterwards, and as I had from them corrected the epochæ of the old satellites, and improved the tables of the new ones, I found that, besides many of these bright points which were completely accounted for by the calculated places of the satellites, there were also many more mentioned in my journal that would not accord with the situation of any of them.

The question then presented itself very naturally, what to make of these protuberant points? To admit two or three more satellites by way of solving such phænomena appeared to me too hazardous an hypothesis; especially as these lucid points, though some of them had a motion, did not seem willing to conform to the criterion I had before used of coming off the ring, and shewing themselves as satellites. And yet a suspicion of at least one more satellite would often return; it was even considerably strengthened when I discovered, by means of re-calculating with great precision the whole series of observations, that in the beginning of the season there had been some few mistakes in the names of the satellites, when the observations of them were entered in the journal. In setting them right, which threw a great light upon the revolution of the 6th, and more especially upon that of the 7th, I found also, that some of the observations which were entered by the name of the 7th satellite could not belong to that, nor to any

The material originally positioned here is too large for reproduction in this reissue. A PDF can be downloaded from the web address given on page iv of this book, by clicking on 'Resources Available'.

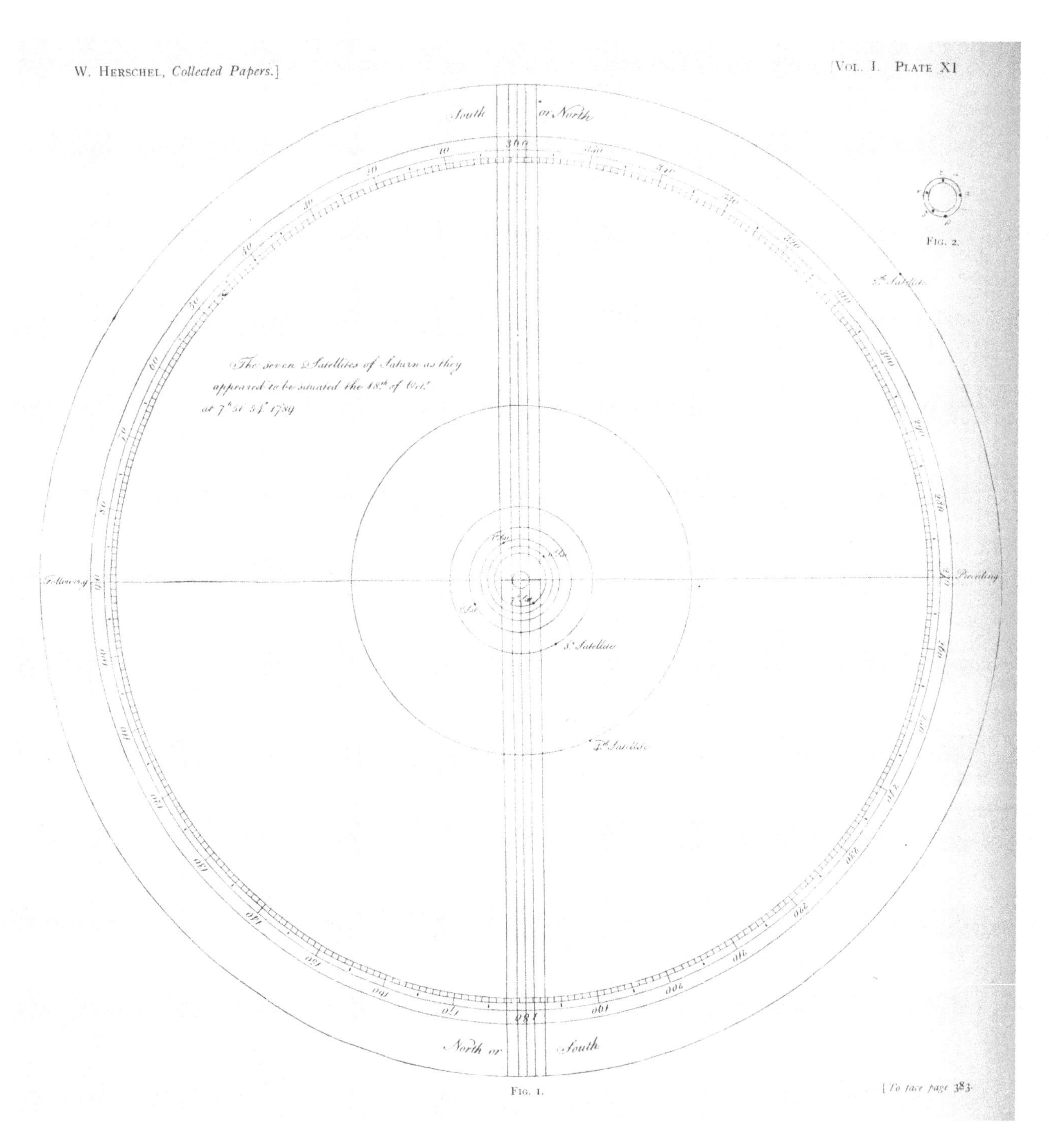

FIG. 2.

FIG. 1.

[To face page 383.

other known one. It remained therefore to be examined whether there might not be sufficient ground to suspect the existence of an eighth satellite.

In this situation of things, I thought it most advisable to draw out the whole series of observations in a paper, beginning at the 5th satellite, and thus gradually through the 4th, 3d, 2d, 1st, 6th, and 7th, to approach towards the center of Saturn; that it might appear at last what observations were left unaccounted for. By this means also it will be seen clearly with how scrupulous an attention the identity of every satellite has been ascertained; and with a view to give the strongest satisfaction in this respect, at least one observation of each has been calculated for each night; and the place thus computed is put down in the notes, that it may be compared with the observed one.

To facilitate this comparison, I have delineated a scheme,* wherein the orbits of the satellites are drawn in their due proportion. A few words will explain the construction and use of this figure, which, notwithstanding its simplicity, is yet amply sufficient to ascertain the accuracy of every observation.

In each of the orbits, by way of marking them, is placed the satellite to which it belongs, as it appeared to be situated the 18th of October, 1789. The graduated circle is of use to find, by means of the tables, the apparent place of a satellite for any given time; or, the apparent situation of the same satellite being given, its real Saturnicentric place may be deduced from it. In the center of the scheme is the planet Saturn, and its ring, expressed by a line which represents the direction of its ansæ; or the ring itself, as it appeared in my telescopes during the months of July, August, September, October, and November, 1789. The five lines which are carried on parallel to each other serve to convey the measure of the planet, and its ring, to the orbits of the satellites, as will be seen in several instances that occur hereafter.

The graduated circle is divided into degrees, and begins to count from that part of every satellite's orbit beyond the planet, which is intercepted by a plane passing from the eye of the observer, at rectangles to the ring, through the center of Saturn. Hence it follows, that the point of zero, or 360 degrees, is the same with the geocentric place of the planet in those four parts of the orbit of the satellite where the eye is in the plane of the ring, and where it appears the most open; and that, in other places, it may be had by solving one spherical triangle. This is to be understood as relating only to the inner satellites; the 5th, or outermost, requiring a different reduction, on account of its deviation from the plane of the ring. Moreover, I am inclined to believe, that the surest way of observing the 5th, is to trust only to measures, taken with micrometers which give the distance and angle of position, except in such cases when the eye is nearly in the plane of this satellite's orbit, where the different reductions may be neglected, without bringing on any considerable inaccuracies. The order of the numbers by which 90 comes to the

* See Plate XI. fig. 1.

left, and 270 to the right, is taken from the motion of the satellites, as they appear to revolve in their orbits, when seen in telescopes of my construction; and which is also the real direction of their motion according to the order of the signs. But the points 360 and 180 must occasionally be changed in their denomination of north or south, according to the real situation of the plane of each satellite's orbit. At present, for instance, when the satellites are at 360, that part of their orbits in which we find them lies to the south of the center of the planet; but about the end of August, 1789, and afterwards, the orbits of the six inner satellites were differently situated; so that the same points then were turned towards the north. I need not remark, that the situation of these points was changed again when the earth passed through the plane of the ring, and that it will change, in the 5th satellite likewise, when we come to be in the plane of its orbit.

The calculations of the places of all the satellites have been made according to tables which are given at the end of this Paper. Their form being very simple, I thought it not amiss to communicate them, for the use of those who may wish to enter into a more particular examination of the following observations; or to follow the satellites in their orbits at any future time. It will be proper to mention, that I have deduced the epochæ of all the seven satellites from my own observations, and they will be found to differ considerably from those which are given by M. DE LA LANDE, in the *Connoissance des Temps* for 1791. But I have not attempted to extend them farther than a few years backwards or forwards, as I am not in possession of any observations that could authorize me to undertake such a work. On the contrary, I am well convinced, that no tables will give us the situation of the satellites accurately, till we have at least established the dimensions of their elliptical orbits, and the motion as well as the situation of their aphelia. The epochæ for 1789, therefore, must be looked upon not as *mean* ones, but such as respect the orbits of these satellites in their situation during the time of the following observations; and the two preceding, and two following years, must be already a little affected with those errors which are the necessary consequence of our not knowing the required elements. I flatter myself, however, that the observations, which are delivered in this Paper, will serve as a beginning to a proper foundation for investigating them. The many conjunctions between the satellites, for instance, will undoubtedly throw some light on the situation and excentricity of their orbits; as it will be found, that the calculated places of these conjunctions require elliptical motions to bring the satellites to such appearances, which, in circular orbits, could not so accurately have taken place. Nor can we ascribe the disagreements to the fault of the observations, since a very few minutes will suffice to determine the time of a conjunction, which never lasts long. For this reason also, I have carefully avoided deducing my epochæ from conjunctions, even with the 6th satellite, which moves so rapidly that, at first sight, we might think those situations favourable.

The mean motion of the five old satellites, as being sufficiently accurate for my present purpose, I have taken from the above-mentioned tables of M. DE LA LANDE; and those of the 6th and 7th, of course, are the result of my own observations.

The geocentric place of Saturn, whose complement is to be added, in order to reduce the Saturnicentric situation of the satellites to the apparent one, I have taken from the *Nautical Almanac* to the nearest minute; and, as I have always confined myself to a literal transcription of the observations from the original journal, all the memorandums which are necessary either to explain them, or to correct mistakes in the names of the satellites, are thrown into notes, that there may be no interruption in the succession of the observations.

*Observations on the fifth satellite of Saturn.**

1789, July 18. 20 h. 20′ (A). The supposed fifth satellite (B) 6° or 7° sp. (C) the ring (D).

July 23. 19 29. The 5th sp. at a great distance (E).

July 28. 22 37. By a figure in the journal, at a great distance np. (F).

Aug. 18. 21 11. The supposed 5th at a great distance 25° np. R. (G).

Aug. 28. 1 28. A line drawn through a large star north of Saturn, and passing between one pretty considerable star nf. and another sf. Saturn, leaves the supposed 5th satellite a little on the following side. By two figures in the journal, the 5th is at a great distance nf. ♄. (H).

Aug. 29. 23 29. The supposed 5th is a very little preceding a line drawn from the large star of last night, through a very small star; and a good deal following a line drawn from the first pretty considerable star of last night, through the same very small star (I).

Aug. 31. 21 3. The 5th I take to be nf. ♄ at a good distance (K).

Sept. 8. 22 37. The 5th about 15° nf. R. and, by a figure, at a great distance (L).

Sept. 11. 20 11. The supposed 5th satellite and two small stars sf. a star *x*, which is sf. ♄, form an exact line.

22 32. The supposed 5th, and the two S st. sf. *x*, form no longer a line; so that is the real fifth satellite.

23 52. The 5th satellite keeps advancing; its situation is 20 or 22° nf. the line of the R. and, by a figure, it is at a considerable distance (M).

* [Iapetus.—ED.]

(A) The time of my observations being sidereal, it is necessary to mention, that this relates only to the hours, minutes, and seconds, the day itself being that which is generally used by astronomers, beginning at noon, and ending the noon following. By this means there can never be a mistake which sidereal hour I mean to point out, as no two such hours can occur in the same astronomical night.

It will also be necessary to remark, that all the times are those shewn by the clock; which, by equal altitudes, has been found to lose very equally at the rate of 0″·4 *per* day; and to be 8′ 51″·5 too fast at midnight the 18th of July, 1789, which is the time on which my observations on Saturn commenced.

(B) The satellite itself not being known, it is here called the supposed fifth.

(C) By six or seven degrees south preceding the line of the ring, is meant, that the satellite in the first place was at the preceding side of the planet; that is, in the semicircle from 180 to 360, which passes through 270 degrees. And in the next, that the situation of its orbit was such as to bring the satellite, at its proper distance, into a line drawn from the center of Saturn, making an angle of 6 or 7 degrees with the line of the ring, and declining towards the south.

(D) The calculated place for 20 h. 20′, shewn by the clock, corrected by −8′ 51″·5, and reduced to 12 h. 22′ 16″ mean time, is 245°·5 which, as no distance is mentioned, leaves it doubtful whether the observation was that of the 5th satellite, or of a fixed star.

(E) By calculation the situation is 268°·1; which agrees well enough for 11 h. 11′ 47″ mean time.

(F) 13 h. 59′ 38″ mean time gives 291°·5, which agrees with the distance and direction; but as the satellite was sp. the observation, which says np. must belong to some small fixed star.

(G) The calculation for 11 h. 11′ 27″ gives 27°·4 or at a good distance sf; therefore this was not the satellite, but a star.

(H) It appears from the calculation for 14 h. 48′ 29″ which gives 74°·2, and also from the following observations, that this was the real 5th satellite; and that, having once obtained its place, I kept it in view all the rest of the season.

(I) 12 h. 45′ 53″. 78°·4. (K) 10 h. 12′ 26″. 87°·2. (L) 11 h. 14′ 47″. 124°·2. (M) 12 h. 17′ 48″. 138°·3.

Sept. 13. 22 17. The 5th sat. of the 11th of Sept. is advanced, and is now north preceding a considerable large star, which was that night sp. ♄. By a figure it is nf. ♄, at a considerable distance (N).

Sept. 14. 20 33. The 5th a little nearer than last night (O).

22 30. The 5th sat. of Sept. 11, observed at 20 h. 11′, has left the place where it was at that time.

Sept. 16. 19 39. The 5th is drawing nearer towards its conjunction.

22 18. Much the same as before.

23 59. About 33° north following the direction of the R. (P).

1 3. The 5th nearly as before.

Sept. 17. 19 48. The 5th sat. of ♄ 30° nf. R. and at the distance of about 3 dia. of R. (Q).

Sept. 18. 21 15. About 2 dia. of R. and near 40° north following (R).

Sept. 20. 23 24. The 5th sat. is within a degree of its conjunction. It is north of ♄, and its motion is retrograde.

23 54. A perpendicular from the 5th sat. to the ring of Saturn, falls towards the following side short of the center by $\frac{1}{8}$ dia. of ♄.

0 19. Distance of the 5th satellite from the parallel of the R. of ♄, 3 rev. 36·7 parts = 1′ 0″·966 central measure.

1 25. The 5th very nearly central.

1 28. With a power of 240, perfectly central. With 300, perfectly central (S).

Sept. 21. 21 15. The 5th sat. is perpendicular to a place half a projection of the ring preceding the edge of it (T).

Sept. 23. 22 51. At a considerable distance np. ♄ (V).

Sept. 24. 19 56. At a good distance np. ♄ (W).

Sept. 25. 19 34. The 5th pursues its track (X).

Oct. 12. 21 18. The supposed 5th forms nearly an isosceles triangle with two preceding stars, the southern one of which is double, consisting of a very considerable star and a small one. By a figure, at a considerable distance, np. ♄ (Y).

Oct. 15. 21 1. The large star of the double star in the figure of the 12th of Oct. is gone from its place, and the supposed 5th of that night is left (Z).

21 8. The real 5th is so bright this evening, and was so the 12th of Oct. that I mistook it on that account for a considerable star; it was then nf. ♄. By three figures to-night it is at a great distance np. ♄. I saw it move to-night; for at 21 h. 1′ it made an angle of 50° on the following side with three stars in a line, sp. ♄. At 1 h. 9′, that angle was less than 40°; and at 1 h. 41′, it was no more than about 35°.

Oct. 16. 20 16. The 5th now precedes a line drawn through the three stars which it followed last night at 21 h. 1′. By five figures, at a great distance sp. ♄ (A).

Oct. 18. 20 18. At a great distance sp.

21 51. At 7 or 8′ distance sp. The same by two figures (B).

Oct. 20. 20 50. By three figures, at a great distance sp. (C).

Oct. 28. 21 1. The 5th about $3\frac{1}{2}$ dia. of ♄ distant, and 45° sp. (D).

Oct. 29. 21 49. The 5th sat. of ♄ is approaching towards its opposition (E).

Oct. 30. 20 53. The 5th sat. is past the opposition a little more than yesterday it wanted of it (F).

Oct. 31. 21 13. The 5th about $3\frac{1}{2}$ diameters of Saturn sf. (G).

	h.	′	″	°		h.	′	″	°		h.	′	″	°
(N)	10	35	13.	147·2.	(O)	8	47	34.	151·5.	(P)	12	5	9.	161·3.
(Q)	7	50	55.	165·1.	(R)	9	13	45.	170·0.					

(S) 13 18 13. 180·0, or directly at rectangles to the ring, to the north. I have used this observation for settling the epocha of this satellite, in which I have made no other allowance than that of the geocentric place of Saturn, as I knew this would answer all my purposes. But when we would obtain the mean motion of this satellite in comparing its present place with other situations at a great distance of time, proper reductions of the geocentric place of ♄ to the orbit of this satellite should be made. This may, however, be done much better when the real situation of its orbit is properly ascertained.

	h.	′	″	°		h.	′	″	°		h.	′	″	°		h.	′	″	°
(T)	9	1	59.	183·8.	(V)	10	29	52.	193·3.	(W)	7	31	25.	197·4.	(X)	7	5	33.	201·9.

(Y) 7 42 32. 280·3. The distance and situation agree well enough, but not the angle, which, by what will appear hereafter from the situation of the nodes of this satellite, should be sp.

	h.	′	″	°		h.	′	″	°		h.	′	″	°		h.	′	″	°
(Z)	7	13	49.	294·0.	(A)	6	25	0.	298·4.	(B)	7	51	54.	307·9.	(C)	6	43	13.	316·8.
(D)	6	22	47.	353·4.	(E)	7	6	44.	358·1.	(F)	6	6	57.	2·5.	(G)	6	22	59.	7·1.

Oct. 31. 21 43. It is very faint; fainter than the first; not much brighter than the sixth (H).

Nov. 2. 23 57. By a figure, the 5th sat. at some distance sf. (I).

Nov. 3. 22 1. At a good distance sf. R. (K).

22 13. The 5th a little following a line drawn through two fixed stars and between them. It is south of the ring.

Nov. 4. 22 17. The 5th at a great distance following, and a very little south; it precedes the line of the two stars which it followed last night (L).

Nov. 7. 22 9. The 5th at a great distance following, and a little north (M).

Nov. 8. 20 46. At a great distance following (N).

Nov. 10. 23 30. As the calculation gives it (O).

Nov. 13. 22 33. On the following side (P).

Nov. 15. 22 33. By a figure, nf. Saturn at a great distance (Q).

Nov. 19. 22 15. Dist. of the 5th sat. 1st measure 8′ 54″·94 but too small.

22 28.	—	—	2d	—	8 58 ·28
22 43.	—	—	3d	—	8 58 ·23
22 54.	—	—	4th	—	8 58 ·85

Mean of the three last measures 8′ 58″·45.

This, when the exact inclination of the orbit is ascertained, must be brought to the greatest elongation, and also reduced to the mean distance of the planet from the sun.

Dec. 2. 0 56. The 5th sat. is in its calculated place (R).

Dec. 5. 0 10. As the calculation gives it (S).

Dec. 16. 23 59. At a great distance preceding (T).

*Observations on the fourth satellite of Saturn.**

July 18. 19 50. The 4th satellite is about 6 or 7° np. R. (A).

July 23. 19 29. About 3½ dia. of ♄ following the body (B).

July 27. 20 27. 4 or 5 dia. following (C).

July 28. 19 40. Near 4 dia. following (D).

Aug. 18. 21 11. Many dia. preceding (E).

Aug. 28. 0 14. About 4½ dia. of ♄ following the body (F).

Aug. 29. 22 18. 3 dia. of ♄ following the body (G).

Aug. 31. 20 48. About 2¼ dia. of ♄ p. the body; a few seconds farther off than the 2d satellite, and a little south of it (H).

Sept. 8. 22 30. About 2½ dia. of ♄ following the body (I).

Sept. 10. 19 42. Following Saturn (K).

(H) From the considerable change in the light of this satellite, we may surmise, that it has a revolution upon its axis; the situation (see note Z and G), which affects the apparent brightness, should however be taken into the account.

	h. ′ ″	°		h. ′ ″	°
(I)	8 58 41.	16·8.	(K)	6 59 4.	21·0.

(L) 7 11 6. 25·2. A few days ago I perceived, that in the former part of these observations I had omitted a pretty essential circumstance, which is an attention to the nodes of the 5th sat. with the ring of Saturn.

(M) 6 h. 51′ 21″. 39°·2. It appears from this and the foregoing observation, that the ascending node of the 5th satellite, with regard to the ring of Saturn, apparently lies between the 25th and 39th degrees, which, reduced to a Saturnicentric position, is about the 19th degree from the point Aries reckoned upon the ring.

	h. ′ ″	°		h. ′ ″	°		h. ′ ″	°		h. ′ ″	°
(N)	5 24 39.	43·5.	(O)	8 0 21.	53·0.	(P)	6 51 44.	66·5.	(Q)	6 43 53.	75·6.
(R)	7 59 46.	152·8.	(S)	7 2 7.	166·2.	(T)	6 7 58.	215·4.			

* [Titan.—Ed.]

(A) 11 52 21. 284·0.

(B) 11 11 47. 36·4. Following or preceding the *body*, denotes that we are to reckon from the nearest part of the circumference, and not from the center; but it is also to be observed, that estimations in diameters, when they exceed one, or one and an half, are not intended as measures, but merely to point out the situation in a very coarse way; so that we are to look upon the calculation as not disagreeing with this estimation, though we should find the satellite considerably farther from the body.

	h. ′ ″	°		h. ′ ″	°		h. ′ ″	°		h. ′ ″	°
(C)	11 53 55.	127·5.	(D)	11 3 7.	149·3.	(E)	11 11 27.	264·7.	(F)	13 34 42.	133·4.
(G)	11 35 5.	154·2.	(H)	9 57 29.	198·0.	(I)	11 7 48.	20·3.	(K)	8 12 24.	62·8.

Sept. 11. 20 26. Following ♄, too far to estimate by the diameter (L).

Sept. 13. 22 17. By a figure, at a distance following (M).

Sept. 14. 20 33. About 1¼ dia. of the ring following (N).

0 42. 1¼ dia. of R. f. the edge.

1 24. 1 dia. of R. f. the ring; exactly in the line.

Sept. 16. 19 37. Not quite 1 dia. of the ring preceding (O).

22 15. 1½ dia. of R. p. the edge, and a little south.

23 59. Near 1½ dia. of R. preceding the edge of the ring.

Sept. 17. 19 48. About 3 dia. of the ring p. the projection (P).

Sept. 18. 21 15. Almost at its greatest distance p. (Q).

Sept. 20. 23 24. At a great distance p, and a little n. (R).

0 40. I can see the 4th satellite of Saturn without an eye-glass in the 20-feet speculum by drawing back the eye about three or four feet.

21 Sept. 21 15. At a good distance preceding, and a little north (S).

Sept. 23. 22 33. The 4th satellite emerged a few seconds ago. It is now in the line of a tangent to that part of Saturn where the projection of the ring comes from the body (T).

22 51. ⅓ of the projection (V) following the body of ♄; and about 2 of its own diameters north of the ring; or not quite half way northwards between the center of ♄ and the northern limb.

Sept. 24. 19 56. About 2 of its own dia. nearer the ring than the 3d sat. and a little more north.

20 48. The 4th advances to its conjunction with the 3d.

22 47. The 4th is past by the 3d. By a figure, it is less than one of its diameters past the 3d satellite, and is more north than the 3d (W).

Sept. 19 34. It pursues its track (X).

Oct. 12. 20 37. At a great distance following (Y).

Oct. 15. 20 54. Many diameters of Saturn f. (Z).

23 20. The 4th sat. at a considerable dist. f.

Oct. 16. 20 16. 2½ dia. of ♄ following the body.

21 59. The colour of the 4th satellite is red, or inclining to red; it approaches towards a conjunction with the 3d.

0 9. The 4th is very nearly in conjunction with the 3d; it is about ¼ of its own dia. nearer to Saturn than the 3d, and near one dia. of the 3d satellite more south than the 3d (A).

Oct. 18. 20 18. 2½ dia. of ♄ p. the body.

21 51. About 2½ dia. of ♄ preceding (B).

Oct. 20. 20 50. At a great distance preceding (C).

Oct. 28. 21 1. At a great distance following (D).

Oct. 29. 21 49. At a good distance sf. (E).

Oct. 30. 20 53. At a great distance following (F).

Oct. 31. 21 13. At a considerable distance following (G).

Nov. 2. 21 6. The 4th sat. is invisible.

21 51. I cannot see the 4th satellite (H).

22 53. The 4th sat. is not visible; I looked for the shadow of it upon ♄, but could not perceive it. The weather a little hazy.

23 21. Upon the dark equatorial belt of ♄, on the f. side, near the edge of the disk, seems to be a small black spot which is darker than the rest of the belt.

23 29. A protuberance on the sp. part of ♄; I suppose it to be the 4th satellite emerging.

	h. ′ ″	°		h. ′ ″	°		h. ′ ″	°		h. ′ ″	°
(L)	8 52 22.	86·1.	(M)	10 35 13.	133·0.	(N)	8 47 34.	154·0.	(O)	7 43 52.	198·3.
(P)	7 50 55.	221·1.	(Q)	9 13 45	245·0.	(R)	11 14 33.	292·2.	(S)	9 1 59.	312·8.
(T)	10 11 55.	359·2.									

(V) The distance from the body of Saturn to the end of the projecting part of the ring, I call the projection, and have made use of it as a measure for estimating.

	h. ′ ″	°		h. ′ ″	°		h. ′ ″	°		h. ′ ″	°
(W)	10 21 57.	22·0.	(X)	7 5 33.	41·6.	(Y)	7 1 39.	66·5.	(Z)	7 6 50.	134·4.
(A)	10 17 22.	160·1.	(B)	7 51 54.	203·0.	(C)	6 43 13.	247·2.	(D)	6 22 47.	67·9.
(E)	7 6 44.	91 2.	(F)	6 6 57.	112·8.	(G)	6 22 59.	135·7.			

(H) The tables I used at the time of this observation being different from my present ones, I expected the 4th satellite to be past its conjunction, and consequently visible again.

Nov. 2. 23 31. The black spot upon the equatorial belt seems to be a little advanced towards the preceding side.

23 38. With 300 the satellite is very nearly detached; the black spot keeps advancing; it is a very little north of the equatorial belt, but part of it is upon the belt.

23 43. The black spot is a little more than ⅛ of the dia. of ♄ advanced from the f. side towards the center.

23 46. The satellite seems to be detached.

23 47. With 300, it is detached; and the black spot keeps advancing.

23 57. The black spot is advanced so as to be ⅓ of its way towards the center; the 4th satellite is near ½ its own dia. clear of the edge.

0 13. The black spot a little more than half way towards the center; it is much darker than the belt.

0 34. The black spot is not arrived to the center yet.

0 53. The black spot is not come to the center, but does not want much of it.

0 57. It is more upon the belt than it was before; that is, more south.

1 6. The black spot is not yet come to the center.

1 10. It is drawn towards the south, so as to be nearly in the middle of the equatorial belt.

1 11. It is not far from a central position.

1 15. It is not come to the center yet.

1 18. The black spot is very near central.

1 21. Very near central.

1 25. Begins to be in the center.

1 30. It is in the center (I).

Nov. 3. 22 1. The 4th sat. about 3 dia. p. the body (K).

Nov. 4. 22 17. At a considerable distance preceding (L).

Nov. 7. 22 9. At a good distance preceding (M).

Nov. 8. 20 46. At a good distance p. (N).

Nov. 10. 21 3. The 4th satellite not yet visible.

21 32. Not yet visible.

22 26. Not yet visible.

23 24. About ½ dia. of ♄ nf. (O).

Nov. 13. 22 33. On the following side (P).

Nov. 15. 22 33. Following ♄ at a good distance (Q).

Nov. 26. 0 28. The 4th satellite is emerged some time past.

0 30. It is nearly in conjunction with the 6th. By a figure, it is half the diameter of the 6th nearer to Saturn than the sixth, and north of it (R).

Nov. 30. 23 36.	Dist. of the 4th satellite		3′ 12″·379.
23 42.	2d measure	—	3 10 ·972.
23 52.	3d —	—	3 10 ·494.
23 59.	4th —	—	3 10 ·579.

Mean of the four measures 3′ 11″·106 (S).

Dec. 2. 0 56. In its calculated place (T).

Dec. 5. 0 10. As the calculation gives it (V).

Dec. 16. 23 59. At a great distance following (W).

0 43. The 4th satellite, with a power of about 500, shews a pretty considerable, visible disk (X).

(I) 10 h. 31′ 25″·5. 184°·8. An extract of these observations being printed in my last Paper, I am to remark, that here the time is uncorrected; but the correction for this evening being − 8′ 8″·7, it will be seen, that in the former Paper − 8′ has been applied to all the times, and − 8′ 9″ to the time of the exact conjunction.

	h. ′ ″	°		h. ′ ″	°		h. ′ ″	°		h. ′ ″	°
(K)	6 59 4.	204·1.	(L)	7 11 6.	226·9.	(M)	6 51 21.	294·4.	(N)	5 24 39.	315·6.
(O)	7 54 22.	3·1.	(P)	6 51 44.	69·9.	(Q)	6 43 53.	115·0.	(R)	7 57 23.	4·4.

(S) The middle of the time to which we may suppose the measures to answer is 23 h. 48′, or 6 h. 59′ 48″ mean time. And by computation the apparent place of the satellite at that time was 93°·778, which is 3°·778 or 3° 46′ 41″·52 past the greatest elongation; therefore its distance, if it had been measured at the greatest elongation, would have been 3 11″·522. This quantity brought to the mean distance of Saturn from the sun, amounts to 3′ 8″·918.

(T) 7 h. 59′ 46″. 139°·8. (V) 7 h. 2′ 7″. 206°·6. (W) 6 h. 7′ 58″ 93°·6.

(X) And from its ruddy colour (see Oct. 16) we may surmise it to have a considerable atmosphere. This satellite, therefore, seems to approach more to the condition of a planet than any of the fourteen known satellites.

*Observations on the third satellite of Saturn.**

July 18. 19 50. The 3d satellite about 1 or 2° sf. R. By a figure, at a considerable distance (A).

July 23. 19 29. Near 2 dia. of ♄ following (B).

July 27. 20 27. About 2½ dia. of ♄ following (C).

July 28. 19 40. The 3d sat. ½ dia. following ♄; it is much larger than the 2d, and a little more north.

22 34. ⅓ part of a dia. following (D).

Aug. 18. 21 11. 1½ dia. of ♄ following the ring (E).

Aug. 28. 0 14. Full 2 dia. f. ♄ (F).

Aug. 29. 22 21. A satellite on the edge of the preceding arm (G).

23 1. The sat. a very little separated from it. I suppose it to be the 3d, on account of its size and brightness.

23 41. The satellite is now fully detached, so as to be near ⅓ of the projection preceding the end of it.

Aug. 31. 20 56. The preceding arm, about the middle, seems to be charged with a satellite; power 157 (H).

21 3. With 300, the same as before.

Sept. 8. 22 30. The 3d sat. about 2½ or 3 dia. of Saturn p. the body (I).

Sept. 10. 19 42. The 3d following Saturn (K).

Sept. 11. 20 26. 1½ dia of ♄ f. the body of ♄.

22 36. 1 dia of ♄ f. the body.

0 16. The 3d a little less than the projection from the f. edge.

0 34. ¾ of the projection following the ring.

1 57. A little less than ½ the projection f. the ring (L).

Sept. 13. 22 2. The 3d, 1¼ dia. of R. preceding the edge of R. (M).

Sept. 14. 20 27. The 3d sat. ½ the projection f. R. (N).

21 55. 1⅓ of the projection f. and a very little north.

0 42. 2 projections following.

1 24. Near two projections following.

Sept. 16. 22 18. (O).

Sept. 17. 19 48. The 3d sat. 1¾ dia. of the ring preceding the projection (P).

Sept. 18. 21 15. ⅓ of the projection, or 1½ dia. of the satellite preceding the edge of the ring.

21 45. ¼ of the projection preceding R.

21 53. The 3d almost touches the R.

21 59. Quite close to the ring, and a little north.

22 7. Not quite so near but that I can still see a small division.

22 20. With 157, I can no longer see a division between the 3d satellite and the R.

22 22. With 300, the sat. is completely joined to the R. but so as to make it appear a little longer, and a very little knotty towards the north (Q).

Sept. 20. 23 27. The 3d sat. 1½ projection f. R. It is within less than the diameter of the 2d satellite preceding the 2d, and a little more south (R).

23 51. The 3d sat. is now more separated from the 2d.

0 45. 1¼ projection f. R.

1 22. 1 projection f. R.

Sept. 21. 21 15. 2½ projections preceding the edge.

22 44. Near 3 projections p. R. (S).

* [Rhea.—Ed.]

	h. ′ ″	°		h. ′ ″	°		h. ′ ″	°		h. ′ ″	°
(A)	11 52 41.	85·5.	(B)	11 11 47.	121·9.	(C)	11 53 55.	83·1.	(D)	13 56 39.	169·6.
(E)	11 11 27.	35·0.	(F)	13 34 42.	120·6.	(G)	11 38 5.	193·9.	(H)	10 5 27.	348·3.
(I)	11 7 48.	269·9.	(K)	8 12 24.	59·7.	(L)	14 22 27.	159·9.	(M)	10 20 15.	306·1.
(N)	8 41 35.	20·4.									

(O) 10 24 26. 185·6. By this it appears that the 3d satellite was invisible; but observations being made on a satellite, by mistake supposed to be the third, they will be found among those of the 6th, to which they belong.

(P) 7h. 50′ 55″. 256°·9. (Q) 10h. 20′ 34″. 345°·0. (R) 11h. 17′ 33″. 147°·6. (S) 10h. 30′ 44″. 224°·8.

Sept. 23. 22 51. The 3d sat. 1 projection f. the edge (T).

Sept. 24. 19 56. 1$\frac{1}{2}$ dia. of R. f. the edge; about 2 diameters of the 4th satellite farther from R. than the 4th, and a little more south.

20 48. It advances towards a conjunction with the 4th.

22 47. It is past by the 4th. By a figure, it is about half its own diameter past the conjunction, and is more south than the 4th (V).

Sept. 25. 19 34. (W).

Oct. 12. 20 37. The 3d sat. about 3$\frac{1}{2}$ dia. of ♄ f. the body (X).

Oct. 15. 20. 47. 1$\frac{1}{4}$ dia. of ♄ p. the body.

21 30. A little more than 1 dia. of ♄ from the body; and its whole diameter north of the line of the projection.

22 25. The 3d sat. will be in conjunction with the 6th, in a very short time, the 3d being still a little preceding.

22 39. The conjunction is so complete now, that I have lost the 6th. The 3d, however, appears to be a little lengthened out towards the south. Distance from the body barely one diameter of ♄; or just one dia. of ♄ including the dia. of the 3d (Y).

23 54. Near two of its own diameters past the conjunction with the 6th.

0 59. The 1st, the 3d, and the 6th, are at equal distances from each other.

1 14. The 3d is nearer to the 1st than to the 6th.

1 35. The 3d approaches to a conjunction with the 1st.

1 45. The 3d is very near its conjunction with the 1st. By a figure, it wants less than $\frac{1}{2}$ a dia. of the 3d.

Oct. 16. 20 16. The 3d, 1$\frac{3}{4}$ dia. of ♄ following the body.

21 59. It draws towards a conjunction with the 4th. The colour of the 3d is inclining to blue.

0 9. The 3d is in conjunction with the 4th, or $\frac{1}{4}$ of the dia. of the 4th satellite past the conjunction; and one of its own dia. more north than the 4th; that is, there is a vacancy between them of one dia. of the 3d (Z).

Oct. 18. 20 18. 1 dia. of ♄ preceding the body.

21 51. 1$\frac{1}{4}$ dia. of ♄ p. the body (A).

Oct. 20. 22 19. With 300, I see the 3d sat. emerging; about $\frac{3}{4}$ of its own dia. is out, at the following side of ♄.

22 44. At the distance of $\frac{3}{4}$ of its dia. following the body (B).

0 8. The f. projection passes over the 3d sat. just so as to clear it.

Oct. 28. 21 1. The 3d about 3 dia. of ♄ p. the body (C).

Oct. 30. 20 53. About 3 dia. of ♄ following the body (D).

Oct. 31. 21 13. $\frac{5}{8}$ dia. of ♄ following the body.

21 57. There is a complete conjunction between the 3d sat. and the 2d; the following arm of the ring passes exactly between them, and points to the 6th. The distance between the 3d and 2d is about $\frac{1}{2}$ the diameter of the 3d satellite, the 2d being to the north, and the 3d to the south (E).

22 0. I can see that the conjunction between the 3d and 2d satellites is past.

23 13. The 3d, $\frac{1}{4}$ dia. of ♄ following the body (F).

23 49. The 3d approaching to a contact with the body, but I can see a division yet.

23 58. A division still visible between the 3d satellite and the body of ♄.

Nov. 2. 21 6. The 3d satellite, 1$\frac{3}{4}$ dia. of ♄ p. the body; just following the 2d, but a little more north (G).

21 44. 1$\frac{3}{4}$ dia. of ♄ preceding the body.

Nov. 3. 22 1. 2$\frac{1}{4}$ dia. of ♄ f. the body (H).

Nov. 4. 22 17. About 2 dia. of ♄ following the body (I).

Nov. 7. From 21 h. 28′ to 23 h 12′ (K).

	h. ′ ″	°		h. ′ ″	°
(T)	10 29 52.	24·3.	(V)	10 21 57.	103·6.
(W)	7 5 33.	172·5.	Hence it appears that the satellite could not be seen this night.		

	h. ′ ″	°		h. ′ ″	°		h. ′ ″	°		h. ′ ″	°
(X)	7 1 39.	88·2.	(Y)	8 51 33.	333·5.	(Z)	10 17 22.	57·9.	(A)	7 51 54.	209·4.
(B)	8 36 54.	11·4.	(C)	6 22 47.	281·8.	(D)	6 6 57.	80·4.	(E)	7 6 51.	163·5.
(F)	8 22 39.	167·7.	(G)	6 8 9.	319·6.	(H)	6 59 4.	42·5.	(I)	7 11 6.	122·5.

(K) From 6 h. 10′ 28″ to 7 h. 54′ 11″. 358°·3 to 4°·0, and consequently invisible.

Nov. 8. 20 46. At a considerable distance following (L).
Nov. 9. 1 2. ⅛ dia. of ♄ following the body; its whole dia. is south of the arm (M).
Nov. 10. 23 30. As the calculation gives it (N).
Nov. 13. 22 33. On the following side of ♄ (O).
Nov. 15. 22 33. At some distance, preceding (P).
Nov. 16. 22 50. (Q).
Nov. 21. 1 54. The 3d, about 1½ dia. of ♄ f. the body (R).
Nov. 25. 1 21. (S).
Nov. 30. 23 47. The 3d sat. about 2½ of its own dia. following the 6th (T)
Dec. 2. 23 36. ¾ of the projection p. the body; its whole diameter is to the south of the arm.
0 22. The 3d, the 1st, and the 6th, nearly at equal distances from each other.
0 50. The 3d and 1st are in conjunction with a little space between them; the 3d being to the south, and the 1st to the north (V).
Dec. 5. 0 10. As the calculation gives it (W).
Dec. 16. 23 59. 1½ dia. of ♄ p. the body (X).

*Observations on the second satellite of Saturn.**

July 18. 19 50. The 2d satellite in the line of the ring p. Saturn; but about 2 or 3° north (A).
July 23. 19 29. ¾ dia. of ♄ preceding (B).
July 27. 20 24. Upon the sf. part of the R. are two small bright points, the largest is to the south, and is nearest to the body of ♄ (C).
20 29. The largest of the knobs is about ¼ dia. of ♄ from the body. Memorandum, I have no doubt, but that the large knob is the 2d satellite; I could nearly see its whole diameter to the south of the ring, but not separated. Clouds came on (D).
July 28. 19 40. The 2d sat. of ♄ ¾ dia. following.
22 34. Almost a dia. f. (E).
Aug. 18. 21 11. ¾ dia. f. R. (F).
Aug. 28. 0 14. Near 2 dia. of ♄ p. the body (G).
Aug. 29. 22 18. About 1 dia. of ♄ p. the body (H).
Aug. 31. 20 48. A few seconds nearer to ♄ than the 4th satellite, and a little more north (I).
Sept. 8. 22 30. The 2d sat. 1¾ dia. of ♄ p. the body (K).
Sept. 10. 19 42. The 2d within 1 or 2 of its own diameters of the edge of the projection (L).
23 2. Invisible.
Sept. 11. 20 26. The 2d, 1¾ dia. of ♄ preceding the body.
22 36. 1¼, or almost 1½ dia. of ♄ preceding the body.
1 34. 1¼ of the projection preceding R.
1 57. 1 projection p. R. (M).
Sept. 13. 22 6. There are two satellites emerging instead of one (N).
22 13. With 300, the nearest, 1½ of its own diameters preceding the projection. This I take to be the 2d satellite (O).
Sept. 14. 20 27. Barely 1 projection p. R. and a little north.
21 55. The 2d, 1 dia. of the satellite p. R. (P).

(L) 5h. 24′ 39″. 75°·5. (M) 9h. 36′ 1″. 169°·1. (N) 8h. 0′ 21″. 243°·5. (O) 6h. 51′ 44″. 118°·8.
(P) 6 43 53. 277·8. (Q) 6 56 55. 358·2. invisible.
(R) 8 40 57. 42·4. (S) 8 52 10. 1·8. therefore invisible. (T) 6 58 48. 33·9.
(V) 7 53 47. 196·3. (W) 7 2 7. 72·4. (X) 6 7 58. 225·5.
* [Dione.—Ed.] (A) 11 52 21. 271·6. (B) 11 11 47. 205·8.
(C) By calculation we find that these two bright points were the 2d and 6th satellites; but at the time of these observations I only took down phænomena as they presented themselves, leaving a solution of them to future considerations. See Note (B) to the 6th satellite.
(D) 11h. 55′ 55″. 16°·0. (E) 13h. 56′ 39″. 158°·7. (F) 11h. 11′ 27″. 26°·8. (G) 13h. 34′ 42″. 276°·0.
(H) 11 35 5. 36·7. (I) 9 57 29. 291·0. (K) 11 7 48. 270·3. (L) 8 12 24. 157·5.
(M) 14 22 27. 322·9.
(N) See the observation of the 1st sat. Sept. 13.
(O) 10h. 31′ 13″. 205°·0. (P) 10h. 9′ 21″. 334°·6.

Sept. 14. 22 23. The 2d is now vanished (Q). With 300, I think there is about ½ dia. of the satellite left.

22 30. There now, certainly, is nothing left of the 2d sat.

Sept. 16. 19 39. The 2d sat. 1¼ of the projection p. R. (R).

22 15. 1 dia. of ♄ from the edge of the R. preceding, and exactly in the line of the R.

23 59. Almost 1 dia. of the R. preceding the edge of it.

1 3. About 1¼ dia. of ♄ from the preceding edge.

Sept. 17. 19 48. (S).

Sept. 18. 21 15. The 2d sat. 1¼ of the projection f. the edge; it is 1 dia. of the 1st satellite nearer to ♄ than the 1st, and a little more south.

22 35. A little less than 1 projection f. the edge.

23 14. About ¾ of the projection f. the edge.

0 12. ¼ of the projection following, and a little south.

0 27. With 300, the 2d sat. 1 of its own dia. f. the R.

0 51. With 157, the 2d sat. close to the R. so that no division can be perceived (T).

0 55. With 300, the satellite touches the R. and is a little south; its whole dia. is still out.

0 58. With 300, about ¾ of the dia. of the 2d sat. may yet be seen.

Sept. 20. 23 27. The 2d within one of its diameters following the 3d, and a little north (V).

23 51. The 2d is now more separated from the 3d.

0 45. 2 projections f. the edge of the ring.

1 22. 2½ projections f. R.

Sept. 21. 21 20. (W).

Sept. 23. 22 51. Almost 1 dia. of the R. f. the edge (X).

Sept. 24. 19 49. The 2d, upon the point of the ring p.; but I can see no vacancy.

19 56. With 300, the same appearance nearly; but the weather is too hazy, and the planet too low to bear it well.

20 45. The 2d sat. begins now to project a little, and is a little south of the ring (Y).

20 48. I can see a division between the 2d sat. and the R.

Sept. 25. 19 34 to 22 h. 38′ (Z).

Oct. 12. 20 37. The 2d sat. 1 full dia. of ♄ following the body (A).

Oct. 15. 20 54. The 2d sat. 1½ dia. of ♄ f. the body (B).

23 20. About 1½ dia. of ♄ following.

Oct. 16. 20 16. The 2d sat. ¾ dia. of ♄, or a little less p. the body (C).

20 36. ¾ dia. of ♄ p. the body.

22 35. 1¼ dia. of ♄ p. the body.

0 11. 1½ dia. of ♄ p. the body.

Oct. 17. 21 30. The 2d sat. 1 projection p. the body of ♄; very hazy weather (D).

Oct. 18. 20 18. The 2d sat. near 2 dia. of ♄ f. the body.

21 51. 1¾ dia. of ♄ f. the body.

0 52. It approaches to a conjunction with the 1st.

1 25. The 2d sat. very nearly in conjunction with the 1st.

1 38. The conjunction is complete (E). By a figure, the 2d is towards the north of the 1st; but they seem to be in contact.

Oct. 20. 21 17. The 2d sat. is emerged some time ago, and is now 1¼ of its own diameters from the body of ♄. I perceived the satellite as a protuberance before 20 h. 50′ (F).

(Q) The word vanished is here probably meant to denote its being gone upon the ring, to the projection of which it was approaching 28′ before.

	h. ′ ″	°		h. ′ ″	°		h. ′ ″	°		h. ′ ″	°
(R)	7 45 48.	224·9.	(S)	7 50 55.	356·8 invisible.	(T)	12 49 10.	155·7.	(V)	11 17 33.	50·6.
(W)	9 6 58.	170·2 consequently invisible.				(X)	10 29 52.	81·0.	(Y)	8 20 17.	200·7.

(Z) From 7 h. 5′ 33″ to 10 h. 9′ 3″, the satellite is not mentioned in my observations; though by calculation it appears, that its situation was from 325°·6 to 342°·3; and that therefore it ought to have been seen. I conclude from this that some particular cause must have rendered it invisible. Most probably it suffered an occultation from the 1st satellite, which was situated in such a manner as nearly to cover it the whole evening; in this case, the observation of the 1st belongs also to the 2d, since their diameters would certainly run together so as, perhaps, if the occultation was not always central, to form only one satellite, of rather a larger diameter than either of them.

	h. ′ ″	°		h. ′ ″	°		h. ′ ″	°
(A)	7 1 39.	42·4.	(B)	7 6 50.	77·7.	(C)	6 25 1.	205·4.
(D)	7 34 53.	343·4.	(E)	11 38 17.	137·2.	(F)	7 10 8.	15·9.

Oct. 28. 20 58. The preceding arm, on the north side, very near to the body, contains a considerable satellite.

21 5. The 2d sat. is close to the body, on the p. side, to the north of the ring (G).

Oct. 29. 21 49. The 2d about 1⅗ dia. of ♄ f. the body (H).

Oct. 30. 20 53. About 2 dia. of ♄ p. the body (I).

Oct. 31. 21 13. The 2d, ¼ dia. of ♄ following the body.

21 57. There is a complete conjunction between the 2d and 3d. The arm passes exactly between them, and points to the 6th. The distance between the 2d and 3d is about ½ the dia. of the 3d; the 2d satellite being to the north, and the 3d to the south (K).

22 0. I can see that the conjunction between the 2d and 3d satellites is past.

23 13. The 2d sat. is past the conjunction with the 6th.

Nov. 2. 21 6. The 2d sat. just preceding the 3d, but a little more south.

21 44. 2 dia. of ♄ p. the body, and a little more south than the 3d satellite (L).

Nov. 3. 22 1. The 2d sat. 2 dia. of ♄ following the body (M).

Nov. 4. 23 42. The 2d sat. ¼ dia. of ♄ p. the body (N).

23 57. The dia. of the 2d satellite is intirely south of the p. arm.

Nov. 7. 21 28. The 2d and 1st satellites, about 1 dia. of ♄, or a little farther, p. R

21 53. The 2d. sat 1⅓ dia. of ♄ p. the body (O).

Nov. 8. 20 46 to 23 h. 40′ (P).

Nov. 10. 23 30. As the calculation gives it (Q).

Nov. 13. 22 33. On the preceding side (R).

Nov. 15. 22 33. The 2d sat. is upon the preceding arm of the ring about half-way; all its dia. is towards the south (S).

22 49. The 2d is not quite to the end of the R. yet, but keeps advancing.

22 56. ¾ of the projection preceding the body, or ¼ wanting to being at the end of the ring.

Nov. 26. 22 27. The 2d is upon the p. arm; its whole dia. is towards the south (T).

0 28. The 2d is emerged some time past (V).

Dec. 2. 22 50. The 2d sat. about 1¾ dia. of ♄ preceding the body (W).

Dec. 5. 0 10. As the calculation gives it (X).

Dec. 16. 23 59. The 2d, about 1¼ or 1⅜ dia. of ♄ p. the body (Y).

*Observations on the first satellite of Saturn.**

July 18. 19 50. The following part of the ring of Saturn, which is a very thin lucid line, ends in a bright point like a very faint satellite (A).

I suppose the bright point on the f. part of the ring to be a very small fixed star (B).

20 14. The bright point on the following part of R. seems to have its whole dia. towards the north; and in all appearance adheres to the line.

0 48. Possibly the bright point on the nf. part of the ring may be one of the satellites, and one of the before supposed satellites may be a small fixed star (C).

(G) 6 h. 26′ 46″. 344°·6. (H) 7 h. 6′ 44″. 119°·8. (I) 6 h. 6′ 57″. 245°·9. (K) 7 h. 6′ 51″. 22°·9.
(L) 6 46 3. 284·1. (M) 6 59 4. 56·9. (N) 8 35 52. 197·3. (O) 6 35 24. 220·9.
(P) 5 24 39. 346·1. The first part of the evening was not very clear, and afterwards, by the calculation, the satellite was invisible.

(Q) 8 h. 0′ 21″. 263°·4. (R) 6 h. 51′ 44″. 291°·8. (S) 6 h. 43′ 53″. 194°·2. (T) 5 h. 54′ 43″. 196°·5.
(V) To emerge was here probably put for coming off the arm. (W) 5 54 6. 265·5. (X) 7 2 7. 306·3.
(Y) 6 h. 7′ 58″. 307°·8. * [Tethys.—ED.]

(A) 11 h. 52′ 21″. 146°·7. This shews, that the bright point was the 1st satellite.

(B) Being the first night of my viewing the satellites this year, their places were unknown. The 6th, which was in view, I took for the 1st satellite; but, the 2d, 3d, and 4th being also before me, there remained only the supposition of some small fixed star to account for the bright point.

(C) The motion of the bright point on the ring led me to the supposition of its being a satellite; and, to make room for one, it occurred, that one of the others might be a star: for still the thought of an unknown satellite did not happen to strike me. I should have made an attempt to calculate the places of the satellites by the manuscript tables of M. DE LA LANDE, which are now printed in the *Connoissance des Temps* for 1791; but as there chanced to be an erratum of one day's motion in the epochæ of all the satellites for 1788, of which I was not aware, I had so little satisfaction from them the year before, that I laid them by as useless, and resolved to investigate the epochæ and revolutions of the satellites from my own observations.

July 18. 21 15. The nf. bright point is advanced towards the body, so that it no longer hangs at the far end of the ring (D).

July 23. 19 29. (E).

July 27. 20 27. One dia. of ♄ f. is a small satellite (F).

July 28. 19 40. The 1st sat. 1½ dia. of ♄ preceding the body (G).

22 34. The 1st sat. as before.

Aug. 18. 21 11. 1½ dia. of ♄ p. R. (H).

Aug. 28. 0 9. 1¼ or 1⅓ projection f. R. or 1 dia. of ♄ from the body (I).

Aug. 29. 22 18. About 1½ dia. of ♄ p. the body (K).

Aug. 31. 20 54. 1½ dia. of ♄ p. the body (L).

Sept. 8. 22 51. (M).

Sept. 10. 22 49 and 23 h. 4′ (N).

Sept. 11. 1 0. The 1st sat. about 2 of its own dia. p. the projection; emerged since I looked last (O).

1 34. Half a projection p. R.

1 57. ¾ of the projection p. R.

Sept. 13. 22 0. There is a satellite emerging from the preceding arm; I take it to be the 1st (P).

22 6. There are two satellites emerging instead of one.

22 13. The one that emerged first, ⅓ of the projection p. R.

Sept. 14. 21 55. The 1st sat. ¾ of the projection f. R. and a very little south.

23 22. With 300, 1 or 1½ of its own dia. farther from the R. than the 6th, and a little more south (Q).

0 42. 1½ projection f. R.

1 24. 1½ projection f. R.

1 46. Much the same as before.

Sept. 16. 19 39. The 1st sat. ½ projection f. R. (R).

22 18. 1½ projection f. the edge, and a very little south.

23 59. About 1 dia. of ♄ f. the edge of the R.

1 3. ¾ dia. of ♄ f. R.

Sept. 17. 19 48. The 1st almost 1 projection p. R. (S).

20 38. 1⅓ projection p. R.

Sept. 18. 21 15. The 1st sat. 1¾ projection f. R. or one of its own dia. following the 2d satellite (T).

22 35. 2 projections following the R.

0 14. 1¼ projection f. R.

Sept. 20. From 23 h. 24′ to 1 h. 28′ (V). Notwithstanding my utmost endeavour, I could not perceive the 1st satellite. From the tables I surmise that it might be under an occultation, or eclipsed by the 3d satellite; I looked for it above two hours. It could be neither in the shadow of Saturn, nor in that of the ring.

(D) 13 h. 17′ 7″. 158°·0. Hence we see, that the satellite had advanced 11 degrees in its orbit towards ♄, since 11 h. 52′ 21″ which agrees with the motion of the bright point.

(E) 11 h. 11′ 47″. 15°·1. Therefore the 1st satellite was not visible.

(F) 11 53 55. 63°·5. It was the first.

(G) 11 3 7. 247°·5. The names of the satellites were by this time ascertained, and I found that the above-mentioned manuscript tables agreed pretty well with my observations this evening.

(H) 11 h. 11′ 27″. 294°·3. (I) 13 h. 49′ 42″. 60°·4. (K) 11 h. 35′ 5″. 236°·0. (L) 10 h. 3′ 29″. 245°·4.

(M) 11 28 45. 342°·9. Consequently the satellite was invisible, or at least might easily be overlooked, so near the body upon the p. arm as it must have been situated; but there was an observation made upon what is called the 1st satellite, which will be reported hereafter. See observations on the 7th satellite, Sept. 8.

(N) From 11 h. 18′ 54″ to 11 h. 33′ 51″. 3°·1 to 5°·1. The 1st satellite was invisible; but two observations were made upon what is called the first, which will be seen in the observations upon the 6th satellite. Such mistakes may easily be made during the time of observation, as a few hours will bring one of the inner satellites in view; but with such accuracy of calculating the precise moment and situation of the satellites, as has now been used, there can be no doubt to which satellite an observation belongs.

(O) 13 h. 25′ 37″. 210°·7. (P) 10 h. 18′ 15″. 207°·4. (Q) 11 h. 36′ 6″. 48°·5.

(R) 7 45 52. 39°·6. (S) 7 50 55. 231°·0. (T) 9 13 45. 72°·8.

(V) From 11 h. 14′ 33″ to 13 h. 18′ 13″. 110°·4 to 126°·6.

Sept. 21. 21 15. The 1st sat. 2 projections p. R. (W).

22 44. 1½ or 1¾ projection p. R.

Sept. 23. 22 51. ¾ projection p. R. and a very little north. It follows the 6th satellite 1 dia. of the 6th (X).

23 55. The 1st sat. almost touches the ring; it may want one of its diameters. Clouds interrupted the observation.

Sept. 24. 19 49. The first sat. 1¼ projection f. R. or about one of its own dia. f. the 6th.

20 45. The 1st sat. 1 full projection f. R.

22 47. Close to the following projection (Y).

Sept. 25. 19 34. The 1st sat. 1 full projection p. the edge of R. and a little north.

20 41. A little more than ½ projection p. R.

22 38. The 1st sat. has half its dia. projecting towards the north from the ring, on the preceding side; its place on the ring is about ⅔ of the projection from the body of ♄. The night is extremely clear (Z).

Oct. 12. 20 37. About ¾ dia. of ♄ p. the body.

21 24. ½ projection of R. preceding the edge, and considerably more north.

22 6. The 1st almost touches the p. projection.

22 24. It very nearly touches the p. projection.

23 26. ¾ of the projection p. the body of ♄; or, as it were, fastened upon the projection, ¼ from the end of it (A).

1 8. I see nothing of the 1st sat.

Oct. 15. 0 52. The 1st sat. ½ projection from the body (B).

0 59. The distance of the 1st from the body is almost, but not quite, equal to the distance of the 1st, 3d, and 6th from each other.

1 14. The first is nearer to the 3d than the 3d to the 6th.

1 35. The 1st and 3d approach to a conjunction.

1 45. The 1st and 3d very near their conjunction.

Oct. 16. 0 11. The 1st satellite ¼ dia. of ♄ f. the body (C).

1 20. ¼ of the projection f. the edge of the R.; the weather remarkably clear. I can see the R. very distinctly, so as to judge with safety of the projection.

Oct. 18. 21 7. The 1st sat. is lately emerged from the body of ♄ on the f. side.

21 12. The emerged sat. 1 of its own dia. f. the body of ♄.

21 32. Above 2 of its own dia. following ♄.

21 51. The 1st sat. ⅜ dia. of ♄ f. the body.

22 36. Very nearly clear of the f. projection (D). By hiding the planet behind the field-bar very carefully, I can see the projection of the R. very well on the f. side. The preceding projection cannot be distinguished so well on account of the satellites (E) that are upon it.

0 52. The 1st sat. approaches to a conjunction with the 2d.

1 25. The 1st and 2d sat. very nearly in conjunction.

1 38. The conjunction is complete.

Oct. 20. 20 5. The 1st sat. ⅛ dia. of ♄ f. the body (F).

20 50. It draws towards a conjunction with the 6th sat. distance 1 dia. of the 1st. The 1st is a little towards the north.

21 26. I can just see a very small division between the 1st and the 6th.

21 51. There is a perfect conjunction between the 1st and 6th.

Oct. 28. 21 1. The 1st sat. 1 dia. of ♄ f. the body.

21 50. It draws towards a conjunction with the 6th, distance 1 full dia. of the 1st (G).

Oct. 29. 21 49. The 1st about ⅞ dia. of ♄ p. the body (H).

Oct. 30. 20 53. The 1st sat. ¾ dia. of ♄ f. the body (I).

Oct. 31. 21 13. ¾ dia. of ♄ p. the body (K).

	h. ′ ″	°		h. ′ ″	°		h. ′ ″	°		h. ′ ″	°
(W)	9 1 59.	283·5.	(X)	10 29 52.	316·7.	(Y)	10 21 57.	146·4.	(Z)	10 9 3.	335·5.
(A)	9 50 11.	336·0.	(B)	11 4 11.	198·0.	(C)	10 19 22.	22·8.	(D)	8 36 47.	30·8
(E)	The 6th and 7th.		(F)	5 58 20.	31·3.	(G)	7 11 39.	127·0.	(H)	7 6 44.	317·1.
(I)	6 6 57.	139·8.	(K)	6 22 59.	332·7.						

Oct. 31. 23 13. ⅜ dia. of ♄ p. the body; its whole dia. seems to be north of the arm.

Nov. 2. 23 26. I suppose the 1st sat. to be upon the f. arm.

0 8. The f. arm contains a lucid point at the distance of ⅛ dia. of ♄ f. the body.

0 34. The 1st sat. almost ¼ dia. of ♄ f. the body (L).

Nov. 3. 22 3. The preceding arm is loaded in two places; at the far end, and about the middle.

23 48. The 1st sat. ½ dia. of ♄ p. the body (M); there seems to be another closely following it (N).

0 10. I can distinguish the two satellites that follow one another upon the arm; the distance between them is ½ dia. of the smallest of them.

Nov. 4. 22 14. The 1st sat. ¼ dia. of ♄ f. the body (O).

22 57. The dia. of the 1st sat. is north of the arm.

Nov. 7. 21 28. The 1st and 2d about 1 dia. of ♄, or a little more, p. the body (P).

21 53. The 1st satellite 1¼ dia. of ♄ p. the body.

Nov. 8. 20 46. ⅞ dia. of ♄ f. the body (Q).

Nov. 10. 23 30. The 1st as the calculation gives it (R).

Nov. 13. 22 33. On the preceding side (S).

Nov. 15. 22 33. About ¾ dia. of ♄ p. the body (T).

Nov. 16. 22 50. Upon the end of the f. arm (V).

Nov. 21. 0 54. ¾ dia. of ♄ f. the body (W).

Dec. 2. 22 49. The 1st sat. about 1 of its own dia. p. the 6th.

23 38. It is past the conjunction with the 6th, which it now follows, and it is a little more north than the 6th.

0 22. The 1st is equally distant from the following 6th, and the preceding 3d (X).

0 50. The 1st and 3d are in conjunction, with a little space between them, the 1st being to the north.

Dec. 5. 0 8 (Y).

Dec. 16. 23 59. The 1st sat. a little more than 1 dia. of ♄ f. the body (Z).

Dec. 24. 0 5. (A).

Dec. 25. 1 36. The 1st sat. is upon the end of the f. arm (B).

*Observations on the sixth satellite of Saturn.**

July 18. 19 50. The *first satellite* of ♄ exactly in a line of the R. preceding (A).

July 27. 20 24. Upon the sf. part of the ring are two small bright points; the largest, to the south, is nearest to the body (B), and the smallest, to the north, is at the farther end (C).

Aug. 28. 23 26. With the 40-feet reflector, I see the five known satellites of Saturn, and also another exactly in a line with the ring, interposed between the 2d satellite and the ring on the preceding side, while the 1st, 3d, 4th, and 5th are on the following one. It has so much the appearance of the other satellites, and ranges so well with them, that I have not a moment's doubt but that it is a sixth satellite. It is less bright than the rest, but seems to have light enough to be seen by my 20-feet telescope.

(L) 9 h. 35′ 35″. 19°·7. (M) 8 h. 45′ 47″. 203°·8. (N) It was the 6th. (O) 7 h. 8′ 6″. 21°·5.

(P) 6 h. 10′ 28″. 226°·0. By the equal distance which is mentioned, it appears, that the 1st and 2d satellites were in conjunction; and this agrees also with the next observation, compared with that of the 2d sat. Nov. 7th.

	h. ′ ″	°		h. ′ ″	°		h. ′ ″	°		h. ′ ″	°
(Q)	5 24 39.	50·8.	(R)	8 0 21.	92·7.	(S)	6 51 44.	295·9.	(T)	6 43 53.	316·2.
(V)	6 56 55.	148·6.	(W)	8 40 57.	35·9.	(X)	7 25 51.	323·5.	(Y)	7 0 7.	172·1 invisible.
(Z)	6 7 58.	102·4.	(A)	5 42 33.	184·3 invisible.	(B)	7 9 23.	26·3.	*	[Enceladus.—Ed.]	

(A) By computation for 11 h. 52′ 21″ we find, that the 6th sat. was 302·4; which is exactly in the place where a satellite called the first was observed; but it appears also from the calculation which has been given in the note A of the 1st satellite, that this observation cannot belong to the real 1st; the 6th satellite therefore was seen this evening without being known; and this explains all the difficulties which occurred with regard to the real 1st satellite. See observations on the 1st satellite, July 18.

(B) It was the 2d satellite. See observation on the 2d satellite, July 27.

(C) 11 h. 50′ 56″. 147°·1, which agrees exactly with the place of the 6th satellite.

Aug. 28. 0 9. 20-feet reflector. The new satellite $\frac{4}{5}$ of the projection of the ring preceding the edge of the R. (D).

0 20. A very small star about 60° sf. ♄, and 1$\frac{1}{4}$ dia. of ♄ distant from the body (E).

1 16. The small star is gradually left behind so as now to make an angle of about 35° sf. ♄; while at the same time the planet has carried along with him the new satellite.

1 24. The 6th satellite $\frac{2}{3}$ of the projection of the R. p. the ring.

1 46. The same small star is now only about 25° sf. ♄.

1 49. The new satellite is now not much more than $\frac{1}{2}$ the projection from the ring.

2 2. Saturn is gone on, in a retrograde order with respect to the small star, and has carried along with him the new discovered satellite.

Sept. 8. 22 30. The new or 6th sat. $\frac{4}{5}$ of the projection of the R. directly preceding (F).

Sept. 10. 22 49. The *first satellite* less than the projection from the following arm; extremely faint (G).

23 4. The *first satellite* the length of the projection following the arm; it is so faint that I cannot expect to see the new satellite (H).

Sept. 14. 21 59. I think I perceive a satellite between the 1st and the following projection close to the ring. 300 leaves it doubtful (I).

22 23. The 6th sat. $\frac{3}{4}$ of the projection f. R.; so close to the 1st that it requires great attention to be distinguished. With 300, the 6th is 1 or 1$\frac{1}{2}$ dia. of the 1st sat. nearer the R. than the 1st, and a little more north than the 1st, that is to say, very exactly in the line of the ring.

23 45. With 460, the 6th sat. is very near one whole projection f. R.

0 42. The 6th sat. 1 full projection f. R. (K).

1 24. The new sat. 1 projection f. R.

1 46. Very nearly, but not quite, 1 projection f. R.

Sept. 16. 22 18. The *third satellite* $\frac{1}{2}$ the projection from the preceding edge of R.

22 25. The *third satellite* is extremely small, and hardly to be seen; but I have no doubt.

23 59. *The third satellite* a little more than $\frac{1}{2}$ the projection preceding the edge.

0 16. *The third* much less than the 1st and 2d, partly owing to its proximity to the planet; but probably there may be an apparent change of magnitude from a revolution upon its axis.

1 3. *The third* about $\frac{3}{4}$ of the projection p. R. (L).

Sept. 17. 19 52. The 6th sat. 1 projection, or rather more f. the projection; extremely faint.

20 38. 1 full projection following (M).

22 55. $\frac{1}{2}$ projection f. and a little south; extremely small.

23 49. $\frac{1}{3}$ projection f. and a little south.

0 58. $\frac{1}{3}$ projection following.

1 46. Near $\frac{3}{4}$ of the projection following (N).

Sept. 21. 21 10. The 6th sat. 1 full projection f. R.; much fainter than the 1st; hazy weather.

21 20. 1 full projection f. the edge, and exactly in the line of the ring. I see it very well; it is less than the 1st.

22 9. 1$\frac{1}{4}$ of the projection f. the edge (O).

22 39. Nearly 1$\frac{1}{4}$ projection f. the edge of the R. exactly in the line of the R.

Sept. 23. 22 51. The 6th sat. 1 projection p. the edge; on 1 of its own diameters p. the 1st (P).

(D) 13 h. 29′ 42″. 294°·2.

(E) This star was immediately taken notice of, to verify the discovery of the 6th satellite.

(F) 11 h. 7′ 48″. 279°·2.

(G) 11 h. 18′ 54″. 86°·8. Which agrees perfectly with the 6th satellite, though it is here by mistake called the 1st.

(H) As I mistook the 6th satellite for the 1st, it was natural enough to find it very faint, and of course to suppose that the night was not clear enough to see the 6th, while at the same time I was making an observation on that very satellite. But it must here be remembered, that the time of its revolution was not yet well ascertained.

(I) The calculation for 10 h. 13′ 20″ gives 46°·0, which shews that the satellite was there.

(K) 12 h. 55′ 53″. 75°·7.

(L) The 6th sat. was this evening mistaken for the 3d; but the calculation for 13 h. 8′ 59″, which gives 243°·7, shews that these observations belong to the 6th; and therefore explains all the difficulty about the supposed change of magnitude of the 3d.

(M) 8 h. 40′ 47″. 97°·6.

(N) The estimation $\frac{3}{4}$ is probably a mistake in writing down, and should have been $\frac{1}{4}$; perhaps also some change in the atmosphere, or other circumstance, may have induced an error of estimation, which, in such minute objects, will now and then take place.

(O) 9 h. 55′ 50″. 82°·5. (P) 10 h. 29′ 52″. 254°·3.

Sept. 24. 19 46. I suspect the new or 6th satellite to be 1 projection f. the edge (Q).
19 49. The 6th is very near 1 projection f. the edge; it precedes the 1st sat. about 1 dia. of the 1st.
20 45. The 6th a little more than ½ projection f. the edge.
Sept. 25. 22 36. The 6th almost ¾ projection f. the edge of the R.
23 42. The 6th sat. very nearly 1 projection f. R. (R).
23 52. The 6th sat. is much larger than the 7th.
Oct. 12. From 22 h. 6′ to 1 h. 8′ (S).

0 58. *The seventh sat.* extremely small upon the point of the preceding projection, and a little towards the south.

1 20. The distance of *the seventh* increases. The satellite seems to be clear of the projection; but I can see no division yet; its whole dia. seems to be to the south; I see it full as well as I saw the 6th, or rather better (T).

1 35. *The seventh* is clear of the projection.

Oct. 15. 20 47. The 6th sat. 1 dia. of ♄ p. the body.

21 34. The 6th about ¾ of the projection p. the edge of R. or very near 1 dia. of ♄ p. the body; just sf. the 3d; it is in the line of the R. (V).

22 25. The 6th sat. will be in conjunction with the 3d in a very short time, the 3d being still a little p.

22 39. The conjunction is so complete now, that I have lost the 6th. The 3d, however, appears to be a little lengthened out towards the south. Distance of the conjoined satellites barely 1 dia. of ♄ from the body; or just 1 dia. of ♄, including the dia. of the 3d sat.

23 59. The 6th near 2 diameters of the 3d satellite past the conjunction.
0 59. The 6th, the 3d, and the first satellites are at equal distances from each other.
1 3. The 6th is 1 dia. of ♄ from the body.
1 39. The 6th nearly 1 dia. of ♄ p. the body.
Oct. 16. 20 16. The 6th sat. ⅔ dia. of ♄ f. the body; too low to be very accurate.
20 36. The 6th one full projection f. the body; extremely faint.
20 50. The 6th one projection f. the body.
21 11. The 6th ¾ of a projection f. the body.
21 55. The 6th ½ projection f. the body, or a little less (W).

22 5. The 6th advances towards a contact with the f. part of the body; I can, however, still look between them.

22 18. I can still see between the planet and the 6th sat.
22 22. The 6th less than its own dia. from the planet.
22 25. It is in contact with the body of ♄.
22 41. I can still perceive the 6th sat.
22 44. The 6th is not quite vanished.
22 47. The satellite is no longer visible.
Oct. 17. 21 30. The 6th sat. ½ projection f. the body, very hazy weather (X).
Oct. 18. 20 40. The 6th sat. is emerging from behind the 3d.

20 46. The 6th sat. which emerged from behind the 3d is a little north of the line of the ring, and of the 3d.

21 36. The 6th is going towards ♄, and is about ¾ dia. of ♄ preceding the body.
22 51. ¾ dia. of ♄ p. the body.
22 26. The 6th about ⅝ dia. of ♄ p. the body.
22 40. The 6th ½ dia. of ♄ p. the body.
23 17. The 6th approaches to a conjunction with the 7th.

23 37. The conjunction of the 6th and 7th satellites is past. The satellites are, however, too near the planet to see exactly how they are placed.

0 12. The 6th a little more than 1 of its own dia. p. the body of ♄ (Y).

(Q) 7 h. 21′ 28″. 122°·6. (R) 11 h. 12′ 52″. 67°·7.

(S) There were five observations made upon the 6th satellite, but they belong to the 7th. There were also three observations made upon the 7th which belong to the 6th, and are here given; we are to observe, that the revolution of the 7th was not yet ascertained, and that, consequently, a mistake of one new satellite for another could easily be made.

(T) 11 h. 43′ 53″. 220°·9. The calculation of its place shews plainly that it was the 6th; and the remark in this observation of its being brighter than the other satellite perfectly agrees with the calculation.

(V) 7 h. 46′ 43″. 246°·0. (W) 8 h. 3′ 44″. 151°·9. (X) 7 h. 34′ 53″. 49°·4. (Y) 10 h. 12′ 31″. 341°·0.

Oct. 20. 20 5. The 6th sat. 1¼ dia. of ♄ f. the body (Z).

20 50. The 6th and 1st satellites are drawing towards a conjunction; distance between them 1 dia. of the 1st sat.

21 26. I can just see a very small division between the 6th and the 1st.

21 51. There is a perfect conjunction between the 6th and the 1st.

22 22. The 6th sat. appears again.

22 43. The 6th is in the middle, between the 1st and 2d satellites.

23 50. I perceive the 6th sat. near the 3d towards the 2d; and on the south of the line that joins the 3d and 2d; but nearer the 3d than the 2d.

Oct. 28. 20 58. The 6th sat. about ½ projection f. the edge of the R.

21 5. The 6th about ¾ dia. of ♄ f. the body (A).

Oct. 29. 21 49. The 6th just f. the 1st (B).

Oct. 30. 20 55. I suspect the 6th on the edge of the p. arm; but moon-light is too strong.

23 44. The 6th sat. ¾ projection p. the edge of the R.

23 55. The 6th sat. ⅞ dia. of ♄ p. the body.

0 42. The 6th sat. 1⅛ dia. of ♄ p. the body (C).

Oct. 31. 21 13. The 6th sat. ⅞ dia. of ♄ p. the body.

21 57. The f. arm of the R. passes between the 3d and 2d satellites, and points to the sixth.

23 57. The 6th full ½ dia. of ♄ following the body (D).

Nov. 2. 21 44. The 6th sat. 1⅛ dia. of ♄ p. the body, and a little north; extremely faint (E).

22 17. The 6th sat. 1 full dia. or $1\frac{1}{16}$ dia. of ♄ p. and a very little north.

22 53. The 6th seems to be still $1\frac{1}{16}$ dia. of ♄, or rather more, p. the body, but the weather is hazy and foggy.

23 15. The 6th sat. 1 dia. of ♄ p. the body.

23 27. The 6th sat. ⅘ of the projection p. the edge or very near 1 dia. of ♄ p. the body.

0 15. The 6th sat. ⅞ dia. of ♄ p. the body.

0 58. The 6th sat. is still clear of the p. arm.

1 16. The 6th sat. ⅝ dia. of ♄ p. the body; a very little p. the edge of the ring, and a little north.

Nov. 3. 22 3. The p. arm is loaded in two places, at the far end and about the middle (F).

23 54. There seems to be a satellite closely following the 1st (G).

0 10. The distance between the two satellites upon the arm is half the dia. of the smallest.

Nov. 4. 22 17. The 6th satellite ⅞ dia. of ♄ f. the body (H).

23 48. The 6th about ¾ or ⅞ dia. of ♄ following the body.

Nov. 7. 21 28. The 6th towards the end upon the f. arm (I).

22 39. The 6th is drawn a little nearer towards ♄.

Nov. 8. 20 46. The 6th about ½ dia. of ♄ f., hazy weather, I do not see it well enough to estimate its distance very exactly.

21 16. The 6th sat. ⅝ dia. of ♄ f. the body.

22 2. The 6th sat. ⅞ dia. of ♄ f. the body.

23 40. The 6th sat. 1⅛ dia. of ♄ f. the body (K).

Nov. 9. 21 42. The 6th sat. ⅝ dia. of ♄ p. the body, and a little north (L).

Nov. 10. 21 33. The p. arm, near the end, seems to contain a sat. probably the 6th.

21 39. The 6th full ⅜ dia. of ♄ p. the body; almost intirely to the south of the arm (M).

22 28. The 6th is clear of the p. arm, and is about ⅝ dia. of ♄ p. the body.

23 27. The 6th almost ⅞ dia. of ♄ p. the body.

0 10. The 6th sat. 1 full dia. of ♄ p. the body.

Nov. 13. 22 33. The 6th sat. on the p. side (N).

Nov. 15. 22 33. The 6th sat. about 1 dia. of ♄ f. the body (O).

(Z) 5h. 58′ 20″. 100°·0. (A) 6h. 26′ 46″. 47°·5.

(B) 7 6 44. 317·5. For the place of the 1st, see the 1st sat. Oct. 29.

(C) 9 55 20. 250·9. (D) 9h. 6′ 32″. 144°·9. (E) 6h. 46′ 3″. 284°·7.

(F) They were the 1st and 6th satellites. See 1st sat. Nov. 3.

(G) 8h. 51′ 46″. 210°·5. (H) 7h. 11′ 6″. 94°·8. (I) 6h. 10′ 28″. 151°·9. (K) 8h. 18′ 11″. 78°·0.

(L) 6 16 34. 318·7. (M) 6 9 39. 220·1. (N) 6 51 44. 296·0. (O) 6 43 53. 100·0.

Nov. 19. 21 55. The 6th sat. $\frac{7}{8}$ dia. of ♄ f. the body (P).
Nov. 21. 0 54. The 6th at a little dist. p. the edge of the ring (Q); cloudy weather.
Nov. 25. 1 21. The 6th about $\frac{3}{4}$ dia. of ♄ following the body (R).
1 27. The 6th about $\frac{1}{2}$ the projection preceding the edge of the projection.
Nov. 26. 22 22. The 6th near 1 dia. of ♄ f. the body (S).
0 30. Very nearly in conjunction with the 4th.
Nov. 30. 23 47. The 6th full $\frac{7}{8}$ dia. of ♄ f. the body; about $2\frac{1}{2}$ dia. of the 3d sat. p. the 3d (T).
Dec. 2. 22 49. The 6th about $\frac{3}{4}$ dia. of ♄ p. the body; about 1 dia. of the 1st f. the 1st.
23 38. The 6th sat. is past its conjunction with the 1st, which it now precedes.
0 22. The 6th, the 1st, and 3d satellites, are nearly at equal distances from each other.
0 52. The 6th sat. nearly 1 dia. of ♄ p. the body (V).
Dec. 5. 0 8. The 6th sat. $\frac{1}{2}$ projection p. the arm (W).
Dec. 15. 0 35. The 6th about $\frac{3}{4}$ dia. of ♄ f. the body, and a little north (X).
Dec. 16. 23 59. The 6th full $\frac{5}{8}$ dia. of ♄ p. the body (Y).
Dec. 24. 0 5. The 6th sat. $1\frac{1}{8}$ dia. of ♄ p. the body; or $1\frac{1}{4}$ projection p. the edge of R. (Z).

*Observations on the seventh satellite of Saturn.**

Sept. 8. 22 51. The *first satellite* $\frac{1}{2}$ the dia. or a little less of the projection sf. R. (A).

Sept. 14. 1 29. A supposed 7th sat. excessively faint, $\frac{1}{2}$ projection p. R. exactly in the line of the R. fainter than the last new one.

1 46. The supposed 7th half a projection p. the R. (B).

Sept. 17. 21 0. A second new satellite excessively faint, $\frac{1}{2}$ projection p. the edge of the R. (C).

22 55. The new, or 7th sat. $\frac{1}{4}$ projection p. R. so excessively small that, if I had not seen it before, it would have been impossible to perceive it now.

23 1. After a more attentive observation and hiding the planet, I see the 7th sat. is not less than $\frac{1}{2}$ projection p. R.

23 31. Forty-feet reflector. I see six satellites at once, and being perfectly assured that the 2d is invisible, it becomes evident that Saturn has seven satellites. This new sat. is excessively small.

Sept. 18. 22 4. The new sat. near $\frac{1}{2}$ projection p. R. and a little south, but so faint that I hardly perceive it (D).

22 36. I cannot perceive the new satellite with the utmost attention (E). Indeed it was so faint before, that I almost entertained a doubt of its reality.

Sept. 25. 23 48. The 7th sat. I believe is between the 6th and the R. or $\frac{1}{3}$ projection f. the edge (F).

23 52. I see it very plainly; it is much smaller than the 6th; I have many times this evening before suspected it, but the weather has been too hazy.

Oct. 12. 22 6. *The sixth sat.* (G) close to the f. projection, and a little north.

	h. ′ ″	°		h. ′ ″	°		h. ′ ″	°		h. ′ ″	°
(P)	5 50 17.	61·1.	(Q)	8 40 57.	257·8.	(R)	8 52 10.	230·7.	(S)	5 49 44.	100·1.
(T)	6 58 48.	83·6.	(V)	7 55 46.	259·4.	(W)	7 0 7.	317·3.	(X)	6 47 48.	62·0.
(Y)	6 7 58.	317·3.	(Z)	5 42 33.	254·2.						

* [Mimas.—Ed.]

(A) Not being acquainted with more than six satellites, and having the 2d, 3d, 4th, 5th, and 6th in view, it was natural enough to call the remaining one, on which this observation is made, the 1st; but from the note M of the 1st sat. it appears, that it could not be in the place where this was seen; and by calculating from the tables of the 7th sat. we have its place for 11 h. 28′ 45″. 106°·0, which agrees exactly with the situation pointed out. From a figure, it appears, that the sat. was extremely small, and less than half its diameter south of the line of the R.

(B) I was now on the look-out for very small stars that were in any situation likely to be satellites of ♄, and always noticed them: for instance, "Sept. 11. 20 h. 42′. A supposed 7th sat. exactly in the line of the R. or a very little south, excessively faint, only to be seen when I hide ♄ by the field bar. Sept. 14. 20 h. 40′. The 7th of Sept. 11. is left in its place." So here this supposed 7th is marked down, and Sept. 16. 20 h. 13′ I find it is said, that "the supposed 7th of the 14th is a small fixed star, left in the place where it was that evening"; but as the configuration of stars which pointed out this supposed 7th was very coarse, and hardly sufficient to determine the place, and as by calculation it appears, that the 7th satellite was in the situation where this observation places it, at 13 h. 59′ 43″, *viz.* 278°·2, it is probable enough, that I saw the real satellite this evening.

(C) 9 h. 2′ 43″. 265°·6. (D) 10 h. 2′ 37″. 303°·5.

(E) From the calculated place 10 h. 34′ 32″. 312°·0, we see, that the satellite was drawn upon the arm, and therefore might easily be overlooked, especially as its revolution was unknown.

(F) 11 h. 18′ 48″. 117°·9.

(G) The satellite is here called the 6th, and we have seen before, in the note (S) of the 6th sat. that the 6th was called the 7th; but the tables of these satellites leave no doubt to which of them the observations belong.

Oct. 12. 22 13. I see *the sixth sat.* very well; but the projection is too faint to estimate the distance by it with any accuracy.

22 24. *The sixth* being nearer to Saturn on the f. side than the 1st on the preceding, must be quite close to the f. projection, or touching it.

Oct. 12. 23 35. *The sixth sat.* 1 projection, or perhaps a little less, f. the body; I see it with great difficulty, but have no doubt (H).

1 8. I see nothing of *the sixth* (I).

Oct. 16. 20 33. The 7th sat. ½ projection f. the body; that is, the sat. is upon the middle of the arm (K).

20 36. The 7th a little more than ¼ projection f. the body; extremely faint.

20 50. The 7th sat. 1 of its own dia. f. the body of ♄.

21 11. The 7th is very nearly in contact with the body.

21 15. I can still perceive the 7th sat. by means of the field bar hiding the planet.

21 55. The 7th is gone.

1 29. I have a strong suspicion of a sat. upon the p. projection not far from the end of it (L).

Oct. 18. 21 25. I am pretty sure the sat. is about 1 of its own dia. p. ♄ (M).

21 26. Very clear. I see the 7th sat. very plainly.

21 35. About 2 of its own dia. from the body of ♄, and a little north of the R.

21 43. The 7th sat. ¾ projection p. the body.

21 51. The 7th sat. ¼ dia. of ♄ p. the body.

22 26. The 7th sat. ¼ dia. of ♄ p. the body.

22 40. The 7th seems still to be where it was.

23 17. The 7th approaches to a conjunction with the 6th.

23 37. The conjunction of the 7th and 6th satellites is past. They are too near the planet to see exactly how they are placed.

0 20. The 7th sat. ⅝, or near ¾, dia. of ♄ p. the body.

0 24. The 7th is clear of the projection.

0 36. The 7th near ½ projection p. the edge of the ring. I see the R. well enough to estimate by it.

0 59. The 7th sat. ¾, or nearly ⅞, dia. of ♄ p. the body.

1 21. The 7th sat. ⅞ dia. of ♄ p. the body (N).

Oct. 20. 21 26. I have a glimmering sight of the 7th sat.

21 56. The 7th is perfectly detached from the p. arm.

23 5. The sat. ¾ of the dia. of ♄ p. the body, or thereabout.

23 37. The 7th extremely faint, near 1 dia. of ♄ p. the body; but the estimation of the distance is not very exactly to be had, as I am obliged to hide the planet when I see the sat. There is a high wind, and the air being dry, the telescope does not act so well as it did 1½ hour ago.

0 8. The 7th sat. ¾ dia. of ♄ p. the body; or ⅓ (or nearly ½) projection p. R. I see the ring very plainly (O).

0 20. The 7th sat. about ¾ dia. of ♄ p. the body.

1 20. The 7th sat. about ⅝ dia. of ♄ p. the body.

Nov. 4. 22 23. The p. arm seems to be loaded about ⅜ dia. of ♄ from the body (P).

Nov. 7. 22 0. The 7th sat. ⅞ dia. of ♄ p. the body, excessively small; but I see it extremely well, and can keep it in view; it is just following the 1st and 2d sat.

22 9. I see the 7th extremely well, notwithstanding its smallness.

22 39. The 7th about ⅞ dia. of ♄ p. the body (Q).

23 12. The 7th nearly the same as at 22 h. 39′, or perhaps a little nearer to ♄.

Nov. 8. 21 17. The 7th is clear of the p. arm; but I do not see it well enough yet to estimate how much.

22 0. The 7th sat. ¾ dia. of ♄ p. the body (R).

(H) The calculated place 9 h. 59′ 10″. 111°·2, gives the satellite farther from the arm than the observation; but as this also mentions the sat. was seen with difficulty, the interval might appear less than it would have done in a very clear view of the sat.

(I) The least change in the atmosphere would make the sat. invisible; and by the tables it also was now very nearly going upon the ring.

	h.	′	″	°		h.	′	″	°		h.	′	″	°		h.	′	″	°
(K)	6	31	59.	144·4.	(L)	11	37	9.	225·3.	(M)	7	25	58.	202·7.	(N)	11	21	20.	265·1.
(O)	10	0	40.	287·9.	(P)	7	17	5.	214·3.	(Q)	7	21	16.	281·4.	(R)	6	38	27.	291·9.

Nov. 8. 23 31. The 7th is upon the p. arm I suppose, for the weather is now very clear, and I see it no longer. There is a small protuberant point on the arm, which I take to be the satellite; but, as it has been cloudy, I have not been able to follow it so as to see it go on since 22 h. 2′.

23 40. I see the 7th upon the arm.

Nov. 10. 21 39. The p. arm is a little gouty, not quite $\frac{3}{8}$ dia. of ♄ preceding the body (s).

Nov. 15. 22 27. The 7th sat. is clear of the f. projection (T).

22 39. The 7th sat. between the 6th and the projection of the R. and a very little to the north.

22 44. The 7th considerably less than the 6th; I see it however very well, notwithstanding the difficulty of its situation.

22 56. By a figure, at a considerable distance following.

Nov. 16. 22 50. The 7th less than 1 dia. of the 1st following the first, and a little north (V).

Nov. 29. 0 38. A small luminous point on the f. arm (W).

Dec. 2. 23 38. I suspect the 7th to be just detached from the f. arm (X).

0 24. I cannot see the 7th, though I tried often for it.

0 56. The 7th sat. is not visible.

From the observations on the seven satellites of Saturn that have been here delivered, and closely compared with their calculated places, it appears evidently that the revolutions of these satellites are so well ascertained, that we may, without hesitation, determine that no phænomenon on the ring of Saturn, in the shape of lucid spot, protuberant point, or latent satellite, can be occasioned by any of them, when, upon computation, we find that the place of the satellite differs from that where such appearances were observed. In consequence of this deduction, I found, that the observations, which will be given presently, could not be explained by any of the known satellites; it remained, therefore, to be examined to what cause to ascribe the appearance of such lucid spots.

The first idea that occurred was that of another satellite, still closer to the ring than the seventh; and if a revolution, slower than about 15 hours and a quarter, could have been found, which would have taken in the most material places in which bright spots were seen, I should have continued of opinion that an eighth satellite, exterior to the ring, did exist, notwithstanding more observations had been wanting to put the matter out of all doubt. But this being impracticable, I examined, in the next place, what would be the result if these supposed satellites, or protuberant points, were attached to the plane or edge of the ring.

As observations, carefully made, should always take the lead of theories, I shall not be concerned if such lucid spots as I am now going to admit, should seem to contradict what has been said in my last Paper, concerning the idea of inequalities, or protuberant points. We may however remark, that a lucid, and apparently protuberant point, may exist without any great inequality in the ring. A vivid light, for instance, will seem to project greatly beyond the limits of the body upon which it is placed. If therefore the luminous places on the ring should be such as proceed from very bright reflecting regions, or, which is more probable, owe their existence to the more fluctuating causes, of inherent fires acting with great

	h.	′	″	°
(S)	6	9	39.	328·5.
(W)	7	53	35.	53·3.
(T)	6	37	54.	85·7.
(X)	6	41	59.	100·0.
(V)	6	56	55.	112·8.

violence, we need not imagine the ring of Saturn to be very uneven or distorted, in order to present us with such appearances as will be related. In this sense of the word, then, we may still oppose the idea of protuberant points, such as would denote immense mountains of elevated surface.

On comparing together several observations, a few trials shew that the brightest and best observed spot agrees to a revolution of 10 h. 32′ 15″·4; and, calculating its distance from the center of Saturn on a supposition of its being a satellite, we find it 17″·227, which brings it upon the ring. It is therefore certain, that unless we should imagine the ring to be sufficiently fluid to permit a satellite to revolve in it, or suppose a notch, groove, or division in the ring, to suffer the satellite to pass along, we ought to admit a revolution of the ring itself.

The density of the ring indeed may be supposed to be very inconsiderable by those who imagine its light to be rather the effect of some shining fluid, like an aurora borealis, than a reflection from some permanent substance; but its disapparition in general, and in my telescopes its faintness when turned edgeways, are in no manner favourable to this idea. When we add also, that this ring casts a deep shadow upon the planet, is very sharply defined both in its outer and inner edge, and in brightness exceeds the planet itself, it seems to be almost proved, that its consistence cannot be less than that of the body of Saturn; and that consequently, no degree of fluidity can be admitted sufficient to permit a revolving body to keep in motion for any considerable time.

A groove might afford a passage, especially as on a former occasion we have already considered the idea of a divided ring. A circumstance also which seems rather to favour this idea is, that, in some observations, a bright spot has been seen to project equally on both sides, as the satellites have been observed to do when they passed behind the ring. But, on the other hand, we ought to consider that the spot has often been observed very near the end of the arms of Saturn's ring, and that the calculated distance is consequently a little too small for such appearances, and ought to be 19 or 20 seconds at least. We should also attend to the size of the spot, which seems to be variable; for it is hardly to be imagined that a satellite, brighter than the sixth, and which could be seen with the moon nearly at the full, should so often escape our notice in its frequent revolutions, unless it varied much in its apparent brightness.

To this we must add another argument drawn from the number of lucid spots, which will not agree with the motion of one satellite only; whereas, by admitting a revolution of the ring itself, in 10 h. 32′ 15″·4; and supposing all the spots to adhere to the ring, and to share in the same periodical return, provided they last long enough to be seen many times, we shall be able to give an easy solution to all the remaining observations.

For instance, let α, β, γ, δ, ϵ, represent five spots on the ring of Saturn, situated as in fig. 2; where the ring is supposed to be divided into 360 degrees, and the spot

α placed at 271°·5; β at 70°·2; γ at 183°·0; δ at 142°·5; and ϵ at 358°·6. Then will the ring, with the spots thus placed, serve as an epocha for the year 1789; by which, with the assistance of a table constructed upon the before-mentioned period of the rotation of the ring, we may calculate their situation for any required time; and to render this calculation perfectly convenient, I have given a table, ready prepared for the purpose, at the end of the other tables.

The following observations have all been previously calculated by the tables of such of the seven satellites as were not already in view, and have been found to belong to neither of them; but in the notes that are given with them they have been again calculated by the table of the rotation of the ring for every time they were observed, on a supposition of their being spots adhering to it.

Observations not accounted for by Satellites.

July 28. 22 31. I now perceive between the nearest sat. and ♄, on the s. side, a small lucid point, like an emerging satellite (A).

22 37. The last discovered point, not quite half-way between the 3d sat. and the body of Saturn; may be it is a 6th sat. By a figure, the greatest part of its diameter is to the north of the ring (B).

Aug. 29. 23 1. The preceding projection contains a small inequality. By a figure, it follows the 3d sat. about $\frac{1}{8}$ of the projection of the ring (C).

Sept. 16. 19 39. I suspect one of the satellites close to the ring following (D).

20 6. I am pretty sure there is a satellite close to the following arm, and a very little to the north. 300 leaves it doubtful (E).

Oct. 15. 20 58. I suspect a satellite upon the preceding projection, not far from the end of it (F).

21 39. I cannot perceive the satellite on the preceding arm suspected at 20 h. 58′ (G).

Oct. 16. 1 29. I was not without a suspicion of another satellite upon the preceding arm, not quite so far advanced as the former (H).

Oct. 18. 20 22. I suspect two satellites upon the p. arm (I).

20 42. I make no doubt but that there is at least one sat. upon the p. arm (K).

21 14. I am in doubt whether it be a sat. upon the p. arm, or the arm itself (L).

21 17. Unless the p. arm be much brighter than the f. one, it must contain a satellite (M).

(A) My surmise of its being an emerging satellite so early as the beginning of the season, when I was still unacquainted with the minute phænomena that offered themselves afterwards, shews plainly, that the lucid point was of a sufficient brightness to deserve notice. The five old satellites were in view, and the 6th and 7th by calculation could not occasion this appearance, the former being at 72°·3; the latter on the opposite side at 299°·4. Supposing this, therefore, to be the spot I have called α, its place for 13 h. 53′ 39″ would be 36°·8; which might make it appear like an emerging satellite.

(B) By this time the spot was at 40°·3, which agrees with the observed situation. As the greatest part of its diameter appeared to be north, we may surmise, that the spot, which must have been of a very considerable size and brightness, was situated on the northern plane of the ring, and within a second or two from the outward edge of it. The ring itself was now so near having its edge directed towards us, that it required no great elevation of the spot to render it visible, notwithstanding it was then in the farthest part of its circuit.

(C) The spot α, at 12 h. 17′ 58″ was 301°·5.

(D) The spot β, at 7 h. 45′ 52″ was 58°·1. This spot was probably also on the northern plane, and on the very edge, but not so considerable as α.

(E) It was now advanced to 73°·4.

(F) The spot γ, at 7 h. 10′ 49″ was 305°·0. Its situation on the ring was probably on the southern plane, and at some considerable distance from the outward edge.

(G) It was now advanced to 328°·3; and therefore could hardly be seen any longer.

(H) The spot β. By calculation, at 11 h. 37′ 9″ it was 309°·2, or just following the 7th satellite, which appeared then upon the preceding arm in the shape of a small bright point.

(I) The spots γ and δ. The former at 6 h. 23′ 9″ was 217°·6, the latter 289°·9. δ is probably a spot upon the northern plane of the ring of a considerable degree of brightness, though but small in its dimensions, and at no great distance from the edge.

(K) The spot γ was by this time at 229°·0, and therefore in a situation to be easily perceived.

(L) The spot γ 247°·2.

(M) The spot γ 248°·9.

Oct. 18. 1 1. I am pretty sure the end of the preceding projection is loaded with two satellites. By a figure, one is placed ¼ projection from the end; the other, ⅓ projection from the body (N).

1 5. I can distinguish one upon the preceding projection very certainly (O).

Oct. 20. 21 26. I suspect the end of the preceding arm to be loaded with a satellite (P).

21 56. The preceding arm is certainly loaded with one or two satellites, or is more knotty than I have ever observed it to be. The weather is very beautiful (Q).

Oct. 30. 20 53. I suppose the 7th to be upon the following arm ¼ dia. of ♄ f. the body (R).

23 55. The 7th sat. ⅜ dia. of ♄ p. the body, or very near to the end of the p. arm, in the shape of a protuberant point (S).

0 1. I see it so well that there is no doubt but that it is a satellite (T).

0 42. The 7th is upon the p. arm, but a little nearer than it was before (V).

0 47. The 7th sat. ⅜ dia. of ♄ p. the body (W).

Oct. 31. 21 13. The 7th sat. ⅜ dia. of ♄ p. the body (X).

21 43. The 7th sat. is brighter than usual; I see it with great ease, notwithstanding the moon is almost at the full. It is brighter now than the 6th (Y).

22 11. The 7th is drawn nearer to the body of ♄. Flying clouds prevent estimations of the distance (Z).

23 13. The 7th is now no longer visible (A).

Nov. 2. 22 14. The 7th sat. ⅜ dia. of ♄ p. the body, it is upon the arm (B).

22 53. The 7th appears to be full ¼ dia. of ♄ p. the body; but hazy weather (C).

23 13. The 7th sat. ¼ dia. of ♄ p. the body (D).

0 8. The following arm contains a lucid point near the end of it (E).

0 16. The preceding arm seems to be loaded with two small points towards the end (F).

Nov. 4. 22 14. The 7th sat. I think, is between the 1st and 6th, but I cannot be sure (G).

22 27. I cannot perceive the 7th sat. where I suspected it (H).

23 47. The f. arm about ¼ dia. of ♄ from the body contains a small lucid point (I).

23 54. I see the point on the f. arm so well that I have not much doubt but that it is a satellite (K).

Nov. 7. 22 9. At the end of the p. arm is a place that is brighter than nearer to the body (L).

23 12. The preceding arm has still the appearance of a small protuberant point towards the south, near the end of the arm (M).

Nov. 8. 23 40. There is a protuberant point on the preceding arm besides the 7th sat.; so that at present I cannot tell whether the satellite be the nearest or farthest of them (N).

Nov. 10. 21 35. The following arm ¼ dia. of ♄ from the body contains a bright point; perhaps an 8th satellite (O).

Nov. 25. 1 21. The p. arm is loaded (P).

Nov. 29. 0 38. There are two small luminous points on the p. arm (Q).

Dec. 5. 0 8. Upon the end of the p. arm appears to be a bright point (R).

0 10. The spot on the preceding arm is rather larger than the 6th satellite (S).

Dec. 16. 0 7. The end of the p. arm seems to be loaded with a satellite (T).

(N) The spot α, at 11 h. 1′ 23″, was at 217°·2; and ϵ was also visible, being at 304°·3. This spot is probably a very small one, on the northern plane of the ring, at some distance from the edge.

(O) The spot α was now at 219°·4, and being very bright could be distinguished easily.

(P) The spot α, at 7 h. 19′ 7″, was at 290°·7.

(Q) The spot α was now at 307°·8; and at the same time the spot β being come on as far as 219°·3, was therefore visible.

(R) The spot δ, at 6 h. 6′ 57″, was at 40°·3.

(S) The spot α, at 9 h. 8′ 28″, was at 272°·4.

(T) It was now at 275°·8.

(V) At 299°·1.

(W) At 302°·0.

(X) The spot α, at 6 h. 22′ 59″, was 278°·4.

(Y) It was now at 295°·5.

(Z) At 305°·7.

(A) At 346°·7.

(B) The spot ϵ, at 7 h. 15′ 58″, was at 235°·6.

(C) At 257°·8.

(D) At 269°·1.

(E) The spot δ, at 84 ·4.

(F) The spot ϵ, at 305°·0.

(G) The spot ϵ, at 7 h. 8′ 7″, was at 71°·0, which agrees with the place, and it might be the supposed satellite.

(H) At 78°·3.

(I) The spot α, at 8 h. 40′ 51″, was at 36°·7.

(K) It was now at 40°·7.

(L) The spot α, at 6 h. 51′ 21″, was at 274°·0.

(M) It was at 309°·8.

(N) The spot δ, at 8 h. 18′ 11″, was at 294°·4.

(O) The spot δ, at 6 h. 5′ 40″, was at 59°·1.

(P) The spot ϵ, at 8 h. 52′ 10″, was at 68°·6.

(Q) The spot β, at 7 h. 53′ 35″, was at 259°·6.

(R) The spot ϵ, at 7 h. 0′ 7″, was at 283°·8.

(S) It was now at 285°·0.

(T) The spot ϵ, at 6 h. 15′ 57″, was at 277°·4.

Dec. 24. 0 7. The p. arm contains a pretty bright point ⅗ towards the end of it (v).
Dec. 25. 23 39. The p. arm very near the end is loaded with a sat. (w).
1 10. The p. arm is loaded very nearly at the far end of it, and a little towards the south (x).
1 37. The bright point is near the far end of the p. arm (y).

The great accordance between the observed places of these spots and the calculated ones, seems to establish the rotation of the ring of Saturn on an axis so as hardly to leave any doubt upon the subject. The time of it, we have already seen, is 10 hours, 32 minutes, and 15·4 seconds. It may be objected, that many of the observations are such as would also agree with other assignable periods, especially when the numbers of spots is so considerable as five; but the most material observations, which are those on the spot α, setting aside all the rest, seem alone to amount to a proof not only of a rotation of the ring, but of the time in which it is performed.

It may be expected, that having now sufficiently examined the whole series of observations of the last new satellites, we can give their periodical times and distances more accurately than before. The times, indeed, are full as well ascertained as we can expect to have them: for on calculating six satellites by my tables back to Aug. 19 d. 12 h. 19′ 56″, 1787, we find their places 341°·1 the 5th; 10°·6 the 4th; 211°·1 the 3d; 158°·9 the 2d; 80°·2 the 1st; and 288°·8 the 6th. And my journal contains the fullest assurance that they were thus situated at the time for which this calculation is made. We may therefore fix the period of the sixth at 1 d. 8 h. 53′ 8″·9. The 7th satellite can only be traced back as far as the 8th of Sept. 1789; so that its revolution will require at least another season to come to some degree of accuracy, till when we shall state it at 22 h. 37′ 22″·9.

The distance of these satellites, deduced from calculation, depends intirely upon the time and distance of the 4th, which is the satellite that has been used. In order to obtain more accuracy in these elements, I have applied myself to measuring the distance of the 4th satellite in those moments which were most favourable for the purpose. It is well known that this subject, on account of the quantity of matter in Saturn, to be deduced from the periodical times and distances of the satellites, is of considerable importance to astronomers; I shall therefore defer a full investigation of it till I can have an opportunity of calculating a great number of measures, not only of the 4th and 5th, but also of the other satellites which I have already by me, and still intend next season to take. Meanwhile, having brought the measures of the 30th of November, which seem to me to be very good ones, to the mean distance of Saturn from the sun, I find they give the distance of the 4th satellite from Saturn 3′ 8″·918. In reducing these measures to the mean distance, I have used the new tables of M. DE LAMBRE for Saturn, and MAYER'S for the sun.

(v) The spot α, at 5 h. 44′ 33″, was at 251°·6.
(w) The spot β, at 5 h. 12′ 42″, was at 244°·8.
(x) It was now at 296°·0.
(y) And now at 311°·4.

Admitting therefore the above quantity as the distance, and 15 d. 22 h. 41′ 13″·4 as the period of the 4th satellite, we compute that the distance of the 6th from the center of Saturn is 36″·7889; and that of the 7th, 28″·6689.

Tables for the seven satellites of Saturn.

Epochs of the mean longitude of the satellites.

	5. sat.	4. sat.	3. sat.	2. sat.	1. sat.	6. sat.	7. sat.
Years.	Deg. dec.	Deg. dec.	Deg. dec.	Deg. dec.	Deg. dec.	Deg. dec.	Deg. dec.
1787	335·91	149·16	87·21	272·18	176·46	269·31	307·07
1788	196·84	132·41	93·86	173·95	131·91	307·48	65·02
1789	53·23	93·09	20·82	304·19	256·66	82·92	161·00
1790	269·63	53·77	307·78	74·43	21·41	218·36	256·98
1791	126·02	14·45	234·74	204·68	146·16	353·81	352·97

Saturnicentric motion of the satellites in months.

	5th.	4th.	3d.	2d.	1st.	6th.	7th.
Months.	Deg. dec.	Deg. dec.	Deg. dec.	Deg. dec.	Deg. dec.	Deg. dec.	Deg. dec
January	000·00	000·00	000·00	000·00	000·00	000·00	000·00
February	140·68	339·89	310·40	117·58	151·64	224·54	320·81
March	267·75	252·05	21·73	200·56	91·18	20·91	215·73
April	48·43	231·95	332·13	318·14	242·81	245·45	176·54
May	184·57	189·26	202·84	304·19	203·75	207·27	115·39
June	325·25	169·16	153·24	61·77	355·39	71·81	76·20
July	101·39	126·47	23·94	47·82	316·33	33·63	15·05
August	242·07	106·37	334·34	165·40	107·96	258·17	335·86
September	22·75	86·26	284·74	282·98	259·60	122·72	296·67
October	158·89	43·58	155·45	269·03	220·54	84·54	235·52
November	299·57	23·47	105·85	26·61	12·17	309·08	196·33
December	75·71	340·78	336·56	12·66	333·11	270·90	135·17

In the months January and February of a bissextile year subtract 1 from the number of days given.

Motion of the satellites in days.							
	5th.	4th.	3d.	2d.	1st.	6th.	7th.
Days.	Deg. dec.	Deg. dec.	Deg. dec.	Deg. dec.	Deg. dec.	Deg. dec.	Deg. dec.
1	4·54	22·58	79·69	131·53	190·70	262·73	21·96
2	9·08	45·15	159·38	263·07	21·40	165·45	43·92
3	13·61	67·73	239·07	34·60	212·09	68·18	65·88
4	18·15	90·31	318·76	166·14	42·79	330·91	87·85
5	22·69	112·89	38·45	297·67	233·49	233·64	109·81
6	27·23	135·46	118·14	69·21	64·19	136·36	131·77
7	31·77	158·04	197·83	200·74	254·89	39·09	153·73
8	36·30	180·62	277·52	332·28	85·58	301·82	175·69
9	40·84	203·19	357·21	103·81	276·28	204·55	197·65
10	45·38	225·77	76·90	235·35	106·98	107·27	219·62
11	49·92	248·35	156·59	6·88	297·68	10·00	241·58
12	54·46	270·93	236·28	138·42	128·38	272·73	263·54
13	58·99	293·50	315·97	269·95	319·07	175·45	285·50
14	63·53	316·08	35·66	41·49	149·77	78·18	307·46
15	68·07	338·66	115·35	173·02	340·47	340·91	329·42
16	72·61	1·24	195·04	304·56	171·17	243·64	351·39
17	77·15	23·81	274·74	76·09	1·87	146·36	13·35
18	81·69	46·39	354·43	207·63	192·56	49·09	35·31
19	86·22	68·97	74·12	339·16	23·26	311·82	57·27
20	90·76	91·54	153·81	110·70	213 96	214·54	79·23
21	95·30	114·12	233·50	242·23	44·66	117·27	101·19
22	99·84	136·70	313·19	13·77	235·35	20·00	123·16
23	104·38	159·28	32·88	145·30	66·05	282·73	145·12
24	108·91	181·85	112·57	276·84	256·75	185·45	167·08
25	113·45	204·43	192·26	48·37	87·45	88·18	189·04
26	117·99	227·01	271·95	179·91	278·15	350·91	211·00
27	122·53	249·58	351·64	311·44	108·84	253·64	232·96
28	127·07	272·16	71·33	82·98	299·54	156·36	254·92
29	131·60	294·74	151·02	214·51	130·24	59·09	276·89
30	136·14	317·32	230·71	346·05	320·94	321·82	298·85
31	140·68	339·89	310·40	117·58	151·64	224·54	320·81

Motion of the satellites in hours.							
	5th.	4th.	3d.	2d.	1st.	6th.	7th.
Hours.	Deg. dec.	Deg. dec.	Deg. dec.	Deg. dec.	Deg. dec.	Deg. dec.	Deg. dec.
1	0·19	0·94	3·32	5·48	7·95	10·95	15·92
2	0·38	1·88	6·64	10·96	15·89	21·89	31·83
3	0·57	2·82	9·96	16·44	23·84	32·84	47·75
4	0·76	3·76	13·28	21·92	31·78	43·79	63·66
5	0·95	4·70	16·60	27·40	39·73	54·73	79·58
6	1·13	5·64	19·92	32·88	47·67	65·68	95·49
7	1·32	6·58	23·24	38·36	55·62	76·63	111·41
8	1·51	7·53	26·56	43·84	63·57	87·58	127·32
9	1·70	8·47	29·88	49·33	71·51	98·52	143·24
10	1·89	9·41	33·20	54·81	79·46	109·47	159·15
11	2·08	10·35	36·52	60·29	87·40	120·42	175·07
12	2·27	11·29	39·84	65·77	95·35	131·36	190·98
13	2·46	12·23	43·17	71·25	103·29	142·31	206·90
14	2·65	13·17	46·49	76·73	111·24	153·26	222·81
15	2·84	14·11	49·81	82·21	119·19	164·20	238·73
16	3·03	15·05	53·13	87·69	127·13	175·15	254·64
17	3·21	15·99	56·45	93·17	135·08	186·10	270·56
18	3·40	16·93	59·77	98·65	143·02	197·05	286·47
19	3·59	17·87	63·09	104·13	150·97	207·99	302·39
20	3·78	18·81	66·41	109·61	158·91	218·94	318·30
21	3·97	19·75	69·73	115·09	166·86	229·89	334·22
22	4·16	20·70	73·05	120·57	174·81	240·83	350·13
23	4·35	21·64	76·37	126·05	182·75	251·78	6·05
24	4·54	22·58	79·69	131·53	190·70	262·73	21·96

Motion of the satellites in minutes.							
	5th.	4th.	3d.	2d.	1st.	6th.	7th.
Min.	Deg. dec.	Deg. dec.	Deg. dec.	Deg. dec.	Deg. dec.	Deg. dec.	Deg. dec.
1	0·00	0·02	0·06	0·09	0·13	0·18	0·27
2	0·01	0·03	0·11	0·18	0·26	0·36	0·53
3	0·01	0·05	0·17	0·27	0·40	0·55	0·80
4	0·01	0·06	0·22	0·37	0·53	0·73	1·06
5	0·02	0·08	0·28	0·46	0·66	0·91	1·33
6	0·02	0·09	0·33	0·55	0·79	1·09	1·59
7	0·02	0·11	0·39	0·64	0·93	1·28	1·86
8	0·03	0·13	0·44	0·73	1·06	1·46	2·12
9	0·03	0·14	0·50	0·82	1·19	1·64	2·39
10	0·03	0·16	0·55	0·91	1·32	1·82	2·65
11	0·04	0·17	0·61	1·00	1·46	2·01	2·92
12	0·04	0·19	0·66	1·10	1·59	2·19	3·18
13	0·04	0·20	0·72	1·19	1·72	2·37	3·45
14	0·05	0·22	0·77	1·28	1·85	2·55	3·71
15	0·05	0·24	0·83	1·37	1·99	2·74	3·98
16	0·05	0·25	0·89	1·46	2·12	2·92	4·24
17	0·06	0·27	0·94	1·55	2·25	3·10	4·51
18	0·06	0·28	1·00	1·64	2·38	3·28	4·78
19	0·06	0·30	1·05	1·73	2·52	3·47	5·04
20	0·07	0·31	1·11	1·83	2·65	3·65	5·31
21	0·07	0·33	1·16	1·92	2·78	3·83	5·57
22	0·07	0·34	1·22	2·01	2·91	4·01	5·84
23	0·08	0·36	1·27	2·10	3·05	4·20	6·10
24	0·08	0·38	1·33	2·19	3·18	4·38	6·37
25	0·08	0·39	1·38	2·28	3·31	4·56	6·63
26	0·09	0·41	1·44	2·37	3·44	4·74	6·90
27	0·09	0·42	1·49	2·47	3·57	4·93	7·16
28	0·09	0·44	1·55	2·56	3·71	5·11	7·43
29	0·10	0·45	1·60	2·65	3·84	5·29	7·69
30	0·10	0·47	1·66	2·74	3·97	5·47	7·96

Motion of the satellites in minutes.							
	5th.	4th.	3d.	2d.	1st.	6th.	7th.
Min.	Deg. dec.	Deg. dec.	Deg. dec.	Deg. dec.	Deg. dec.	Deg. dec.	Deg. dec.
31	0·10	0·49	1·72	2·83	4·10	5·66	8·22
32	0·11	0·50	1·77	2·92	4·24	5·84	8·49
33	0·11	0·52	1·83	3·01	4·37	6·02	8·75
34	0·11	0·53	1·88	3·10	4·50	6·20	9·02
35	0·12	0·55	1·94	3·20	4·63	6·39	9·29
36	0·12	0·56	1·99	3·29	4·77	6·57	9·55
37	0·12	0·58	2·05	3·38	4·90	6·75	9·82
38	0·13	0·60	2·10	3·47	5·03	6·93	10·08
39	0·13	0·61	2·16	3·56	5·16	7·12	10·35
40	0·13	0·63	2·21	3·65	5·30	7·30	10·61
41	0·14	0·64	2·27	3·74	5·43	7·48	10·88
42	0·14	0·66	2·32	3·83	5·56	7·66	11·14
43	0·14	0·67	2·38	3·93	5·69	7·85	11·41
44	0·15	0·69	2·43	4·02	5·83	8·03	11·67
45	0·15	0·71	2·49	4·11	5·96	8·21	11·94
46	0·15	0·72	2·55	4·20	6·09	8·39	12·20
47	0·16	0·74	2·60	4·29	6·22	8·58	12·47
48	0·16	0·75	2·66	4·38	6·36	8·76	12·73
49	0·16	0·77	2·71	4·47	6·49	8·94	13·00
50	0·17	0·78	2·77	4·57	6·62	9·12	13·27
51	0·17	0·80	2·82	4·66	6·75	9·30	13·53
52	0·17	0·82	2·88	4·75	6·88	9·49	13·80
53	0·17	0·83	2·93	4·84	7·02	9·67	14·06
54	0·18	0·85	2·99	4·93	7·15	9·85	14·33
55	0·18	0·86	3·04	5·02	7·28	10·03	14·59
56	0·18	0·88	3·10	5·11	7·41	10·22	14·86
57	0·19	0·89	3·15	5·20	7·55	10·40	15·12
58	0·19	0·91	3·21	5·30	7·68	10·58	15·39
59	0·19	0·93	3·27	5·39	7·81	10·76	15·65
60	0·20	0·94	3·32	5·48	7·94	10·95	15·92

Table of the rotation of the ring of Saturn.

Epochs for 1789.

Spot α	271·5
β	183·0
γ	70·2
δ	142·5
ϵ	358·6

Motion of the spots in Months.

Months.	Deg. dec.
January	000·00
February	217·52
March	135·28
April	352·80
May	110·40
June	327·92
July	85·52
August	303·04
September	160·56
October	278·16
November	135·68
December	253·28

Motion of the spots in days, hours, and minutes.

Days.	Deg. dec.	Hours.	Deg. dec.	Min.	Deg. dec.	Min.	Deg. dec.
1	99·92	1	34·16	1	0·57	31	17·65
2	199·84	2	68·33	2	1·14	32	18·22
3	299·76	3	102·49	3	1·71	33	18·79
4	39·68	4	136·65	4	2·28	34	19·36
5	139·60	5	170·81	5	2·85	35	19·93
6	239·52	6	204·98	6	3·42	36	20·50
7	339·44	7	239·14	7	3·99	37	21·07
8	79·36	8	273·30	8	4·56	38	21·64
9	179·28	9	307·46	9	5·12	39	22·21
10	279·20	10	341·63	10	5·69	40	22·78
11	19·12	11	15·79	11	6·26	41	23·35
12	119·04	12	49·95	12	6·83	42	23·91
13	218·96	13	84·11	13	7·40	43	24·48
14	318·88	14	118·28	14	7·97	44	25·05
15	58·80	15	152·44	15	8·54	45	25·62
16	158·72	16	186·60	16	9·11	46	26·19
17	258·64	17	220·76	17	9·68	47	26·76
18	358·56	18	254·93	18	10·25	48	27·33
19	98·48	19	289·09	19	10·82	49	27·90
20	198·40	20	323·25	20	11·39	50	28·47
21	298·32	21	357·41	21	11·96	51	29·04
22	38·24	22	31·59	22	12·53	52	29·61
23	138·16	23	65·75	23	13·10	53	30·18
24	238·08	24	99·91	24	13·67	54	30·75
25	338·00			25	14·24	55	31·32
26	77·92			26	14·80	56	31·89
27	177·84			27	15·37	57	32·46
28	277·76			28	15·94	58	33·03
29	17·68			29	16·51	59	33·59
30	117·60			30	17·08	60	34·16
31	217·52						

Example of the use of the tables.

Let it be required to calculate the apparent place of the seven satellites for 1789, Oct. 18, 7 h. 51′ 54″, to the nearest minute of time and to tenths of a degree.

	5th.	4th.	3d.	2d.	1st.	6th.	7th.
1789	53·23	93·09	20·82	304·19	256·66	82·92	161·00
Oct.	158·89	43·58	155·45	269·03	220·54	84·54	235·52
18	81·69	46·39	354·43	207·63	192·56	49·09	35·31
7	1·32	6·58	23·24	38·36	55·62	76·63	111·41
52	0·17	0·82	2·88	4·75	6·88	9·49	13·80
* ♄	12·58	12·58	12·58	12·58	12·58	12·58	12·58
	307·9	203·0	209·4	116·5	24·8	315·3	209·6

The situation of the spot α calculated for July 28, 13 h. 53′ 39″ ; β for Sept. 16, 7 h. 45′ 48″ ; ε for Nov. 2, 7 h. 15′ 58″.

1789, α	271·5	β	183·0	ε	358·6
July	85·52	Sept.	160·56	Nov.	135·68
28	277·76	16	158·72	2	199·84
13	84·11	7	239·14	7	239·14
54	30·75	46	26·19	16	9·11
♄	7·20	♄	10·45	♄	13·18
	36·8		58·1		235·6

* The quantity marked ♄ 12°·58, which is applied to every one of the satellites, is the complement of 11ˢ 17° 25′, or geocentric place of Saturn, taken from the *Nautical Almanac*, for midnight of the required day, and to the nearest minute, which is sufficiently exact. This complement, or 12° 35′ in conformity with the tables, is reduced to decimals of a degree 12°·58.

XXVI.

On Nebulous Stars, properly so called.

[*Phil. Trans.*, vol. lxxxi., 1791, pp. 71–88.]

Read February 10, 1791.

In one of my late examinations of a space in the heavens, which I had not reviewed before, I discovered *a star of about the 8th magnitude, surrounded with a faintly luminous atmosphere, of a considerable extent.* The phænomenon was so striking that I could not help reflecting upon the circumstances that attended it, which appeared to me to be of a very instructive nature, and such as may lead to inferences which will throw a considerable light on some points relating to the construction of the heavens.

Cloudy or nebulous stars have been mentioned by several astronomers; but this name ought not to be applied to the objects which they have pointed out as such; for, on examination, they proved to be either mere clusters of stars, plainly to be distinguished with my large instruments, or such nebulous appearances as might be reasonably supposed to be occasioned by a multitude of stars at a vast distance. The milky way itself, as I have shewn in some former Papers, consists intirely of stars, and by imperceptible degrees I have been led on from the most evident congeries of stars to other groups in which the lucid points were smaller, but still very plainly to be seen; and from them to such wherein they could but barely be suspected, till I arrived at last to spots in which no trace of a star was to be discerned. But then the gradations to these latter were by such well-connected steps as left no room for doubt but that all these phænomena were equally occasioned by stars, variously dispersed in the immense expanse of the universe.

When I pursued these researches, I was in the situation of a natural philosopher who follows the various species of animals and insects from the height of their perfection down to the lowest ebb of life; when, arriving at the vegetable kingdom, he can scarcely point out to us the precise boundary where the animal ceases and the plant begins; and may even go so far as to suspect them not to be essentially different. But recollecting himself, he compares, for instance, one of the human species to a tree, and all doubt upon the subject vanishes before him. In the same manner we pass through gentle steps from a coarse cluster of stars, such as the Pleiades, the Præsepe, the milky way, the cluster in the Crab, the nebula in

Hercules, that near the preceding hip of Bootes,* the 17th, 38th, 41st of the 7th class of my Catalogues,† the 10th, 20th, 35th of the 6th class,‡ the 33d, 48th, 213th of the 1st,§ the 12th, 150th, 756th of the 2d,‖ and the 18th, 140th, 725th of the 3d,¶ without any hesitation, till we find ourselves brought to an object such as the nebula in Orion, where we are still inclined to remain in the once adopted idea, of stars exceedingly remote, and inconceivably crowded, as being the occasion of that remarkable appearance. It seems, therefore, to require a more dissimilar object to set us right again. A glance like that of the naturalist, who casts his eye from the perfect animal to the perfect vegetable, is wanting to remove the veil from the mind of the astronomer. The object I have mentioned above, is the phænomenon that was wanting for this purpose. View, for instance, the 19th cluster of my 6th class,** and afterwards cast your eye on this cloudy star,†† and the result will be no less decisive than that of the naturalist we have alluded to. Our judgement, I may venture to say, will be, that *the nebulosity about the star is not of a starry nature.*

But, that we may not be too precipitate in these new decisions, let us enter more at large into the various grounds which induced us formerly to surmise, that every visible object, in the extended and distant heavens, was of the starry kind, and collate them with those which now offer themselves for the contrary opinion.

It has been observed, on a former occasion, that all the smaller parts of other great systems, such as the planets, their rings and satellites, the comets, and such other bodies of the like nature as may belong to them, can never be perceived by

* R.A. 13 h. 27′ 40″. P.D. 60° 2′. The places of all the objects mentioned in this Paper are not brought to the present time, but given as they were calculated from the best observations I have made of them; the change in their situation arising from the lapse of a few years is too trifling to be any hindrance to our finding them very easily. [N.G.C. 5272 = M. 3.]

			h. ′ ″		° ′	
† VII.	17.	R.A.	7 9 45.	P.D.	114 34.	[N.G.C. 2362.
	38.		6 53 16.		88 37.	2324.
	41.		22 20 20.		38 47.	7296.
‡ VI.	10.		16 14 22.		115 32.	6144.
	20.		0 42 4.		117 46.	288.
	35.		0 19 44.		29 41.	136.
§ I.	33.		11 57 26.		78 25.	4124.
	48.		17 10 46.		107 36.	6356.
	213.		12 17 59.		44 45.	4449.
‖ II.	12.		12 32 37.		72 23.	4651.
	150.		14 19 53.		81 43.	5645.
	756.		14 51 42.		35 22.	5820.
¶ III.	18.		12 41 7.		84 51.	4505.
	140.		1 5 8.		92 0.	6064.
	725.		12 6 57.		43 14.	4242.
** VI.	19.		15 5 2.		110 14.	5897.
†† [IV.	69].		3 56 48.		59 50.	1514.]

us, on account of the faintness of light reflected from small, opaque objects; in my present remarks, therefore, all these are to be intirely set aside.

A well connected series of objects, such as we have mentioned above, has led us to infer, that all nebulæ consist of stars. This being admitted, we were authorized to extend our analogical way of reasoning a little farther. Many of the nebulæ had no other appearance than that whitish cloudiness, on the blue ground upon which they seemed to be projected; and why the same cause should not be assigned to explain the most extensive nebulosities, as well as those that amounted only to a few minutes of a degree in size, did not appear. It could not be inconsistent to call up a telescopic milky way, at an immense distance, to account for such phænomena; and if any part of the nebulosity seemed detached from the rest, or contained a visible star or two, the probability of seeing a few near stars, apparently scattered over the far distant regions of myriads of sidereal collections, rendered nebulous by their distance, would also clear up these singularities.

In order to be more easily understood in my remarks on the comparative disposition of the heavenly bodies, I shall mention some of the particulars which introduced the ideas of *connection* and *disjunction*: for these, being properly founded upon an examination of objects that may be reviewed at any time, will be of considerable importance to the validity of what we may advance with regard to my lately discovered nebulous stars.

On June the 27th, 1786, I saw a beautiful cluster of very small stars of various sizes, about 15′ in diameter, and very rich of stars.* On viewing this object, it is impossible to withhold our assent to the idea which occurs, that these stars are connected so far one with another as to be gathered together, within a certain space, of little extent, when compared to the vast expanse of the heavens. As this phænomenon has been repeatedly seen in a thousand cases, I may justly lay great stress on the idea of such stars being connected.

In the year 1779, the 9th of September, I discovered a very small star near ϵ Bootis.† The question here occurring, whether it had any connection with ϵ or not, was determined in the negative; for, considering the number of stars scattered in a variety of places, it is very far from being uncommon, that a star at a great distance should happen to be nearly in a line drawn from the sun through ϵ, and thus constitute the observed double star.

The 7th of September, 1782, when I first saw the planetary nebula near ν Aquarii,‡ I pronounced it to be a system whose parts were connected together. Without entering into any kind of calculation, it is evident, that a certain equal degree of light within a very small space, joined to the particular shape this object presents to us, which is nearly round, and even in its deviation consistent with

* R.A. 18 h. 20′ 2″. P.D. 107° 3′. [N.G.C. 6645 = VI. 23.]

† *Phil. Trans.* Vol. LXXII. p. 115. Catalogue of Double Stars, I. 1. [Above, p. 60.]

‡ R.A. 20 h. 52′ 36″. P.D. 102° 12′. [N.G.C. 7009 = IV. 1.]

regularity, being a little elliptical, ought naturally to give us the idea of a conjunction in the things that produce it. And a considerable addition to this argument may be derived from a repetition of the same phænomenon, in nine or ten more of a similar construction.

When I examined the cluster of stars, following the head of the great dog,* I found on the 19th of March, 1786, that there was within this cluster a round, resolvable nebula, of about two minutes in diameter, and nearly of an equal degree of light throughout.† Here, considering that the cluster was free from nebulosity in other parts, and that many such clusters, as well as many such nebulæ, exist in divers parts of the heavens, it appeared to me very probable, that the nebula was unconnected with the cluster; and that a similar reason would as easily account for this appearance as it had resolved the phænomenon of the double star near ε Bootis; that is, a casual situation of our sun and the two other objects nearly in a line. And though it may be rather more remarkable, that this should happen with two compound systems, which are not by far so numerous as single stars, we have, to make up for this singularity, a much larger space in which it may take place, the cluster being of a very considerable extent.

On the 15th of February, 1786, I discovered that one of my planetary nebulæ,‡ had a spot in the center, which was more luminous than the rest, and with long attention, a very bright, round, well defined center became visible. I remained not a single moment in doubt, but that the bright center was connected with the rest of the apparent disk.

In the year 1785, the 6th of October, I found a very bright, round nebula, of about 1½ minute in diameter.§ It has a large, bright nucleus in the middle, which is undoubtedly connected with the luminous parts about it. And though we must confess, that if this phænomenon, and many more of the same nature, recorded in my catalogues of nebulæ, consist of clustering stars, we find ourselves involved in some difficulty to account for the extraordinary condensation of them about the center; yet the idea of a connection between the outward parts and these very condensed ones within is by no means lessened on that account.

There is a telescopic milky way, which I have traced out in the heavens in many sweeps made from the year 1783 to 1789.‖ It takes up a space of more than 60 square degrees of the heavens, and there are thousands of stars scattered over it: among others, four that form a trapezium, and are situated in the well known nebula of Orion, which is included in the above extent. All these stars, as well as the four I have mentioned, I take to be intirely unconnected with

		h. ′ ″		° ′		
*	R.A.	7 32 1.	P.D.	104 18.	[N.G.C.	2437 = M. 46.
†		7 32 5.		104 15.		2438 = IV. 39.
‡		17 58 25.		23 22.		6543 = IV. 37.
§		3 30 35.		109 15.		1407 = I. 107.]

‖ R.A. from 5 h. 15′ 8″ to 5 h. 39′ 1″. P.D. from 87° 46′ to 98° 10′. [See above, p. 369.—Ed.]

the nebulosity which involves them in appearance. Among them is also *d* Orionis, a cloudy star, improperly so called by former astronomers; but it does not seem to be connected with the milkiness any more than the rest.

I come now to some other phænomena, that, from their singularity, merit undoubtedly a very full discussion. Among the reasons which induced us to embrace the opinion, that all very faint milky nebulosity ought to be ascribed to an assemblage of stars is, that we could not easily assign any other cause of sufficient importance for such luminous appearances, to reach us at the immense distance we must suppose ourselves to be from them. But if an argument of considerable force should now be brought forward, to shew the existence of a luminous matter, in a state of modification very different from the construction of a sun or star, all objections, drawn from our incapacity of accounting for new phænomena upon old principles, will lose their validity.

Hitherto I have been shewing, by various instances in objects whose places are given, in what manner we may form the ideas of connection and its contrary by an attentive inspection of them only: I will now relate a series of observations, with remarks upon them as they are delivered, from which I shall afterwards draw a few simple conclusions, that seem to be of considerable importance.

To distinguish the observations from the remarks, the former are given in italics, and the date annexed is that on which the objects were discovered; but the descriptions are extracted from all the observations that have been made upon them.

October 16, 1784. *A star of about the 9th magnitude, surrounded by a milky nebulosity, or chevelure, of about 3 minutes in diameter. The nebulosity is very faint, and a little extended or elliptical, the extent being not far from the meridian, or a little from north preceding to south following. The chevelure involves a small star, which is about 1½ minute north of the cloudy star; other stars of equal magnitude are perfectly free from this appearance.**

My present judgement concerning this remarkable object is, that the nebulosity belongs to the star which is situated in its center. The small one, on the contrary, which is mentioned as involved, being one of many that are profusely scattered over this rich neighbourhood, I suppose to be quite unconnected with this phænomenon. A circle of three minutes in diameter is sufficiently large to admit another small star, without any bias to the judgement I form concerning the one in question.

It must appear singular, that such an object should not have immediately suggested all the remarks contained in this Paper; but about things that appear new we ought not to form opinions too hastily, and my observations on the construction of the heavens were then but entered upon. In this case, therefore, it was the safest way to lay down a rule not to reason upon the phænomena that

* R.A. 5 h. 57′ 4″. P.D. 96° 22′. [N.G.C. 2170 = IV. 19.]

might offer themselves, till I should be in possession of a sufficient stock of materials to guide my researches.

October 16, 1784. *A small star of about the 11th or 12th magnitude, very faintly affected with milky nebulosity; other stars of the same magnitude are perfectly free from this appearance.*

February 23, 1786. 5 *or* 6 *small stars within the space of* 3 *or* 4′, *all very faintly affected in the same manner, and the nebulosity suspected to be a little stronger about each star.* But a third observation rather opposes this increase of the faintly luminous appearance.*

Here the connection between the stars and the nebulosity is not so evident as to amount to conviction; for which reason we shall pass on to the next.

January the 6th, 1785. *A bright star with a considerable milky chevelure; a little extended,* 4 *or* 5′ *in length, and near* 4′ *broad; it loses itself insensibly. Other stars of equal magnitude are perfectly free from this chevelure.*†

The connection between the star and the chevelure cannot be doubted, from the insensible gradation of its luminous appearance, decreasing as it receded from the center.

January 31, 1785. *A pretty considerable star, with a very faint, and very small, irregular, milky chevelure; other stars of the same size are perfectly free from such appearance.*‡

I can have no doubt of the connection between the star and its chevelure.

October 5, 1785. *A star with a strong bur all around.* A second observation calls it *a very bright nucleus, with a milky nebulosity, of no great extent.* A third suspects *the milkiness to belong to more of the same, which is diffused over the whole sweep in that place*; but a fourth says, that *the milky nebulosity is much stronger than what the nebulous ground, on which the star is placed, intitles it to.*§

The connection, therefore, between the nebulosity and the star is evident.

January 1, 1786. *A star surrounded with milky chevelure; the star is not central.* A second observation calls it *affected with a very faint, and extensive, milky chevelure.* A third only mentions *a star affected with milky chevelure.*‖

As by the word chevelure I always denoted something relating to a center, the connection cannot be doubted.

February 24, 1786. *A considerable star, very faintly affected with milky chevelure.* A second observation, much the same.¶

* R.A. 6 h. 0′ 33″. P.D. 96° 13′. [N.G.C. 2185 = IV. 20. Feb. 24, 1786, *the stars are not connected with it, though I suspected it last night.*]

† R.A. 5 h. 30′ 53″. P.D. 92° 21′. [N.G.C. 2023 = IV. 24.]

		h. ′ ″		° ′		
‡	R.A.	6 54 27.	P.D.	100 53.	[N.G.C.	2327 = IV. 25.
§		5 25 57.		96 52.		1999 = IV. 33.
‖		5 35 56.		89 50.		2071 = IV. 36.
¶		5 59 4.		96 19.		2182 = IV. 38.]

November 28, 1786. *A star involved in milky chevelure.**

January 17, 1787. *A star with a pretty strong milky nebulosity, equally dispersed all around; the star is of about the 9th magnitude.* A memorandum to the observation says, that, having but just begun, *I suspected the glass to be covered with damp, or the eye out of order; but yet a star of the 10th or 11th magnitude, just north of it, was free from the same appearance.* A second observation calls it *one of the most remarkable phænomena I ever have seen, and like my northern planetary nebula in its growing state.*†

The connection between the star and the milky nebulosity is without all doubt.

November 3, 1787. *A bright star with faint nebulosity.* A second observation mentions *the star to be of the 9th magnitude, and the faint nebulosity of very little extent.*‡

June 11, 1787. *Suspected, stellar.* By a second observation it is verified, and called *a very small star involved in extremely faint nebulosity.*§

November 25, 1788. *A star of about the 9th magnitude, surrounded with very faint milky nebulosity; other stars of the same size are perfectly free from that appearance. Less than 1′ in diameter. The star is either not round or double.*‖

March 23, 1789. *A bright, considerably well defined nucleus, with a very faint, small, round chevelure.*¶

The connection admits of no doubt; but the object is not perhaps of the same nature with those which I call cloudy stars.

April 14, 1789. *A considerably bright, round nebula; having a large place in the middle of nearly an equal brightness, but less bright towards the margin.***

This seems rather to approach to the planetary sort.

March 5, 1790. *A pretty considerable star of the 9th or 10th magnitude, visibly affected with very faint nebulosity of little extent, all around. A power of* 300 *shewed the nebulosity of greater extent.*††

The connection is not to be doubted.

March 19, 1790. *A very bright nucleus, with a small, very faint chevelure, exactly round. In a low situation, where the chevelure could hardly be seen, this object would put on the appearance of an ill-defined, planetary nebula, of* 6, 8, *or* 10″ *diameter.*‡‡

November 13, 1790. *A most singular phænomenon! A star of about the 8th*

		R.A. h ′ ″		P.D. ° ′	N.G.C.
*	R.A.	5 57 4.	P.D.	96 15.	[N.G.C. . . . =IV. 44. See above, p. 368.]
†		7 16 28.		68 39.	2392 =IV. 45.
‡		23 11 26.		30 0.	7635 =IV. 52.
§		17 1 51.		47 26.	6301 =IV. 57.
‖		0 1 57.		18 41.	40 =IV. 58.
¶		11 12 25.		50 17.	3658 =IV. 59.
**		11 45 12.		33 43.	3982 =IV. 62.
††		6 58 40.		90 29.	2346 =IV. 65.
‡‡		9 27 22.		30 11.	2950 =IV. 68.]

*magnitude, with a faint luminous atmosphere, of a circular form, and of about 3′ in diameter. The star is perfectly in the center, and the atmosphere is so diluted, faint, and equal throughout, that there can be no surmise of its consisting of stars; nor can there be a doubt of the evident connection between the atmosphere and the star. Another star not much less in brightness, and in the same field with the above, was perfectly free from any such appearance.**

This last object is so decisive in every particular, that we need not hesitate to admit it as a pattern, from which we are authorized to draw the following important consequences.

Supposing the connection between the star and its surrounding nebulosity to be allowed, we argue, that one of the two following cases must necessarily be admitted. In the first place, if the nebulosity consist of stars that are very remote, which appear nebulous on account of the small angles their mutual distances subtend at the eye, whereby they will not only, as it were, run into one another, but also appear extremely faint and diluted; then, what must be the enormous size of the central point, which outshines all the rest in so superlative a degree as to admit of no comparison? In the next place, if the star be no bigger than common, how very small and compressed must be those other luminous points that are the occasion of the nebulosity which surrounds the central one? As, by the former supposition, the luminous central point must far exceed the standard of what we call a star, so, in the latter, the shining matter about the center will be much too small to come under the same denomination; we therefore either have a central body which is not a star, or have a star which is involved in a shining fluid, of a nature totally unknown to us.

I can adopt no other sentiment than the latter, since the probability is certainly not for the existence of so enormous a body as would be required to shine like a star of the 8th magnitude, at a distance sufficiently great to cause a vast system of stars to put on the appearance of a very diluted, milky nebulosity.

But what a field of novelty is here opened to our conceptions! A shining fluid, of a brightness sufficient to reach us from the remote regions of a star of the 8th, 9th, 10th, 11th, or 12th magnitude, and of an extent so considerable as to take up 3, 4, 5, or 6 minutes in diameter! Can we compare it to the coruscations of the electrical fluid in the aurora borealis? Or to the more magnificent cone of the zodiacal light as we see it in spring or autumn? The latter, notwithstanding I have observed it to reach at least 90 degrees from the sun, is yet of so little extent and brightness as probably not to be perceived even by the inhabitants of Saturn or the Georgian planet, and must be utterly invisible at the remoteness of the nearest fixed star.

More extensive views may be derived from this proof of the existence of a shining matter. Perhaps it has been too hastily surmised that all milky nebulosity,

* R.A. 3h. 56′ 48″. P.D. 59° 50′. [N.G.C. 1514 =IV. 69.]

of which there is so much in the heavens, is owing to starlight only. These nebulous stars may serve as a clue to unravel other mysterious phænomena. If the shining fluid that surrounds them is not so essentially connected with these nebulous stars but that it can also exist without them, which seems to be sufficiently probable, and will be examined hereafter, we may with great facility explain that very extensive, telescopic nebulosity, which, as I mentioned before, is expanded over more than sixty degrees of the heavens, about the constellation of Orion; a luminous matter accounting much better for it than clustering stars at a distance. In this case we may also pretty nearly guess at its situation, which must commence somewhere about the range of the stars of the 7th magnitude, or a little farther from us, and extend unequally in some places perhaps to the regions of those of the 9th, 10th, 11th, and 12th. The foundation for this surmise is, that, not unlikely, some of the stars that happen to be situated in a more condensed part of it, or that perhaps by their own attraction draw together some quantity of this fluid greater than what they are intitled to by their situation in it, will, of course, assume the appearance of cloudy stars; and many of those I have named are either in this stratum of luminous matter, or very near it.

We have said above, that in nebulous stars the existence of the shining fluid does not seem to be so essentially connected with the central points that it might not also exist without them. For this opinion we may assign several reasons. One of them is the great resemblance between the chevelure of these stars and the diffused extensive nebulosity mentioned before, which renders it highly probable that they are of the same nature. Now, if this be admitted, the separate existence of the luminous matter, or its independance on a central star, is fully proved. We may also judge, very confidently, that the light of this shining fluid is no kind of reflection from the star in the center; for, as we have already observed, reflected light could never reach us at the great distance we are from such objects. Besides, how impenetrable would be an atmosphere of a sufficient density to reflect so great a quantity of light? And yet we observe, that the outward parts of the chevelure are nearly as bright as those that are close to the star; so that this supposed atmosphere ought to give no obstruction to the passage of the central rays. If, therefore, this matter is self-luminous, it seems more fit to produce a star by its condensation than to depend on the star for its existence.

Many other diffused nebulosities, besides that about the constellation of Orion, have been observed or suspected; but some of them are probably very distant, and run out far into space. For instance, about 5 minutes in time preceding ξ Cygni, I suspect as much of it as covers near four square degrees; and much about the same quantity 44′ preceding the 125 Tauri. A space of almost 8 square degrees, 6′ preceding α Trianguli, seems to be tinged with milky nebulosity. Three minutes preceding the 46 Eridani, strong, milky nebulosity is expanded over more than two square degrees. 54′ preceding the 13th Canum venaticorum, and again

48′ preceding the same star, I found the field of view affected with whitish nebulosity throughout the whole breadth of the sweep, which was 2° 39′. 4′ following the 57 Cygni, a considerable space is filled with faint, milky nebulosity, which is pretty bright in some places, and contains the 37th nebula of my Vth class, in the brightest part of it. In the neighbourhood of the 44th Piscium, very faint nebulosity appears to be diffused over more than 9 square degrees of the heavens. Now, all these phænomena, as we have already seen, will admit of a much easier explanation by a luminous fluid than by stars at an immense distance.

The nature of planetary nebulæ, which has hitherto been involved in much darkness, may now be explained with some degree of satisfaction, since the uniform and very considerable brightness of their apparent disk accords remarkably well with a much condensed, luminous fluid; whereas to suppose them to consist of clustering stars will not so completely account for the milkiness or soft tint of their light, to produce which it would be required that the condensation of the stars should be carried to an almost inconceivable degree of accumulation. The surmise of the regeneration of stars, by means of planetary nebulæ, expressed in a former Paper, will become more probable, as all the luminous matter contained in one of them, when gathered together into a body of the size of a star, would have nearly such a quantity of light as we find the planetary nebulæ to give. To prove this experimentally, we may view them with a telescope that does not magnify sufficiently to shew their extent, by which means we shall gather all their light together into a point, when they will be found to assume the appearance of small stars; that is, of stars at the distance of those which we call of the 8th, 9th, or 10th magnitude. Indeed this idea is greatly supported by the discovery of a well defined, lucid point, resembling a star, in the center of one of them: for the argument which has been used, in the case of nebulous stars, to shew the probability of the existence of a luminous matter, which rested upon the disparity between a bright point and its surrounding shining fluid, may here be alleged with equal justice. If the point be a generating star, the further accumulation of the already much condensed, luminous matter, may complete it in time.

How far the light that is perpetually emitted from millions of suns may be concerned in this shining fluid, it might be presumptuous to attempt to determine; but, notwithstanding the unconceivable subtilty of the particles of light, when the number of the emitting bodies is almost infinitely great, and the time of the continual emission indefinitely long, the quantity of emitted particles may well become adequate to the constitution of a shining fluid, or luminous matter, provided a cause can be found that may retain them from flying off, or reunite them. But such a cause cannot be difficult to guess at, when we know that light is so easily reflected, refracted, inflected, and deflected; and that, in the immense range of its course, it must pass through innumerable systems, where it cannot but frequently meet with many obstacles to its rectilinear progression. Not to mention the great counteraction of the united attractive force of whole sidereal systems, which must be continually exerting their power upon the particles while they are endeavouring to fly off. However, we shall lay no stress upon a surmise of this kind, as the means of verifying it are wanting: nor is it of any immediate consequence to us to know the origin of the luminous matter. Let it suffice, that its existence is rendered evident, by means of nebulous stars.

I hope it will be found, that in what has been said I have not launched out into hypothetical reasonings; and that facts have all along been kept sufficiently in view. But, in order to give every one a fair opportunity to follow me in the reflections I have been led into, the place of every object from which I have argued has been purposely added, that the validity of what I have advanced might be put to the proof by those who are inclined, and furnished with the necessary instruments to undertake an attentive and repeated inspection of the same phænomena.

W. HERSCHEL.

Slough, Jan. 1, 1791.

XXVII.

On the Ring of Saturn, and the Rotation of the fifth Satellite upon its Axis.

[*Phil. Trans.*, 1792, pp. 1–22.]

Read December 15, 1791.

It is well known to Astronomers that the ring of Saturn becomes alternately enlightened on one of its sides, and that this change of illumination takes place when the planet passes through the node of the ring. This happened in October, 1789, when the southern plane, which had been in the dark for about fifteen years, became visible to us: an event to which I have looked forwards with considerable impatience. In the year 1790, the position of the ring was still too oblique to permit me to examine it well enough to form a proper judgment of its appearance, but lately I have been able to view it to greater advantage, with every one of my telescopes.

In a former paper,* where I ventured to hint at a division of the ring of Saturn, it was highly necessary to express that surmise with proper doubts concerning the reality of so wonderful a construction; but my late views of its southern plane, assisted by some conclusions drawn from the discovery of the quick rotation of the ring, have enabled me to speak decisively on this subject. My suspicion of a divided or double ring arose chiefly from the following circumstances.

In the first place, the black belt, during the time of about ten years that I observed it, on the northern plane, was subject to no kind of change; but remained always, permanently, of the same breadth and colour. With regard to its breadth, it is true that I could only judge of that part of it which goes across the body of the planet, by the rules of perspective, which made me suppose it to be as broad there as it was on the two sides; yet now, as we know that the ring revolves in about ten hours and a half, it is very certain, that the apparently narrow part, across the body, and that which was hidden behind the planet, in the course of an evening, when I have been observing Saturn for many hours together, must have been exposed to view in their full breadth, upon the sides of the ring; and that, if there had been any difference, I must have perceived it; especially as I was continually on the look out for such phænomena, by way of ascertaining, if possible, the rotation of the ring.†

* *Phil. Trans.* Vol. LXXX. page 4 [above, p. 371].

† When I say that the black division was always of the same breadth, I do not mean to exclude very small variations, not only of the breadth of the black mark, but of the ring itself, which I have occasionally observed, and which it may be necessary, hereafter, to communicate at full length; but these, almost imperceptible, differences might arise from causes that are foreign to our present purpose.

In the next place, the colour of this dark belt was also uniformly the same, whenever I observed it under equally favourable circumstances; and being so well defined on both its borders, and, in every part of the revolving ring, presenting us with the same view of colour, breadth, and sharpness of its outlines, no kind of hypothesis but a division of the ring, through which the open heavens may be seen, will answer the conditions of this phænomenon. It remained therefore only to ascertain, whether the southern plane would present us with the same aspect. And since I have lately had a great number of fine views of the ring of Saturn, I shall here deliver as many of the observations as will be sufficient to throw light enough on the subject, to enable us to decide the question, whether this ring be double or single?

Observations on the Ring of Saturn.

Sept. 7, 1790. 20-feet reflector. No dark division can as yet be seen upon the ring of Saturn; but it is hardly open enough to expect it to be visible.

Aug. 5, 1791. 20-feet reflector. The black list, on this side of the ring of Saturn, is exactly in the same relative place where I saw it on the northern plane.

Sept. 25, 1791. 20-feet reflector. The black division goes all around the ring, as far as I can trace it, exactly in the same place where I used to see it on the north side.

Oct. 13, 1791. 10-feet reflector. The black division upon the southern plane of Saturn's ring, is in the same place, of the same breadth, and at the same distance from the outer edge, that I have always seen it on the northern plane. With a power of 400, I see it very distinctly; it is of the same kind of colour as the space between the ring and the body, but not so dark.

Oct. 24, 1791. 7-feet reflector. With a new, machine-polished, most excellent speculum, I see that the division on the ring of Saturn, and the open spaces between the ring and the body, are equally dark, and of the same colour with the heavens about the planet.

20-feet reflector. The black division upon the ring is as dark as the heavens. It is equally broad on both sides of the ring. I see it very steadily, and can trace it a good way towards Saturn; both on the part of the ring which is turned towards us, and on that which lies the other way. I trace it as far as the place where a line, perpendicular to the direction of the ring, would touch the inside of the ring, or the outside of the open space between the ring and the body of the planet.

40-feet reflector. I see the division on the ring of Saturn of the same colour as the surrounding heavens. It is of an equal breadth on both sides, and I can trace it a great way towards the body of Saturn.

20-feet reflector. With a power of 600, I can trace the division very nearly as far as the place, where a perpendicular to the direction of the ring, would divide the open space between the planet and the ring, into two equal parts.

From these observations, added to what has been given in some former papers, I think myself authorized now to say, that the planet Saturn has two concentric rings, of unequal dimensions and breadth, situated in one plane, which is probably not much inclined to the equator of the planet. These rings are at a considerable distance from each other, the smallest being much less in diameter at the outside, than the largest is at the inside.

The dimension of the two rings and the intermediate space are nearly in the following proportion to each other.

Inside diameter of the smallest ring	5900 parts.
Outside diameter - - - -	7510
Inside diameter of the largest ring	7740
Outside diameter - - - -	8300
Breadth of the inner ring - -	805
Breadth of the outer ring - -	280
Breadth of the vacant space - -	115

Admitting, with M. DE LA LANDE, that the breadth of the whole ring, as formerly supposed to consist of one entire mass, is near one third of the diameter of Saturn, it follows that the vacant space between the two rings, according to the above statement, amounts to near 2513 miles.

In giving these proportions, which are merely taken from very accurate representations of the phænomena that offered themselves, I do not mean to be scrupulously exact, but reserve a greater accuracy for a future opportunity; when a micrometer, which I have lately applied to the 40-feet telescope, will assist me to have recourse to proper measures.

It may be remarked, that this opening in the ring must be of considerable service to the planet, in reducing the space that is eclipsed by the shadow of the ring to a much smaller compass; both on account of the direct light it lets through, and because there will be a strong reverberation of the rays of the sun between the two opposite edges. Moreover, if these rings should be surrounded by some atmosphere, which is highly probable, the refractions that will take place upon the edges will still contribute to lessen the darkness which the shadow of an undivided ring would have occasioned.

As we have now admitted Saturn to have two rings entirely detached from each other, so as plainly to permit us to see the open heavens through the vacancy between them; and as in my former paper I have given the revolution of the ring, which was then supposed to be all in one united mass, it will be necessary to examine, whether both rings partake in the same revolution, or to which the period which has been assigned belongs?

To decide this point, we must recur to the observations of the spots by which the rotation of the ring was determined. The spot called α,* for instance, which

* See *Phil. Trans.* Vol. LXXX. page 481 [above, p. 404].

has been observed to revolve with great regularity through upwards of 300 periods, between the 28th of July and the 24th of December, 1789, was certainly situated pretty near the outer edge. The spot β, as may be gathered from the observation of the 16th of September, and 25th of December, was most likely on the very edge itself; nor could the spot δ be very far from it. This, without considering the situation of γ and ϵ, is quite sufficient to determine us to assign the period we have given to belong to the large, thin and narrow, outward ring.

The spots γ and ϵ were probably at some distance from the outward edge of the outer ring; but this distance might possibly not exceed that of the inside edge of the same ring. We may however admit them to have adhered to the inner ring, whose rotation is perhaps not very different from that of the outer one; or we may examine whether these two spots may not perhaps agree to some other supposed revolution of the inner ring; but then the observations that are given of them will hardly be sufficient for establishing the time of that ring's rotation with accuracy, though they undoubtedly must amount to a proof that it also revolves with great velocity on its axis.

That there should be a small difference in the periods of the rotation of the two rings, is highly probable from their different dimensions; and now, that the rotation is known, the division of it into two parts seems to be a very natural consequence of its construction. For, when the extreme thinness is taken into consideration, we find by KEPLER's law, of the periods of revolving bodies placed at different distances, that it would be very wonderful for so thin, and so broad a plane, to have adhesion enough to keep together; and that consequently this ring in its divided state, supposing the rotation of the parts to favour the construction, is more permanent than it would be otherwise. This however is only mentioned as a collateral circumstance, and by no means intended either as a proof of the division, or the different rotation of the two parts of the ring. For, notwithstanding we cannot but set the highest value upon the excellent theories that have been lately delivered in the Memoirs of a learned Society, of which I also have the honour to be a member, we must refer entirely to observation for the necessary data on which to found our subsequent computations.

The memoir to which I allude* refers to observations of many divisions of the ring of Saturn. This must lead us to consider the question, whether the construction of this ring is of a nature so as permanently to remain in its present state? or whether it be liable to continual and frequent changes, in such a manner as in the course of not many years, to be seen subdivided into narrow slips, and then again as united into one or two circular planes only? Now, without entering into a discussion, the mind seems to revolt, even at first sight, against an idea of the chaotic state in which so large a mass as the ring of Saturn must needs be, if phænomena like these can be admitted. Nor ought we to indulge a suspicion of this

* See *Histoire de l'Académie Royale des Sciences de Paris*, 1787, page 249 [by Laplace.—ED.].

being a reality, unless repeated and well-confirmed observations had proved, beyond a doubt, that this ring was actually in so fluctuating a condition. Let us therefore examine what facts we have to guide us in this inquiry.

After looking over all my observations upon Saturn, since the year 1774 to the present time, I can find only four where any other black division upon the ring is mentioned than the one which I have constantly observed, and from which I have deduced the actual division of the ring into two very unequal portions. These observations are as follows.

June 19, 1780. 10^h 15′ mean time. With a new 7-feet speculum, having an aperture of 6·4 inches, with also a much improved small speculum, and a power of about 200. I see a second black list upon the ring of Saturn, close to the inner side, on the preceding arm of the ring. See figure 1 [Plate XII.].

June 20, 1780. 10^h 10′. I see the same double list on the preceding side of the ring.

June 21, 1780. 10^h 1′. Small 20-feet, Newtonian reflector, power 200. I see the second black list on Saturn's ring. It is closer to the inside than the other is to the outside; but it is only visible on the preceding side of the ring. See figure 2.

June 26, 1780. 9^h 34′. Small 20-feet, Newtonian reflector; aperture confined to 7 inches. The 2d black list, on the preceding side of the ring of Saturn, is visible.

June 29, 1780. 10^h 19′. Saturn's belts are very clear. I see but one black list upon the ring. The shadow of the planet is visible upon the side of the ring, as well as upon the small northern part that projects beyond the planet. See fig. 3.

Nov. 21, 1791. 0^h 28′ sid. time. 40-feet reflector, power 370. There is no other black division visible upon the ring of Saturn but the one near the outer edge.

It must be confessed that Saturn was in the very best situation for viewing the plane of the ring, when the first four observations were made; and that consequently they may be looked upon as a strong evidence for another division. But hitherto I have set them aside as wanting more confirmation, not only because I could never perceive the same dark line on the following side of the ring as well as on the preceding side; nor since I could not find it on the 29th of June, 1780, as we have seen above; but chiefly, because I have not been able, with any of my best instruments, to see it again at all. We also find by the observation of the 21st of November, 1791, which has been added, that the southern plane, as yet, presents us with no other division than the capital one, which I have observed these thirteen years, on both sides of the ring. However, if the opening should be very narrow, and the rings eccentric, it is possible that a dark line might by this means become visible on one side only. Moreover, these objects may be so minute, that no other time than when the plane of the ring is exposed as much as it can possibly be, will do to ascertain such phænomena. This will happen again about the year 1796, when we may hope to have a satisfactory view of it, with our large instruments.

It remains now to consider the observations that have been made by M. CASSINI, Mr. SHORT, and Mr. HADLEY.

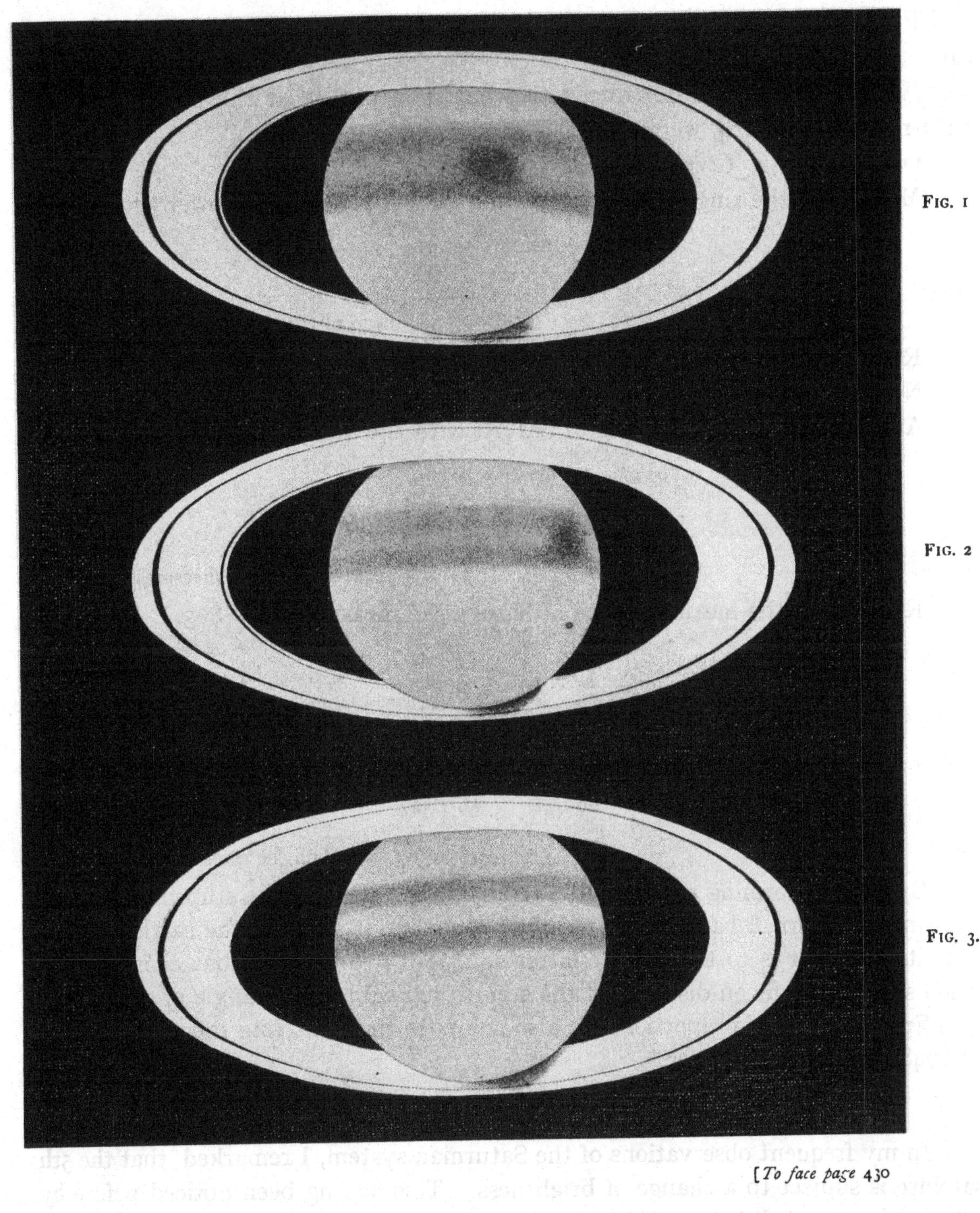

FIG. 1

FIG. 2

FIG. 3.

[*To face page* 430

Proper allowance was made for the wires being tangents to the outside of the ring.

When this measure is reduced to what it would be at the mean distance of Saturn from the earth, we have 46″·832.

Oct. 24, 1791. *Correction of the 40-feet clock* +25″·4.

Measure of the ring of Saturn with the 40-feet reflector. Power 370.

1h 3′	1st measure	53″·914
	2d ———	53 ·260
		53″·587 mean of the two measures.

Reduced to the mean distance of Saturn, the measure is 47″·241.

Nov. 21, 1791. *Correction of the 40-feet clock* −7″·8.

Another measure of the ring of Saturn with the 40-feet reflector. Power 370.

0h 48′	1st measure	50″·627
	2d ———	50 ·042
	3d ———	50 ·808
		50″·492 mean of the three measures.

Reduced to the mean distance of Saturn the measure is 45″·803.

Oct. 24	47″·241
Nov. 21	45 ·803
	46″·522 mean of the measures with the 40-feet reflector.
40-feet	46″·522
20-feet	46 ·832
	46″·677 mean of all the measures.

By way of forming more easily a comparative idea of the stupendous size of this ring of Saturn, I have calculated the proportion it bears to the earth, and find that its diameter is to that of the latter as 25·8914 to 1; and that consequently, when seen at the mean distance of the sun, it will subtend an angle of 7′ 25″·332.

From the above proportions we also compute that this ring must be upwards of 204883 miles in diameter.*

On the Rotation of the fifth Satellite of Saturn,† *on its Axis.*

In my frequent observations of the Saturnian system, I remarked, that the 5th satellite is subject to a change of brightness. This having been noticed before by other observers, I did not at first pay so much attention to it as I soon afterwards

* In this calculation I have used for the earth's diameter the sum of the longer and shorter semi-axes which are given in Mr. Dalby's Paper, published in the last volume of the *Philosophical Transactions*. If we compute the vacant space between the two rings immediately from the above dimensions of the outward ring, we shall have 2839 miles; and this will certainly be more accurate than the result which has been drawn from the proportion of the breadth of the ring to the diameter of Saturn.

† [Iapetus.—Ed.]

found this circumstance deserved. When I saw this satellite always assume the same brightness in the same part of its orbit, and perceived that its change was regular and periodical, it occurred to me very naturally, that the cause of this phænomenon could be no other than a rotation upon its axis. It became necessary therefore to find out a method to determine the time of this rotation.

In order to investigate this, I pursued the satellite with great attention, and marked all its changes of apparent brightness. The result of many observations is as follows. The light of the satellite is in full splendour during the time it runs through that part of its orbit which is between 68 and 129 degrees past the inferior conjunction. In passing through this arch it does not fall above one magnitude short of the brightness of the 4th satellite.* On the contrary, from about 7 degrees past the opposition till towards the inferior conjunction, it is not only less bright than the 3d,† but hardly, if at all, exceeds the 2d,‡ or even the 1st satellite §; provided the latter be then about its greatest elongation, where its light is least impeded by the brightness of the planet. Upon the whole, the alteration seems to amount to what among the fixed stars, and with the naked eye, would be called a change from the 5th to the 2d, and from the 2d to the 5th magnitude.

Having thus observed this satellite, for many of its revolutions round the primary planet, to lose and regain its light regularly, it is evident that the time of its rotation on its axis cannot differ much from that of its revolution round Saturn. I think myself sufficiently authorized to make this conclusion, notwithstanding it may have happened sometimes that the light of the satellite has suffered an occasional change, of short duration, from other causes; for the same reason that we should certainly allow those who first saw the spots in the sun to be in the right to assign the period of its rotation *nearly*, when they perceived that the same spot made several revolutions, notwithstanding that spot might afterwards vanish. But I may go farther, and ascertain upon sufficient grounds, that this satellite turns once upon its axis, exactly in the time it performs one revolution round its primary planet. This degree of accuracy is obtained by taking in the observations of M. CASSINI, which are related in the *Mémoires de l'Académie des Sciences*, 1705, page 121; where we find it mentioned, that "the 5th satellite of Saturn disappears regularly for about one half of its revolution, when it is to the east of Saturn." The same memoir contains also a conjecture of this satellite's rotation upon its axis; but this surmise is contradicted as premature, in 1707, page 96; where we find the following paragraph. "M. CASSINI gives an example of the danger there is in these sort of determinations, that are made too hastily. The 5th satellite of Saturn, of which we have said, in the History of 1705, page 121, that it grew invisible, in the eastern half of the circle it describes about Saturn, began, in the month of Sept. 1705, to be there visible, as well as in the western half, where it always was so. Hence the conjectures which we have related cease to be well founded."

* [Titan.—ED.] † [Rhea.—ED.] ‡ [Dione.—ED.] § [Tethys.—ED.]

Now, without determining whether the satellite, from some cause or other, ceased to change its brightness, or whether its phænomena were not sufficiently followed to come to a proper conclusion, I think that with the assistance of observations at so great a distance of time as those of M. CASSINI, I may sufficiently establish the period of this satellite's rotation. For since I have traced the regular, and periodical change of light, through more than ten revolutions, and find them, in all appearance, to be contemporary with its return about Saturn, it leads us directly to a strong presumption that its rotation upon its axis, like that of our moon, strictly coincides with its revolution round its primary planet; and the observations of M. CASSINI completely confirm this conclusion. For, had he seen the satellite brightest in any other part of its orbit, our observations would not have agreed together; but since the year 1705, the satellite has made about 397 revolutions; and yet the phænomena described by CASSINI answer now as exactly to my own observations, as the spots in our moon, viewed in CASSINI's time, answer to those we now observe.

If it should be objected, that the 5th satellite of Saturn has not been continually observed, and that consequently these appearances might either not happen at all, or fall upon different places in its orbit; I answer, that a period of more than ten revolutions, which I have included, is already a strong argument that no such change has taken place; for if the satellite had but made a single rotation upon its axis more or less than it has made revolutions round Saturn, the change must amount to nearly one degree *per* revolution; that is, to about ten degrees during the time of my taking notice of it; which is a quantity I think I might have perceived. However, to remove all doubt, we have some valuable observations of M. BERNARD, who in the year 1787 also found the 5th satellite of Saturn subject to the same change of light that M. CASSINI had observed.* Now, by joining those to mine, we have a short period of near 20 revolutions that agree together, so as to preclude all doubt of any intermediate change; and therefore we cannot be liable to err, when we extend this period to all the 397 revolutions since CASSINI's time, and by that means ascertain that the 5th satellite of Saturn turns upon its axis, once in 79 days, 7 hours, and 47 minutes.

I cannot help reflecting, with some pleasure, on the discovery of an analogy, which shews that a certain, uniform plan is carried on among the secondaries of our solar system; and we may conjecture, that probably most of the moons of all the planets are governed by the same law; especially if it be founded on such a construction of the figure of the secondaries, as makes them more ponderous towards their primary planets. For, if even the 5th satellite of Saturn, which is at so great a distance from its planet, is affected by such a law, of course the other satellites are not very likely to have escaped its influence.

From the considerable change in the brightness of the 5th satellite of Saturn,

* See *Mémoires de l'Académie*, 1786, page 378.

we may be certain that some part of its surface, and this by far the largest, reflects much less light than the rest; and, from the points of its orbit in which it appears brightest to us, we conclude that neither the darkest nor brightest side of the satellite is turned towards the planet, but partly one and partly the other; though probably rather less of the bright side.

The great regularity of this change of brightness seems to point out another resemblance of this satellite with our moon. It is well known that we see the spots of the moon pretty nearly of the same brightness, so as not to be overcast in a very strong degree by dense clouds to disfigure them, and therefore have great reason to surmise that her atmosphere is extremely rare; which indeed we also know from other principles: In like manner, on account of the uninterrupted changes in the brightness of the 5th satellite of Saturn, we may suppose that it also partakes of a similar fate with respect to its atmosphere, which is probably as rare as that of our moon.

On the Distance of the fifth Satellite.

The distance of the 5th satellite from Saturn is allowed to be the most proper for obtaining a true measure of the quantity of matter contained in the planet; for which reason I have taken many measures of it with the 20-feet reflector. I give them at full length, that the validity of them may appear in its proper light.

Sept. 25, 1791. *Correction of the clock* −2′ 19″·5 *for midnight.*

Distance of the 5th satellite of Saturn from the centre of the planet; measured with the 20-feet reflector, and a magnifying power of 157.

23ʰ 4′	1st measure	8′ 55″·684
23 19	2d ———	8 53 ·175
23 33	3d ———	8 59 ·179
23 47	4th ———	8 52 ·123
23 55	5th ———	8 56 ·361
0 2	6th ———	8 55 ·797
	Mean of the six measures,	8′ 55″·5

Sept. 26, 1791. *Correction of the clock* −2′ 19″·8.

23ʰ 15′	1st measure	9′ 3″·745
23 25	2d ———	9 2 ·758
23 31	3d ———	9 7 ·014
23 38	4th ———	9 6 ·592
23 42	5th ———	9 8 ·001
23 45	6th ———	9 6 ·479
	Mean of the six measures,	9′ 5″·8

Sept. 27, 1791. *Correction of the clock* −2′ 20″·0.

22ʰ 53′	1st measure	9′ 20″·656
22 57	2d ———	9 20 ·359
23 1	3d ———	9 20 ·149
23 5	4th ———	9 20 ·641
23 11	5th ———	9 20 ·064
23 16	6th ———	9 20 ·840
	Mean of the six measures,	9′ 20″·5

Sept. 28, 1791. *Correction of the clock* −2′ 20″·2.

21ʰ 38′	1st measure	9′ 27″·759
21 44	2d ———	9 29 ·957
21 46	3d ———	9 27 ·815
21 50	4th ———	9 28 ·238
21 53	5th ———	9 28 ·970
21 58	6th ———	9 27 ·420

Mean of the six measures, 9′ 28″·4

Sept. 29, 1791. *Correction of the clock* −2′ 20″·4.

0ʰ 17′	1st measure	9′ 37″·060
0 22	2d ———	9 36 ·552
0 26	3d ———	9 36 ·270
0 29	4th ———	9 36 ·862
0 33	5th ———	9 37 ·765
0 36	6th ———	9 37 ·060

Mean of the six measures, 9′ 36″·9

Sept. 30, 1791. *Correction of the clock* −2′ 20″·6.

20ʰ 25′	1st measure	9′ 39″·258
20 29	2d ———	9 38 ·441
20 34	3d ———	9 37 ·596
20 38	4th ———	9 37 ·032
20 40	5th ———	9 40 ·949
20 42	6th ———	9 37 ·793

Mean of the six measures, 9′ 38″·5

Supposing the satellite now to be not far from its greatest elongation, I measured the declination between the centre of Saturn, and the 5th satellite; causing one to pass along one wire, while the other followed upon the other wire.

22ʰ 32′	1st measure	1′ 41″·889
22 47	2d ———	1 45 ·609

Mean of the two measures, 1′ 43″·749

Not being satisfied with the considerable disagreement, I took another measure with the utmost precaution and care; as the apparent curvature of the wires at so great a distance, required more than common attention.

0ʰ 52′ very exact. 1′ 43″·354

Mean between this and the former mean, 1′ 43″·55 south of the parallel of Saturn.

The satellite not being perhaps arrived at its greatest elongation, I took six other measures of its distance.

1ʰ 5′	1st measure	9′ 41″·907
1 11	2d ———	9 38 ·723
1 14	3d ———	9 38 ·159
1 18	4th ———	9 41 ·203
1 23	5th ———	9 40 ·385
1 26	6th ———	9 41 ·935

Mean of the six measures, 9′ 40″·4

In the last six measures of the 5th satellite, I used a method a very little different from that which I employed before, and which is probably more accurate. I used to observe, when the two wires were nearly brought to their proper distance, the moment of intersection of the satellite; and the instant it was hid behind the wire, cast my eye on Saturn, which should be bisected when the measure is justly taken. But this change of attention cannot be made without some very small loss of time. To correct this defect I took alternately the bisection of Saturn, and cast my eye upon the satellite; and the bisection of the satellite, casting the eye upon Saturn. As the latter way gives the interval too small, the former gives it too large, and between both the true measure may be obtained. I do not, however, suppose, that the error of the former method can amount to so much as a single second of space; as, knowing the loss of time, I always used the utmost precaution; and repeated the examination of a measure perhaps 20 times before I let it pass.

Oct. 1, 1791. *Correction of the clock* −2′ 20″·0

22h 25′	1st measure	9′ 43″·767
22 27	2d ———	9 44 ·444
22 30	3d ———	9 43 ·007
22 32	4th ———	9 42 ·499
22 34	5th ———	9 40 ·554
22 36	6th ———	9 43 ·965
	Mean of the six measures,	9′ 43″·0

It grew cloudy, so that no measures later in the night could be obtained; nor could I get another sight of Saturn till October the 7th, when the satellite was far advanced in its orbit, on its return towards the planet.

Supposing the satellite to have been very nearly at its greatest elongation, when the last six measures were taken, I have reduced them to the mean distance of Saturn, where they give 8′ 31″·97.

I forbear making deductions from this result, with respect to the quantity of matter contained in the planet, as, possibly, the orbit of the satellite may be considerably elliptical; in which case measures taken in opposite parts of that orbit will be required, before we can make a strict application of the laws of centripetal forces.

XXVIII.

Miscellaneous Observations.

[*Phil. Trans.*, 1792, pp. 23-27.]

Read December 22, 1791.

Account of a Comet.

LAST Thursday evening, the 15th of December, about half after eight o'clock, while I was taken up with observing Saturn, my sister looked over the heavens, and discovered a pretty large, telescopic comet, in the breast of Lacerta. I viewed it in my seven-feet reflector, and with that instrument settled its place and rate of moving. At 9^h 42′ 4″·8 true mean time, it preceded a small telescopic star 11″·3 in time, and was 2′ 41″ south of the same. The place of this star I have since determined with sufficient accuracy, that it may be found again by those who wish to settle it more exactly. It follows the 2d of FLAMSTEED's stars in the constellation of Lacerta, 1′ 41″·5 in time; and is 45′ 40″·8 more south than the same. The apparent motion of the comet on Thursday evening was direct, and at the rate of about three minutes of time in right ascension, and a little more than two degrees in polar distance *per* day; from which we may suppose that we shall keep it some time in view. Last night I examined it with a twenty-feet reflector, and found it to consist of a great light, pretty regularly scattered about a condensed small part of five or six seconds in diameter; which resembled a kind of nucleus, but had not the least appearance of a solid body. Beside the scattered, and gradually diminishing light, which reached nearly to a distance of three minutes every way beyond the bright centre, there was also a faintly extended, ill defined, pretty broad ray, of about 15 minutes in length, directed towards the north following part of the heaven, which might be called the tail of the comet.

Its place for the same night (Dec. 16th) was determined by a five-feet Newtonian *Sweeper*, carrying an equilateral triangle in the focus of the eye-glass, not so large but that the three intersections, made by the wires at the three angles, may be distinctly perceived. At 5^h 49′ 40″·6 it preceded the 6th Lacerta 4′ 58″·5 in time, and was 52′ 14″·5 more north than that star.

On the periodical Appearance of o Ceti.

The changeable star in the neck of the Whale, *o* Ceti, continues its variations as usual, but with some considerable irregularities of brightness.

In the year 1779, as we have seen,* it excelled α Arietis so far as almost to rival Aldebaran; and continued in that state a full month.

In 1780, its greatest brightness was only like that of δ Ceti.

In the year 1781, it did not come up to the brightness of δ.

In 1782, this star increased to the size of β Ceti, and continued bright for more than twenty days.

In 1783, it did not only vanish to the naked eye, as usual, but disappeared so completely, that I could not find it with a telescope, which permitted not a star of the 10th magnitude to escape me. When it increased again, it did not amount to the brightness of δ.

In 1784, I saw it only of the 8th magnitude in a twenty-feet reflector, but as I did not continue to observe it regularly, it might possibly change as usual.

In 1789, it arrived to the brightness of α Piscium, or rather excelled it.

In 1790, the greatest brightness was almost equal to that of α Ceti.

In the present year, I have seen it only of the magnitude of γ Ceti nearly; or between γ and δ; but, as bad weather has occasioned many interruptions, it may possibly have been larger.

The period of 333 days, assigned by BOUILLAUD, does not agree with present observations compared to those of FABRICIUS made on the 13th of August, 1596, when this star was in its greatest lustre. M. CASSINI also found, that his observations, in the beginning of August, 1703, when the star was brightest, did not agree with the interval of 333 days; and therefore, supposing the star to have changed 117 times since the epoch of FABRICIUS, he gave it a period of 334 days. This will, however, not agree with the present time of the changes; and it appears now that M. CASSINI ought to have assumed 118 instead of 117 variations; which would have pointed out a period of 331 days, and some hours.

That this is, probably, very near the real time of the star's variation, will be seen when we admit it to have undergone 214 changes between the 13th of August, 1596, and the 21st of October, 1790; by which long interval we obtain the period of 331 days, 10 hours, 19 minutes. It will, indeed, be necessary, in order to reconcile all observations, to admit of some occasional deviations in the appearance of the star, amounting almost to a month; but that this is no more than we may allow, is pretty evident from the variations I have taken notice of within the last 14 years; besides, a period of 334 days could not be admitted without totally giving up all regularity in the returning appearance of the star.

I have taken the epoch of the 21st of October, 1790, as one of the best ascertained, modern appearances I have been able to obtain; and believe it to be more proper for settling the period, than that which might be deduced from a brilliant blaze of the star, such as took place in 1779, owing to causes that are not regular, and therefore may be apprehended to disturb the general order of the change.

* *Phil. Trans.* Vol. LXX. page 338. [Above, p. 3.]

On the Disappearance of the 55th Herculis.

Among the changes that happen in the sidereal heavens we enumerate the loss of stars; but, notwithstanding the real destruction of an heavenly body may not be impossible, we have some reasons to think that the disappearance of a star is probably owing to causes which are of the same nature with those that act upon periodical stars, when they occasion their temporary occultations. This subject, however, being of great extent and consequence, we shall not enter into it at present, but only relate a recent instance of the kind.

Two stars of the 5th magnitude, whose places we find inserted in all our best catalogues, were to be seen in the neck of Hercules. They are the 54th and 55th of FLAMSTEED'S, in that constellation. In the year 1781, the 10th of October, I examined them both, and marked down their colour, *red.* The 11th of April, 1782, I looked at them again, and noted my having seen them distinctly, with a power of 460; and that they were single stars.

The 24th of last May, I missed one of the two, and examining the spot again the 25th, and many times afterwards, found that one of them was not to be seen. The situation of the stars is such that, not having fixed instruments, I could not well determine which of the two was the lost one. I therefore requested the favour of my much esteemed friend, the astronomer royal, to ascertain the remaining star; and it appears from Dr. MASKELYNE'S answer to my letter, that the 55th Herculis is the one which we have lost.*

Remarkable Phænomena in an Eclipse of the Moon.

The 22d of October, 1790, when the moon was totally eclipsed, I viewed the disk of it with a twenty-feet reflector, carrying a magnifying power of 360. In several parts of it I perceived many bright, red, luminous points. Most of them were small and round. The brightness of the moon, notwithstanding the great defalcation of light occasioned by the eclipse, would not permit me to view it long enough to take the places of these points. They were, indeed, very numerous; as I suppose that I saw, at least, one hundred and fifty of them. Their light did not much exceed that of Mons Porphyrites HEVELII.

We know too little of the surface of the moon to venture at a surmise of the cause from whence the great brightness, similarity, and remarkable colour of these points could arise.

Slough, Dec. 17, 1791.

* [55 Herculis was not a separate star but only a repeated observation of 54 Herculis. See Baily's edition of Flamsteed's Catalogue, p. 610 and the Memoir of C. H. F. Peters, p. 82.—ED.]

XXIX.

Observations on the Planet Venus.

[*Phil. Trans.*, 1793, pp. 201–219.]

Read June 13, 1793.

THE planet Venus is an object that has long engaged my particular attention. A series of observations upon it, which I began in April, 1777, has been continued down to the present time.

My first view, when I engaged in the pursuit, was to ascertain the diurnal rotation of this planet; which, from the contradictory accounts of CASSINI and BIANCHINI, the former of which states it at 23 hours, while the latter makes it 24 days, appeared to me to remain unknown, as to its real duration: for the observations of these gentlemen, how widely different soever with regard to time, can leave no doubt but that this planet actually has a motion on its axis.

The next object was the atmosphere of Venus; of the existence of which also, after a few months' observations, I could not entertain the least doubt.

The investigation of the real diameter, was the third object I had in view.

To which may be added, in the last place, an attention to the construction of the planet, with regard to permanent appearances; such as might be occasioned by, or ascribed to, seas, continents, or mountains.

The result of my observations would have been communicated long ago, if I had not still flattered myself with the hopes of some better success, concerning the diurnal motion of Venus; which, on account of the density of the atmosphere of this planet, has still eluded my constant attention, as far as concerns its period and direction. Even at this present time, I should hesitate to give the following extract from my journals, if it did not seem incumbent upon me to examine by what accident I came to overlook mountains in this planet, which are said to be "*of such enormous height, as to exceed four, five, and even six times the perpendicular elevation of Cimboraco, the highest of our mountains!*" *

The same paper, which contains the lines I have quoted, gives us likewise many extraordinary relations, equally wonderful; such as hints of the various and singular properties of the atmosphere of Saturn.† A ragged margin in Venus, resembling the uneven border of the moon, as it appears to a power magnifying from 1 to 4.‡ One cusp of Venus appearing pointed, and the other blunt, owing

* See *Phil. Trans.* for 1792, Part II. page 337 [Schröter.—ED.]. † Ibidem, p. 309. ‡ p. 310.

to the shadow of some mountain.* Flat spherical forms conspicuous on Saturn.† All which being things of which I have never taken any notice, it will not be amiss to shew, by what follows, that neither want of attention, nor a deficiency of instruments, could occasion my not perceiving *these mountains of more than 23 miles in height;* ‡ *this jagged border of Venus; and these flat spherical forms on Saturn.*

Indeed with regard to Saturn, I cannot hesitate a single moment to say, that, had any such things as flat spherical forms existed, they could not possibly have escaped my notice, in the numberless observations with 7, 10, 20, and 40-feet reflectors, which I have so often directed to that planet. However, if the gentleman who has seen the mountains in Venus, has made observations on flat spherical forms on Saturn, it is to be regretted that he has not attended to the revolution of this planet on its axis, which could not remain an hour unknown to him when he saw these forms.

Last night,§ for instance, I saw two small dark spots on Jupiter; I shall not call them flat spherical forms, because their flatness, as well as their sphericity, must be hypothetical; moreover, these two terms seem to me to contradict each other. These were evidently removed, in less than an hour, in such a manner as to point out, very nearly, the direction and quantity of the rotation of this planet.

Before I remark on the rest of the extraordinary relations above-mentioned, I will give a short extract of my observations on Venus, with such deductions as it seems to me that we are authorised to make from them.

Observations.

April 17, 1777. The disk of Venus was exceedingly well defined, distinct, and bright, but no spot was visible by which I could judge of her diurnal motion. The same telescope shews the spots on Mars extremely well. 7-feet reflector.

April 26, 1777. The disk well defined, and bright, but no spot. 10-feet reflector.

February 21, 1780. No spot on the disk of Venus; diameter 15″·9, mean of three measures.

May 2, 1780. No spot; power 449; diameter 17″·2.

May 28, 1780. No spot; power 268 and 449; diameter 22″·8.

May 29, 1780. I viewed Venus with a 20-feet Newtonian reflector; power 447. The edge of the disk was so sharp and well defined, that there can be no wish to see it better. There was no spot of any kind.

I could see no projections of any mountains, though the phase of Venus is now such as would be most favourable for shewing them.

June 19, 1780. There is, on Venus, a bluish, darkish spot, *adc*; and another,

* See *Phil. Trans.* for 1792, Part II. p. 312. † Ibidem, p. 336.

‡ The height of Chimbo-raço, according to Mr. CONDAMINE, is 3200 French toises; and the English mile, by Mr. DE LA LANDE, measures 830. If the mountains in Venus exceed Chimbo-raço six times in perpendicular elevation, they must be more than 23 miles in height.

§ May 31, 1793.

which is rather bright, *ced*; they meet in an angle at *c*, the place of which is about one-third of the diameter of Venus from the cusp *a*. See fig. 1.

June 21, 23, 24, 25, 26, 28, 29, 30, and July 3, 1780. Continued observations were made upon these, and other faint spots, and drawings of them annexed. The instrument I used was a 20-feet Newtonian reflector, furnished with no less than five different object specula, some of which were in the highest perfection of figure and polish; the power generally 300 and 450. But the result of them would not give me the time of the rotation of Venus. For the spots assumed often the appearances of optical deceptions, such as might arise from prismatic affections; and I was always very unwilling to lay any stress upon the motion of spots, that either were extremely faint and changeable, or whose situation could not be precisely ascertained.

However, that Venus has a motion on an axis cannot be doubted from these observations; and that she has an atmosphere is as evident, from the changes I took notice of, which surely cannot be upon the solid body of the planet.

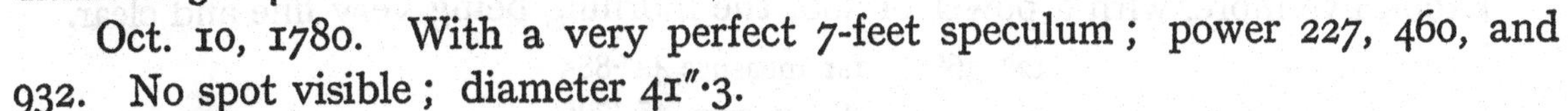

Fig. 1.

Sept. 18, 1780. No spot on Venus; diameter $38''\cdot4$.

Oct. 10, 1780. With a very perfect 7-feet speculum; power 227, 460, and 932. No spot visible; diameter $41''\cdot3$.

Oct. 11, 1780. No spot; diameter $27''\cdot8$.

Oct. 20, 21, 23, 1780. No spot visible.

April 17, 1783. 10-feet reflector; power 324. I see some darkish spots on Venus. 7-feet reflector; power 227. The same appearances; but in neither of the instruments are they determined enough to serve for the purpose of finding the rotation.

May 21, 1783. 10-feet reflector; a new speculum; power 250.

$7^h\ 30'$. No spot visible.

$8^h\ 30'$. There seems to be an ill defined spot.

$9^h\ 15'$. No motion can be perceived that may be depended upon, though the figure seems rather advancing towards the centre.

May 30, 31, and June 1, 6, 1783. Spots were observed with 10 and 20-feet reflectors, and also motion perceived in them. Continued observations were recorded; and a great many figures delineated.

Dec. 3, 1783. With 460, and 932. No spot. No kind of protuberance, or indenture in the line which terminates the illumination, that might denote a mountain.

Feb. 13, 1785. No spot. A new 10-feet Newtonian reflector.

April 8, 1788. No spot on Venus; but she is still at too great a distance for such observations.

Nov. 30, 1789. No satellite visible. If she has one, it must be less in appearance than a star of the 8th or 9th magnitude; power 300.

Dec. 2, 1789. No spot.; power 157, 300, and 460.

May 23, 1791. 40-feet reflector. The light of Venus is so brilliant that it becomes very uneasy for the eye to bear it long. There is no spot on the disk.

I had prepared my apparatus for a regular succession of observations with this instrument, having turned it towards the west, and put on the round-motion to keep the planet in view; but found that the great advantage of this telescope, which is its superior light, was, on this occasion, not only unnecessary, but rather an inconvenience.

Nov. 24, 1791. Correction of the clock, −46″·7.

I took measures of the diameter of Venus with the 20-feet reflector; power 157.

12^h 18′	1st measure	45″·486
	2d ———	46 ·142
	3d ———	45 ·514
	4th ———	45 ·814
	5th ———	46 ·033
	6th ———	46 ·252
	Mean of the six measures	45″·874

I took five more, with a power of 300, the morning being very fine and clear.

12^h 36′	1st measure	44″·885
	2d ———	45 ·705
	3d ———	45 ·104
	4th ———	45 ·322
	5th ———	45 ·842
	Mean of the five measures	45″·372
	Mean of the two sets	45″·623

These measures were taken with a speculum that has been lately re-polished, and therefore required new tables for casting them up. Such tables were made by the following transits.

Nov. 25, 1791. Transits of equatorial stars, taken to determine the value of the micrometer, which is divided into revolutions of sixty parts each.

First set, 23″·0 23·0 23·0 23·0 23·2 23·1 23·1 23·0 23·1 23·1 = 23″·06 = 21 revolutions; correction +7·2 parts, for zero and concave wires.

Second set, 16″·8 16·6 16·4 16·5 16·7 16·6 16·5 16·8 16·4 16·5 = 16·58 = 15 revolutions 1·3 parts. Correction +7·2.

By the first set, 1 part	= 0″·272964
Second set - -	273748
Mean of the two sets	0″·273356

In the first set, the micrometer was opened to 21 revolutions; and ten equatorial stars were observed to pass from one wire to the other. The opening was afterwards changed, and ten other stars were again observed to pass over the wires; after which the micrometer was read off, and found to be 15 revolutions and 1·3 parts.

Feb. 4, 1793. Correction of the clock, $-1' 28''\cdot0$.

$2^h 55'$. 7-feet reflector; power 172. The air is very clear, and I see Venus very well defined; but cannot perceive any inequality on the edge of the planet that might denote a mountain; though the situation is favourable, being a little more enlightened than what we may call her last quarter. With 215, I had a very distinct view for a long time; but cannot perceive any inequality on the line which divides light from darkness.

With 287, I perceive no mountains: with 430, very distinct, I perceive no mountains. The terminating line is not so sharply defined as the circumference; but no inequality is visible.

With the same power, I see on Saturn the equatorial belt, the shadow of the ring on Saturn, the shadow of Saturn on the ring, the division of the ring, etc.

I do not find any spot on Venus; so that there is no possibility to assign its diurnal motion.

March 3, 1793. Correction of the clock, $-2' 0''\cdot6$.

$6^h 30'$. 7-feet reflector; I observed Venus with many powers, but could perceive no spot by which its diurnal motion might be ascertained.

April 3, 1793. Correction of the clock, $-2' 43''\cdot9$.

$9^h 9'$. 7-feet reflector; power 215. The evening remarkably fine. There is no spot upon the disk of Venus, by which its rotation might be ascertained. The horns are equally sharp. There is nothing that has the appearance of a mountain, like what we see in the moon. With 287, very well defined, appearances are the same. With 430, not the least appearance of any mountains.

April 4, 1793. Correction of the clock, $-2' 45''\cdot3$.

$9^h 8'$. There is no spot upon the disk of Venus. The horns are perfectly alike.

Not the least appearance like the mountains of the moon. With 287, and 430, very distinct.

April 5, 1793. Correction of the clock, $-2' 46''\cdot7$.

$8^h 25'$. 7-feet reflector; power 215, 287, and 430. There are no spots upon Venus, by which its diurnal motion could be ascertained. The horns are exactly alike; and no inequality, like the mountains of the moon, is visible.

April 6, 1793. Correction of the clock, $-2' 48''\cdot1$.

$9^h 29'$. With the 7-feet reflector; power 430. There is no kind of spot visible in any part of the disk. The two horns are exactly alike; and no appearance of mountains can be perceived.

April 7, 1793. Correction of the clock, −2′ 49″·6.

9^h 8′. With the 7-feet reflector; power 215, 287, 430, and 860. I can see no spot upon the disk. Both horns are perfectly alike. Nothing resembling the mountains upon the moon can be perceived. I see it beautifully well, and sharply defined.

April 8, 1793. Correction of the clock, −2′ 51″·0.

9^h 2′. With the 10-feet reflector; power 300, and 400. There is no spot upon Venus. The shape of the two horns is perfectly alike, and no appearance of mountains can be perceived. The illumination of the horns is also perfectly alike.

April 9, 1793. Correction of the clock, −2′ 52″·1.

8^h 45′. With the 10-feet reflector; power 300. No spot upon Venus. Both horns perfectly alike. No appearance of mountains.

The light of Venus is brighter all around the limb, than on that part which divides the enlightened, from the unenlightened part of the disk. With 400, appearances are the same.

9^h 16′. The bright part, on the limb of Venus, is like a bright bead, of nearly an equal breadth all around.

April 16, 1793. Correction of the clock, −2′ 59″·5.

10^h 3′. 7-feet reflector, with different powers. No spot upon the disk. No mountains visible. Both horns alike.

A luminous margin, as usual, all around the limb.

April 20, 1793. Correction of the clock, −3′ 3″·8.

10^h 0′. 7-feet reflector; power 172, 215, 287, 430, and 860. No spot upon the disk. Both horns exactly alike. Not the least appearance of any mountains.

With 287, there is a narrow luminous border all around the limb, and the light afterwards diminishes pretty suddenly, and suffers no considerable diminution as we go towards the line which terminates the enlightened part of the disk. It is however less bright near the terminating line than farther from it. With powers lower than 287, the narrow luminous border cannot be so well distinguished.

April 22, 1793. Correction of the clock, −3′ 5″·9.

9^h 30′. 7-feet reflector; power 430. Very distinct. No spot. No appearance of mountains. Both horns perfectly alike.

With 860, 1290, and 1720, not the least appearance of mountains. Even the last power is considerably distinct.

10^h 20′. With 430, the luminous margin, compared to the light adjoining to it, may be expressed by, *suddenly much brighter all around the limb.*

April 28, 1793. Correction of the clock, −3′ 12″·3.

12^h 0′. 7-feet reflector; power 215. No spot. Both horns perfectly alike. No appearance of mountains.

April 29, 1793. Correction of the clock, $-3' \ 13''{\cdot}4$.

$10^h \ 30'$. 7-feet reflector; power 215. No spot. Both horns perfectly alike. Not the least appearance of any mountains.

With 287 and 430. Both horns equally sharp: no mountains visible.

May 1, 1793. Correction of the clock, $-3' \ 15''{\cdot}5$.

$10^h \ 45'$. With the 10-feet reflector; power 300. No spot. Both horns perfectly alike, and very sharp. Not the least appearance of any mountains.

With 600, very distinct. Both horns extremely sharp, and alike. No mountains.

With 400, the same appearances.

May 5, 1793. Correction of the clock, $-3' \ 19''{\cdot}8$.

$11^h \ 27'$. 7-feet reflector; power 215, 287, and 430. Both horns perfectly alike. No spot. Not the least appearance of any mountains.

May 12, 1793. Correction of the clock, $-3' \ 27''{\cdot}3$.

$11^h \ 10'$. 7-feet reflector; power 215. Beautifully distinct. No spot visible; indeed the crescent is so slender, that we cannot expect to see any spots upon the disk.

Not the least appearance of any mountains, or inequality on the border.

The slender part of the crescent appears often knotty, but this is evidently a deception arising from undulations in the air; for, with proper attention, the knots may be perceived to change place. Little scratches in the great, or small speculum, may also occasion seeming irregularities; but, with proper attention, all such deceptions may be easily detected. Both horns perfectly alike.

With 287, 430, and 860, all that has been mentioned before is perfectly verified, and confirmed.

$11^h \ 43'$. I tried also the lower powers of 172, and 115; but they are inferior, in effect, to 215, 287, and 430; and not adequate to the delicacy and power required in such observations.

I have often taken notice, and again this evening, that the illuminated part of Venus is more than a semi-circle. Whether the excess of the sun's diameter alone will account for this, or how far we are to take the twilight of the atmosphere of Venus into consideration, I have hitherto deferred investigating, as my disk-micrometer wants a moveable parallel, in order to be adjustable, by observation, to the quantity of the horns which is enlightened beyond an hemisphere.

May 13, 1793. Correction of the clock, $-3' \ 28''{\cdot}4$.

$11^h \ 45'$. 7-feet reflector; power 115, 172, 215, 287, and 430. Both horns perfectly alike. No appearance of mountains.

The points of the horns appear more blunt than they were last night, and are not drawn out to so slender a point; but this is evidently a deception, owing to the indifference of the night; for great sharpness, and distinct vision, are wanting in every other object I am looking at.

May 18, 1793. Correction of the clock, $-3' \ 33''\cdot 7$.

$12^h \ 28'$. 7-feet reflector; power 287. Both horns perfectly alike. No appearance of mountains. No spot. But, at the present altitude of Venus, it is impossible to make any observations that require delicacy, and demand very distinct vision with high powers.

May 19, 1793. Correction of the clock, $-3' \ 34''\cdot 7$.

$11^h \ 45'$. 7-feet reflector; power 287. Both horns perfectly alike, in shape and illumination. Not the least appearance of any mountains. The horns are exceedingly slender.

$12^h \ 0'$. I do not see any diminution of light on the edge of the horns, but what may be accounted for from their slenderness; being brought to very fine points, that lose themselves by their minuteness.

I saw it in great perfection, with a newly polished, plain speculum, which excels my former one in sharpness.

May 20, 1793. Correction of the clock, $-3' \ 35''\cdot 8$.

$12^h \ 20'$. No spot or unevenness in the light of Venus upon either cusp, or in any other part, that could in the least make me suspect a mountain.

I measured the diameter of Venus, and projection of the cusps beyond an hemisphere, by my disk-micrometer. This was not done by an illumination, as described in the apparatus, (*Phil. Trans.* Vol. LXXIII. p. 4) * when I used it for a nocturnal planet; for, day-light being sufficiently strong, there was no occasion to light the lamps. On the measuring disk were drawn concentric circles; and also a diameter, having several lines parallel to it, in one of the semicircles. If there had been time, I should have prepared a straight edge, *be*, moveable parallel to the diameter *ad*.· See fig. 2.

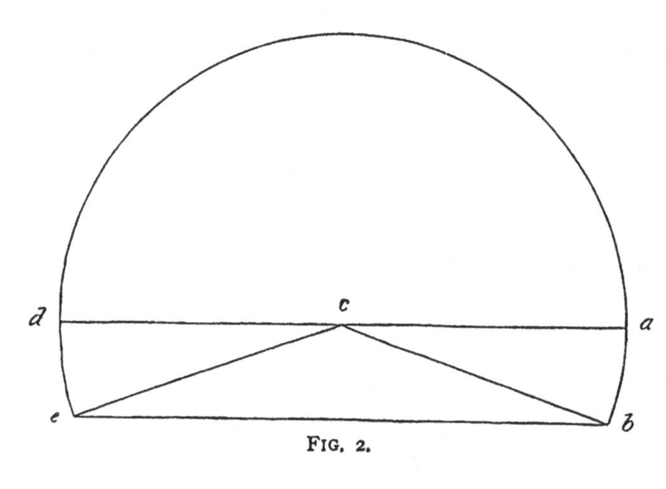

FIG. 2.

First measure, with the double eye-glass; power about 90. Diameter of Venus 2390. Projection 500. But the power is too low to be accurate.

Second measure; power 215. Diameter of Venus 4800. Projection 620. Here the projection is probably as much too small as the former was too large; but the planet is too low for repeating the measures. A mean of both may, perhaps, not be far from the truth; which gives, diameter 3595; projection 560.

Here 1797·5 being radius, and 560 sine, we find the angle *acb*, or *dce*, equal to $18° \ 9' \ 8''\cdot 2$.

* [Above, p. 102.]

A few very evident results may be drawn from the foregoing observations.

With regard to the rotation of Venus on an axis, it appears that we may be assured of this planet's having a diurnal motion, and though the real time of it is still subject to considerable doubts, it can hardly be so slow as 24 days. Its direction, or rather the position of the axis of Venus, is involved in still greater uncertainty.

The atmosphere of Venus is probably very considerable; which appears not only from the changes that have been observed in the faint spots on its surface, but may also be inferred from the illumination of the cusps, when this planet is near its inferior conjunction; where the enlightened ends of the horns reach far beyond a semicircle. I must here take notice, that the author we have before quoted on this subject, has the merit of being the first who has pointed out this inference, but he has overlooked the penumbra arising from the diameter of the sun; * which has certainly a considerable share in the effect of the extended illumination, and in his angle of 15° 19′ will amount to 1° 11′ 47″·6. His measures are also defective; as probably the mirror of his 7-feet reflector, which was a very excellent one, was by that time considerably tarnished, and had lost much of the light necessary to shew the extent of the cusps in their full brilliancy.

I do not give the calculations I have made of the extent of the twilight of Venus, because my measures were not so satisfactory to myself as I wish them to be; nor so near the conjunction as we may hereafter obtain them; neither were they sufficiently repeated. My computations, however, when compared to those given in the paper on the atmosphere of Venus, shew sufficiently that it is of much greater extent, or refractive power, than has been computed in that paper. Those calculations indeed are so full of inaccuracies, that it would be necessary to go over them again, in order to compare them strictly with my own, for which at present there is no leisure.

I ought also to take notice here, that the same author, it seems, has taken measures of the horns of Venus by an instrument, which, in his publications, he calls a *projection table*, and describes as his own; † of which, however, those who do not know its construction may have a very perfect idea, when they read the descriptions of my lamp, disk, and periphery micrometers, joined to what I have mentioned above, of using the disk-micrometer without lamps when day-light is

* He mentions it upon another occasion, and says in a note, p. 313, that "*this whole penumbra, which, according to the greatest apparent diameter of Venus, extends from* 59 *to* 60″, (for what reason he fixes upon these quantities does not appear) "*measures, in the direction perpendicular to the line of the cusps, only* 0″·36." But if, according to him, the apparent diameter of the sun be 44′, (which is less than it ought to be) the penumbra must certainly extend likewise upon the surface of Venus over 44′ of a great circle; and, in the situation which he mentions, that is, perpendicular to the line of the cusps at the time of the greatest elongation, and when the apparent diameter of Venus is 60″, (as he makes it) it must measure 0″·384.

† See *Beiträge zu den neuesten astronomischen Entdeckungen*, p. 210. And *Selenotopographische Fragmente*, p. 63.

sufficiently strong; or even with an illumination in front, where the object is bright enough to allow of it, such as the moon, &c.

I remember drawing the picture of a cottage by it, in the year 1776, which was at three or four miles distance; and going afterwards to compare the parts of it with the building, found them very justly delineated.

I have also many times had the honour of shewing my friends the accuracy of the method of applying one eye to the telescope, and the other to the projected picture of the object in view; by desiring them to make two points, with a pin, upon a card fixed up at a convenient place, where it might be viewed in my telescope; and this being done, I took the distance of these points from the picture I saw projected, in a pair of proportional compasses, one side of which was to the other as the distance of the object, divided by the distance of the image, to the magnifying power of the telescope; and giving the compasses to my friends, they generally found that the proportional ends of them exactly fitted the points they had made on the card. All which experiments are only so many different ways of using the lamp-micrometer.

As to the mountains in Venus, I may venture to say that no eye, which is not considerably better than mine, or assisted by much better instruments, will ever get a sight of them; though, from the analogy that obtains between the only two planetary globes we can compare, (the moon and the earth) there is little doubt but that this planet also has inequalities on its surface, which may be, for what we can say to the contrary, very considerable.

The real diameter of Venus, I should think, may be inferred with great confidence, from the measures I took with the 20-feet reflector, in the morning of the 24th of November, 1791; which, when reduced to the mean distance of the earth, give 18″·79 for the apparent diameter of this planet.

This result is rather remarkable, as it seems to prove that Venus is a little larger than the earth, instead of being a little less, as has been supposed; yet, upon the nicest scrutiny, I cannot find fault with the measures. The planet was put between the two wires of the micrometer, which were outward tangents; and they were, after each measure, shut, so as to meet with the same edge, and in the same place where the planet was measured. In this situation the proper deduction, for not being central measures, was pointed out by the index plate. The transits of the 25th were corrected for a small concavity of the wires, which being pretty thick and stubborn, were not strained sufficiently to make them quite straight, the amount of which was also ascertained by an examination of the division where the wires closed at the ends, and where they closed in the centre. The zero was, with equal precaution, referred to a point at an equal distance from the contact of the wires on each side; for they are at liberty to pass over each other, without occasioning any derangement. The *shake*, or *play*, of the screw is less than 3-tenths of a division.

The two planets, however, are so nearly of an equal size, that it would be necessary to repeat our measures of the diameter of Venus, in the most favourable circumstances, and with micrometers adjusted to the utmost degree of precision, in order to decide with perfect confidence that she is, as appears most likely, larger than the earth.

The remarkable phænomenon of the bright margin of Venus, I find, has not been noticed by the author we have referred to: on the contrary, it is said, page 310, "*this light appears strongest at the outward limb abc, from whence it decreases gradually, and in a regular progression, towards the interior edge, or terminator.*" But the luminous border, as I have described it, in the observations of the 9th, 16th, 20th, and 22d of April, does not in the least agree with the above representation.

With regard to the cause of this appearance, I believe that I may venture to ascribe it to the atmosphere of Venus, which, like our own, is probably replete with matter that reflects and refracts light copiously in all directions. Therefore on the border, where we have an oblique view of it, there will of consequence be an increase of this luminous appearance. I suppose the bright belts, and polar regions of Jupiter, for instance, which have a greater light than the faint streaks, or yellow belts, on that planet, to be the parts where its atmosphere is most filled with clouds, while the latter are probably those regions which are free from them, and admit the sun to shine on the planet; by which means we have the reflection of the real surface, which I take to be generally less luminous.

If this conjecture be well founded, we see the reason why spots on Venus are so seldom to be perceived. For, this planet having a dense atmosphere, its real surface will commonly be enveloped by it, so as not to present us with any variety of appearances. This also points out the reason why the spots, when any such there are, appear generally of a darker colour than the rest of the body.

An Account of the Discovery of a Comet. In a Letter from Miss CAROLINE HERSCHEL, *to* JOSEPH PLANTA, *Esq., Sec. R.S.*

[*Phil. Trans.*, 1794, p. 1.]

Read November 7, 1793.

SIR,—Last night I discovered a comet near 1 (δ) Ophiuchi, but clouds covering the part of the heavens where it was, its place could not be obtained. My brother has just now (7 o'clock) determined its situation, as follows. The comet precedes the 1st (δ) Ophiuchi 6′ 34″ in time, and is 1° 25′ more north than that star.—I remain, sir, &c.,

CAR. HERSCHEL.

Slough,
Tuesday, Oct. 8, 1793.

XXX.

Observations of a quintuple Belt on the Planet Saturn.

[*Phil. Trans.*, 1794, pp. 28–32.]

Read December 19, 1793.

EVERY analogy that can be traced in the appearance of the planets, seems to throw some additional light on what we know of them already. In some of my former papers I have established the spheroidical form of the planet Saturn, and pointed out the motion of a spot on its disk. From the first of these may be inferred a considerable rotation on its axis; while the latter goes a step farther, and shews that it has such a motion. My late observations seem to hint to us, that the period in which it revolves is, probably, not of a long duration.

They are as follows:

Nov. 11, 1793. 3^h 35′, 7-feet reflector, power 287.

Close to the ring of Saturn, where it passes across the body of the planet, is the shadow of the ring; very narrow, and black. See [Plate XIII.] fig. 1.

Immediately south of the shadow is a bright, uniform, and broad belt.

Close to this bright belt is a broad, darker belt; which is divided by two narrow, white streaks; so that by this means it becomes to be five belts; namely, three dark, and two bright ones; the colour of the dark belt is yellowish.

The space from the quintuple belt towards the south pole of the planet which is in view, is of a pale, whitish colour; less bright than the white equatorial belt, and much less so than the ring.

The globular form of Saturn is very visible, so that it has by no means the appearance of a flat disk.

Nov. 13, 3^h 30′. The quintuple belt on Saturn is as it was Nov. 11. I saw it three hours ago, and several times since, without any visible change.

Nov. 19, 3^h 14′. The southern belt of Saturn is still divided into five. The evening is not clear enough to observe changes in it, if there were any.

Nov. 22, 2^h 32′. The quintuple belt on Saturn remains still the same; power 287.

With 430, I see the same very distinctly, but the small divisions have hardly light enough when so much magnified.

I viewed the same belt with four different object specula. One of them shewed the divisions uncommonly well.

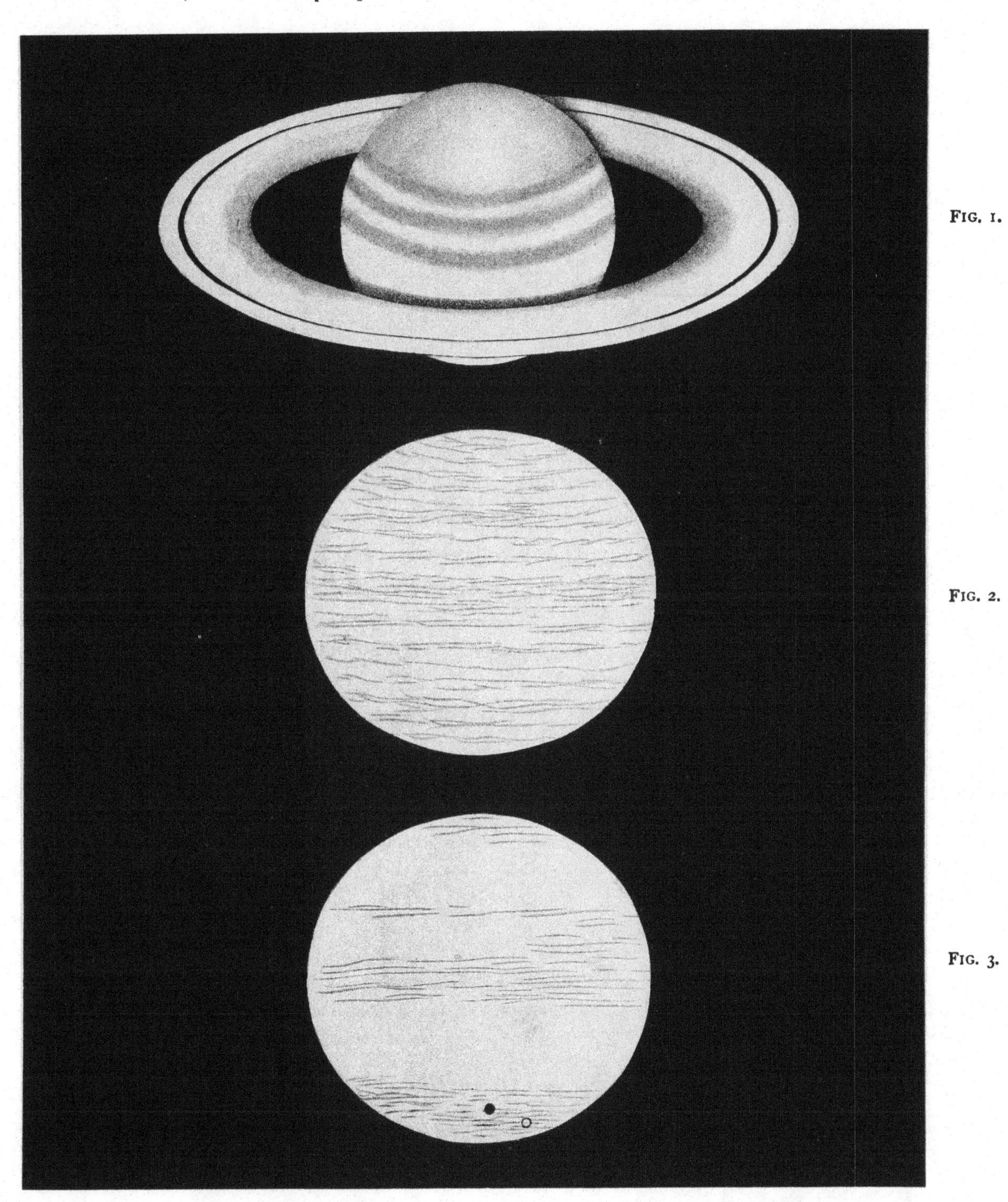

FIG. 1.

FIG. 2.

FIG. 3.

[*To face page* 452.

Dec. 3, 0^h 35′. 7-feet reflector; power 287. The quintuple belt upon Saturn remains as it was Nov. 22.

I tried several double and plano-concave eye-glasses, but found them all defective in figure except one, and that being of one inch focal length, the power was too low to expect seeing these belts well with it.

The smallness of the field of view, with astronomical objects is not so disagreeable as it is generally supposed to be; for the eye may have a motion before the lens, and by that means a small luminous object, when all the rest of the field is dark, and while the telescope remains in the same situation, may be seen for as long a time, passing through the field of a concave eye-glass, as it can in a convex one; whereas with the latter, it is well known that such a motion of the eye can be of no use.

2^h 36′. 20-feet reflector; power 157, 300, 480. I see the quintuple belt very well.

We know that the planet Jupiter has many belts. Some remarkable instances of their being very numerous are recorded in my journal, one of which is accompanied with a figure. The observations are as follows:

May 28, 1780. Jupiter's belts are curved; and there are a multitude of them all over the body of the planet. See fig. 2.

Jan. 18, 1790. I viewed Jupiter with the 40-feet reflector. There were two very dark, broad belts, divided by an equatorial zone or space, the colour of which was of a yellow cast. Next to the dark belts, on each side, towards the poles, were bright and dark small belts, alternately placed, and continued almost up to the poles, both ways.

In taking out fig. 2 from my journal, I perceive one so very unlike it just before, that I am induced to give it here, though rather foreign to my present purpose. It contains, however, an observation which it will not be amiss to record.

April 6, 1780. I had a fine view of Jupiter, and saw, as soon as I looked into the telescope, without having any previous notice of it, the shadow of the 3d satellite, and the satellite itself, upon the lower part of the disk. See fig. 3. The shadow was so black and well defined, that I attempted to measure it, and found its diameter by the micrometer 1″·562.

This measure of the shadow should be checked by the following observation.

March 15, 1792. 11^h 54′. With the 20-feet reflector, and a power of 800, I estimate the apparent diameter of the largest of Jupiter's satellites to be less than one-fourth of the diameter of the GEORGIAN planet, which I have just been viewing. With 1200, it seems also to be less, in the same proportion. With 2400, I can plainly perceive the disk of the satellite. With 4800, the apparent diameter of the largest of the satellites is less than one-quarter of that of the GEORGIAN planet.

The analogy alluded to in the first paragraph of this paper, refers to the numerous parallel belts which we have noticed, in the above given observations, on the disks of Jupiter and Saturn.

That belts are immediately connected with the rotation of the planets will hardly be denied, when those of Jupiter are so well known always to lie in the direction of its equatorial motion. Since, then, it appears that the belts of Saturn are very numerous, like those of Jupiter, and are also placed in the direction of the longest diameter of the planet, it may not be without some reason that we infer the period of the rotation of the former to be short, like that of the latter.

The planet Mars, in all my observations, never presented itself with any parallel belts, nor do we observe such phænomena on the disk of Venus. The first is known to have a rotation much slower than Jupiter; * and the latter, according to the accounts of CASSINI and BIANCHINI, is certainly not one that moves quickly upon its axis.

However, I do not mean to enter into the strength of an argument for a quick rotation of Saturn, that may be drawn from the condition of its belts. The circumstance of a quintuple belt, is adduced here with no other view, than merely to point out an analogy in the condition of the two largest planets of our system; and from thence to infer, that every conclusion on the atmosphere and rotation of the one, drawn from the appearance of its belts, will equally apply to the other.

WM. HERSCHEL.

Slough, near Windsor,
Dec. 14, 1793.

* See *Phil. Trans.* Vol. LXXI. Part I. page 134. [Above, p. 27.]

XXXI.

Account of some Particulars observed during the late Eclipse of the Sun. [Sept. 5, 1793.]

[*Phil. Trans.*, 1794, pp. 39–42.]

Read January 9, 1794.

It will be proper to remark that my attention, in observing this eclipse, was not directed to the time of the several particulars which are usually noticed in phænomena of this kind; such as the beginning, the end, and the digits eclipsed. I was very well assured that the care of other astronomers would render my endeavours, in that respect, perfectly unnecessary. The only view I had was, to avail myself of the power and distinctness of my telescopes, in order to see whether any appearances would arise that might deserve to be recorded; and the following particulars will, at least, serve to point out the way for similar observations to be made in other eclipses, where different circumstances may chance to afford an opportunity for gathering some addition to our knowledge, with regard to the nature and condition of the moon, or of the sun, and perhaps of both these heavenly bodies.

Sept. 5, 1793. 8^h 40′ 3″ by the clock.* My attention being directed to the place where I supposed the first impression would be made, I perceived two mountains of the moon enter the disk of the sun, as delineated at *a*, *b*, fig. 1 [Plate XIV.]. The time of their beginning to appear, when I saw them first, might be one or two seconds past.

9^h 5′. 7-feet reflector; power 287. The internal, luminous angle made on the sun, by the intersection of the limb of the moon, which is now but little more than a rectangle, is perfectly sharp up to the very point. It is not in the least disfigured by the refraction of the lunar atmosphere. The present shape of the angle, however, is not favourable for shewing the effects of that atmosphere.

9^h 17′. The luminous angles of the sun's preceding and following limbs, which are now acute, remain perfectly sharp. One of them, indeed, was disfigured, a little while ago, by the entrance of a mountain of the moon, but is now restored to its sharpness.

10^h 5′. I delineated the appearance of the limb of the moon upon the sun, and found its mountains as in fig. 2. At *a* was a large *table mountain*, as it may be

* By account, my sidereal time-piece was about 5′ 1″·7 too forward; but, as no transits had been lately taken, there may be an error of some seconds.

called, from its flat appearance; at *b* and *c* were elevated, pointed rocks. Their appearance changing pretty fast, no great accuracy can be expected in their expressed relative situation.

I suppose the height of the most elevated of these mountains not to exceed a mile and an half; for, on drawing several of them upon the segment of a large circle, so as to look like what they appeared when projected upon the sun, I found them to be from the 1500dth to the 2000dth part of the diameter of that circle. Then, putting the moon's diameter, as M. DE LA LANDE states it, at 782 French leagues, or 2151 English miles, we find the 1500dth part of this to be less than one mile and an half for the highest; and the 2000dth part, not quite one mile and a tenth for the lowest.

I attended all this time to the appearance of the sharp limb *abc* of the sun, fig. 3, and suspected, sometimes, a little bending of the cusps outwards, as expressed at *b* in fig. 4; but upon long, and attentive inspection, I could not satisfy myself of its reality. If there was a bending, it did probably not amount to one second of a degree; for, having formerly been much in the habit of measuring the moon's mountains,* the quantity of one second, on its disk, was still familiar enough to me to estimate it pretty exactly.

$10^h\ 15'$. I looked out with the natural eye for the planet Venus, and soon perceived her. In the telescope, with 287, she appeared very sharp and well defined, and was a little gibbous.

It may seem, perhaps, extraordinary that in the trial above mentioned, the eye should be able to ascertain the proportion of a quantity so little as the fifteen hundredth, or two thousandth part of the diameter of the moon; but the experiment may be easily repeated in the following manner:

Upon a line, six or eight inches long, drawn on a sheet of paper, make several small marks, representing mountains on the projected circumference of a large globe. The paper being then placed in a proper light and situation, withdraw the eye to the distance of 7, 8, or 9 feet, and take notice which of the marks appear of the same size, and distinctness, with the mountains they represent. Then, from the known angular magnitude of the moon, calculate its diameter, at the distance of your situation; this, multiplied by the power of the telescope, gives the diameter of a circle, to the circumference of which belongs the line, upon which are placed the marks above described. Now, measure the elevation of these marks above that line, and you will obtain the proportion they bear to the diameter of the circle.

* In the years 1779, 1780, and 1781. I did not measure, I suppose, less than an hundred mountains of the moon, in which I used three different methods: the projection of the tops of these mountains beyond the enlightened part of the disk; the length of their shadow on the surface of the moon; and their perpendicular projection on the full edge of the moon's limb. Some of these observations are contained in a former paper (see *Phil. Trans.* Vol. LXX. Part II. page 507 [above, p. 5]); but most of them remain uncalculated in my journal, till some proper opportunity.

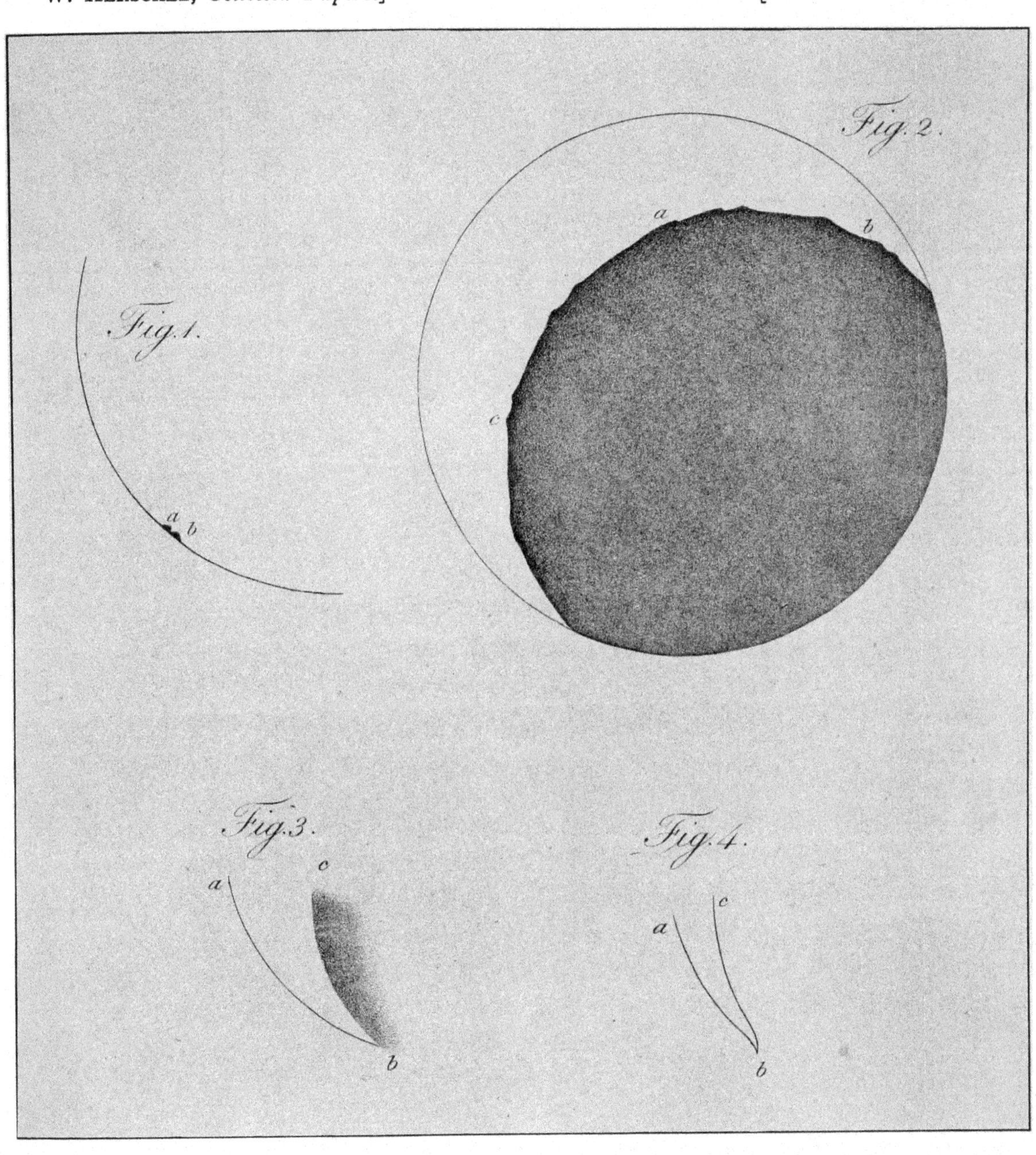

[*To face page* 456.

In my experiment, I found that I could plainly see some small protuberances at 9 feet distance, which were no higher than the 50th part of an inch. Then putting the diameter of the moon at 30′, we have the sum of the logarithms of the tangent of 30′, of the power 287, and of the 50ths of an inch contained in 9 feet; which, taken from the logarithm of the diameter of the moon in miles, gives the logarithm of ·16. By which we find, that so small a mountain as the $\frac{16}{100}$dth, or not much more than the sixth part of a mile, may be perceived and estimated, by the telescope and power that was used upon this occasion; and that, consequently, the estimation of mountains, near a mile and an half high, must become a very easy task.

W. HERSCHEL.

Slough, near Windsor,
Dec. 30, 1793.

XXXII.

On the Rotation of the Planet Saturn upon its Axis.

[*Phil. Trans.*, 1794, pp. 48–66.]

Read January 23, 1794.

In a late paper on the multiplicity of the regular belts of the planet Saturn, I pointed out an analogy, which might lead us to surmise that it had a pretty quick rotation upon its axis; I can at present announce the reality of that rotation. The following series of observations, in which Saturn has been traced through one hundred and fifty-four revolutions of its equator, will sufficiently confirm it.

The changes in the belts of Jupiter, it is well known, are so frequent, that we find some difficulty to make our observations of them agree to within 3, 4, or 5 minutes of time; but the belts on Saturn, which I have been lately observing, seem to have undergone no very material change, during the course of the two last months; so that we may hope the period of the rotation of this planet, which will be assigned in this paper, may be looked upon as having a considerable degree of exactness.

Before we can enter into particulars, it will be necessary to give the series of observations upon which my computations have been founded. It is not sufficient to extract only those parts of them which have served for calculating the period; as the value of astronomical observations consists in having them entire; every circumstance, as it occurred, is of consequence, and facts being stubborn things, we cannot decide upon them properly till they have been entirely laid open to our view, and sufficiently scrutinized. For this purpose the observations are all extracted from the journal, in the regular order in which they were made; and here I must remark, that I purposely avoided any calculations, or even surmises, of the length of a rotation, while the observations were making; in order to be perfectly free from every bias that might mislead the eye. In this I succeeded so well, that, when I began to calculate, I mistook not less than 4 hours and $\frac{3}{4}$ in the first supposition I made; which, happening to agree extraordinarily well with four of the most pointed observations, it misled me so far, that I was very near rejecting the whole series as inconsistent, and began to think the changes in the belts to have been so frequent, and irregular, as not to fall under any kind of calculation. It will, however, soon appear that this has not been the case, and that, on the contrary, there has been more steadiness and regularity in the belts, than might well have been expected in such kind of appearances.

Observations on the Belts of Saturn.*

Nov. 11, 1793. 3^h 35′. (*Correction of the clock* −7′ 27″·1).† Seven-feet reflector; power 287; new specula, uncommonly distinct.‡ Close to the ring of Saturn, where it passes the body of the planet, is the shadow of the ring; very narrow, and black.

Immediately south of the shadow is a bright, uniform, and broad belt.

Close to this belt is a broad, darker belt, which is divided by two narrow, white streaks; so that, by this means, it becomes to be five belts; namely, three dark, and two bright ones; the colour of the dark belt is yellowish.§ (A)

The space from the quintuple belt towards the south pole of the planet which is in view, is of a pale whitish colour; less bright than the white equatorial belt, and much less so than the ring.

The globular form of Saturn is very visible, so that it has, by no means, the appearance of a flat disk.

Nov. 13. 3^h 30′. (*Cor.* −7′ 29″·5.) The quintuple belt on Saturn is as it was Nov. 11. I saw it three hours ago, and several times since, without any visible change. (B)

Nov. 19. 3^h 14′. (*Cor.* −7′ 36″·8.) The southern belt of Saturn is still divided into five; the evening is not clear enough to observe changes in it, if there were any. (C)

Nov. 22. 2^h 32′. (*Cor.* −7′ 40″·4.) The quintuple belt on Saturn remains still the same; power 287. (D)

With 430, I see the same very distinctly; but the small divisions have hardly light enough, when so much magnified.

I viewed the same belt with four different object specula. One of them shewed the divisions uncommonly well.

Dec. 3. 0^h 35′. (*Cor.* −7′ 53″·8.) The quintuple belt upon Saturn remains as it was Nov. 22. (E)

* A few of these observations have been lately given; but as they are essential to the series, I thought it better to repeat them here, than to refer to my former paper.

† My time is kept by one of SHELTON's clocks, set now and then by equal altitudes, taken with a 12-inch BIRD's quadrant; and checked by the passage of a set of stars over the wire of a four-feet telescope, firmly fixed to the wall of my house. By calling the correction minus, I denote, in this case, that the clock is 7′ 27″·1 too fast for true sidereal time.

‡ In the course of these observations, I made 10 new object specula, and 14 small plain ones, for my 7-feet reflector; having already found, that with this instrument I had light sufficient to see the belts of Saturn completely well; and that, here, the maximum of distinctness might be much easier obtained, than where large apertures are concerned.

§ The letters (A) (B) (C), (*a*) (*b*) (*c*), &c. as they occur, refer to calculations which will be given hereafter.

$2^h\ 36'$. 20-feet reflector; power 157; 300; 480. I see the quintuple belt very well.*

Dec. 4. $23^h\ 22'$. (*Cor.* $-7'\ 55''$ *by a transit.*) 10-feet reflector. The quintuple belt is on Saturn, as it was last night. (F)

$4^h\ 57'$. 7-feet. The quintuple belt is uncommonly distinct.† The three narrow dark belts are of an equal breadth over the whole disk; the two bright belts which divide them, are also of an equal breadth throughout, but a little narrower than the dark ones. (G)

$5^h\ 58'$. I see the quintuple belt so clearly defined throughout all the 3 dark, and 2 bright belts, that I am apt to guess that the side which presents itself to me now is not the same which I saw in the first part of the evening; but it is not easily possible to determine whether the air might not be less clear then. I saw all other phænomena on Saturn extremely well, many times, between 1^h and 3^h; but not the belts so well as now. (H)

$6^h\ 36'$. The belt remains as free from interruptions as it was at $5^h\ 58'$. (I)

$6^h\ 52'$. I see the planet not so well defined now, as I did in the first part of the evening; being at present nearer the horizon; but I see the belts better than I did at that time. (K)

Dec. 6. $22^h\ 28'$. (*Cor.* $-7'\ 57''\cdot 2$.) I see the quintuple belt very distinctly. (L)

$22^h\ 55'$. I see the quintuple belt as readily as I see the rest of the appearances on Saturn. (M) I took care to bend my head so as to receive the picture of the belt in the same direction upon the retina, as I did December 4, at $5^h\ 58'$.‡

$23^h\ 55'$. I now see the quintuple belt full as well as I did Dec. 4, at $5^h\ 58'$. (N)

$1^h\ 25'$. That part of the quintuple belt which is now on the meridian, or centre of Saturn, is much less separated and defined than what was there at $23^h\ 55'$. (O)

$2^h\ 26'$. The divisions in the quintuple belt are not grown distincter, but rather more confused than they were before; I can scarcely perceive them. (P)

$3^h\ 28'$. The uppermost of the small dark belts, in that part which is on the meridian, is very faint, and the most north is rather darker than before. (Q)

* I found that the strong light of this instrument was too great a fatigue for the eye, which cannot bear to look at a very luminous object for a long time together. For this reason, I chiefly used the 7-feet reflector; and in future, all the observations, not expressly marked otherwise, are to be understood as having been made with that instrument; bearing an eye-glass of 3-tenths of an inch focal length. My object specula are generally from 84 to 88 inches in focus, and, therefore, give a power from 280 to 293. The favourite one gave 287. I had another reason for chiefly confining myself to one instrument, and one power; which was, that every circumstance being as much as possible the same, a change in the object I viewed might be the sooner perceived.

† When I found the divisions between the small belts so remarkably distinct, I began to suspect that they might not be, all around the planet, perfectly uniform in brightness and figure; and therefore now described the phænomena that occurred more minutely and carefully.

‡ This was a precaution that occurred to me, as there was a possibility that the vertical diameter of the retina might be more or less sensible than the horizontal one; but I had no reason afterwards to suppose that any such difference really exists.

4^h 28′. The most north of the two belts is darker, and a little broader than the most south. (R)

5^h 53′. Saturn is remarkably distinct; much more so than Dec. 4, at 5^h 58′; but the quintuple belt is less distinct than it was that evening; it has also undoubtedly a different appearance. The northern belt is the darkest and broadest, that next to it is less dark; and the third, or southmost belt, is faint, and hardly to be seen; the narrow white belts that separate them are contracted, and but just visible. (S)

Observation upon the Double Ring of Saturn.*

The outer ring is less bright than the inner ring. The inner ring is very bright close to the dividing space; and, at about half its breadth, it begins to change colour, gradually growing fainter; and just upon the inner edge, it is almost of the colour of the dark part of the quintuple belt.

7^h 52′. This is evidently another part of the planet than what I saw in the beginning of the evening. (T)

Dec. 9. 5^h 33′. (*Cor.* −8′ 0″·4.) The quintuple belt is extremely distinct, but not so much so, in proportion to the rest of the appearances, as I might expect. (U)

6^h 9′. The southmost dark belt is very faint; the northmost is the strongest and broadest; the bright divisions are very small, and difficult to be seen; I can, however, trace them all along. (V)

Dec. 11. 1^h 25′. (*Cor.* −8′ 2″·6.) I see the quintuple belt. The southmost belt is extremely faint; that to the north is the darkest and broadest; the middlemost is nearly as dark, but not quite so broad. (W) The air is much disturbed by wind, and flying haziness.

5^h 5′. The quintuple belt is very distinct. The southmost belt is less faint than it was at 1^h 25′; but the wind is too high, and the air too disturbed, to examine it minutely. (X)

Dec. 13 23^h 40′. (*Cor.* −8′ 4″·7.) I see the divisions of the quintuple belt very well; but there is a dry wind, and the telescope will not shew objects with that degree of distinctness which it usually does, when moisture is discharged from the air, by the precipitation of dew. (Y)

0^h 46′. I see the quintuple belt very well.

Remark on the Shadows of Saturn and its Ring.

On the south following part of the ring, close to the body of the planet, is the shadow of the body.

The shadow of the ring upon the body of the planet close to the ring, is not parallel to the ring at the two extremes, but a little broader there, than in the middle; the ends turning towards the south.

* This observation is foreign to the present purpose, but as it is new, and but short, I would not omit it; and for the same reason, two or three more are retained hereafter.

2^h 4′. The bright divisions between the belts are very narrow. The southmost dark belt is not much less faint than the northmost. (Z)

2^h 51′. The southmost dark belt on the preceding side, which at 23^h 40′ I thought was a little more south than the inside of the ring, now falls short of it. The broad bright belt also seems to be narrower now, than it was at that time.*

3^h 2′. The broad bright belt is as broad as the next three belts and a half, of the quintuple belt: that is, not quite so broad as the quintuple belt, without the southmost narrow dark belt.

4^h 11′. The broad white belt is of the breadth of the three adjoining belts.

5^h 11′. Appearances seem to be the same as they were an hour ago. The evening is indifferent. (*a*)

Dec. 16. 0^h 43′. (*Cor.* −8′ 8″·0 *by a transit.*) I see the quintuple belt extremely well.

1^h 3′. The most south of the three dark belts is full up to the inner edge of the preceding side of the ring; or rather a little above (or south of) it. The following side is less south, with regard to the ring, than the preceding; so that the quintuple belt is not parallel with that part of the ring, which crosses the body.

2^h 21′. The southmost belt is very faint; the middle and northern ones are much darker, and seem to be broader and closer, than I have at other times seen them. (*b*)

With 215. This power is too small to distinguish the divisions of the quintuple belt sufficiently.

3^h 12′. There seems to be no material alteration, but I do not see the divisions of the belt so distinctly as from the appearance of the other phænomena I ought to do. (*c*)

4^h 5′. The southmost belt meets the inner edge of the preceding part of the ring very nearly, and the quintuple belt seems to be parallel to the northern part of the ring, or nearly so.

4^h 16′. 10-feet reflector; power 300. The quintuple belt is parallel to the northern part of the ring, which is turned towards us.

5^h 35′. 7-feet. The southmost belt is brighter than it was at 2^h 21′. (*d*)

Dec. 18. 23^h 54′. (*Cor.* −8′ 9″·8.) I see the quintuple belt. (*e*)

1^h 19′. All the parts of the quintuple belt seem to be very uniform. The southmost on the preceding side reaches full up to the ring, where it passes behind the body. (*f*)

I viewed the planet with eight different object specula, they all shewed the quintuple belt very well.

Dec. 19. 1^h 41′. (*Cor.* −8′ 10″·6.) I have a very beautiful view of Saturn.

* Suspecting that the situation and direction of the belts might not be uniform all around the planet, I began to be more careful in describing the particulars relating to those circumstances; but found, soon after, that no additional light could be gathered from an attention to them.

I should suppose this part of Saturn to be that, where the quintuple belt is not so distinct as it is on some other parts; though I see it very well; yet from the extraordinary distinctness of the other phænomena, I think I ought to see it still better. (g)

The five old Satellites.

$1^h\ 56'$. 10-*feet reflector; with a power of* 60 *only, I see all the five old satellites.**

$4^h\ 15'$. 7-feet reflector. The southmost belt is faint. The quintuple belt is a little broader on the preceding side, than on the following. (h)

Dec. 23. $3^h\ 46'$. (*Cor.* $-8'\ 14''\cdot2$.) I see the quintuple belt very distinctly. The northmost of the dark belts is the broadest and darkest; the southmost is very faint; they are parallel to the ring, and the southmost just comes up to the inner edge of the ring. (i)

Dec. 29. $1^h\ 10'$. (*Cor.* $-8'\ 19''\cdot5$.) The quintuple belt remains on Saturn as it used to be; the southmost of the dark belts is faint. (k)

Jan. 1, 1794. $2^h\ 2'$. (*Cor.* $-8'\ 22''\cdot1$.) I see the quintuple belt very well; the southmost belt is not much fainter than the northmost. (l)

Observations of the South Pole of Saturn, and the Shadow of the Ring.

$3^h\ 40'$. *The south polar regions of Saturn are a little brighter in proportion to the bright equatorial belt, than they used to be; they are almost as bright as that belt.*

The shadow of the ring upon Saturn is perfectly black, like the shadow of Saturn upon the ring. The shadow of the ring upon Saturn, on each side, is bent a little southwards; so that the apparent curve it makes departs a little from the ring.

$5^h\ 18'$. The three dark belts of the quintuple belt seem to be very close, but the air is tremulous; I can however see them divided. (m)

Jan. 2. $23^h\ 53'$. (*Cor.* $-8'\ 23''\cdot0$ *by a transit.*) The quintuple belt is very distinct. The southmost belt is almost as dark as the other two. (n)

$1^h\ 52'$. The southmost belt is very nearly, if not quite as dark and distinct as the northmost. The air is very fine, and all the phænomena on Saturn are beautifully distinct. (o)

Jan. 4. $1^h\ 36'$. (*Cor.* $-8'\ 23''\cdot8$.) The quintuple belt is not so distinct as, from the appearance of the rest of the phænomena on Saturn, one might expect to see it. All the three dark belts are fainter than I have often seen them; and the southmost of them is much fainter than the northmost. (p)

$2^h\ 35'$. The quintuple belt is still faint. The southmost belt much fainter than the northmost; the latter is not only stronger, but also broader than the former. (q)

Jan. 7. $1^h\ 4'$. (*Cor.* $-8'\ 25''\cdot0$ *by a transit.*) The three dark belts of the

* This observation was made to try with how low a power they might be seen.

quintuple belt seem not to differ much in colour and breadth, but the evening is very indifferent. (*r*)

1^h $53'$. The air is a little clearer. The southmost is very little (if at all) less dark than the northmost; they are all very faint. (*s*)

3^h $18'$. As well as the night will permit to see, I judge the dark belts to be pretty equal in colour and size, the northmost, however, is still a little darker, and broader than the other two. (*t*)

3^h $44'$. The dark belts seem to be as equal as I have seen them at any time. I see them very well. (*u*)

Trial of Concave Eye Glasses.

I tried five new concave eye glasses, but they all proved defective in figure; with one of them, power 360, I saw the quintuple belt pretty well. With regard to the field of view they are full as convenient as convex glasses. (*v*)

5^h $50'$. The three dark belts are still nearly alike, and uniformly divided by the bright ones. (*w*)

Jan. 16. 2^h $52'$. (*Cor.* $-8'$ $28''\cdot 6$.) I suppose this to be the part of the quintuple belt, which is nearly uniform; the southmost one however is not quite so dark as the northmost. (*x*)

4^h $20'$. The belts seem to be equal and uniform throughout. (*y*)

5^h $5'$. The belts seem to be uniform; the southmost however is the faintest. (*z*)

Determination of the Period.

I shall now enter upon the method, which has been used to determine the rotation of the planet, from these observations.

Let K, T, Z, P, in the annexed figure, [see Plate XV.] represent the quintuple belt on the southern hemisphere of Saturn; the different parts of which are diversified as expressed by the different tints of the belts: those at A E K N I H M L G being uniform, while others at T B C D V, and R Q Y F have the southmost of the small dark belts very faint, and the northmost pretty strongly marked Let α be the south pole of Saturn, and let the circle 90, 180, 270, 360 represent its equator, divided into degrees; so that α 180, α 210, α 240, α 270, &c. are meridians of Saturn, which, as the planet turns upon its axis, from west to east, will successively pass over the line $\alpha\beta\gamma$, representing that meridian on Saturn which passes through the earth.

Then the eye of the observer being placed in the line $\alpha\beta\gamma$, at a great distance, and the hemisphere of Saturn, which is here projected on the plane of the equator, being in an oblique position, will only see the semicircle $\delta\beta\epsilon$. But on account of the great inclination of the arches $\zeta\delta$, $\eta\epsilon$, to the visual ray $\gamma\beta\alpha$, the eye will not perceive minute divisions, or marks, till they come within the limits $\zeta\eta$, and even

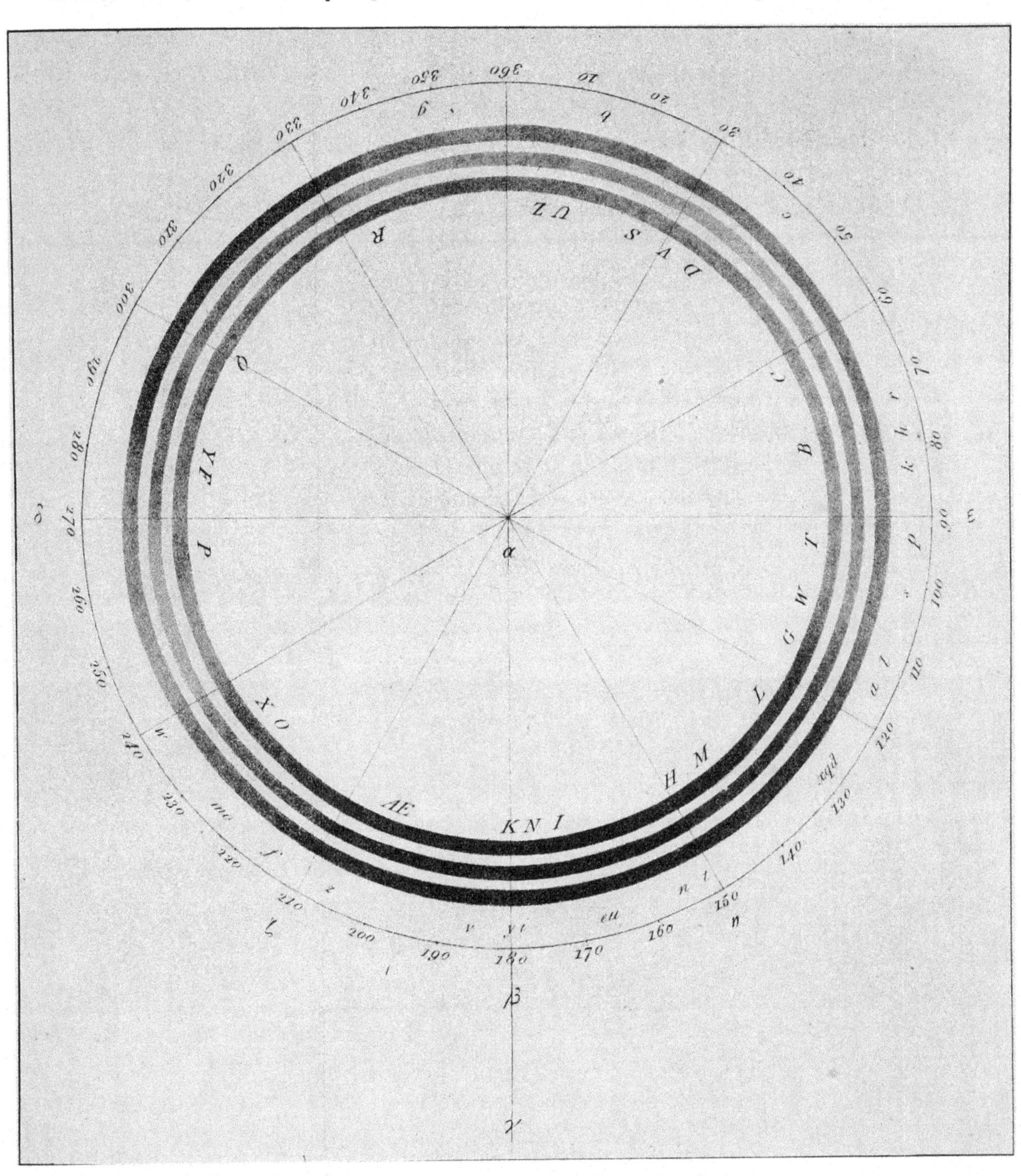

[*To face page* 464.

then will nowhere judge so well of their brightness and figure as when they draw near the situation β.

The divisions on the equator are to serve us to point out those places of the quintuple belt, which, by future calculations of the motion of this equator, will be shewn to have been on the meridian, $\alpha\beta\gamma$, at any given time; and the numbers are placed in the reverse order of the rotation, that the calculated motion of the belt may immediately point out the place which is come to the meridian. Thus, if the point K at 180 has moved 53 degrees forward, the situation x, on the belt, may be concluded to be on the meridian; because it is at 180 + 53, that is, the 233d degree.

Assuming for an epoch the observation of Dec. 4. 4^h 57′, where the different small belts, that make up the quintuple belt, are described as quite uniform; let it be placed at the 180th degree of the divided circle, where it will be fully exposed to the view of the observer.

I now select a few observations that are strongly marked, and as far distant from each other as can be found, by way of trying in what rotation they will agree. Such are the following two: Dec. 4. 6^h 52′; Jan. 7. 3^h 44′, for the places where the belts were uniform; and other two, Dec. 6. 7^h 52′, Jan. 4. 2^h 35′, for places where they appeared unequal. The sidereal times being corrected, and brought to true mean time, we have from Dec. 4. 13^h 46′ 51″, to Jan. 7. 8^h 25′ 11″ an interval of 33 days 18^h 38′ 20″; in which time let us suppose 79 revolutions to have been made. This will give 10^h 15′ 40″ for the time of each revolution.

In the next place, we have from Dec. 6. 14^h 38′ 47″ to Jan. 4. 7^h 28′ 12″ an interval of 28 days 16^h 49′ 25″; and allotting 67 revolutions to this, we obtain 10^h 16′ 51″ for the time of the rotation. These periods being independent of each other, and the observations having been made upon different parts of the belt, agree very well together. But now, some intermediate places are wanted, by way of trying, whether the period thus determined will accord with the rest of the observations; and for this purpose I select Dec. 18. 1^h 19′, Jan. 2. 1^h 52′; and bringing them also to true mean time, we have,

From Dec. 4. 13^h 46′ 51″ to Dec. 18. 7^h 19′ 28″ an interval of 13 days 17^h 32′ 37″. Then, supposing 32 revolutions to have been made, we obtain a period of 10^h 17′ 54″. Also,

From Dec. 4. 13^h 46′ 51″ to Jan. 2. 6^h 53′ 11″ is an interval of 28 days 17^h 6′ 20″; and admitting 67 revolutions of the belts, the period will be 10^h 17′ 6″.

These trials of intermediate times agreeing with the former sufficiently well, there can remain no doubt about the true quantity of the period in general. I therefore take a mean of the two first determinations, which gives 10^h 16′ 15″·5 for the approximate rotation of Saturn upon its axis.

It now becomes necessary to construct tables for a general calculation of all the observations. For, if these should contain descriptions contradicting the calculated appearances of the quintuple belt, our assigned period could not be looked upon as sufficiently established; on the contrary, if the calculated and observed appearances are found to agree, we may rest satisfied that the rotatory motion of this planet, which has so long eluded our strictest attention, is at length obtained.

In consequence of a few trials, which were made after the 7th of January, by tables constructed upon this mean period, I found that some small correction was required; and obtaining another very good observation on the 16th of the same month, it gave me an interval which included one hundred revolutions of the equator of Saturn. Now, making the proper deduction for the planet's retrograde motion during the time that passed between the first and last observation, we have from Dec. 4. $13^h\ 46'\ 51''$ to Jan. 16. $8^h\ 25'\ 39''$ an interval of 42 days $18^h\ 38'\ 48''$, in which the equator of Saturn moved over 35998·87 degrees, from which we compute a period of $10^h\ 16'\ 0''\cdot44$.

The following tables have been constructed upon this last period, and in the use of them the complement of the geocentric longitude of Saturn is always to be added, as has been explained in the tables of the satellites of that planet. *Phil. Trans.* Vol. LXXX. part II. page 495.*

Tables for the Equatorial Motion of Saturn.		
Epochs of the Longitude of the uniform Part of the Belts.		
1793 November	-	284°·18
1793 December	-	330 ·62
1794 January	-	138 ·61

* [Above, p. 414]

Motion of the Equator of Saturn, in Days, Hours, Minutes, and Seconds.

Days.	Deg. dec.	Hours.	Deg. dec.	Min.	Deg. dec.	Min.	Deg. dec.	Sec.	dec.	Sec.	dec.
1	121·55	1	35·06	1	·58	31	18·12	1	·01	31	·30
2	243·10	2	70·13	2	1·17	32	18·70	2	·02	32	·31
3	4·64	3	105·19	3	1·75	33	19·29	3	·03	33	·32
4	126·19	4	140·26	4	2·34	34	19·87	4	·04	34	·33
5	247·74	5	175·32	5	2·92	35	20·45	5	·05	35	·34
6	9·29	6	210·39	6	3·51	36	21·04	6	·06	36	·35
7	130·84	7	245·45	7	4·09	37	21·62	7	·07	37	·36
8	252·38	8	280·52	8	4·68	38	22·21	8	·08	38	·37
9	13·93	9	315·58	9	5·26	39	22·79	9	·09	39	·38
10	135·48	10	350·65	10	5·84	40	23·38	10	·10	40	·39
11	257·03	11	25·71	11	6·43	41	23·96	11	·11	41	·40
12	18·58	12	60·77	12	7·01	42	24·54	12	·12	42	·41
13	140·12	13	95·84	13	7·60	43	25·13	13	·13	43	·42
14	261·67	14	130·90	14	8·18	44	25·71	14	·14	44	·43
15	23·22	15	165·97	15	8·77	45	26·30	15	·15	45	·44
16	144·77	16	201·03	16	9·35	46	26·88	16	·16	46	·45
17	266·32	17	236·10	17	9·93	47	27·47	17	·16	47	·46
18	27·86	18	271·16	18	10·52	48	28·05	18	·17	48	·47
19	149·41	19	306·23	19	11·10	49	28·64	19	·18	49	·48
20	270·96	20	341·29	20	11·69	50	29·22	20	·19	50	·49
21	32·51	21	16·35	21	12·27	51	29·80	21	·20	51	·49
22	154·06	22	51·42	22	12·86	52	30·39	22	·21	52	·50
23	275·60	23	86·48	23	13·44	53	30·97	23	·22	53	·51
24	37·15	24	121·55	24	14·03	54	31·56	24	·23	54	·52
25	158·70			25	14·61	55	32·14	25	·24	55	·53
26	280·25			26	15·19	56	32·73	26	·25	56	·54
27	41·80			27	15·78	57	33·31	27	·26	57	·55
28	163·34			28	16·36	58	33·90	28	·27	58	·56
29	284·89			29	16·95	59	34·48	29	·28	59	·57
30	46·44			30	17·53	60	35·06	30	·29	60	·58
31	167·99										

The following 50 positions have been calculated by the above tables, and belong to the several observations which are marked with the tabular letters of references, as has been explained in the fourth note [page 459]. They have been computed to two places of decimals, but are only put down in degrees, as being sufficiently near for comparing them with the descriptions belonging to those parts of the belt which they point out.

Table of calculated Positions.

A 201	K 180	T 94	*c* 43	*m* 225	*v* 186
B 77	L 125	U 10	*d* 127	*n* 155	*w* 239
C 63	M 141	V 31	*e* 167	*o* 224	*x* 129
D 37	N 176	W 105	*f* 216	*p* 93	*y* 180
E 201	O 228	X 233	*g* 348	*q* 128	*z* 206
F 278	P 264	Y 282	*h* 78	*r* 73	
G 113	Q 300	Z 6	*i* 178	*s* 101	
H 149	R 335	*a* 115	*k* 83	*t* 151	
I 171	S 24	*b* 14	*l* 111	*u* 166	

I may venture to say, that there are not many of these calculated observations which do not very forcibly concur in proving the assigned revolution to be properly stated. I shall however only mention a few of the leading ones, and leave the rest to be looked over at leisure by those who wish to examine the subject minutely.

In observations previous to the 4th of December, no particular attention had been given to the minutiæ of the belts; but we may suppose them, on the 11th of November, to have been considerably distinct and uniform, to occasion their being noticed; and this the calculation verifies. For the position A points out 201 degrees, as the situation which was on the meridian at the time of observation, and by the figure we find it to be a very marked place.

A strong evidence of the rotation is the position H 149, observed Dec. 4, contrasted with the place F 278, which had been viewed in the early part of the same evening. The calculation here completely supports the suspicion which is expressed in the observation, "that the side which presented itself then was not the same which had been seen in the beginning of the evening."

The observations of Dec. 6, are of the most decisive nature; as will clearly appear by viewing the calculated positions L M N O P Q R S T, and comparing them with the descriptions belonging to them, that have been given in the observations. For, here the revolving belts were successively seen, in all their various tints, and the last position T was marked down as leaving no doubt of the evident rotation. By the calculation it appears that the belts had moved over 329 degrees, in the course of this evening's observation.

When the positions *c*, *d*, are compared, which were observed Dec. 16, we see that the southmost belt had acquired an additional brightness, as the observation expresses. It may not be amiss to remark upon this occasion, that brightness relates to clearness, distinctness, and easiness of vision; in opposition to faintness and confused outlines; therefore, the belt being brighter here, denotes its being more strongly marked by a deeper tint of dusky yellow, and by clearer divisions; so as to be easier perceived.

Dec. 19 furnishes a good instance of the exactness of our period; as the calculated position *g* perfectly justifies the surmise which is expressed in the observation.

Last of all, the place *y*, being truly pointed out by computation for Jan. 16, after a series of an hundred revolutions since the 4th of December, must concur in supporting our assigned period.

I shall only add one general remark, which is, that if we lengthen the time of the rotation but 2 minutes, it will throw the last observation back above 116 degrees; and if we diminish it by 2 minutes, there will arise an excess of more than 117; and, in either case, the calculations and observations would be totally at variance: from which we may conclude that our period must be exact to much less than 2 minutes, either way. Indeed, what alterations may have taken place in the belts themselves, it is impossible to determine. That there have been some, we may admit, and rather suppose, but we have no particular reason to suspect them to have been very considerable. And, after we have shewn that a proper motion, in the spots of the belts, of 116 degrees one way, or of 117 the other, would only occasion an error of 2 minutes in time, we need not hesitate to fix the rotation of the planet Saturn upon its axis at $10^h\ 16'\ 0''\cdot4$.

WM. HERSCHEL.

Slough, near Windsor,
Jan. 22, 1794.

XXXIII.

On the Nature and Construction of the Sun and fixed Stars.

[*Phil. Trans.*, 1795, pp. 46–72.]

Read December 18, 1794.

AMONG the celestial bodies the sun is certainly the first which should attract our notice. It is a fountain of light that illuminates the world! it is the cause of that heat which maintains the productive power of nature, and makes the earth a fit habitation for man! it is the central body of the planetary system; and what renders a knowledge of its nature still more interesting to us is, that the numberless stars which compose the universe, appear, by the strictest analogy, to be similar bodies. Their innate light is so intense, that it reaches the eye of the observer from the remotest regions of space, and forcibly claims his notice.

Now, if we are convinced that an inquiry into the nature and properties of the sun is highly worthy of our notice, we may also with great satisfaction reflect on the considerable progress that has already been made in our knowledge of this eminent body. It would require a long detail to enumerate all the various discoveries which have been made on this subject; I shall, therefore, content myself with giving only the most capital of them.

Sir ISAAC NEWTON has shewn that the sun, by its attractive power, retains the planets of our system in their orbits. He has also pointed out the method whereby the quantity of matter it contains may be accurately determined. Dr. BRADLEY has assigned the velocity of the solar light with a degree of precision exceeding our utmost expectation. GALILEO, SCHEINER, HEVELIUS, CASSINI, and others, have ascertained the rotation of the sun upon its axis, and determined the position of its equator. By means of the transit of Venus over the disc of the sun, our mathematicians have calculated its distance from the earth; its real diameter and magnitude; the density of the matter of which it is composed; and the fall of heavy bodies on its surface.

From the particulars here enumerated, it is sufficiently obvious, that we have already a very clear idea of the vast importance, and powerful influence of the sun on its planetary system. And if we add to this the beneficent effects we feel on this globe from the diffusion of the solar rays; and consider that, by well traced analogies, the same effects have been proved to take place on other planets of this system; I should not wonder if we were induced to think that nothing remained to be added in order to complete our knowledge: and yet it will not be difficult to shew that we are still very ignorant, at least with regard to the internal construction of the sun. The various conjectures, which have been formed on this subject, are evident marks of the uncertainty under which we have hitherto laboured.

The dark spots in the sun, for instance, have been supposed to be solid bodies revolving very near its surface. They have been conjectured to be the smoke of volcanoes, or the scum floating upon an ocean of fluid matter. They have also been taken for clouds. They were explained to be opaque masses, swimming in the fluid matter of the sun; dipping down occasionally. It has been supposed that a fiery liquid surrounded the sun, and that, by its ebbing and flowing, the highest parts of it were occasionally uncovered, and appeared under the shape of dark spots; and that, by the return of this fiery liquid, they were again covered, and in that manner successively assumed different phases. The sun itself has been called a globe of fire, though perhaps metaphorically. The waste it would undergo by a gradual consumption, on the supposition of its being ignited, has been ingeniously calculated. And in the same point of view its immense power of heating the bodies of such comets as draw very near to it has been assigned.

The bright spots, or faculæ, have been called clouds of light, and luminous vapours. The light of the sun itself has been supposed to be directly invisible, and not to be perceived unless by reflection; though the proofs, which are brought in support of that opinion, seem to me to amount to no more than, what is sufficiently evident, that we cannot see when rays of light do not enter the eye.

But it is time to profit by the many valuable observations that we are now in possession of. A list of successive eminent astronomers may be named, from GALILEO down to the present time, who have furnished us with materials for examination.

In supporting the ideas I shall propose in this paper, with regard to the physical construction of the sun, I have availed myself of the labours of all these astronomers, but have been induced thereto only by my own actual observation of the solar phænomena; which, besides verifying those particulars that had been already observed, gave me such views of the solar regions as led to the foundation of a very rational system. For, having the advantage of former observations, my latest reviews of the body of the sun were immediately directed to the most essential points; and the work was by this means facilitated, and contracted into a pretty narrow compass.

The following is a short extract of my observations on the sun, to which I have joined the consequences I now believe myself entitled to draw from them. When all the reasonings on the several phænomena are put together, and a few additional arguments, taken from analogy, which I shall also add, are properly considered, it will be found that a general conclusion may be made which seems to throw a considerable light upon our present subject.

In the year 1779, there was a spot on the sun which was large enough to be seen with the naked eye. By a view of it with a 7-feet reflector, charged with a very high power, it appeared to be divided into two parts. The largest of the two, on the 19th of April, measured 1′ 8″·06 in diameter; which is equal, in length, to more than 31 thousand miles. Both together must certainly have extended above 50 thousand.

The idea of its being occasioned by a volcanic explosion, violently driving away a fiery fluid, which on its return would gradually fill up the vacancy, and thus restore the sun, in that place, to its former splendour, ought to be rejected on many accounts. To mention only one, the great extent of the spot is very unfavourable to that supposition. Indeed a much less violent and less pernicious cause may be assigned, to account for all the appearances of the spot. When we see a dark belt near the equator of the planet Jupiter, we do not recur to earthquakes and volcanoes for its origin. An atmosphere, with its natural changes, will explain such belts. Our spot in the sun may be accounted for on the same principles. The earth is surrounded by an atmosphere, composed of various elastic fluids. The sun also has its atmosphere, and if some of the fluids which enter into its composition should be of a shining brilliancy, in the manner that will be explained hereafter, while others are merely transparent, any temporary cause which may remove the lucid fluid will permit us to see the body of the sun through the transparent ones. If an observer were placed on the moon, he would see the solid body of our earth only in those places where the transparent fluids of our atmosphere would permit him. In others, the opaque vapours would reflect the light of the sun, without permitting his view to penetrate to the surface of our globe. He would probably also find that our planet had occasionally some shining fluids in its atmosphere; as, not unlikely, some of our northern lights might not escape his notice, if they happened in the unenlightened part of the earth, and were seen by him in his long dark night. Nay, we have pretty good reason to believe, that probably all the planets emit light in some degree; for the illumination which remains on the moon in a total eclipse cannot be entirely ascribed to the light which may reach it by the refraction of the earth's atmosphere. For instance in the eclipse of the moon, which happened October 22, 1790, the rays of the sun refracted by the atmosphere of the earth towards the moon, admitting the mean horizontal refraction to be 30′ 50″·8, would meet in a focus above 189 thousand miles beyond the moon; so that consequently there could be no illumination from rays refracted by our atmosphere. It is, however, not improbable, that about the polar regions of the earth there may be refraction enough to bring some of the solar rays to a shorter focus. The distance of the moon at the time of the eclipse would require a refraction of 54′ 6″ equal to its horizontal parallax at that time, to bring them to a focus so as to throw light on the moon.

The unenlightened part of the planet Venus has also been seen by different persons, and not having a satellite, those regions that are turned from the sun cannot possibly shine by a borrowed light; so that this faint illumination must denote some phosphoric quality of the atmosphere of Venus.

In the instance of our large spot on the sun, I concluded from appearances that I viewed the real solid body of the sun itself, of which we rarely see more than its shining atmosphere.

In the year 1783, I observed a fine large spot, and followed it up to the edge of the sun's limb. Here I took notice that the spot was plainly depressed below the surface of the sun; and that it had very broad shelving sides. I also suspected some part, at least, of the shelving sides to be elevated above the surface of the sun; and observed that, contrary to what usually happens, the margin of that side of the spot, which was farthest from the limb, was the broadest.

The luminous shelving sides of a spot may be explained by a gentle and gradual removal of the shining fluid, which permits us to see the globe of the sun. As to the uncommon appearance of the broadest margin being on that side of the spot which was farthest from the limb when the spot came near the edge of it, we may surmise that the sun has inequalities on its surface, which may possibly be the cause of it. For, when mountainous countries are exposed, if it should chance that the highest parts of the landscape are situated so as to be near that side of the margin, or penumbra of the spot, which is towards the limb, it may partly intercept our view of it, when the spot is seen very obliquely. This would require elevations at least five or six hundred miles high; but considering the great attraction exerted by the sun upon bodies at its surface, and the slow revolution it has upon its axis, we may readily admit inequalities to that amount. From the centrifugal force at the sun's equator, and the weight of bodies at its surface, I compute that the power of throwing down a mountain by the exertion of the former, balanced by the superior force of keeping it in its situation of the latter, is near six and a half times less on the sun than on our equatorial regions; and as an elevation similar to one of three miles on the earth would not be less than 334 miles on the sun, there can be no doubt but that a mountain much higher would stand very firmly. The little density of the solar body seems also to be in favour of the height of its mountains; for, *cæteris paribus*, dense bodies will sooner come to their level than rare ones. The difference in the vanishing of the shelving side, instead of explaining it by mountains, may also, and perhaps more satisfactorily, be accounted for from the real difference of the extent, the arrangement, the height, and the intensity of the shining fluid, added to the occasional changes that may happen in these particulars, during the time in which the spot approaches to the edge of the disc. However, by admitting large mountains on the surface of the sun, we shall account for the different opinions of two eminent astronomers; one of whom believed the spots depressed below the sun, while the other supposed them elevated above it. For it is not improbable that some of the solar mountains may be high enough occasionally to project above the shining elastic fluid, when, by some agitation or other cause, it is not of the usual height; and this opinion is much strengthened by the return of some remarkable spots, which served CASSINI to ascertain the period of the sun's rotation. A very high country, or chain of mountains, may oftener become visible, by the removal of the obstructing fluid, than the lower regions, on account of its not being so deeply covered with it.

In the year 1791, I examined a large spot in the sun, and found it evidently depressed below the level of the surface; about the dark part was a broad margin, or plane of considerable extent, less bright than the sun, and also lower than its surface. This plane seemed to rise, with shelving sides, up to the place where it joined the level of the surface.

In confirmation of these appearances, I carefully remarked that the disc of the sun was visibly convex; and the reason of my attention to this particular, was my being already long acquainted with a certain optical deception, that takes place now and then when we view the moon; which is, that all the elevated spots on its surface will seem to be cavities, and all cavities will assume the shape of mountains. But then, at the same time the moon, instead of having the convex appearance of a globe, will seem to be a large concave portion of an hollow sphere. As soon as, by the force of imagination, you drive away the fallacious appearance of a concave moon, you restore the mountains to their protuberance, and sink the cavities again below the level of the surface. Now, when I saw the spot lower than the shining matter of the sun, and an extended plane, also depressed, with shelving sides rising up to the level, I also found that the sun was convex, and appeared in its natural globular state. Hence I conclude that there could be no deception in those appearances.

How very ill would this observation agree with the ideas of solid bodies bobbing up and down in a fiery liquid? with the smoke of volcanoes, or scum upon an ocean? And how easily it is explained upon our foregoing theory. The removal of the shining atmosphere, which permits us to see the sun, must naturally be attended with a gradual diminution on its borders; an instance of a similar kind we have daily before us, when through the opening of a cloud we see the sky, which generally is attended by a surrounding haziness of some short extent; and seldom transits, from a perfect clearness, at once to the greatest obscurity.

Aug. 26, 1792. I examined the sun with several powers, from 90 to 500. It appears evidently that the black spots are the opaque ground, or body of the sun; and that the luminous part is an atmosphere, which, being interrupted or broken, gives us a transient glimpse of the sun itself. My 7-feet reflector, which is in high perfection, represents the spots, as it always used to do, much depressed below the surface of the luminous part.

Sept. 2, 1792. I saw two spots in the sun with the naked eye. In the telescope I found they were clusters of spots, with many scattered ones besides. Every one of them was certainly below the surface of the luminous disc.

Sept. 8, 1792. Having made a small speculum, merely brought to a perfect figure upon hones, without polish, I found, that by stifling a great part of the solar rays, my object speculum would bear a greater aperture; and thus enabled me to see with more comfort, and less danger. The surface of the sun was unequal; many parts of it being elevated, and others depressed. This is here to be understood of the shining surface only, as the real body of the sun can probably be seldom seen, otherwise than in its black spots.

It may not be impossible, as light is a transparent fluid, that the sun's real surface also may now and then be perceived; as we see the shape of the wick of a candle through its flame, or the contents of a furnace in the midst of the brightest glare of it; but this, I should suppose, will only happen where the lucid matter of the sun is not very accumulated.

Sept. 9, 1792. I found one of the dark spots in the sun drawn pretty near the preceding edge. In its neighbourhood I saw a great number of elevated bright places, making various figures: I shall call them faculæ, with HEVELIUS; but without assigning to this term any other meaning than what it will hereafter appear ought to be given to it. I see these faculæ extended, on the preceding side, over about one-sixth part of the sun; but so far from resembling torches, they appear to me like the shrivelled elevations upon a dried apple, extended in length, and most of them are joined together, making waves, or waving lines.

By some good views in the afternoon, I find that the rest of the surface of the sun does not contain any faculæ, except a few on the following, and equatorial part of the sun. Towards the north and south I see no faculæ; there is all over the sun a great unevenness in the surface, which has the appearance of a mixture of small points of an unequal light; but they are evidently an unevenness or roughness of high and low parts.

Sept. 11, 1792. The faculæ, in the preceding part of the sun, are much gone out of the disc, and those in the following are come on. A dark spot also is come on with them.

Sept. 13, 1792. There are a great number of faculæ on the equatorial part of the sun, towards the preceding and following parts. I cannot see any towards the poles; but a roughness is visible every where.

Sept. 16, 1792. The sun contains many large faculæ, on the following side of its equator, and also several on the preceding side. I perceive none about the poles. They seem generally to accompany the spots, and probably, as the faculæ certainly are elevations, a great number of them may occasion neighbouring depressions: that is to say, dark spots.

The faculæ being elevations, very satisfactorily explains the reason why they disappear towards the middle of the sun, and re-appear on the other margin; for, about the place where we lose them, they begin to be edge-ways to our view; and if between the faculæ should lie dark spots, they will most frequently break out in the middle of the sun, because they are no longer covered by the side views of these faculæ.

Sept. 22, 1792. There are not many faculæ in the sun, and but few spots; the whole disc, however, is very much marked with roughness, like an orange. Some of the lowest parts of the inequalities are blackish.

Sept. 23, 1792. The following side of the sun contains many faculæ, near the limb. They take up an arch of about 50 degrees. There are, likewise, some on the preceding side. The north and south is rough as usual; but differently disposed. The faculæ are ridges of elevations above the rough surface.

Feb. 23, 1794. By an experiment I have just now tried, I find it confirmed that the sun cannot be so distinctly viewed with a small aperture and faint darkening glasses, as with a large aperture and stronger ones; this latter is the method I always use.

One of the black spots on the preceding margin, which was greatly below the surface of the sun, had, next to it, a protuberant lump of shining matter, a little brighter than the rest of the sun.

About all the spots, the shining matter seems to have been disturbed; and is uneven, lumpy, and zig-zagged in an irregular manner.

I call the spots black, not that they are entirely so, but merely to distinguish them; for there is not one of them, to-day, which is not partly, or entirely, covered over with whitish and unequally bright nebulosity, or cloudiness. This, in many of them, comes near to an extinction of the spot; and in others, seems to bring on a subdivision.

Sept. 28, 1794. There is a dark spot in the sun on the following side. It is certainly depressed below the shining atmosphere, and has shelving sides of shining matter, which rise up higher than the general surface, and are brightest at the top. The preceding shelving side is rendered almost invisible, by the overhanging of the preceding elevations; while the following is very well exposed: the spot being apparently such in figure as denotes a circular form, viewed in an oblique direction.

Near the following margin are many bright elevations, close to visible depressions. The depressed parts are less bright than the common surface.

The penumbra, as it is called, about this spot, is a considerable plane, of less brightness than the common surface, and seems to be as much depressed below that surface as the spot is below the plane.

Hence, if the brightness of the sun is occasioned by the lucid atmosphere, the intensity of the brightness must be less where it is depressed; for light, being transparent, must be the more intense the more it is deep.

Oct. 12, 1794. The whole surface of the sun is diversified by inequality in the elevation of the shining atmosphere. The lowest parts are every where darkest; and every little pit has the appearance of a more or less dark spot.

A dark spot, which is on the preceding side, is surrounded by very great inequalities in the elevation of the lucid atmosphere ; and its depression below the same is bounded by an immediate rising of very bright light.

Oct. 13, 1794. The spot in the sun I observed yesterday is drawn so near the margin, that the elevated side of the following part of it hides all the black ground, and still leaves the cavity visible, so that the depression of the black spots, and the elevation of the faculæ, are equally evident.

It will now be easy to bring the result of these observations into a very narrow compass. That the sun has a very extensive atmosphere cannot be doubted ; and that this atmosphere consists of various elastic fluids, that are more or less lucid and transparent, and of which the lucid one is that which furnishes us with light, seems also to be fully established by all the phænomena of its spots, of the faculæ, and of the lucid surface itself. There is no kind of variety in these appearances but what may be accounted for with the greatest facility, from the continual agitation which we may easily conceive must take place in the regions of such extensive elastic fluids.

It will be necessary, however, to be a little more particular, as to the manner in which I suppose the lucid fluid of the sun to be generated in its atmosphere. An analogy that may be drawn from the generation of clouds in our own atmosphere, seems to be a very proper one, and full of instruction. Our clouds are probably decompositions of some of the elastic fluids of the atmosphere itself, when such natural causes, as in this grand chemical laboratory are generally at work, act upon them; we may therefore admit that in the very extensive atmosphere of the sun, from causes of the same nature, similar phænomena will take place ; but with this difference, that the continual and very extensive decompositions of the elastic fluids of the sun are of a phosphoric nature, and attended with lucid appearances, by giving out light.

If it should be objected, that such violent and unremitting decompositions would exhaust the sun, we may recur again to our analogy, which will furnish us with the following reflections. The extent of our own atmosphere, we see, is still preserved, notwithstanding the copious decompositions of its fluids, in clouds and falling rain ; in flashes of lightning, in meteors, and other luminous phænomena ; because there are fresh supplies of elastic vapours, continually ascending to make good the waste occasioned by those decompositions. But it may be urged, that the case with the decomposition of the elastic fluids in the solar atmosphere would be very different, since light is emitted, and does not return to the sun, as clouds do to the earth when they descend in showers of rain. To which I answer, that in the decomposition of phosphoric fluids every other ingredient but light may also return to the body of the sun. And that the emission of light must waste the sun, is not a difficulty that can be opposed to our hypothesis. For as it is an evident fact that the sun does emit light, the same objection, if it could be one, would equally militate against every other assignable way to account for the phænomenon.

There are moreover considerations that may lessen the pressure of this alleged difficulty. We know the exceeding subtilty of light to be such, that in ages of time its emanation from the sun cannot very sensibly lessen the size of this great body. To this may be added, that, very possibly, there may also be ways of restoration to compensate for what is lost by the emission of light; though the manner in which this can be brought about should not appear to us. Many of the operations of nature are carried on in her great laboratory, which we cannot comprehend; but now and then we see some of the tools with which she is at work. We need not wonder that their construction should be so singular as to induce us to confess our ignorance of the method of employing them, but we may rest assured that they are not a mere *lusus naturæ*. I allude to the great number of small telescopic comets that have been observed; and to the far greater number still that are probably much too small for being noticed by our most diligent searchers after them. Those six, for instance, which my sister has discovered, I can from examination affirm had not the least appearance of any solid nucleus, and seemed to be mere collections of vapours condensed about a centre. Five more, that I have also observed, were nearly of the same nature. This throws a mystery over their destination, which seems to place them in the allegorical view of tools, probably designed for some salutary purposes to be wrought by them; and, whether the restoration of what is lost to the sun by the emission of light, the possibility of which we have been mentioning above, may not be one of these purposes, I shall not presume to determine. The motion of the comet discovered by Mr. MESSIER in June, 1770, plainly indicated how much its orbit was liable to be changed, by the perturbations of the planets; from which, and the little agreement that can be found between the elements of the orbits of all the comets that have been observed, it appears clearly that they may be directed to carry their salutary influence to any part of the heavens.

My hypothesis, however, as before observed, does not lay me under any obligation to explain how the sun can sustain the waste of light, nor to shew that it will sustain it for ever; and I should also remark that, as in the analogy of generating clouds I merely allude to their production as owing to a decomposition of some of the elastic fluids of our atmosphere, that analogy, which firmly rests upon the fact, will not be less to my purpose to whatever cause these clouds may owe their origin. It is the same with the lucid clouds, if I may so call them, of the sun. They plainly exist, because we see them; the manner of their being generated may remain an hypothesis; and mine, till a better can be proposed, may stand good; but whether it does or not, the consequences I am going to draw from what has been said will not be affected by it.

Before I proceed, I shall only point out, that according to the above theory, a dark spot in the sun is a place in its atmosphere which happens to be free from luminous decompositions; and that faculæ are, on the contrary, more copious mixtures of such fluids as decompose each other. The penumbra which attends the spots, being generally depressed more or less to about half way between the solid body of the sun and the upper part of those regions in which luminous decompositions take place, must of course be fainter than other parts. No spot favourable for taking measures having lately been on the sun, I can only judge, from former appearances, that the regions in which the luminous solar clouds are formed, adding thereto the elevation of the faculæ, cannot be less than 1843, nor much more than 2765 miles in depth. It is true that in our atmosphere the extent of the clouds is limited to a very narrow compass; but we ought rather to compare the solar ones to the luminous decompositions which take place in our *aurora borealis*, or luminous arches, which extend much farther than the cloudy regions. The density of the luminous solar clouds, though very great, may not be exceedingly more so than that of our *aurora borealis*. For, if we consider what would be the brilliancy of a space two or three thousand miles deep, filled with such coruscations as we see now and then in our atmosphere, their apparent intensity, when viewed at the distance of the sun, might not be much inferior to that of the lucid solar fluid.

From the luminous atmosphere of the sun I proceed to its opaque body, which by calculation from the power it exerts upon the planets we know to be of great solidity; and from the phænomena of the dark spots, many of which, probably on account of their high situations, have been repeatedly seen, and otherwise denote inequalities in their level, we surmise that its surface is diversified with mountains and vallies.

What has been said enables us to come to some very important conclusions, by remarking, that this way of considering the sun and its atmosphere removes the great dissimilarity we have hitherto been used to find between its condition and that of the rest of the great bodies of the solar system.

The sun, viewed in this light, appears to be nothing else than a very eminent, large, and lucid planet, evidently the first, or in strictness of speaking, the only primary one of our system; all others being truly secondary to it. Its similarity to the other globes of the solar system with regard to its solidity, its atmosphere, and its diversified surface; the rotation upon its axis, and the fall of heavy bodies, leads us on to suppose that it is most probably also inhabited, like the rest of the planets, by beings whose organs are adapted to the peculiar circumstances of that vast globe.

Whatever fanciful poets might say, in making the sun the abode of blessed spirits, or angry moralists devise, in pointing it out as a fit place for the punishment of the wicked, it does not appear that they had any other foundation for their assertions than mere opinion and vague surmise; but now I think myself authorized, *upon astronomical principles*, to propose the sun as an inhabitable world, and am persuaded that the foregoing observations, with the conclusions I have drawn from them, are fully sufficient to answer every objection that may be made against it.

It may, however, not be amiss to remove a certain difficulty, which arises from the effect of the sun's rays upon our globe. The heat which is here, at the distance of 95 millions of miles, produced by these rays, is so considerable, that it may be objected, that the surface of the globe of the sun itself must be scorched up beyond all conception.

This may be very substantially answered by many proofs drawn from natural philosophy, which shew that heat is produced by the sun's rays only when they act upon a calorific medium ; they are the cause of the production of heat, by uniting with the matter of fire, which is contained in the substances that are heated : as the collision of flint and steel will inflame a magazine of gunpowder, by putting all the latent fire it contains into action. But an instance or two of the manner in which the solar rays produce their effect, will bring this home to our most common experience.

On the tops of mountains of a sufficient height, at an altitude where clouds can very seldom reach, to shelter them from the direct rays of the sun, we always find regions of ice and snow. Now if the solar rays themselves conveyed all the heat we find on this globe, it ought to be hottest where their course is least interrupted. Again, our aëronauts all confirm the coldness of the upper regions of the atmosphere ; and since, therefore, even on our earth the heat of any situation depends upon the aptness of the medium to yield to the impression of the solar rays, we have only to admit, that on the sun itself the elastic fluids composing its atmosphere, and the matter on its surface, are of such a nature as not to be capable of any excessive affection from its own rays ; and, indeed, this seems to be proved by the copious emission of them ; for if the elastic fluids of the atmosphere, or the matter contained on the surface of the sun, were of such a nature as to admit of an easy, chemical combination with its rays, their emission would be much impeded.

Another well known fact is, that the solar focus of the largest lens, thrown into the air, will occasion no sensible heat in the place where it has been kept for a considerable time, although its power of exciting combustion, when proper bodies are exposed, should be sufficient to fuse the most refractory substances.*

It will not be necessary to mention other objections, as I can think of none that may be made, but what a proper consideration of the foregoing observations will easily remove ; such as may be urged from the dissimilarity between the luminous atmosphere of the sun and that of our globe will be touched upon hereafter, when I consider the objections that may be assigned against the moon's being an inhabitable satellite.

I shall now endeavour, by analogical reasonings, to support the ideas I have suggested concerning the construction and purposes of the sun ; in order to which, it will be necessary to begin with such arguments as the nature of the case will admit,

* The subject of light and heat has been very ably discussed by Mr. DE LUC, in his excellent work, *Idées sur la Météorologie*, Tome I. part 2, chap. 2, section 2, *De la Nature du Feu* ; and Tome II. part 3, chap. 6, section 2, *Des Rapports de la Lumière avec la Chaleur dans l'Atmosphère.*

to shew that our moon is probably inhabited. This satellite is of all the heavenly bodies the nearest, and therefore most within the reach of our telescopes. Accordingly we find, by repeated inspection, that we can with perfect confidence give the following account of it.

It is a secondary planet, of a considerable size; the surface of which is diversified, like that of the earth, by mountains and vallies. Its situation, with respect to the sun, is much like that of the earth; and, by a rotation on its axis, it enjoys an agreeable variety of seasons, and of day and night. To the moon, our globe will appear to be a very capital satellite; undergoing the same regular changes of illuminations as the moon does to the earth. The sun, the planets, and the starry constellations of the heavens, will rise and set there as they do here; and heavy bodies will fall on the moon as they do on the earth. There seems only to be wanting, in order to complete the analogy, that it should be inhabited like the earth.

To this it may be objected, that we perceive no large seas in the moon; that its atmosphere (the existence of which has even been doubted by many) is extremely rare, and unfit for the purposes of animal life; that its climates, its seasons, and the length of its days, totally differ from ours; that without dense clouds (which the moon has not), there can be no rain; perhaps no rivers, no lakes. In short, that, notwithstanding the similarity which has been pointed out, there seems to be a decided difference in the two planets we have compared.

My answer to this will be, that that very difference which is now objected, will rather strengthen the force of my argument than lessen its value: we find, even upon our globe, that there is the most striking difference in the situation of the creatures that live upon it. While man walks upon the ground, the birds fly in the air, and fishes swim in water; we can certainly not object to the conveniences afforded by the moon, if those that are to inhabit its regions are fitted to their conditions as well as we on this globe are to ours. An absolute, or total sameness, seems rather to denote imperfections, such as nature never exposes to our view; and, on this account, I believe the analogies that have been mentioned fully sufficient to establish the high probability of the moon's being inhabited like the earth.

To proceed, we will now suppose an inhabitant of the moon, who has not properly considered such analogical reasonings as might induce him to surmise that our earth is inhabited, were to give it as his opinion that the use of that great body, which he sees in his neighbourhood, is to carry about his little globe, that it may be properly exposed to the light of the sun, so as to enjoy an agreeable and useful variety of illumination, as well as to give it light by reflection from the sun, when direct daylight cannot be had. Suppose also that the inhabitants of the satellites of Jupiter, Saturn, and the Georgian planet, were to look upon the primary ones, to which they belong, as mere attractive centres, to keep together their orbits, to direct their revolution round the sun, and to supply them with reflected light in the absence of direct illumination. Ought we not to condemn their ignorance, as proceeding from want of attention and proper reflection? It is very true that the earth, and those other planets that have satellites about them, perform all the offices that have been named, for the inhabitants of these little globes; but to us, who live upon one of these planets, their reasonings cannot but appear very defective; when we see what a magnificent dwelling place the earth affords to numberless intelligent beings.

These considerations ought to make the inhabitants of the planets wiser than we have supposed those of their satellites to be. We surely ought not, like them, to say "the sun (that immense globe, whose body would much more than fill the whole orbit of the moon) is merely an attractive centre to us." From experience we can affirm, that the performance of the most salutary offices to inferior planets, is not inconsistent with the dignity of superior purposes; and, in consequence of such analogical reasonings, assisted by telescopic views, which plainly favour the same opinion, we need not hesitate to admit that the sun is richly stored with inhabitants.

This way of considering the sun is of the utmost importance in its consequences. That stars are suns can hardly admit of a doubt. Their immense distance would perfectly exclude them from our view, if the light they send us were not of the solar kind. Besides, the analogy may be traced much farther. The sun turns on its axis. So does the star Algol. So do the stars called β Lyræ, δ Cephei, η Antinoi, ο Ceti, and many more; most probably all. From what other cause can we so probably account for their periodical changes? Again, our sun has spots on its surface. So has the star Algol; and so have the stars already named; and probably every star in the heavens. On our sun these spots are changeable. So they are on the star ο Ceti; as evidently appears from the irregularity of its changeable lustre, which is often broken in upon by accidental changes, while the general period continues unaltered. The same little deviations have been observed in other periodical stars, and ought to be ascribed to the same cause. But if stars are suns, and suns are inhabitable, we see at once what an extensive field for animation opens itself to our view.

It is true that analogy may induce us to conclude, that since stars appear to be suns, and suns, according to the common opinion, are bodies that serve to enlighten, warm, and sustain a system of planets, we may have an idea of numberless globes that serve for the habitation of living creatures. But if these suns themselves are primary planets, we may see some thousands of them with our own eyes; and millions by the help of telescopes; when at the same time, the same analogical reasoning still remains in full force, with regard to the planets which these suns may support.

In this place I may, however, take notice that, from other considerations, the idea of suns or stars being *merely* the supporters of systems of planets, is not absolutely to be admitted as a general one. Among the great number of very compressed clusters of stars, I have given in my catalogues, there are some which open a different view of the heavens to us. The stars in them are so very close together, that, notwithstanding the great distance at which we may suppose the cluster itself to be, it will hardly be possible to assign any sufficient mutual distance to the stars composing the cluster, to leave room for crowding in those planets, for whose support these stars have been, or might be, supposed to exist. It should seem, therefore, highly probable that they exist for themselves; and are, in fact, only very capital, *lucid*, primary planets, connected together in one great system of mutual support.

As in this argument I do not proceed upon conjectures, but have actual observations in view, I shall mention an instance in the clusters, No. 26, 28, and 35, VI. class, of my catalogue of nebulæ, and clusters of stars. (See *Phil. Trans.* Vol. LXXIX. Part II. p. 251.)* The stars in them are so crowded, that I cannot conjecture them to be at a greater apparent distance from each other than five seconds; even after a proper allowance for such stars, as on a supposition of a globular form of the cluster, will interfere with one another, has been made. Now, if we would leave as much room between each of these stars as there is between the sun and Sirius, we must place these clusters 42104 times as far from us as that star is from the sun. But in order to bring down the lustre of Sirius to that of an equal star placed at such a distance, I ought to reduce the aperture of my 20-feet telescope to less than the two-and-twenty hundredth part of an inch; when certainly I could no longer expect to see any star at all.

The same remark may be made, with regard to the number of very close double stars; whose apparent diameters being alike, and not very small, do not indicate any very great mutual distance. From which, however, must be deducted all those where the different distances may be compensated by the real difference in their respective magnitudes.

To what has been said may be added, that in some parts of the milky way, where yet the stars are not very small, they are so crowded, that in the year 1792, Aug. 22, I found by the gages that, in 41 minutes of time, no less than 258 thousand of them had passed through the field of view of my telescope.†

It seems, therefore, upon the whole not improbable that, in many cases, stars are united in such close systems as not to leave much room for the orbits of planets

* [Above, pp. 360–361.]

† The star-gages ran thus:

From $19^h\ 35'$ to $19^h\ 51'$	600 stars in the field
19 51 — 19 57	440
19 57 — 20 12	360
20 12 — 20 16	260

The breadth of the sweep was 2° 35′, the diameter of the field 15′, and the mean polar distance 73° 54′. Then let F be the diameter of the field of view, S the number of stars in each field, B the breadth of the sweep, plus F, T the length of the sweep expressed in minutes of space, ϕ the sine of the mean polar distance, C the constant fraction ·7854, and the stars in these four successive short sweeps will be found by the expression $\frac{BTS\phi}{F^2\,C}$ equal to 133095 + 36601 + 74866 + 14419 or in all 258981.

or comets; and that consequently, upon this account also, many stars, unless we would make them mere useless brilliant points, may themselves be lucid planets, perhaps unattended by satellites.

POSTSCRIPT.

The following observations, which were made with an improved apparatus, and under the most favourable circumstances, should be added to those which have been given. They are decisive with regard to one of the conditions of the lucid matter of the sun.

Nov. 26, 1794. Eight spots in the sun, and several subdivisions of them, are all equally depressed.

The sun is mottled every where.

The mottled appearance of the sun is owing to an inequality in the level of the surface.

The sun is equally mottled at its poles and at its equator; but the mottled appearances may be seen better about the middle of the disc than towards the circumference, on account of the sun's spherical form.

The unevenness arising from the elevation and depression of the mottled appearance on the surface of the sun, seems, in many places, to amount to as much, or to nearly as much as the depression of the penumbræ of the spots below the upper part of the shining substance; without including faculæ, which are protuberant.

The lucid substance of the sun is neither a liquid, nor an elastic fluid; as is evident from its not instantly filling up the cavities of the spots, and of the unevenness of the mottled parts. It exists, therefore, in the manner of lucid clouds swimming in the transparent atmosphere of the sun; or rather, of luminous decompositions taking place within that atmosphere.

XXXIV.

Description of a Forty-feet Reflecting Telescope.

[*Phil. Trans.*, 1795, pp. 347–409.]

Read June 11, 1795.

THE uncommon size of my forty-feet reflecting telescope will render a description of it not unacceptable to lovers of astronomy. I shall therefore endeavour to give as complete an idea of its construction as the limited compass of this paper will permit, and hope that, with the assistance of the annexed drawings, the mechanism of it will be sufficiently intelligible to such as have been in the habit of viewing machines and mechanical works.

It will be necessary to mention a few circumstances that led the way to the construction of this large instrument, in the execution of which two very material requisites were necessary: namely, the support of a very considerable expence, and a competent experience and practice in mechanical and optical operations.

When I resided at Bath I had long been acquainted with the theory of optics and mechanics, and wanted only that experience which is so necessary in the practical part of these sciences. This I acquired by degrees at that place, where in my leisure hours, by way of amusement, I made for myself several 2-feet, 5-feet, 7-feet, 10-feet, and 20-feet NEWTONIAN telescopes; besides others of the GREGORIAN form, of 8 inches, 12 inches, 18 inches, 2 feet, 3 feet, 5 feet, and 10 feet focal length. My way of doing these instruments at that time, when the direct method of giving the figure of any of the conic sections to specula was still unknown to me, was, to have many mirrors of each sort cast, and to finish them all as well as I could; then to select by trial the best of them, which I preserved; the rest were put by to be repolished. In this manner I made not less than 200, 7-feet; 150, 10-feet; and about 80, 20-feet mirrors; not to mention those of the GREGORIAN form, or of the construction of Dr. SMITH's reflecting microscope, of which I also made a great number.

My mechanical amusements went hand in hand with the optical ones. The number of stands I invented for these telescopes it would not be easy to assign. I contrived and delineated them of different forms, and executed the most promising of the designs. To these labours we owe my 7-feet NEWTONIAN telescope-stand, which was brought to its present convenient construction about 17 years ago; a description and engraving of which I intend to take some future opportunity of presenting to the Royal Society. In the year 1781 I began also to construct a 30-feet aërial reflector; and after having invented and executed a stand for it, I cast the mirror, which was moulded up so as to come out 36 inches in diameter. The composition of my metal being a little too brittle, it cracked in the cooling. I cast it a second time, but here the furnace, which I had built in my house for the purpose, gave way, and the metal ran into the fire.

These accidents put a temporary stop to my design, and as the discovery of the GEORGIAN planet soon after introduced me to the patronage of our most gracious King, the great work I had in view was for a while postponed.

In the year 1783 I finished a very good 20-feet reflector with a large aperture, and mounted it upon the plan of my present telescope. After two years' observation with it, the great advantage of such apertures appeared so clearly to me, that I recurred to my former intention of increasing them still farther; and being now sufficiently provided with experience in the work I wished to undertake, the President of our Royal Society, who is always ready to promote useful undertakings, had the goodness to lay my design before the King. His Majesty was graciously pleased to approve of it, and with his usual liberality to support it with his royal bounty.

In consequence of this arrangement I began to construct the 40-feet telescope, which is the subject of this paper, about the latter end of the year 1785. The woodwork of the stand, and machines for giving the required motions to the instrument, were immediately put in hand, and forwarded with all convenient expedition. In the whole of the apparatus none but common workmen were employed, for I made drawings of every part of it, by which it was easy to execute the work, as I constantly inspected and directed every person's labour; though sometimes there were not less than 40 different workmen employed at the same time.

While the stand of the telescope was preparing I also began the construction of the great mirror, of which I inspected the casting, grinding, and polishing; and the work was in this manner carried on with no other interruption than what was occasioned by the removal of all the apparatus and materials from Clay-hall, where I then lived, to my present situation at Slough.

Here soon after my arrival, I began to lay the foundation, upon which by degrees the whole structure was raised as it now stands; and the speculum being highly polished and put into the tube, I had the first view through it on Feb. 19, 1787. I do not however date the completing of the instrument till much later; for the first speculum, by a mismanagement of the person who cast it, came out thinner on the centre of the back than was intended, and on account of its weakness would not permit a good figure to be given to it. A second mirror was cast Jan. 26, 1788; but it cracked in cooling. Feb. 16, we recast it with particular attention to the shape of the back, and it proved to be of a proper degree of strength. Oct. 24, it was brought to a pretty good figure and polish, and I observed the planet Saturn

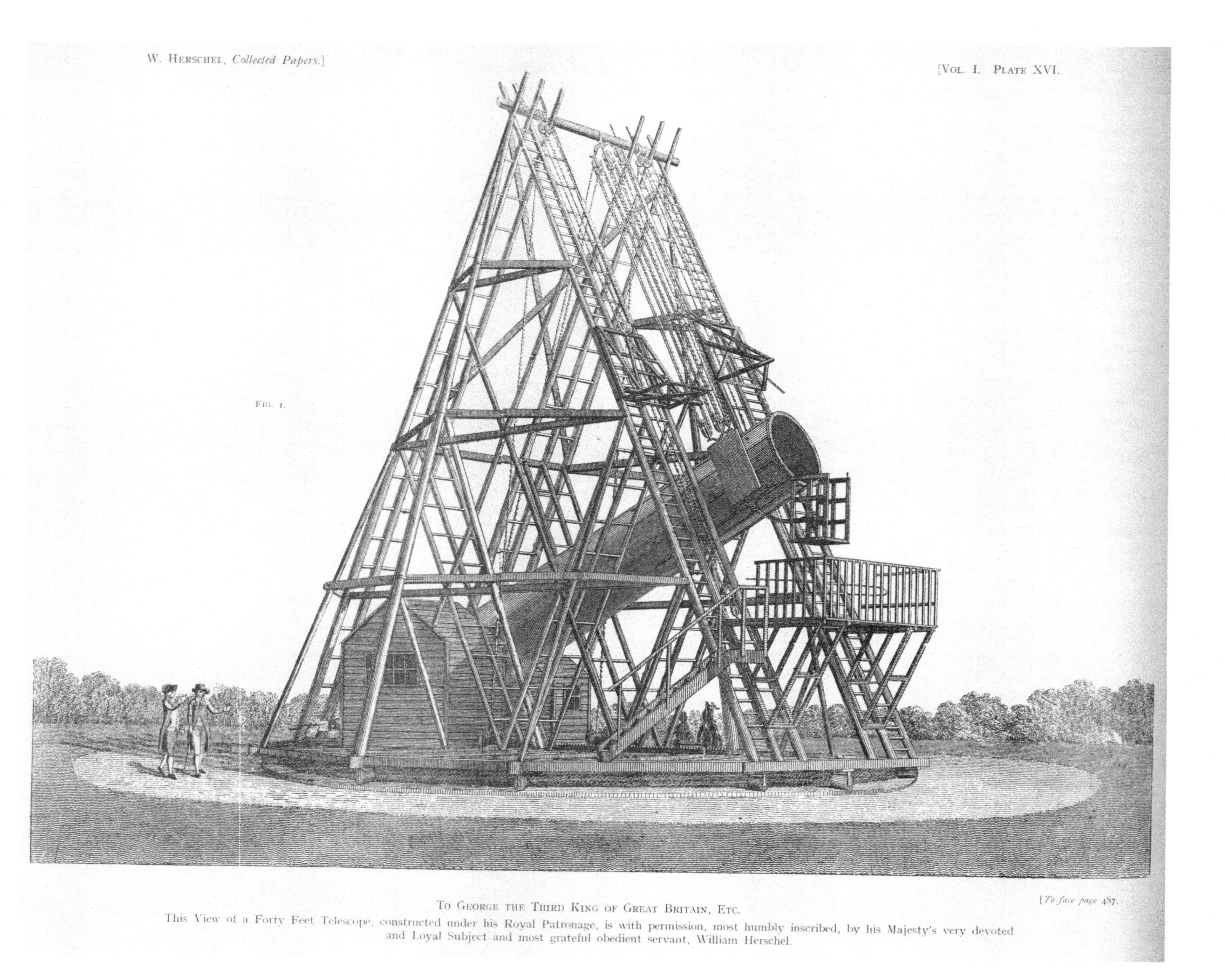

[*To face page* 487.

To George the Third King of Great Britain, Etc.

This View of a Forty Feet Telescope, constructed under his Royal Patronage, is with permission, most humbly inscribed, by his Majesty's very devoted and Loyal Subject and most grateful obedient servant, William Herschel.

with it. But not being satisfied, I continued to work upon it till Aug. 27, 1789, when it was tried upon the fixed stars, and I found it to give a pretty sharp image. Large stars were a little affected with scattered light, owing to many remaining scratches in the mirror.

Aug. the 28th, 1789. Having brought the telescope to the parallel of Saturn, I discovered a sixth satellite of that planet; and also saw the spots upon Saturn, better than I had ever seen them before, so that I may date the finishing of the 40-feet telescope from that time.

Description of the Instrument. [Plates XVI. and XVII.]

Fig. 1 represents a view of the telescope in a meridional situation, as it appears when seen from a convenient distance by a person placed towards the south-west of it.

The foundation in the ground consists of two concentric circular brick walls, the outermost of which is 42 feet in diameter, and the inside one 21 feet; these measures are reckoned from the centre of one wall to the centre of the other. They are 2 feet 6 inches deep under ground; two feet 3 inches broad at the bottom, and 1 foot two inches at the top; and are capped with paving stones about 3 inches thick, and 12¾ broad. Fig. 2 represents a section of one of them.

These walls were brought to an horizontal plane by means of a beam turning upon a pivot fixed in the centre of the circle, which had a roller under it at the end. Upon this beam and over the roller was fixed a spirit level, to point out any defect in the walls; and by correcting every inequality that could be perceived, they were by degrees brought to be so uniformly horizontal that the beam would roll about every where upon them without occasioning any alteration in the bubble of the spirit level.

The timber of the groundwork (see fig. 3), in the construction of which it was necessary to join strength to lightness, is put together in the following manner: three principal beams, AA, BB, CC, are extended from south to north, when the telescope is in a meridional situation. They are 43 feet 2 inches long, 6 inches broad, and 6 inches thick. The distances of the centre of the two outside ones is 17 feet. Within one foot of the ends of them are bolted down the cross beams DD, EE; which serve as a foundation to the two sets of ladders. These cross beams are 19 feet 2 inches long, 12 inches broad, and 6 inches thick; and by way of additional strength two more, FF, GG, of the same breadth and thickness, are bolted against the sides of the former, resting also upon the longitudinal beams, but these are rounded off at the ends, as marked in the figure.

The firmness of the foundation in the direction from south to north being thus secured, it became equally necessary to provide for strength in the support from east to west. For this purpose there are three latitudinal cross beams, HH, II, KK, bolted down upon the former longitudinal ones. That which crosses the centre is 45 feet 2 inches long, and 12 inches broad ; its thickness, like that of all the ground timber, being 6 inches. The other two are about 39 feet 9 inches long, and 6 inches broad. They project beyond the circular foundation wall about 8 inches, while the middle one projects 12. The use of these cross beams is to receive six supporters upon their respective ends, at the places which are marked with an ellipsis ; the supporting beams which stand upon them being round and inclined towards the ladders, which they are to keep steady in the east and west direction.

Under each end of the principal beams at AA, BB, CC, HH, II, KK, is

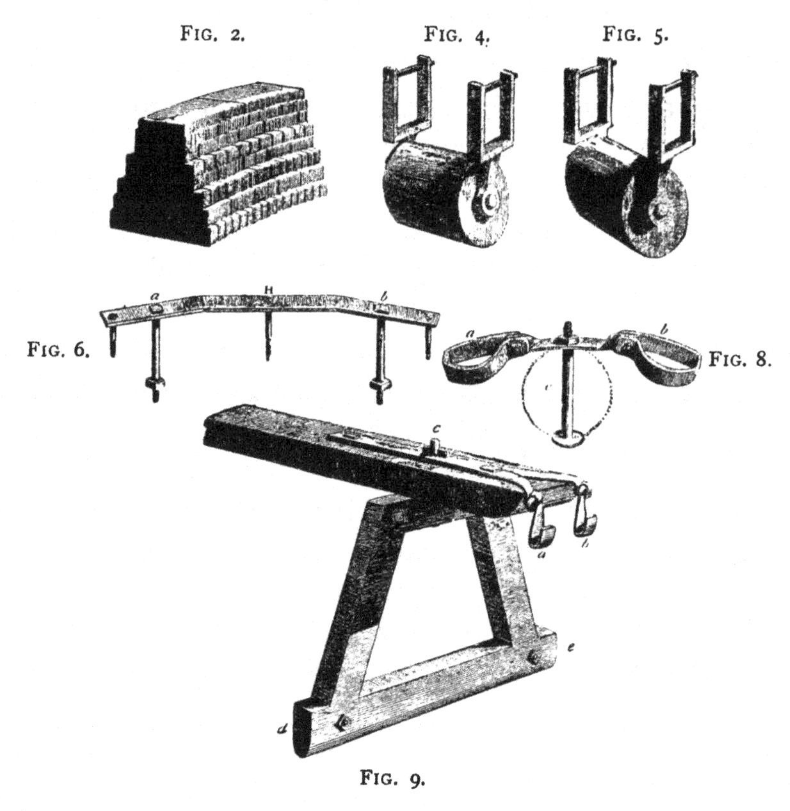

Fig. 9.

placed a roller which rests upon the outer foundation wall. The three latter of these beams being placed higher than the former, have a piece of a proper thickness under the ends to bring the bottom of them to the level with the former. The rollers are set in iron frames, and bolted to the beams, so as to be directed to the centre of their motion. They are 8 inches long, and 6 in diameter. The construction of those which are under beams that come from the centre, is expressed in fig. 4 ; but the irons which hold the rollers under beams in other directions are more or less eccentric, as may be seen in fig. 5.

No other fastening in the whole machinery of the wood-work has been admitted but screw bolts ; as tenons of any kind, in an apparatus continually to be exposed to the open air, will bring on a premature decay, by lodging wet. In order to obtain steadiness, however, the cross timbers of the frame, in all places where they are bolted together, are let in, and receive each other about $\frac{3}{4}$ of an inch, which

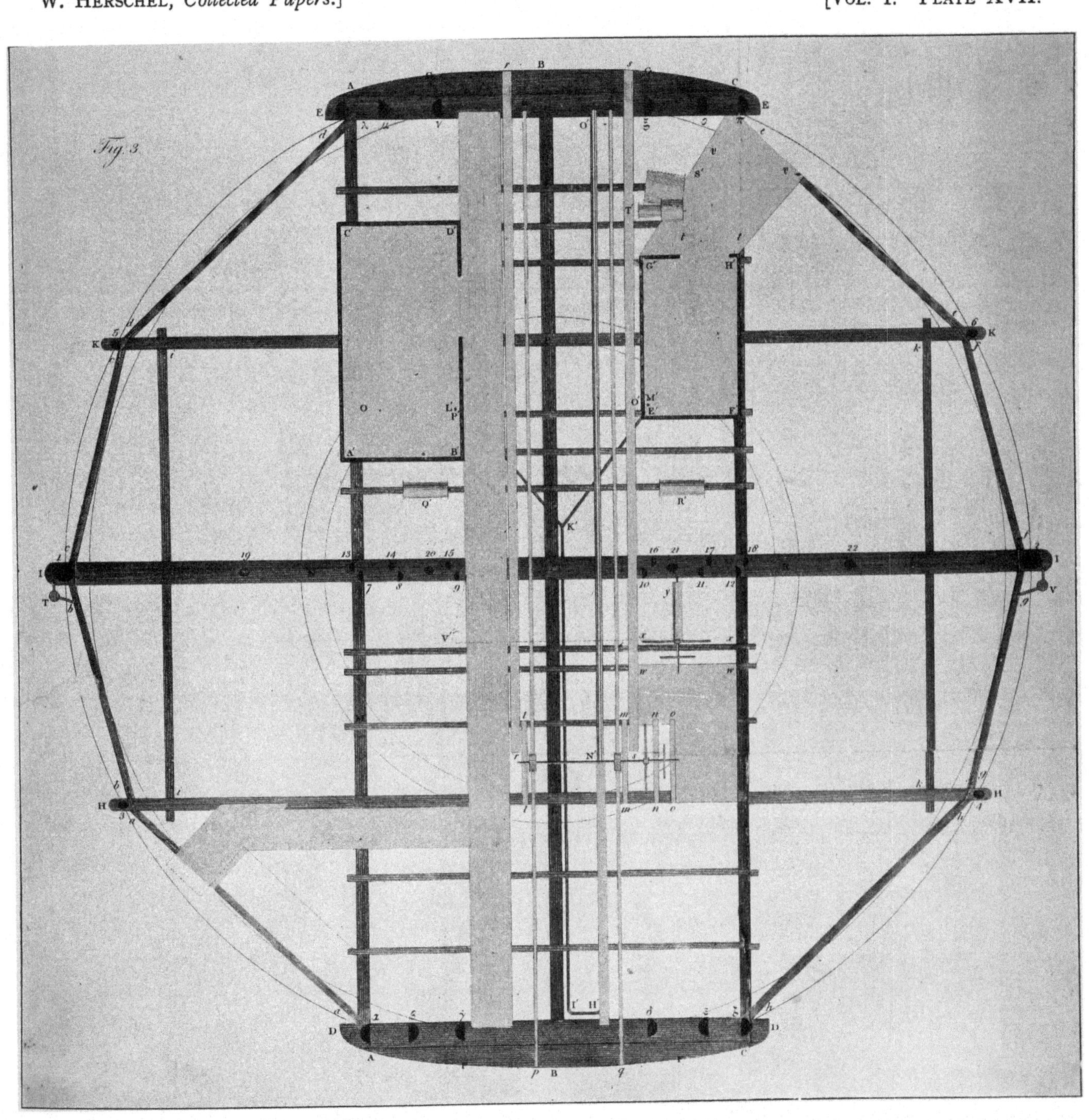

[*To face page* 488.

makes an entering of 1½ inch into each other, and produces the required firmness without any material weakening of the timber.

The twelve rollers whose place has been pointed out, would not have been sufficient to support the length of the beams to which they are fastened, the shortest of which, as we have seen, being near 40 feet long. Eight additional rollers, therefore, sustain the ground timber half way towards the centre at L M N O P Q R S. They are, like the former, directed to the pivot upon which this frame moves, and rest upon the inner foundation wall.

In the centre is a large post of oak, framed together with braces under ground, and walled fast with brick-work so as to make it steady. The two central beams BB, II, cross each other over this post; and a strong iron pin, or pivot, goes through them both into a socket within the centre of the post, so as to permit the whole of the foundation timber to turn freely upon this centre, when a proper force is applied for that purpose.

Although by means of the 20 rollers, and this support in the centre, the bottom frame of the stand to the telescope be firmly supported, it may notwithstanding be easily seen that there was occasion for some additional braces, in order to keep each beam in its proper situation. For this reason 8 pieces, *aa*, *bb*, *cc*, *dd*, *ee*, *ff*, *gg*, *hh*, of a proper length, 4 inches broad and 6 deep, are applied near the end of the beams against the sides of them; these are held together by irons that are bent, as in fig. 6. One of them, for instance, is bolted down, at *a* and *b*, fig. 3, upon the braces that hold the beam H in its place; upon which it is also screwed down, and makes a firm joint of the three pieces. A small entrance of *a* and *b* into H takes off the weight from the iron, and keeps the braces in their places, when bolted together.

Two other braces, *ii*, *kk*, are added. Their use is evident in the horizontal motion of the telescope; for that being effected, as will be seen hereafter by means of the strong centre beam II, the connection of the whole frame with this beam is completed by the pieces *ii*, AA, BB, CC, *kk*.

Before I proceed to explain any other part of the work in this figure, it will be necessary to describe the construction of the ladders and their braces.

Fig. 7 represents the front set of ladders. $\alpha\ \beta\ \gamma\ \delta\ \epsilon\ \zeta$ are six tapering halves of three large poles, or rather small masts, cut through the middle. Before the masts were cut they measured between 11 and 12 inches at the bottom; but when they had been sawed through, the pieces were flattened on the front and back, so as to be reduced to 8 inches at the bottom, and 5½ at the top, while the other dimension, or thickness, was left to its full extent. By trimming up and making them pretty equal, however, they became also reduced to about 5½ and 5 inches in that direction.

The length of the ladders is 49 feet 2 inches, and their construction is as follows. The top of each step is 9 inches from that of the one below it; and, beginning

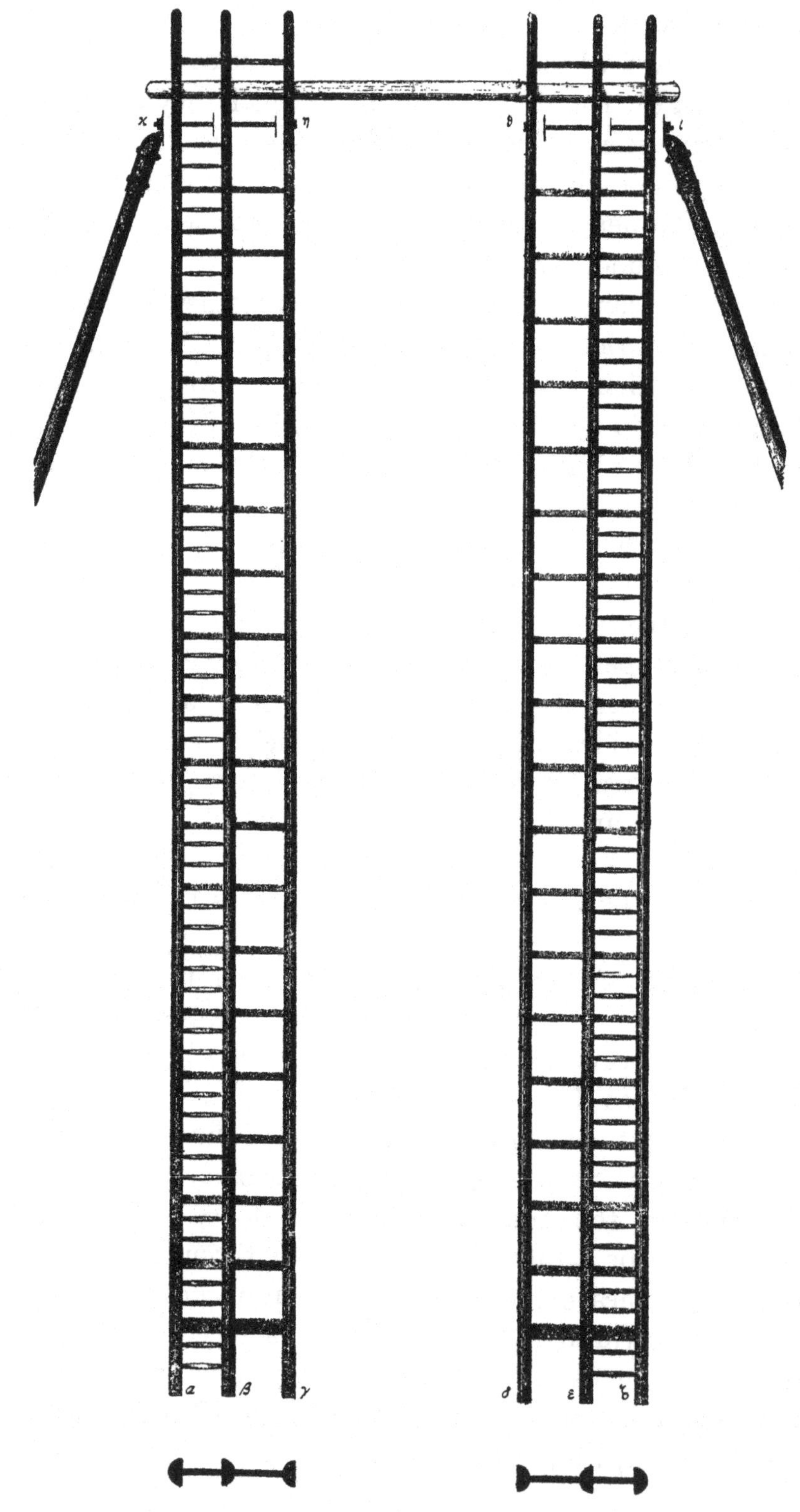

FIG. 7.

12 inches from the bottom, there are two rounds and one flat placed alternately, as far as 40 rounds, and 19 flats. In the place of the 20th flat is the centre of the meeting of the front and back sets of the ladders; above this is another flat, with a termination of 16 inches at the top.

The timber of the sides being tapering, a similar diminution of the flats and rounds has been attended to, especially as their size, in proportion to the sides, is far above what is generally used in building ladders. The flats and rounds are all made of solid English split oak.

The lowest rounds are 2 inches thick in the middle, and $1\frac{1}{2}$ where they enter the sides. At the 21st round the thickness is $1\frac{3}{4}$ in the middle, and $1\frac{3}{8}$ at the shoulder. About the 31st round the thickness is $1\frac{1}{2}$ in the middle, and $1\frac{1}{8}$ at the shoulder, and this size is nearly preserved up to the end. Those parts of the rounds which enter the sides of the ladders have all been turned in a lathe, and are about $\frac{1}{8}$ of an inch tapering, in order to fill the holes properly, which were also made a little tapering so as perfectly to answer the size of the rounds.

The lowest flat, for a particular purpose in the erection of the ladders, which will be explained hereafter, is $4\frac{1}{2}$ inches by 2. The next, as far as the 10th, are $3\frac{1}{8}$ by $1\frac{5}{8}$; from the 11th to the 16th they are $2\frac{3}{8}$ by $1\frac{1}{4}$; and from the 17th to the last $2\frac{1}{2}$ by $1\frac{1}{8}$ inches.

The two outside divisions of the ladders serve for mounting into the gallery, and therefore contain rounds as well as flats. The distance of the sides, the flat parts of which, as in common ladders, are put facing each other, is 18 inches, and remains the same up to the end. But the two inside divisions, which have no rounds, are placed with the flat face outwards, and the distance between these faces being 2 feet 8 inches up to the top, the parallelism of these divisions is preserved outside, while that of the mounting ladders is continued within. The reason of this arrangement is, that the brackets which support the moveable gallery rest upon the inside frames of the ladders. These go upon 24 rollers, as will be seen hereafter, and 12 of them confining the gallery sideways, while the other 12 support it, the parallelism was of course required where it is placed. The mounting ladders are made parallel within, that a moveable chair, intended to be made if required, might be drawn up with a person seated in it, to prevent the fatigue of mounting, or take up in safety any one who chanced to be afraid of ascending an open ladder.

The back set is constructed like the front; and, the ladders being of the same length, the only difference is that no rounds have been put into them. The flats have been preserved on a double account; first, that the connection of all the side timber might be firm and strong; and secondly, that every part of the frame might be accessible. For by means of these flats we have steps of 27 inches, which may be ascended with tolerable ease, when occasion requires.

The method of joining the front and back at the top, is by passing one set of ladders through the other so as to embrace it; the backs, therefore, which go outside, are placed a little farther asunder than the fronts; and the same pins pass through them both, at $\eta\kappa$ and $\theta\iota$, where a section of the back ladders is shewn, with the pins going through them. The last flat was put into the ladders after they had been erected and secured together.

The method of setting up and bracing the ladders was as follows:

When the eight principal beams of the groundwork A B C D E H I K, fig. 3, had been put together, the eastern front ladder $\delta\epsilon\zeta$, fig. 7, with its back $\xi o\pi$ (not expressed in the figure) bolted to it, was laid down edgeways, with the ends δ and ξ opposite to the same letters in fig. 3, and the centre θ towards the west. In this situation the ends were properly secured to the foundation beams, that they might not slip. The centre θ was then raised up to about 10 or 12 feet from the ground, and a tackle was fastened on one side to the crossing of the ladders at ι, and on the other, about 2 feet from the ground, to a tree at a convenient distance in the east. By assisting a little at first, in lifting the centre, our tackle soon got hold of the ladders, and drew them up. A rope had been provided to prevent their going farther than the perpendicular; and being secured in that position, the tackle was now fastened to the other ladder at η; but instead of making use again of our tree, the corresponding tackle was secured on the top of the first ladder at θ; by which means we easily drew up the second. Both sets of ladders stood now upon the ground, within the frame, and with the front legs, $\alpha\beta\gamma\delta\epsilon\zeta$, nearly opposite to the same letters on the front beam; while the legs of the back stood opposite the letters $\lambda\mu\nu\xi o\pi$ on the back beam.

We now proceeded to put on the middle top cross-beam, which is placed above the two sets of ladders in the angle made by their crossing each other. It is expressed by points in fig. 7, and may be seen in its place, fig. 1. The method of keeping it there, and securing the proper distance of the ladders by this beam, which is of a cylindrical form, is as follows: twelve iron loops, shaped to the ends of the ladders, with arms to them like lamp-irons, and a hole at the end of each arm, are slipped down upon the ends of the ladders, till two and two of them, as *a b*, fig. 8, meet in the middle of the cross-beam *c*, which is about 8 inches in diameter. Here a screw bolt, coming up through the beam, passes into the holes of the two irons, where all is screwed firmly together. By this means no holes are made to weaken the tapering ends of the ladders, and the centre beam takes firmly hold of every one of them; so that were even the pins η and θ pulled out, the ladders would still remain firmly kept together.

Before, however, the ladders were screwed to the centre cross-beam they were lifted up into their places upon the front and back foundation beams DD, EE. This was done by a strong lever-beam, about 25 feet long (see fig. 9), with two moveable iron claws, *a b*, at the end; which took hold in two places, equally distant from the middle division of the lowest flat of the ladders: this flat having been made, as has been noticed before, sufficiently strong for sustaining the whole weight of a set of ladders. Thus they were lifted one by one into their proper places, and supported till they could be shaped with their lower ends to fit upon their respective bearings, and were in the same manner brought to the required parallel situation: this kind of lever affording the means of giving some small motion to the weight it sustains, not only upon the pivot *c*, but also on the support *de*, which is rounded off at the bottom.

When the ladders had been properly adjusted to their places, we proceeded to support them immediately by two capital side braces. These consist of two whole masts, of nearly the same dimensions with those which were sawed through for making the ladders: the upper end of each was mounted with an iron loop *a*, two claws *b c*, and ring *d*, which were put on with bolts, *e f*, as expressed in fig. 10. The poles being drawn up, the loops were put upon the centre pins $\iota\,\kappa$, fig. 7, and keyed on; while the lower ends of the poles were lifted into their places, on the cross foundation beam II, fig. 3, and fitted upon the elliptical marks at the ends of them at 1, 2.

Before the ladders and side braces were fixed down, a line with an hundred weight at the end, immersed in a tub of water, was hung upon the centre at θ; and being viewed from a considerable distance, was made to range with the flat side $\delta\theta$ of the ladder. The bottom of every end of the bracing poles and ladders being finally adjusted by this plumb-line, they were all screwed down by strap-bolts, as delineated in fig. 11. The top centre cross-beam was now also screwed to the loops on the ends of the ladders, which, as we have mentioned before, see fig. 8, had been already prepared.

The most essential part of the stand being now erected, we proceeded to brace and support it finally. Four small ladders, but without rounds, 22 feet 9 inches long, having been made, were erected to support the large ones half way. They consist of three half-poles each, placed at the same distance from one another as the large ones, and have their faces turned like them. These meet the former at the 10th flat with their upper ends; while the lower parts of them rest upon the middle foundation beam in fig. 3, at 7, 8, 9, 10, 11, 12; 13, 14, 15, 16, 17, 18. They are screwed both at the top and bottom with flat corner-irons as in fig. 12, and their situation may be seen in fig. 1.

In the next place four less poles were now added to support the ladders sideways. They stand upon the beams HH, KK, at 3, 4, 5, 6; and are fastened against the ladders at the 10th flat, or half way up. The upper end is secured with an iron, as in fig. 10, through the loop of which passes a strap-bolt that holds it at the same time with the triangular brace, which will be described, to the ladders. The lower end is strap-bolted down upon the beams, as in fig. 11.

The two long side-bracing poles are supported each by two less poles, which meet them at one half and one quarter of their length from the bottom; or at a height opposite to the 5th and 10th flats of the ladders. Those poles which meet the great ones in the middle are placed with their lower ends upon the middle beam at 19 and 20; and the shortest rest at 21 and 22. They add very materially to the steadiness of the frame in the east and west, or lateral direction; at the bottom they are also fastened by strap-bolts as in fig. 11, and at the top by loops, as in fig. 10. They may also be seen in fig. 1.

The next braces we are to describe are those of the sides of the ladders, and these it will be seen by fig. 13 are of so simple a nature, that a bare inspection of their representation will be sufficient. The size of the horizontal pieces is 6 inches by 3½; but those which are parallel to the ladders, and are of no other use than to keep the rest in their places, or as it were brace-bracers, are only 3½ by 2½.

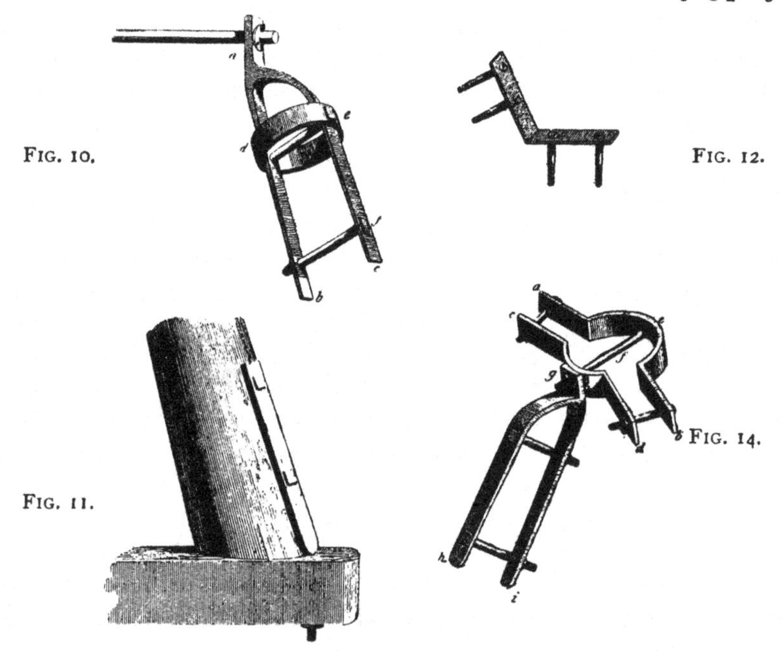

FIG. 10. FIG. 12. FIG. 11. FIG. 14.

Besides these there are three sets of braces, which serve to confine the poles to their stations. The highest set meets the side brace of fig. 13 at the 15th flat. The next meets the middle brace at the 10th flat, and both these make with the here mentioned side braces a triangle, in the vertex of which is inclosed the large pole that is braced by them. At the 5th flat a third set of braces, which incloses the two small poles as well as the large one, is carried round with four divisions. In order not to weaken the great pole by many holes, the braces secure it by a double iron strap, *abcd*, fig. 14; and the small supporting poles which rest upon 19, 20, 21, and 22, fig. 2, are at the same time joined by a single screw-bolt, *ef*, which passes through the loop *ghi* at the end of them, and through the straps which hold the braces to the great pole; see fig. 1.

The back of the ladders is bound together by a large cross, *ab*, *cd*, from the 10th flat to the middle braces, and by two horizontal pieces, *ef*, *gh*, as represented upon a small scale in fig. 15. The cross is bolted in twelve places to the ladders, and the horizontal pieces in six places each. The size of these braces is 6 inches by 4;

but the lowest horizontal beam, which is used for a point of suspension to lift the mirror in and out of the tube, is 6 inches and a half by $5\frac{1}{2}$; and the bolts that hold it to the ladders are also very substantial.

The front of the ladders, it is very evident, would admit of no brace, and is left entirely open for the tube of the telescope to range in. It receives, however, some confinement from the moveable gallery, which is always hung across the front, in the place where observations are to be made.

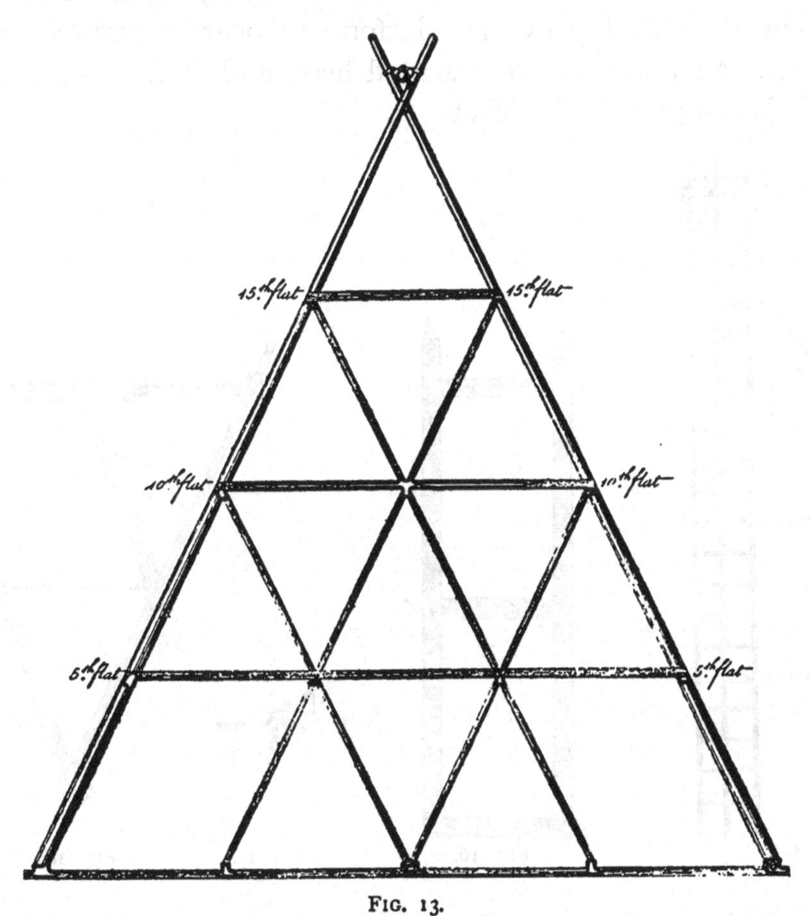

FIG. 13.

This gallery is next to be described. It consists of three separate parts: two double side brackets with a small platform upon them, and a middle passage. The whole of it when joined together is properly railed in at the front by wooden palisades; and on the inside by light iron-capped bars. Each of the brackets by which the gallery is supported consists of three frames; a parallelogram for the bottom, with two triangular sides erected on the former, and held together by a narrow platform on the top. Fig. 16 represents the bottom frame. Its length, *ag*, is 8 feet 10 inches, and breadth, *gh*, 2 feet 8 inches. It is made of yellow fir deal, 4 inches broad, and 2 inches thick. Six sets of brass rollers, in iron frames, constructed as represented in fig. 17, by means of the two small screw-bolts, *i k*, are screwed under the frame, fig. 16, at *c d e f g h*; so that when this frame comes to be placed upon the front of the ladders at $\beta\gamma$, or $\delta\epsilon$, fig. 3, the 6 rollers in the direction *m*, fig. 17, will sustain the frame, while the other six, in the direction *l*, by embracing the flat sides of the ladders, which as has been described are turned outwards for this purpose, will prevent the frame from slipping off sideways, when it rolls up and down the ladders.

Two such triangles as delineated in fig. 18, the wood of which is $3\frac{1}{2}$ by $2\frac{1}{2}$ inches, are fastened to the frame, fig. 16; one, with the side *np* upon *cg*, the other upon *dh*. These being joined at the top by the platform of boards, screwed down to the supporters *q r s*, of that which is represented here, and of the corresponding one, which rests on *dh*, complete the bracket.

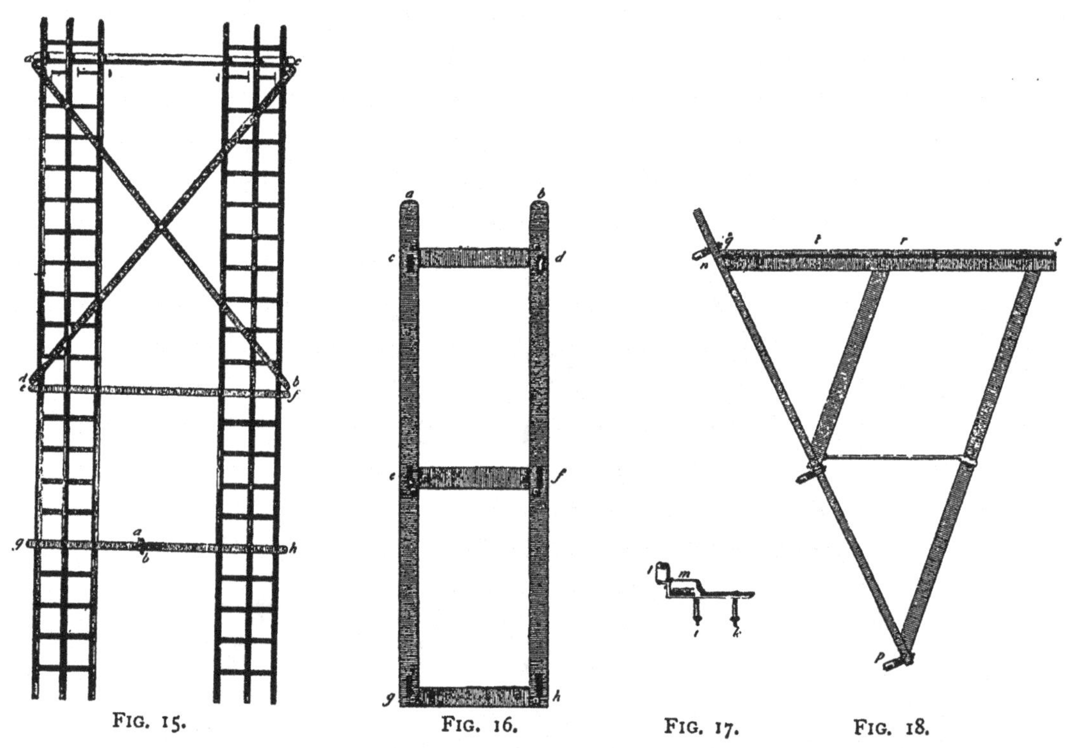

Fig. 15. Fig. 16. Fig. 17. Fig. 18.

Upon the platform are fixed palisades commencing at *t*, 19 inches from *q*; which turn the corner at the front *s*, and are continued so as to meet the middle platform of the gallery. The palisades over *trs* are strengthened and rendered steady, by a seat which is fastened against them, and supported from the floor by slight iron bars.

The other double bracket, with its platform, palisades, and seat, which runs upon the right side of the front ladders, is in every respect the same as that on the left, except that the entrance is here upon the right side, instead of being at the left in the former.

The whole gallery together, the floor of which is represented in fig. 19, takes up a space of 13 feet 6 inches broad, by 6 feet $1\frac{1}{2}$ inches in depth; the middle platform, however, is cut away so as to leave sufficient room for the tube to come forward in high altitudes. At *ac* and *bd* it is 4 feet 3 inches, but at *ef* and *gh*, a space of 4 feet 10 inches long, it is only 2 feet deep. The front, *cfhd*, contains palisades, which meet those of the left bracket *trsc* at *c*, and the similar ones of the right at *dik*. These palisades are 3 feet 2 inches high. The light iron rails on the inside pass along the edge, *laegbm*, and are only 2 feet 3½ inches in height.

The first requisite in this gallery being that it should be drawn up to any required altitude, it became necessary to connect the two double brackets and the middle platform in such a manner as to bear some little derangement in their level, arising from the inequality of the motion of the side brackets. With a view to this end, the method of uniting the parts is as follows. The dotted lines 1, 2, &c. shew the place of the joists which support the floor of the platform. At the

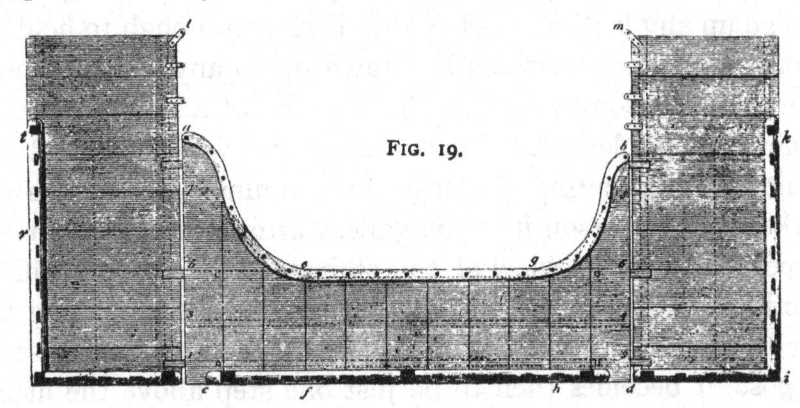

Fig. 19.

Fig. 20.

ends, 1, 2, 5, 6, 7, 8, of these joists are six iron hooks, shaped as in fig. 20; they are bolted and screwed with the end *no* under the bottom of the joists, and rise to the level of them with the arms *p*, leaving the hooks *q* projecting. These enter into six proper openings made in the side brackets; three in each: they leave a space of about ¾ of an inch between the two brackets and the middle platform, which permits a small irregularity in the level of the three parts to take place without injury to either of them. The hooks sink down into the floor of the sides so as to be level with the surface; and go over the inside of the supporting triangles, fig. 18; which, for the sake of additional strength, and to prevent their being galled by friction, are lined with an iron plate at the inside, in all their length.

The light iron rails joining the bars of the inside, which are along the margin *laegba*, fig. 19, are left moveable at the bottom, in the places *la* and *mb*; where they run down into loops; by which means they admit of being a little displaced.

The contrivance to make the junction of the front and side palisades moveable is by means of a front bar. This being slipped upon pins at the end of the rails belonging to the sides, a hole at each end of the bar, lined with an iron plate about 2 inches long, through which the pins pass, permits the bar to be drawn either way. There are moreover at the ends of the rails, which are fixed to the platform, two iron hooks; which, though they bind the rails to the front bar, still permit it to go up or down a little way, as occasion may require. By this means a deviation from the level, amounting to six or eight inches, will occasion no injury to the wood-work. The greatest security against such a derangement of the platform, however, will be explained hereafter, when we come to the mechanism by which it is moved.

There is a small staircase by which we may ascend into the gallery, without being obliged to go up any ladder; and as that is strong enough to hold a company of several persons, and can afterwards be drawn up to any altitude, observations may be made with great conveniency: the activity of an astronomer, however, will seldom require this indulgence. The readiness with which I ascend the ladders, has even prevented my executing the projected running chair, which may easily be added, to take a single person into the gallery after it has been already drawn up to its destined situation. A view of the staircase in fig. 1 will suffice to point out its construction. I ought only to observe, that in the engraving the gallery is placed higher than where it will join the staircase properly, but that when it is lowered on purpose, it becomes then to be just one step above the little landing-place of the staircase, and the palisades of the former unite with the railing of the latter.

The next piece to be described, is the tube of the telescope. This, though very simple in its form, which is cylindrical, was attended with great difficulties in its construction. No one will wonder at this who considers the size of the tube, and the materials of which it is made.

Its length is 39 feet 4 inches; it measures 4 feet 10 inches in diameter, and every part of it is of iron. Upon a moderate computation, the weight of a wooden tube must have exceeded an iron one at least 3000 pounds; and its durability would have been far inferior to this of iron.

The body of the tube is made of rolled, or sheet iron, which has been joined together without rivets; by a kind of seaming, well known to those who make iron funnels for stoves. It is represented by fig. 21, where the two sheets of iron are left a little open at *ab* to shew the construction, but which being properly compressed will become very nearly flat: the whole outside was thus put together in all its length and breadth, so as to make one sheet of near 40 feet long, and 15 feet 4 inches broad. The tools, forms, and machines, we were obliged to make for the construction of the tube were very numerous. For instance, in the formation of this large sheet, a kind of table was built for its support, which grew in size as the sheet advanced, till when finished, it was as large as the whole of it. In the formation of the sheet, cramping irons, seaming bars, setting tools, and claw-screws, such as are represented in the figures 22, 23, 24, 25, and 26, were made in great number, to confine and stretch the parts as they were seamed together. The small single sheets of which this large one is composed, are 3 feet 10 inches long, and about 23½ inches broad. Their thickness is less than the 36th part of an inch; or, what will be a more precise measure, a square foot of it weighs about fourteen ounces. They are joined so, that the middle of a whole one always butts against the seam of the preceding two, in the manner of brick-work, where joints are crossed by bricks above and below.

When the whole sheet was formed, which was done in a convenient barn not far from my house, the sides were cut perfectly parallel, and afterwards bent over at the ends in contrary directions, as in fig. 21, to be ready to receive each other. A number of broad hooks, such as were proper for grasping the sides of the sheet, with loops at the other end for cords to go through, see fig. 22, were now prepared with their necessary tackle.

Twelve pulleys were fastened about 11 feet high, on moveable beams, that might be drawn together; six on each side. The sheet was now taken up, by occasioning all the corded hooks to be drawn at the same time, and while it was kept suspended our large table was taken to pieces. Another kind of support was now put under the middle of the sheet to receive it. The form of this was that of an hollow segment, or quarter of a cylinder, cut lengthways, to the extent of a few feet more than the length of the intended tube; and the concavity of which was formed by the same radius as that of the tube.

The sheet being let down, it rested upon the hollow gutter; for so we may call the machine that was placed under it. Six moveable segments of a whole cylinder, or circular arches, about 3 feet wide each, which had been prepared, were now brought upon the sheet and placed at proper distances from each other. By these the sheet was pressed down upon the foundation, so that no injury could be done by walking upon it. The beams which held the pulleys were now brought close together; which being done, we hung the pulleys of one upon the hooks of the other beam, so as by that means to cross the cords which held the sheet. In this operation we slackened only one of the cords at a time, the rest being sufficient to keep the whole up.

The beams were now again separated, and the cramping hooks by the crossing of the cords drew the two sides of the sheet together.

Here I must take notice, that the circular inside supports, which resembled the machines upon which arches of brick-work are built, were cut in two in the middle, as in fig. 27; some part of the circumference being taken out, that when they were laid down upon each other they might not fill the tube. Four long wedges, *a b c d*, in opposite directions, were confined two and two in the notches *ef*, *gh*; and similar ones at the back. By driving them in very equally, the upper half of the arches might be forced up so as to swell to the full extent of the tube.

When all this was properly arranged, and the arches lowered, the two sides of the sheet were gradually brought to take hold of each other. As we proceeded, the wedges within the arches were forced in successively, till at last, with much care and considerable difficulty, the two sides completely embraced one another, and were kept stretched by the swelled inside arches.

Another circular arch, closed in with boards all around, well rounded off, and only about 2 feet 3 inches long, had a vacancy at the top into which we could

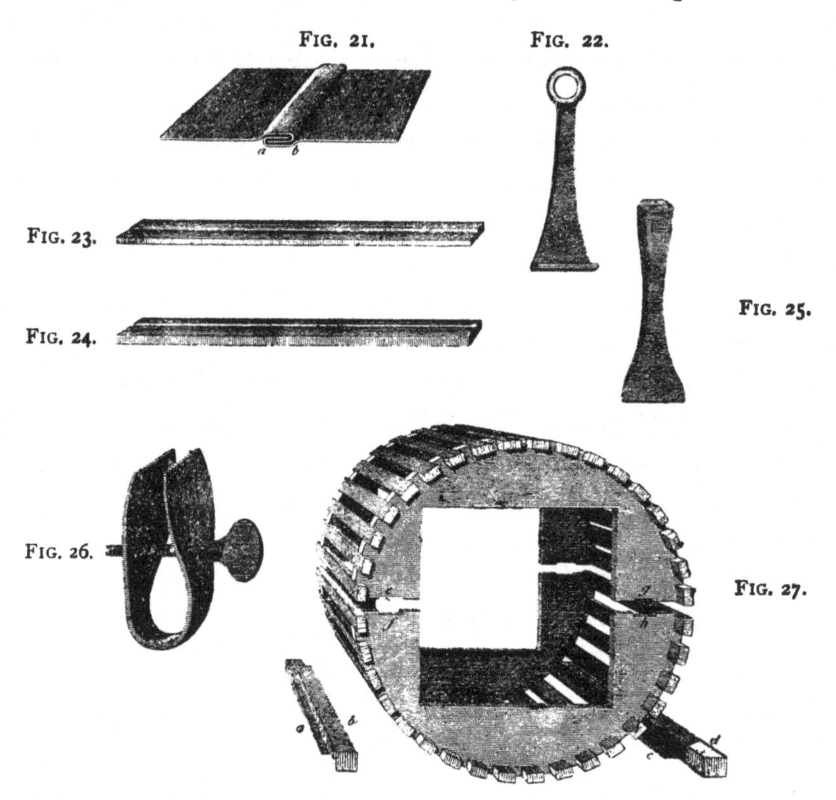

introduce the iron seaming bars, fig. 23, for indenting, and 24, for closing up the long seam of the two sides. This arch also had its stretchers for swelling it up, and served at the same time, as soon as the seam was properly closed, to beat with mallets the whole sheet all around upon its well-finished outside, in order to take away any accidental bulge which it might have received in the long preparations it had undergone, till it came to the present state.

The same arch, as soon as any portion of the tube had been done, was removed to another place, and the whole was by this means completely seamed up.

The theory upon which the strength of so thin a cylinder of iron is founded, is, that the sides of it must unavoidably support it, provided you can secure the cylindrical form of the tube.

It appeared to me the most practical way to obtain this end by the following contrivance. By a few experiments I found that a slip of sheet iron, a little thicker than that of the tube, and doubled to an angle of about 40 degrees, as in fig. 28, might afterwards be made circular, as in fig. 29. The deepest we could conveniently bend, and such as I supposed would answer the end, was when the sides *a b* were about 2½ inches broad. They were shaped red hot upon a concave tool, which had the required curvature and angle of the slips. The pieces were long enough to form a complete quadrant of the circle, with the ends sufficiently projecting to be seamed together.

Before they were joined the sides received another bending, as in fig. 30,

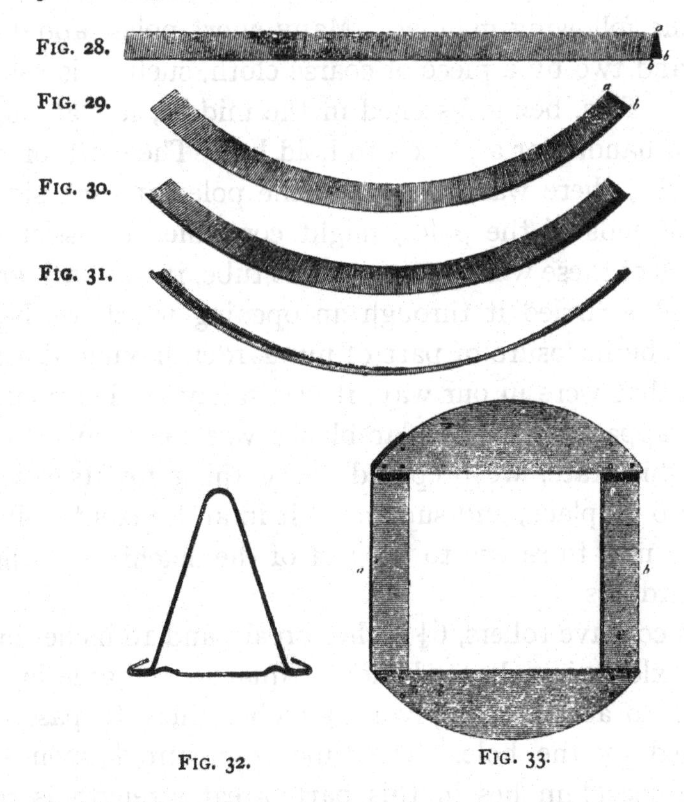

FIG. 28. FIG. 29. FIG. 30. FIG. 31.

FIG. 32. FIG. 33.

which was given them by tools of a proper convexity. A back was next prepared, consisting of a slip of iron turned up at both sides, and also bent to the circle, as in fig. 31. Last of all, the four quadrants having been put together, and a back put round them, the whole was firmly seamed together, so as to resemble a hollow triangular bar made into a hoop or ring, of a proper diameter to go closely into the tube, so as to keep it extended, and braced to the cylindrical form. A section of the ring with the bottom seam not quite pressed down, in order to shew it better, is represented in fig. 32.

One of these rings was put into the middle of every one of the small sheets, which brought them to about 23 inches from each other. They were carried in edgeways, and afterwards turned about and forced into their respective places. In order to get them in, as they were all obliged to go in from one side, there was substituted, in the room of the circular arches, a kind of temporary props, like fig. 33, that could be easily removed, one at a time, and were narrow enough at *ab* to pass through the hoops while they advanced; and as soon as a ring was in its proper place, no further support became necessary.

In this manner we secured the cylindrical form of the tube; and as soon as this was accomplished, we had every thing removed from within and without, and began to give the tube three or four good coats of paint, inside as well as outside; in order to secure it against the damp air, to which it was soon to be exposed.

As the tube was now much lighter than it would be hereafter, we transported it into my garden in the following manner. Many short poles, about 5 feet long each, were joined two and two by a piece of coarse cloth, such as is used for sacks, about 7 feet long each. This, being fastened in the middle, left at each end part of the pole to serve as a handle for a person to hold by. The cloth of one of these being put under the tube, there was left one of the poles at each side, and four men taking hold of the ends of the poles, might conveniently assist in carrying the tube. When six sets of these were put under the tube, it was with great facility lifted up by 24 men, who carried it through an opening which we had made at one end of the barn. The inclosure of part of my garden having also been taken down, with some trees that were in our way, it was safely landed upon my grass-plot; where a proper apparatus of circular blocks was put under to receive it. While it remained in this state, we prepared every thing for its reception, and afterwards moved it into its place, and supported it in an horizontal situation.

It will be necessary now to return to the rest of the machinery, which by this time was in great forwardness.

Two solid cast-iron concave rollers, 6¼ inches broad, and 10 inches in diameter, are mounted upon an axle or iron bar, 2½ inches square; the axle in the middle being swelled out so as to admit of a pivot 2¼ inches thick to pass through it, without being weakened by the hole. The tube is mounted upon this at the lower end, and as the speculum lies in this part, great strength is requisite to support it firmly, as also an extensive connection of this strong part with the length of the tube. The speculum likewise is to be put in here, and when the telescope is in use the cover of the speculum is to be taken off, and afterwards to be put on again; for which reason a convenient door or opening must be had. The line of collimation of the mirror also requires an adjustment at this end of the tube; and a small side motion is required upon the pivot of the axle, which must not only be perfectly smooth, but equally firm and steady. All these exigencies have been provided for in the following manner.

Fig. 34 represents the back of the tube, closed up by six iron bars, *a b c d e f*, which cross each other. The middle bar is 4 inches broad, and 1⅛ thick at *e*, but is swelled so as to measure 5 broad, and 1½ thick at *l*, where it is turned at rect-angles, and passes under the bottom of the tube. In this bar is a square hole, through which a pivot, or pin, passes from the inside of the tube, where it is confined by a square head, into the hole of the axle, AB, under which at the bottom it is keyed fast at C; with proper washers between the joints to allow of a very smooth motion.

The bar, *cg*, is of the same strength with *el*, and passes over it at *e*. It is bent at rectangles at *c* and *g*, so as to pass along the sides of the tube. The two bars, *dm*, *fk*, fastened upon *cg*, and afterwards turned down to the back, are 2 inches broad, and $\frac{5}{8}$ thick; and are also bent at rectangles at *m* and *k*, so as to go under the tube: the remaining two bars, *bh*, *ai*, cross the other three bars, with proper offsets; and are bent at rectangles on both sides, that they may turn round the end of the tube, to go along the sides of it. At the crossing of these bars they are fastened by screws, which pass through the upper bar, and are lodged in the lower one. The same screws pass on through a moderate plate of sheet iron which closes the back, and is held by nuts upon them within the tube. The eight returning bars, at *a b c g h i k*, extend only to about 6 inches along the tube; but they are immediately received by other eight bars of the same size, which are screwed upon them. These bars are made tapering, so as only to measure $1\frac{1}{2}$ inch broad, and $\frac{3}{8}$ thick at the ends; and they are 9 feet 8 inches long. The middle bar is turned over about 16 inches, and made tapering; and the bar which meets it is laid under it, and also made tapering to answer the former. The pivot goes through both, and they form, as it were, only one bar; this is soon reduced gradually, and at the end measures $1\frac{3}{4}$ inches by $\frac{1}{2}$; its length being the same as the rest.

FIG. 34.

The segment *cng* is cut off to leave an opening, which is 2 feet broad at the sides. A cover of the same shape with the piece cut out of the tube, is laid upon the place, to overlap the opening properly. But this would not have been sufficient: for, after observations at night, this cover, though close enough to preserve the inside of the tube from damp or wet, would itself be covered with dew or condensed vapour. And by taking it off in order to secure the mirror, many drops of water would unavoidably fall upon it from this wet cover. To prevent this, an outward cover has been applied, which completely preserves the inner one from moisture.

The tube being much too weak, in this place, for the support of the mirror, a piece consisting of three sheets of iron, 2 feet 4 inches broad, $\frac{1}{8}$ thick, and dovetailed together so as to be long enough to reach from *c* around *b a m l k i h* to *g*, was added to its thickness within; upon this again, an iron bar, 6 inches broad, and $\frac{5}{8}$ thick, was bent round close to the end, from *b* to *h* only; and another bar, $2\frac{1}{8}$ inches broad, and $\frac{1}{2}$ thick, made into a complete circle, was added to support that end of the tube which had been cut, to make the entrance. All these pieces were well secured, by screw-bolts passing through the nine long outside bars *a b c g h i k l m*, next through the tube, then through the strong sheet, and at last through the broad strap and circular bars, upon which they were screwed down with nuts at the inside. The more advanced parts of the long bars were secured also by screw-bolts passing through the tube, and through circular straps of hoop iron, about $2\frac{3}{4}$ inches broad, and $\frac{1}{8}$ thick; one of these being put into every sheet of the tube as far as the bars went.

As we had now secured what I call *the point of support*, it was no less necessary to form a strong *point of suspension*. This was obtained by grasping the tube with a system of bars similar to that which has been employed at the bottom.

Ten bars, equally divided around the circumference, about 10 feet 4 inches long, are placed longitudinally so as to have one of them at the top, and an opposite one at the bottom. Every one of these has six screw-bolts, which pass through the bar and the tube, and also through complete circles of hoop iron, which is of the same breadth and thickness as has been mentioned before. The bars also, except the highest and lowest, are of the size of those which have been used about the point of support. They are also, like them, chamfered at the sides, and begin to lessen in breadth and thickness about 4 feet from the front, to the same dimensions with the former. The lowest bar is a little stouter in its dimensions, but otherwise exactly the same.

The middle bar at the top is strongest about the point of suspension, where it is 4 inches broad, and 1 inch thick. In this place it is crossed by another bar, which is a segment of a circle, and embraces the middle one, and two other bars at each side. This crossing bar is $3\frac{1}{2}$ inches broad, and 1 inch thick in the centre; chamfered or sloped at the sides, and reduced in thickness towards the ends. It passes over the middle bar with a proper offset, and its two ends terminate upon the two farthest bars; but the bars next to the middle, on each side, are made to pass over it. The middle bar receives a loop, by which the telescope is suspended, the centre of which is 3 feet 8 inches from the mouth of the tube. The loop is made of iron, 4 inches broad, and 1 inch thick; doubled together, and the ends of it opened again, so as to cross the circular bar, and to rest upon the strong middle one, to which it is fastened with four large screw-bolts. These pass through the bar into the tube, where they are well secured with substantial nuts. The long middle bar is reduced gradually after the place of the loop, the ends of which extend about 18 inches, till it comes at last to the breadth of $2\frac{1}{4}$ inches, and thickness $\frac{3}{8}$.

All the ten bars are secured with six screw-bolts each, which pass through the tube, and through iron hoops, four of which are of the same dimensions with those which are used about the point of support. The hoop which is under the suspension is 8 inches broad, and a little thicker than the rest. The front hoop is of a different construction: its thickness is about two-tenths of an inch, and being bent at rectangles, that part which is held down to the tube, and to the ten bars, keeps it steady, while that in the other direction serves as a ring, both to strengthen and confine the aperture. It projects about three inches all around, and leaves an opening of 4 feet 4 inches to the mouth of the telescope.

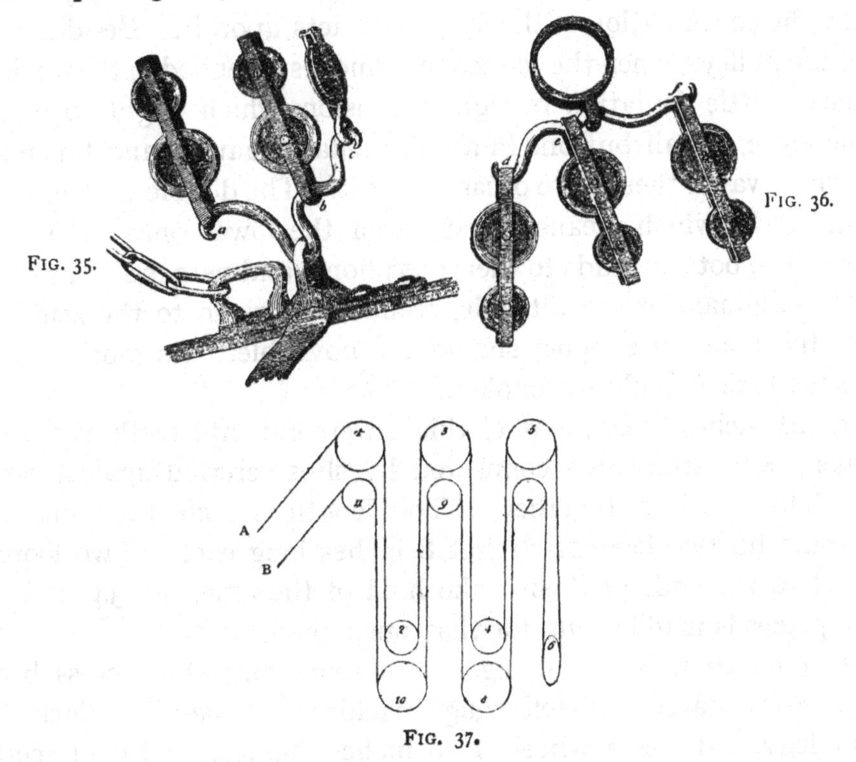

Fig. 35.

Fig. 36.

Fig. 37.

The loop of suspension stands across the tube, and receives a round bar of iron, shaped as in fig. 35, which is left at liberty to take its own position. To the places *a b* are hung two double pulleys, and at *c*, a single one; all turning upwards to meet the upper set of pulleys.

On the top of the stand, and round the centre beam, passes a ring of iron, 4 inches broad, and 1 inch thick, which contains a loop resembling that on the point of suspension at the telescope. This also receives a round iron bar, bent as in fig. 36, and supports three double pulleys at *d e f*.

Nothing can obstruct the motion of a tackle more than the friction of the ropes against each other; and as the utmost ease was required in the action of my pulleys, it was particularly necessary to guard against a defect of that kind. Another inconvenience was to be avoided, still more pernicious than the friction of the tackle. When pulleys are set, two, three, or four in a row, side by side, they will incline one way when the weight is drawn up, and another when it is let down. This may easily occasion an accident, which in the case of my large telescope must have been exceedingly troublesome, and probably in the end proved fatal; for by the side inclination of the set, a rope will sometimes slip out of its place; especially as my ropes are well soaked in melted tallow to preserve them from moisture. This in summer will occasion dust to settle upon them, and sometimes fill up a channel of the pulleys, so that the least deviation from the perpendicular may throw a rope out of its place. Should this happen in the night, when it might not be immediately perceived, the rope would soon be injured, or even cut through, by the continuation of the force that acts upon it. Besides, this irregular motion of the pulleys, when the telescope is finely suspended in the meridian, will tend to produce a little deviation in right ascension, which ought to be avoided. My pulleys, therefore, are all but one in a meridional situation, and this might also be turned the same way if there were occasion for it. The double pulleys are placed under each other; by which means the stress of the lower ones at the top, and the upper ones at the bottom, adds to their meridional and perpendicular steadiness.

In order to command every altitude, from the horizon to the zenith, it was necessary that the point of support should be moveable. Its motion is effected by a mechanism which I shall now explain.

Eight bars, $2\frac{1}{2}$ inches broad, and $1\frac{1}{8}$ thick, were cut into teeth at the distance of $1\frac{1}{2}$ inch each; and afterwards connected by slips screwed against both sides of the places where two butt together. Their length is such, that four and four being joined make up two bars of 29 feet 8 inches long each. Two loops which are screw-bolted to the ends of them, take hold of the axle, fig. 34, at D and E, which in those places is made round for that purpose.

Upon the foundation beams in fig. 3 are fixed four short cross beams, at *ll mm nn oo*; these carry the following machine. A handle which turns a pinion of eight leaves, drives a wheel of 20 inches diameter, with 51 teeth; the axle of the wheel contains a pinion of 12 leaves, driving a wheel 3 feet in diameter, of 88 teeth. On the axle of this latter are fixed, upon a long bar, two lantern pinions of twelve leaves, at a distance of 3 feet 9 inches from each other, and these are confined down to work in the two long cut bars, which pass under them at that distance in iron notches, to prevent their receding sideways. The long bars are supported by narrow slips of timber, *pp*, *qq*, which are extended from the front to the back; as otherwise the weight of these bars would bend them down so as to render them unfit for action. The slips are covered with sheet iron, that they may not be injured by friction. The front ends of the bars are furnished with claws, which keep them in their places upon the slips.

Two supporters of oak, 29 feet 8 inches long each, 6 inches broad, and 4 thick, are extended from *rs* near the pinions, to *rs* at the back. These are made convex at the top so as to fit the concavity of the iron rollers A B, fig. 34. They are also covered with pretty thick sheet iron, to prevent their being worn by the motion of the weight which is to go upon them. The distance from the centre of one to that of the other is 5 feet 4½ inches.

These things being arranged as has been described, it appears clearly that, when the handle of the first pinion is turned, the system of wheels and pinions in the machine will draw the bars, and consequently the point of support of the tube, forward into any required situation ; and return it back to its former place, by turning the same handle a different way.

At S′, fig. 3, near the platform of boards, *tt vv*, is placed a barrel, 19 inches in diameter, and 17 broad, with high sides to confine the long rope which draws up the point of suspension of the telescope by means of the pulleys that have been described. On one side of the barrel is a wheel, 2 feet 3 inches in diameter, with 91 teeth ; and a handle with a pinion of 4 leaves gives motion to it, when the telescope is to be lifted up or let down.

The method of stringing the pulleys is expressed in fig. 37. The rope A, coming from the great barrel, passes successively over the pulleys 1, 2, 3, 4, 5, 6, 7, 8, 9, 10, 11 ; and from B goes to another barrel, T′, fig. 3, which is also near the platform *tt vv*, the use of which will be explained hereafter.

By the assistance of these two motions, the telescope may be set to any altitude, up to the very zenith ; and in order to have the direction of it at command, a foot quadrant of Mr. BIRD's is fixed at the west side of the tube, near the end of it, inclosed in an iron case ; upon the top of which is also planted a finder, or night-glass, about 21 inches long, with cross wires in the focus. The divisions of the quadrant are indicated by a spirit-level, instead of a plumb-line.

The axle, which turns the first pinion of the mechanism for moving the point of support, carries a pallet. This gives motion to a small wheel with studs, contained in a machine fixed to the frame of the great wheel-work, and inclosed in a little box. The wheel with the studs carries a perpetual screw, which moves a central wheel, upon the axis of which is fixed an index-hand, that passes over a graduated plate of 140 divisions. Each of these divisions answers to four turns of the handle ; and they are large enough that a 4th part of one of them may be distinguished. In this manner the hand will point out how many turns of the handle have been made to move the telescope from its most backward point of support to the most forward.

I call this machine the *bar-index*. It is of eminent use in giving us immediately, by means of a table made for that purpose, the place of the point of support for any given altitude or zenith distance of the quadrant.

In order to come at every part of the heavens, the vertical motion of the telescope requires the addition of the horizontal one. This has been obtained by another very simple mechanism. We have already seen that the bottom frame rests upon 20 concentric rollers, and is moveable upon a pivot.

At *ww xx y*, fig. 3, is a machine in every respect like that which has been described as giving motion to the point of support, except that instead of a bar with two lantern pinions, the great wheel here carries an iron barrel, 2 feet 8 inches long, and 5 inches in diameter. Near the ends of the great cross-beam II, are planted two pulleys; one at T, the other at V. Round the outer circular wall is a gravel-walk, 12 feet broad; and on a grass-plot close to the margin of this walk are eight posts of oak, in large frames, firmly buried in the ground, at equal distances, so as only to shew their heads sufficiently to admit an iron ring and pulley to be hung upon them as occasion may require; the middle beam also carries an iron loop at each end.

A strong rope is now thrown round one of the spokes of the wheel, next to the barrel, which passes with one of its ends under the bottom of it, while the other remains at the top. As soon as the handle puts the wheel-work in motion, the barrel will draw both ends of the rope, but in contrary directions. One of the ends is then to be led to the pulley on the great beam at T, while the other is made to pass over that at V; but in a contrary direction. Upon the nearest post at rectangles to the great beam, towards the south, for instance, is hung a ring with a pulley to it; while, at the same time, a similar ring and pulley is fastened to another post, near the opposite end of the beam, but situated towards the north. The ends of the rope are now returned through these pulleys, and with iron hooks, which are fastened to them, are hung in the loops at their respective ends of the middle beam.

As soon as the ropes are sufficiently and equally stretched, the telescope will begin its horizontal motion, which may be continued as long as the same posts will be conveniently situated. In order to go on with the motion, the ropes are to be slackened, and the rings being then hung upon the two next posts, we may continue at pleasure to turn the telescope to any part of the heavens that may be required. The arrangement is represented upon a small scale in fig. 38.

It should be noticed, that the ends of the rope must be equally stretched, for which reason a mark ought to be made in the place, which is to be thrown round the spoke of the wheel. The fastenings of the pulleys also, which are joined to the rings that are thrown over the posts, ought to have an adjustment by links and hooks, to be either lengthened or shortened at pleasure.

With the assistance of the motions that have now been described, I have in the year 1789, many times taken up Saturn, 2 or 3 hours before its meridian passage, and kept it in view with the greatest facility, till 2 or 3 hours after the passage; a single person being able, very conveniently, to continue both the horizontal and vertical motions, at the command of the observer. In this however ought to be included an assisting third motion, which I am in the next place to explain.

We have seen that in fixing the ladders they were set at 8 feet 2 inches distance in front, in order to permit the telescope to have a side motion, without displacing the whole apparatus, which is designed for a meridional situation.

Every celestial object, when it passes the meridian, is then in its most favourable situation for being viewed, on account of the greater purity of the atmosphere in high altitudes. The advantage also of being able to direct the instrument, by means of the quadrant, to the spot in which we are to view the object, is considerable, in so large an instrument as the 40-feet telescope. With unknown objects, it is likewise of the greatest consequence to be enabled, by a meridional situation, to ascertain their place. But, as a single passage through the field of view, especially with my examinations of the heavens in zones, would not have been sufficient to satisfy the curiosity of an observer, when a new object presented itself, it became necessary to contrive a method to lengthen this interval. The tube, therefore, as we have seen, is made to rest with the point of support in a pivot, which permits it to be turned sideways.

Its diameter being 4 feet 10 inches, and that part which is generally opposite the ladders that confine it in front being about 35 feet from the pivot, it appears that a motion of 3 feet 4 inches may be had, which to the radius of 35 feet gives upwards of five degrees of a great circle.

Several abatements must be made on account of the disposition of the apparatus that gives this side motion, and the shortness of the ropes in high altitudes; but there remains, notwithstanding, a sufficient quantity of this lateral motion to answer the purpose of viewing, pretty minutely, every object that passes the meridian.

Before I can give the particulars of this side motion, some other things must be explained. The point of support rests in a pivot; but this alone could not have given steadiness to a tube of 4 feet 10 inches in diameter, loaded with the weight of the strengthening bars, and speculum, which rest upon it. Two moveable supporters have therefore been provided at $p\ q$, fig. 34. They consist of two solid brass rollers, 3 inches thick, and 4½ in diameter; set in strong frames firmly united to the sides of the tube, and resting upon the flat face of the square axle AB, which carries the pivot in the centre. The middle of these rollers is applied about 2 feet 2 inches from the centre of the pivot; and being set so as to lose none of the motion which they may have upon the axle, we find that there is room for full as much angular motion of the rollers upon the axle, as there is for the tube between the sides of the ladders; and indeed more than can be wanted, as 10 minutes of time are generally sufficient for viewing any object.

The method of observing with this telescope is by what I have called the front view; and the size of the instrument being such as would permit its being loaded with a seat, there is a very convenient one fixed to the end of it. The foot-board or floor, is 3 feet broad, and 2 feet 2½ inches deep. The seat is moveable from the height of 1 foot 7 inches to 2 feet 7 inches, not so much for the accommodation of different observers, as for the alteration which is required at different altitudes, and which amounts to nearly 12 inches. One half of the seat falls down, to open an entrance at the back; and being inclosed at the front and sides, a bar which shuts up the back after the observer is in his place, secures the whole in such a manner as to render it perfectly safe and convenient.

There are two strong iron quadrants with teeth, at the sides of the seat, in which run two pinions fixed upon a bar, with a ratchet and handle at the end of it. By turning that handle, the seat is easily brought to an horizontal position,

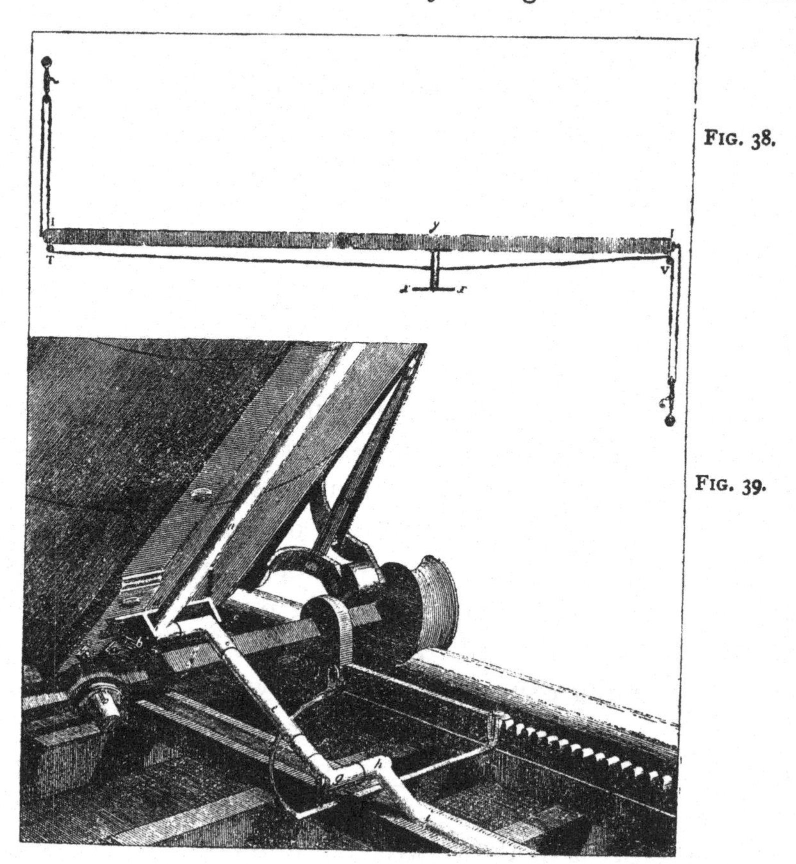

Fig. 38.

Fig. 39.

before the observer enters it; or restored to it, when any considerable alteration in the altitude of the telescope renders a change necessary.

The focus of the object mirror, by its proper adjustment, is brought down to about 4 inches from the lower side of the mouth of the tube, and comes forward into the air. By this arrangement, there is room given for that part of the head, which is above the eye, not to interfere much with the rays that go from the object to the mirror; the aperture of the speculum being 4 feet, while the diameter of the tube is 4 feet 10 inches; especially as we suppose a night observer will prefer some kind of warm cap to a hat, the rim of which might obstruct a few of the entering rays.

A long coarse screw-bar is confined in a collet, which takes on and off, and may readily be put to the inclosing right side of the seat, so as to present the observer with a short handle. The other end of this bar passes into a nut, which, like the collet, moves upon a double swivel, so as to admit of every motion. The nut is planted upon a machine which will be described hereafter, and may be drawn up to any altitude, so as to bring the nut upon a level with the swivel of the handle. Upon turning the handle, the observer will screw himself, the seat, and the telescope, from the ladder; and may thus follow the object he wishes to pursue in its course, for as many minutes as may be convenient. If, indeed, he is inclined to give up the meridian for some time, he may order the whole frame to be moved by the great round motion, which ought to be in readiness; and may even keep his object in view, as I have often done, by screwing the telescope backwards as fast as the round motion advances it. Then screwing himself forward again, he may repeat these successive motions as long as he pleases.

In those observations, which I have called sweeps (from the method of oscillating or sweeping over an arch, which at first I had adopted in the way of right ascension, but which in the year 1783 I reduced to a systematical method of sweeping over zones of polar distance), several conveniences are required: the principal of them are as follows.

An assistant, provided with an apparatus for writing down observations; with catalogues of stars, atlasses, and other resources of that kind.

A small apartment, as near to the observer as possible, in which this apparatus, with candles and other conveniences, may be inclosed.

A sidereal time-piece.

A right ascension apparatus.

A polar distance apparatus.

A polar distance clock.

A zoned catalogue of the stars.

And a ready communication between the observer and assistant, both ways.

There is also wanting, a person to give the required motions for sweeping the zones of the heavens.

A micrometer-motion to perform the sweeps.

A zone-piece, to point out the required limits of the intended zones.

A small apartment to inclose these motions, and the candle which is required for the workman.

And a ready communication, for the observer to direct the workman in the required motions.

All these conveniences were gradually brought to perfection with my 20-feet telescope; but here, they were at once, and with great advantage, designed and executed in their most improved state.

A′ B′ C′ D′, fig. 3, is the floor of the observatory, 8 feet 5 inches by 5 feet 5. It is of a proper height, and has a double window towards the west, with a shutter to be used at night. Fig. 1 gives a sufficient view of it.

E′ F′ G′ H′, is the floor of the working room, which is 6 feet 6 by 4 feet 5; and has two small windows, one to the south, the other to the east. Its height is considerably less than that of the observatory; and a view of this may also be seen in fig. 1.

The distance between the observatory and the end of the telescope, is evidently too far for a conversation in the open air, between the observer and assistant; especially as the latter, on account of his candles, must be inclosed; and ought not to leave his post at the time-piece and writing-desk. Add to this, that when the observer is elevated 30 or 40 feet above the assistant, a moderate breeze will carry away the sound of his voice very forcibly. A speaking-pipe was therefore necessary, to convey the communications of the observer to their destination.

At the opening of the telescope, near the place of the eye-glass, is the end of a tin pipe, into which at the time of observation a mouth-piece may be put, which can be adapted, by drawing out, or turning sideways, so as conveniently to come to the mouth of the observer, while his eye is at the glass. This pipe is 1½ inch in diameter, and runs down to the bottom of the telescope, to which it is held by proper hooks, that go into the tube, and are screwed fast at the inside. When it is arrived as near to the axle AB, fig. 34, as convenient, it goes into a turning joint; thence into a drawing tube, and out of this into another turning joint; from which it proceeds by a set of sliding tubes towards the front of the foundation timber.

The mechanism of the first turning joint and short sliding tube, as well as the next turning joint, is executed in brass, as represented in fig. 39. The tube *a* is the continuation of the pipe which comes down from the observer; at *b* and *c* it is turned about in an angle, but the part *b* and *c* consists of a double brass tube, one of which may be turned within the other. *bde* is an arm which has two pivots, one at *b*, the other at *c*; the part *d* is put through, and pinned to a fastening at the tube, where it is also permitted to turn about if required. When the telescope is lifted up, the pipe *abc* turns upon the pivot *b*, and within the pipe *bce*; which also turns upon the pivot *c*; so that *abc* may come at rectangles to *bce*, when the telescope is turned up to the zenith. At the same time *ce* sliding in *fg*, will be drawn out, since *bc* is not in the axis of the vertical motion, which lies in AB, fig. 34, but turns in a small arch about it. The point *c* will not only be drawn back, but will also be lifted up, and therefore a second turning joint, *hg*, becomes necessary, which is of the same nature with the first. From *hi* the pipes are continued in three joints of 9 feet 6 inches long each. These slide into one another as far as is required, and, all together, into a fourth pipe, when the telescope is advanced to the place where it rests in zenith observations. The fourth pipe, which is the largest, goes to the end of the frame H′, fig. 3, where it turns towards I′; and is there again made to return to K′. At this last place it divides itself into two branches, one going into the observatory, to L′, where it rises up through the floor; the other going into the work-room to M′, where it also ascends through the floor, up to the level of the workman's ear, who stands just by the place where it terminates in the usual shape of speaking pipes. Notwithstanding the passage of the sound through a pipe with many inflections, and not less than 115 feet in length, I find that it requires no particular exertion to be very well understood; and that the communication is quite sufficient for the purpose; though undoubtedly some advantage might have been gained if brass sliding tubes had been used throughout the whole length. Under the long pipes that slide into one another, is a semicircular gutter, extended from N′ to O′, which keeps the pipes in their place, as they are carried along by the motion of the telescope, when the point of support is advanced or drawn back; and the large gathering pipe is inclosed in a box, N′ H′, to secure it from accidents.

The right ascension apparatus is constructed thus. Against the sides of the tube, and 2 feet 6 inches from the mouth of it, are fixed the centres of two rubbing plates, 3 feet 10 inches long, 2 feet 1 inch broad, and near 2 tenths of an inch thick. These plates are fastened to the long bars of the tube, nearest the top and bottom, by six arms each; and screwed on so as to be perpendicular to the horizon. The plate on the west is fixed, but that on the east is adjustable, in order to be kept perfectly vertical on every part of its surface. One of these is visible against the tube, in fig. 1. An iron roller, 1 inch thick, and 2½ in diameter, is set in a strong frame, in such a manner as to allow the claw, which holds it, to be set to any direction; where it can be afterwards fastened by a large horned nut. This roller is mounted upon a frame, see fig. 40, that may be drawn up to any altitude, and lies upon the whole set of ladders on the east; where it rolls up and down on six sets of brass rollers, *a b c d e f*, which are constructed as in fig. 17. This machine consists of a bottom frame, and a bar, *gh*, at rectangles to it, which, when the frame lies upon its rollers on the ladders, stands also at rectangles to them, on the lowest part of the frame: it is braced so as to make the greatest resistance from east to west. The bar carries the iron roller, which may be shifted to two different situations; *g* almost down to the ladders, and *h* more elevated. The latter is used in high altitudes. The iron roller, standing out, is then turned so about as to be in the direction of the length of the eastern rubbing iron; in which situation it is fixed by the horned nut. The telescope is then brought forward, or backward, by the bar machine, till the rubbing iron comes to be opposite the roller.

Upon one of the braces of this same frame at *i*, is also planted the nut belonging to the lateral screw motion, which has been described; and its long bar goes always with this machine, when it is disengaged from the observing chair, and is laid back at *k* into a secure resting place.

On the opposite set of ladders is another machine, which carries a large spring-bolt, on the end of which is mounted an iron roller, exactly like that which has been described. This is also adjustable to any direction. The bolt is contrived in such a manner as to come out of the frame, in which it runs, with a pressure of 34 pounds; and it exerts, very nearly, the same force during the time in which it goes through every part of the space it describes. The construction of the springs is expressed in fig. 41: *a b c* are two iron bars, 5 feet 6 inches long; jointed at *b*, like a pair of compasses, and fastened on a pivot at *a*, which remains immoveable upon the frame, while the other end is also fastened on a pivot, fixed to the bolt, which carries the roller *f*. The bolt is 7 feet 1 inch long, and 3 inches square.

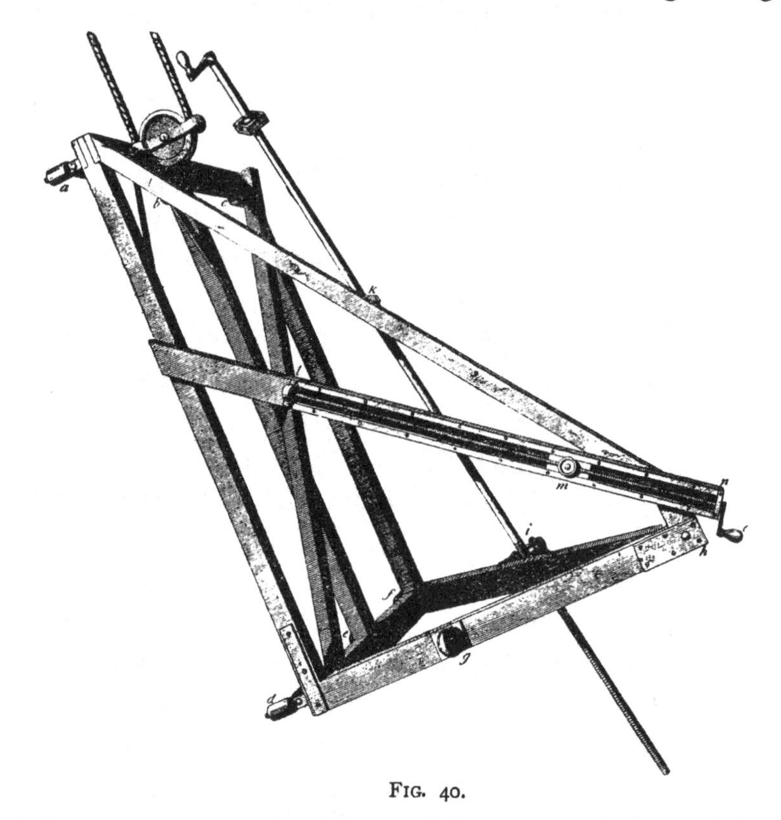

Fig. 40.

It runs in two sets of four brass rollers each, at *g* and *h*, which embrace it completely, and prevent friction as much as possible. The joint *b* is sustained by a brass roller, which runs on the iron plate *ik*. Two tapering steel bars, or springs, *lm*, *no*, are fastened against the lower ends *a c* of the iron bars; one of them is convex at *m*, the other concave at *o*; and they exert their force against each other at *mo*, where the convex one rests in the concavity of the other. There is an adjustment of the flap which carries the bolt, see fig. 1, by which it may be raised up, so as to become exactly opposite to the roller on the east, when that is raised to its highest position.

It will now be easily perceived, that when the eastern rubbing plate, in its well adjusted vertical position, is pressed against the right ascension roller, by a roller exactly opposite, and with a force sufficient to keep it firmly poised against that roller, a vertical motion may be given to the telescope, in which the same meridional situation will be preserved. Accordingly, I find that the right ascension of unknown objects, deduced from known ones, observed by the same instrument, and in the same zone, is capable of great precision ; and this construction will therefore answer all the ends that were proposed. For it would not be doing justice to the telescope to require of it all the accuracy of a transit instrument.

The spring-bolt, as I call this latter machine, is brought to any required situation by a rope fastened to the middle cross-beam of the stand, which comes down, and goes through a pulley placed upon the machine ; in its return to the top, it passes over a second pulley, and then goes down to a barrel with a wheel and pinion, on the ground timber at Q′.

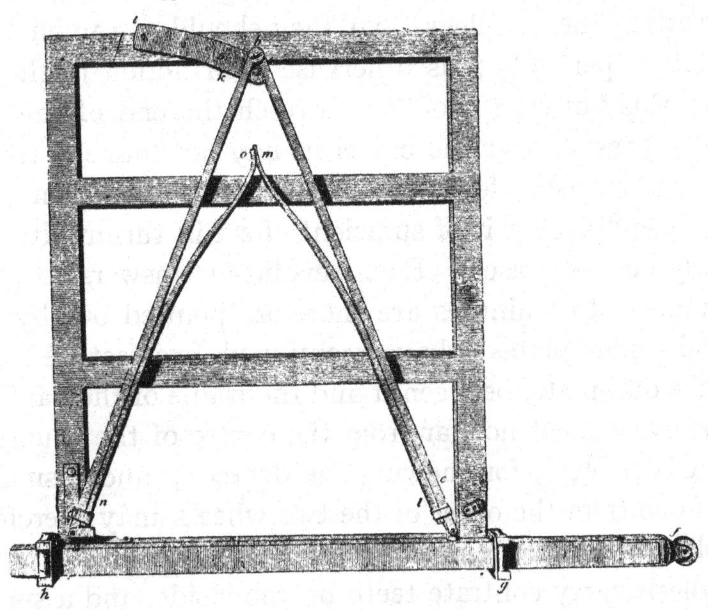

FIG. 41.

The polar distance machine, as I call the opposite one, on account of its chief use, which remains still to be explained, is drawn up and down in a similar manner, by the handle of a pinion, wheel, and barrel placed at R′.

In the observatory is placed a valuable sidereal time-piece, made by Mr. SHELTON, for which I am obliged to my astronomical friend Mr. AUBERT, as a gift that will always be highly esteemed. Close to it, and of the same height, is a polar distance piece, which has a dial-plate of the same dimensions with the time-piece ; and is also divided into sixty parts on the outside ; but these are to express minutes of space. Every tenth is marked with large figures, but every single one is also denoted with its proper figure, in a smaller character. The degrees are shewn in a square opening under the centre, and change backwards and forwards as the telescope rises or falls. This piece may be made to shew polar distance, zenith distance, declination, or altitude, by setting it differently ; but in conformity with FLAMSTEED's British catalogue of stars, I have generally adopted polar distance.

The construction of this piece is very simple. It contains only one barrel, for the weight and line, which gives motion to the work; and two small index wheels. The line is conducted from the polar distance machine into the observatory at the bottom of the polar distance clock, where it rises up, and passes over the barrel. By making this revolve, it moves the hand upon the axle of it, which points out the minutes upon the dial-plate. The hand is made adjustable in the usual way of the minute hand of common clocks, by going upon a pipe, kept firm by springs. The line is of considerable length; but the case of the clock being no larger than that of the time-piece, a set of neat and very thin pulleys, four and four, are used to draw the end, after its having crossed the barrel. It is necessary to mind, in setting these pulleys, that they should run upon very thin pivots, and clear one another perfectly; as otherwise their action might not be adequate to the purpose; this however is only to stretch the end of the line freely and sufficiently, that in passing over the barrel it may not make it turn about irregularly. There will be no occasion for a revolution of the line upon the barrel, as I have found a mere passage over it of sufficient effect in turning it; for the hands must all be properly counterpoised. Each revolution answers to one degree of change in polar distance; the minutes are therefore pointed out by the hand it carries. The two small index-plates I have mentioned, are fastened upon pivots against the back of the dial-plate, between it and the frame of the barrel. They are placed so, that their edges meet not far from the centre of the square hole, I have mentioned, in the dial-plate, for shewing the degrees; and a small square portion, a little more of one than the other of the two wheels, may therefore be seen, in front of the dial-plate, through the opening in it.

These wheels carry contrate teeth on the inside, and a small dial-plate on the back. The face of the dial-plate of the wheel which presents itself at the right, carries the units of the degrees; 1, 2, 3, 4, 5, 6, 7, 8, 9, 0; while that on the left has a blank which remains till the 0 of the first appears. Upon the axle of the barrel, close to the frame-plate on the outside, is fixed a long counterpoised contrate pallet; which at every revolution sweeps over one of the teeth of the first wheel, of which there are ten. The shape of the pallet must be like the barb of an arrow; but more obtuse, that it may take as much time in entering very obliquely into the teeth as possible, to avoid a sudden shock. The movement will even then be found to be quite quick enough, for shewing almost instantly the proper degree of polar distance. But to counteract the sudden stroke of the long pallet, there is over each wheel a small lever, see *ab*, fig. 42, that rests with its end between the two uppermost teeth *c d*; and its shape is that of a very obtuse angle, such as 160 degrees. The point of the angle sinking down between the two teeth, by its slope both ways, prevents their overshooting. The lever is held down with a very weak spring, *ef*, the point of which touches the lever at *e*, near the place of its pivot. This method will even throw back the figure upon the dial, if it should have been overshot a little. Care must be taken to let all this work be light, that no great force may be required in the long pallet to move it.

The first wheel in turning about carries a short pallet, of a shape similar to the long one. This must be placed low enough to let the long pallet pass freely, and high enough to clear the spring-lever in going over it. The pallet, on the appearance of the 0, strikes a tooth of the second wheel, and brings the figure 1 into view, which with the other forms 10. The second dial-plate has a blank, and the figures 1, 2, 3, 4, 5, 6, 7, 8, 9, 10, 11, 12, engraved upon its face, and presents thirteen teeth to the pallet on the first wheel, by which the blank and figures are successively brought into view, along with the succession of the units on the dial-plate of the first wheel.

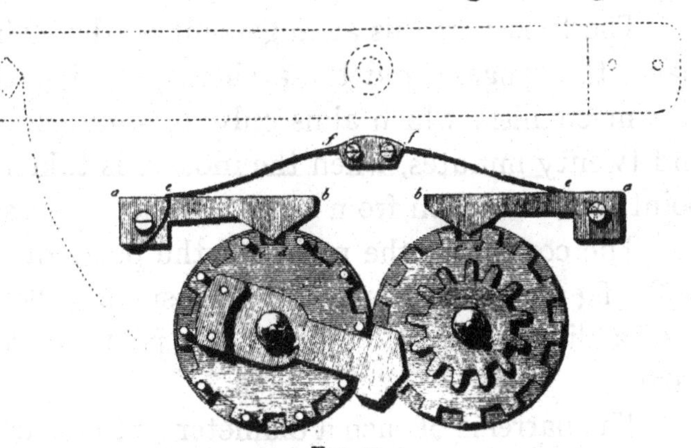

FIG. 42.

In this manner the degrees are shewn from 0 to 129, which includes the whole range of north polar distance in this latitude; while at the same time they are properly subdivided into minutes. A more minute division was not thought necessary with this instrument, and indeed ought not to be aimed at.

The cord which gives motion to the polar distance clock, is rendered a just representative, or true index, of the angular movement of the telescope in the following manner.

On the machine which holds the right ascension roller, is an arm *lmn*, fig. 40, in an oblique direction, upon which is fastened a brass slider, 3 feet 1 inch long. A coarse screw passes from one end of it to the other, and is confined between its shoulders *l n*. At *o* there is a handle, by which the screw being turned, a small sliding plate *m*, which carries a pulley, is drawn backwards or forwards at pleasure, along the whole range of the slider.

On the telescope, near the bottom of the front edge of the eastern rubbing-plate, is a small square bar with a loop upon it, which is adjustable, so that it may be occasionally brought a little nearer to the mouth of the telescope, or removed farther from it. The end of the polar distance cord is fastened to the loop upon this bar, where it remains when the polar distance clock is not in use. By this means the weight which stretches it in all its length from the telescope to where it is suspended in the clock-case, is kept always equally exerted, and no relaxation of the cord, which ought to be avoided, can take place.

When the polar distance clock is to be used, the cord is lifted into the pulley of the slider at *m*, and now goes from thence to its destination as before. The right ascension roller resting against the rubbing plate, the pulley of the slider is near at hand, and the cord may easily be lifted into it. The handle *o* is now to be turned till the cord, which goes from the loop at the telescope to the pulley upon the slider, comes to cover a certain white line or mark upon the side of the tube. This line when it is first made, must be placed so as to be vertical when the radius of motion of the loop is a little more than one degree of elevation above the horizon.

The theory of this arrangement is, that when a motion in polar distance takes place, the tangent and the arch may be looked upon as equal for a few degrees, in a mechanism which aims only at minutes. And, indeed, as far as two degrees and twenty minutes, when the motion is taken equally both ways of the adjusting point, the deviation from truth will not even amount to quite one second.

The cord from the pulley of the polar distance machine passes straight away to O′, fig. 3, where it is bent over a small pulley to one just close to it, which leads it in a direct line to P′, under the polar distance clock, where it rises up to the barrel.

The barrel is of such a diameter as to answer as nearly as possible to the length of the cord which is drawn by the motion of the telescope over one degree of polar distance; but as the utmost accuracy could not have been obtained in the make of the barrel, the loop at the telescope which draws the end of the cord, as we have described, may be slipped backwards or forward upon its bar, which will either lengthen or shorten the radius of its motion, and occasion its drawing more or less of the cord.

As there is a good quadrant upon the telescope, there remains nothing else to obtain a just position of this loop than to compare the indication of the polar distance piece with that of the quadrant; and when the former is regulated to a perfect agreement with the latter, we may safely rely upon the truth of its report.

The time and polar distance pieces are placed so that the assistant sits before them at a table, with the speaking-pipe rising between them; and in this manner observations may be written down very conveniently. The place of new objects also may directly be noted, as their right ascension and polar distance is before the assistant upon the table, where nothing is required but to read them off, on the signal of the observer.

By a catalogue in zones the assistant may guide the observer, who is with his back to the objects he views, and who ought to have notice given him of such stars as have their places well settled, in order to deduce from their appearance the situations of other objects that may occur in the course of a sweep. In the year 1783, when I began this kind of observations, no catalogue of stars in zones had ever been published; I therefore gave a pattern to my indefatigable assistant, CAROLINA HERSCHEL, who brought all the British catalogue into zones of one degree each, from the 45th degree of north polar distance down to the horizon, and reduced the right ascension of the stars in it to time, in order to facilitate observations by the clock. This catalogue was afterwards completed from the same degree up to the pole in zones of 5 degrees each; and the variation in right ascension from one degree of change in longitude, was also reduced to time, for every star in the catalogue. To this were added computed tables for carrying back present observations to the time of that catalogue; which method I preferred to bringing the stars it contains forward to the present time, on account of conforming with the construction of the *Atlas Cœlestis*, which was of great service.

The evident use of such a catalogue must undoubtedly soon have been perceived by every person who was acquainted with the method I used for sweeping the heavens; and as the same is practicable, not only with my telescopes, but likewise with transit instruments, and mural quadrants, we are now much indebted to the Rev. Mr. WOLLASTON, who in the year 1789 favoured the astronomical world with a work of nearly a similar construction with that which I was in the habit of using; but much enlarged, and enriched with stars taken from the best authors; and moreover reduced to the time of the year 1790.

We now seem only to want an atlas on the same construction, upon a scale equally extensive, and plentifully stored with well ascertained objects.

The micrometer-motion which is required for sweeping the heavens, and indeed for viewing the planets or other objects, is obtained by means of the end of the rope B, fig. 37, which draws up the telescope. This goes down to a barrel, S′, fig. 3, 12 inches long, and 4 in diameter, joined upon the same axle with another barrel, 12 inches long, and 12 in diameter. A smaller rope goes from the largest barrel into the working-room, where it is fastened to the top of a thin vertical spindle, 2 feet 6 inches long, and 3 inches in diameter, at *a*, fig. 43. Another rope of equal size is fastened to the bottom of the same spindle at *b*; and when by turning the handle *cd* the rope *ae* is wound upon the spindle one way, the rope *bf* is wound off the contrary way. This second rope *bf* goes out of the work-room over a pulley, which leads it upwards to the top of the middle cross beam of the ladders, where it descends over another pulley, by a weight with shifters which is suspended to the end of it. In this manner a balance is obtained between the stress of the ropes *ae* and *bf*, which leaves the spindle at rest in any position where it may chance to stand, and considerably eases the labour of the workman, who turns this handle a certain number of times one way, and then the same number of times back again. By such a motion of the handle the telescope is alternately depressed and elevated; and this being continued for as long a time as the observer chooses, enables him to review the heavens as they pass by the telescope.

By the arrangement of the barrels, it is easy to see that the motion is sufficiently divided, as many turns of the handle are wanted to pass over a small space of the heavens. The method of barrels and ropes is to be preferred to wheel-work, on account of the smoothness as well as silence of the motion, both which in observations of this kind are highly necessary. It is true that the great stress which lies on the ropes of the micrometer-motions wears them out very fast, and they must therefore be carefully watched, and often renewed; but this ought to be no objection where the end to be obtained is of such consequence.

It would not only be troublesome to the workman, but often bring on mistakes, were he to count the turns of the handle, which perhaps for hours together he is moving; a zone-clock, therefore, has been contrived to release him from that care. This is a machine which is placed upon a table just by the workman. It strikes a bell when he is no longer to turn one way; that is, when the telescope is come to one of the limits of the zone, which if it be after going down, is called

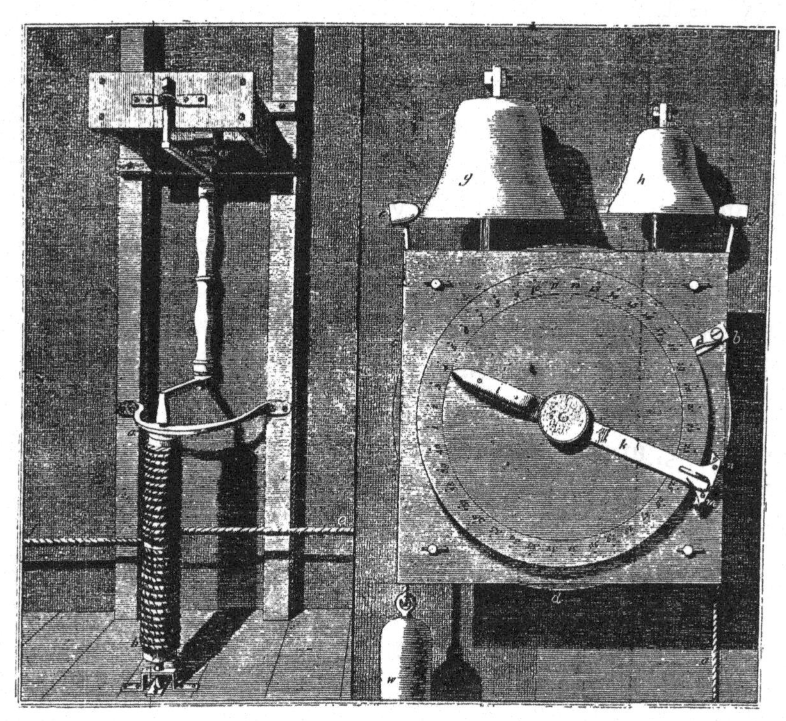

FIG. 43. FIG. 44.

the bottom bell; and it strikes another bell when he has made the same number of turns in a contrary direction. The telescope is by this motion restored to its former situation, and this second notice is called the top bell; which marks out the other limit of the zone. These bells not only give notice to the workman when he is to change, but their different sound indicates the position of the telescope, and prevents mistakes.

An additional precaution has been used, to make the bells repeat their stroke, the very next turn, if by some mistake the workman should have been inattentive to the first notice. In a long continuation of uniform intervals of sound, we may become so used to them as hardly to perceive them at all; but the coming in of an additional sound will immediately rouze the proper attention. Another very necessary use which I have often made of a second or third bell, is to extend the zone, either towards the north or south, for some time, when notice has been given of a star that was a little above or below the sweep; for in some parts of the heavens known stars are scarce, and it becomes necessary to take in all those that may be come at.

The construction of the zone-clock is very simple and convenient. The end of the axle which holds the double barrel, fig. 3, must be left projecting at T′. Upon this a small hollow cylindrical pipe is placed, which holds the end of the cord that is to move the zone-clock. The pipe must be guarded at both ends like a clock barrel, to keep on the cord, but remain open at the end which goes upon the axle, upon which it must fit upon a square so as to keep firm. It should be about $1\frac{3}{4}$ inches long, and 1 in diameter.

From this piece the cord is made to pass to the work-room, where it rises up into the clock at *a*, fig. 44. It then passes over a large narrow barrel, *bcd*, and by means of a weight *w* at its end, descends when the handle of the micrometer-motion turns the spindle and double barrel with which the pipe that holds this cord is connected. At *cb* are two levers that, in the usual way, occasion the hammers *e f* to strike the bells *g h*, when the pins quit the levers which they have lifted up. But these levers have spring joints, so as to permit the pins to pass back again without disordering the work. The pins which move the lever *c* are fastened to the barrel *bcd*. The lever *b* must be brought out so as to be before the front of the first frame-plate, and close to a dial-plate, which is to contain about 40 numbers. The dial-plate must be pretty thick, and be fixed upon a hollow arbor. The axle of the barrel, which should be strong, must be long enough to come through the hollow arbor, and project a little way to receive a milled nut upon its end, which must have a screw upon it. The arbor which carries the dial-plate is then to be pinned fast upon the axle, and an adjustable hand being put upon the projecting arbor, a collet is slipped over it, and the milled nut screws it down, in any position that is to be given to it.

The adjustable hand is made of a piece of springy iron, or steel, formed as represented at *ik*; but broader than clock hands usually are. It must have a pretty large circle in the middle, with a hole wide enough to go upon the plate-arbor. The end *k* of the hand must project beyond the dial-plate a little way, so as to permit two screws, *m n*, to pass by it into a brass plate, with a small piece between, to allow some motion up and down to the hand. The plate which is fixed to the hand by the screws *m n*, returns under the dial-plate sufficiently to carry three pins that are to lift the lever *b*, when they come to the proper situation, in the same manner as those on the barrel lift the lever *c*. The dial-plate, close to the margin, should have as many small holes, to receive a pin, as there are numbers marked upon it; and in the hand, answering to the holes, must be fixed a steady pin to fall into any one of them, when the hand comes to be placed over it. There must be a small handle near the end of the hand, by which it may be lifted up, and moved into any situation that shall be required; and care must be taken to have both ends properly counterpoised.

In order to set the zone-piece to the breadth of any particular sweep, as for instance two degrees, we make the workman begin at the striking of the top bell, and while he turns the handle till the quadrant or polar distance-piece points out a change of two degrees, we keep the hand of the zone-clock lifted up, that the pin may be out of the holes upon the dial-plate; for which purpose also the nut in the centre must be unscrewed a little to permit it to pass freely. When the telescope has descended two degrees the workman must stop the handle. We then lift the hand to the place where the first pin strikes the lever of the bottom bell. Here we let the pin drop into its proper hole, and screw fast the central nut. When this is done, the workman may turn backwards and forwards from bell to bell, and the telescope will perform the required motion of two degrees.

The work of the zone-piece is arranged in such a manner as to make the numbers on the dial-plate answer to turns of the working handle: this however, though convenient, is not absolutely necessary. The number of turns to a degree varies a little in different altitudes; but by trial a table may be made, which will shew with sufficient accuracy the figure on the dial-plate to which the hand must point, that the zone-piece may give any required breadth to a sweep, at any certain polar distance.

By means of the speaking-pipe the workman may be directed to begin, to stop, to go fast, or slow. And these, with a very few other orders, will be all that are wanted; which being known to him and to the assistant, will occasion no mistake, notwithstanding the pipes which go into the two apartments are united.

The ropes that come from the gallery, each bracket of which is separately drawn up, go through a double pulley, hung to the top cross beam, and a double pulley fastened to the upper end of the gallery bracket; after this they pass over a single pulley at the top, down to two barrels placed under the back of the ladders, one on each side. Each barrel is moved by a handle, on the axle of which is fixed a pinion of four leaves: this works in a wheel, on one side of the barrel, of 61 teeth, and 18 inches in diameter.

The barrels are 25½ inches long, and 12 inches in diameter, that the rope may not be doubled often, which might hurt the uniformity of drawing up the gallery. They are made exactly alike, and draw an equal length of rope at every stroke of the handle; but as one of the persons who draw the gallery might go on quicker than the other, each of the handles strikes a bell at every turn, going up as well as going down; the different tone of the bells easily shews, by sounding in regular alternate succession, when the gallery is properly moved; which therefore may be safely done in the dark. The mechanism of the bell-work at each handle is in a little box, to keep it dry, but sufficiently open at the side to throw out the sound.

A single bell being suspended, as in fig. 45, upon a plate of iron, at *a*, there is a cock, *bc*, planted upon it; between which, at *d* and *e*, are inserted two axles continued outwards. On the outside of the cock, and upon these axles, are two inverted hammers suspended, with lever arms, *f g*. These are made with spring joints, like those that have been described in the zone-clock. The axle which moves the barrel has a pallet upon it, and the plate with the bell apparatus being presented at rectangles to the axle, so as to make the pallet play in the notch of the plate between *f* and *g*, where the lever arms meet, it will make the bell give a stroke, either with one hammer going up, or with the other coming down.

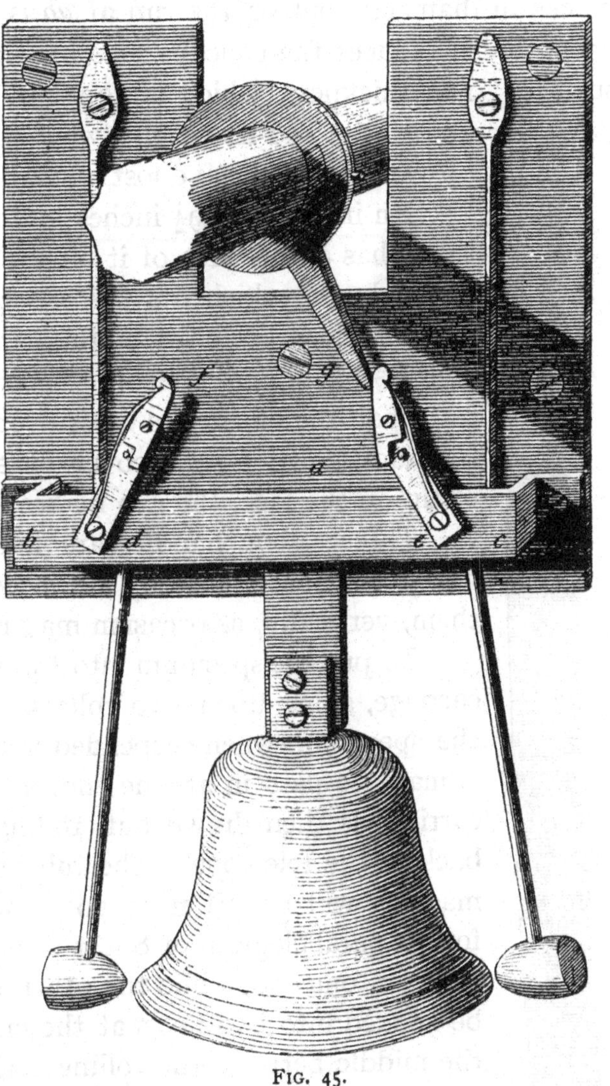

FIG. 45.

It is necessary to preserve the pliability of the ropes, for which reason no tar has been used with any of them that are about the telescope. To preserve them, however, they are passed through very hot melted tallow, and kept a sufficient time immersed in it, that they may be thoroughly penetrated. In this state they will last a considerable time, especially when care is taken not to relax them often. The gallery being suspended by ropes in this state, it would be unpleasant to trust entirely to them. Each bracket therefore is furnished with four strong broad iron hooks, two of which take hold of one of the flats of the division $\beta\gamma$, fig. 7, while on the opposite side two more take hold of the corresponding flat of $\delta\epsilon$. When the gallery has been drawn up to the required altitude, the hooks are let down, and the ropes slackened a little, so as to permit it to hang in the hooks. The other two hooks on each side serve for an elevation between the flats half way from one to the other. They are upon the same centre with the former, and fall back as the others do when the gallery is to go down.

For the safety of the tube also, there is a strong chain, which will sustain it, in case the ropes by which it is suspended should give way. This is fastened into a loop near the point of suspension. The other end of it is hooked upon a flat, and passes round one of the side beams of the ladder at a certain elevation above the telescope, and is sufficiently long to permit the tube to move a few inches more than is necessary. By this means a fall can never be considerable: if the ropes were to break in the worst part of a sweep of $2\frac{1}{2}$ degrees broad, the telescope would hardly descend two feet.

The construction of the great mirror is as in fig. 46. The metal itself is $49\frac{1}{2}$ inches in diameter, but on the rim at *ab* is an offset of $\frac{3}{4}$ inch broad, and 1 inch deep, which reduces the concave face of it to a diameter of 48 inches of polished surface. The thickness, which is equal in every part of it, remains now about $3\frac{1}{2}$ inches; and its weight, when it came from the cast, was 2118 pounds, of which it must have lost a small quantity in polishing.

FIG. 46.

An iron ring, $49\frac{1}{2}$ inches in diameter within, 4 inches broad, and $1\frac{1}{8}$ thick, has at the face of it on the inside a strong bead or rim added to its thickness, which fits the offset in the speculum, but is not quite so deep as that. A cross of the same substance of iron as the ring, goes over its back, and when the speculum is placed into the ring, so as to rest upon the offset, the cross over the back confines it in the ring. By the addition of a thin cover of sheet iron on the back, and another of tin on the face, the rim makes a complete case for the mirror.

Three strong handles are fixed against the sides of the ring, by which the speculum may be lifted horizontally, or using only one of them, vertically, as occasion may require.

To put the speculum into the tube, there is provided a small narrow carriage, going upon two rollers. It has upright sides, between which the speculum, when suspended vertically by a crane in the laboratory, is made to pass in at one end, and being let down, is bolted in. The carriage is then drawn out, rolling upon planks, till it comes near the back of the telescope. The tube must be put back as far as the bar-machine will permit it to go. Two beams connected together so as to form a parallelogram of 8 feet 6 inches long, and 2 feet broad, are sloped away on one end, while the other contains two hooks, by which it may be hooked into two holes at the end of the foundation timber, fig. 3, in the middle between the rolling beams *r s*. This affords a passage of an easy ascent to the speculum carriage, which must now be brought into a proper position for rolling up. When this is done, the carriage is to be tied to the axle of the point of support, AB, fig. 34, and by turning the bar-machine handle, the speculum with its carriage will be drawn up to the foundation beams EE, AA, fig. 3, which are 16 inches above the foundation wall. By the time that the carriage comes near to the top, there will be room for six 3-inch planks that are provided, to be laid one after another upon the rolling beams *r s*, which will form a platform of 5 feet 10 inches by 5 feet 5, for the reception of the carriage. But these planks must not be put down till the telescope has been first brought back, and fixed again close to the carriage, which must be sustained in its place while this is doing. Then, advancing again, the platform is laid down, board by board, till completed, while at the same time the carriage will be drawn upon it.

As soon as that is safely landed, a strong rope is to be hooked into a loop, fixed upon the beam at *a*, fig. 15. This going down to a pulley with a swivel hook at the bottom, which is put through one of the three handles of the speculum, returns to a pulley hung upon the hook *b*. From that pulley it goes forward to a leading pulley at V′, upon the foundation timber, fig. 3. This directs the end of the rope to the barrel, which serves for the great round motion of the whole telescope. When the handle of that machine is moved, the speculum will be lifted up in its carriage, which being eased, must now be turned about while the mirror is yet partly resting upon it, so as to become parallel with the back of the tube, and close to it. As soon as the mirror is fairly suspended, the carriage must be unbolted, and drawn sideways from under it. At the same time the platform must be gradually removed, that the tube may be brought back by the bar-motion, whenever the mirror is high enough to pass over the back of it. Then letting down gently the round motion handle, and guiding the mirror properly, it is to be placed upon a small hollow square, with a sloping back, which is planted under its support. The height of the square frame is such as will bring the centre of the mirror into the centre of the tube; and the sloping back receives it in going down, and throws it from the back of the tube, just as much as is required to make the adjustment at the top act properly.

When the mirror is in its place, two loops which are prepared are to be screwed fast to it. They contain the collets that receive the adjusting screws from the back, through the strong upper bar *cg*, fig. 34, and as soon as these are fastened the pulleys may be unhooked, and all the apparatus that has been used removed. The six planks are then to be laid upon the same rolling bars at *no*, where a passage across the work is wanted, and where they may remain till zenith sweeps require them to be moved.

The method of preserving the speculum from damp is by having a flat cover of tin soldered upon a rim of iron, about $1\frac{3}{4}$ inches broad, and $\frac{1}{8}$ thick, the diameter of which is equal to the iron ring which holds the speculum. Upon the flat part of the rim is cemented, all around, some close-grained cloth of an equal breadth with the rim. The cover has two handles near the upper end, and under them two flaps that project about an inch and are six inches broad. When the cover is hung or laid upon the speculum, so that the two flaps are close to the ring which incloses it, the rim of the cover, as far as it is lined with cloth, will rest against the edge of the iron ring, and fit it all around very closely.

In six places are painted white marks which divide the circumference equally; and six claw-screws are provided, of the shape that is represented in fig. 47. These are applied to the six marked places. The end *a* being put over the iron ring *b* to take hold of the back, the screw *c* is then fastened so as to press upon the outside of the cover and rim, till the lining of it is brought into close contact with the iron ring.

To take off and put on again the cover, a small ladder is provided, which being set at the outside against the back of the tube, the person who is to uncover it goes up, and descends into the tube by means of a board with steps. This board goes across the mirror in a parallel direction with it, and being narrow, does not interfere with the work of loosening the screws to take them off. When they are removed, the person comes out of the tube the same way, still leaving the speculum covered, but when at the top of the ladder brings out the inside board-steps. The two handles of the cover now present themselves at the back, so that two persons can easily lift it off, without suffering it to touch the mirror in any place. It must then immediately be carried into the observatory, and remain there till the mirror is to be covered again; but first of all the inner and outer cover of the tube ought to be carefully closed up.

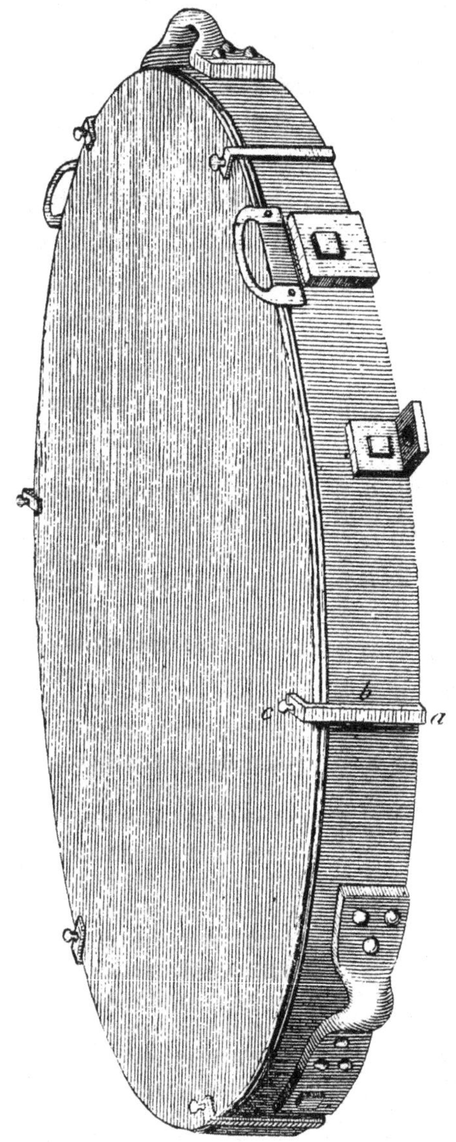

FIG. 47.

When the speculum is to be covered again, great care is required to see that no drops of dew may fall from the outer cover of the tube upon the inner one; or at least that these may not find their way to the mirror; and to let the first object be to put its own cover upon it before any thing be done about fixing it there.

In very high observations the tube will not fall down again readily, and in the zenith, by its great weight added to that of the mirror, will even tilt backwards. A counterpoise therefore is applied by a meridional post, 7 feet high, well fastened by a frame in the ground; and placed about 20 yards from the front of the telescope. To this is fastened at the top on the back an arm, which carries a pulley, and at the bottom on the front, a barrel moved by a wheel and pinion. A rope with a weight fastened to the end of it, goes over the pulley on the post, and towards the mouth of the telescope. At the end of the tube on each side is a loop, into which a chain is hung with both ends. It is long enough to go round the seat to a considerable distance, and holds a pulley in the middle, over which the rope from the weight is made to pass back again to the barrel at the post. Here it may be drawn up till the weight is lifted sufficiently to keep the telescope steady, and to make it fall again, when its own motions lower it. In zenith sweeps 300 weight are required for that purpose, but one hundred of that quantity is in shifters that may be taken off in lower altitudes, when less is sufficient.

A similar post and apparatus is fixed about the same distance from the telescope towards the north, to be used when the instrument is turned about for high observations in the northern meridian.

Another inconvenience to be removed in very high altitudes, is that the long bars, which bring the point of support forwards, begin to project beyond their supporters. When this comes to take place, a light iron frame with two small wheels, or rather rollers, is pinned to the ends of them. This not only keeps them together, but also supports them sufficiently as far as they are to come out.

A slider, upon an adjustable foundation, is planted at the mouth of the telescope so as to be directed towards the centre of the mirror. It carries a brass tube, into which all the single eye-glasses, or micrometers, are made to slide. When they are nearly brought to the focus, a milled head under the end of the tube turns a bar, the motion of which adjusts them completely.

The focus of the great mirror is directed to its proper place, by putting two plates with springs upon the rim that limits the aperture of the tube, into two places which are marked. Then a cap with a small hole being put into the sliding tube, an assistant with a proper handle must screw in or out one or other of the adjusting screws at the back of the mirror, till the plates upon the aperture in front of the telescope become both visible; for they are contrived so that when the mirror is not properly adjusted, either one or both will vanish. At the same time these plates, by their situation, serve to inform us which of the screws, whether that to the right or that to the left, is in fault, by which means the adjustment becomes a very easy operation.

WM. HERSCHEL.

Slough, near Windsor,
May 18, 1795.

Account of the Discovery of a new Comet. By Miss CAROLINE HERSCHEL. *In a Letter to Sir* JOSEPH BANKS, *Bart. K.B. P.R.S.*

[*Phil. Trans.*, 1796, pp. 131–132.]

Read November 12, 1795.

Slough, November 8, 1795.

SIR,—Last night, in sweeping over a part of the heavens with my 5-feet reflector, I met with a telescopic comet. To point out its situation I transcribe my Brother's observations upon it from his Journal.

November 7, 1795.

0^h 33′ Sidereal time. Place of the comet 2° 20′ *np*. 37 (γ) Cygni, in a line continued from 66 (υ) through γ nearly. It is just visible to the naked eye.

0^h 44′. It is in a line between two small stars at a considerable distance from each other, to which are perpendicular the two extreme stars of three other stars, that form a small arch approaching to a straight line.

0^h 49′. The comet precedes the point in the line where the perpendicular of the arch crosses the line of the two stars one-fifth of the distance of the bisecting point from the preceding star.

1^h 25′. The comet is visibly moved from the place where it was 0^h 49′.

The direction of its motion seems to be towards the south preceding side, and it is about 5 or 6′ removed from its former place.

1^h 51′. The diameter of the comet is about 5′. It has no kind of nucleus, and has the appearance of an ill-defined haziness, which is rather strongest about the middle.

2^h 16′. The comet is about 2° 38′ *np* γ, in a line continued from 69 through 37 Cygni.

3^h 37′. The comet is about 2° 50′ *np* γ, in a line continued from 65 through 37 Cygni, or, perhaps more accurate, in a line from 70 continued through 37 Cygni.

It will probably pass between the head of the Swan and the constellation of the Lyre, in its descent towards the sun.

The direction of its motion is retrograde.

Place of the comet deduced from the above.

	h ′		h ′ ″		° ′ ″
Nov. 7.	0 33	R.A.	20 3 48	P.D.	49 17 18
	3 37		20 0 58		49 37 18

As the appearance of one of these objects is almost become a novelty, I flatter myself that this intelligence will not be uninteresting to astronomers.—I have the honour to be, &c.

CAROLINA HERSCHEL.

XXXV.

Additional Observations on the Comet. By WILLIAM HERSCHEL, *LL.D. F.R.S.*

[*Phil. Trans.*, 1796, pp. 133–134.]

November 8, 1795.

0^h 10′. The comet is about 42′ north of 22 Cygni, in a line continued from 21 (η) through 22 nearly; it is not quite come to the line.

It is exactly in a line with 22 and a north following star 1° 34′ from 22 towards 21.

0^h 31′. Distance of the comet from 19 Cygni, 1° 10′. From 22 Cygni, about 42′. From 25 Cygni, 2° 10′. From 15 Cygni, exactly 3°.

2^h 27′. The comet is 36′ from 22 Cygni; its motion has been very nearly in the line pointed out before. It will however not pass over 22, but go by it towards 19 Cygni, having left the line pointed out, a little on the following side.

November 9, 1795.

20^h 45′. The comet is about 17 or 18′ from 15 Cygni.

21^h 59′. The comet is now centrally upon a small star north following 15 Cygni. It is a small telescopic star of about the 11th or 12th magnitude, and is double, very unequal, the smallest of the two being much smaller than the largest.

With a power of 287 I can see the smallest of the two stars perfectly well; this shews how little density there is in the comet, which is evidently nothing but what may be called a collection of vapours.

Angle of the small north following star with respect to 15 Cygni, 71° 55′ north following.

Position of the small star belonging to the double star, a few degrees south following.

The north following star must be of the 11th or 12th magnitude at least, for it is not visible in my achromatic finder, and its smaller companion therefore is an extremely small star indeed.

The double star is about 5 or 6′ from 15 Cygni.

November 10, 1795.

21^h 55′. The comet is about 1° 5′ north of, and a little following, 8 Cygni, and exactly 1° 30′ south following 4 Cygni.

2^h 16′. The comet is about 40′ from 8 Cygni, in the line between 8 and 4, but rather past the line.

It is about 1° 26′ from 4 Cygni.

XXXVI.

On the Method of observing the Changes that happen to the fixed Stars; with some Remarks on the Stability of the Light of our Sun. To which is added, a Catalogue of comparative Brightness, for ascertaining the Permanency of the Lustre of Stars.

[*Phil. Trans.*, 1796, pp. 166–226.]

Read February 25, 1796.

THE earliest observers of the stars have taken notice of their different degrees of brilliancy, and, by way of expressing their ideas to others, have classed them into magnitudes. Brightness and size among the stars were taken as synonymous terms, and may still be used as such with sufficient truth, notwithstanding the latter, it seems, can only be looked upon as the consequence of the former. The brightest stars were called of the first magnitude; the next of the second; and those of an inferior lustre of the third, fourth, and fifth magnitudes; and so on.

Among the stars of the first two or three classes there seems to be some natural limit which confines them to a particular order. If we suppose the stars to be about the size of our sun, and at nearly an equal distance from us and from each other, those which form the first inclosure about us will appear brighter than the rest, and there can be only a small number of them. This hypothesis is nearly confirmed by observation, as may be seen by looking over a globe, and applying a pair of compasses opened to 60 degrees, which should be the angle subtended by the stars of the first magnitude, if they were all scattered equally. For it will be found that the distances from Lyra to Arcturus; from Arcturus to Regulus; from Regulus to Sirius; from Sirius to β Navis; from Elgeuse to Canopus; from Canopus to α Centauri; from α Centauri to Achernar; from Achernar to α Crucis; from Procyon to Canopus; from Fomalhaut to Altair; and from Altair to Antares, agree sufficiently well with this hypothesis. It must also be remembered that a perfect equality in the mutual angular distribution of the stars that form the first inclosure, is a thing that is mathematically impossible, and therefore not to be looked for. This would authorize us to take in other intervals, such as from Arcturus to Antares; from Elgeuse to Regulus; from Achernar to Rigel; from Rigel to Capella; from Capella to Sirius; from Regulus to Spica; from Spica to α Crucis; and from Rigel to Castor; all which concur, in a great measure, to support the same hypothesis. But as the distribution and real magnitude of stars is not my present subject, what has been mentioned will be sufficient.

A second layer of stars will be more extensive; for the superficies of the celestial regions allotted for the situation of these successive stars exceeds the former in the ratio of 4 to 1. And on looking over the collection of stars which astronomers have pointed out as belonging to the second class, we find that their number is proportionally larger.

A similar way of considering the stars of the third order might be applied, if it did not already appear, from what has been said of the two former orders, when strictly compared with the state of the heavens, that such kind of limits can be of no real use in the classification of stars. The hypothesis of an equality and an equal distribution of stars to which we have referred, is too far from being strictly true to be laid down as an unerring guide in this research. The stars of the first and second class, when scrupulously examined, evidently prove that if we would be accurate, we must admit them, in some degree at least, to be either of different sizes, or placed at different distances. Both varieties undoubtedly take place. This consideration alone is fully sufficient to shew, that how much truth soever there may be in the hypothesis of an equal distribution and equality of stars, when considered in a general view, it can be of no service in a case where great accuracy is required.

Since therefore it appears that in the classification of stars into magnitudes, there either is no natural standard at all, or at least none that can be satisfactory; it follows, that astronomers who have classed them thus, have referred their size or lustre to some imaginary idea of brightness. The great number of stars, indeed, which have been placed into every particular class, may assist us to form a kind of confused type in our minds, by which we may be enabled to arrange others; but how doubtful this must ever remain, we may see from the circumstance of the intermediate expressions that have been introduced.

1·2 m * for instance, denotes that a star so marked is between the first and second magnitude. 2·1 m signifies the same thing, with an intimation that the star so distinguished is nearly of the second magnitude, but partakes still something of the lustre of a star of the first order. With stars of the first, second, and third classes there may be some necessity to introduce such subdivisions; but how very vague must be the expressions 5m, 5·6m, 6·5m, 6m! In vain have I endeavoured to find a criterion for a star of any one of these magnitudes. On looking over, for instance, the stars of the fifth order, I found that in the list of other stars which ought to be less bright because they were marked 5·6m, 6·5m, or 6m, there were many that exceeded the former in brightness, while among those that are put down 5·4m, 4·5m, or even 4m, which ought to be more bright, I found several of a lustre not equal to some of this fifth magnitude, which I was desirous to ascertain. I may therefore justly call the method that has been hitherto in use to point out the lustre of stars, a reference to an imaginary standard.

The inconvenience arising from this unknown, or at least ill ascertained type

* I use the letter m in a short way to express the magnitude of the stars.

to which we are to refer, is such, that now our most careful observations labour under the greatest disadvantage. If any dependence could be placed upon the method of magnitudes, it would follow, that no less than eleven stars in the constellation of the Lion, namely, $\beta\, \sigma\, \pi\, \xi$ A *b c d* 54 48 72, had all undergone a change in their lustre since FLAMSTEED's time. For if the idea of magnitudes had been a clear one, our author, who marked β 1·2m, and γ 2m, ought to be understood to mean that β is larger than γ; but we now find that actually γ is larger than β. Every one of the eleven stars I have pointed out may be reduced to the same contradiction; and as the subject is of some consequence, I shall give a few other instances of them.

σ by FLAMSTEED is 4·5m, $\iota\, \phi\, \upsilon\, \lambda\, \kappa\, \pi\, \xi$ are all marked 4m, and therefore ought to be larger; but σ is larger than any of them.

π is marked 4m; *d* 6·5m, χ and *e* 4·5m, *c* and 72 5m; therefore π should be larger than all the former; but it is less.

ξ is marked 4m; but there are twelve stars, namely, σ *b* 54 A *d* χ *e c* 72 27 48 69, all marked in various manners less than that star, yet they all exceed it in magnitude.

Not to proceed any farther with particulars, we ought to account for this by allowing that FLAMSTEED did not compare the stars to each other, but referred each of them separately to its own imaginary standard of magnitude. This is the real source of all such contradictions, which therefore cannot be charged to our author. As we should, however, take it for granted, that the magnitudes were affixed to the stars with as much care as the nature of an unsettled standard would allow, a short inquiry into the extent of the confidence we may place upon the method of magnitudes will be of considerable use.

We have observed that in this method the brightness of stars is referred to unsettled standards; but admitting that a pretty general though coarse idea may be formed of these magnitudes, it may be granted that a mistake of a whole order in the first class cannot be supposed. The difference between a star of the first and second magnitude is so palpable that it excludes all suspicion of taking one for the other.

When subdivisions are introduced, the case becomes doubtful. 1·2m may easily pass for 2·1m. But though these two notations should not be sufficiently clear to be distinguished from each other, yet I am inclined to believe that the former may be precise enough to point out a difference from 2m, and the latter from 2·3m.

With the next order of stars the difference is much less striking; but yet 2m will convey an idea which may be pretty well distinguished from 3m. 2·3m, however, cannot be sufficiently kept apart from 3·2m, or either of these expressions from 3m, or from 2m. Perhaps the former may be distinguished from 3·4, and the latter from 4m.

The following step from 3m to 4m, or indeed from 3·4m to 4·5m, is less decisive than from 2 to 3m.

Again, if a star had changed from 4m to 5m, or from 4·5m to 5·6m since FLAMSTEED's time, we could hardly entertain more than a very slight suspicion of the alteration. From 4 to 5·6m, or from 4·5 to 6m, would be a pretty considerable step, and might serve as a foundation for an argument.

A change from 5m to 6m is such as no stress could be laid upon; and such are the changes from 5·6 to 6·7m, and from 6 to 7m. In all these inferior orders less than an alteration of a magnitude and an half could hardly deserve attention.

Here we have supposed all references to be made to the same author; for when other astronomers are consulted the uncertainty is much increased. A star which in FLAMSTEED's catalogue stands 1·2m, may be found 2m in another author: 2m in the former may be rated 2·3m, or even 3m by the latter. Of course 3m and 4m may be written for the magnitude of the same star by different persons. 4 and 5m as well as 5 and 6m are frequently interchanged, and no stress can be laid upon such nominal differences in different catalogues. We can hardly allow less than half a magnitude in the higher orders, and a whole one in the inferior classes, for this uncertainty.

To apply what has been said: suppose there should be some inducement to believe a certain star, such as β Leonis, to have changed its lustre. Now having no real, existing type of comparison, we can only refer to the general, imaginary one; and here the rules we have laid down will be of considerable service. The magnitude of this star given by FLAMSTEED is 1·2m; but as we have shewn that there is some ground to admit that 1·2m, even in this coarse way of reference, may be distinguished from what the same author seems to have taken for 2m, we conclude that the star has probably lost some of its former brightness. Again, he gives β 1·2m, and γ 2m. This notation may be taken to imply, though indirectly, that β is larger than γ; which not being the case, we have an additional reason to suspect a change. DE LA CAILLE puts down β 2m. Now the difference between the notation 1·2m of FLAMSTEED and 2m of the latter author can add nothing to the force of the argument for a change, as we have observed before, that a considerable allowance must be made for nominal varieties in different authors. Nor can we draw any support from the magnitude itself, because the star will pass very well for one of that order, when compared with other stars which are marked 2m by the same author. But when DE LA CAILLE marks β 2m, and γ 3m, we may then conclude that he estimated β to be larger than γ, though we do not know that he compared these two stars together; because a whole magnitude in the second class, as we have said, cannot well be mistaken, coarse as is the type to which the reference is made. Upon the whole, therefore, we conclude that β Leonis is now less brilliant than it was formerly.

In this manner, with proper circumspection, we may get at some certainty, even by the method of magnitudes; the imperfection of it, however, in other cases is very obvious. σ Leonis, for instance, being marked by FLAMSTEED 4·5m, the star itself will in every respect pass for one of that magnitude, when compared to a mental standard taken from other stars of the same author. Nor can its being brighter than stars which have a magnitude of a superior lustre affixed to them, do more than raise a considerable suspicion of a change. But as this subject will occur again hereafter, and as it must be sufficiently apparent that the present method of expressing the brightness of the stars is very defective, we now proceed to propose a different one.

I place each star, instead of giving its magnitude, into a short series, constructed upon the order of brightness of the nearest proper stars. For instance, to express the lustre of D, I say CDE. By this short notation, instead of referring the star D to an imaginary uncertain standard, I refer it to a precise, and determined, existing one. C is a star that has a greater lustre than D; and E is another of less brghtness than D. Both C and E are neighbouring stars, chosen in such a manner that I may see them at the same time with D, and therefore may be able to compare them properly. The lustre of C is in the same manner ascertained by BCD; that of B by ABC; and also the brightness of E by DEF; and that of F by EFG.

That this is the most natural, as well as the most effectual way to express the brightness of a star, and by that means to detect any change that may happen in its lustre; will appear, when we consider what is requisite to ascertain such a change. We can certainly not wish for a more decisive evidence, than to be assured, by actual inspection, that a certain star is now no longer more or less bright than such other stars to which it has been formerly compared; provided we are at the same time assured that those other stars remain still in their former unaltered lustre. But if the star D will no longer stand in its former order CDE, it must have undergone a change; and if that order is now to be expressed by CED, the star has lost some part of its lustre; if on the contrary, it ought now to be denoted by DCE, its brightness must have had some addition. Then, if we should doubt the stability of C and E, we have recourse to the orders BCD, and DEF, which express their lustre; or even to ABC, and EFG, which continue the series both ways. Now having before us the series BCDEF, or if necessary even the more extended one ABCDEFG, it will be impossible to mistake a change of brightness in D, when every member of the series is found in its proper order, except D.

Here I have used the letters of the alphabet merely to explain my way of fixing the order of brightness of the stars. In the journal or catalogue itself, which gives this order of brightness, each star must bear its own proper name, or number. For instance, the brightness of the star δ Leonis may be expressed by $\beta\,\delta\,\epsilon$ Leonis, or better by 94 – 68 – 17 Leonis; these being the numbers which the three above stars bear in the British catalogue of fixed stars.

Perhaps it may be thought that the known introduction of letters, added to the magnitudes of the stars, seems to be that very method which I now recommend, as different from what has already been used. And certainly if letters had been annexed to stars with a strict view to their order of brightness, they would now be of considerable service; but the intention of the astronomers who lettered the stars seems only to have been to give them a name, whereby to call them more readily, than by the descriptive method of pointing out their situation. It was indeed natural enough to give the name α to the brightest star, on account of its being the most remarkable in a constellation; and we may admit that with a few of the most conspicuous stars the letters $\alpha\beta\gamma$ would present themselves in succession; but whoever compares all the letters of the Greek and English alphabet that have been used, with the numerical magnitudes annexed to the same stars, will immediately give up all thoughts of intended order. In the constellation of Andromeda, which happens to lie before me, I find the following arrangement: $\delta o \mu \epsilon$, $\theta \pi \xi$, $\lambda \upsilon \upsilon \lambda$,* and *d b c*. In that of Hercules $\epsilon\delta$, $\xi\lambda\kappa$, $\pi\theta$, $\rho\mu$, $\sigma\nu$, τo, and *h* A *e b k q c m* Z.

It will be needless to point out the irregularities which take place in every other constellation; they go indeed so far, that it would be wrong to call them irregularities, because certainly no order could be intended in the arrangement of the letters. A doubt has even arisen whether any succession of brightness might be argued from the very first, second, or third letters of the alphabet; and when we find them arranged thus: $\beta\alpha$ Cassiopeæ, $\beta\alpha$ Cancri, $\gamma\beta$ Aquilæ, $\beta\zeta$ Canis minoris, 41 γ Arietis, we can hardly think it safe to regard the order of letters as of the least consequence. To which may be added, that in many constellations $\alpha\beta\gamma$ are all marked to be of the same magnitude, in which case again the order of the letters can bring no information. And therefore, even in those cases where the order of the letters agrees with the different magnitudes assigned to them, the knowledge we can have of the former state of the heavens must be derived from the magnitudes, and cannot be from the letters.

It may in the next place be remarked, that if not the letters, at least the numerical magnitudes affixed to the stars by astronomers, point out an order of brightness; and therefore contain my method already established. A succession of the marks 1, 2, 3, 4, 5, &c. and other intermediate notations, which are to be found in the British and other catalogues, give us a long list of stars that are (or should be) in a regular order of brightness, from a star of the first magnitude down to one of the eighth or ninth.

That these marks, denoting the magnitudes of the stars, are of some use every astronomer will readily perceive; but if we would apply them to the purpose of detecting a change in the lustre of some suspected star, the defect of this method will easily appear, and has already been shewn in the instance of σ Leonis. It was hinted before that the subject would recur again, I shall therefore mention

* [Flamsteed gives the letter λ to two stars, 16 and 52 Andromedæ, and υ both to 50 and to 51 Andromedæ.—ED.]

two other instances, in the first of which the common notation is sufficiently expressive. It will be so in all cases where a very considerable change takes place. Thus, β Persei being marked 2·3m, and ρ of the same constellation 4m, there could be no doubt of a change in the light of Algol when it was found to be not brighter than ρ. But let us in the next place take an observation recorded in my journal.

"May 12, 1782. β Lyræ is much less than γ."

Now, examining the British catalogue, we find β 3m, and γ 3m. Had the method of orders been adopted by FLAMSTEED, we should at once have pronounced this star to be changeable. For it would have been $\beta\gamma$ in his time, and $\gamma\beta$ at the time of observation; but since we have shewn that no inference can be drawn from the order of the letters, we have only the magnitudes to refer to. And here again the deviation of β from its usual brightness not being so considerable, but that a star such as it appeared to be at the time of observation might pass for one of the third magnitude, we are left in the dark; notwithstanding which a few years after, this star was actually found to be not only changeable, but periodical.*

M. DE LA LANDE in mentioning the change of δ Ursæ majoris arranges the seven bright stars of that constellation as they appeared to him; and remarks that sometimes γ and ϵ should stand before β, and sometimes after it. Here we have something like an order of seven remarkable stars; but as it happens, the stars themselves are not favourable to the formation of a regular series. Mr. PIGOTT and Mr. GOODERICKE also compared the stars whose changes they were examining to other neighbouring stars that were proper to be estimated with them, and were in a manner forced to lay aside the method of magnitudes.† These instances contribute to support the arguments I have used, to shew that another method of ascertaining the lustre of the stars is required, while at the same time they sufficiently indicate that the comparative brightness of stars is the only safe one to which we can have recourse.

It will be necessary now to enter into a full display of my proposed method; for simple as it is in its principle, it is not only difficult but very laborious in its progress. I began to put it into execution about 14 years ago; but other very interesting astronomical pursuits have broken in upon the regular continuation of it. By relating the difficulties or inconveniences as they happened, it will appear that my present notation, as well as method of arranging the observations, are liable to the fewest objections.

The general disposition of the stars is in constellations. This order is to be preferred to that of right ascension, or polar distance, because the stars being to be compared to the nearest proper stars that can be found, the constellations themselves will generally answer that purpose better than other selections.

* *Phil. Trans.* Vol. LXXVI. Part I. page 197 [PIGOTT].

† *Phil. Trans.* Vol. LXXV. Part I. page 127 and 154.

My first design was to draw each whole constellation into one series. Accordingly I began July 16, 1781, to arrange the stars in Ophiuchus thus:

"Order of the stars in Ophiuchus; $\alpha\beta\delta\zeta\eta\kappa\gamma\epsilon$."

This way of placing the stars agrees so far with my present one, that any star, such as κ for instance, may be taken, and the expression of its lustre will be had by $\eta\kappa\gamma$. And as FLAMSTEED marks the magnitudes of these stars 3m 4m 3m, my arrangement does not agree with his. If we should now suspect κ to have changed its lustre, recourse may be had to another star on both sides, which gives $\zeta\eta\kappa\gamma\epsilon$. The magnitudes of FLAMSTEED are 3m 3m 4m 3m 3·4m, where κ again seems to be placed in a situation to which it is not intitled.

A defect of this arrangement, which was not immediately perceived, is that in taking the stars of a constellation we have not always a proper connection of the steps of the series that may be formed of them: there being too much difference in the lustre of some of the stars, and too little in others.

Other inconveniences will also arise from the multiplicity of the members of a general series, and the trouble of arranging them when they are nearly equal. To get over these difficulties I marked the stars that differed much in lustre by magnitudes or degrees of difference; in which I assumed three different sorts of each; namely, 1′ 1″ 1‴ 2′ 2″ 2‴, &c. For instance,

"May 12, 1783. Order of the stars in Bootes;
"$\alpha 1'$ $\epsilon 2''$ $\eta 2'''$ $\gamma\beta\delta 3'$ $\rho 3''$ $\zeta 3'''$ $\pi 4$."

That this is not recurring to the old method of magnitudes, will appear when we consider that the stars are strictly compared. The series $\alpha\epsilon\eta\gamma\beta\delta\rho\zeta\pi$ remains established, but the difference in the gradation of brightness between the members of the series is added to it. At first this seemed to answer the intended purpose; for $\alpha\epsilon\eta$ not being sufficiently distinguished, the addition 1′ to α, and 2″ to ϵ, shewed that α was very much brighter than ϵ, while 2‴ added to η denoted only a very small difference between this and ϵ. The difficulty which immediately after arose in the choice of the magnitudes, however, soon convinced me that the fallacy of them would still have some influence upon the arrangements. The same evening I marked the stars in Leo thus:

"Order of the stars in Leo;
"$\alpha 1'''$ $\gamma 2'$ $\beta 2'$ $\delta\epsilon$ $\zeta\theta\eta$ μo $\rho\nu\sigma$."

Here I parcelled them together in the order of brightness, but could not find a convenient way to denote the different degrees by using any derivation from magnitudes; therefore I contented myself with placing those close together that agreed nearly with each other, and kept a little distance between those that differed rather more. This might perhaps have answered the required end, if the confusion which would arise from the distance of letters had not proved a great objection.

And that it would unavoidably bring on mistakes we may see by the other constellations which were arranged that evening.

"Draco $\gamma\eta\beta$ $\delta\zeta\iota$ $\theta\lambda\alpha\kappa$ ξ
"Cygnus α $\gamma\epsilon$ $\beta\delta\zeta$ θ
"Hercules $\beta\zeta\alpha$ $\delta\eta\pi$ $\gamma\epsilon\mu$ r* changed."

August 16, 1783, being upon the same subject of assigning comparative magnitudes, I introduced lines to shew the intended distances of the letters, with a view to prevent mistakes that might be made in transcribing them, and expressed the order as follows:

"Order of the stars in Auriga;
"α —— $\gamma\beta$ —— $\iota\theta$ —— $\epsilon\eta\zeta$ —— $\upsilon\pi\tau$"

The marks denoted that all the stars were in succession, but that the distance between those which are separated by lines was greater than that between the rest. When stars occurred that were nearly equal, I placed them under each other, thus:

"Order of the stars in Ursa Minor, $\alpha\beta$ —— γ —— $\begin{matrix}\delta\text{"} \\ \epsilon \\ \zeta\end{matrix}$

But in this expression there is the inconvenience of its breaking in upon the lines above and below.

Another cause of disorder arose from the stars which are not lettered. For here we are obliged to use numbers in lieu of them; and these, unless properly separated, will run into one another, and occasion mistakes.

In the next place, the letters themselves became troublesome; for a star cannot be found so readily in a catalogue or in an atlas by a letter, as it may be by a number.

The inconveniences attending the above different ways of notation having now been sufficiently pointed out, it remains only to lay down the method upon which, after many trials, I have fixed, in order to avoid them.

Setting aside the letters entirely I use only numbers in all my observations, and these numbers are such as I have added with red ink both to the edition of 1725 of the British catalogue, and to the *Atlas Cœlestis* taken from that catalogue, and printed in 1729. When I use other stars than what are contained in the British catalogue, the authors who have given them, and their numbers in the catalogues from whence they are taken, are particularly mentioned.

In the choice of the stars which are to express the lustre of any particular

* I called it r changed, because this star, which in my edition of 1725 is marked 3 m, is only of the 5th magnitude. At that time I ascribed the difference to a change in the star; but I have since found that there is an error in the edition of 1725 which is not in that of 1712, where the star is marked as it ought to be, of the 5th magnitude. [5 Herculis, in Flamsteed's MS. obs. of the 6th mag.—Ed.]

one, my first view is directed to a perfect equality. When two stars are perfectly alike in brightness, so that by looking often and a long while at them, I either cannot tell which is the brightest, or occasionally think one the largest, and sometimes, not long after, give the preference to the other, I put down their numbers together, only separated by a point. For instance, 30 . 24 Leonis. However, it can happen but very seldom that the equality in the lustre of two neighbouring stars is so perfect as not to leave an inclination to prefer one to the other; therefore I place that first which may probably be the largest, even though I do not particularly judge it to be so. But this preference is never to be understood to extend so far as to make it improper to change the order of the two stars; and the expression 24 . 30 Leonis will be equally good with the former. When a third star is concerned, such as 30 . 24 . 77 Leonis, the order of them ought not to be changed; notwithstanding an equality between each member of the series has been strictly ascertained. The reason of this is obvious. For by the order in which they are placed, it appears that 30 has been deemed equal to 24, and 24 equal to 77; but it is not affirmed that 30 has been compared to 77. There will be a great probability that these two last stars do not differ sensibly or materially; but since actual comparison is what we are to go by, the order in which the stars are given must remain.

When two stars are so nearly alike in their lustre that they may be almost called equal, and even now and then leave us doubtful to which to give the preference, but when upon a longer inspection of them we always return to decide it in favour of the same, I separate the numbers that denote these stars by a comma. For instance, 41 , 94 Leonis. This expression can certainly not be changed to 94 , 41 Leonis; much less can the order of three such stars, as 20 , 40 , 39 Libræ, admit of a different arrangement. If ever the state of the heavens should be such as to require a different order in these numbers, we need not hesitate a moment to declare a change in the brightness of one or more of the stars that are contained in the series to have taken place.

When two stars differ but very little in brightness, but so that even a doubt cannot arise to which the preference ought to be given, I separate the numbers by which they are to be found in the catalogue by a short line. For instance, 17 – 70 Leonis; or 68 – 17 – 70 Leonis. If, in the former instance, a breaking in upon the order is to be looked upon as a proof that at least one of the stars has undergone a change in its lustre, much more must that change be evident in this case, where the stars are separated by lines instead of commas.

When two stars differ so much in brightness that one or two other stars might be put between them, and still leave sufficient room for distinction, they become partly unfit for standards by which the lustre of other stars can be ascertained. But as proper intermediate stars sometimes cannot conveniently be had, we are often obliged to retain them; and in that case I distinguish them by a line and comma –, or by two lines, as 32 – – 41 Leonis. A difference which exceeds those that are expressed by the above marks, I denote by a broken line, thus – – –, for instance 16 – – – 29 Bootis. It would be very easy to give a more extensive signification to lines by adding cross marks to them, such as, + ++ +++ ++++ &c.; but in estimations that are to ascertain the brightness of stars, such expressions would rather throw us back again to look for imaginary differences, resembling those which have been rejected in the old system of magnitudes. On the contrary, the marks I have introduced admit of so precise a definition, that they cannot possibly be mistaken: a point denoting equality of lustre: a comma indicating the least perceptible difference: a short line to mark a decided but small superiority: a line and comma, or double line, to express a considerable and striking excess of brightness; and a broken line to mark any other superiority which is to be looked upon as of no use in estimations that are intended for the purpose of detecting changes.

In a foregoing paragraph we have said that this method of ascertaining the lustre of the stars was difficult and laborious. The difficulty consists in avoiding the various causes of error that may bias our judgment in assigning the comparative brightness of the stars: the different altitudes at which we view them: the state and situation of the moon: the time of the night with regard to twilight: the uncertainty of flying clouds: the twinkling and continual change of star-light, to whatever cause it may be owing; I mean such changes as last but few moments, or at most but a few minutes: a return into the dark after having been writing by candle-light: the zodiacal light: aurora borealis: and dew or damp upon the glasses or specula when a telescope is used. All these, it must be confessed, are real difficulties, which it requires much attention and perseverance to get the better of.

That the method is also laborious may be easily conceived; for each star must at least have two other stars to be compared with, and even these will often be found not to be sufficient. To look out for such proper objects, and then to make the necessary comparisons for every star in the heavens, can be no easy task, especially when we remember the difficulties I have enumerated, to which every single estimation of comparative brightness is subject. This ought, however, not to discourage us from a work which has in view the investigation of a point of great importance; and as I have already made a considerable progress, I shall give the result of my labour in small catalogues, of which I have joined one at the end of this paper.

That these investigations are of the importance we have ascribed to them, will appear when we call to our remembrance the great number of alterations of stars that we are certain have happened within the last two centuries, and the much greater number that we have reason to suspect to have taken place. If we consider how little attention has formerly been paid to this subject, and that most of the observations we have are of a very late date, it would perhaps not appear extraordinary were we to admit the number of alterations that have probably happened to different stars to be a hundred; this compared with the number of stars that have been examined, with a view to ascertain their changes, which we can hardly rate at three thousand, will give us a proportion of 1 to 30. But we are very certain that had a number of observers applied themselves to the same subject, which is of such a nature as to require the attentive scrutiny of many diligent persons at the same time, many more discoveries might probably have been made of changeable and periodical stars, whose variations are too small to strike a general observer. In the application we shall make of this subject however, a proportion, such as 1 to 30, or even 1 to 300, is sufficiently striking to draw our attention.

By observations such as this paper has been calculated to promote and facilitate, we are enabled to resolve a problem not only of great consequence, but in which we are all immediately concerned. Who, for instance, would not wish to know what degree of permanency we ought to ascribe to the lustre of our sun? Not only the stability of our climates, but the very existence of the whole animal and vegetable creation itself is involved in the question. Where can we hope to receive information upon this subject but from astronomical observations? If it be allowed to admit the similarity of stars with our sun as a point established, how necessary will it be to take notice of the fate of our neighbouring *suns*, in order to guess at that of our own! That *star* which among the multitude we have dignified by the name of *sun*, to-morrow may slowly begin to undergo a gradual decay of brightness, like β Leonis, α Ceti, α Draconis, δ Ursæ majoris, and many other diminishing stars that will be mentioned in my catalogues. It may suddenly increase, like the wonderful star in the back of Cassiopea's chair, and the no less remarkable one in the foot of Serpentarius; or gradually come on like β Geminorum, β Ceti, ζ Sagittarii, and many other increasing stars, for which I also refer to my catalogues. And lastly, it may turn into a periodical one of 25 days duration, as Algol is one of 3 days, δ Cephei of 5, β Lyræ of 6, η Antinoi of 7 days, and as many others are of various periods.

Now, if by a proper attention to this subject, and by frequently comparing the real state of the heavens with such catalogues of brightness as mine, it should be found that all, or many of the stars which we now have reason to suspect to be changeable, are indeed subject to an alteration in their lustre, it will much lessen the confidence we have hitherto placed upon the permanency of the equal emission of light of our sun. Many phænomena in natural history seem to point out some past changes in our climates. Perhaps the easiest way of accounting for them may be to surmise that our sun has been formerly sometimes more and sometimes less bright than it is at present. At all events, it will be highly presumptuous to lay any great stress upon the stability of the present order of things;

and many hitherto unaccountable varieties that happen in our seasons, such as a general severity or mildness of uncommon winters or burning summers, may possibly meet with an easy solution in the real inequality of the sun's rays.

A method of ascertaining the quantity or intenseness of solar light might be contrived by some photometer or instrument properly constructed, which ought probably to be placed upon some high and insulated mountain, where the influence of various causes that affect heat and cold, though not entirely removed, would be considerably lessened. Perhaps the thermometer alone might be sufficient. For though the lustre of the sun should be the chief object of this research, yet, as the effect of light in producing expansion in mercury seems to be intimately connected with the quantity of the incident solar rays, it may be admitted that all conclusions drawn from their action upon the thermometer will apply to the investigation of the brilliancy of the sun. And here the forms laid down by Mr MAYER, in his little treatise *De Variationibus Thermometri accuratius definiendis,** may be of considerable service to distinguish the regular causes of the change of the thermometer from the adventitious ones, among which I place the probable instability of the sun's lustre.

Introductory Remarks and Explanations of the Arrangement and Characters used in the following Catalogue.

This catalogue contains nine constellations, which are arranged in alphabetical order. I have called the present collection the first catalogue. The rest of the constellations, which are pretty far advanced, will be given in successive small catalogues as soon as time will permit to complete them.

Each page † is divided into four columns, the first of which gives the number of the stars in the British catalogue of Mr. FLAMSTEED, as they stand arranged in the edition of 1725.

The second column contains the letters which have been affixed to the stars.

The third column gives the magnitude assigned to the stars by FLAMSTEED in the British catalogue; and

The fourth contains my determination of the comparative brightness of each star, by a reference to proper standards.

All numbers used in the fourth column refer to the stars of the same constellation in which they occur, except when they are marked by the name of some other constellation; and in that case the alteration so introduced extends only to the single number which is marked, and which then refers to the constellation affixed to the number.

The numbers at the head of the notes, which will be found at the end of the catalogue, refer to the stars in the same constellation to which the notes belong.

* *Tobiæ Mayeri opera inedita*, I.

† [In this edition each page is subdivided by a vertical thick line.]

They contain particulars which it will be useful to know for those who wish to review that constellation.

To each star which I could not find in the heavens, and which, upon examining FLAMSTEED'S observations, appeared never to have been seen by him, I have put down " Does not exist." To such as I could not find in the heavens, but which nevertheless appeared to have been observed by FLAMSTEED, I have put down " Lost." This is to be understood only to mean that the star in question was not to be seen when I looked for it, but that possibly at some future time, if it be a changeable or periodical star, it may come to be visible again.

The observations in the notes, distinguished by marks of quotation, " " are taken from my own journals.

Errors in FLAMSTEED'S catalogue, or in the *Atlas Cœlestis*, are pointed out at the end of the constellations in which they occur, that they may be corrected.

Simple Characters.

' The least perceptible difference less bright.
. Equality.
, The least perceptible difference more bright.
– A very small difference more bright.
–, A small difference more bright.
– – A considerable difference more bright.
– – – Any great difference more bright in general.

Compound Characters, expressing the wavering of Star-light.

' From the least perceptible difference less bright to equality.
; From equality to the least perceptible difference more bright.
–, From a very small difference more bright, to the least perceptible difference.
=, From –, to – &c.
; The wavering expressed by the passing of the light from a state of the least perceptible difference less bright to equality, and to the least perceptible difference more bright.
–; The wavering expressed by the changes from – to , and to . or from . to , and to –

General Characters.

= Perfect equality.
< Less, but undetermined.
> Larger, but undetermined.

All the observations contained in this catalogue have been made in very fine nights, where no suspicion of any whitish haziness or thin clouds can be admitted that might have deceived me.

The compound expressions which occur in the catalogue are not such as have arisen from want of attention, but on the contrary from more than common and long inspection.

Whoever looks a long while at two stars which are equal, A and B for instance, will find that he is not always pleased with the expression A . B, but would incline rather to put them down A , B when A seems to have the preference, or A ‘ B when the advantage is on the side of B. Since, therefore, these three expressions A ‘ B A . B A , B seem equally to belong to the stars, my compound character A ;‘ B is in that instance an useful one, which includes them all. This may seem to be a doubtful expression, but it is in fact a very positive one, amounting to A = B. For had the stars not been perfectly equal, the same causes which bring on these little waverings in the appearance of stars, whatever they are, would have operated so as perhaps to produce the comparative wavering lustres expressed by A ; B and A ∸ B or A ‘∸ B which denotes the union of the three expressions A . B and A , B and A – B. But if this had been the case, we could certainly not admit A = B.

Sometimes, when I was not willing to put down these compound marks, I have cast my eyes upon the ground, and after a few moments lifted them quickly up to the stars A B, and instantly decided which of the expressions ought to be used : this being repeated perhaps a dozen or more times, I took that expression for the most proper one which would occur oftener than any other in these transitory glances.

All observations upon stars of any considerable magnitude have been made with the naked eye. I was unwilling to introduce the fallacies, or at least the difficulties that occur in the use of a telescope, owing to various causes that need not be mentioned, where I could possibly do without it. In numberless instances, however, the telescope has been recurred to, notwithstanding the stars under examination were not so small but that I saw them very well with the naked eye ; for in very fine nights, and in high situations, all the stars of the sixth, and most of the seventh magnitude, are sufficiently visible. But when small stars were situated very near each other, or very near brighter ones, it became necessary to remove the objection arising from the light of one star either overpowering or blending with that of the other.

Care has been taken in observations with the naked eye not to fix upon a star as a standard which has another close to it ; for the united light of the two stars would certainly cause deceptions. And stars that stand in this predicament of course have been referred to others with the assistance of a telescope.

The largest stars, and in general all such as had no convenient stars in the same constellation to be compared with them, have their lustre ascertained by such as I could find in the neighbouring part of the heavens.

Whenever I use the expression of magnitude, which though not of so nice and critical distinction as would be required for the purpose of my catalogue, is still a very useful one for general purposes, I have endeavoured to conform my mental standard to the notation of FLAMSTEED.

The most remarkable expressions of brightness which are contradictory to FLAMSTEED'S magnitudes, are pointed out in the notes annexed to the constellations. They are pretty numerous, and with many stars so considerable, that we have great reason to suspect changes in their lustre since FLAMSTEED'S time. It is to be noticed, that in collating my observations of brightness with FLAMSTEED'S magnitudes, I have not only taken those which are in the British catalogue, but also those that are to be found in the *Observationes Fixarum.* The very extraordinary disagreement between the former and the latter ought not to pass unnoticed. Were it not for what FLAMSTEED says in his *Prolegomena,* when he mentions the arrangement of the catalogue, " Undecima columna indicat cujus magnitudinis stellam esse arbitratus sum quando eam observatam habui," I should entirely reject the magnitudes of the catalogue as being without authority to support them. Nor can I conceive how such a remarkable disagreement could escape the author's notice, or remain unperceived by astronomers till this time, if the lustre of the stars in general had not been looked upon as a thing of no material consequence.

To shew what the difference is to which I allude, let us cast an eye upon the 9 constellations which are contained in the following catalogue of brightness.

In Aquarius there are 108 stars. To 49 of these no magnitudes can be found in FLAMSTEED'S observations; of 38 the magnitudes annexed to them agree with those of the catalogue; and of 21 they disagree with them.

In Aquila there are 71 stars. 39 are not observed; 16 agree; 16 disagree.

In Capricornus are 51 stars. 22 not observed; 17 agree; 12 disagree.

In Cygnus are 81 stars. 47 not observed; 21 agree; 13 disagree.

In Delphinus are 18 stars. 11 are not observed; 3 agree, and 4 disagree.

In Equuleus are 10 stars. 5 are not observed; 3 agree, and 2 disagree.

In Hercules are 113 stars. 10 are not observed; 54 agree, and 49 disagree.

In Pegasus are 89 stars. 22 are not observed; 37 agree, and 30 disagree.

In Sagitta are 18 stars. 3 are not observed; 13 agree, and 2 disagree.

To this may be added, that the disagreement in several stars is so considerable as to amount to two magnitudes; in many to one and an half, and in still more to one magnitude: not only with stars of a small size, but with some of the brightest in the constellation. I do not include α Cygni, which is marked 2m in the catalogue, and in the observation 7m, as that must certainly be a mistake; but cannot help regretting that a work to which every astronomer has been taught to look up as the first authority, should have been sent to the press with so many errors, that we hardly know how far to give our confidence to what is laid down in it.

I. *Catalogue of the comparative Brightness of the Stars.*

Lustre of the stars in Aquarius.

1		6	70 Aquilæ , 1	46	ρ	5 . 6	46 , 30
2	ε	5 . 4	2 - 23 2 - - 6	47		5 . 6	47 , 41 59 , 47 . 68
3		5	3 - 5	48	γ	3	62 . 48 - 52
4		6	5 , 4	49		5	41 - 49
5		6	3 - 5 , 4	50		6	45 . 50 , 56
6	μ	4 . 5	13 , 6 - 7 6 , 18 2 - - 6 - 7	51		6	63 , 51 , 44
7		6	6 - 7 - - 8 18 , 7	52	π	5	48 - 52
8		6 . 7	7 - - 8 , 9	53		6	42 , 53
9		6	8 , 9	54		6	58 . 54
10		6	11 , 10	55	ζ	4	76 . 55 , 62
11		6	12 , 11 , 10	56		6	50 , 56 , 61
12		6	12 , 11	57	σ	5	33 , 57 43 , 57
13	ν	5	23 . 13 , 6	58		6	58 . 54 74 - 58 - 64
14		6	17 - 14	59	υ	5	66 -, 59 , 47
15		6	21 . 15 , 16	60		6	30 - 60
16		6	15 , 16 , 20	61		6	56 , 61 40 , 61
17		6	19 , 17 - 14	62	η	4	55 , 62 . 48
18		6	6 , 18 , 7	63	κ	5	63 , 51
19		6	19 , 17	64		6	58 - 64 . 65
20		6	16 , 20	65		6	64 . 65 . 75
21		6	21 . 15	66	g^1	6	66 -, 59
22	β	3	34 ¦ 22 , 49 Capricorni	67		6	67 . 78
23	ξ	6	2 - 23 . 13	68	g^2	6	47 . 68
24		6	26 - 24	69	τ^1	5	7 -, 69 69 -, 77
25	d	6	25 . 27	70		6	70 . 74
26		6	27 , 26 - 24	71	τ^2	6 . 5	73 , 71 , 43 71 -, 69
27		6	25 . 27 , 26	72		6	Does not exist.
28		6	32 , 28 28 , 30	73	λ	4	73 , 33 73 , 71 73 , 88
29		6	35 , 29	74		6	70 . 74 - 58
30		6	46 , 30 30 - 60 28 , 30 30 - - 36	75		7	65 . 75
31	ο	5	31 - 32	76	δ	3	76 . 55
32		6	31 - 32 , 28	77		6	69 -, 77
33	ι	4	73 , 33 , 57 33 . 23 Capricorni	78		6	67 . 78 81 , 78 80 - 78 . 80
34	α	3	34 . 88 Pegasi 34 ¦ 22	79		2 . 1	8 Pegasi , 79 , 44 Pegasi
35		6 . 5	41 , 35 , 29	80		7	80 - 78 . 80
36		6	30 - - 36 37 , 36	81		7	82 , 81 , 78
37		6	45 . 37 37 , 36	82		7	82 , 81
38	e	6	38 , 42	83	h^1	6	83 , 92
39		6	42 . 39 , 45	84	h^2	7	87 - 84
40		7 . 8	45 . 40 40 , 61	85	h^3	6	92 - - 85 , 87
41		6	47 , 41 , 35 41 - 49	86	c^1	6	18 Piscis aust . 86 99 , 86 . 89
42		7	38 , 42 , 45 42 . 39 42 , 53	87	h^4	6	85 , 87 - 84
43	θ	4	71 , 43 , 57	88	c^2	4	88 - 18 Piscis aust 73 , 88 -, 98
44		6	51 , 44	89	c^3	5 . 6	86 . 89 . 101 89 , 104
45		6	42 , 45 39 , 45 . 40 45 . 37 45 . 50	90	φ	5	93 . 90 - 92 91 , 90 . 93
				91	ψ^1	5	91 , 90
				92	χ	6	90 - 92 , 96 83 , 92 - - 85

Lustre of the stars in Aquarius.

93	ψ^2	5	93 . 90 90 . 93	101	b^4	5	89 . 101 – 100
94		6	94 , 95	102	ω^1	5	105 , 102
95	ψ^3	5	94 , 95 , 97	103	A^1	5	104 . 103 . 106
96		6 . 7	92 , 96	104	A^2		89 , 104 . 103
97		6	95 , 97	105	ω^2	5	105 , 102
98	b^1	5	88 -, 98 , 99	106	A^3	5	103 . 106 . 107
99	b^2	5	98 , 99 , 86	107	A^4	6	106 . 107 . 108
100	b^3	5	101 – 100	108	A^5	6	107 . 108

Lustre of the stars in Aquila.

1	*m*	4	16 , 1 , 12	38	μ	4	41 – 38 . 44 38 , 32 38 -, 59
2	*o*	5	2 . 3				67 ; 38
3	*n*	5	2 . 3 , 9	39	κ	3 . 4	39 , 26 39 , 37
4		5	9 , 4 , 5	40		6	Does not exist.
5		6	4 , 5	41	ι	3 . 4	55 – 41 – 38 41 – 32 41 -, 71
6	*l*	4	6 . 12 6 . 63 Serpentis				41 , 55
7		} 6	8 , 7	42		6	36 , 42 , 45 62 , 42 . 66 58 , 42
8			8 , 7	43		6	Does not exist.
9	*k*	5 . 4	3 , 9 12 , 9 , 14 9 , 4	44	σ	5	38 . 44 32 , 44 59 , 44 . 54
10		6	11 , 10	45		6	42 , 45 36 – 45
11		6	18 -, 11 , 10	46		6	61 , 46 47 , 46 , 48
12	*i*	4	1 , 12 6 . 12 , 9 12 . 63 Ser-	47	χ	6	47 . 52 47 , 46
			pentis	48	ψ	6	46 , 48
13	ϵ	3 . 4	13 – 18	49	υ	6	63 . 49
14	*g*	6	9 , 14 , 15	50	γ	3	53 – – – 50 , 17 50 , 34 Sagit-
15	*h*	6	14 , 15				tarii
16	λ	3	16 , 30 16 , 65 16 , 1	51		5	37 , 51 56 . 51
17	ζ	3	50 , 17 – 65	52	π	6	47 . 52 . 61
18		6	13 – 18 , 11 18 , 19 18 – 11	53	α	1 . 2	53 – – – 50 21 Scorpii – – 53
19		6	18 , 19 , 22 21 ; 19				– – 50 Cygni
20		5 . 6	26 . 20 37 . 20	54	*o*	6 . 5	44 . 54 54 , 63 59 – 54
21		5	21 ; 19 23 . 21	55	η	3 . 4	30 , 55 . 60 55 – 41 41 , 55
22		6	19 , 22	56		5	57 – – 56 . 51
23		7	23 . 24 23 . 21 23 – 24	57		6	57 – – 56
24		7	23 . 24 23 – 24	58		6	62 , 58 , 42 58 – 66
25	ω^1	6	28 . 25 25 -, 29	59	ξ	5	38 -, 59 , 44 59 – 54
26	*f*	6	39 , 26 . 20	60	β	3 . 4	30 , 60 . 55
27	*d*	6	32 – 27 27 – 35	61	ϕ	6	52 . 61 , 46
28	A	6	31 , 28 . 25	62		6	62 , 58 66 . 62 , 64
29	ω^2	7	25 -, 29	63	τ	6	54 , 63 . 49
30	δ	3	65 , 30 , 55 16 , 30 65 ; 30	64		6	66 , 64 62 , 64
31	*b*	6	31 , 28	65	θ	3	17 – 65 , 30 16 , 65 65 ; 30
32	ν	5	41 – 32 – 27 38 , 32 , 44	66		5 . 6	42 . 66 66 , 64 66 . 62 58 – 66
33		6	Does not exist.	67	ρ	5	67 ; 38
34		6	Does not exist.	68		6	69 , 68 69 – 68
35	*c*	6	27 – 35	69		5	70 , 69 , 68 70 , 69 – 68
36	*e*	6	36 , 42 36 – 45	70		5	71 – 70 , 69 70 , 1 Aquarii
37	*k*	6	37 . 20 39 , 37 , 51	71		4	71 – 70 41 -, 71

Lustre of the stars in Capricornus.

1		6	2 , 1 . 3
2	ξ	6	2 , 1
3		6	1 . 3
4		6	7 , 4
5	α^1	4	6 - 5 -, 8
6	α^2	3	6 - 5
7	σ	obs.	10 , 7 , 12 15 . 7 , 4
8	ν	6	5 -, 8 , 11
9	β	3	49 - 9 - - 40
10	π	obs.	11 , 10 , 12 10 , 7 10 , 15
11	ρ	6	8 , 11 , 10
12	ο	obs.	10 , 12 7 , 12 15 , 12
13		6	15 - 13 14 = -, 13
14	τ	6	14 , 15 14 = -, 13
15	ν	6	10 , 15 , 12 15 . 7 14 , 15 - 13
16	ψ	5	16 , 18
17		6	24 -, 17
18	ω	6	16 , 18 ; 24
19		6 . 7	23 - - 19 , 21
20		6 . 7	20 . 21 22 - 20 . 25
21		6	19 , 21 20 . 21
22	η	5	23 -, 22 - 20 22 - 24
23	θ	5	23 - - 19 23 -, 22 33 Aquarii . 23
24	A	6	18 ; 24 -, 17 22 - 24 , 25
25	χ^1	6	20 . 25 -, 26 24 , 25
26	χ^2	6	25 -, 26 . 27
27	χ^3	6	26 . 27
28	φ	6	36 - 28 , 33
29		6	32 - 29 - 30
30		6	32 - - 30 - 31 29 - 30
31		7	30 - 31
32	ι	5	32 - - 30 32 - 29
33		6	36 , 33 , 35 41 , 33 28 , 33 , 35 34 Sagittarii . 33
34	ζ	5	34 , 39 34 - - 36
35		6	33 , 35
36	b	6	39 - 36 , 43 36 , 33 34 - - 36 - 28
37		6	43 -, 37 , 38
38		6	37 , 38
39	ε	4	34 , 39 - 36
40	γ	4	9 - - 40
41		6	41 , 33
42	d^1	6	42 - 44 42 -, 44 42 -, 51 42 . 48 51 ; 42 -, 48
43	κ	5	36 , 43 -, 37
44	d^2	6	42 - 44 . 45 42 -, 44 , 45 44 ; 45
45		6	44 ; 45 44 , 45 44 . 45
46	c^1	6	46 -, 47
47	c^2	6	46 -, 47
48	λ	5	42 . 48 , 51 48 - - 50 42 -, 48
49	δ	3	49 - 9 22 Aquarii , 49
50		6	48 - - 50
51	μ	5	42 ; 51 48 , 51 51 ; 42

Lustre of the stars in Cygnus.

1	κ	4	1 . 10 1 -, 10 10 -; 1 10 . 1 1 - 32
2		5	12 ; 2 -, 9
3		6	8 Vulpeculæ . 3 3 Vulpeculæ - - 3
4		6	14 . 4 12 , 4 15 , 4 - 11 8 - 4 , 11
5			Does not exist. A small star near the place 9 - - 5
6	β	3 . 4	53 , 6 , 18 18 ; 6 53 -, 6 6 -, 14 Lyræ
7		6	16 , 7
8		6	17 . 8 8 ; 15 8 , 12 8 , 14 8 - 4 21 - 8 - 17
9			12 -, 9 12 - 9
10	ι	6	1 . 10 10 . 1 10 -; 1 65 -; 10
11		6	4 , 11
12	φ	5	12 ; 2 8 , 12 , 4 12 - 9
13	θ	4	32 , 13 33 , 13 , 20 33 , 13 , 23
14		6	8 , 14 . 4
15		6	15 , 22 8 ; 15 , 4 15 , 25
16	c^1	6	16 , 7
17	χ	5	8 - 17 21 - - 17 . 8 21 = -, 17
18	δ	3 . 4	6 , 18 , 64 53 , 18 53 -, 18 - 64 53 , 18 , 64 8 ; 6 18 -, 64
19		6	22 , 19 25 , 19
20		5 . 6	24 . 20 . 26 13 , 20
21	η	6	21 . 41 21 - - 17 21 = -, 17 21 - 8
22		6	25 . 22 22 , 25 15 , 22 , 19
23		6	13 , 23
24	ψ	5	24 . 20

Lustre of the stars in Cygnus.

No.	Letter	Mag.	Order of lustre
25		6	15 , 25 . 22 22 , 25 , 19
26	c^2	6	20 . 26
27	b^1	5	36 – 27
28	b^2	5	34 ; 28 , 36 28 . 35 28 . 34
29	b^3	6	29 . 34 34 . 29
30	o^1	4	32 – 30 32 -, 30
31	o^2	5	31 , 32
32		5 . 6	32 . 33 32 , 13 31 , 32 – 30 1 – 32 31 – 32 -, 30
33			32 . 33 , 13
34		6	29 . 34 ; 28 28 . 34 34 . 29 34 , 36 34 , 40
35	m	6	28 . 35 39 – 35
36		6	28 , 36 – 27 34 , 36
37	γ	3	37 , 53 37 – 53 37 -, 53 37 -; 8 Peg. 37 – 8 Peg. 37 ; 8 Pegasi 5 Cephei -, 37
38			Does not exist, or is lost.
39	h	6	41 – 39 – 35 47 – 39 – 35
40		6	34 , 40 40 , 42
41	i	4	41 , 52 21 . 41 41 – 39
42		6	40 , 42 , 44
43	w^1	5	46 , 43
44		6	42 , 44
45	w^2	5	45 , 46
46	w^3	5	45 , 46 , 43
47	l	6	47 – 39
48		6	49 – 48 ; 48 49 -, 48 ; 48 49 – – 48 , 48
49		6	49 – 48 ; 48 49 -, 48 ; 48 49 – – 48 , 48
50	α	2	50 – – – 37 53 Aquilæ -; 50 53 Aquilæ – – 50
51		6	56 , 51
52	k	6	41 , 52
53	ε	3	37 , 53 , 6 37 – 53 , 18 37 – 53 -, 18 37 -, 53 -; 6
54	λ	4	67 . 54
55		6	56 . 55 – 59 55 . 63 63 . 55 , 59
56		6	57 , 56 . 55 57 , 56 , 51
57		6	57 , 56
58	ν	4	62 , 58 . 67
59	f^1	5 . 6	55 – 59 , 60 63 , 59 55 , 59 59 , 68
60		6	59 , 60
61		6	70 . 61 , 69
62	ξ	4	65 , 62 , 58
63	f^2	6	55 . 63 , 59 63 . 55 68 , 63
64	ζ	3	18 , 64 18 -; 64
65	τ	4	65 , 62 65 – 66 65 -; 10
66	υ	5	66 , 78 65 – 66 , 67
67	σ	4	58 . 67 . 54 67 ; 78 66 , 67
68	A	6	59 , 68 , 63
69		6	69 . 70 70 – 69 79 , 69 61 , 69
70		6	69 . 70 72 , 70 – 69 70 . 61
71	g	6	80 . 71
72		6	72 , 70 74 , 72
73	ρ	4	73 – 81
74		6	74 , 72 74 – 77 75 . 74 , 72
75		6	75 . 74
76		6	77 , 76
77		6	74 – 77 , 76
78		3 . 4	66 , 78 67 ; 78 – 14 Pegasi
79		6	79 , 69
80	π^1	4	81 – 80 . 71
81	π^2	5	73 – 81 – 80

Lustre of the stars in Delphinus.

No.	Letter	Mag.	Order of lustre
1		6	8 , 1 , 10
2	ε	3	12 , 2 , 11
3	η	6	5 . 3 , 8
4	ζ	5	11 – 4 . 7
5	ι	6	7 , 5 . 3
6	β	3	6 , 9
7	κ	6	4 . 7 , 5
8	θ	6	3 , 8 , 1
9	α	3	6 , 9 . 12
10		6	1 , 10
11	δ	3 . 4	2 , 11 – 4
12	γ	3	9 . 12 , 2
13		5	13 – 14 1 Equulei – 13
14		6	13 – 14
15		6	18 , 15
16		6	17 , 16 , 18
17		6	17 , 16
18		6	16 , 18 , 15

Lustre of the stars in Equuleus.

1		5	10 , 1 – 13 Delphini	6		6	5 – 6
2		6	4 , 2	7	δ	4	5 . 7 , 10
3		6	3 . 4	8	α	4	8 – – 5
4		6	3 . 4 , 2	9		6	10 – 9
5	γ	4	8 – – 5 . 7 5 – 6	10	β	4	7 , 10 , 1 10 – 9

Lustre of the stars in Hercules.

1	χ	6	1 –, 4 1 , 30 30 . 1	36	} m	6	37 – – 36 38 – 36
2		6	4 , 2 –, 14	37			37 – – 36 37 – 38 37 , 45
3		5	3 – 9	38		6	37 – 38 41 , 38 – 36
4		6	1 –, 4 , 2	39		5	39 – 50
5	r	5	7 – – 5 – 16	40	ξ	3	40 – 27 27 ⸵ 40 27 ; 40 27 , 40 27 -̦ 40 37 Serpentis – 40 –, 64
6	υ	6	35 , 6 – – 14 35 – 6 , 30 6 , 52 11 , 6 11 -̦ 6				
7	κ	5	7 – – 5	41		6	45 – – 41 47 , 41 , 38 41 , 33
8	q	5 . 6	16 , 8	42		5	52 , 42 , 30
9		6	3 – 9 , 43 Serpentis	43	i	5 . 6	45 , 43 , 47
10		5	10 – 17 10 – 19	44	η	3	65 , 44 , 22 44 , 86 44 – 22 65 – 44 – 58
11	φ	6	11 . 35 – 6 11 – – – 14 35 . 11 , 6 11 -̦ 6 22 , 11				
12		6	21 – – 12 – 15	45	e	5	45 – – 41 37 , 45 , 43
13		5 . 6	15 . 13	46		7	48 – 46
14		7	6 – – 14 11 – – – 14 2 –, 14	47	k	5	43 , 47 , 41
15		6	12 – 15 . 13	48		6	50 – 48 , 32 48 – 46
16		6	5 – 16 , 8	49		6	60 – – 49
17		6	10 – 17 , 18 19 , 17 , 18	50		5	39 – 50 – 48 53 – – 50
18		7	17 , 18	51		5	53 . 51 – – 56
19		6	10 – 19 , 17	52		5 . 6	6 , 52 , 30 52 , 42
20	γ	3	64 –, 20 22 , 20 . 58 22 – 20	53		5	53 – – 50 53 . 51
21	ο	6	21 – – 12 28 – 21	54		5 {	One of these two does not exist.
22	τ	4	44 , 22 , 20 22 , 85 44 – 22 – 20 22 , 11	55			60 , 54 or 55
				56		6	51 – – 56 . 57
				57		6	56 . 57
23		5	20 Coronæ – 23 23 – 26	58	ε	3	20 . 58 , 35 58 , 91 44 – 58 103 – 58 – 76
24	w	6	24 -̦ 29 24 , 60 24 – 29 24 –, 29	59	d	6	68 , 59 , 61 59 , 25
25		5	30 – 25 30 –, 25 59 , 25	60		6	24 , 6 ; 29 60 – – 49 60 , 66 60 , 54 or 55
26		7 . 6	23 – 26 26 – 31				
27	β	3	27 –, 64 40 – 27 27 ; 40 27 – – 64 27 ⸵ 40 27 , 40	61	c	6	59 , 61
				62		6	71 – – 62
28		6	28 – 21	63		6	72 . 63 63 , 78
29	h	4	24 -̦ 29 60 ; 29 24 – 29 24 –, 29	64	α	3	64 – 27 Ophiu 27 Ophiu – 64 64 , 65 65 –, 64 27 –, 64 –, 20 64 ⸵ 67
30	g	5	1 , 30 30 – 25 6 , 30 52 , 30 42 , 30 , 34 30 . 1 30 –, 25				
31		7	26 – 31	65	δ	4	65 – 64 67 , 65 64 , 65 – 44
32		6	48 , 32	66	ω	6	60 , 66 , 37 Ophiuchi
33		6	41 , 33	67	π	3 . 4	67 , 65 67 , 27 Ophiuchi
34		6	30 , 34	68	u	5	70 . 68 , 72 68 , 90 68 , 59 69 – 68
35	σ	4	58 , 35 85 , 35 11 , 35 , 6 35 . 11	69	e	4 . 5	76 – 69 69 – 68 94 , 69 – 99

Lustre of the stars in Hercules.

70		4	70 - - 62 70 . 68 70 - 73	93		5	78 , 93
71		5	Does not exist.	94	ν	5	75 , 94 , 69 103 - 94 . 100 109 , 94
72	w	6	68 , 72 . 63 90 , 72	95		4	102 , 95 , 101 95 . 98
73		6	70 - 73	96		5	101 , 96 . 97 101 . 96 - 98
74		6	77 , 74 . 88	97		5.6	96 . 97 98 , 97
75	ρ	4	75 , 76 91 , 75 75 , 94	98		5	95 . 98 96 - 98 , 97
76	λ	4.5	75 , 76 - 69 58 - 76	99	b	5	69 - 99 - 104 106 . 99 100 , 99 107 , 99 . 108
77	κ	6	82 . 77 , 74	100	i	6	94 . 100 , 99
78		6.7	63 , 78 78 , 93	101		5	95 , 101 . 96 95 , 101 : 96
79		6	79 - 83 89 , 79	102		4.5	102 , 95
80		4	Does not exist.	103	o	4	109 . 103 92 . 103 - 94 92 ; 103 - 58
81		4	Does not exist.	104	A	4.5	99 - 104
82	γ	6	82 . 77	105		5	106 - 105
83		7	79 - 83 . 84	106		5.6	106 . 99
84		7	83 . 84	107	t	6	107 , 99
85	ι	4	22 , 85 , 35 14 Lyræ , 85	108		6	99 . 108
86	μ	4	44 , 86 , 92 86 - 92	109		4	92 . 109 . 103 92 , 109 , 94 109 , 111
87		6	87 - 89	110		4.5	111 , 110 , 113
88	Z	6	74 . 88	111		4	109 , 111 , 110
89		6	87 - 89 , 79	112		5	113 , 112
90	f	6	68 , 90 , 72	113		5	110 , 113 , 112
91	θ	4	58 , 91 , 75 92 . 91				
92	ξ	4	86 , 92 . 103 103 . 92 . 91 92 . 109 92 , 109 86 - 92 ; 103				

Lustre of the stars in Pegasus.

1	e	4	24 - 1 , 10 1 . 9	19		6	18 . 19 19 , 18 22 - 19
2	f	4.5	10 - 2 , 16 9 - 2 . 13 2 - 13	20		6	21 , 20
3		6	3 , 4	21		5	17 . 21 , 20 17 , 21 , 20
4		6	3 , 4 . 7	22	ν	5	46 , 22 - 18 22 , 35 22 - 19 31 , 22
5		6.7	13 - 5 12 , 5 13 , 5	23		6	38 , 23
6		6	16 , 6 - 11	24	ι	4	24 - 1
7		6	4 . 7	25		6.7	25 - 28
8	ε	3	8 - 54 37 Cygni - 8 21 Andr ; 8 , 54 8 , 53 8 ; 16 Ceti 8 , 24 Pisc austr	26	θ	4	42 , 26 - 46
				27		5	29 - 27
9	g	4.5	1 . 9 - 2	28		6.7	25 - 28
10	κ	4	1 , 10 - 2	29	π	4.5	29 - 27
11		6	6 - 11	30		6	35 , 30 31 , 30 - 36 31 =' 30
12		6	16 , 12 , 5	31		4.5	31 , 30 31 =' 30 31 , 22 50 . 31 , 49
13		6	2 . 13 - 5 13 , 17 13 , 5 2 - 13	32		6	43 , 32 , 38 43 . 32
14		6	16 , 14 , 15 78 Cygni - 14	33		6.7	39 . 33
15		6	14 , 15	34		6	37 , 34
16		6	2 , 16 16 , 6 16 , 12 16 , 4	35		6	22 , 35 , 37 35 , 30
17		6	13 , 17 . 21 17 , 21	36		6.7	30 - 36
18		5	22 - 18 . 19 19 , 18	37		6	35 , 37 , 34

Lustre of the stars in Pegasus.

38		6	32 , 38 , 23
39		6 . 7	41 , 39 , 45 39 . 33
40		6	40 , 41
41		6 . 7	40 , 41 , 39
42	ζ	3	48 , 42 , 26
43	o	5	47 – 43 , 32 56 , 43 , 32
44	η	3	53 ,̄ 44 ,̄ 48 53 , 44 , 88 54 , 44 ; 88 16 Ceti ⁝ 44
45		6 . 7	39 , 45
46	ξ	5	26 – 46 , 22 46 – 50
47	λ	4	48 , 47 – 43 47 , 68 47 –, 68 48 , 47 – 68
48	μ	4	44 ,̄ 48 , 47 88 – – 48 , 42
49	σ	6	50 , 49 , 52 31 , 49
50	ρ	6	46 – 50 , 49 50 . 31
51		6	56 – 51 , 60
52		6	49 , 52
53	β	2	54 , 53 ,̄ 44 54 . 53 , 44 8 , 53 . 54
54	α	2	8 – 54 54 , 53 8 , 54 . 53 53 . 54 , 44
55	l	5	55 , 59
56		5 . 6	62 – 56 – 51 56 , 64 56 . 72 56 , 71 62 – 56 56 . 43 78 ; 56
57	m	6	58 , 57
58	n	6	59 , 58 , 57
59	p	6 . 5	55 , 59 , 58
60		6	51 , 60 , 61
61		6	60 , 61
62	τ	6	68 , 62 – 56 62 , 78 84 , 62 62 – 56
63		6	67 – 63 . 73
64		6	56 , 64 . 67
65		6	69 , 65
66		6	70 – 66 , 86
67		6 . 7	64 . 67 – 63
68	υ	6	47 , 68 , 62 68 , 70 47 –, 68 47 – 68
69		6	71 , 69 , 65
70	q	5 . 6	68 , 70 – 66 70 , 82
71	y	6	56 , 71 , 69 71 , 85
72		6	56 . 72
73		6	63 . 73
74		7	75 , 74 , 76 75 ,̄ 74 , 76
75	S	6	81 , 75 , 74 75 ,̄ 74
76		6	74 , 76
77		6	82 , 77 – 80
78		5 . 6	62 , 78 – 79 62 , 78 ; 56
79		6	78 – 79
80		6	77 – 80
81	φ	6	89 , 81 , 75
82		6	70 , 82 , 77
83	r	6	87 , 83 23 Piscium – 83
84	ψ	6	84 . 62 84 , 89
85		6	71 , 85 87 . 85
86		5 . 6	66 , 86
87	u	6	87 . 85 87 , 83
88	γ	2	44 , 88 44 ; 88 – – 48 34 Aquarii . 88 88 ; 6 Arietis
89	χ	6	84 , 89 , 81

Lustre of the stars in Sagitta.

1		6	1 Vulpeculæ – 1 2 Vulpeculæ , 1
2		6	2 , 3
3		6	2 , 3
4	ε	5	6 –, 4
5	α	5	5 . 6 7 –, 5 . 6 5 – 9 Vulpeculæ
6	β	4	5 . 6 –, 4 5 . 6 , 8
7	δ	4 . 5	7 –, 5 12 – 7
8	ζ	6	6 , 8 – 9 8 . 16 9 Vulpec , 8 12 Vulp , 8
9		6	8 – 9
10		6	11 , 10 . 15
11		6	11 , 10
12	γ	4	12 – 7
13	χ	6	15 . 13
14	y	6	14 , 15
15	Z	6	14 , 15 . 13 10 . 15
16	η	6	8 . 16 –, 17
17	θ	6	16 –, 17 – 18
18		6	17 – 18

*Notes to Aquarius.**

2 "August 2, 1788. 20-feet reflector 2 (ε) 4·3m FL. 5·4m." The difference amounts to one whole magnitude. In FLAMSTEED's observations no magnitude is mentioned.

6 Is less than 13, and very little brighter than 18. The former is contrary to the catalogue, and the latter inconsistent with the magnitude assigned to 18. None of these stars have any magnitude in FLAMSTEED's observations.

8 Is larger than 9, contrary to the catalogue. In the observations 8 is 6m, but 9 has no magnitude.

13 Is less than 23, and is larger than 6; both are contrary to the catalogue. There are no magnitudes of either of these stars in FLAMSTEED's observations.

23 Is larger than 13, contrary to the catalogue, and from the expression 2–23 (see 2) it appears that 23 is undervalued by FLAMSTEED, or has changed its lustre. FLAMSTEED's observations give no magnitude of 23.

34 Is equal to 88 Pegasi, which the catalogue has 2m. See 88 Pegasi.

35 Is less than 41, contrary to the catalogue, "Oct. 13, 1786, 5·6m." There is no magnitude of 35 in FLAMSTEED's observations.

40 Is larger than 61, contrary to the catalogue. This is a considerable deviation, amounting to 1½m. In the observations 40 is 7m, 61 6m.

41 Is larger than 49 and 35, contrary to the catalogue. It is also contrary to the observations. "Oct. 13, 1786, 41 6·5m."

42 Is larger than 45, 39 and 53, contrary to the catalogue. The observations give 6m to 53.

43 Is less than 71. See 71. There is no magnitude to either of these stars in FLAMSTEED's observations.

48 Is less than 62, contrary to the catalogue; and is now probably less bright than it was formerly. 48 being but little brighter than 52 confirms the same. There is no observation of 48, but 62 is 5m.

59 Is less bright than 66, contrary to the catalogue. The observations give 59 6m.

71 Is brighter than 69, contrary to the catalogue. These stars are so near each other that a change must be evident, unless FLAMSTEED should have made a mistake in writing down their magnitudes. 71, 43 confirms the same conclusion. In the observations neither 69 nor 71 has a magnitude assigned.

72 There is no observation of FLAMSTEED upon this star.

78 Is less than 81, contrary to the catalogue; and in the observations it is 5m.

79 Is less than 8 Pegasi, contrary to the catalogue. The difference between 2·1m and 3m would be striking, if the lowness of the situation of 79 did not render its real magnitude very uncertain. In my estimation no allowance is made for that low situation. In the observations there is no magnitude to either of these stars.

80 There are two stars, the smallest of which agrees best with the place of 80 in Atlas, but neither of them seems to accord completely in relative situation with 81 and 82. In one of my sweeps a star, supposed to be 80, was taken with the following deduction; "Sept. 12, 1785. This star requires a correction of −1′ 13″ in time of R.A., and −6′ in P.D."

84 Is larger than 87, contrary to the catalogue. In the observations they are both 8m.

85 Is much less than 92, which does not agree with the magnitudes of the catalogue. In the observations it is marked 8m.

86 Is larger than 89, contrary to the catalogue. There is no magnitude to either of these stars in the observations.

88 "Oct. 13, 1786, 20-feet reflector, 4·3m." FLAMSTEED's observations give it 4m.

89 Is larger than 101 and 104, contrary to the catalogue. In the observation 104 is 6m.

94 Is larger than 95, contrary to the catalogue. In the observations they are both 5m.

96 "Sept. 12, 1785, 6m." In FLAMSTEED's observations it is also marked 6m.

Catalogue.

80 Requires −18′ in R.A., and −6′ in P.D.
The P.D. of 96 requires −8′.

* [For additional Notes see below, p. 575.—ED.]

Atlas.

The R.A. of 30 requires +1°.
72 must be out.
80 requires −18′ in R.A., and − 6′ in P.D.
84 requires −10′ in R.A., and −12′ in P.D.
85 requires −28′ in R.A.
96 requires − 8′ in P.D.

Notes to Aquila.

"July 23, 1781. Order of magnitude $\alpha\,\gamma\,\xi\,\theta\,\delta\,\eta\,\beta\,\epsilon$."

3 Is larger than 9, contrary to the catalogue. In FLAMSTEED's observations both are marked 6m.

6 and 12 are both larger than 63 Serpentis; but that star is placed among the changeable ones. See *Phil Trans.* Vol. LXXVI. page 211. FLAMSTEED's observations give 3m to 6.

13 "Sept. 3, 1784; 20-feet reflector, 13 (ϵ) 5·6 FL. 3·4m, but strong twilight." It is not much larger than either 11, 18, or 19, so that we may be pretty certain it must have lost some of its lustre since the time of FLAMSTEED. In his observations it is marked 4m.

20 Is less than 26 and 37, contrary to the catalogue. In the observations 20 is marked 5m.

21 Is less than 23, contrary to the catalogue. But in the observations 23 is marked 5m. The error therefore is probably in the catalogue.

24 The star I estimate is one of two small ones.

33 There is no observation of this star in FLAMSTEED's work.

34 This star was never observed by FLAMSTEED.

37 Is larger than 20 and 51, contrary to the catalogue. The latter is marked 6m in the observations.

38 Is less than 67, contrary to the catalogue.

39 Is not much larger than 26 and 37, which will not agree with 3·4m of the catalogue; but in FLAMSTEED's observations it is put down only 5m.

40 There is no observation of this star in FLAMSTEED's work.

43 There is no observation of this star in FLAMSTEED's work.

55 This star is periodical. The time of its period as given by Mr. PIGOTT, the discoverer, is $7^{d}\,4^{h}\,15'$. See *Phil. Trans.* Vol. LXXV. page 127.

56 Is much less than 57, contrary to the catalogue; but in the observations 56 is only marked 6m.

66 Is less than 42 and 58, contrary to the catalogue, and it is moreover marked 5m in the observations.

Atlas.

The R.A. of 23 requires −20′.
29 Should be 30′ from 25, and 48′ from 28, on the south following side of the two stars.
The stars 33, 34, 40, and 43 should be out.

Notes to Capricornus.

"Sept. 27, 1782. Order of magnitude $\delta\,\beta\,\alpha\,\gamma$."

6 in 1780 was less than $\beta\,\gamma\,\delta\,\zeta\,\eta\,\theta\,\iota$.

13 Is not equal to 14 as the catalogue gives it. In FLAMSTEED's observations 14 is without magnitude assigned to it.

19 and 20 are larger than 21, contrary to the catalogue. In the observations 19 is marked 6m.

34 Is larger than 39, contrary to the catalogue. Neither of them has any magnitude given with them in FLAMSTEED's observations.

36 Is larger than 43, contrary to the catalogue. It has either been under-rated, or gained additional lustre since FLAMSTEED's time. Neither of the stars has any magnitude in his observations.

42 Is larger than 48, contrary to the catalogue. The latter has no magnitude in the observations, and the former is marked once of the 5th and once of the 6th, which may be put down 5·6m.

Catalogue.

The letter *e* should be added to 26. FLAMSTEED has used it in his observations, page 75.

Atlas.

31 requires about +22′ in R.A.

Notes to Cygnus.

" May 12, 1783. Crder of magnitude $\alpha \quad \gamma \epsilon \beta \delta \quad \zeta \quad \theta$."

5 There is no observation of this star by FLAMSTEED. Page 67 a star was observed without time, but by page 71 and 122 it appears that the defective observation belongs to 2. There is a star 8 or 9m, about 50′ from 2, 1° 20′ from 9, and 1° 30′ from 6; and calling that star 5, its brightness may be expressed by 9 -- 5.

10 " Sept. 15, 1783. 10 is at least 4m. It is larger than 13." If the authority of the catalogue be good, there can be no doubt of a change since FLAMSTEED's time; but in his observations there is no magnitude to this star.

12 Is less than 8, contrary to the catalogue. " Sept. 7, 1784, 12 (ϕ) 6m." In FLAMSTEED's observations there is no magnitude to either of the stars.

13 Is less than 32, contrary to the catalogue. But in the observations 13 has no magnitude.

17 Is less than 21, contrary to the catalogue. But in the observations neither of the stars has any magnitude.

18 Is larger than 64, contrary to the catalogue. But in FLAMSTEED's observations neither of the stars has any magnitude.

21 Is larger than 41, contrary to the catalogue. But 21 is without magnitude in the observations.

23. The expression 13 , 23 does not agree with the catalogue. But 13 has no magnitude in the observations.

27 Is less than 36, contrary to the catalogue; but in FLAMSTEED's observations are no magnitudes of these stars.

30 Is less than 32, contrary to the catalogue; it is also contrary to the observations, which give 30 5m and 32 6·5m.

31 Is larger than 30, contrary to the catalogue. It is also contrary to the magnitudes given in the observations " Sept. 27, 1788; 20-feet reflector 30 (1st o) 5m, FL. 4m. 31 (2d o) 4m, FL. 5m."

34 Is a changeable star. Its period perhaps is about 18 years. See *Phil. Trans.* Vol. LXXVI. page 201.

38 In FLAMSTEED's observations, page 75, a star was taken without R.A., marked " quæ præcedit ω." The time of this observation however is sufficiently determined by the 37 before it, and 45 and 46 just after; but there is no star visible in the space pointed out that can possibly be taken for 38. " Sept. 22, 1783, 38 lost. There is not a star of the 7, 8, 9, or 10th magnitude near the place." It therefore does not exist, or rather is lost.

41 Is less than 21, and not much larger than 52, which is contrary to the catalogue. " It is less than 4m." In FLAMSTEED's observations it is marked 4m, but 21 and 52 are without magnitudes.

48. " Sept. 5, 1784; I could not see this star, but instead of it found in the neighbourhood 2 stars of the 7th magnitude within 5 or 6′ of each other." " Nov. 15, 1795. If one of the stars be 48, its magnitude is over-rated, and must be about 7·8m. That of the two which is nearest to 49 is the largest."

59 Is less than 55, 56 and 63, contrary to the catalogue. It is also contrary to the magnitudes given in the observations: 63 is without magnitude.

66 Is larger than 78, contrary to the catalogue. See 78. " Sept. 13, 1784; 20-feet reflector 66 (v) 4m, FL. 5m. It is larger than 54 (λ), contrary to the catalogue." Neither 66 nor 54 has any magnitude in FLAMSTEED's observations.

71 Is equal to 80, contrary to the catalogue. Neither of them has any magnitude given with them in the observations.

78 Is less than 66 and 67. " It is much too small for 3·4m." In FLAMSTEED's observations I find it marked 6m.

81 Is larger than 80, contrary to the catalogue. But in the observations there is no magnitude to either of the stars. " Sept. 27, 1788; 20-feet reflector, 81 ($2^d\ \pi$) 3·4m, FL. 5m." It is either undervalued in the catalogue, or grown brighter since FLAMSTEED's time.

P. The changeable star in the neck of the swan. Its period is 396 days 21 hours. See *Phil. Trans.* Vol. LXXVI. page 200. Its present lustre is 17 -- P.

Atlas.

14 requires $+1^\circ$ in P.D.
5 should be out.

Notes to Delphinus.

"Aug. 14, 1781. Order of magnitude $\beta\alpha\ \delta$ / $\gamma\ \epsilon$"

9 In the catalogue it is marked 3m; in FLAMSTEED'S observations it is 6m. My expression 6, 9 · 12 agrees best with the catalogue.

13 "Aug. 7, 1785, 6m." In FLAMSTEED'S observations it is also marked 6m.

Atlas.

12 Should be placed about 52′ more south on plate 23. It is right on plate 25.

Notes to Equuleus.

"Aug. 13, 1781. Order of magnitude α γ/δ β"

6 In the catalogue we have 4m; in the observations FLAMSTEED has once marked it 6m, and once 8m. If there be any accuracy in these various notations, the star must certainly be changeable.

Notes to Hercules

May 12, 1783. Order of magnitude $\beta\zeta\alpha\quad\delta\eta\pi\quad\gamma\epsilon\mu$."

5 Is much less than 7. My edition has this star 3m; that of 1712 has it 5m. FLAMSTEED'S observations give 6m, which agrees best with 7 -- 5 as I give its present lustre.

8 Is less than 16, contrary to the catalogue. But in the observations this star has been marked twice 7m, twice 6m, and once 5m.

11 Is larger than 6 and 35, contrary to the catalogue. But in the observations we have this star given twice 4m, and once 3m. It is therefore undervalued in the catalogue, or is subject to changes in its lustre.

13 Is less than 15, contrary to the catalogue. The observations give them both 6m. "May 25, 1795, 13 and 15 are both smaller than FL. gives them, and are about 7·8m."

20 Is less than 22, contrary to the catalogue. In the observations they are both 4m.

22 "May 12, 1787. 22 (τ) 3m, FL. 4m."

23 Is not much larger than 26, contrary to the catalogue. The observations give 23 6m, and 26 7m.

25 "May 16, 1787. 25 7·6m FL. 5m." In the observations this star is also 5m.

27 By my observations the light of this star seems to be subject to change. FLAMSTEED'S observations give it twice 3m, and once 2m.

29 Is less than 24 and 60, contrary to the catalogue. In the observations 24 is marked 6 and 5m; 60 is given 5m, 6m, and 4m; and 29 is put down five times 5m, once 6m, and once 4m. Very possibly this star may be changeable.

30 Is larger than 1 and 52, contrary to the catalogue. In the observations 30 is given three times 5m; 1 twice 5m, and twice 4m; and 52 twice 5m.

37 Is larger than 45, contrary to the catalogue. But in the observations we have 37 twice 6m, once 5m; and 45 twice 6m, and twice 5m.

40. From the expressions I have given of the brightness of this star, we have great reason to suppose it to be changeable. FLAMSTEED'S observations give it 3m.

47 Is less than 43, contrary to the catalogue. The observations, however, give 47 three times 6m, and only once 5m.

52 Is larger than 42, contrary to the catalogue. In the observations both are twice marked 5m.

54 or 55. FLAMSTEED observed but one of these stars, once 4m, once 5m, and once 6m.

58 Is less than 103, and not much larger than 76, contrary to the catalogue. It is also contrary to the magnitudes of the observations.

62 " Is less than it is marked. I suppose it to be 7 or 7 . 8m." FLAMSTEED's observations give it 6m.

64 From my expression of brightness it appears that this star is changeable, and I may venture to announce it periodical. A series of observations upon it will be given when the period of the changes shall have been more fully ascertained. FLAMSTEED has but one observation of its magnitude, which is 5m.

65 This star is probably changeable, but its connected reference to neighbouring changeable stars has hitherto rendered it difficult to come at the truth. In FLAMSTEED's observations it is three times 3m, and twice 4m.

67 This star is probably changeable. FLAMSTEED's observations give it twice 3m.

69 Is less than 94, contrary to the catalogue; but the magnitudes in the observations are favourable to my notation.

78 Is larger than 93, contrary to the catalogue. The latter has no magnitude in the observations, and the former is marked 6m.

95 Is less than 102, contrary to the catalogue. The observations give 95 twice 4m, and 102 once 4m, and once 5m.

99 Is less than 100, 106 and 107, contrary to the catalogue; and also to the magnitudes of the observations. It is larger than 104, which is doubly inconsistent with the catalogue, and yet the observations also give to 104 a larger magnitude.

105 Is less than 106, contrary to the catalogue. The observations give once 3m, and once 6m. " July 17, 1785; 20-feet reflector, 105 7·6m FL. 5m is visibly less than 106."

Catalogue.

In the edition of 1725, 5 (*r*) should be 5m.

Atlas.

The P.D. of 2 requires +34′.
The R.A. of 4 requires +16′ and the P.D. −34′.
The R.A. of 110, 111, 112 and 113 requires +5°.
55, 71, 80 and 81 should be out.

Notes to Pegasus.

2 The expressions 2 , 16 and 2 . 13 shew that this star is over-rated in the catalogue. In FLAMSTEED's observations it stands 6m.

8. Is larger than 53 and 54, contrary to the catalogue. In the observations are no magnitudes of these stars.

18 Is not sufficiently distinguished from 19 to agree with the magnitudes of the catalogue. In FLAMSTEED's observations 18 is marked once 5m, once 6m, and once 7m; and 19 is 7m.

20 " Oct. 19, 1784, 7m." In FLAMSTEED's observations it stands 6m.

21 Is less than 17, contrary to the catalogue. It is also contrary to the magnitudes given in the observations. If there be any accuracy in the magnitudes of the catalogue and of the observations, we ought to look upon this star as changeable; for the latter give it once 3m, and once 6m, while the former has 5m.

27 " Sept. 6, 1784, 6m." In FLAMSTEED's observations there is no magnitude of this star.

31 Is no larger than 50, contrary to the catalogue, " Sept. 5, 1784, 6m," and " Oct. 19, 1784, 5·6m." FLAMSTEED's observations give it twice 5m, and 50 also 5m.

32 " Sept. 8, 1784, 6 . 5m." In FLAMSTEED's observations it stands once 4m, and once 5m.

42 Is less than 48, contrary to the catalogue. But in the observations there is no magnitude to 48. " Sept. 19, 1784, 48 (μ) 4·3m."

43 Is less than 56, contrary to the catalogue. It is also contrary to the magnitudes in the observations, where 43 is 4m, 56 5m.

47 " Sept. 19, 1784, 4 . 3m." FLAMSTEED's observations give no magnitude.

62 Is larger than 56 and 78, contrary to the catalogue. In the observations 56 is 5m.

63 Is less than 67, contrary to the catalogue. The observations give no magnitude of these stars.

68 Is larger than 70, contrary to the catalogue. In the observations 68 is without magnitude, and 70 is 5m and 6m.

76 Is less than 74, contrary to the catalogue. In the observations both stars are marked 6m.

79 " Sept. 8, 1784, 6.7m." The observations give no magnitude.

86 Is less than 66, contrary to the catalogue, and contrary to the observations, where the former is marked 5m, the latter 5m, and twice 6m.

88 Is less than 44, contrary to the catalogue. There are no magnitudes of these stars in the observations. It is also less than 34 Aquarii, which FLAMSTEED has observed 3m, and hardly larger than 6 Arietis, which he has also observed 3m. Therefore, if the catalogue may be trusted where this star is 2m, it must have lost some of its former lustre. But I rather suppose that this star, as well as 53 and 54, have been overvalued in the catalogue.

Catalogue and Atlas.

The letter *t*, which FLAMSTEED has annexed to 31 in his observations page 57 and 130 should be added.

Notes to Sagitta.

" Sept. 7, 1781. Order of magnitude $\gamma\,\delta\,{\alpha \atop \beta}$"

7 Is larger than 5 and 6, contrary to the catalogue. By the order of magnitude, it appears that 14 years ago it was also larger. In FLAMSTEED's observations 5, 6 and 7 are marked 4m.

WM. HERSCHEL.

Slough, near Windsor,
Jan. 1, 1796.

XXXVII.

On the periodical Star α Herculis; with Remarks tending to establish the rotatory Motion of the Stars on their Axes. To which is added a Second Catalogue of the comparative Brightness of the Stars.

[*Phil. Trans.*, 1796, pp. 452–482.]

Read June 9, 1796.

In my first catalogue of the comparative brightness of the stars, I announced α Herculis as a periodical star. The precision of the characters introduced in that catalogue is such, that the smallest alteration in the lustre of the stars may be discovered, by a proper attention to their expressions: the variation in the light of α Herculis is, however, pretty considerable, and cannot easily be mistaken, when strictly compared to a proper standard. The star most conveniently situated for this purpose is κ Ophiuchi; and as I have had no reason, during the time of my observations, to doubt the uniformity of its lustre, I have made use of it in the following comparisons; which seem to be sufficiently decisive, with regard to the periodical variations of the light of α Herculis.

Other stars besides κ Ophiuchi have also been consulted; but the unsteadiness of their light would draw me into difficulties, which at present it will be proper to avoid. For this reason I only give the following table, which will be found to contain at least four regular changes of the alternate increase and decay of the apparent lustre of our new periodical star, deduced from a comparison of its brightness with that of κ Ophiuchi.

In this short table, the number 64 refers to the stars in the constellation of Hercules, and 27 to those in the constellation of Ophiuchus. The characters that are used to denote their relative lustre, have been explained among the introductory remarks to the first catalogue of the comparative brightness of the stars.

In order now from the table to obtain the time of the period, we shall first take all the successive observations from the 16th of September till the 28th of November. They shew very clearly that the star has completely gone through all its changes. For, admitting a maximum of the light of α Herculis to have been the 16th of September, we find a minimum the 25th of October; and a second maximum about the 28th of November. The period, therefore, is of somewhat more than two months' duration.

Table of the Variation of Light observed in α (Fl. 64) Herculis, compared to κ (Fl. 27) Ophiuchi.

Date	Comparison	Date	Comparison
1795. May 18	64=27	1795. November 20	64 ; 27
22	64=27	26	64 , 27
July 6	27 ; 64	28	64 -, 27
August 17	27 – 64	December 15	27 ; 64
18	27 – 64	1796. January 9	64 ; 27
September 16	64 – 27	March 5	64 ; 27
17	64 – 27	9	64 ⁝ 27
18	64 , 27	10	64 ⁝ 27
20	64 ; 27	28	27 , 64
21	64 -, 27	April 4	27 -, 64
22	64 , 27	5	27 -, 64
October 2	64 ; 27	9	27 =, 64
6	64 ; 27	May 2	64 -; 27
7	64 ; 27	4	64 -, 27
9	64 . 27	5	64 -, 27
11	64 ⁝ 27	10	64 =, 27
15	27 , 64	11	64 =, 27
16	27 , 64	14	64 =, 27
17	27 , 64	16	64 =, 27
25	27 -; 64	17	64 =, 27
26	27 ; 64	18	64 – 27
28	27 , 64	19	64 =, 27
30	27 ; 64	22	64 – 27
November 3	27 , 64	24	64 -, 27
6	64 . 27	25	64 -, 27
9	64 ⁝ 27	26	64 -, 27
10	64 . 27	27	64 , 27

But as changeable stars are subject to temporary inequalities, which will render a determination of the length of a period, from a single series of changes, liable to considerable errors, we shall now take the assistance of the most distant observations. By an inspection of the table, we find again the first maximum to have been about the 16th of September, 1795; and the fourth the 14th of May, 1796. This being an interval of 241 days, in which four successive changes have been gone through, we obtain about 60 days and a quarter for the duration of the period.

In confirmation of this computation, the table shews that our periodical star was very faint in August, 1795; bright about the middle of September;

faint towards the end of October; bright the latter part of November; faint in December; bright in January, 1796; not observed in February; bright in March; faint in April; and lastly, bright again in May. This is just what should have happened according to the above determination, which, as we have seen, gives a period of eight weeks, four days and a quarter. Greater accuracy can only be obtained by future observations.

On the Rotatory Motion of the Stars on their Axes.

I shall now add a few remarks on the subject of the rotation of the fixed stars on their axes. This motion has been lately mentioned in a paper, where I could not have an opportunity to enter into the reasons why it ought to be admitted.* The discovery of the period of α Herculis furnishes me with an opportunity to say a few words upon the subject, as every addition to the list of periodical stars increases our knowledge of the construction of the celestial bodies. Not so much because now one star more is known to be subject to periodical changes in its lustre; for this would indeed be of no great consequence. But we ought not to be satisfied with merely inrolling this circumstance among the list of facts we are acquainted with. The rotatory motion of stars upon their axes is a capital feature in their resemblance to the sun. It appears to me now, that we cannot refuse to admit such a motion, and that indeed it may be as evidently proved as the diurnal motion of the earth.

Dark spots, or large portions of the surface, less luminous than the rest, turned alternately in certain directions, either towards or from us, will account for all the phænomena of periodical changes in the lustre of the stars, so satisfactorily, that we certainly need not look out for any other cause. Let us, however, take a review of any objections that might be made.

The periods in the change of the lustre of Algol, β Lyræ, δ Cephei, and η Antinoi, are short; being only 3, 5, 6, and 7 days respectively: those of ο Ceti, the changeable star in Hydra, and that in the neck of the Swan are long, amounting to 331, 394, and 497 days. Will not a doubt arise whether the same cause can be admitted to explain indiscriminately phænomena that are so different in their duration?

To this it may be answered, that the whole force of the objection is founded upon our very limited acquaintance with the state of the heavens. Hitherto we have only had seven stars whose periodical changes have been determined. No wonder then that proper connections between their different periods were wanting. But let us now place α Herculis among the list, which is not less than 60 days in performing one return of its changes. Here we find immediately, that the step from the rotation of α Herculis to that of ο Ceti, is far less considerable than that from the period of Algol to the rotation of α Herculis; and thus a link in the chain is now supplied, which removes the objection that arose from the vacancy.

* *Phil. Trans.* for the year 1795, Part I. page 68. [Above, p. 482.]

There is, however, another instance of a slow rotatory motion; and it is doubly instructive upon this occasion. In one of my former papers it has been shewn, that the 5th satellite of Saturn revolves on its axis in 79 days; this not only shews that very slow rotatory motions take place among the celestial bodies; but from the arguments that were brought to prove its rotation, which I believe no astronomer will oppose, we are led to apply the same reasoning to similar appearances among the fixed stars. A variation of light, owing to the alternate exposition of a more or less bright hemisphere of this periodical satellite, plainly indicates that the similar phænomenon of a changeable star arises from the various lustre of the different parts of its surface, successively turned to us by its rotatory motion.

The rotations of the sun and moon, and of several of the planets, become visible in a telescope by means of the spots on their surfaces; the remote situation and smallness of the 5th satellite of Saturn leave us without this assistance; but what we can no longer perceive with our best optical instruments, we now supply by rational arguments. The change in the light of the satellite proves the rotation; and the rotation once admitted, proves the existence of spots, or less luminous regions on its surface, which at setting off were only hypothetical. In the same manner a still more extended similarity between the sun and the stars offers itself, by the spots that now must also be admitted to take place on their surfaces, as well as on that of the sun.

To return to the difficulty which has been started, it may be further urged, that there are some reasons to surmise that the 34 Cygni is a periodical star of 18 years return; * and that other stars seem very slowly to diminish their lustre, and may probably recover it hereafter.

In answer to this, I remark that it will not be necessary to remove objections to the rotatory motion of the stars, inferred from their very slowly changeable lustre, till they come properly supported by well ascertained facts. Many causes in the physical construction of the stars may occasion an accidental and gradual increase or decay of brightness, not subject to any regularity in its duration. But when settled periods can be ascertained, notwithstanding they should be of the most extended duration, it will not be difficult to find other causes to explain them, without giving up the rotatory motion. When the biography of the stars, if I may be allowed the expression, is arrived to such perfection as to present us with a complete relation of all the incidents that have happened to the most eminent of them, we may then possibly not only be still more assured of their rotatory motion, but also perceive that they have other movements, such as nutations or changes in the inclination of their axes; which, added to bodies much flattened by quick rotatory motions, or surrounded by rings like Saturn, will easily account for many new phænomena that may then offer themselves to our extended views.

* *Phil. Trans.* for the year 1786, Part I. page 201.

Memorandum relating to the following Catalogue.

It was my intention to have continued to remark all the deviations of brightness, here assigned to the stars, from the magnitudes which are given by FLAMSTEED, either in his observations, or in the British catalogue ; but I now find this author so little consistent, that it appears to be of no use to refer to his determinations. In the constellation of Aries are no less than 31 stars, in which the magnitudes of his catalogue differ from those we find in his observations ; and these are generally the stars of the greatest brightness. The difference is also very considerable : thus, 41 Arietis is observed 6 m, the catalogue gives 3 m ; 50 is observed 7 m, the catalogue gives 5 m ; 35 is 5 m, catalogue 4 m ; 33 is 6 m, catalogue 5 m ; 38 is 6 m, catalogue 7 m.

The notes to these and the remaining constellations, therefore, will now be confined chiefly to my own observations, and to the correction of errors that have fallen under my notice.

II. *Catalogue of the comparative Brightness of the Stars.*

Lustre of the stars in Aries.

1		7.6	Does not exist.	25		7	64 Ceti – 25
2		7.6	2 , 8 2 ; 8	26		6.7	29 , 26 , 27
3		6	4 , 3	27		6.7	26 , 27
4		7.6	8 – 4 , 3	28		6	Does not exist, or is lost
5	γ	4	41 . 5 5 – 3 Aurigæ 92 Ceti , 5 41 – 5 5 ⁝ 37 Cassiop 99 Pisc – 5, 110 Piscium	29		6.7	29 , 26
				30		7	35 , 30 – 33 16 Trianguli, 30
				31		5.6	38 – 31 , 24
6	β	3	6 – 4 Trianguli 88 Pegasi ; 6	32	ν	6	48 =, 32 ,̄ 34
7		tel.	14 – 7	33		5	30 – 33 35 , 33 33 –, 16 Trianguli
8	ι	6	17 . 8 . 22 2 ; 8 – 4				
9	λ	5	9 , 12 9 ,̄ 12 9 , 14	34	μ	6	32 ,̄ 34
10		6.7	14 , 10 – 21	35		4	39 , 35 , 33 35 , 30 35 . 57
11		6	20 , 11 . 16	36		7	53 ; 36 . 50
12	κ	6.5	9 , 12 9 ,̄ 12 12 – 17	37	ο	6	43 . 37
13	α	2	57 Andro ; 13 13 , 23 Aurigæ 43 Andromedæ ,̄ 13	38		7	38 – 31
				39		4	39 , 35
14		6	14 , 10 9 , 14 – 7	40		6	46 . 40 , 45
15		6	22 . 15 –, 18	41		3	41 . 5 41 – 5
16		8	11 . 16	42	π	6	42 ,̄ 46 42 , 43
17	η	6	17 . 8 12 – 17 , 22	43	σ	6	42 , 43 . 37
18		7	15 –, 18 . 23	44	ρ^1	6	54 , 44
19		6.7	20 , 19	45	ρ^2	6.7	40 , 45 , 53
20		6	21 . 20 , 11 20 , 19	46	ρ^3	6.7	42 ,̄ 46 . 40
21		7	10 – 21 . 20	47		6.7	63 . 47 , 65
22	θ^1	6.5	8 . 22 17 , 22 . 15	48	ε	5	58 . 48 . 63 48 , 57 48 =, 32
23	θ^2	7	18 . 23	49		7	52 ,̄ 49 . 51
24	ξ	6	31 , 24 65 Ceti 24, 64 Ceti	50		7	36 . 50 . 54

Lustre of the stars in Aries.

51		7	49 . 51	59		7	62 – 59 , 60
52		6	52 −, 49 52 , 56	60		7	59 , 60 . 64
53		7	45 , 53 ; 36	61	τ^1	7	58 , 61 , 63
54		6 . 7	50 . 54 , 44	62		6	62 – 59
55		7	56 . 55	63	τ^2	6	48 . 63 61 , 63 , 65 63 . 47
56		6 . 7	52 , 56 . 55	64		6	60 . 64 – 66
57	δ	4	35 . 57 . 58 48 , 57 , 58	65		7	63 , 65 47 , 65
58	ζ	5	57 . 58 . 48 57 , 58 , 61	66		7	64 – 66 7 Tauri . 66

Lustre of the stars in Canis major.

1	ζ	3	1 . 24	17	π^2	6	15 – 17 . 12
2	β	2	21 , 2 , 25	18	μ	4	20 , 18 . 11
3	λ	4	3 . 13 3 ; 8 Columbæ	19	π^3	6	19 . 15
4	ξ^1	5	4 , 5	20	ι	4	23 , 20 14 , 20 , 18
5	ξ^2	5	4 , 5	21	ϵ	3 . 2	21 , 2 21 – 2
6	ν^1	5	8 – 6	22		4	22 , 16
7	ν^2	5	7 , 8 7 , 23	23	γ	3	7 , 23 , 20
8	ν^3	5	7 , 8 – 6	24	o^2	5 . 4	1 . 24
9	α	1	9 – – – all 1 m	25	δ	2 . 3	2 , 25 , 31
10	κ^1	6	13 – 10	26		7	27 , 26
11		5	18 . 11	27		7	28 – 27 , 26
12	ρ	6	17 . 12	28		5	16 , 28 – 27 28 – 30
13	κ^2	5	3 . 13 – 10	29		5	30 . 29
14	θ	5	14 , 20	30		5	28 – 30 . 29
15	π^1	6	19 . 15 – 17	31	η	3 . 2	25 , 31 15 Navis , 31
16	o^1	5	22 , 16 , 28				

Lustre of the stars in Canis minor.

1		7 . 6	6 , 1 7 , 1 11 , 1	9	δ^3	6	7 , 9 . 8
2	ϵ	6	4 , 2 , 5	10	α	1 . 2	19 Orionis −, 10 58 Orionis . 10 19 Orionis –, 10
3	β	3	3 – – 4				
4	γ	6	3 – – 4 , 2 4 . 6	11	π	6	6 , 11 , 1
5	η	6	2 , 5	12		5 . 6	Does not exist.
6	o	6	4 . 6 , 1 6 , 11	13	ζ	5	13 Navis , 13 – 14 13 Navis – 13
7	δ^1	6	7 , 1 7 , 9	14		6	13 – 14
8	δ^2	6	9 . 8				

Lustre of the stars in Cassiopea.

1	e	6	3 , 1 – – 2	6		6	4 – 6 . 9
2		7	1 – – 2	7	ρ	6	7 . 3
3		6	7 . 3 , 1 7 . 3 . 8	8	σ	6	3 . 8 – 5
4	d	5	5 , 4 – 6	9		6	6 . 9 . 10
5	τ	5	8 – 5 , 4 5 – 12 5 , 14	10		6	9 . 10

Lustre of the stars in Cassiopea.

11	β	3.2	27, 11 27.11
12		6	5 − 12 − 13 14, 12
13		6	12 − 13.16
14	λ	5	17 − − 14 33, 14 5, 14, 12
15	κ	4	15, 33 -
16		6	13.16
17	ζ	4	24, 17 − − 14 24 − 17
18	α	3	18; 27 18.27
19	ξ	6	19 − 25
20	π	6	22, 20
21		6	21; 23
22	o	6	22, 20
23		6	21; 23
24	η	4	24, 17 24 − − 33 24 − 17
25	ν	5	19 − 25
26	υ^1	7	28; 26
27	γ	3	18; 27, 11 18.27.11
28	υ^2	6	28; 26
29		6	Does not exist.
30	μ	5	33 =; 30
31		6	36 − 31, 43 46, 31
32		6	32 − 35
33	θ	4	24 − − 33, 14 33 =; 30 15, 33
34		6	44 − 34
35		7	32 − 35
36	ψ	5.6	36 − 31 36, 46
37	δ	3	5 Arietis ¦ 37 − − 45
38		6	42 − 38 40, 38
39	χ	6	39, 44
40	h	6	42, 40, 38
41		6	Does not exist.
42	g	6	48 − 42 − 38 42, 40
43	c	6	31, 43
44		6	39, 44 − 34
45	ε	3	37 − − 45
46	d	6	36, 46, 31
47		5	47, 49
48	e	5	50 − 48 − − 54 48 − 42
49		6	47, 49
50	f	4.5	50 − 48
51		6	Does not exist, or is lost.
52		7	53 −; 52 53, 52 − 55
53		7	53 −; 52 53, 52
54		6	48 − − 54
55		6	52 − 55

Lustre of the stars in Cetus.

1		6	6, 1 9, 1
2	g	4.5	2, 7
3		6	17, 3, 19
4		6	4.5
5		6	4.5
6	f	5	7 − 6, 1
7	h	5	2, 7 − 6
8	ι	3	8.31
9		6	9, 1
10		6	12 − 10, 11 12, 10 44 Pis- cium − 10
11		6	10, 11
12		6	13, 12 − 10 13, 12.15 13 − 12, 10
13		6	13, 12 20, 13, 12 13 − 12
14		6	Does not exist.
15		6	12.15
16	β	3	8 Pegasi; 16 ¦ 44 Pegasi
17	ϕ^1	5	17, 19 17, 3
18		6	23, 18
19	ϕ^2	5	17, 19 − 22 3, 19
20		6	20, 13 20 − 26
21		6	27.21
22	ϕ^3	5	19 − 22.23
23	ϕ^4	5	22.23, 18
24		6	Does not exist.
25		6	25.37
26		6	20 − 26, 33
27		6	28, 27.30 27.21
28		6	37, 28, 27
29		6	35, 29
30		6	27.30, 32
31	η	3	8.31.45
32		6	30, 32.41 32.44
33		6	26, 33, 35
34		6	34, 38
35		6	33, 35, 29
36		6	36.41
37		5	25.37, 28
38		6	34, 38.39
39		6	38.39, 40
40		6	39, 40.42
41		6	32.41 36.41
42		6	40.42.43
43		6	42.43
44		6	32.44

Lustre of the stars in Cetus.

45	θ	3	31 . 45 45 , 52
46		5	50 , 46 48 . 46 53 − 46
47		6	47 , 49
48		6	48 . 46
49		6	47 , 49 . 50
50		6	49 . 50 , 46
51		6	110 Piscium − 51 , 98 Piscium
52	τ	3 . 4	45 , 52 . 55
53	χ	5	55 − − 53 − 46
54		6	64 , 54
55	ζ	3	52 . 55 − − 53
56	υ¹	4	59 − 56 , 57
57		5	56 , 57 57 . 74
58		6	58 , 66
59	υ²	4 . 5	59 − 56
60		6	60 − 61
61		7	60 − 61 , 63
62		6 . 7	63 − 62 71 . 62
63		6	66 , 63 − 62 61 , 63
64		6	65 − − 64 , 54 65 −, 64 24 Arietis , 64 − 25 Arietis
65	ξ¹	6	65 − − 64 73 . 65 −, 64 65 ,̄ 91 65 , 24 Arietis
66		6	66 , 63 58 , 66 , 63
67		6	80 , 67
68	ο	3 . 2	68 < 10 m 68 = 13 Arietis 68 > 13 Arietis 87 Tauri > 68
69		6	69 . 70
70		6	69 . 70 , 75
71		6	71 . 62 81 , 71 − 79
72	ρ	4	83 . 72
73	ξ²	5 . 4	87 , 73 . 65 73 − 91
74		6	57 . 74
75		5 . 6	70 , 75 , 84
76	σ	4	89 − 76 . 83
77		6	77 . 80
78	ν	4 . 5	96 , 78 , 97 88 , 78 − 85
79		6	71 − 79
80		6	77 . 80 , 67
81		6	81 , 71
82	δ	3	86 − 82
83	ε	3	90 , 83 76 . 83 . 72
84		6	75 , 84
85		6	78 − 85 88 −, 85
86	γ	3	86 − 82
87	μ	4	87 , 73
88		6	88 , 78 88 −, 85
89	π	4 . 3	89 , 90 89 − 76
90		8	89 , 90 , 83
91	λ	4	73 − 91 65 ,̄ 91
92	α	2	92 , 4 Arietis
93		6	96 − 93 , 97
94		6	96 , 94 − 95
95		6	94 − 95
96	κ¹	5	96 , 78 96 , 94 96 − 93
97	κ²	4	78 , 97 93 , 97

Lustre of the stars in Corvus.

1	α	4	7 − 1 − 8
2	ε	4	9 , 2 . 7
3		6	5 − 3 −, 6
4	γ	3	4 , 9
5	ζ	5	8 − 5 − 3
6		6	3 −, 6
7	δ	3	9 , 7 − 1 2 . 7 − 1
8	η	5	1 − 8 − 5
9	β	3	4 , 9 , 7 4 , 9 , 2

Lustre of the stars in Eridanus.

1	τ¹	4	11 , 1 − 2
2	τ²	4	1 − 2 , 4
3	η	3	18 . 3
4		6	2 , 4 , 6 15 ,̄ 4
5		6	5 −, 7
6		6	4 , 6
7		6	5 −, 7
8	ρ¹	6	10 . 8
9	ρ²	5	13 − 9 , 10 13 − 9 − 14
10	ρ³	4	9 , 10 . 8
11		3 . 4	16 , 11 , 1 16 , 11 . 27
12		3	19 , 12
13	ζ	3	13 − 9
14		6	9 − 14
15		6	15 ,̄ 4 15 − 20
16		4	16 , 11
17		4 . 5	17 − 22
18	ε	3 . 4	23 , 18 − 26 23 , 18 . 3

Lustre of the stars in Eridanus.

19		4	27 , 19 , 12	44		5 . 6	49 . 44
20		5 . 6	15 – 20	45		5 . 6	45 . 49
21		6	22 ; 21	46		5	46 . 47
22		5 . 6	17 – 22 ; 21	47		4	46 . 47 , 56
23	δ	3 . 4	34 –, 23 , 18 53 , 23	48	ν	4	48 . 57 48 –, 51
24		5	32 , 24 ; 25 24 – 25	49		5 . 6	45 . 49 . 44
25		6	24 ; 25 24 – 25	50	υ^1	4	50 . 52 52 , 50
26	π	4	18 – 26 , 32	51	c	4	57 –, 51 48 –, 51
27		4	11 . 27 , 19 27 – 33	52	υ^2	3	50 . 52 52 , 50
28		6 . 5	36 , 28	53		3 . 4	53 , 23
29		6 . 7	30 , 29	54		3 . 4	54 – 60
30		5 . 6	32 – 30 , 29	55		6	56 , 55
31		5 . 6	Does not exist.	56		6	47 , 56 , 55 63 – 56
32		4 . 5	26 , 32 – 30 26 , 32 –, 35 32 , 24	57	μ	4	48 . 57 –, 51
				58		5 . 6	60 , 58 . 59
33		4 . 5	27 – 33 . 36	59		6	58 . 59
34	γ	2	67 , 34 –, 23	60		6	54 – 60 , 58
35		5	32 –, 35	61	ω	5	69 , 61 , 65
36		4	33 . 36 , 28	62	b	6	65 –, 62
37		6	40 , 37	63		6	64 – 63 – 56
38	ο	3 . 4	38 , 40	64		6	64 – 63
39	A	5	40 . 39 , 42	65	ψ	5	61 , 65 –, 62
40	d	5	38 , 40 – 37 40 . 39	66		6	68 . 66
41		4 . 3	41 , 43	67	h	3	67 , 34 67 – – 68
42	ξ	4 . 3	39 , 42	68		6	67 – – 68 . 66
43		5	41 , 43	69	λ	4	69 , 61

Lustre of the stars in Gemini.

1	H	5	18 – 1 – 139 Tauri	21		6 . 7	Does not exist.
2		8	5 – 2 . 4	22		7	23 , 22
3		8	3 , 5	23		5	20 , 23 19 , 23 , 22 20 – 23
4		7	2 . 4 . 6	24	γ	2 . 3	50 Orionis , 24 . 37 Aurigæ 66 –, 24
5		7	3 , 5 – 2				
6		7	4 . 6	25		7	54 Aurigæ – 25
7	η	4 . 5	13 , 7 – 31 13 ; 7 –, 31	26		5	26 . 61
8		7 . 8	8 . 9	27	ε	3	27 ⁝ 13 13 , 27
9		7	8 . 9 , 10	28		6	53 Aurigæ . 28 , 54 Aurigæ 36 , 28
10		8	9 , 10 . 11				
11		8	10 . 11 . 12	29		6 . 7	Does not exist.
12		8	11 . 12	30	ξ^1	6	31 – 30 . 38
13	μ	3	27 ⁝ 13 , 7 13 , 27 13 ; 7	31	ξ^2	5 . 4	7 – 31 – 30 7 –, 31
14		7 . 8	16 . 14 . 15	32		6	35 , 32
15		7	14 . 15 , 17	33		6 . 7	35 . 33
16		7	16 . 14	34	θ	4	77 , 34 – 43
17		8	15 , 17	35		6	38 , 35 , 32 35 . 33
18	ν	4	69 . 18 – 1	36	d	6	36 , 28 36 , 37
19		7	20 , 19 , 23	37		6	36 , 37 , 39
20		7 . 8	20 , 23 20 , 19 20 – 23	38	e	6	30 . 38 – 35

Lustre of the stars in Gemini.

39		6 . 7	37 , 39 . 40
40		6	39 . 40
41		6	45 – 41 45 , 41 , 50
42	ω¹	6	42 –, 44
43	ξ	3 . 4	34 – 43
44	ω²	6 . 7	42 –, 44
45	o	6	51 – 45 – 41 68 , 45 , 41
46	τ	5	60 – 46 69 , 46 62 – 46
47		6	48 , 47 , 52
48	m	6	57 – 48 , 47
49		7	52 – 49 58 , 49
50		6	41 , 50
51		6 . 5	51 – 45 51 , 68
52	n	7 . 6	47 , 52 – 49 52 , 58
53		6 . 7	59 . 53
54	λ	5	54 ; 55
55	δ	3	55 . 77 54 ; 55
56	q	6 . 7	56 ; 63
57	A	5 . 6	57 – 48
58		7 . 8	52 , 58 , 49
59		6	64 – 59 59 . 53
60	ι	4 . 5	60 – 65 60 – 46 60 , 62
61	r	6	26 . 61 63 , 61
62	s	5	60 , 62 , 46
63	p	6	56 ; 63 , 61
64	b¹	6	65 . 64 – 59
65	b²	6	60 – 65 . 64
66	α	1	78 –, 66 –, 24 66 . 34 Aurigæ
67		7 . 8	68 –, 67
68	k	6	68 –, 67 51 , 68 , 45 68 , 81
69	υ	5	69 . 18 69 , 46
70		5	80 – 70
71	o	5	71 . 80
72		6	Does not exist.
73		6 . 7	Does not exist.
74	f	6	81 , 74 – 85
75	σ	5	75 , 83
76	c	6	83 – 76
77	κ	4 . 5	55 . 77 , 34
78	β	2	78 –, 66 78, 67 Virginis 87 Tauri – – 78
79		7	82 . 79 , 84
80	π	5	71 . 80 – 70
81	g	6	68 , 81, 74
82		6	85 – 82 , 84 82 . 79
83	φ	5	83 – 76 75 , 83
84		5	82 , 84 79 , 84
85	l	6	74 – 85 – 82

Lustre of the stars in Leo.

1	κ	4	4 –̦ 1 1 –, 22
2	ω	5	6 – 2 –̦ 3
3		6	2 –̦ 3
4	λ	4	4 –̦ 1
5	ξ	4	10 , 5 – 6
6	h	6	5 – 6 – 2
7		6	16 – 7 , 11 8 , 7
8		6	16 , 8 , 7
9		6	13 , 9 – 12
10		5	10 , 5
11		6	7 , 11
12		7 . 8	9 – 12
13		6	13 , 9
14	o	3 . 4	14 , 47 14 , 30
15	f	6	22 . 15
16	ψ	6	16 – 7 16 , 8
17	ε	3	68 – – 17 – 70 17 – 36 17 – 30 68 –, 17 –, 36 17 –, 70
18		6	18 , 23
19		7	23 . 19 , 21
20		6	22 –, 20
21		7	19 , 21
22	g	6	22 . 15 1 –, 22 –, 20
23		6	18 , 23 . 19
24	μ	3 . 4	47 , 24
25		6 . 7	Does not exist.
26		7	Does not exist.
27	ν	4 . 5	29 – 27
28		7	Does not exist.
29	π	4	29 – 27 31 – 29
30	η	3 . 4	70 30 17 – 30 36 . 30 14 . 30
31	A	5	31 – 29 31 – 40
32	α	1	78 Geminorum –, 32 , 66 Geminorum
33		6 . 7	Lost or 34 – – 33
34		7	42 – 34
35		6	39 ; 35
36	ζ	3	17 – 36 . 70 70 . 36 . 30 36 – 46 Leonis minoris 17 –, 36
37		6	37 – 42
38		6	Does not exist.
39		6	39 ; 35
40		6	31 – 40
41	γ	2	41 , 94 41 – 94 41 –, 94

Lustre of the stars in Leo.							
42		6	37 – 42 – 34	69		5.6	61 , 69 – 66
43		6	44 , 43	70	θ	3	17 – 70 36 . 70 . 30 17 –, 70 . 36
44		5.6	44 , 45 49 , 44 , 43	71		6	Does not exist.
45		6	44 , 45 46 , 45	72		5	54 –, 72 –, 67
46	*i*	6	46 – 47 Leonis minoris 46 , 45 46 – 50 46 . 51	73		6	73 ; 81
47	ρ	4	14 , 47 , 24	74	*n*	4	91 . 74 –, 87
48		6	48 , 49	75		6	84 , 75 , 79
49		6	48 , 49 , 44	76		7	79 – 76 65 , 76
50		7.6	46 – 50	77	σ	4.5	77 , 78
51	*m*	6	46 . 50	78	ι	4	77 , 78 – 91
52	K	6	41 Leonis minoris – 52 53 , 52	79		5.6	75 , 79 – 76 79 , 65
53	*l*	6	53 , 52	80		6	89 – 80 – 83
54		4.5	60 . 54 –, 72	81		6	73 ; 81 86 – 81 , 90
55		6.5	55 – 57 55 , 62	82		7.8	83 , 82
56		6.7	59 –, 56	83		8	80 – 83 , 82
57		6	55 – 57 62 , 57	84	τ	4	84 , 75
58	*d*	6.5	63 – 58 ; 59	85		6	85 , 88 90 , 85
59	*c*	5	58 ; 59 59 –, 56	86		6	86 – 81
60	*b*	5	60 . 54	87	*e*	4.5	74 –, 87
61		5	61 , 69	88		6	85 , 88
62	*g*	6	55 , 62 , 57	89		6	89 – 80
63	χ	4.5	63 – 58	90		6	81 , 90 , 85
64		6	67 – 64	91	υ	4	78 – 91 . 74
65		6	79 , 65 , 76	92		6	95 , 92
66		6	69 – 66	93		4	93 –, 95
67		6	72 –, 67 – 64	94	β	1.2	41 , 94 41 – 94 41 –, 94 – 68
68	δ	2.3	94 – 68 – – 17 68 –, 17	95	*o*	6	93 –, 95 , 92

Notes to Aries.

1 There is no observation of this star in FLAMSTEED's works. Perhaps one of the observations of 104 Piscium was placed 8 degrees too much north, which would produce the first Arietis.*

2 There are two stars, the largest of which I take to be the 2d Arietis, and have estimated its brightness.

28 There is an observation in the second volume of the *Historia Cœlestis*, page 181; and in all appearance it is a good one. But as we find no where the 26th and 28th observed together, it is probable that only one of them existed in FLAMSTEED's time. The observation, Dec. 10, 1692, being corrected –1′ of time in R.A., will agree with the four observations on page 81, 181, 273, and 283.

Catalogue and Atlas.

1 and 28 should be out.

Atlas.

A small star under 41 towards 16 Trianguli should be out.

Notes to Canis major.

" Order of magnitude Nov. 17, 1784. ε 2m β 2.3m δ 3.2m η 3m o 4m."

2 " Is 3m, January 9, 1796."

23 " Is hardly 5m, January 9, 1796."

* [See below, p. 576.—ED.]

Notes to Canis minor.

3 " Is 4m, January 9, 1796."

10 The comparative brightness at first sight is 10 – 19 Orionis ; but the light of Rigel being more brilliant than that of Procyon, a continuance of its impression, when we look a good while, occasions an increase ; and it becomes 19 Orionis -, 10.

12 FLAMSTEED never observed this star.

Catalogue and Atlas.

12 should be out.

Notes to Cassiopea.

" Order of magnitude May 12, 1783. $\beta\,\alpha\,\gamma\,2''$m $\delta\,\epsilon$."

8 There are two stars, nearly equal.

29 There is no observation of FLAMSTEED of this star.

41 There is no observation of this star by FLAMSTEED.

51 FLAMSTEED observed this star Dec. 16, 1691. The observation, however, is marked defective in zenith-distance. There is a star, about 9 or 10m, near the place where 51 should be.

55 That which I have estimated is the largest of several small stars about the place of 55.

Catalogue and Atlas.

29, 41, and 51 should be out.

52 and 53 are not laid down in Atlas as they are in the heavens; and on examining FLAMSTEED'S observation on page 208, it appears that the catalogue is erroneous; for by that observation 52 is the most north, and 53 the most south, contrary to the catalogue.

Notes to Cetus.

14 There is no observation of this star by FLAMSTEED.*

24 There is no observation of this star by FLAMSTEED.

51 Is 106 Piscium.

68 A periodical star of a great range in its lustre, as appears from the expressions I have given of its comparative brightness. Its period is 331 days, 10 hours, 19 minutes. See *Phil. Trans.* 1792, p. 25 [above, p. 438]. Different authors, however, vary a little in their determinations of the length of this period ; for which we may account by admitting the star to be subject to considerable alterations in the emission of light, from some parts of its surface, which being more copious sometimes in one place, and sometimes in another at some small distance, will give a different result to the observations of the time of its maximum ; while, notwithstanding, the general period of its changes will not be considerably affected by it. We have a similar instance in the rotation of Jupiter, which seems to vary on account of the little stability of its spots. See *Phil. Trans.* 1781, Part I. page 123 to 126 [above, pp. 20–22].

90 Is 1 Eridani.

Catalogue and Atlas.

14 and 24 should be out.

54 " requires + 1' 8" of time in R.A. Sept. 4, 1786."

Atlas.

18 " Is not laid down very accurately in Atlas, Sept. 22, 1795."

Notes to Corvus.

" Order of magnitude May 12, 1783. $\gamma\,\delta\,\beta\,\alpha$."

The order, by the present catalogue, does not agree with that in 1783. It is no longer $\delta\,\beta$, but $\beta\,\delta$; but the constellation is so low that precision is difficult.

* [See below, p. 577.—ED.]

Notes to Eridanus.

1 Is 90 Ceti.
31 FLAMSTEED has no observation of this star.

Catalogue and Atlas.

31 must be out.
40 " requires + 6′ in P.D., January 31, 1786."

Notes to Gemini.

21 There is an observation by FLAMSTEED on page 294, but with a small correction, minus, in time it will agree with 20 ; and my double star IV. 46 marked 21 : : is 20 Geminorum.

29 There is no observation by FLAMSTEED upon this star.

54 Seems to be increasing. There is an interval of 9 months between the two observations of my catalogue. Mr. BODE supposes the star to be changeable. See *Astronomisches Jahrbuch*, 1788, page 255. And 1793, page 201.

72 and 73 There are no observations of these stars by FLAMSTEED.

78 β appears to be of a deeper colour than it was a good many years ago. "I should now (Jan. 8, 1796) place it among the red, or ruddy stars ; which formerly I did not use to do."

Catalogue and Atlas.

21, 29, 72, and 73 should be out.

Atlas.

17 requires −15′ in P.D.

Notes to Leo.

" Order of magnitude May 12, 1783. α 1‴ m γ 2′ m β 2′ m δ ϵ ζ θ η μ ο ρ ν σ."

25 There is no observation of this star by FLAMSTEED ; but if −14′ 30″ in R.A. and +2° 25′ 25″ in P.D. be applied, according to the edition of 1712, the place will then agree with 10 Sextantis. There are two very small stars near the place where 25 is put in Atlas ; one of them is the star V. 63, which in my catalogue of double stars is called 25 Leonis.

26 There is an observation of this star by FLAMSTEED on page 299 ; but it is defective.

28 There is an observation without time on page 299, marked " *Leonis*," taken after the transit of Mars, which has probably occasioned the insertion of 28 into the British catalogue ; but the observed star must have been 11 Sextantis.

38 In the observation, March 10, 1691 ; the number 821,66 is cast up 6 degrees too little, which produced this star ; when the error is corrected, we find that the star observed was 37.

41 This is the double star in my second catalogue I. 28, and by a series of observations it appears now that the distance between the two stars is considerably increased, since the year 1782. I mention this circumstance here, as it will probably explain some apparent increase of brightness, that seems to have taken place in this star ; for although the same quantity of light, when it is spread over more space, must appear less intense than it will do when it remains in a more confined state, it may nevertheless, by an increase of apparent magnitude, become entitled to be ranged upon a par with a brighter star.

71 FLAMSTEED never observed this star.*

94 From the expressions of this catalogue, it is evident that the star is less now than it was 13 years ago.

420 of MAYER's catalogue is not visible.

Catalogue and Atlas.

25, 26, 28, 38, and 71 should be out.

WM. HERSCHEL.

Slough, near Windsor,
June 1, 1796.

* [See Baily, p. 579 ; probably = LL21660.—ED.]

XXXVIII.

A Third Catalogue of the comparative Brightness of the Stars; with an introductory Account of an Index to Mr. FLAMSTEED'S *Observations of the fixed Stars contained in the second Volume of the* Historia Cœlestis. *To which are added, several useful Results derived from that Index.*

[*Phil. Trans.*, 1797, pp. 293–324.]

Read May 18, 1797.

IN my earliest reviews of the heavens, I was much surprised to find many of the stars of the British catalogue missing. Taking it for granted that this catalogue was faultless, I supposed them to be ost. The deviation of many stars from the magnitude assigned to them in that catalogue, for the same reason, I looked upon as changes in the lustre of the stars. Soon after, however, I perceived that these conclusions had been premature, and wished it were possible to find some method that might serve to direct us from the stars in the British catalogue to the original observations which have served as a foundation to it. The labour and time required for making a proper index, withheld me continually from undertaking the construction of it: but when I began to put the method of comparative brightness in practice, with a view to form a general catalogue, I found the indispensable necessity of having this index recur so forcibly, that I recommended it to my Sister to undertake the arduous task. At my request, and according to a plan which I laid down, she began the work about twenty months ago, and has lately finished it.

The index has been made in the following manner. Every observation upon the fixed stars contained in the second volume of the *Historia Cœlestis* was examined first, by casting up again all the numbers of the screws, in order to detect any error that might have been committed in reading off the zenith-distance by diagonal lines. The result of the computation being then corrected by the quantity given at the head of the column, and refraction being allowed for, was next compared with the column of the correct zenith-distance as a check.

Every star was now computed by a known preceding or following star; and its place according to the result of the computation laid down in the *Atlas Cœlestis*, by means of proportional compasses. This was necessary, in order to ascertain the observed star: for the observations contain but little information on the subject; most of the small stars being without names, letters, or descriptions. The many errors in the names of the constellations affixed to the stars, and in the letters by which they are denoted, also demanded a more scrupulous attention; so that only their relative situation, examined by calculation, could ascertain what the stars really were which had been observed.

Every observed star being now ascertained, its number in the British catalogue was added in the margin at the end of the line of the observation; and a book with all the constellations and number of the stars of the same catalogue, with large blank spaces to each of them, being provided, an entry of the page where FLAMSTEED's observation is to be found, was made in its proper place.

If the star observed was not in the British catalogue, it was marked as such in the margin of the observations; and being provided with another book of constellations and numbers, it was entered into the blank space belonging to some known preceding or following star, by which its place had been settled. The Greek and English letters used by FLAMSTEED, whether they were such as had been introduced before, or which he thought it expedient to add to them at the time of observation, were also entered into their proper places; and to complete the whole, the magnitude affixed to the stars was likewise joined to the entry made in the blank spaces of the index.

I have been so far particular in giving the method by which the index has been constructed, that it may appear what confidence ought to be given to the conclusions which will be drawn from its report.

About three or four examples of its use, will completely shew how the results, which will be mentioned, have been obtained.

Suppose I wish to be informed of the particulars relating to the 13th Arietis. Then by the index I am referred, in the column allotted for that star, to 77 observations; and find that FLAMSTEED used the letter α 72 times, and that in two places he calls it a star of the 2d magnitude; the rest of the observations being without any estimation of its brightness.

If it be required to know FLAMSTEED's observations upon the 34th Tauri, which star is supposed to have been the Georgian planet, mistaken by FLAMSTEED for a small fixed star; * we find in our index, that on page 86, December 13, 1690, a star of the 6th magnitude was observed, which answers to the place of the 34th Tauri in the British catalogue; and that no other observation of the same star occurs in the second volume. In my catalogue of comparative brightness, the 34th Tauri is put down among the lost stars, it being no longer to be seen in the place where it was observed by FLAMSTEED.

If in my review of the heavens I cannot find 38 Leonis, and examine this index, I am at once informed that FLAMSTEED never observed such a star; and that of consequence it has been inserted in the British catalogue by some mistake or other. In many cases, these mistakes may be easily traced, as has been shewn with regard

* See *Astronomisches Jahrbuch* for 1789, page 202.

to this star in my second catalogue of comparative brightness. See the note to 38 Leonis.

When we wish to examine 90 Ceti in the heavens, and cannot find it, we are informed by our index, that 90 Ceti is the same star with 1 Eridani; and that, consequently, we are not to look out for two different stars.

We may now proceed to give some general results that are to be obtained from an inspection of our index. They are as follows.

111 Stars inserted in the British catalogue have never been observed by FLAMSTEED. This will explain why so many stars in the heavens seem to have been lost.

There are 39 stars in the same catalogue that want considerable corrections in right-ascension or polar-distance. In many it amounts to several degrees.

54 stars more, besides the 39 that are taken from the erroneous stars in the catalogue, want corrections in the *Atlas Cœlestis*; several of them also of many degrees.

42 stars are put down, which must be reduced to 21; each going by two names in different constellations.

371 stars, completely observed both in right-ascension and zenith-distance, have been totally overlooked.

35 more, which have one of the two, either right-ascension or polar-distance doubtful, have been omitted.

86 with only the polar-distance, and 13 with only the right-ascension, have also been unnoticed.

About 50 more that are pointed out by pretty clear descriptions, are likewise neglected; so that upon the whole between five and six hundred stars observed by FLAMSTEED have been overlooked when the British catalogue was framed.

These additional stars will make a considerable catalogue, which is already drawn up and nearly finished by Miss HERSCHEL, who is in hopes that it may prove a valuable acquisition to astronomers.

Neither the index to FLAMSTEED's observations, nor the catalogue of omitted stars, were finished when my former two catalogues of comparative brightness were given; I shall therefore now select a few notes to be added to those which are at the end of these catalogues. They will contain such additional light as I have been enabled to gather from this newly acquired assistance.*

* [Caroline Herschel's "Catalogue of Stars taken from Mr. Flamsteed's observations and not inserted in the British Catalogue, with an Index to point out every observation &c.," folio, London, 1798, contains two prefaces by W. Herschel describing the use of the catalogue and the index.—ED.]

Additional Notes to the Stars in the First Catalogue of the comparative Brightness of the Stars.

Aquarius.

25 Is the same star with 6 Pegasi. There are but two observations upon it. The first is on page 57; FLAMSTEED calls it "*in constellatione Pegasi sub capite.*" The second, on page 71, is described "*in constellatione Aquarii trianguli in capite præcedens et borealis.*" Here we see that the double insertion in the catalogue is owing to the star's having been called by different names in the observations. See also Mr. WOLLASTON's catalogue, zone 88°.

27 Is the same with 11 Pegasi. There are three observations: the first places the star in the constellation of Pegasus, the two latter in that of Aquarius. See also Mr. WOLLASTON's catalogue for this star, and others of the same kind.

65 Has not been observed by FLAMSTEED; notwithstanding which we find it inserted in my first catalogue, where its relative brightness is given. It should be considered that, in the first place, several stars of which there are no observations in the second volume of FLAMSTEED's works, and which are, nevertheless, inserted in the British catalogue, such for instance as θ and ι Draconis, are well known to exist in the heavens. Now whether they were put into the catalogue from observations that are not in the second volume, or taken from other catalogues, it so happens that observations of them cannot be found. Therefore the want of a former observation by FLAMSTEED, is not sufficient to prove that a star does not exist. In the next place it should be recollected, that the method used to ascertain the stars in estimating their brightness, is not so accurate, as to point out with great precision the absolute situation of a star; and that, consequently, another star which happens to be not far from the place where the catalogue points out the star we look for, may be taken for it; especially when there are no neighbouring stars of the British catalogue that may induce us to exert uncommon attention in ascertaining the identity of such a star. MAYER, however, has an observation of 65 Aquarii in his zodiacal catalogue, No. 932, which puts the existence of the star out of doubt.

72 As the star neither was observed by FLAMSTEED, nor does exist, we cannot admit the remark which Mr. WOLLASTON in his catalogue, zone 95°, has upon MAYER's 939 star; where he supposes an error in declination of 3 degrees to have been committed, on a supposition of its being FLAMSTEED's 72.

80 Requires + 2′ in time in R.A., and therefore is not the star I have given, which requires − 1′ 35″.

104 Which is without R.A. in the British catalogue, has three complete observations, page 8, 70, and 331.

Aquila.

29 Is without R.A. There is but one observation of FLAMSTEED, page 53, which has no time. The R.A. is given by M. DE LA LANDE, in Mr. BODE's *Jahrbuch* for 1796, page 163.

33 and 34 Which do not exist, were probably inserted by a mistake of one hour in the time of one of the observations on the two stars 68 and 69. In the zenith-distance, page 71 of FLAMSTEED's observation of 69 Aquilæ, for 53° read 55°.

40 and 43 Which do not exist, were probably also inserted by the same mistake of one hour in the R.A. of 70 and 71.

Capricornus.

1 and 2 Should be ξ^1 ξ^2. FLAMSTEED calls them so in his observations, and MAYER has also adopted the same letters in his catalogue, No. 821 and 822.

Cygnus.

5 Is without R.A. in the British catalogue; but the star has not been observed by FLAMSTEED.

9 Is without R.A.; FLAMSTEED, however, has a complete observation of it, page 67.

24 Has no R.A. The time observed by FLAMSTEED is only doubtful in the seconds. Its R.A. has been given in Mr. BODE's *Jahrbuch* for 1797, page 163.

33 Has no R.A. FLAMSTEED never observed this star; but it is 3 Cephei Hevelii.

38 Has no R.A. in the British catalogue; but as the defective and only observation of FLAMSTEED on page 75, which might be supposed to belong to 38, will agree better with 43, it follows that he never observed 38.

68 Has no R.A. There is a complete observation by FLAMSTEED, page 75.

78 Has no time in FLAMSTEED's observations. It is No. 146 in DE LA CAILLE's catalogue.

79 Has no R.A. FLAMSTEED has but one observation, which is without time. Mr. BODE gives it in his *Jahrbuch* for 1797, page 163.

Hercules.

24 Is the same with 51 Serpentis.

28 Is the same with 11 Ophiuchi.

54 There is no observation of this star. The zenith-distance of 55 was taken twice April 8, 1703 (instances of which we find in several other stars), which occasioned its being inserted as two stars.

63 There is no observation of this star, nor does it exist. The star of which the brightness is given in my catalogue, is at some distance from the place assigned in the British catalogue. FLAMSTEED observed a star, page 444, which will be No. 269 in Miss HERSCHEL's manuscript catalogue. This, with an error in the calculation of the P.D., probably occasioned the insertion of 63. And if this be the star, the P.D. of the British catalogue must be corrected $+ 3°$.

71 Has never been observed by FLAMSTEED, nor does it exist. A small error in the calculation of one of the four observations of 70, may have produced it.

80 and 81 Were never observed. The two stars ν 24 and 25 Draconis, miscalled ι in FLAMSTEED's observations, page 55 and 175, with an error of P.D., accounts for the insertion of these stars. See Mr. BODE's *Jahrbuch* for 1787, page 194.

93 The P.D. is marked : : (doubtful), in the British catalogue; but the observation of FLAMSTEED, page 520, is complete.

Pegasus.

6 Is the same star with 25 Aquarii.

11 Is the same star with 27 Aquarii.

Additional Notes to the Stars in the Second Catalogue of the comparative Brightness of the Stars.

Aries.

1 There is an observation of a star by FLAMSTEED, which being calculated with an error of 10′ of time in R.A., would produce 1 Arietis; we may therefore correct the British catalogue R.A. $+10'$, and the star will be found to exist. In Miss HERSCHEL's manuscript catalogue it is No. 143.

2 Is the same star with 107 Piscium.

38 is the same star with 88 Ceti. In three observations, page 85, 285, and 485, FLAMSTEED has called it Arietis; and on page 481 he has called it Ceti. See also Mr. BODE's *Jahrbuch* for 1793, page 200.

50 By FLAMSTEED's observation, page 273, the catalogue requires $-1'$ in time of R.A.

Cassiopea.

3 The place in the catalogue by two observations of FLAMSTEED requires $+ 5'\frac{1}{4}$ of time in R.A., and $+ 7'$ of P.D.

8 Is marked : : but has four complete observations on page 140, 144, 145, and 147.

29 There is an observation of FLAMSTEED on page 144 which has produced this star, but the time of it requires a correction of $+ 6'$; and it will then belong to 32. That this correction should be used, will appear when we compare this observation with another on page 213. In both places a star which is not inserted in the British catalogue, but which is No. 384 of Miss HERSCHEL's manuscript catalogue, was taken at the same time. On page 144 it is "Duarum infra γ, versus polum, borealis. Simul fere transit, austrea;" and on page 213 we have "post transitum" for the new star, and "cum priore" for 32; and in both places the zenith-distance perfectly shews that they were the same stars: the 32d and a star south of it. And they are now both in the places where FLAMSTEED has observed them.

30 FLAMSTEED has no observation of this star. It is μ 21 Cassiopeæ Hevelii.

33 FLAMSTEED observed no R.A. of this star. It is θ 23 Cassiopeæ Hevelii.

34 Is wrong in the catalogue. By two observations of FLAMSTEED, page 144 and 521, it requires a mean correction of − 9′ of time in R.A. In this case my double star III. 23 will no longer be ϕ 34 Cassiopeæ, but a star 9′ of time preceding ϕ; for it exists in the place where 34 is put in Atlas, according to the erroneous catalogue, and is rather larger than FLAMSTEED's star ϕ.

35 The R.A. is marked : : The single observation, page 207, has the time marked *circiter*, being probably set down to the nearest minute only; and by the same observation the P.D. requires + 20′

47 Is also marked : : but has one complete observation, page 149.

51 The observation of FLAMSTEED which produced this star should be corrected + 1 hour. This makes it 37 Cassiopeæ Hevelii.

52 and 53 By FLAMSTEED's observation page 208, should be the reverse in P.D. of what they are.

Cetus.

14 If we correct the British catalogue + 3° in P.D., it will become a star observed by FLAMSTEED, which is No. 312 in Miss HERSCHEL's manuscript catalogue.

26 FLAMSTEED has no observation of this star; but we find it in DE LA CAILLE's zodiacal catalogue, No. 10.

51 Is the same with 106 Piscium. FLAMSTEED has 23 observations of the star, and has always called it ν, except once on page 482, where it is without letter, and where the constellation is marked Aquarii; now, as there was immediately following an observation of 54 Ceti, and Aquarius was evidently wrong, the star has been put in Cetus.

58 By FLAMSTEED's observation, page 358, the R.A. in the British catalogue requires a correction of − 3′ in time.

74 FLAMSTEED has no observation of this star, nor can I find it in any other catalogue. The place of it is so distant from other stars of the British catalogue, that my estimation of brightness may belong to some star not far from the situation assigned, and that the star of the British catalogue may not exist.

88 Is the same with 38 Arietis. See Mr. BODE's *Jahrbuch* for 1793, page 200.

Eridanus.

44 In the British catalogue is marked : : The single observation of FLAMSTEED, page 153, is perfect, all but a difference of 5′ between the zenith-distance by the diagonal lines and by the screw.

45 Marked : : has a complete observation, page 153.

68 Marked : : has a complete observation, page 146.

Gemini.

50 There is no observation on this star. The star I have given is at a considerable distance from the place assigned by the British catalogue, so that in fact the star of the catalogue does not exist. It has been inserted in the British catalogue by a mistake in the calculation of a star which is about 1° 49′ more south. This will be No. 139 in Miss HERSCHEL's manuscript catalogue, and it is probably the real intended 50 of FLAMSTEED. The expression of its brightness 41 , 50 of my catalogue will do very well for it.

70 and 71 By FLAMSTEED's observations should be called π^1 and π^2. TYCHO and HEVELIUS also call 71 π.*

72 and 73 Have been inserted by a mistake in 64 and 65. See Mr. BODE's *Jahrbuch* for 1788, page 175.

76 FLAMSTEED has no observation of this star. It is, however, MAYER's No. 310.

80 Is not π, but according to FLAMSTEED's observation *quæ sequitur* π ; and has no letter

Leo.

10 Is the same with 1 Sextantis.

25 This star does not exist in the place where the British catalogue gives it ; but if we admit that it has been inserted by a mistake in the calculation of 10 Sextantis, it may be taken into the constellation of Leo, as a star inserted in two constellations ; and it will then be " 25 is the same with 10 Sextantis." †

* [71 Geminorum is not in Tycho's Catalogue.—ED.]

† [See Baily's note to No. 1310.—ED.]

26 In FLAMSTEED'S observations, page 299, the *strias cochleæ* give 26′ less than the *lineas diagonales*. The former are right; therefore the British catalogue must be corrected P.D. – 26′.

28 FLAMSTEED has no observation of this star. It was probably inserted by a mistake in calculating an imperfect observation of 11 Sextantis. If this be allowed, we then must say "28 is the same with 11 Sextantis."

66 FLAMSTEED has no observation of this star. There is a small star near the place where the British catalogue has given it, of which I have expressed the brightness; but as its situation is not exactly where it ought to be, my catalogue should have, "does not exist."

67 Is the same with 53 Leonis minoris.

71 May have been inserted by a mistake in one of the three observations of 73; putting the star north of θ instead of south.

III. *Catalogue of the comparative Brightness of the Stars.*

Lustre of the stars in Andromeda.

1	o	3.4	15 - 1 - 16
2		6	20 - 2 , 4
3		6	8 . 3
4		6	2 , 4 , 6
5		6	11 -, 5
6		6.7	4 , 6
7		5.6	7 - 8
8		6	7 - 8 , 11 8 . 3
9		6	10 . 9
10		6.7	13 - 10 . 9
11		6	8 , 11 -, 5
12		6	15 . 12 , 13
13		6	12 , 13 - 10
14		6	14 , 15
15		6	14 , 15 . 12
16	λ	4	16 - 17 1 - 16
17	ι	4	16 - 17 , 19 19 -, 17
18		6	20 . 18
19	κ	4	17 , 19 - 20 19 -, 17
20	ψ	5.6	19 - 20 20 - 2 22 - 20 . 18 22 - 20 - 23
21	α	2	21 , 43 21 -, 8 Pegasi 21 -, 43 21 - 43
22		5	22 - 20
23		6	20 - 23 , 26
24	θ	4.5	25 , 24 - 27
25	σ	5	25 , 24
26		6	23 , 26
27	ρ	5	24 - 27
28		6	29 - 28 32 . 28 -, 40
29	π	4.5	30 . 29 - 28 29 , 35
30	ε	4	37 - 30 . 29
31	δ	3	4 Trianguli =, 31 -, 2 Trianguli
32		6	35 - 32 . 28 32 - 39
33		Neb.	is a Nebula
34	ζ	4	35 , 34 , 38

35	ν	4	29 , 35 - 32 35 , 34 35 - 48 50 -, 35 , 53
36		6	38 -, 36
37	μ	4.3	37 - 30 37 - 50
38	η	4.5	34 , 38 -, 36
39		6	32 - 39
40		6	28 -, 40
41	d	5	42 -, 41 . 45
42	φ	5	54 ; 42 -, 41
43	β	2	21 , 43 . 57 21 -, 43 ; 57 21 - 43 , 57 43 -, 13 Ari 43 - 13 Ari 43 =, 57
44		6	45 . 44
45		5.6	41 . 45 . 44 45 . 47
46		4.5	48 , 46 , 49
47		6	45 . 47
48		5	35 - 48 , 46
49	ξ	5	46 , 49
50	υ	6.5	37 - 50 -, 35
51	υ	5	51 - 1
52	λ	6	53 , 52 . 55
53	τ	5	35 , 53 , 52 58 , 53 - 56 53 , 60
54	φ	4	54 ; 42
55		Neb.	52 . 55
56		6	53 - 56 . 59 60 , 56
57	γ	2.3	43 . 57 57 ; 13 Arietis 43 ; 57 43 , 57 43 =, 57
58		6	58 , 53
59		6	56 . 59
60	b	6	53 , 60 , 56
61		6	63 , 61 66 , 61
62	c	6	65 , 62
63		6	64 . 63 , 61 6 Persei , 63
64		6	65 - 64 . 63
65		5	65 - 64 65 , 62 65 , 6 Persei
66		6.7	66 ; 61

Lustre of the stars in Bootes.

1		6	7 , 1 6 – 1 , 2
2		6	1 , 2 . 10
3		6	11 . 3
4	τ	4	5 -̦ 4 – 6
5	υ	4	5 -̦ 4 30 –, 5 , 35
6		5 . 6	6 , 7 4 – 6 – 1
7		7	6 , 7 , 1 7 – 26
8	η	3	8 , 27 79 Virginis ; 8 8 – 27 36 – 8
9		5	12 -̦ 9 – 11
10	e	7	2 . 10
11		7 . 6	9 – 11 . 3
12	d	5	28 ; 12 -̦ 9
13		6	13 – 24
14		6	18 , 14 , 15
15		6	14 , 15
16	α	1	16 – – 3 Lyræ
17	κ	4	21 . 17
18		6	20 , 18 , 14
19	λ	4	19 . 23
20		5	20 , 18 20 ; 22
21	ι	4	23 , 21 . 17
22	f	5	20 ; 22
23	θ	4	19 . 23 , 21
24	g	6 . 7	13 – 24
25	ρ	4	25 – – 51
26		7	7 – 26 34 – – 26
27	γ	3	8 , 27 – 49 27 –, 49 8 – 27 27 –, 42
28	σ	5	51 – 28 28 ; 12
29	π	4 . 3	35 , 29
30	ξ	3	30 –, 5
31		5	35 – 31 – 32
32		6	31 – 32
33	h^1	6	39 . 33 – 38
34		6	34 – – 26
35	o	4 . 5	5 , 35 , 29 37 . 35 – 31
36	ε	3	5 Coronæ – 36 – 8
37	ξ	4	37 . 35
38	h^2	6	33 – 38
39		6	47 . 39 39 . 33
40		6 . 7	47 – 40
41	ω	5	45 ; 41 – 46 41 , 48 41 . 50
42	β	3	49 , 42 42 -̦ 49 27 –, 42 . 49 42 ⁝ 49
43	ψ	5	43 – 45
44		6	44 , 47
45	c	5	43 – 45 ; 41
46	b	6	41 – 46 48 , 46
47	k	5	44 , 47 . 39 47 – 40
48	χ	5	41 , 48 , 46
49	δ	3	27 –, 49 42 –, 49 42 . 49 42 , 49 42 ⁝ 49 27 – – 49
50		5	41 . 50
51	μ	4	25 – – 51 – 28 4 Coronæ , 51 , 7 Coronæ
52	ν^1	6	53 ; 52 , 54
53	ν^2	6	53 ; 52
54	φ	6	52 , 54

Lustre of the stars in Cancer.

1		6	5 . 1
2	ω^1	6	9 , 2 , 4 14 , 2 , 4
3		6	16 –, 3 , 5 8 – 3 , 12
4	ω^2	6	2 , 4 . 13
5		6	3 , 5 . 1
6	χ	5	6 – 14 6 – 15 6 -̦ 18
7		8	9 , 7
8		6	8 – 3
9	μ^1	7	10 – 9 , 2 9 , 7
10	μ^2	5	10 – 9
11		6	14 , 11 15 – 11
12		6	3 , 12
13	ψ^1	6 . 7	4 . 13
14	ψ^2	4	14 , 2 6 – 14 , 11
15	ψ^3	5	6 – 15 – 11
16	ξ	5 . 6	43 . 16 –, 3
17	β	4 . 3	17 -̦ 47 17 , 48
18	χ	6	6 -̦ 18 , 23
19	λ	6	19 – 30 , 28
20	d^1	6	31 , 20 , 25
21		6	37 , 21 , 34 29 – 21
22	ϕ^1	6 . 7	23 , 22
23	ϕ^2	6	18 , 23 , 22
24	υ^1	6	32 . 24
25	d^2	6	20 , 25
26	ϕ^3	6	Does not exist.
27		6	27 ; 29
28	υ^2	6 . 7	30 , 28 , 32
29		6 . 7	27 ; 29 – 21
30	υ^3	6	19 – 30 , 28
31	θ	6 . 5	31 , 20 31 . 33
32	υ^4	7 . 8	28 , 32 . 24

Lustre of the stars in Cancer.

33	η	6 . 7	31 . 33	58	ρ^3	6	48 − − 58 − 55 58 ; 75 58 ; 57
34		6	21 , 34 . 36	59	σ^2	5 . 6	64 . 59 . 66
35		7	42 ; 35 . 38	60	α^1	4 . 5	45 . 60 , 50
36	c^1	6	34 . 36	61		6	46 . 61
37	c^2	6	49 − 37 , 21	62	o^1	6	63 . 62 −, 54
38	o	8	42 ; 38 . 40 35 . 38	63	o^2	6	63 . 62
39		6	39 , 41	64	σ^3	6	51 . 64 . 59
40		6	38 . 40	65	α^2	4	65 . 47
41	ε	7	39 , 41 . 42	66	σ^4	6	59 . 66
42	c	7 . 8	41 . 42 ; 38 42 ; 35	67	ρ^4	6 . 7	67 . 55
43	γ	4	43 . 16 47 −, 43	68		6	81 − 68 , 71 68 . 78 68 ; 80
44		6	20 − − − 44 4 − − 44	69	ν	6	69 ⁝ 77
45	A^1	6	76 , 45 . 60	70	ρ^5	6 . 7	55 , 70
46		6	55 ; 46 . 61	71		7	68 , 71 78 , 71
47	δ	4	17 ˉ, 47 −, 43 65 . 47 − − 76 48 , 47	72	τ	6 . 7	57 , 72
48	ι^1	5	17 , 48 , 47 48 − − 58	73		6	Does not exist.
49	b	6	49 − 37	74		6	Does not exist.
50	A^2	6	60 , 50	75		6 . 7	58 , 75
51	σ^1	6	51 . 64	76	κ	4 . 5	47 − − 76 , 45
52		6	54 − 52	77	ξ	5 . 6	69 ⁝ 77 −, 79
53	ρ^1	6	55 ; 53	78		6	68 . 78 , 71 83 , 78 80 ˉ, 78 , 71
54		7	62 −, 54 − 52 82 , 54 , 81	79		8	77 −, 79
55	ρ^2	6	58 − 55 ; 53 67 . 55 , 70 57 − 55 ; 46	80		7	80 − 83 68 ; 80 ˉ, 78
56	ρ^3	6	Does not exist.	81	π	7	54 , 81 81 − 68 81 , 83
57	ι^2	5 . 6	58 ; 57 − 55 57 , 72	82		6	82 , 54
				83		6	81 , 83 80 − 83 , 78

Lustre of the stars in Centaurus.

1	i	4 . 5	3 . 1 . 5	4	h	4 . 5	4 , 3
2	g	4 . 5	5 − 2	5	θ	2 . 3	1 . 5 − 2
3	k	4 . 5	4 , 3 . 1				

Lustre of the stars in Cepheus.

1	κ	5	1 . 17	12		7	9 − 12
2	θ	5	3 −, 2	13	μ	6	13 , 14
3	η	4	32 , 3 −, 2 21 ⁝ 3 32 ⁝ 3	14		6	13 . 14 −, 15
4		6	6 − 4 . 7	15	ν	7 . 6	14 −, 15 − 15
5	α	3	5 −, 37 Cygni 5 . 37 Cassiopeæ	16		5 . 6	24 , 16 , 78 Draconis
6		6	6 − 4	17	ξ	5	1 . 17 , 33 10 . 17 − 9 23 , 17 − − 30
7		6	4 . 7	18			19 , 18 ; 20
8	β	3	35 −, 8 −, 32	19		6	22 . 19 , 18
9		6	17 − 9 − 12 11 , 9	20		6	18 ; 20
10		5	10 . 17	21	ζ	4 . 5	21 ⁝ 3
11		5	11 , 9				

Lustre of the stars in Cepheus.

22	λ	6	22 . 19	29	ρ	6	28 – 29
23	ε	4	23 , 17	30		6	17 – – 30 –, 26
24		5 . 6	24 , 16	31		6	34 – 31
25		7	26 –, 25	32	ι	4	8 –, 32 . 3 32 ⁝ 3
26		6	30 –, 26 –, 25	33	π	5	17 , 33
27	δ	4 . 5	32 , 27 27 . 23 21 – 27 – 23	34	ο	5	34 – 31
			21 =, 27	35	γ	3	35 –, 8
28		6	28 – 29				

Lustre of the stars in Corona Borealis.

1	ο	6	2 – – 1	12	λ	5	12 , 9
2	η	5	2 – – 1	13	ε	4 . 5	3 – 13 ; 4
3	β	4	8 ; 3 – 13	14	ι	5 . 6	14 , 19
4	θ	4 . 5	13 ; 4 – 10 4 , 7	15	ρ	6	17 , 15
5	α	2 . 3	55 Ophiuchi , 5 5 – 36 Bootis	16	τ	6	16 – 17
6	μ	5	11 – 6 , 9	17	σ	6	16 – 17 17 , 15
7	ζ	4	4 , 7 , 10	18	υ	6	19 –, 18
8	γ	4	8 ; 3	19	ξ	5	14 , 19 –, 18
9	π	5	6 , 9 12 , 9	20	ν^1	5	20 =, 21
10	δ	4	4 – 10 7 , 10	21	ν^2	5	20 =, 21
11	κ	5	11 – 6				

Lustre of the stars in Lacerta.

1		5	7 , 1 , 8 1 , 1 Hevelii . 6	9		6	3 , 9
2		5	7 –, 2 , 5	10		6	8 . 10 , 12
3		4 . 5	4 . 3 , 9	11		5	6 , 11 , 15
4		5	5 . 4 . 3	12		6	10 , 12
5		4 . 5	7 – 5 2 , 5 . 4	13		6	15 . 13 , 14
6		5	7 – 6 , 11 1 Hevelii . 6	14		6	13 , 14 . 16
7		4	7 – 5 7 –, 2 7 – 6 7 , 1	15		5	11 , 15 . 13
8		6	1 , 8 . 10	16		6	14 . 16

Lustre of the stars in Lepus.

1		9	7 – 1 10 . 1 . 12	11	α	3	11 – 9
2	ε	4	5 , 2 , 13	12		6	1 . 12
3	ι	5	3 . 6	13	γ	3 . 4	2 , 13 , 15
4	κ	5	6 , 4 , 7 4 , 8	14	ζ	4	5 , 14 , 16
5	μ	4	9 , 5 , 2 5 , 14	15	δ	4 . 3	13 , 15
6	λ	4 . 5	3 . 6 , 4	16	η	4	14 , 16 – 18
7	ν	5 . 6	4 , 7 8 ; 7 – 1	17		6	18 , 17 –, 19
8		6	4 , 8 ; 7	18	θ	4	16 – 18 , 17
9	β	3	11 – 9 , 5	19		6	17 –, 19
10		6	10 . 1				

Lustre of the stars in Navis.

1		6	686 De la Caille – 1 – 12	13		4	13 , 13 Canis min. 13 – 13 Canis min.
2		6	5 . 2 . 10	14		6	16 – 14 16 –, 14
3	τ	4 . 5	3 , 11	15	ι	3	15 , 31 Canis maj. 15 – – 7
4		6	4 , 9 4 , 6	16		5	11 . 16 – 14 16 –, 14
5		6	9 . 5 . 2	17		6	20 – 17 –, 18
6		5	4 , 6 . 9	18		6	20 , 18 –, 22
7	ξ	3 . 4	15 – – 7 = ⁏ 11	19		4 . 5	19 , 20
8		5 . 6	10 , 8	20		5 . 6	19 , 20 – 21 19 , 20 , 18 20 –, 21 20 – 17
9		4	4 , 9 . 5 6 . 9	21		6	20 – 21 20 –, 21
10		6	2 . 10 , 8	22		6	18 –, 22
11	e	4	7 = ⁏ 11 – 12 11 . 16 11 –, 12 3 , 11 ⁏ 665 De la Caille.				
12		6	11 – 12 11 –, 12 – 1				

Lustre of the stars in Orion.

1		4	1 –, 3 1 – 8	31		6	22 . 31 , 27
2	π^1	4	3 – – 2 – 7	32	A	5	32 , 47
3		4	1 –, 3 8 , 3 – – 2 3 , 9	33	n	6	38 , 33
4	o^1	4 . 5	9 – 4 11 , 4 , 15 4 – – 96 Tauri 4 , 97 Tauri.	34	δ	2	50 – 34 ; 53 50 –, 34 53 ⁝ 34
5		6	10 –, 5	35		6	15 – 35
6	g	6	7 – 6 , 14	36	υ	4	29 , 36 – 49
7	π^2	6	2 – 7 – 6	37	ϕ^1	5	40 – 37 61 , 37
8	Z	4	1 – 8 , 3 8 –, 10	38		6	23 , 38 , 33
9	o^2	4 . 5	3 – 9 – 4	39	λ	4	39 – 40
10		4 . 5	8 –, 10 –, 5	40	ϕ^2	5	39 – 40 – 37 40 , 61
11	y^1	5	11 , 4	41	θ^1	6	41 . 43
12		6	Does not exist.	42	c^1	5	42 , 45
13		6	16 – 13 18 – 13	43	θ^2	4	41 . 43
14	i	5	6 , 14 ; 16	44	ι	3 . 4	44 – 28
15	y^2	5	4 , 15 – 35	45	c^2	5	42 , 45
16	h	6	14 ; 16 – 13 16 . 18	46	ϵ	2	46 , 50 – 34 24 – 46 46 – 30 Hydræ 46 – 50 24 –, 46
17	ρ^1	4 . 5	25 , 17 – 21	47	ω	5	32 , 47 , 25
18		6 . 5	16 . 18 – 13	48	σ	4	28 , 48
19	β	1	19 ⁏ 10 Canis min. 19 = ⁏ 87 Tauri 19 –, 10 Canis min.	49	d	5	36 – 49 49 – 55
20	τ	4	20 , 29 28 – 20 =, 29	50	ζ	2	50 , 24 Gemin 46 , 50 – 34 46 – 50 –, 34 50 – 24 Gemin
21		6	17 – 21	51	b	5	56 ; 51 ; 52
22		5	22 , 27 22 . 31 22 – 11 Monocerotis	52		6	51 ; 52 , 60
23	m	6	30 ⁏ 23 , 38	53	κ	3	35 ; 53 29 ; 53 30 Hyd – 53 53 ⁝ 34
24	γ	2	112 Tauri –, 24 – 46 24 –, 46	54	χ^1	5	54 – – 57 54 – 62
25	ψ^1	5	25 , 17 47 , 25 . 30	55		6	49 – 55
26		6	Does not exist.	56		6	56 ; 51
27	ρ^2	6	22 , 27 31 , 27	57	χ^2	5	54 – – 57 68 , 57
28	η	3	44 – 28 , 48 28 – 20	58	α	1	58 . 10 Canis min. 58 – – 87 Tauri 58 –, 10 Canis min.
29	e	5	20 , 29 , 36 20 =, 29 ; 53	59		6	60 –, 59
30	ψ^2	5	25 . 30 ⁏ 23				

Lustre of the stars in Orion.

60		6	52 , 60 –, 59	70	ξ	4 . 5	67 ; 70 –, 74 70 – 75 67 , 70 – 69
61	μ	4	40 , 61 , 37	71		6	71 , 68
62	χ^3	6	54 – 62 – 64	72	f^2	6	69 . 72
63		6	66 . 63 66 ; 63	73	k^1	6	74 ; 73
64	χ^4	6	62 – 64	74	k^2	6	70 –, 74 ; 73 75 , 74
65	χ^5	5 . 6	Does not exist.	75	l	6	70 – 75 , 74
66		6	66 . 63 66 ; 63	76		6	Does not exist.
67	ν	4 . 5	67 ; 70 67 , 70	77		6	77 – 78
68		6	71 , 68 , 57	78		6	77 – 78
69	f^1	6	70 – 69 . 72				

Notes to Andromeda.

1 By three observations of FLAMSTEED, page 130, 138, and 140, the polar-distance in the edition of 1725 requires + 9°.

40 Is the same with 69 Piscium. FLAMSTEED observed it five times; twice among the stars of the constellation Pisces, and three times among those of Andromeda. See pp. 14, 134, 139, 149, and 210.

61 M. DE LA LANDE says is lost. See Mr. BODE'S *Jahrbuch* for 1794, page 97; but as the star is now in its place, it may perhaps be changeable, and ought to be looked after.

Notes to Bootes.

47 The R.A. in the British catalogue is only given to the nearest degree, and Mr. BODE and Mr. WOLLASTON, in their catalogues, have left it out; but FLAMSTEED has four complete observations of it, on pp. 166, 168, 414, and 415, and the star is called *k* in all of them.

Notes to Cancer.

26 Was not observed by FLAMSTEED. An observation on page 297 has occasioned the insertion of this star; but by correcting the time – 1′, it will agree with two other observations of 22 Cancri on pages 21 and 26. See Mr. BODE'S *Jahrbuch* for 1788, page 172.

56 This star has not been observed by FLAMSTEED, nor does it exist. Page 25 FLAMSTEED observed 55 Cancri with a memorandum, " *Hæc habet comitem sequentem ad austrum* ; " which has probably occasioned the insertion of this star; but he had not then observed all the ρ's, and might possibly mean to point out ρ 53; which he afterwards observed on page 27. The stars are so near together that he might easily mistake *sequens* for *præcedens ad austrum.* FLAMSTEED in his observations calls 58 3d ρ, 67 4th ρ, and 70 5th ρ; this shews that there is no authority for six ρ's. See Mr. BODE'S account of the same star in his *Jahrbuch* for 1788, page 171.

71 " April 5, 1796. 71 Cancri is 15′ nearer to 78 and 15′ farther from 68 than it is placed in Atlas."

73 and 74 Have not been observed by FLAMSTEED, nor do they exist. How they came to be inserted, does not appear to be satisfactorily accounted for by Mr. BODE in his *Jahrbuch* for 1788, page 172. He gives us four observations of 62 and 63 Cancri; but FLAMSTEED has thirteen, and they are all perfect except the last on page 564.*

Notes to Cepheus.

15 " October 25, 1796. 15 Cephei consists of two stars. Both taken together for one, by the naked eye, give 14 . 15. In the telescope they are 14 –, 15 – 15."

18 Has no time in FLAMSTEED'S observations. " March 26, 1797. 18 is a very little preceding 19. It is 1½ degree from 17. The stars 18, 20 and 19 are in a line which bends a little at 18 towards the preceding side."

* [See Baily's notes to Nos. 1261 and 1262.—ED.]

Notes to Corona Borealis.

21 In the British catalogue requires a correction of − 28′ 21″ in time of R.A. and − 14′ 55″ in P.D. In the place where it is marked in Atlas, according to the erroneous catalogue, is no star; but very unaccountably it is also marked in its right place in the same Atlas. FLAMSTEED has four complete observations of it on pp. 167, 445, 477, and 478. Mr. WOLLASTON not being acquainted with the existence of 21 Coronæ in its right place, supposes zone 55°, that I have made a mistake in calling my double star VI. 18, very unequal; but in his corrections he gives us the place of a star, as he calls it "near ν," which is the real second ν of FLAMSTEED; who very particularly describes it on page 167, "*Duarum ad ν sequens et clarior*;" and this is the double star I have given in my catalogue as 21 Coronæ.

Notes to Navis.

1 There is no observation of this star; but in Miss HERSCHEL's manuscript catalogue, No. 92, is a star 2° more south, which has probably been calculated wrong, and has given occasion for its insertion; correcting, therefore, the P.D. of 1 Navis + 2°, the expression of its brightness is as I have given it.

17 There is no observation of this star; but if we correct the P.D. + 3°, it will then agree with No. 238 in Miss HERSCHEL's manuscript catalogue.

21 By FLAMSTEED's observation page 431, the P.D. of the British catalogue requires + 18′.

Notes to Orion.

12 FLAMSTEED never observed this star. It does not appear how it came to be inserted in the British catalogue.*

26 FLAMSTEED never observed this star. An error of 20′ in P.D. in the calculation of one of the four observations of 25 Orionis, may have occasioned the insertion of it.

35 Is marked : : in the British catalogue; but FLAMSTEED has seven complete observations of this star; therefore the marks : : should be out.

63 There is no observation of this star; but supposing an error of + 2′ 14″ of time in R.A., and of + 0′ 22″ in P.D., it will then agree with No. 33 of Miss HERSCHEL's manuscript catalogue. I have taken the comparative brightness of that star, supposing it to be 63.

64 and 65 Have no observation by FLAMSTEED; but their insertion has been accounted for by Mr. BODE in his *Jahrbuch* for 1793, page 195. He mentions FLAMSTEED's two observations on page 17 and 94. There is a third on page 292, which confirms what Mr. BODE says. The 64 of which I give the brightness, is not far from the place assigned to it in the British catalogue. It is No. 1 in Miss HERSCHEL's manuscript catalogue.

76 There is no observation of this star. A mistake of 41′ in P.D. in calculating one of the four observations of 8 Monocerotis, might occasion its insertion.

WM. HERSCHEL.

Slough, near Windsor,
April 12, 1797.

* [See Baily's note to No. 660, 13 Orionis.—ED.]

XXXIX.

Observations of the changeable Brightness of the Satellites of Jupiter, and of the Variation in their apparent Magnitudes; with a Determination of the Time of their rotatory Motions on their Axes. To which is added, a Measure of the Diameter of the Second Satellite, and an Estimate of the comparative Size of all the Four.

[*Phil. Trans.*, 1797, pp. 332–351.]

Read June 1, 1797.

It may be easily supposed when I made observations on the brightness of the 5th satellite of Saturn, by way of determining its rotation upon its axis, and found that these observations proved successful, that I should also turn my thoughts to the rest of the satellites, not only of Saturn, but likewise of Jupiter, and of the Georgian planet. Accordingly I have from time to time, when other pursuits would permit, attended to every circumstance that could forward the discovery of the rotation of the secondary planets; especially as there did not seem to lie much difficulty in the way. For since I have determined, by observation, that the 5th satellite of Saturn is in its rotation subject to the same law that our moon obeys, it seems to be natural to conclude that all the secondary planets, or satellites, may probably stand in the same predicament with the two I have mentioned; consequently a few observations that coincide with this proposed theory, will go a good way towards a confirmation of it.

I had another point in view when I made the observations which are contained in this paper. It was an attempt to avail myself of the abundant light and high powers of my various telescopes, to examine the nature and construction of the bodies of the satellites themselves, and of their real magnitudes. Here phænomena occurred that will perhaps be thought to be remarkable, and even inconsistent or contradictory. So far from attempting to lessen the force of such animadversions, I shall be the first to point out difficulties, in order that future observations may be made to resolve them.

Perhaps it would have been better to delay the communication of these observations, till I had continued them long enough to be able to account for things which at present must be left doubtful. But as in final conclusions to be drawn from astronomical observations we ought to take care not to be precipitate; so on the other hand I am perhaps too scrupulous in satisfying myself, and should probably require the observations of several years before I could venture to be decisive. It will also be seen by the dates of the first observations, that a further delay in the communication cannot be adviseable; since much information may possibly be gained by throwing open, to other observers, the road it will be eligible to take for a satisfactory investigation of the subject; especially as we have reason to congratulate ourselves on the spirit of observation, and increase of large instruments, that seem to have taken place in various parts of Europe.

I shall now transcribe the observations from my journals. They are as follows.

OBSERVATIONS.

A remarkable Conjunction of two Satellites of Jupiter.

May 14, 1790. 11^h 30′ 10″; correct sidereal time. The 2d and 3d satellites of Jupiter are so closely in conjunction, that with a 7-feet reflector, charged with a magnifying power of 350, I cannot see a division between them.

11^h 34′ 10″. The shadow of the 1st satellite is still upon the disc of the planet.

Intenseness of Light and Colour of the Satellites.

July 19, 1794 17^h 12′ 47″. 7-feet reflector. The 1st satellite of Jupiter is of a very intense bright, white, shining light. It is brighter than the 2d or 4th. I speak only of the light, and not of the size. The colour of the 4th satellite is inclining to red. In brightness it is very nearly, but not quite equal to the 2d. I make no allowance for its being farther from the bright disc of Jupiter than the 2d. 10-feet reflector, power 170; the 3d satellite is just gone upon the body; before it went on, it appeared to me to be smaller than usual. The 2d satellite is of a dull, ash-colour; not in the extreme, but rather inclining to that tint.

July 21, 1794. 16^h 56′ 45″. 10-feet reflector; power 170. The 3d satellite of Jupiter is round, large, and well defined. It is very bright, and its light is very white. The 4th satellite is also round, large, and well defined. I estimate its magnitude in proportion to that of the 3d satellite to be as 4 to 5. Its light is not white, but inclined to orange.

Brightness and Diameter distinguished.

July 26, 1794. 17^h 14′ 41″. 10-feet reflector; power 170. The 4th satellite is very dim. It is of a pale, dusky, reddish colour. The 2d satellite is of a bright, white colour. The 3d satellite is very bright, and white. The 1st satellite is very brilliant, and white.

17^h 22′ 41″. *The Magnitudes with* 240. The 3d satellite is the largest. The 2d satellite is the smallest.

With 300. The 4th satellite is a very little larger than the 2d, though less bright. The 1st satellite is larger than either the 4th or 2d. With 400, the order of the magnitudes is 3 1 4 2. With the same power, the order of the light is 3 1 2 4. Now and then it appeared to me doubtful whether the 4th satellite was larger than the 2d; and as their light is of an unequal intensity, it is difficult without much attention to be decisive about the magnitudes.

Diameter of the second Satellite by entering on the Disc of the Planet.

July 28, 1794. 17^h 25′ 40″. 10-feet reflector; power 170. The 2d satellite is nearly in contact with the following limb of Jupiter.

17^h 29′ 40″. It seems to be very near the contact. With 300, very near the contact.

17^h 30′ 40″. It seems to be in contact. It is brighter than that part of Jupiter where it enters.

17^h 31′ 40″. It is more than half entered.

$17^h\ 33'\ 40''$. It seems to be nearly quite entered. Its superior brightness makes it seem protuberant.

$17^h\ 34'\ 40''$. It is certainly quite entered.

$17^h\ 35'\ 25''$. I see a little of the disc of Jupiter on the outside of the satellite, equal to about $\frac{1}{4}$ of its diameter.

$17^h\ 39'\ 40''$. The 3d satellite is very bright, and of its usual colour. The 4th satellite is faint, and also of its usual colour. The 1st satellite is very bright, and the light of it is of its usual intenseness.

The Magnitudes with 300. The diameter of the 4th seems to be to that of the 3d, as 2 to 3; or perhaps more exactly, as 3 to 5. The diameter of the 4th satellite exceeds that of the 1st a very little.

With 400. With this power the diameter of the 4th satellite certainly exceeds that of the 1st. The diameter of the 4th, is to that of the 3d, as 3 to 5.

July 30, 1794. $19^h\ 1'\ 37''$. 10-feet reflector; power 300. The 4th satellite of Jupiter is a little larger than the 1st. It is of its usual colour. The 2d is less than the 1st. The 3d is larger than the 4th.

July 31; 1794. $17^h\ 18'\ 38''$. 10-feet reflector;| power 170. The four satellites of Jupiter are very favourably placed for my purpose. The 1st is less bright than the 2d; it is a very little larger than the 2d: the difference in the size is but barely visible. The light of the 2d is very intense and white. The light of the 3d is very intense and bright. The light of the 4th is dull; and seems to be inferior to the usual proportion it bears to the other satellites.

$18^h\ 38'\ 38''$. *With* 300. The 4th satellite is larger than the 1st. The 2d satellite is a little larger than the 1st, or at least equal to it. The 3d is undoubtedly the largest. The order of the magnitudes therefore is, 3 4 2 1. My Brother, ALEXANDER HERSCHEL, looked at the satellites, and estimated the order of their magnitudes exactly the same; though he was not present when I made the foregoing estimations.

August 1, 1794. $17^h\ 38'\ 37''$. 10-feet reflector; power 170. The light of all the four satellites is very brilliant, the evening being very fine.

With 300. The northmost and farthest of the two satellites which are in conjunction, is the smallest: I suppose it to be the 2d. The southmost and nearest of the two satellites in conjunction, is the next in size: I suppose it to be the 1st. The 4th satellite is a little larger than the largest of the two satellites which are in conjunction; but the difference is only visible with a great deal of attention. The 3d satellite is much larger than the 4th.

August 9, 1794. $17^h\ 56'\ 32''$. 10-feet reflector; power 170. The light of the 1st satellite is very intense and white. The light of the 2d satellite is also pretty intense and white. The light of the 3d satellite is neither so intense nor so white as that of the 1st. The light of the 4th is dull and of a ruddy tinge. With 300, and 400, the second is the least, and the 3d is the largest. I am in doubt whether the 4th or the 1st is largest; with 600, I suppose the 1st to be larger than the 4th.

September 30, 1795. $20^h\ 15'\ 17''$. 7-feet reflector; power 210. Order of the magnitudes of the satellites of Jupiter 3 – 2 . 1 , 4. Power 110. 3 – 2 , 1 . 4. With 460, 3 – 2 , 1 , 4.*

October 2, 1795. $20^h\ 18'\ 22''$. 7-feet reflector; power 287. Jupiter's satellites 3 – – 2 – 1 , 4. The 2d and 3d satellites are not yet in conjunction.

$20^h\ 43'\ 22''$. The conjunction between the 3d and 2d satellites is past. The distance between them is now one diameter of the 3d.

August 18, 1796. $18^h\ 47'\ 21''$. 7-feet reflector; power 287. The 4th satellite is less bright than the 1st; notwithstanding the latter is so near the planet as to have its light overpowered by Jupiter, while the 4th is at a great distance. I mean light or brightness, not magnitude. The 1st is very bright.

September 15, 1796. $19^h\ 25'\ 25''$. 10-feet reflector; power 300. The 2d satellite of Jupiter is a little less than the 1st. The 3d is much larger than any of the rest. Power 600. The difference in the magnitude of the 1st and 2d satellites, with this power, is pretty considerable.

September 21, 1796. $19^h\ 24'\ 5''$. 10-feet reflector; power 600. The shadow of the 1st satellite is upon one of the dark belts of Jupiter. In order to use very high powers with this telescope, I tried

* Here, in order to denote the different magnitudes of the satellites, I used the notation which has been explained in my *First Catalogue of the comparative Brightness of the Stars.* See *Phil. Trans.* for the year 1796, page 189. [Above, p. 543.]

it upon the double star ζ Aquarii with 1200. The air is very tremulous, but I see now and then the two stars of this double star very well defined. With the same power, the satellites of Jupiter are very large, but not so well defined as the above star.

The Brightness of the Satellites compared to the Belts and Disc of the Planet.

The 1st satellite, which is lately come off the southern belt, is nearly of the same brightness with that belt; power 600. With 400, it is nearly as bright as the brighter part of the planet, or rather a mean between the belt and the planet.

The 2d satellite is considerably bright; its colour is whiter than that of the 1st; it is however not so white as the colour of the bright part of Jupiter.

The colour of the 4th satellite is as dingy as that of the belt; very much less bright and less white than that of the 2d.

The brightness of the 3d satellite is not intense; its colour, however, is white, though not so white as the bright part of the planet.

September 24, 1796. 20^h 55′ 24″. 10-feet reflector; power 600. The 1st satellite of Jupiter is very bright, and of a white colour; it is also very large. The 2d satellite is faint and bluish; its light is not much brighter than that of the belt. The 3d satellite is pretty bright; its light is whitish. It seems to be comparatively less than it ought to be; or rather, its apparent smallness is owing to the uncommon largeness of the 1st. The 1st satellite, with 200, compared to the 3d, is proportionally larger than I have seen it before.

September 30, 1796. 20^h 8′ 4″. 10-feet reflector; power 600. The satellites of Jupiter are well defined, and the night is beautiful. The 3d satellite, in proportion to the 1st, is much larger than it was September 24. I ascribe the change to an apparent diminution of the 1st.

20^h 30′ 4″. The 1st satellite is evidently less in proportion to the 3d, than it was September 24. The 2d satellite is considerably bright; its light is whitish; much brighter than the belt, but not so bright as the bright part of the disc. Its magnitude is less than that of the 4th; but its light is considerably superior. The 3d satellite is remarkably well defined. Its light is considerably brighter than that of the belts. The magnitude of the 1st satellite exceeds that of the 2d. It is nearly equal to that of the 4th.

22^h 58′ 4″. Appearances as before.

October 15, 1796. 21^h 23′ 42″. 10-feet reflector; power 600. The 2d satellite is uncommonly bright; its apparent magnitude is also larger than usual. The 4th satellite is very faint; it is not brighter than the belt, but is of a bluish, ruddy colour. The apparent magnitude of the 2d satellite, after long looking, is very nearly equal to that of the 1st; but at first sight it seems to be larger, owing to its superior brightness. The apparent diameter of the 2d satellite is certainly larger than that of the 4th.

23^h 55′ 42″. The light of the 1st satellite, compared to that of the 2d, is considerably increased since the last observation. It is now nearly as bright as the 2d.

October 16, 1796. 0^h 23′ 49″. 10-feet reflector; power 600. The 1st, 2d, and 3d satellites of Jupiter seem all considerably bright. The 3d is much larger than the 1st, and the 1st a little larger than the 2d. The intensity of the light seems to be pretty equal in all the three; that of the 2d, however, is perhaps a little stronger than that of the 1st; for, notwithstanding its apparent less diameter, it seems to make as strong an impression as the 1st.

October 25, 1796. 21^h 44′ 48″. 10-feet reflector; power 600. The 1st satellite of Jupiter, compared to the 3d, is small. The 3d satellite is bright and large. The 2d is brighter than the 1st. Compared to its usual brightness and magnitude, it is very bright and small. The 1st satellite, compared to its usual brightness and magnitude, is faint and small.

The air is so tremulous that the power of 600 is too high, and the necessary uniformity required in these observations will not permit a lower to be used. Perhaps one of 400 might be more generally employed; and it may be proper to use it constantly.

November 3, 1796. 23^h 55′ 47″. 10-feet reflector; power 600. The 4th satellite of Jupiter is large and bright. The 3d satellite is large and bright. The 1st satellite is pretty small, and not very bright. The 2d satellite is small, and considerably bright. The brightness and magnitude of each satellite refer to its own usual brightness and magnitude.

Before we can proceed to draw any conclusions from these observations, we ought to take notice of many causes of deception, and of various difficulties that attend the investigation of the brightness of the satellites.

The difference in the state of the atmosphere between two nights of observation cannot influence much our estimation of the brightness of a satellite, provided we adopt the method of comparative estimations. If we endeavour by much practice to fix in our mind a general ideal standard of the brightness of each satellite, we shall find the state of the atmosphere in different nights very much disposed to deceive us; but if we learn to acquire a readiness of judging of the comparative brightness of each satellite with respect to the other three, we may arrive at much more precision, since the different disposition of the air will nearly affect all the satellites alike. But here, as we get rid of one cause of deception, we fall under the penalty of another. The situation of those very satellites to which we are to refer the light of the satellite under estimation, being changeable, permits us no longer to trust to their standard, without a full scrutiny of the causes that may have produced an alteration in them.

In the foregoing observations it will also be seen, that I attempted to compare the intenseness of the light of the satellites with the different brightness of the disc of Jupiter; but these endeavours will always fail, on account of the little assurance we can have that the parts of the disc, setting aside its quick rotation, will remain for any time of the same lustre.

A very material difficulty arises from the magnifying power we use in our estimations. If it be a low one, such as for instance 180 (for a lower should not even be attempted), then we run the risk of being disappointed in bright nights by the sparkling of the brilliant light of the satellites. Besides, we cannot then see the bodies of them, and judge of their comparative magnitude, with the same power that we view their light. If we choose a high magnifier, we shall be often disappointed in the state of the atmosphere, which will of course occasion an interruption in the series of our observations, of which the regular continuance is of the greatest consequence. If we change our power according to the state of the atmosphere, we introduce a far worse cause of confusion; for it will be next to impossible to acquire, for each magnifying power, an ideal standard of comparative brightness to which we can trust with confidence.

If the magnitudes are not attended to, and carefully contradistinguished from the intenseness of light, we shall run into considerable error, by saying that a satellite is large, when we mean to express that it is bright. It is so common to call stars that are less bright than others, small, that we must be careful to avoid such ambiguities, when the condition of the satellites is under investigation. Nor is it possible to throw the size and light into one general idea, and take the first *coup d'œil* in looking at them, to decide about the general impression this compound may make. When our attention is forcibly drawn by a considerable power to the apparent size of the satellite we are looking at, its brightness can no longer be taken in that general way, but must be abstracted from size.

Let us now see what use may be drawn from the observations I have given.

It appears in the first place very obviously, that considerable changes take place in the brightness of the satellites. This is no more than might be expected. A variegated globe, whether terraqueous like the earth, or containing regions of soil of an unequal tint, like that side of the moon which is under our inspection, cannot, in its rotation, present us with always the same quantity of light reflected from its surface.

In the next place, the same observations point out what we could hardly expect to have met with; namely, a considerable change in the apparent magnitude of the satellites. Each of them having been at different times the standard to which another was referred, we cannot refuse to admit a change so well established, singular as it may appear.

The first of these inferences proves that the satellites have a rotatory motion upon their axes, of the same duration with their periodical revolutions about the primary planet.

The second either shews that the bodies of the satellites are not spherical, but of such forms as they have assumed by their quick periodical and slow contemporary, rotatory motions, and which forms in future may become a subject for mathematical investigation; or it may denote, in case geometrical researches should not countenance a sufficient deviation from the spherical form, that some part of the discs of these satellites reflects hardly any light, and therefore in certain situations of the satellite makes it appear of a smaller magnitude than in others.

Here then we see evidently that a considerable field for speculation, as well as observation, is opened to our view; and almost every attempt to enter upon the work must seem premature, for want of more extended observations. However, from those that have been given, such as they are, I will shew how far we may be authorized to say, that the satellites revolve on their axes in the same time that they perform a periodical revolution about the planet.

I shall take the usual method of throwing the observations of each satellite into a graduated circle. The zero of the degrees into which I suppose it divided, is in all observations assumed to be in the place of the geocentric opposition.

In order to bring these observations to the circle, the places of the satellites have been calculated from my own tables of the mean motion in degrees, and according to epochs continually assumed from the geocentric conjunctions pointed out in the configurations of the *Nautical Almanac*; and the nearest of these conjunctions have been always used. This method is fully sufficient for the purpose, as greater precision in the calculation is not required.

The observations extend from July 19, 1794, to November 3, 1796; and therefore include a period which takes in 470 rotations of the 1st satellite; 234 of the 2d; 116 of the 3d; and 50 of the 4th: that is, provided we admit that these rotations are performed in the same time the satellites revolve in their orbits.

In the following table are the calculated places of the satellites; the correct sidereal times, given with the observations, having been turned into mean time.

Table of the Positions of the four Satellites of Jupiter at the Time of the Observations.

Time of Observ.	I	II	III	IV	Time of Observ.	I	II	III	IV
1794. July 19d 9h 21′	127°	346°	179°	46°	1796. Aug. 18d 8h 21′	115°	°	°	191°
July 21. 8. 57			278	89	Sept. 15. 7. 44	36	328	198	
July 26. 8. 56	124	333	169	205	Sept. 21. 7. 19	172	214	138	210
July 28. 8. 59	171	176	270	248	Sept. 24. 8. 38	74	163	305	275
July 30. 10. 27	231	25	13	292	Sept. 30. 7. 27	206	46	244	36
July 31. 8. 40	59	118	60	312	Oct. 15. 7. 44	28	130		5
Aug. 1. 8. 56	265	221	111	334	Oct. 15. 10. 15	49			
Aug. 9. 8. 42	83	310	152	138	Oct. 16. 10. 39	256	243	334	
					Oct. 25. 7. 25	261	72	59	
1795. Sept. 30. 7. 37	294	62	219	100	Nov. 3. 9. 0	306	270	151	58
Oct. 2. 7. 32	341	264	319	143					

It will be necessary now to explain in what manner, with the assistance of this table, the observations of the brightness and magnitudes of the satellites have been reduced to the expressions they bear in the four circles of the figures contained in figs. 1–4. By way of uniformity I judged it would be best to reduce the estimations of magnitude to those of brightness; as it may be justly supposed that when a satellite is at any given time larger in proportion to another than it was at another time, it will also be brighter than it was at that other time, due regard being had to the light of the satellite to which its magnitude has been compared. To manage the space allotted to the figure advantageously, I have used the abbreviations formerly employed in my catalogue of Nebulæ, *v* B, *c* B, B, *p* B, *p* F, F, *c* F, *v* F, for all the gradations of light that are necessary to express the brightness of the satellites at the time of observation. It will be easily remembered that B and F mean bright and faint; and *p*, *c*, *v*, stand for pretty, considerably, and very.

Now, when the observation mentions the brightness of the satellite, I place it in the figure as it is given. In that of the first, for instance, July 19, 1794, we find the satellite called very bright; I therefore put down in fig. 1 at 127 degrees, *v* B. But where the brightness is not expressed, I have recourse

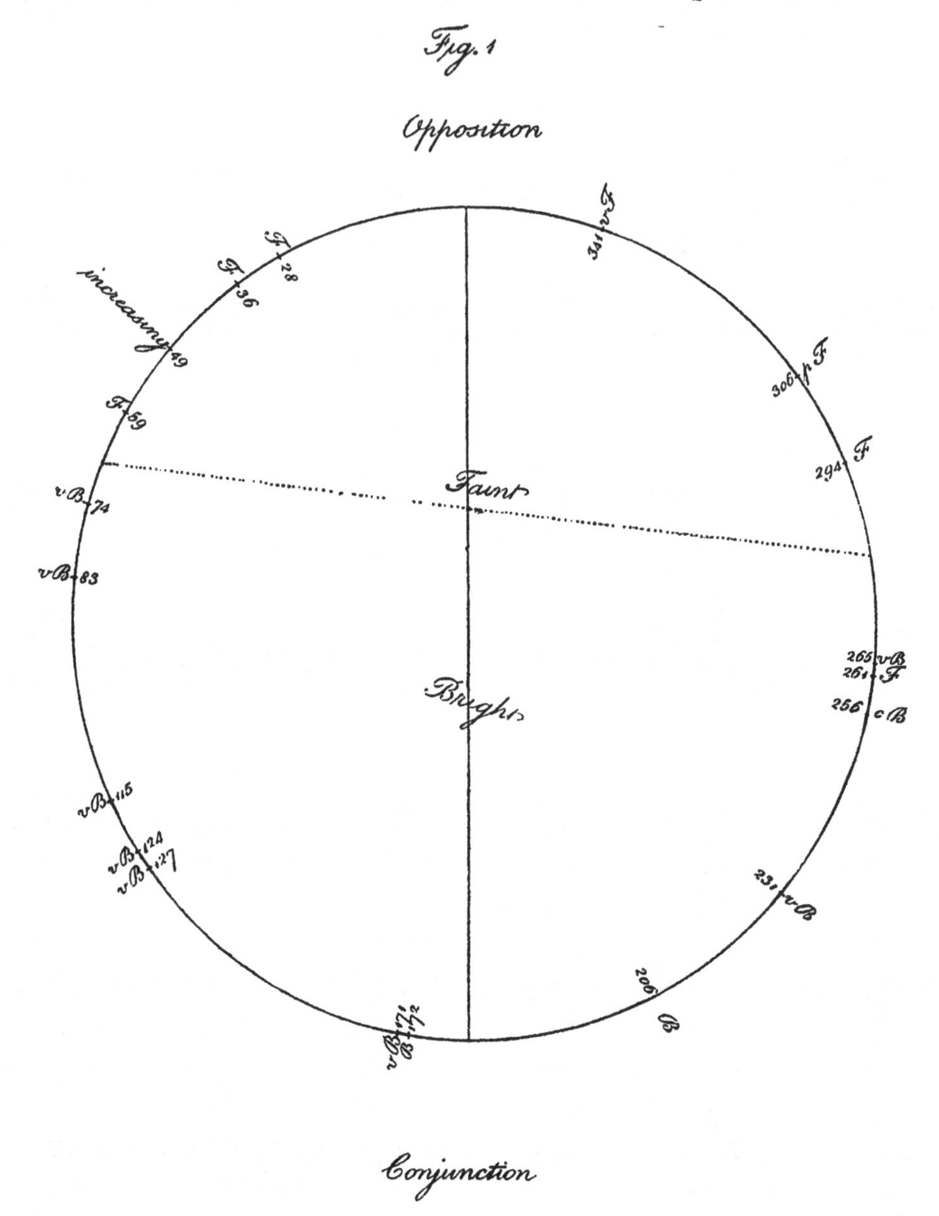

to the comparative magnitude, if that can be had. By fig. 3 it appears that the 2d satellite is less subject to a change of brightness than either the 1st or 4th: it becomes, for that reason, a pretty good standard for the light of these other satellites. Therefore, in the observation of October 2, 1795, for instance, where the 1st satellite is described as undoubtedly less than the 2d, I put down very faint, or *v* F, at 341 degrees of the circle in fig. 1; for in the observation of July 19, before mentioned, when the satellite was called very bright, it was at the same time described as undoubtedly larger than the 2d. In this case, as regard must be had to the relative state of the satellite we refer to, the four figures I have given will assist us in determining the condition of the light of the satellite we wish to admit as a standard.

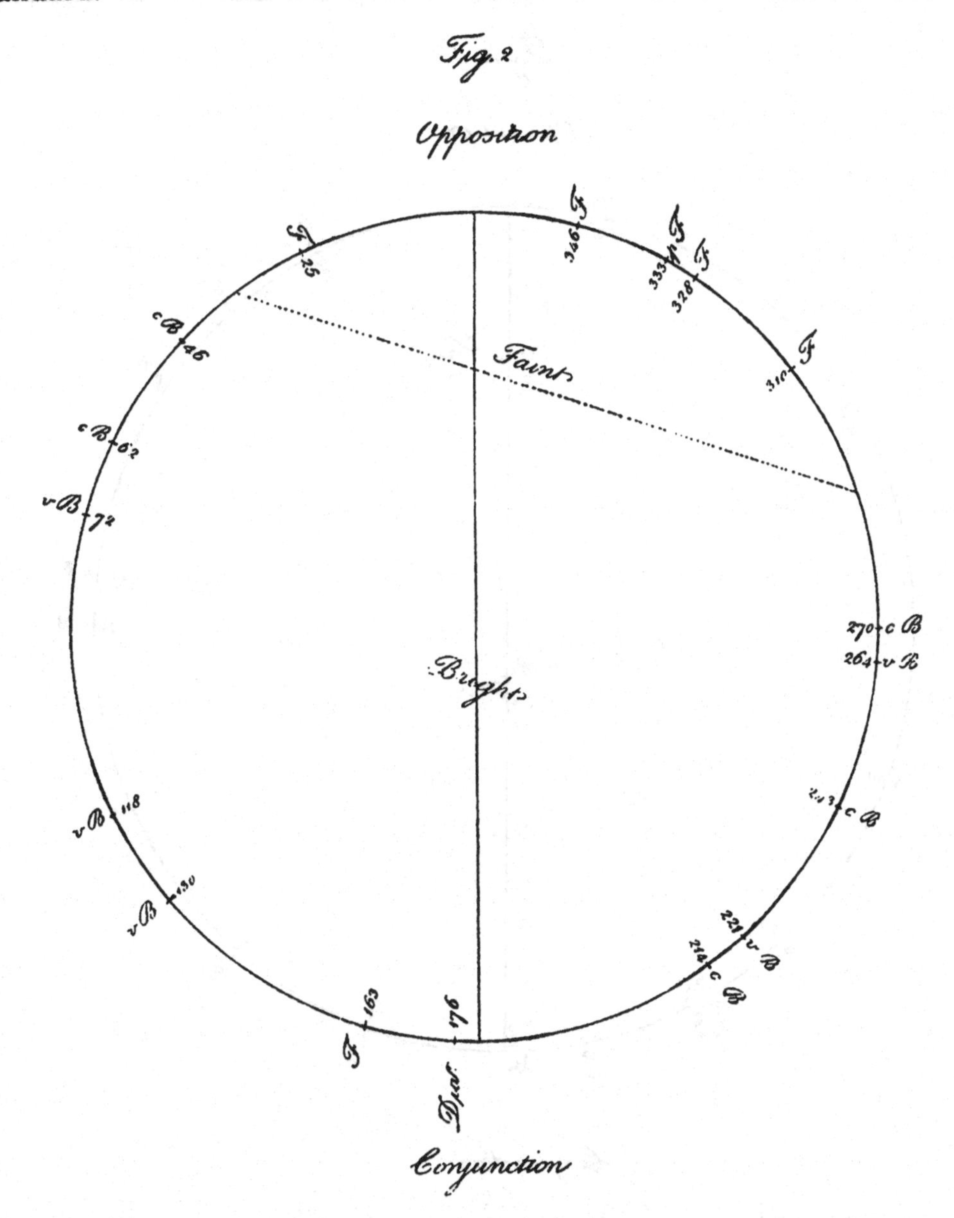

In reducing the 2d satellite to the circle, I have generally used a reference to the magnitude of the 1st, where marks of brightness were wanting; and sometimes also to the magnitude of the 4th, and even of the 3d.

The 3d satellite can hardly be ever compared to any but the 2d in magnitude; and this only in its degree of excess.

The magnitude of the 4th satellite has been generally compared with that of the 1st; and also sometimes with that of the 2d.

To make an application of the contents of the figures, will now require little more than a bare inspection of them.

The 1st satellite appears evidently to have a rotation upon its axis that agrees with its revolution in its orbit. It cannot be supposed that, in the course of 470

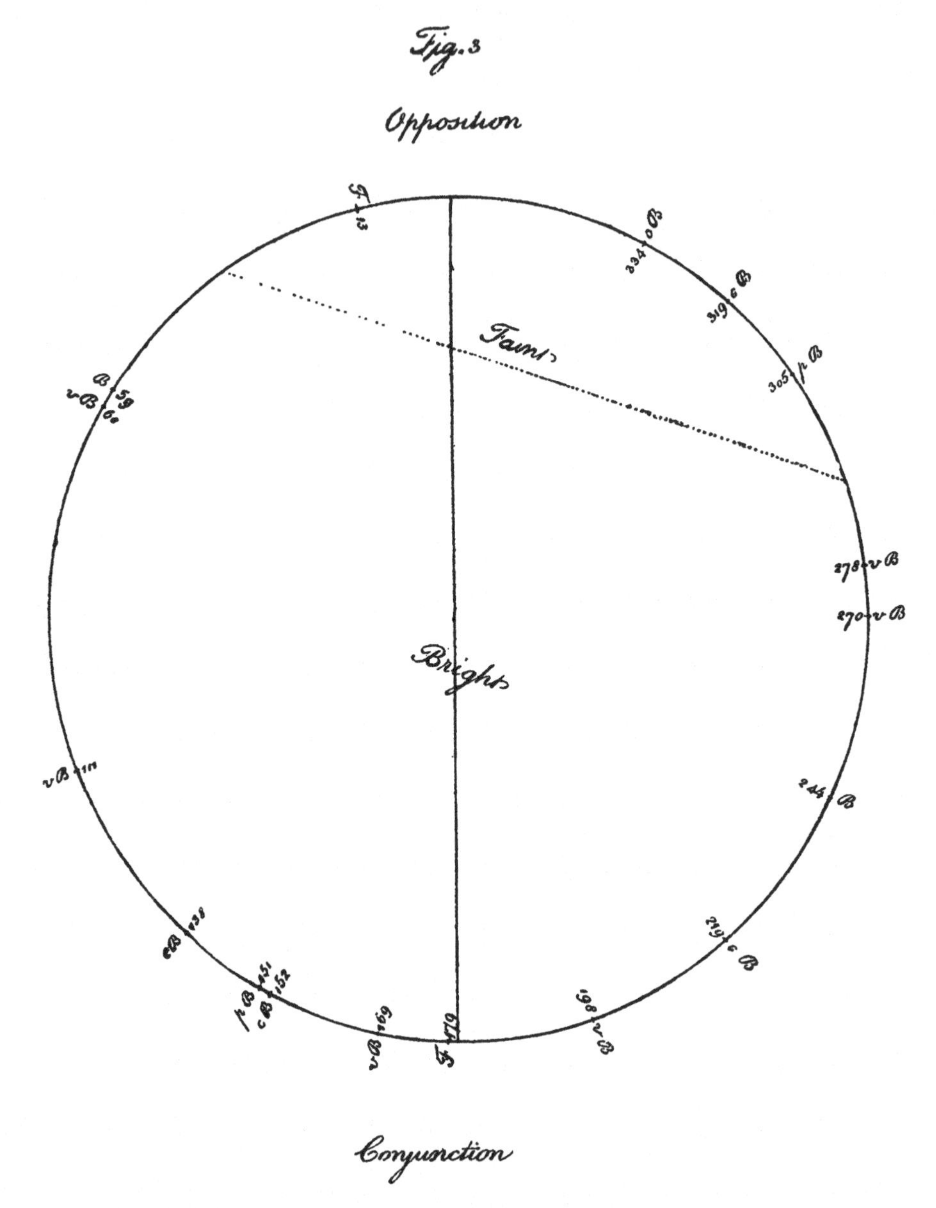

revolutions, all the bright observations could have ranged themselves in one half of the orbit, while the faint ones were withdrawn to the other. The satellite appears in the middle of the duration of its brightness, when it is nearly half way between its greatest eastern elongation and the nearest part of its orbit; or when advancing towards its conjunction. I have pointed out this circumstance by a division with dotted lines, and the words bright and faint, inserted within the circle, fig. 1. This satellite, therefore, revolves on its axis in $1^d\ 18^h\ 26'\cdot6$,

The 2d satellite, though much less subject to change, on account, as we may suppose, of having only a small region on its body which reflects less light than the rest; has, nevertheless, its rotation directed by the same law with the 1st. It will hardly be necessary to take notice of a single deviation which occurs at 163

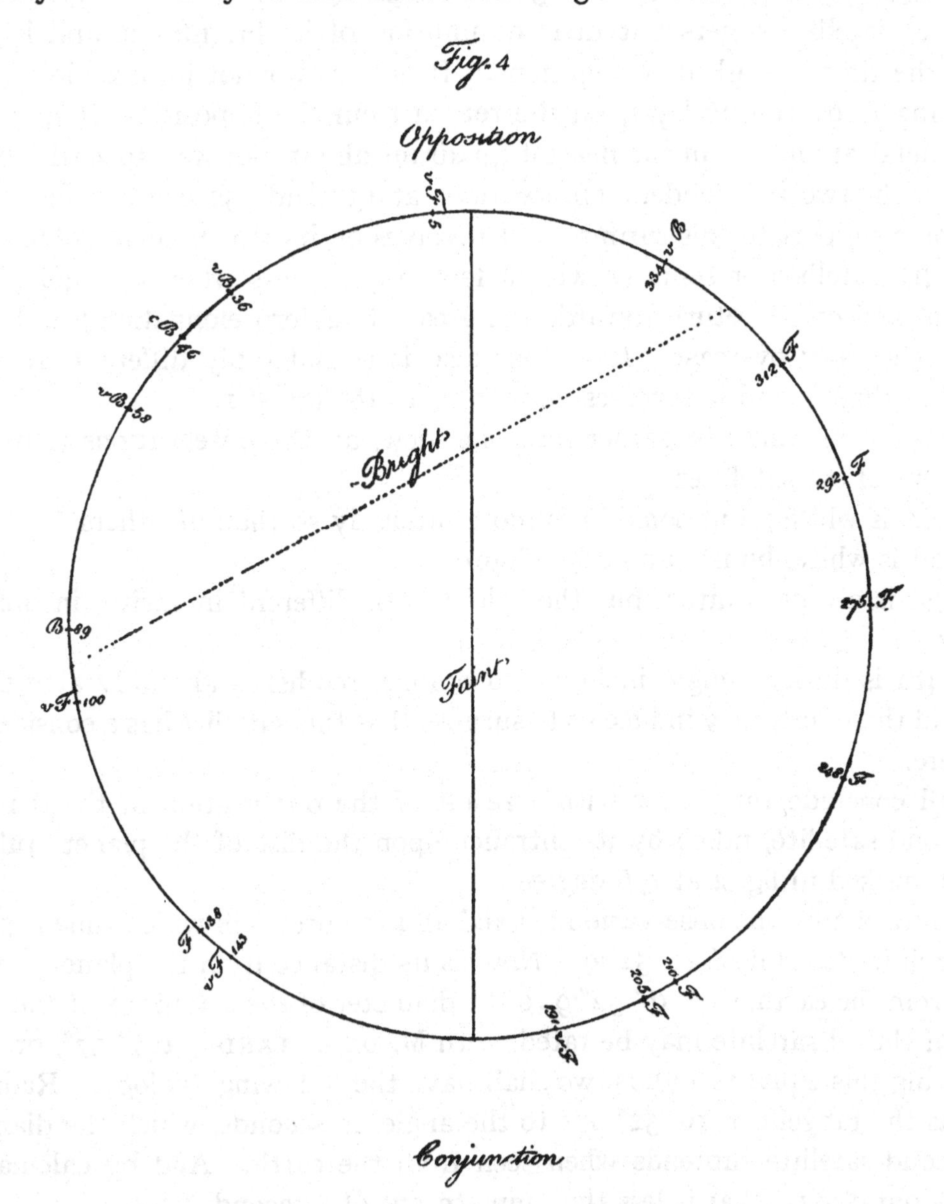

degrees, fig. 2; as from the proximity of the satellite to the conjunction, a mistake in the estimation may easily take place. I generally made it a rule not to make allowance for the influence of the superior light of the planet; but it seems that we can hardly abstract sufficiently on such occasions. Two similar cases occur, in fig. 3, at 179; and fig 4 at 5 degrees.

It is indeed not impossible but that occasional changes, on the bodies of the satellites themselves, may occasion some temporary irregularity of their apparent brightness: it will, however, not be necessary to make such an hypothesis, till we have better authority for it. The brightest side of this satellite is turned towards us when it is between the greatest eastern elongation and the conjunction. It revolves, consequently, on its axis in $3^d\ 18^h\ 17'\cdot 9$.

The 3d satellite suffers but little diminution of its brightness, and is in full lustre at the time of both its elongations. It is however not impossible but that, after having recovered its light, on the return from the opposition, it may suffer a second defalcation of it in the nearest quadrant about half way towards the conjunction. The two independent observations at 151 and 152 degrees, fig. 3, seem to give some support to this surmise. It revolves on its axis in $7^d\ 3^h\ 59'\cdot 6$.

The 4th satellite presents us with a few bright views when it is going to its opposition, and on its return towards the greatest eastern elongation; but otherwise it is generally overcast. Its colour also is considerably different from that of the other three; and it revolves on its axis in $16^d\ 18^h\ 5'\cdot 1$.

It will not be amiss to gather into one view, all the observations that relate to the colour of the satellites.

The 1st is white; but sometimes more intensely so than at others.

The 2d is white, bluish, and ash-coloured.

The 3d is always white; but the colour is of different intensity, in different situations.

The 4th is dusky, dingy, inclining to orange, reddish and ruddy at different times; and these tints may induce us to surmise that this satellite has a considerable atmosphere.

I shall conclude this paper with a result of the observation of the diameter of the second satellite, taken by its entrance upon the disc of the planet, July 28, 1794, and marked in fig. 2 at 176 degrees.

The duration by the observation is fixed at 4 minutes; in which time it passes over an arch in its orbit of $16'\ 52''\cdot 9$. Now as its distance from the planet is to its distance from the earth, so is $16'\ 52''\cdot 9$ to the diameter of the satellite; or the mean distance of the 2d satellite may be rated, with M. DE LA LANDE, at $2'\ 57''$, or $177''$. Then putting this equal to radius, we shall have the following analogy. Radius is to $177''$, as the tangent of $16'\ 52''\cdot 9$ is to the angle, in seconds, which the diameter of the second satellite subtends when seen from the earth. And by calculation, this comes out $0''\cdot 87$; that is less than nine-tenths of a second.

I have not been scrupulously accurate in this calculation, as the real distance of Jupiter at the time of observation should have been computed, whereas I have contented myself with the mean distance. Nor am I very confident that the angle of the greatest elongation, admitted to be $2'\ 57''$, is quite accurate; but I judged it unnecessary to be more particular, because the time of my observation in the beginning of the transit upon the disc, I find, was only taken down in whole minutes of the clock. The end, however, is more accurately determined, by the observation which was made 45″ after the immersion ; when a part of the disc, equal to about $\frac{1}{4}$ of the diameter of the satellite, is said to be visible. It seems that observations of this kind, made with very good telescopes, charged with high powers, are capable of great precision. For the remark that a margin of Jupiter, equal to about $\frac{1}{4}$ of the diameter of the satellite, became visible in 45″ of time, adds great support to the accuracy of the observation of the foregoing 4 minutes : and, at all events, it is evidently proved, from the whole of the entrance upon the disc, that the diameter of this satellite is less, by one half at least, than what from the result of the measures of former observers it has been supposed to be.

A method has also been used, of deducing the diameter of the satellites from the time they employ to immerge into the shadow of the planet ; but this must be very fallacious, and ought not to be used.

I should not pass unnoticed the apparent magnitude of the satellites. The expressions that have been given of them may be collected into the following narrow compass.

1,4,2   4;1   3-,4;1-2   4,2,1   3--4;1;2   1,4,2   3-2,1,4
3--2-1,4   1-2   4.1-2   1;2-4   3--1,2   2-1

From which we may conclude, that the 3d satellite is considerably larger than any of the rest ; that the 1st is a little larger than the 2d, and nearly of the size of the 4th ; and that the 2d is a little smaller than the 1st and 4th, or the smallest of them all.

WM. HERSCHEL.

Slough, near Windsor,
April 30, 1797.

END OF VOL. I.

PRINTED BY NEILL AND CO., LTD., EDINBURGH

Printed in the United States